The 7th International Conference on Time Series and Forecasting

The 7th International Conference on Time Series and Forecasting

Editors

Ignacio Rojas
Fernando Rojas
Luis Javier Herrera
Hector Pomares

MDPI • Basel • Beijing • Wuhan • Barcelona • Belgrade • Manchester • Tokyo • Cluj • Tianjin

Editors
Ignacio Rojas
University of Granada
Spain

Fernando Rojas
University of Granada
Spain

Luis Javier Herrera
University of Granada
Spain

Hector Pomares
University of Granada
Spain

Editorial Office
MDPI
St. Alban-Anlage 66
4052 Basel, Switzerland

This is a reprint of articles from the Special Issue published online in the open access journal *Proceedings* (ISSN 2504-3900) (available at: https://www.mdpi.com/2673-4591/5/1).

For citation purposes, cite each article independently as indicated on the article page online and as indicated below:

LastName, A.A.; LastName, B.B.; LastName, C.C. Article Title. *Journal Name* **Year**, *Volume Number*, Page Range.

ISBN 978-3-0365-1731-5 (Hbk)
ISBN 978-3-0365-1732-2 (PDF)

Contents

About the Editors

Ignacio Rojasis a full professor at the University of Granada, Spain. His field of research focuses on the study of complex multidimensional systems using intelligent systems, supported by high-performance computing platforms, focused on solving real problems in various fields, such as bioinformatics, biomedicine, and time series prediction, among others. As a result of his research, more than 270 contributions have been added to the ISI Web of Knowledge database, of which more than 153 are classified as articles, and more than 120 of his contributions are collected in journals indexed in the Journal Citation Reports of the Institute for Scientific Information (ISI). In addition to publication in indexed journals, he has presented his scientific contributions at more than 125 international conferences related to his field of research, has directed 25 doctoral theses, and has organized various international conferences (a total of 23 international and 5 national conferences), workshops and special sessions. Throughout his entire research career, he has participated in 33 projects. He has been a visiting researcher, mainly at the University of Dortmund (Germany), University of California at Berkeley (USA), and University of Applied Sciences Muenster (Germany). As for university management, he was the Deputy Director of the Library Infrastructure and Economic Management of the Higher Technical School of Computer Engineering and Telecommunications of the University of Granada from 2004 to 2008, and since 2013, he has been the Director of the Research Center in Information and Communication Technologies (CITIC-UGR) from the University of Granada (http://citic.ugr.es/).

Fernando Rojas works at the Department of Computer Architecture and Computer Technology at the University of Granada as an Associate Professor. The research field focuses on signal processing, artificial intelligence techniques for optimization (evolutionary computing, fuzzy logic, neural networks, etc.) and the study of computer architectures for parallel processing. As a result of the research carried out, he has published more than 40 articles collected in journals indexed in the Institute for Scientific Information (ISI). In addition to publication in indexed journals, he has participated in numerous conferences related to his field of research, resulting in the publication of 67 contributions. Rojas has participated in various research projects (13 in total), which were respectively funded by the Ministry of Science and Technology, Ministry of Education and Science and the Ministry of Innovation, Science and Business of the Junta de Andalucía. His work has also contributed to the transfer of knowledge to the productive sector by participating in three research contracts of special relevance in this area. He has carried out research and teaching stays in centers in Ireland, Cuba and Germany. He also has management experience. Rojas was the coordinator of the Official University Master's Degree in Computer and Network Engineering from March 14, 2013 to September 30, 2014. Since October 1, 2014, he has been secretary of the Official University Master's Degree in Data Science and Computer Engineering. Since March 18, 2018, he has been secretary of the Department of Computer Architecture and Technology at the University of Granada.

Luis Javier Herrera works at the Department of Computer Architecture and Computer Technology at the University of Granada as an Associate Professor. His research field covers the study of machine learning techniques (fuzzy logic, deep learning, genetic algorithms, etc.), and their optimization and application over a wide range of scientific problems related to classification, approximation and time series prediction, sometimes requiring high-performance computing systems. These applications include relevant problems in biomedicine, bioinformatics, biochemistry, physics, and time series prediction. As a result of his research, he has published 40 journal papers. He has participated in several international conferences related to my research scope. He has been an editor of proceedings of national and international conferences, and he is a co-author of a number of book chapters in national and international editorials. He has also supervised three PhD theses, producing relevant journal publications. He has participated in several research projects (Spanish Ministry and the regional Junta de Andalucía Government funding programs), among others. He has been the main researcher in other competitive calls, such as CEI-BIOTIC Campus. He has also contributed to the knowledge transfer to the productive sector through his participation in three contracts. Finally, he has performed three research stays at research centers in Germany, Belgium and United Kingdom, with satisfactory results, and reflected in publications in conferences proceedings.

Hector Pomares (MSc in Electrical Engineering in 1995, MSc in Physics in 1997, PhD from the University of Granada in 2000, all of them with honors) is currently a full professor at the University of Granada. He has published more than 60 articles in the most prestigious scientific journals and contributed with more than 150 papers in international conferences. He has been a visiting researcher at the University of Dortmund (Germany), University of California at Berkeley (USA), University of Texas A&M (USA), University of Applied Sciences Muenster (Germany), Technical University of Graz (Austria) and University of Amsterdam (Netherlands). At the present time, he is the director of the Doctoral Program in Information & Communication Technologies at the University of Granada.

Preface to "The 7th International Conference on Time Series and Forecasting"

The ITISE 2021 (7th International conference on Time Series and Forecasting) seeks to provide a discussion forum for scientists, engineers, educators, and students about the latest ideas and realizations in the foundations, theory, models, and applications for interdisciplinary and multidisciplinary research, encompassing disciplines of computer science, mathematics, statistics, forecaster, econometric, etc., related to the field of time series analysis and forecasting.

The aims of ITISE 2021 is to create a friendly environment that could lead to the establishment or strengthening of scientific collaborations and exchanges among attendees, and, therefore, ITISE 2021 solicits high-quality original research papers (including significant work-in-progress) on any aspect time series analysis and forecasting, in order to motivate the generation, and use of knowledge and new computational techniques and methods on forecasting in a wide range of fields.

Ignacio Rojas, Fernando Rojas, Luis Javier Herrera, Hector Pomares
Editors

engineering proceedings

MDPI

Proceedings

Forecasting and Analysis Tools for Regional Industries' Dynamics †

Valeriy Semenychev [1] and Anastasiya Korobetskaya [2,*]

1 Department of Mathematical Methods in Economics, Samara University, 34, Moskovskoye Shosse, 443086 Samara, Russia; 505tot@mail.ru
2 System Integrator "Webzavod", 156, Galaktionovskaya Str., 443001 Samara, Russia
* Correspondence: kornast@yandex.ru

† Presented at the 7th International Conference on Time Series and Forecasting, Gran Canaria, Spain, 19–21 July 2021.

Abstract: The article is devoted to the author's approach and tools for regional industries' modeling, analysis and forecasting, following the general idea of splitting time series into four components: trend, cycles, seasonal component, and residuals. However, the authors introduce new approaches, models, metrics, and identification algorithms, and the components' interaction structures, having included the analysis of 12 industries in 82 regions of Russia. The models and forecast accuracy were tested on 3–12 month forecasts, thus proving their high accuracy. Therefore, the article proposes not only new systematic econometric tools but a methodology for decision making, developed to provide stable and adequate characteristics of complex non-linear evolutionary dynamics of Russian regions.

Keywords: time series; regional economy; forecasting; modeling; medians; wavelets; Russia

Citation: Semenychev, V.; Korobetskaya, A. Forecasting and Analysis Tools for Regional Industries' Dynamics. *Eng. Proc.* **2021**, *5*, 1. https://doi.org/10.3390/engproc2021005001

Published: 24 June 2021

1. Introduction

Regional industries, having both spatial and temporal dimensions, represent a complex meso-economic object of analysis [1]. On the one hand, they represent inter-related socio-economic systems, and their dynamics should be coherent with each other and the country. On the other hand, the regions' development level and their connectivity may vary a lot.

The common equilibrium approach is based on a mechanistic view of economic systems and is focused on a return to the equilibrium state. However, regional economies tend to shift to an evolutionary approach based on long-term forecasting [2] and the regions' abilities to change their economic structures [3]. Researchers also point out that regional differences in repeatability tend to become a significant factor in effective economic policy decisions [4].

The research aims at designing tools for modeling and forecasting, in the context of time series concerning regional industries' dynamics. The acquired results should be adequate and accurate to provide a facility for decision making and to support sustainable evolutionary development.

2. Data

As a statistical database, we are using an official data source provided through the Unified Interdepartmental Information and Statistical System (EMISS) by the Russian Federal State Statistic Service. The EMISS database possesses operational monthly data on the production level of each economic industry (real (volume) growth rate, percent) for each subject (region) of the Russian Federation. The industries are classified hierarchically by the All-Russian Classifier of Types of Economic Activity (OKVED2). We have chosen the twelve most important industries in terms of presence in different regions:

1. Extraction of minerals.
2. Crude oil and natural gas extraction.

3. Mining of metal ores.
4. Manufacturing.
5. Food production.
6. Petroleum production.
7. Chemical production.
8. Pharmaceuticals and materials used for medical purposes.
9. Rubber and plastic production.
10. Metallurgy.
11. Computers, electronics and optical production.
12. Automotive industry. Production of motor vehicles, trailers and semi-trailers.

The data reflect the situation in 82 regions of Russia (except Crimea, Sevastopol, and the Republic of Chechnya due to lack of statistics), as well as for Russia as a whole. However, some industries may be not presented in particular regions, so, in total, there are about 750 time series.

The analysis embraces the period from January 2005 to August 2020. This time interval seems interesting as it covers important periods and milestones within the Russian economy's evolution. It reveals the economic growth in the noughties of the XXI century, the crisis of 2008–2009, subsequent recession and recovery, the growth of political turbulence, the adoption of economic sanctions against Russia, and the beginning and extension of the COVID-19 pandemic.

3. Research Methods

3.1. Approach

To reach the research goal, we defined some approaches to be used in our algorithms and models. Basically, we are following the idea of splitting time series into trend, cycles, seasonal fluctuations, and residuals (stochastic component). However, we are trying to review the details and characteristics of economic objects at the meso-level.

Firstly, the meso-economic systems are non-linear and show evolutionary development. So, the simple models such as linear or exponential trends are applicable only for short periods of time. For longer periods, perspective changes inside the region, as well as its interrelations with other regions, affect the dynamics and should be appropriately reflected in the models. We provide a complex of different trends, possessing extremums, inflection points, asymptotes, asymmetry, and thus, the ability to adapt to such volatility.

Another approach is using points of structural change [5] to reveal the moments of time where dynamics change drastically and cannot anymore be described by the same model.

The other important factor is the model residuals. Traditionally, the distribution law of the residuals is supposed to be normal (Gaussian). However, real economic systems are rarely normally or even log-normally distributed. The practice shows very different asymmetric and heavy-tailed distributions. We examined all the above-mentioned industries and regions and discovered a wide variety of distributions. So, choosing some particular distribution law or even a fixed set of laws seems inappropriate. It is better to use tools that are more robust and do not depend upon the distribution law.

One of the basic and robust metrics is the median. Instead of choosing the one "best" model, we identify many different models, and at each moment of time use the median of all their fitted values. Thus, we eliminate one of the hardest problems—structural identification of the model. Using the median effectively filters inadequate models and provides sustainable fits.

To find the median, it is better to take the maximum possible fits for each point. Using the bootstrap procedure [6], it is possible to identify a few models with the same structure (formula) but different parameters. The bootstrap procedure is a common approach to increase small sample sizes.

The other important point is the criterion used to identify models' parameters. The most common approach is using the least squares. However, it is very sensitive to outliers presented in heavy-tailed distributions. The least absolute deviations method is seen as more reliable [7]:

$$\sum_{t=1}^{n} \left|Y_t - \hat{Y}_t\right| \rightarrow min, \quad (1)$$

where Y_t is the original time series, t is time (ordering indices from 1 to n), $\hat{Y}_t$ is the model's fitted values.

On the other hand, for multiplicative residuals, the least absolute percentage deviations seem more correct:

$$\sum_{t=1}^{n} \left|\frac{Y_t - \hat{Y}_t}{Y_t}\right| \rightarrow min. \quad (2)$$

Unfortunately, the choice between the additive and multiplicative residuals' structure is not obvious. The mixed additive–multiplicative structures can also be present. So, we defined combined measures to minimize both:

$$\frac{1}{\overline{Y}} \sum_{t=1}^{n} \left|Y_t - \hat{Y}_t\right| + \sum_{t=1}^{n} \left|\frac{Y_t - \hat{Y}_t}{Y_t}\right| \rightarrow min. \quad (3)$$

The criterion (1) includes two parts to minimize both additive and multiplicative residuals, but the first part is divided by the time series average $\overline{Y}$ (which is constant and does not change extremum position), to underline the parts' comparability. The same effect may be achieved by multiplying the second part by $\overline{Y}$, but we prefer to use relative values.

It should be also mentioned that all the models and algorithms described below were implemented in the R language using both the authors' program code and open-source libraries.

3.2. Models

The most common models for time series structures appear as additive (2) and multiplicative (4):

$$Y_t = T_t + C_t + S_t + \varepsilon_t, \quad (4)$$

$$Y_t = T_t(1 + C_t)(1 + S_t)(1 + \varepsilon_t), \quad (5)$$

where T_t—trend values; C_t—cyclical component values; S_t—seasonal component levels; ε_t—stochastic component.

It is also reasonable to consider mixed additive–multiplicative structures:

$$Y_t = (T_t + C_t)S_t + \varepsilon_t, \quad (6)$$

$$Y_t = T_t(1 + C_t) + S_t + \varepsilon_t. \quad (7)$$

The authors' complex of trend models currently includes linear, generalized exponential, power trends, four cumulative logistic (S-shaped) and four impulse logistic (bell-shaped) trends with different asymmetry settings:

$$T_t = C_0 + A_0 t, \quad (8)$$

$$T_t = C_0 + A_0 t^{\alpha}, \quad (9)$$

$$T_t = C_0 + A_0 e^{\alpha t}, \quad (10)$$

$$T_t = C_0 + \frac{A_0}{1 + e^{-\alpha(t-t_0)}}, \quad (11)$$

$$T_t = C_0 + A_0 arctg(\alpha(t - t_0)), \quad (12)$$

$$T_t = C_0 + A_0 exp(-exp(-\alpha(t - t_0))), \quad (13)$$

$$T_t = C_0 + A_0(1 + exp(-\alpha(t - t_0)))^{\sigma}, \quad (14)$$

$$T_t = C_0 + A_0 exp\left(-\alpha(t-t_0)^2\right), \quad (15)$$

$$T_t = C_0 + \frac{A_0}{1+\alpha(t-t_0)^2}, \quad (16)$$

$$T_t = C_0 + \frac{A_0}{1+\alpha(t-t_0)^2} \cdot \frac{1}{1+exp(-\sigma(t-t_0))}, \quad (17)$$

$$T_t = C_0 + \frac{A_0}{1+(\sigma(t)(t-t_0))^2}, \sigma(t) = \frac{1}{1+exp(-\sigma(t-t_0))}, \quad (18)$$

All the trend models (8)–(18) use the unified naming of parameters where: C_0 is the vertical shift constant and asymptotic level (if any), A_0 is the trend amplitude (vertical scale), α is the growth/decline velocity (horizontal scale), t_0 is the horizontal shift (inflection point for S-shaped trends, extremum point for bell-shaped trends), σ is the asymmetry coefficient. The models differ by their shape, growth velocity, and skewness (symmetric, fixed asymmetry or free asymmetry).

For each dynamic series, all the trend models are identified through the total sample length and can grouped by means of structural changes (the points may be different for each model). Thus, we have up to 22 fitted values for each point in the time series.

As for cycles, we used two general approaches to modeling. The first approach is based on the E. Slutsky [8] hypothesis that any fluctuations could be presented as a sum of a few sinus functions with non-proportional frequencies:

$$C_t = \sum_{i=1}^{\infty} A_i \sin(\omega_i t + \varphi_i), \quad (19)$$

where A_i is the ith sinus amplitude, ω_i is the sinus frequency and φ_i is the sinus phase.

This approach is effective for modeling as it gives a well-smoothed model of the cycles. However, there is no guarantee that the amplitudes, phases, and frequencies that optimally described dynamics in the past will remain the same in the future. So, the extrapolation of such a model is simple but unproven. Thus, we turned our attention to wavelet transformation [9–11]. The wavelet transformation is used widely in signal processing to eliminate signal noise, but is now adopted in economics and other sciences for time series smoothing and forecasting.

Wavelets are seen as functions used to identify local non-periodical fluctuations and monitor their changes through time periods. The time series are decomposed on a few levels of so-called wavelet and scaling coefficients. These components may vary from high-frequency ones (representing the "noise") to lower-frequency components representing local cycles. The low-frequency components of wavelet decomposition can be easily modeled and forecasted with ARMA models and reversely transformed back to provide a smoothed model and forecast.

The variety of wavelet functions' families is wide. In this study, we used the most generalized discrete transformation from the wavelet families: Haar, Daubechies, etc. (in total, 42 wavelet functions).

3.3. Identification Algorithm

Based on the below-mentioned principles, an algorithm to identify time series models is designed with the following sequence of steps:

1. Preprocessing of the initial time series, removing random outliers and using *R*'s standard library, then replacing them with median smoothed values.
2. Determining the structure of seasonal fluctuations (additive or multiplicative).
3. Detecting seasonal fluctuations using the STL function, which returns the smoothed trend, seasonal fluctuations, and random residuals based on LOESS smoothing. For multiplicative structures, the logarithms are used.
4. Deseasonalization (removing seasonal fluctuations from the initial series).
5. Determining the structure of cyclic fluctuations.

6. Building the median trend without structural shifts and without bootstrapping. To do this, all available types of trends are selected using criterion (1), and the median value from all trend estimates is taken at each point in the time series.
7. Detrending (removing a trend from a series).
8. Fitting cyclic fluctuations to the detrended and deseasonalized data.
9. Removal of cyclical fluctuations from the deseasonalized data.
10. Repeating step 6 but with structural shifts.
11. Repeating steps 7–9 for the newly fitted trend values.
12. Plotting the median trend with both the structural changes and bootstrapped values.
13. Repeating steps 7–9 for the newly fitted trend values.

The resulting estimates of dynamics and their components are used for modeling and forecasting. When studying the Russian regions' dynamics, regional models are built independently of each other, and general trends are revealed upon the modeling results.

The following methods are applied in the algorithm:

- LOESS smoothing to define seasonal coefficients as provided in the *stl* function in the *stats* package [12];
- The Breusch–Pagan test on heteroscedasticity to separate additive and multiplicative structures (using the *bptest* function in the *lmtest* package) [13];
- The probabilistic simulated annealing algorithm [14] for finding the global minimum area and initial estimates of model parameters;
- The *RPROP* algorithm [15,16], which is used to minimize errors in training neural networks;
- The minimization algorithm implemented in the standard *nlm* function [17];
- Wavelet transformation using the *wavelets* package;
- The ARIMA-models identification algorithm using the *forecast* package.

4. Results

The identification algorithm was applied for all analyzed time series. The results are shown in Figure 1.

The top part of the chart shows the original data (black points), fitted values (grey solid line), median trend fits (black dashed line) and the fits of all of the trends (dotted grey lines). The middle part of the chart demonstrates the median cycles model. The bottom part of the chart shows seasonal fluctuations. The titles of the middle and bottom of the chart appear as structures ('mult.' is abbreviation for multiplicative).

The example demonstrates a median-declining S-shaped trend, and the "cloud" of all the trends, depicting possible distributions of fits for estimates at each point. The cycles clearly show a decline in 2008–2009 (global crisis), 2014–2015 (economic sanctions against Russia) and 2020 (pandemic). Seasonality achieves its peak in December, and shows slow growth through the given period. This is one "typical" example of the dynamics, but for other regions and industries it varies drastically.

Our research goal was, however, not only to obtain the forecasts and models but also to measure their accuracy. To achieve this, we split the time series into two parts: one to identify the model (working sample) and the other to measure forecast accuracy (test sample). At the regional level, short-term and middle term forecasts appear as the most useful. So, we tested the models on 3, 6, 9 and 12-months forecasts. We also varied the forecasting year from 2018 to 2020 to generalize the conclusions, and thus, verify the models' overall accuracy.

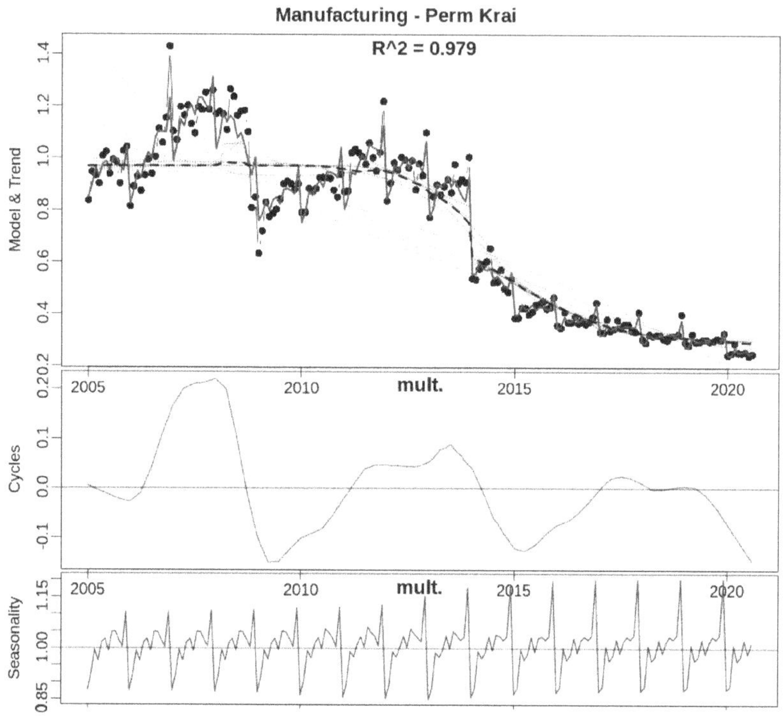

Figure 1. Modeling results for the manufacturing in Perm Krai.

This study uses two common measures of accuracy. The determination coefficient is used to measure the modeling accuracy:

$$R^2 = 1 - \frac{\sum(Y_t - \hat{Y}_t)^2}{\sum(Y_t - \overline{Y_t})^2} \tag{20}$$

The Theil's coefficient is used to measure the forecast error:

$$U_2 = \sqrt{\frac{\sum(Y_t - \hat{Y}_t)^2}{\sum Y_t + \sum \hat{Y}_t}} \times 100\% \tag{21}$$

For high accuracy, R^2 is supposed to be above 0.7 and U_2 below 30%.

Table 1 shows the median estimates of R^2 and U_2 among all regions, separated by industry. The industries are enumerated as mentioned in Section 2.

Judging by the table, the forecast accuracy is generally high. R^2 estimates are above 0.7, and U_2 are below 20%, for most industries except for the pharmaceutical, electronic and computer production, and automotive industries. These industries are highly volatile at the meso-economic level in Russia, especially the electronic and computer production industry, which is highly subsidized by the state and depends on government support. More stable industries such as mineral extraction, manufacturing, the chemical industry and metallurgy demonstrate low forecast errors. The predictability of the industries' futures could be assessed as their stability indicator.

Table 1. Median estimates for models and forecasts on test samples.

Year	2018					2019					2020				2021
Forecast Depth, Months	-	3	6	9	12	-	3	6	9	12	-	3	6	8	-
Industry No	R^2	U_2, %				R^2	U_2, %				R^2	U_2, %			R^2
1	0.888	4.7	6.4	8.4	10.5	0.908	3.9	6.6	9.0	10.3	0.910	5.0	9.1	11.5	0.894
2	0.966	2.3	4.1	4.9	6.1	0.973	1.3	2.4	3.2	3.5	0.976	2.2	6.7	7.8	0.956
3	0.897	4.4	5.1	6.6	7.4	0.911	3.6	5.2	8.3	9.3	0.915	6.2	7.6	7.5	0.917
4	0.881	4.3	5.2	7.0	7.7	0.883	4.0	5.8	6.9	9.8	0.891	4.8	9.2	11.6	0.892
5	0.891	3.9	4.8	6.0	7.1	0.893	3.5	4.8	5.9	6.8	0.896	4.5	5.9	7.5	0.905
6	0.772	3.1	5.9	7.9	8.6	0.797	3.9	8.6	9.3	9.4	0.798	4.2	10.1	10.2	0.791
7	0.851	4.6	6.9	9.5	10.6	0.856	5.0	7.0	10.4	12.1	0.862	7.2	11.3	11.7	0.853
8	0.712	12.3	17.9	18.5	18.0	0.723	8.8	19.0	21.1	25.9	0.696	17.3	28.8	29.8	0.725
9	0.916	6.8	9.1	9.7	11.7	0.916	5.4	8.2	11.1	11.9	0.916	6.5	11.3	13.0	0.914
10	0.846	8.5	10.7	12.0	13.4	0.851	6.5	9.9	12.0	14.8	0.858	8.8	12.6	14.2	0.868
11	0.650	14.0	21.1	26.6	33.7	0.581	12.2	20.2	26.8	34.7	0.582	23.2	36.0	38.6	0.626
12	0.864	10.4	16.0	22.8	26.3	0.859	9.4	15.0	17.8	23.6	0.877	13.4	26.0	28.5	0.873

5. Conclusions

The key findings of the research are as follows:

1. The approaches used to analyze and forecast regional industries' dynamics are justified. They include time series decomposition, the median approach, increases in the models' variety, using weighted additive–multiplicative criterion, and applying wavelet transformation for cycles.
2. The complex of models and algorithms is designed, upgraded and applied in the form of a program code in the R language.
3. The designed tools are applied to 12 industries in 82 Russian regions. Decompositions and forecasts are obtained for each time series. The median trend model shows general tendencies (growth, decline and bell) and structural change points. Cycle models define cycle stages and reversion points (peaks, troughs and zero-points). Seasonal models describe calendar effects and their changes through years.
4. The results' accuracy is proven by short-term and mid-term forecasts (3–12 months), even including the pandemic period.

This paper mostly demonstrates individual series analysis. However, more significant results may be achieved by comparing different regions, both between each other and with Russia as a whole. In our previous research, we showed that cycles and trends in the regions are not synchronous [18] but they may be clustered in terms of model type and the values of parameters.

We plan to continue to develop tools by increasing the trends' variety, using bootstrapping at all algorithm steps, improving calculation methods, adding interval forecasts, and analyzing regions' interactions and neighborhoods.

Author Contributions: Conceptualization, V.S. and A.K.; methodology, V.S. and A.K.; software, A.K.; validation, A.K.; formal analysis, A.K.; investigation, V.S. and A.K.; resources, V.S. and A.K.; data curation, A.K.; writing—original draft preparation, V.S. and A.K.; writing—review and editing, A.K.; visualization, A.K.; supervision, V.S.; project administration, V.S.; funding acquisition, V.S. All authors have read and agreed to the published version of the manuscript.

Funding: The reported study was funded by RFBR, project number 20-010-00549.

Institutional Review Board Statement: Not applicable.

Informed Consent Statement: Not applicable.

Data Availability Statement: Not applicable.

Conflicts of Interest: The authors declare no conflict of interest. The funders had no role in the design of the study; in the collection, analyses, or interpretation of data; in the writing of the manuscript, or in the decision to publish the results.

References

1. Semenychev, V.K.; Korobetskaya, A.A. Tools for Estimation of "Deterministic Chaos" of Economic Sectoral Mesodynamic. In *Economic Systems in the New Era: Stable Systems in an Unstable World. IES 2020*; Ashmarina, S.I., Horák, J., Vrbka, J., Šuleř, P., Eds.; Lecture Notes in Networks and Systems; Springer: Cham, Switzerland, 2021; Volume 160. [CrossRef]
2. Boschma, R. Towards an Evolutionary Perspective on Regional Resilience. *Reg. Stud.* **2015**, *49*, 733–751. [CrossRef]
3. Simmiea, J.; Martin, R. The economic resilience of regions: Towards an evolutionary approach. *Camb. J. Reg. Econ. Soc.* **2010**, *3*, 27–43. [CrossRef]
4. Beraja, M.; Hurst, E.; Ospina, J. The Aggregate Implications of Regional Business Cycles. *Econometrica* **2019**, *87*, 1789–1833. [CrossRef]
5. Perron, P. Structural change, econometrics of. In *Macroeconometrics and Time Series Analysis. The New Palgrave Economics Collection*; Durlauf, S.N., Blume, L.E., Eds.; Palgrave Macmillan: London, UK, 2010. [CrossRef]
6. Cavaliere, G.; Taylor, A. Bootstrap Unit Root Tests for Time Series with Nonstationary Volatility. *Econom. Theory* **2008**, *24*, 43–71. [CrossRef]
7. Tyrsin, A.N. Robust construction of regression models based on the generalized least absolute deviations method. *J. Math. Sci.* **2006**, *139*, 6634–6642. [CrossRef]
8. Slutsky, E.E. Slozhenie sluchajnyh prichin kak istochnik ciklicheskih processov. *Voprosy kon"yunktury* [Addition of random causes as a source of cyclic processes. *Market issues.*] **1927**, *3*, 34–64. (In Russian)
9. Morettin, P.A. Wavelets in Statistics. *Rev. Inst. Math. Stat. Univ. Sao Paulo* **1997**, *3*, 211–272.
10. Percival, D.B.; Walden, A.T. *Wavelet Methods for Time Series Analysis*; Cambridge University Press: London, UK, 2000.
11. Raihan, S.M.; Wen, Y.; Zeng, B. Joint Time-Frequency Distributions for Business Cycle Analysis. In *Wavelet Analysis and Its Applications. WAA 2001*; Tang, Y.Y., Yuen, P.C., Li, C., Wickerhauser, V., Eds.; Lecture Notes in Computer Science; Springer: Berlin, Germany, 2001; Volume 2251. [CrossRef]
12. Cleveland, R.B.; Cleveland, W.S.; McRae, J.E.; Terpenning, I. STL: A Seasonal-Trend Decomposition Procedure Based on Loess. *J. Off. Stat.* **1990**, *6*, 3–73.
13. Breusch, T.S.; Pagan, A.R. A Simple Test for Heteroscedasticity and Random Coefficient Variation. *Econometrica* **1979**, *47*, 1287–1294. [CrossRef]
14. Xiang, Y.; Gubian, S.; Suomela, B.; Hoeng, J. Generalized Simulated Annealing for Efficient Global Optimization: The GenSA Package for R. *R J.* **2013**, *5*, 13–28. [CrossRef]
15. Igel, C.; Huesken, M. Empirical evaluation of the improved Rprop learning algorithms. *Neurocomputing* **2003**, *50*, 105–123. [CrossRef]
16. Riedmiller, M. Advanced supervised learning in multilayer perceptrons—From backpropagation to adaptive learning techniques. *Comput. Stand. Interfaces* **1994**, *16*, 265–278. [CrossRef]
17. Dennis, J.E.; Schnabel, R.B. *Numerical Methods for Unconstrained Optimization and Nonlinear Equations*; Prentice-Hall: Englewood Cliffs, NJ, USA, 1983.
18. Khmeleva, G.A.; Semenychev, V.K.; Korobetskaya, A.A.; Kozhukhova, V.N.; Agaeva, L.K.; Burets, Y.S.; Egorova, K.S.; Zemtsov, S.P.; Koroleva, E.N.; Chertopyatov, D.A. *Rossiyskie Regiony v Usloviyakh Sanktsiy: Vozmozhnosti Operezhayushchego Razvitiya Ekonomiki na Osnove Innovatsiy [Russian Regions in the Conditions of Sanctions: The Possibility of Priority Development of the Economy Based on Innovation]*; Khmeleva, G.A., Ed.; Pub. Sam. State Univ.: Samara, Russia, 2019. (In Russian)

engineering proceedings

MDPI

Proceedings

An Advanced Markov Switching Approach for the Modelling of Consultation Rate Data †

Emmanouil-Nektarios Kalligeris [1,‡], Alex Karagrigoriou [1,‡] and Christina Parpoula [2,*]

1 Lab of Statistics and Data Analysis, Department of Statistics and Actuarial-Financial Mathematics, University of the Aegean, 83200 Samos, Greece; ekalligeris@aegean.gr (E.-N.K.); alex.karagrigoriou@aegean.gr (A.K.)
2 Department of Psychology, Panteion University of Social and Political Sciences, 17671 Athens, Greece
* Correspondence: chparpoula@panteion.gr
† Presented at the 7th International conference on Time Series and Forecasting, Gran Canaria, Spain, 19–21 July 2021.
‡ These authors contributed equally to this work.

Abstract: Regime switching in conjunction with penalized likelihood techniques could be a robust tool concerning the modelling of dynamic behaviours of consultation rate data. To that end, in this work we propose a methodology that combines the aforementioned techniques, and its performance and capabilities are tested through a real application.

Keywords: elastic net; dynamic system; markov switching; penalized likelihood; regimes; regularization methods; two-state modelling; variable selection

Citation: Kalligeris, E.-N.; Karagrigoriou, A.; Parpoula, C. An Advanced Markov Switching Approach for the Modelling of Consultation Rate Data. *Eng. Proc.* **2021**, *5*, 2. https://doi.org/10.3390/engproc2021005002

Published: 24 June 2021

1. Introduction

Dynamic behaviours in time events are always quite complex, and their modelling is often a challenging task. The level of difficulty is accelerated in cases where the dynamics of a system cannot be satisfactorily described by linear models, but more perplexed non-linear functions are required. Classical time series approaches are not capable of capturing complex functional behaviours. Even advanced models recently proposed are not flexible enough and as a result are not easily adjusted to handle more general non-linear schemes [1–4].

It is also frequently observed that the behaviour changes its general pattern in different regions of the time space. Such changes may affect either the mean or the variation or both. A breakthrough in this field took place 40 years ago with the proposal of Markov regime models and the switching regressions [5]. Such models allow a great degree of flexibility, and as a result they could be implemented to capture complex dynamic behaviours. The unobserved state variable associated with such models is an attractive feature directly related to the switching mechanism of the underlying modelling approach. The resulting advanced models rely on the Markovian property, which is an easily handled issue in terms of inferential statistics. Note finally that censoring [6] or semi-Markov approaches [7] may also be considered in such frameworks.

In this work, within the switching framework, we introduce the classical likelihood combined with a penalty term controlled by a properly chosen tuning parameter. In other words, the switching modelling technique is combined with the so-called penalized likelihood with the parameter estimation being dealt via the Expectation-Maximization algorithm [8]. Depending on the phenomenon under investigation, a proper switching model can be used. Two states often suffice to describe the classical dynamic behaviour of incidence data or epidemics, with one state representing the normal stage of the phenomenon and the other the outbreak stage. In such a case, the frequency (usually of daily or weekly data) changes (increases) considerably (and in some cases dramatically) when the system enters into the second stage. Such a change is considered statistically significant, and therefore the unobserved state variable ignites the switch. In addition, possible covariates may affect the variable of interest, which is denoted by y_t and represents either the frequency or the associated rate.

2. The Modelling

The model used in this work is the 2-state switching model of conditional mean, the general form of which is given by

$$x_t = \mu_{s_t} + \sum_{i=1}^{p} \phi_{is_t} x_{t-i} + \epsilon_t, \tag{1}$$

where μ_{s_t} is a switching intercept, ϕ_{is_t}, $i = 1, ..., p$, are autoregressive (AR) switching coefficients, s_t represents the state variable that takes the values 1 (normal or typical state), and 2 (the extreme or outbreak state) and ϵ_t are *i.i.d.* random variables with zero mean and variance σ_ϵ^2.

If k covariates are allowed to enter into the model, (1) extends to

$$x_t = \mu_{s_t} + \sum_{i=1}^{p} \phi_{is_t} x_{t-i} + \sum_{j=1}^{k} \theta_{js_t} W_j + \epsilon_t, \tag{2}$$

where θ_{js_t} the coefficient associated with the W_j covariate.

As i is clear from the presentation of the above model, a different set of parameters is involved for each state considered. It is important to state that the set of covariates involved in each state may or may not be the same.

3. The Algorithmic Procedure

The approach we choose to follow for modelling phenomena that exhibit a dynamic switching behaviour consists of three steps, which are briefly discussed in this section.

3.1. Step 1-The Change Point Detection

The detection of a change point in a time series and in general in events over time constitutes an integral part of time series analysis, since their identification is directly related to a distributional change. Such changes, even light ones, should cause alarm due to the fact that they may alter the data generating process in such a way that the process under investigation may fail to fulfill the purpose for which it is intended. Applications can be found in most scientific fields from finance and business to engineering, biosciences, climatology, geosciences etc. [9–11].

The proposed methodology requires a preliminary analysis to identify a set of possible change points. It should be noted that such analysis involves only the response variable, and no covariates are involved. The method to be used may be the classical method of change-point identification [12].

For the change point detection, an offline algorithm is used to examine the entire set of observations in a single step to recognize where the change occurred. The online approach could be chosen instead, as long as a certain number of new data are available for the algorithm to function properly and satisfactorily.

To check the performance of the selected change points, we have used the mean absolute error (MSE) according to which predicted and actual values are compared. The general expression is given by

$$MSE = \frac{1}{T} \sum_{i=1}^{T} (\hat{x}_t - x_t)^2.$$

3.2. Step 2-The Variable Identification

For the identification of the statistically significant covariates, a kind of model selection technique can be implemented. In this work, we propose the use of computationally advanced regularization methods, such as Lasso, Ridge or Elastic-Net with the latter considered to be a generalization of the former ones that overcomes their disadvantages. For the interested reader, a number of articles investigate the interrelation of time series and regularization techniques [13–15].

The generalized regularization method used in this work is given by

$$SSE + 2T\lambda\left[\alpha\left(\sum_{i=1}^{p}\sum_{j=1}^{k}(|\phi_{is_t}| + |\theta_{js_t}|)\right)\right.$$
$$\left. + \left(\frac{1-\alpha}{2}\right)\left(\sum_{i=1}^{p}\sum_{j=1}^{k}(\phi_{is_t}^2 + \theta_{js_t}^2)\right)\right], \quad (3)$$

where SSE is the sum of squared errors or any other loss function chosen by the researcher, T the sample size, $\alpha \in [0,1]$, and λ the tuning parameters that result the penalty in the loss function. Note that α balances the amount of emphasis given to minimize the loss function versus minimizing the sum of squared coefficients and/or the sum of absolute coefficients.

Observe that the above generalized regularization method is reduced to

- The Lasso method for $a = 1$;
- The Ridge method for $a = 0$; and
- To Elastic-Net for $a \in (0,1)$.

Note that a proper weighted version of (3) can be used if it is needed, for instance, to resolve a heteroscedasticity issue. In such a case, (3) takes the general form

$$SSE + 2T\lambda\left[\alpha\left(\sum_{i=1}^{p}\sum_{j=1}^{k}(|w_i^{\phi}\phi_{is_t}| + w_j^{\theta}|\theta_{js_t}|)\right)\right.$$
$$\left. + \left(\frac{1-\alpha}{2}\right)\left(\sum_{i=1}^{p}\sum_{j=1}^{k}(w_i^{\phi}\phi_{is_t}^2 + w_j^{\theta}\theta_{js_t}^2)\right)\right], \quad (4)$$

where w_i^{ϕ} and w_j^{θ} appropriate weights, $i = 1, \ldots, p$ and $j = 1, \ldots, k$.

3.3. Step 3—The Switching

The selected model for each state is obtained together with the parameter estimates and the associated standard errors.

Note that in practice, we do not know and we do not observe the state s_t, but we could infer it from the observed data. Indeed, although the state variable s_t is an unobserved variable, the process y_t is observed. To make an inference about s_t, we need to make an assumption about the process s_t, which usually is assumed to follow a first order Markov chain. For the 2-state case, the probabilities of transition are also obtained. Thus, the transition probability to state j at time t, given that the process was in state i at the time point $t-1$, is given by $P(s_t = j|s_{t-1} = i) = p_{ij}$, $i, j = 1, 2$.

4. An Application on Epidemiology

Using a data set of 105 weekly influenza-like-illness (ILI) consultation rate data for Greece for the period 2014–2016, including a series of meteorological and climatological covariates such as temperature and wind, we were able to identify an ideal tuning parameter λ and the best value of the index α in the sense that the minimum mean squared error is achieved. Figure 1 shows that the ideal value of λ is around one (1), while the best value of

α is between 0.5 and 1. A more detailed analysis reveals that $\lambda = 1$ and $\alpha = 0.729$, with the corresponding value of MSE being equal to 0.0734.

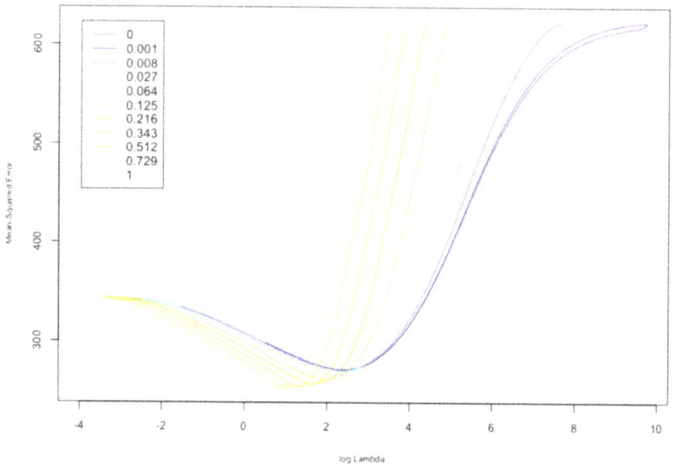

Figure 1. Behaviour for various values of a as opposed to different values of $\log \lambda$ and MSE.

Under this setting, the regimes of the data and consequently the form of the selected models are identified together with the estimates of the parameters involved.

The models obtained with the use of the MSwM R-package [16] are as follows (with three decimal points):

Regime 1-typical period/state

$$\begin{aligned}\hat{y}_t = {} & 146.480 - 0.375t - 37.488sin\left(\frac{2\pi t}{n}\right) - 2.682cos\left(\frac{2\pi t}{n}\right) \\ & - 44.002sin\left(\frac{4\pi t}{n}\right) - 23.945cos\left(\frac{4\pi t}{n}\right) + 8.330sin\left(\frac{8\pi t}{n}\right) - 14.244cos\left(\frac{8\pi t}{n}\right) \\ & + 0.035\mathrm{T1} + 15.326\mathrm{T2} - 11.081\mathrm{T3} + 10.042\mathrm{WF} - 0.595\hat{y}_{t-1}.\end{aligned} \tag{5}$$

Regime 2-Outbreak period/state

$$\begin{aligned}\hat{y}_t = {} & 19.552 - 0.005t + 18.933sin\left(\frac{2\pi t}{n}\right) - 10.652cos\left(\frac{2\pi t}{n}\right) \\ & - 5.119sin\left(\frac{4\pi t}{n}\right) - 2.244cos\left(\frac{4\pi t}{n}\right) + 0.971sin\left(\frac{8\pi t}{n}\right) - 0.313cos\left(\frac{8\pi t}{n}\right) \\ & - 0.236\mathrm{T1} + 4.006\mathrm{T2} - 4.139\mathrm{T3} + 1.657\mathrm{WF} + 0.697\hat{y}_{t-1}.\end{aligned} \tag{6}$$

It is worth mentioning that the same set of covariances are found to be significant for both regimes together with a first-order autoregressive, a first degree trend polynomial (linear trend) and a periodic (seasonal) part. The covariates chosen are the minimum, mean, and median temperature, denoted, respectively, by $T1$, $T2$, and $T3$ and the mean of the wind force denoted by WF, implying that the influenza is closely connected to meteorological/climatological factors like the temperature and the wind.

References

1. Kalligeris E.N.; Karagrigoriou, A.; Parpoula, C. On Mixed PARMA Modeling of Epidemiological Time Series Data. *Commun. Stat. Case Stud. Data Anal.* **2019**, *6*, 36–49. [CrossRef]
2. Pelat, C., Boëlle, P.Y.; Cowling, B.J.; Carrat, F.; Flahault, A.; Ansart, S.; Valleron, A.J. Online Detection and Quantification of Epidemics. *BMC Med. Inform. Decis. Mak.* **2007**, *5*, 29.
3. Tong, H. Nonlinear Time Series Analysis Since 1990: Some Personal Reflections. *Acta Math. Appl. Sin.* **2002**, *18*, 177. [CrossRef]
4. Kalligeris, E.N.; Karagrigoriou, A.; Parpoula, C. Periodic-Type Auto-Regressive Moving Average Modeling with Covariates for Time-Series Incidence Data via Changepoint Detection. *Stat. Meth. Med. Res.* **2020**, *29*, 1639–1649. [CrossRef]
5. Lindgren, G. Markov Regime Models for Mixed Distributions and Switching Regressions. *Scand. J. Stat.* **1978**, *5*, 81–91.
6. Huber, C. Efficient Regression Estimation Under General Censoring and Truncation. In *Mathematical and Statistical Models and Methods in Reliability*; Statistics for Industry and Technology; Birkhäuser: Boston, MA, USA, 2010; Volume 12, pp. 235–241.
7. Barbu, V.S.; Karagrigoriou, A.; Makrides, A. Semi Markov Modelling for Multi State Systems. *Methodol. Comput. Appl. Prolab.* **2017**, *19*, 1011–1028. [CrossRef]
8. Green, P.J. On Use of the EM Algorithm for Penalized Likelihood Estimation. *J. R. Stat. Soc. Ser. B Stat. Methodol.* **1990**, *52*, 443–452. [CrossRef]
9. Karagrigoriou, A.; Makrides, A.; Tsapanos, T.; Vougiouka, G. Earthquake Forecasting Based on Multi State System Methodology. *Methodol. Comput. Appl. Probab.* **2016**, *18*, 547–561. [CrossRef]
10. Votsi, I.; Limnios, N.; Tsaklidis, G.; Papadimitriou, E. Hidden Markov Models Revealing the Stress Field Underlying the Earthquake Generation. *Phys. A* **2013**, *392*, 2868–2885. [CrossRef]
11. Shaby, B.A.; Reich, B.; Cooley, D.; Kaufman, C.G. A Markov Switching Model for Heat Waves. *Ann. Appl. Stat.* **2016**, *10*, 74–93. [CrossRef]
12. Page, E.S.: Continuous Inspection Schemes. *Biometrika* **1954**, *41*, 100–115. [CrossRef]
13. Nardi, Y.; Rinaldo, A. Autoregressive Process Modeling via the Lasso Procedure. *J. Multivar. Anal.* **2011**, *102*, 528–549. [CrossRef]
14. Chen, K.; Chan, K.S. Subset ARMA Selection via the Adaptive Lasso. *Stat. Interface.* **2011**, *4*, 197–205. [CrossRef]
15. Medeiros, C.M.; Eduardo, M. L1-Regularization of High-Dimensional Time-Series Models with Non–Gaussian and Heteroskedastic Errors. *J. Econom.* **2016**, *191*, 255–271. [CrossRef]
16. Sanchez-Espigares, J.A.; Lopez-Moreno, A. MSwM: Fitting Markov Switching Models. CRAN 2018. R Package Version 14. Available online: https://CRANR-projectorg/package=MSwM (accessed on 25 May 2021).

engineering proceedings

MDPI

Abstract

Cycles and Uncertainty: Applications in the Tourist Accommodation Market †

Miguel Ángel Ruiz Reina

Department of Theory and Economic History (Staff of Fundamentals), PhD Program in Economics and Business, s/n, University of Málaga, Plaza del Ejido, 29013 Málaga, Spain; ruizreina@uma.es

† Presented at the 7th International Conference on Time Series and Forecasting, Gran Canaria, Spain, 19–21 July 2021.

Abstract: In the socio-economic field, it is not surprising that decision-making is based on asymmetric information. Economic agents make decisions to forecast in primary and secondary industries related to the tourism sector. This study aims to provide knowledge in situations of asymmetric information with increasing randomness using time series for tourism accommodation markets. We are trying to solve the question of how consumers exchange their preferences for tourist accommodation between tourist apartments and hotel accommodation in Spain. The emergence of the sharing economy concept has emerged as a competitor to the traditional hotel accommodation in the tourist market. To do this, we will develop a theoretical framework to measure situations of uncertainty and their temporal evolution. Information Theory (IT) is the central axis of the study, particularly the concept of entropy. The Shannon entropy (SE) concept is a static measure of information. This work proposes to model the temporal arrangement of SE to discover the behaviors of the systems. The study in the domain of time and frequency allows us to understand the cycles of uncertainty between systems. To apply the theoretical framework, we will work with data from official Spanish sources for tourist accommodation from January 2008 to December 2019. The results of the empirical analysis show the decision changes of economic agents according to a seasonal pattern. Consumers have new accommodation options, and the answer we get from this work is that consumers have different preferences depending on seasonality. The use of SE allows us to make better predictions compared to SARIMA models, the traditional modelling of seasonal dummy variables, and VAR models. The results of the Matrix U1 Theil verify this hypothesis. The theoretical framework and empirical analysis find an answer to asymmetric information. The implications of this work contribute to the field of social sciences related to the tourism sector, in particular to thermodynamics, statistical mechanics, and IT. The modelling of uncertainty allows for the forecasting and control of accommodation tourist markets in random situations. The applications of this study can be tested in other areas of the economy such as finance, transportation, or investment.

Keywords: randomness; forecasting; Information Theory

Citation: Ruiz Reina, M.Á. Cycles and Uncertainty: Applications in the Tourist Accommodation Market . *Eng. Proc.* **2021**, *5*, 3. https://doi.org/10.3390/engproc2021005003

Published: 24 June 2021

Funding: This research received no external funding.

Institutional Review Board Statement: Not applicable.

Informed Consent Statement: Not applicable.

Data Availability Statement: The datasets generated during and analysed during the current study are available in the repository (https://drive.google.com/file/d/1TP63q-1IFW0_oRJHXBDBFJxMPANOTc9b/view?usp=sharing, accessed on 23 June 2021) [1–7].

Acknowledgments: The author wishes to acknowledge the support given by the University of Malaga. PhD. Program in Economics and Business, effective from 16 July 2013. Especially to Associate Professor Antonio Caparrós Ruiz from the University of Malaga for reviewing this work. Group of research: "SEJ157-INIDICADORES SOCIALES".

Conflicts of Interest: The authors declare no conflict of interest.

References

1. Bakker, M.; Twining-Ward, L. *Tourism and the Sharing Economy: Policy and Potential of Sustainable Peer-to-Peer Accommodation*; World Bank: Washington, DC, USA, 2018.
2. Delgado-Bonal, A.; Marshak, A. Approximate entropy and sample entropy: A comprehensive tutorial. *Entropy* **2019**, 541. [CrossRef] [PubMed]
3. Ruiz-Reina, M.Á. Entropy of Tourism: The unseen side of tourism accommodation. In Proceedings of the International Conference on Applied Research in Business, Management and Economics, Barcelona, Spain, 12–14 December 2019; Available online: https://www.dpublication.com/wp-content/uploads/2019/12/424.pdf (accessed on 23 June 2021).
4. Harvey, A. Chapter 7 Forecasting with Unobserved Components Time Series Models. *Handb. Econ. Forecast.* **2006**. [CrossRef]
5. Box, G.E.P.; Jenkins, G.M.; Reinsel, G.C. *Time Series Analysis: Forecasting and Control*, 4th ed.; John Wiley & Sons: Hoboken, NJ, USA, 2013.
6. Lütkepohl, H. *New Introduction to Multiple Time Series Analysis*; Springer Science & Business Media: Berlin/Heidelberg, Germany, 2005.
7. Reina, M.Á.R. Big Data: Forecasting and Control for Tourism Demand. In *Theory and Applications of Time Series Analysis. ITISE 2019*; Valenzuela, R.I.O., Rojas, F., Herrera, L.J., Pomares, H., Eds.; Springer: Cham, Switzerland, 2020; pp. 273–286.

engineering proceedings

MDPI

Proceedings

Airbnb Host Scaling, Seasonal Patterns, and Competition †

Ruggero Sainaghi [1,*] and Rodolfo Baggio [2,3]

1 Department of Business, Law, Economics, and Consumer Behavior, IULM University, 20147 Milan, Italy
2 Master in Economics and Tourism, Bocconi University, 20136 Milan, Italy; rodolfo.baggio@unibocconi.it
3 National Research Tomsk Polytechnic University, 634050 Tomsk, Russia
* Correspondence: ruggero.sainaghi@iulm.it

† Presented at the 7th International conference on Time Series and Forecasting, Gran Canaria, Spain, 19–21 July 2021.

Abstract: This paper explores the scaling (size) effect in the seasonal patterns, a proxy for competitive threats, of Airbnb's host providers, with the aim of understanding possible similarities and differences. This explorative study uses the city of Milan (Italy) as a case and daily occupancy data from Airbnb listings for four completed years (2015–2018). A mutual information-based technique was applied to assess possible synchronizations in the seasonal patterns. Empirical findings show progressive dissimilarities when moving from single to multiple listings, thus indicating a differentiation correlated to the presence of managed listings. There are fewer differences during the seasonal periods more centered around leisure clients and they are higher when considering business travelers. The evidence supports the scaling effect and its ability to reduce the competitive threat among different hosts.

Keywords: host scaling; seasonal patterns; competition; synchronization; Airbnb; Milan

check for updates

Citation: Sainaghi, R.; Baggio, R. Airbnb Host Scaling, Seasonal Patterns, and Competition. *Eng. Proc.* **2021**, *5*, 4. https://doi.org/10.3390/engproc2021005004

Published: 24 June 2021

Publisher's Note: MDPI stays neutral with regard to jurisdictional claims in published maps and institutional affiliations.

1. Introduction

This paper explores the scaling (size) effect, focusing on Airbnb's host providers, with the aim of understanding the similarities and differences in seasonal patterns. Since the launch of this commercial peer-to-peer accommodation platform in 2007 [1], Airbnb has attracted academic debate, especially in the last few years [2]. Airbnb is a web platform that rents idle assets, called listings (typically rooms, apartments, and houses) that are owned by hosts, to travelers or guests [3].

There are few studies investigating the supply side (host) [3]. Previous papers centered on listing performance focused on the scaling effect [4]. In fact, many studies have distinguished between the host managing only one listing (usually called "mom-and-pop" hosts or simply single-listing hosts) or more than one (usually defined as "professional", "commercial," or multi-listing hosts) [5]. Generally speaking, the two types of host (single versus multiple) depict different results, as discussed in more detail in the next section.

However, knowledge about the managerial differences among these two groups is very limited, despite the ability of the scaling effect to deeply change the hosts' business model [6]. Furthermore, the large majority of these studies simply distinguish between single and multi-listing hosts, without any additional segmentation. Scaling, as usual in managerial studies [7], in this paper refers to the number of listings managed by one host. The higher the number of listings, the higher the scaling effect and the opposite. To contribute to reducing the current gap in the commercial peer-to-peer accommodation platform literature, the present article explores the ability of scaling to change the seasonal patterns to measure the degree of similarity and differences among Airbnb hosts. These similarities and differences are used as a proxy for the competition threat among different (in size) Airbnb listings (as later discussed in detail in the Methodology). According to Butler, the definition of seasonality is "a temporal imbalance in the phenomenon of tourism, (which) may be expressed in terms of dimensions of such elements as numbers of visitors, expenditure of visitors, traffic on highways and other forms of transportation, employment, and admissions to attractions" [8] (p. 332). As clarified in the methodology, the seasonality is measured considering the daily data and developing a longitudinal (four-year) approach. The higher the similarity of seasonal patterns, the higher the potential competition among the Airbnb hosts; the opposite in the case of different seasonal patterns [9]. The current literature has developed studies exploring the potential disruptive innovation generated by peer-to-peer accommodation platforms on hotels [10], but any study has explored the competition among Airbnb listings and, in particular, the role of scaling. Therefore, this study's research question focuses on seasonal patterns and the competition threat among Airbnb hosts.

Research questions: can the scaling effect change the seasonal patterns of Airbnb hosts? Does the scaling effect reduce or increase the competition among Airbnb hosts?

2. Literature Review

This section is structured in three parts. The first section analyzes the results suggested in previous studies about Airbnb hosts focusing on the scaling effect. The second section explores the seasonal patterns of Milan. The third section (based on the previous two) formulates the hypotheses tested in the empirical findings.

2.1. Host Scaling Effect

Peer-to-peer accommodation platform literature, despite the rising number of contributions [11], is in its infancy, and many research areas are less investigated [12]. One of them is the qualitative description of the host business model and the scaling effect [13]. For this reason, this paper has analyzed some adjacent but different supply research streams and in particular the impact studies on one hand and the determinants of listings results and pricing strategies on the other.

The impact literature is centered on the effect of Airbnb on hotels [10], tourism destinations [14], and local stakeholders [15], with prevailing attention on housing and long-term rentals [16]. Although the effects on hotels are contradictory [17], the social transformation generated by commercial peer-to-peer accommodation platforms is usually described as relevant. For this reason, the impact studies include a growing area of inquiry exploring the regulation of peer-to-peer accommodation platforms [18]. Despite the importance of the impact research, the focus is usually on the whole effect of the hosts; therefore, the topic of this article (the host scaling effect) is not developed.

A second supply-side research stream has analyzed the determinants of listings results and the pricing strategies [19]. As anticipated in the introduction, this second area of inquiry usually considers the host size as an independent variable that can influence, respectively, the listing results or the pricing strategies. These two distinct sub-topics (performance and pricing) are now discussed. In both groups, many papers (as later presented) distinguish between single- and multiple-listing hosts, also called commercial or professional hosts. The latter (multiple) includes the hosts managing two or more listings.

The determinants of results represent a small research stream that explores the determinants or antecedents of listing performance [20,21], usually operationalized using review volume or rating [22] or, more rarely, occupancy [20]. It is quite a small area of inquiry and is separated in this article from the second, wider, research stream that focuses on pricing strategy. Some studies have considered the number of listings managed by a host as a relevant independent variable. Xie and Mao have found a trade-off between host quality and the quantity of her/his listings. In particular, "as the number of listings managed by a host increases, the performance effects of host quality diminish" [21] (p. 2240). Gunter has investigated the conditions improving the likelihood of obtaining the superhost badge [22]. The author has found four variables, one of them being the status of "commercial" Airbnb host.

Moving to the second sub-topic (pricing strategy), many previous studies have considered the distinction between single-unit and multi-unit hosts [23]. Multi-unit hosts, generally speaking, are described as being more proficient in using a dynamic pricing strategy [24] and in achieving a higher price or revenue (variously operationalized) than a single-unit host [25]. In the theoretical model created by Chen, Zhang and Liu [26], the adoption of a flexible pricing strategy leads to higher performance in a large market, but the accommodation quality is not better. The study of Gibbs et al., reveals that the host's experience positively influences the adoption of dynamic pricing [27]. Realistically, a multi-listing host has more opportunity to improve her/his experience than a single unit host because the host manages a higher number of transactions [28] and, for this reason, they have more skills to manage also new start-ups [29]. In another study, professional hosts achieve a higher price per night. Similarly, professional hosts receive higher rates in rural Switzerland, intermediate (between rural and urban areas), and in cities [30]. These findings are confirmed both using a random and a quantile estimation model [31]. Other authors demonstrate a negative correlation between multi-listing and price but a positive correlation between professional hosts and revenue per month [32].

However, there are some exceptions. For example, [24] focused on five metropolitan areas in Canada. Professional hosts show a positive and significant coefficient with the dependent variable (price), considering all the cases, but the coefficient is negative and insignificant in the case of Calgary. Similarly, in the study of [33], two cities show negative and significant correlations with price, while Madrid shows a positive (but not significant) coefficient. In Hong Kong, multi-listing hosts book at lower price their capacity [34].

These contradictory results can be explained using at least six different arguments. First, multi-listing hosting reduces social interaction with the guests, which is called reciprocity [35], and this can generate a drop in price [36]. Second, the correlation coefficient that tied the multi-listing and rates together is usually small and can therefore suddenly change from being slightly positive to slightly negative [33]. Third, the studies employ different frameworks (i.e., hedonic models, regression, quantile analysis, and artificial neural networks), which can generate different results [37]. Fourth, the relationship can change [38] in different destination contexts [39], also considering the diverse destination positioning and governance [40]. Fifth, the studies use a diverse set of control variables that can influence the relationship between host size and price [31]. Finally, different studies use diverse dependent variables, and sometimes the relationships change [32].

However, as previously stated, the large majority of studies reveal a positive correlation between multi-listings and rates. Curiously, the vast majority of the analyzed studies, with the exception of [32], have operationalized the scaling effect only by distinguishing between single and multiple hosts, without any additional segmentation. In other words, a host managing two listings is considered similar to a host renting 50 or more listings.

2.2. Milan Seasonal Patterns

This section explores the Milan seasonal patterns. The city is the economic capital of Italy and previous studies have identified three main market segments attracted by Milan: (i) business, (ii) trade-fair, and (iii) leisure [41]. Each segment is characterized by a clear seasonality [42]. During weekdays, the business target is prevalent, while during weekends the leisure is the main market segment [43]. Some previous articles introduced the distinction between "working days" and holiday (or "non-working days") [44]. The first group (working days) includes, in line with the study of [45], the weekdays not affected by religious (such as Christmas and Easter) or civil holidays (as Republic Day or Labor Day). The opposite is for holidays, that include the weekends and all the religious and civil holiday periods. During holidays, the leisure clients are prevalent, while the business is the core target of working days [46]. Finally, Milan is a leading European city for exhibitions. When the local trade-fair center (Fiera Milano) organizes some top events, the hotels achieve top performance in both occupancy and revenue. For this reason, this study included these top events that are mainly business-to-business exhibitions able to attract a large international audience.

2.3. Hypotheses Development

In this section, some previous insights related to both the number of listings managed by a host and Milan's seasonal patterns are considered as formulating two different groups of hypotheses, which guide the empirical analysis.

The first group focuses on the scaling effect and explores the degree of synchronization (the similarity) of the five groups of hosts. The precise meaning of synchronization is described in the methodology section. However, in order to perceive the meaning of the proposed hypotheses, a qualitative explanation is anticipated. The synchronization (as the name itself suggests) evaluates the similarities (differences) in time series [47]. This paper explores whether the scaling effect is able or not to change the synchronization degree among different (in size) groups of hosts. Put differently, do small and big hosts show similar series or does the scale differentiate them?

The hosts are segmented into five groups. As discussed (Section 2.1), previous studies usually distinguish only between single and multiple listings. However, some recent studies adopted a more fine-grained classification for multiple listings [48]. In line with these last papers, the current article distinguishes between: (i) a mom-and-pop host (single listing); (ii) a host renting two listings; (iii) a host selling three listings; (iv) a host managing four to 10 listings; (v) a host renting more than 10 listings. The five groups represent three different scaling effects. Logically, a host managing one to three listings can organize her/his business without (or by limiting) the employment of external workers. Four is assumed as the threshold for moving from a personal to a more professional business model, where professional means the involvement of external collaborators [48]. Finally, as suggested in another study [32], more than 10 can represent new, important scaling, which can favor more specialization and professionalization in the main business functions (selling, housekeeping, customer relationship management, and information technology). In this paper, the scaling effect is considered and can, therefore, improve the host's knowledge and managerial skills. For this reason, the following two hypotheses are formulated.

Hypothesis 1. *A rise in the number of listings progressively reduces the synchronization degree among the five groups.*

Hypothesis 2. *A rise in the number of listings progressively reduces the synchronization degree between each group and the overall (sample) mean.*

The second set of hypotheses focuses on Milan's seasonal patterns. As previously explained, Milan shows a strong demand fluctuation between the holidays, the weekends, and the days without trade-fair events compared with working days, midweek and days with trade-fair events [9]. Many previous studies agree that Airbnb listings are mainly specialized and categorized as leisure [49]. Therefore, Airbnb listings are expected to be more proficient when leisure clients are more relevant (holidays and weekends) as well as when the city hosts trade-fair events (many trade-fair guests combine business and leisure). In another words, when the key target of Airbnb is prevalent (leisure) the differences among the five groups of hosts (based on their size) are less nuanced. By contrast, when the key target of the city is business, reasonably smaller hosts are less skilled to serve this target and, therefore, show different seasonal patterns (and therefore less synchronization degree) than bigger (scaled) hosts. Therefore, the following three hypotheses are proposed.

Hypothesis 3. *The synchronization degree between the five groups and the total is higher during: 3A—the holiday period than the working period; 3B—the weekend period than the midweek period; and 3C—the trade-fair period than the non-trade-fair period.*

The scaling effect should progressively increase the multi-hosts' ability to serve the three main Milan segments (leisure, business, and trade-fair guests) differently. In fact, these diverse targets have different needs, seasonal patterns, performance metrics and require diverse host's skills and services. Therefore, in the six seasonal periods (holiday and working; weekend and midweek; trade-fair and non-trade-fair events), the scaling effect is expected to show progressive desynchronization patterns compared with the five groups. The following six hypotheses are formulated.

Hypothesis 4. *The synchronization degree among the five groups progressively reduces during: 4A—the holiday period; 4B—the working period; 4C—the weekend period; 4D—the midweek period; 4E—the trade-fair period; and 4F—the non-trade-fair period.*

3. Methodology

3.1. The Sample

This study has chosen the city of Milan due to its prevalent focus on business and trade-fair clients on one side but in association with its non-marginal presence of leisure travelers on the other. Previous papers that explore the effects of Airbnbs in Europe are mainly focused on large leisure cities, such as Barcelona [50] and Venice [51], or mixed leisure and business destinations, such as Madrid [52], Paris [53], London [54], and Berlin [55].

To explore the scaling and seasonal patterns of Airbnb listings, AirDNA data were used by the research team, which cover the period of 2014 (from November) to June 2019. Therefore, the data include four completed years (2015–2018) and support a longitudinal analysis, in line with some previous studies [56]. Many previous papers have used AirDNA data [4]. To test the hypotheses, daily data were used, as in other similar studies [57]. AirDNA considers the available and sold listings as well as the price for each day and for each listing. The sample includes all Milan's population represented by more than 50,000 listings.

3.2. The Host Segmentation

As anticipated in the section dedicated to the hypotheses' development, the 31,000 listings in Milan are classified into five groups and consider the number of rented listings (one, two, or three, from four to 10, or more than 10). The segmentation is based on the difference skills and competences required to manage the increasing number of listings and the business organizational complexity and it is in line with some previous studies [48]. In this section, additional quantitative data are considered to test the validity of this segmentation.

Figure 1 reports the host and listing distribution, which shows a clear power-law pattern. The graph illustrates the long tail with a strong concentration on the right side of the horizontal axis. Essentially, a handful of hosts manages a wide number of listings.

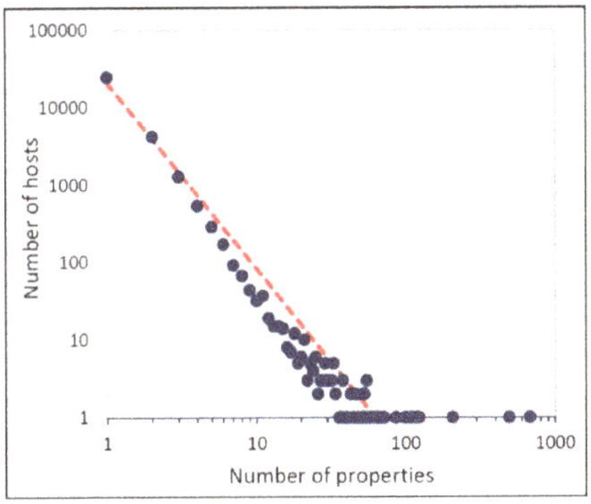

Figure 1. Host and listing distribution (number of hosts as a function of number of properties).

Table 1 reports the descriptive statistics of the five groups; it is structured in five different vertical sections.

Table 1. Descriptive statistics of the five groups.

2015–2018											
	Absolute Measures					Clusters' Weight			Unitary Values		
Clusters	Host	Listings (list)	Available Days (Av_D) (/000)	Book Days (Bo_D) (/000)	Revenue (Rev) (mil.)	Rev %	List %	Host %	List per host	Av_D per list	Bo_D per list
P1	24,535	24,535	9702	1950	194	35%	48%	78%	1	395	79
P2	4215	8430	3476	746	72	13%	17%	13%	2	412	89
P3	1289	3867	1666	383	48	7%	8%	4%	3	431	99
P10	1238	6539	2812	709	87	15%	13%	4%	5.3	430	109
P > 10	242	7536	2737	878	125	29%	15%	1%	31.1	363	117
PAll	**31,519**	**50,907**	**20,393**	**4666**	**526**	**100%**	**100%**	**100%**	**1.6**	**401**	**92**

	Performance				Performance Scaling		
Clusters	Rev per list	ADR	Occ. (*)	RevPAN	Var. ADR	Var. occ.	Var. RevPAN
P1	7922	100	20.10%	20			
P2	8520	96	21.50%	21	−3.50%	6.80%	3.10%
P3	12,352	125	23.00%	29	29.80%	7.00%	38.80%
P10	13,342	123	25.20%	31	−1.50%	9.90%	8.20%
P > 10	16,530	142	32.10%	46	15.40%	27.20%	46.70%
PAll	**10,328**	**113**	**22.90%**	**26**			

Legend: (*) occupancy here is calculated as the ratio between book days over available days.

The first depicts the absolute metrics, and the second shows the relative measures. The first cluster includes 78% of the hosts, but only 48% of the listings, which generate 35% of the total revenue. Focusing on this latter figure (revenue), there is a good division among the remaining four groups: The second cluster is 13%, the third is 7%, the fourth is 15%, and the last is 29%. The small size of the third cluster in terms of listings (8%), hosts (4%), and revenue (7%) appears coherent to the managerial description. In fact, this group reasonably represents the breaking point of the "personal" business model, which is centered around the work and the competencies of the host. The unitary values (third column) show the rising ability of the bigger hosts to book a higher number of days, moving form 79 (cluster 1) to 117 days (cluster 5). Generally speaking, the scaling generates approximately an augment of 10 additional booked days moving from one cluster to the following. The penultimate column depicts the operating performance indices as occupancy (booked days divided by available days), the average daily rate (ADR, revenue divided by booked days), and the revenue per available night (RevPAN, revenue across available days). Focusing on the revenue per available night, the scaling effect is associated with a progressive rise of this value. The last column reports the variation of the performance metrics moving from the first to the last cluster. The revenue per available night shows an impressive growth, rising to 3.1% (from the first to the second group), 38.8% (from the second to the third), 8.2% (from the third to the fourth), and 46.7% (from the fourth to the fifth), respectively.

3.3. *The Method*

As anticipated, this paper evaluates the synchronization degree comparing the five groups of hosts, in order to perceive the similarities and differences. The method developed by Cazelles has been adopted [47]. It requires three different steps, which are introduced and described below [58].

The first phase transforms the series (values) in a set of symbols, by comparing each value with its neighbors'. As reported in Figure 2, there are some possible cases: (i) trough point, (ii) peak point, (iii) increase, (iv) decrease, (v) stability. These five trends are then observed comparing couple groups of different hosts (in terms of size).

Formally, these five situations are identified using the following relationships:

(i) through: $x(t+1) < x(t)\ x(t+2)$ or $x(t+1) < x(t+2)\ x(t)$;
(ii) peak: $x(t) < x(t+2)\ x(t+1)$ or $x(t+2) < x(t)\ x(t+1)$ or $x(t+2)\ x(t) < x(t+1)$;

(iii) increase: x(t) x(t + 1) < x(t + 2);
(iv) decrease: x(t + 2) x(t + 1) < x(t);
(v) stability: x(t) = x(t + 1) = x(t + 2).

An example of the five situations is reported in Figure 2.

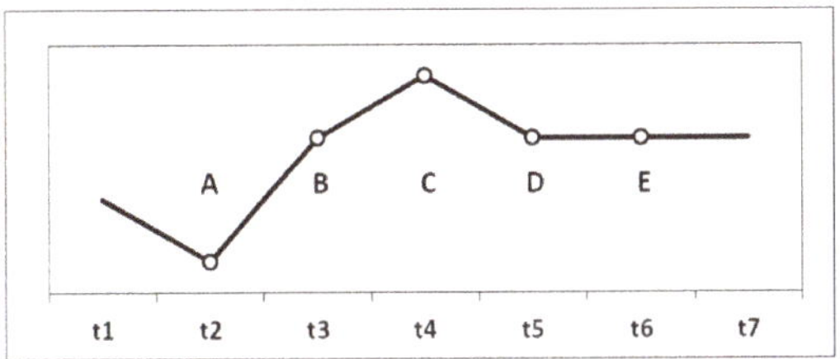

Figure 2. The transformation of a time series in symbols (A: through, B: peak, C: increase, D: decrease, E: stability). Source: adapted from [48].

The second phase is the heart of the analysis and calculates the mutual information degree. It is a quantitative method that compares two series and evaluates the degree of similarity (synchronization) or dissimilarity (de-synchronization). "Given the series *X* and *Y*, the mutual information *I(X,Y)* is calculated as:

$$I(X,Y) = H(X) + H(Y) - H(X,Y) \quad (1)$$

where H( ) is the entropy of each series:

$$H(X) = -\sum p(x_i)log_2(p(x_i)) \quad (2)$$

and *H(X,Y)* is the joint entropy of the two series:

$$H(X,Y) = -\sum\sum p(x_i, y_i)log_2(p(x_i, y_i)) \quad (3)$$

We then normalize the mutual information using:

$$U(X,Y) = I(X,Y)/(H(x) + H(Y)) \quad (4)$$

thus *U(X,Y)* is in the interval [0, 1].

It is easy to demonstrate that if X and Y are independent random variables, then:

$$H(X, Y) = H(X) + H(Y). \quad (5)$$

therefore, the "mutual information is zero" [48] (p. 5). To calculate these quantities, Python scripts adapted from at https://github.com/people3k/pop-solar-sync (last accessed April, 2021) were used.

The last phase evaluates the statistical significance of the values *U(X,Y)*. In line with previous studies [47] 500 surrogate pairs of series were created, based on a Markov process (with a one time-step memory), that preserve the structure of the series [48]. Finally, the five groups of hosts were compared to the corresponding surrogate series and a *t*-test was used for evaluating the statistical significance.

4. Findings

The findings are structured in two sub-sections. The first tests the hypotheses focused on the scaling effect, while the second explores the seasonal patterns.

4.1. Scaling Effect

Table 2 reports the findings related to the first (synchronization degree among clusters) and the second hypothesis (synchronization degree between each cluster and the overall sample). In both hypotheses, the rise of listings is expected to reduce the synchronization degree. As explained in the methodology section, the synchronization is measured by the mutual information. The higher the value of the mutual information score, the higher the similarity and vice versa. A ratio of 0.40 identifies a good similarity (or synchronization), while a value lower than 0.20 depicts a strong dissimilarity or desynchronization [59–61]. The following columns are based on the comparison between the different clusters and the 500 random series.

Table 2. Mutual information for the five clusters.

Clusters	P1	P2	P3	P10	P > 10	PAll	Mean
P1	1						
P2	0.498	1					0.498
P3	0.429	0.377	1				0.403
P10	0.363	0.367	0.337	1			0.356
P > 10	0.334	0.308	0.276	0.343	1		0.315
PAll	0.665	0.560	0.499	0.454	0.395	1	0.515

The evidence reported in the first five columns (from P1 to P > 10) is used to test the first hypothesis. The mutual information of the first cluster (P1) shows a progressive reduction comparing the single host with cluster P2 (0.498), P3 (0.429), P10 (0.363), and P > 10 (0.334). The other column shows the same pattern. Therefore, the evidence confirms the first hypothesis that the synchronization among the different clusters reduces as the the number of managed listings rises.

Moving to the second hypothesis, reading the values of the last line is sufficient. In fact, the values show a very strong synchronization for the first cluster (0.665), but the mutual information progressively reduces, moving to P2 (0.560), P3 (0.499), P10 (0.454), and P > 10 (0.395). The second hypothesis is confirmed, and it implicitly confirms that the overall sample (PAll) is largely influenced by the first three clusters, which together represent the 95% of hosts, 72% of listings, and 55% of total revenue.

4.2. Seasonal Patterns

The analysis explores the seasonal patterns characterizing the chosen destination. The hypothesis focuses on the synchronization degree between the five groups and the total (PAll) comparing the opposed seasonal patterns: holiday and working; weekend and midweek; trade fair and non-trade fair. The values are reported in Table 3. The values should be read while comparing each vertical couple for each cluster. If the synchronization degree reduces (for each cluster and for each of the opposed seasonal periods), then the three hypotheses are confirmed. Focusing on Hypothesis 3A, the cluster P1 moves from 0.649 (holiday) to 0.643 (working); P2 from 0.553 to 0.505; P3 from 0.521 to 0.438; P10 from 0.434 to 0.412; and P > 10 from 0.403 to 0.331. The values confirm Hypothesis 3A, which means that each cluster is more synchronized with the overall sample during the holiday period rather than the working days (when business is dominant). This evidence confirms the prevalent specialization of Airbnb listings for leisure rather than business guests. These results can be extended to the second (Hypothesis 3B) and third (Hypothesis 3C) seasonal periods. During the last seasonal period (trade-fair), when Milan hosts some events, the synchronization degree registers as the highest value in all clusters (for example, is 0.860 for P1).

Table 3. Mutual information between clusters and overall sample during opposed seasonal periods.

	P1	P2	P3	P10	P > 10
PAll					
Hypothesis 3A					
Holiday	0.649	0.533	0.521	0.434	0.403
Working	0.643	0.505	0.438	0.412	0.331
Hypothesis 3B					
Weekend	0.721	0.573	0.533	0.472	0.415
Midweek	0.596	0.477	0.413	0.396	0.330
Hypothesis 3C					
Trade-fair	0.860	0.801	0.676	0.508	0.513
Non-trade-fair	0.654	0.549	0.490	0.465	0.396

Finally, Table 4 reports the relationships among the five clusters during the different seasonal periods in order to test the last six hypotheses, according to which the synchronization decreases when the scaling effect rises. Table 4 contains six panels—one for each seasonal period. Hypothesis 4A focuses on holiday. Reading the table by column, cluster P1 shows a progressive reduction in the mutual information moving from top (0.47) to down (0.35). Generally speaking, all the values show this trend with very few exceptions (three out of 60) identified by the squared cells in Table 4. Therefore, the evidence largely confirms the six hypotheses.

Table 4. Mutual information for seasonal period.

Holiday (4.A)	P1	P2	P3	P10	P > 10
P1	1				
P2	0.47	1			
P3	0.42	0.34	1		
P10	0.35	**0.36**	0.34	1	
P > 10	0.35	0.28	0.27	0.29	1

Weekend (4.C)	P1	P2	P3	P10	P > 10
P1	1				
P2	0.48	1			
P3	0.49	0.39	1		
P10	0.4	**0.44**	0.38	1	
P > 10	0.36	0.33	0.3	0.31	1

Trade fair (4.E)	P1	P2	P3	P10	P > 10
P1	1				
P2	0.75	1			
P3	0.58	0.68	1		
P10	0.43	0.51	0.55	1	
P > 10	0.44	0.47	0.54	0.6	1

Non trade fair (4.F)	P1	P2	P3	P10	P > 10
P1	1				
P2	0.48	1			
P3	0.42	0.36	1		
P10	0.37	**0.37**	0.34	1	
P > 10	0.34	0.31	0.27	0.33	1

Working (4.B)	P1	P2	P3	P10	P > 10
P1	1				
P2	0.45	1			
P3	0.37	0.36	1		
P10	0.32	0.29	0.27	1	
P > 10	0.26	0.24	0.2	0.28	1

Midweek (4.D)	P1	P2	P3	P10	P > 10
P1	1				
P2	0.41	1			
P3	0.34	0.3	1		
P10	0.3	0.27	0.26	1	
P > 10	0.25	0.23	0.19	0.26	1

Legend: squared bold values = increase

5. Discussion

The research question of this paper focuses on the relationship between the scaling effect and the seasonal patterns. The latter are used as a proxy for the competition among different (in terms of size) Airbnb hosts. Based on the findings previously shown, this section discusses the main results. Focusing on the overall (annual) seasonal patterns (Hypotheses 1 and 2), the data confirm that the scaling effect increases the dissimilarities between the hosts managing a few and many listings, respectively. Realistically, this evidence supports a progressive competitive reduction among big and small hosts.

The second set of Hypotheses 3 and 4 move from the whole (annual) patterns to the specific seasonal periods characterizing the destination under study. Generally speaking, the synchronization degree is higher during the holiday and weekend periods, confirming the prevalent specialization of Airbnb listings for leisure clients. However, the mutual information degree registers an important drop moving from single to multi-listing hosts, suggesting, also in this case, different (or partially different) business models. By contrast, during the working and especially midweek periods, the synchronization is lower, and the dissimilarities augment when comparing mom-and-pop hosts with large multi-listing providers. Therefore, the progressive reduction in the competition appears more relevant. These results can be extended to the trade-fair and non-trade fair seasonal periods.

6. Conclusions

The conclusions are articulated in four sections: Some theoretical, as well as practical, implications are traced, future research avenues are proposed, and some study limitations are identified.

6.1. Theoretical Implications

As discussed in the introduction and in the literature review, the current studies largely distinguish only between single- and multi-listing hosts, ignoring the magnitude of the scaling effect. The findings proposed in this study depict a very different picture, showing a progressive differentiation in the seasonal patterns correlated to the rise in the managed listings. The results, therefore, can significantly change the theoretical knowledge in this field and can re-orient future studies, especially in the sub-field of competition, the determinants of listing performance, and pricing strategies.

Second, the synchronization degree among the different (in scale) hosts is not homogenous but changes according to the different seasonal patterns of the Milan destination. The higher the specialization in leisure segments, the higher the competition among the different hosts; the higher the specialization in the business segment, the lower the similarities and, therefore, the competitive pressure.

Finally, this paper introduces important innovations to analyze the competition among Airbnb listings. The first innovation is to clearly identify the main destination market segments (in the case of Milan, leisure, business, and trade-fair guests) and the corresponding seasonal patterns. The second methodological innovation is the use of mutual information to perceive and measure the degree of synchronization between the series, variously articulated to measure the scaling effect and the identified seasonal periods. This approach can open new research opportunities in other destination contexts.

6.2. Practical Implications

This paper sheds new light on the competition threat among Airbnb listings considering their scale. In particular, the findings clearly suggest a progressive reduction in the similarity of seasonal patterns when the size of the host, measured by her/his listings, rises. Therefore, the results support identifying different groups in the Airbnb arena that have a diverse competitive threat according to the specific seasonal period. Therefore, a single host, according to her/his scaling effect, can create a different competitive set. Furthermore, the competitive intensity varies according to the specific seasonal period and appears to be higher when the attracted segments are leisure, which reduces in the case of business guests.

6.3. Research Avenues

The findings reported open many new research opportunities. Some of them are discussed in this sub-section. It is surprising that the current peer-to-peer accommodation platform literature has completely omitted any studies qualitatively exploring the business models of mom-and-pop and multi-listing hosts. Future research must cover this gap, identifying the advantages and disadvantages of moving from a limited size to a bigger scale. This qualitative research can shed light on the main resources and competences that can be stretched as the scale rises.

A second interesting area of inquiry can explore the competitive threat between Airbnb listings and hotels. In particular, based on the current findings, this research area can explore if the listing scaling augments the synchronization degree between Airbnb and hotels, thereby increasing the competition and the substitution threat. Furthermore, the competition intensity can be articulated considering the different seasonal periods characterizing the destination under study.

A third research avenue focuses on the studies exploring the determinants of performance and pricing strategies. As analytically discussed in the literature review, the research standard, with very few exceptions, is to segment the hosts into single- or multi-listings. Based on the current results, future research should investigate more analytical segmentation and consider different relevant seasonal periods. In fact, the determinants of the results and rates can change consistently.

Finally, future studies can employ new methods for testing the hypotheses (as Bayesian Hypothesis Testing) or SARIMA (Seasonal Auto Regressive Integrated Moving Average) for the seasonal patterns.

6.4. Limitations

This is an explorative paper, which is in line with similar previous studies focused on competitive threats [62]. It is centered around a single case study. The findings' generalization is partially limited. However, the paper adopts a longitudinal approach, creating a consistent temporal pattern. A future research agenda is called for to verify whether, within a multi-destination study, the evidence reported is confirmed.

Author Contributions: R.S. wrote the introduction, the literature review and the findings and was responsible of formal analysis; methodology; project administration; supervision; visualization; roles/writing. R.B. developed the statistical analysis and wrote the methodology and was responsible of data curation; funding acquisition; investigation; software; validation. Both authors discussed and contributed to the conclusions. Both authors have read and agreed to the published version of the manuscript.

Funding: R.B. was partially supported by the Ministry of Education and Science of the Russian Federation in the framework of the Competitiveness Enhancement Program of the Tomsk Polytechnic University.

Institutional Review Board Statement: Not applicable.

Informed Consent Statement: Not applicable.

Data Availability Statement: Data are available through the companies cited in the methodology section (see Section 3.1).

Conflicts of Interest: The authors declare no conflict of interest. The funders had no role in the design of the study; in the collection, analyses, or interpretation of data; in the writing of the manuscript; or in the decision to publish the results.

References

1. Sainaghi, R.; Köseoglu, M.A.; d'Angella, F.; Mehraliyev, F. Sharing economy: A co-citation analysis. *Curr. Issues Tour.* **2019**. [CrossRef]
2. Altinay, L.; Taheri, B. Emerging themes and theories in the sharing economy: A critical note for hospitality and tourism. *Int. J. Contemp. Hosp. Manag.* **2019**, *31*, 180–193. [CrossRef]
3. Sainaghi, R. Determinants of price and revenue for peer-to-peer hosts. The state of the art. *Int. J. Contemp. Hosp. Manag.* **2020**. [CrossRef]
4. Garau-Vadell, J.B.; Gutiérrez-Taño, D.; Díaz-Armas, R. Residents' Support for P2P Accommodation in Mass Tourism Destinations. *J. Travel Res.* **2019**, *58*, 549–565. [CrossRef]
5. Koh, Y.; Belarmino, A.; Kim, M.G. Good fences make good revenue: An examination of revenue management practices at peer-to-peer accommodations. *Tour. Econ.* **2019**. [CrossRef]
6. Xie, K.L.; Kwok, L.; Heo, C.Y. Are Neighbors Friends or Foes? Assessing Airbnb Listings' Agglomeration Effect in New York City. *Cornell Hosp. Q.* **2019**. [CrossRef]
7. Kim, J.; Tang, L.R.; Wang, X. The uniqueness of entrepreneurship in the sharing accommodation sector: Developing a scale of entrepreneurial capital. *Int. J. Hosp. Manag.* **2020**, *84*, 102321. [CrossRef]
8. Bocken, N.M.; Fil, A.; Prabhu, J. Scaling up social businesses in developing markets. *J. Clean. Prod.* **2016**, *139*, 295–308. [CrossRef]
9. Butler, R.W. Seasonality in tourism: Issues and problems. In *Tourism: The State of the Art*; Wiley: Hoboken, NJ, USA, 1994; pp. 332–340.

10. Sainaghi, R.; Mauri, A.; d'Angella, F. Decomposing seasonality in an urban destination: The case of Milan. *Curr. Issues Tour.* **2018**. [CrossRef]
11. Guttentag, D. Airbnb: Disruptive innovation and the rise of an informal tourism accommodation sector. *Curr. Issues Tour.* **2018**, *12*, 1192–1217. [CrossRef]
12. Dolnicar, S. A review of research into paid online peer-to-peer accommodation: Launching the Annals of Tourism Research curated collection on peer-to-peer accommodation. *Ann. Tour. Res.* **2019**, *75*, 248–264. [CrossRef]
13. Sainaghi, R.; Baggio, R. Clusters of topics and research designs in peer-to-peer accommodation platforms. *Int. J. Hosp. Manag.* **2020**, *88*, 102393. [CrossRef]
14. Prayag, G.; Ozanne, L.K. A systematic review of peer-to-peer (P2P) accommodation sharing research from 2010 to 2016: Progress and prospects from the multi-level perspective. *J. Hosp. Mark. Manag.* **2018**, *27*, 649–678. [CrossRef]
15. Gant, A.C. Holiday rentals: The new gentrification battlefront. *Sociol. Res. Online* **2016**, *21*, 1–9. [CrossRef]
16. Fang, B.; Ye, Q.; Law, R. Effect of sharing economy on tourism industry employment. *Ann. Tour. Res.* **2016**, *57*, 264–267. [CrossRef]
17. Vinogradov, E.; Leick, B.; Kivedal, B.K. An agent-based modelling approach to housing market regulations and Airbnb-induced tourism. *Tour. Manag.* **2020**, *77*, 104004. [CrossRef]
18. Akbar, Y.H.; Tracogna, A. The sharing economy and the future of the hotel industry: Transaction cost theory and platform economics. *Int. J. Hosp. Manag.* **2018**, *71*, 91–101. [CrossRef]
19. Ključnikov, A.; Krajčík, V.; Vincúrová, Z. International Sharing Economy: The Case of AirBnB in the Czech Republic. *Econ. Sociol.* **2018**, *11*, 126–137. [CrossRef]
20. Sainaghi, R.; Abrate, G.; Mauri, A. Price and RevPAR determinants of airbnb listings: Convergent and divergent evidence. *Int. J. Hosp. Manag.* **2021**, *92*, 102709. [CrossRef]
21. Sainaghi, R. The current state of academic research into peer-to-peer accommodation platforms. *Int. J. Hosp. Manag.* **2020**, *89*, 102555. [CrossRef]
22. Liang, S.; Schuckert, M.; Law, R.; Chen, C.C. Be a "Superhost": The importance of badge systems for peer-to-peer rental accommodations. *Tour. Manag.* **2017**, *60*, 454–465. [CrossRef]
23. Gunter, U.; Önder, I. Determinants of Airbnb demand in Vienna and their implications for the traditional accommodation industry. *Tour. Econ.* **2018**, *24*, 270–293. [CrossRef]
24. Xie, K.; Mao, Z. The impacts of quality and quantity attributes of Airbnb hosts on listing performance. *Int. J. Contemp. Hosp. Manag.* **2017**, *29*, 2240–2260. [CrossRef]
25. Gunter, U. What makes an Airbnb host a superhost? Empirical evidence from San Francisco and the Bay Area. *Tour. Manag.* **2018**, *66*, 26–37. [CrossRef]
26. Li, J.; Moreno, A.; Zhang, D.J. Agent behavior in the sharing economy: Evidence from Airbnb. *Ross Sch. Bus. Work. Pap. Ser.* **2015**, *1298*, 1–33. [CrossRef]
27. Gibbs, C.; Guttentag, D.; Gretzel, U.; Morton, J.; Goodwill, A. Pricing in the sharing economy: A hedonic pricing model applied to Airbnb listings. *J. Travel Tour. Mark.* **2018**, *35*, 46–56. [CrossRef]
28. Oskam, J.; van der Rest, J.P.; Telkam, B. What's mine is yours—But at what price? Dynamic pricing behavior as an indicator of Airbnb host professionalization. *J. Revenue Pricing Manag.* **2018**, *17*, 311–328. [CrossRef]
29. Chen, Y.; Zhang, R.; Liu, B. Fixed, flexible, and dynamics pricing decisions of Airbnb mode with social learning. *Tour. Econ.* **2020**. [CrossRef]
30. Gibbs, C.; Guttentag, D.; Gretzel, U.; Yao, L.; Morton, J. Use of dynamic pricing strategies by Airbnb hosts. *Int. J. Contemp. Hosp. Manag.* **2018**, *30*, 2–20. [CrossRef]
31. Kwok, L.; Xie, K.L. Pricing strategies on Airbnb: Are multi-unit hosts revenue pros? *Int. J. Hosp. Manag.* **2019**, *82*, 252–259. [CrossRef]
32. Campopiano, G.; Minola, T.; Sainaghi, R. Students Climbing the Entrepreneurial Ladder: Family Social Capital and Environment-related Motives in Hospitality and Tourism. *Int. J. Contemp. Hosp. Manag.* **2016**, *28*, 1115–1136. [CrossRef]
33. Magno, F.; Cassia, F.; Ugolini, M.M. Accommodation prices on Airbnb: Effects of host experience and market demand. *TQM J.* **2018**, *30*, 608–620. [CrossRef]
34. Falk, M.; Larpin, B.; Scaglione, M. The role of specific attributes in determining prices of Airbnb listings in rural and urban locations. *Int. J. Hosp. Manag.* **2019**, *83*, 132–140. [CrossRef]
35. Deboosere, R.; Kerrigan, D.J.; Wachsmuth, D.; El-Geneidy, A. Location, location and professionalization: A multilevel hedonic analysis of Airbnb listing prices and revenue. *Reg. Stud. Reg. Sci.* **2019**, *6*, 143–156. [CrossRef]
36. Tong, B.; Gunter, U. Hedonic pricing and the sharing economy: How profile characteristics affect Airbnb accommodation prices in Barcelona, Madrid, and Seville. *Curr. Issues Tour.* **2020**. [CrossRef]
37. Cai, Y.; Zhou, Y.; Scott, N. Price determinants of Airbnb listings: Evidence from Hong Kong. *Tour. Anal.* **2019**, *24*, 227–242. [CrossRef]
38. Proserpio, D.; Xu, W.; Zervas, G. You get what you give: Theory and evidence of reciprocity in the sharing economy. *Quant. Mark. Econ.* **2018**, *16*, 371–407. [CrossRef]
39. Chen, Y.; Xie, K. Consumer valuation of Airbnb listings: A hedonic pricing approach. *Int. J. Contemp. Hosp. Manag.* **2017**, *29*, 2405–2424. [CrossRef]

40. Izquierdo, L.M.; Egorova, G.; Rovira, A.P.; Ferrando, A.M. Exploring the use of artificial intelligence in price maximisation in the tourism sector: Its application in the case of Airbnb in the Valencian Community. *J. Reg. Res.* **2018**, *42*, 113–128.
41. Chattopadhyay, M.; Mitra, S.K. Do airbnb host listing attributes influence room pricing homogenously? *Int. J. Hosp. Manag.* **2019**, *81*, 54–64. [CrossRef]
42. Baggio, R.; Sainaghi, R. Mapping time series into networks as a tool to assess the complex dynamics of tourism systems. *Tour. Manag.* **2016**, *54*, 23–33. [CrossRef]
43. d'Angella, F.; de Carlo, M.; Sainaghi, R. Archetypes of destination governance: A comparison of international destinations. *Tour. Rev.* **2010**, *65*, 61–73. [CrossRef]
44. Sainaghi, R.; Mauri, A. The Milan World Expo 2015: Hospitality operating performance and seasonality effects. *Int. J. Hosp. Manag.* **2018**, *72*, 32–46. [CrossRef]
45. Baggio, R.; Sainaghi, R. Complex and chaotic tourism systems: Towards a quantitative approach. *Int. J. Contemp. Hosp. Manag.* **2011**, *23*, 840–861. [CrossRef]
46. Sainaghi, R.; Baggio, R. The effects generated by events on destination dynamics and topology. *Curr. Issues Tour.* **2019**. [CrossRef]
47. Sainaghi, R.; Canali, S. Commercial mix, seasonality and daily hotel performance: The case of Milan. In *Strategic Management Engineering: Enterprise, Environment and Crisis*; Sichuan University Press: Chengdu, China, 2009.
48. Sainaghi, R.; Mauri, A.; Ivanov, S.; D'Angella, F. Mega events and seasonality: The case of the Milan World Expo 2015. *Int. J. Contemp. Hosp. Manag.* **2019**, *31*, 61–86. [CrossRef]
49. Sainaghi, R.; Canali, S. Exploring the effects of destination's positioning on hotels' performance: The Milan case. *Tour. Int. Multidiscip. J. Tour.* **2011**, *6*, 121–138.
50. Cazelles, B. Symbolic dynamics for identifying similarity between rhythms of ecological time series. *Ecol. Lett.* **2004**, *7*, 755–763. [CrossRef]
51. Sainaghi, R.; Baggio, R. Substitution threat between Airbnb and hotels: Myth or reality? *Ann. Tour. Res.* **2020**, *83*, 102959. [CrossRef]
52. Nofre, J.; Giordano, E.; Eldridge, A.; Martins, J.C.; Sequera, J. Tourism, nightlife and planning: Challenges and opportunities for community liveability in La Barceloneta. *Tour. Geogr.* **2018**, *20*, 377–396. [CrossRef]
53. Oxoli, D.; Prestifilippo, G.; Bertocchi, D. Enabling spatial autocorrelation mapping in QGIS: The hotspot analysis Plugin. *Geoing. Ambient. Min.* **2017**, *151*, 45–50.
54. Garcia-Ayllon, S. Urban Transformations as an Indicator of Unsustainability in the P2P Mass Tourism Phenomenon: The Airbnb Case in Spain through Three Case Studies. *Sustainability* **2018**, *10*, 2933. [CrossRef]
55. Heo, C.Y.; Blal, I.; Choi, M. What is happening in Paris? Airbnb, hotels, and the Parisian market: A case study. *Tour. Manag.* **2019**, *70*, 78–88. [CrossRef]
56. Ferreri, M.; Sanyal, R. Platform economies and urban planning: Airbnb and regulated deregulation in London. *Urban Stud.* **2018**, *55*, 3353–3368. [CrossRef]
57. Schäfer, P.; Braun, N. Misuse through short-term rentals on the Berlin housing market. *Int. J. Hous. Mark. Anal.* **2016**, *9*, 287–311. [CrossRef]
58. Sainaghi, R.; Baggio, R. Complexity traits and dynamics of tourism destinations. *Tour. Manag.* **2017**, *63*, 368–382. [CrossRef]
59. Sainaghi, R.; Baggio, R. Destination Events, Stability, and Turning Points of Development. *J. Travel Res.* **2019**. [CrossRef]
60. Freeman, J.; Baggio, J.A.; Robinson, E.; Byers, D.A.; Gayo, E.; Finley, J.B.; Meyer, J.A.; Kelly, R.L.; Anderies, J.M. Synchronization of energy consumption by human societies throughout the Holocene. *Proc. Natl. Acad. Sci. USA* **2018**, *115*, 9962–9967. [CrossRef]
61. Latham, P.E.; Roudi, Y. Mutual information. *Scholarpedia* **2009**, *4*, 1658. [CrossRef]
62. Zervas, G.; Proserpio, D.; Byers, J.W. The rise of the sharing economy: Estimating the impact of Airbnb on the hotel industry. *J. Mark. Res.* **2017**, *54*, 687–705. [CrossRef]

engineering proceedings

MDPI

Proceedings

The Use of Satellite TIR Time Series for Thermal Anomalies' Detection on Natural and Urban Areas †

Malvina Silvestri [1,*], Federico Rabuffi [1], Massimo Musacchio [1], Sergio Teggi [2] and Maria Fabrizia Buongiorno [1]

1 Istituto Nazionale di Geofisica e Vulcanologia (INGV), 00143 Rome, Italy; federico.rabuffi@ingv.it (F.R.); massimo.musacchio@ingv.it (M.M.); fabrizia.buongiorno@ingv.it (M.F.B.)

2 Department of Engineering "Enzo Ferrari", University of Modena and Reggio Emilia (UNIMORE), 41125 Modena, Italy; sergio.teggi@unimore.it

* Correspondence: malvina.silvestri@ingv.it

† Presented at the 7th International Conference on Time Series and Forecasting, Gran Canaria, Spain, 19–21 July 2021.

Abstract: In this work, the land surface temperature time series derived using Thermal InfraRed (TIR) satellite data offers the possibility to detect thermal anomalies by using the PCA method. This approach produces very detailed maps of thermal anomalies, both in geothermal areas and in urban areas. Tests were conducted on the following three Italian sites: *Solfatara-Campi Flegrei* (Naples), *Parco delle Biancane* (Grosseto) and *Modena* city.

Keywords: thermal anomaly; time series analysis; geothermal site; urban heat island

check for updates

Citation: Silvestri, M.; Rabuffi, F.; Musacchio, M.; Teggi, S.; Buongiorno, M.F. The Use of Satellite TIR Time Series for Thermal Anomalies' Detection on Natural and Urban Areas . *Eng. Proc.* **2021**, *5*, 5. https://doi.org/10.3390/engproc2021005005

Published: 24 June 2021

1. Introduction

Thermal anomalies, i.e., areas where the surface temperature has a value significantly different from the background, are potentially related to the underground energy sources or to land use and coverage variations in urban areas where the urban heat island (UHI) phenomenon can be observed. Current satellite missions, providing imagery in the Thermal InfraRed (TIR) spectral region at 90–100 m of spatial resolution, provide the potential to estimate the land surface temperature (LST) and highlight the main surface thermal anomalies [1–6]. In this work, two case studies were carried out. The first case study is the detection of thermal anomalies on geothermal active areas (volcanic or not). The second focuses on the detection of UHIs [7,8]. Both the studies are based on the remote sensing LST time series. In both the case studies, the thermal anomalies' detection is also inspected using the principal component analysis (PCA) of the LST time series.

2. Materials and Methods

All the analyses are based on the following two types of data: Advanced Spaceborne Thermal Emission and Reflection Radiometer (ASTER) and the Landsat 8 satellite data, with a 90-m and 100-m pixel spatial resolution, respectively, in TIR channels (8–12 μm) and a temporal resolution of 16 days.

In the first study, nighttime ASTER and TIRS/Landsat 8 time series have been processed using the following two different methodologies: the Temperature and Emissivity Separation (TES, [9]) algorithm for ASTER and the Single Channel Algorithm (SCA, inverting radiative transfer equation, already tested in [10,11]) for Landsat 8. Two LST time series have been obtained and the results are cross-compared and validated with ground measurements. TES and SCA are well-known methodologies and have been used to evaluate LST on the following two different test sites with different geological features: the volcanic area of *Solfatara-Campi Flegrei* (near Naples, Italy) and the geothermal area of *Parco delle Biancane* (near Grosseto, Italy).

The second case study has been addressed to the characterization of the UHI of the city of *Modena* (Italy). The analysis is based on nighttime TIRS/Landsat 8 image time series processed using the SCA methodology.

The availability of a substantial number of these satellite data for the three test sites (as reported in Table 1) offered the possibility of obtaining three LST time series over a long period (Figures 1–3), thus allowing an accurate analysis of thermal anomalies.

Table 1. LST processed for three test sites.

Data	Number of Processed Images		
	Solfatara-Campi Flefrei	*Parco delle Biancane*	*Modena City*
LST-ASTER	40	20	NA
LST-Landsat 8	55	40	43
TOTAL LST	95	60	43

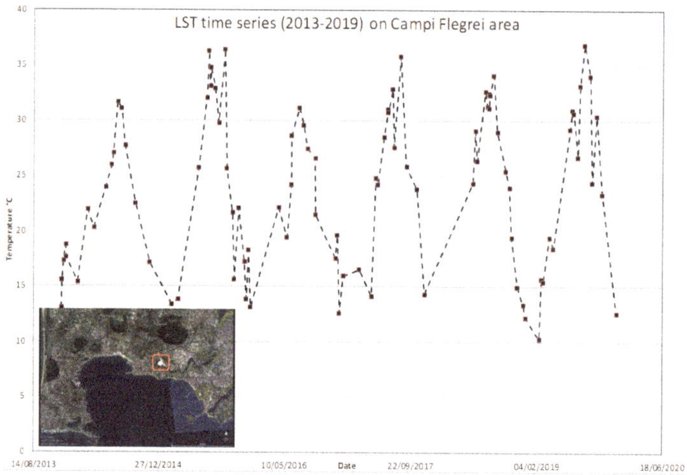

Figure 1. LST Time series on *Solfatara-Campi Flegrei* test site. Plot referred to is the red box area.

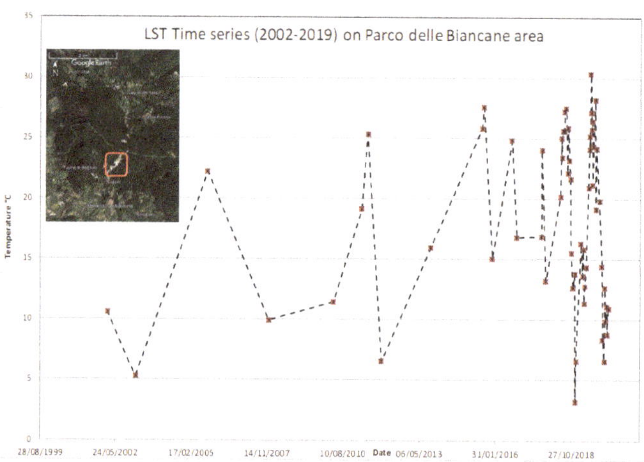

Figure 2. LST Time series on *Parco delle Biancane* test site. Plot referred to is the red box area.

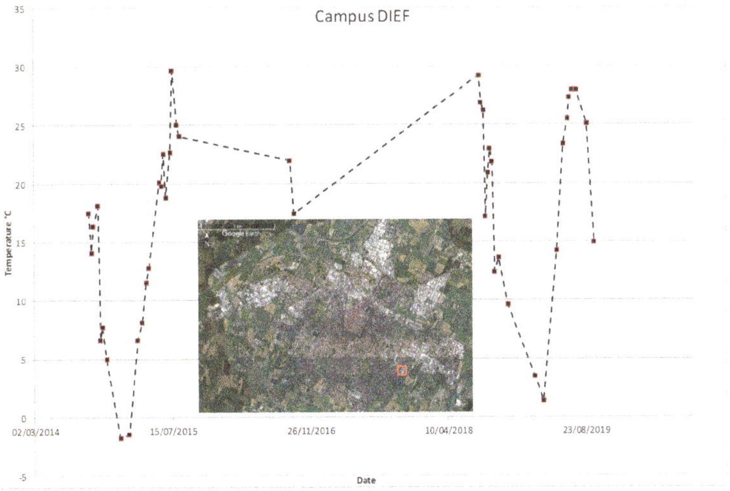

Figure 3. LST Time series on *Modena* city (Campus-DIEF) test site. Plot referred to is the red box area.

The relative accuracy of the LST estimates can be assessed in comparison to the ground measures provided by the ground network that operates independently of the satellites. Alternatively, the accuracy can be estimated by a cross-validation between the products obtained with different LST retrieval algorithms and/or for different sensors, even if largely complicated by the spatial scale mismatch between the satellite sensors and/or the ground-based sensors. In fact, the areas observed by ground radiometers usually cover small areas, whereas satellite measurements in the thermal infrared typically cover between 1 and 100 km^2. In this case, the following data have been used for the validation steps:

- for *Solfatara-Campi Flegrei*, the ground measurements collected by permanent thermal cameras installed in *Solfatara* volcano. The images have been reprojected to have the same point of view as the satellite;
- for *Parco delle Biancane*, the TIR images acquired using thermal cameras mounted on drones and collected during three separated field campaigns synchronized with satellite passages;
- for *Modena*, weather stations at the 4 stations around Modena city.

3. Results and Discussion

In both case studies (natural and urban areas), the thermal anomalies' detection is inspected using the principal component analysis (PCA) on the LST time series obtained by processing the satellite data. In this work, analysis on the use of PCA has demonstrated the possibility of detecting thermal anomalies in studied sites, neglecting the seasonality effect present in long LST time series. The use of nighttime data has been considered to remove the "noise" due to the solar irradiation that is strong during the day.

PCA allowed the extraction of the dominant patterns within the time series as the detection of thermal anomalies, offering a good and easy way to produce very detailed maps of thermal anomalies, both in geothermal areas and in urban areas (UHI). Thermal anomalies were detected by considering the first two PCs and selecting three sets of pixels from the clusters used as endmembers for the maximum likelihood classification. The position of the cluster, which has been selected considering the PC1 and PC2, is approximatively the same in the three scatter plots (Figure 4). The thermal anomaly points (geothermal or UHI), background points (rural areas for UHI) and water points (sea, lakes, rivers) are grouped in distinct sectors of the scatter plots. This leads to the conclusion that thermal anomalies can also be individuated using this combination of PC components of the time series of temperature images derived from satellite imagery.

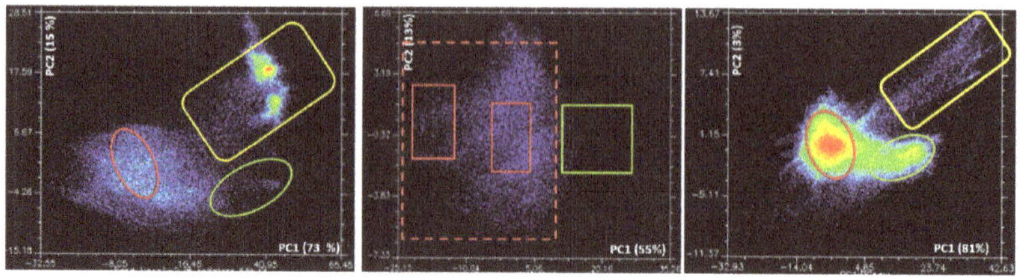

Figure 4. PC1 vs. PC2 cluster: *Campi Flegrei* (left), *Parco delle Biancane* (middle), *Modena* City (right). Yellow line represents the water cluster, red line the background (land, rural areas) and green line represents the thermal anomaly or warm area due to UHI.

In Figure 5, the main warm areas are the *Solfatara* volcano and the lakes that have a temperature greater than the land during the night.

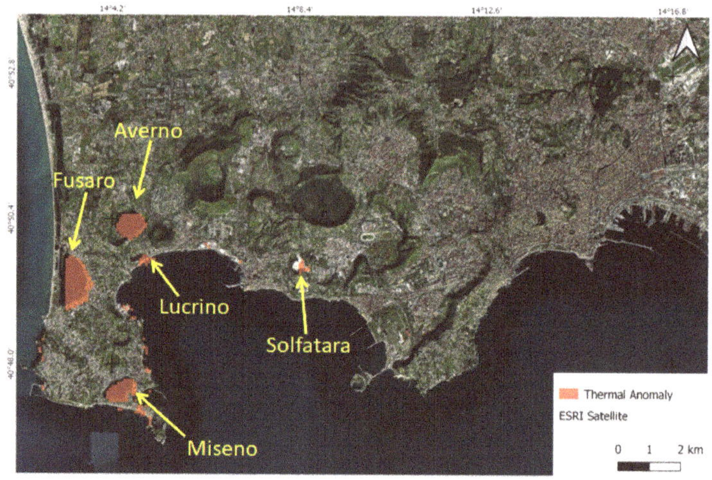

Figure 5. The red areas represent the results obtained for Landsat 8 data in the *Solfatara-Campi Flegrei* area.

A comparison with a different methodology is also presented, confirming that satellite data can be a very powerful tool to study surface thermal anomalies quantitatively.

In fact, the result obtained by using PCA (Figure 5) is in agreement with the one obtained using a different methodology, as described in [11]. In particular, in [11], the process of removing the seasonal component of temperature time series is considered. The existence of a thermally anomalous area at the *Campi Flegrei* site is analyzed by considering the land surface Median Temperature values greater than a Threshold Value (MTTV) on a de-seasonalized time series. In [11], the threshold values of $+1\sigma$ (16.36 °C for ASTER and 17.01 °C for Landsat 8), $+1.5\sigma$ (17.19 °C for ASTER and 18.02 °C for Landsat 8) and $+2\sigma$ (18.03 °C for ASTER and 19.04 °C for Landsat 8) allowed us to obtain the results showed in Figure 6. The use of PCA confirms that the process of removing seasonality, applied in MTTV (Figure 6), is not necessary. The spatial distribution of the thermal anomalies detected using PCA and MTTV is coincident, as shown in Figures 4 and 6.

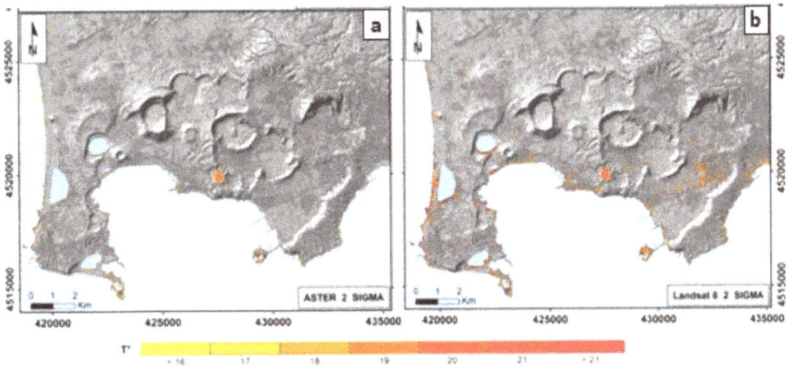

Figure 6. Maps of MTTV of temperature satellite frame with temperature threshold at +2σ (**a**,**b**); modified after [11].

A similar approach to that taken in *Solfatara-Campi Flegrei*, using PCA as showed in Figure 4, has been adopted for the *Parco delle Biancane* area, where validation data are lacking. An example of the result obtained with Landsat 8 for *Parco delle Biancane* is shown in Figure 7. The red areas cover *Parco delle Biancane*'s geothermal areas that have a temperature greater than the land during the night. Moreover, the *Valle Secolo* and *Nuova San Martino* Enel Green Power central are also detected. Even though there are several geothermal centrals in the area analyzed, the two power plants detected as "thermal anomalies", together with the *Parco delle Biancane* area, are, indeed, those with a high rated power.

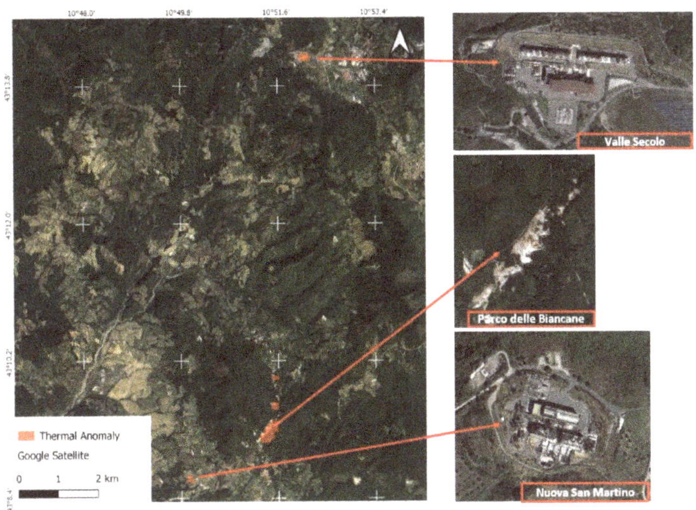

Figure 7. Landsat 8—The red polygons represent the results in the *Parco delle Biancane* area.

Moreover, concerning a UHI, PCA was able to separate the statistics for the rural environment, built areas and water surfaces without additional information on land cover (e.g., the classification obtained using VIS/NIR imagery or other land cover databases). In Figure 8, the results of the UHI phenomenon in *Modena* city are shown. In particular, in Figure 8, green represents water (warm) and red represents built (warm) pixels. The warm areas are included inside the "0" isoline that represents the line that separates warm and cold surfaces (15.4 °C has been assumed as the reference). Similarly, the isoline corresponding to +2 °C of difference (higher than the reference) has been added and marked as "2" (dashed line).

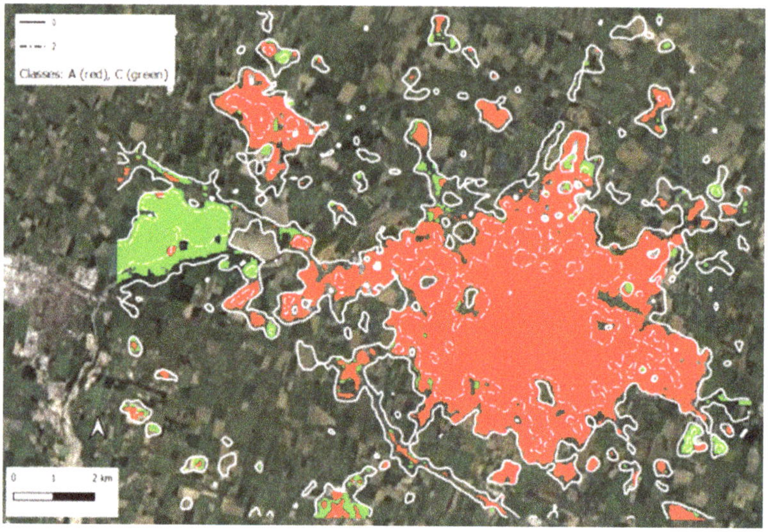

Figure 8. *Modena* city test site: the classes representing warm areas are reported in green and red.

4. Conclusions

The results of these studies furnished some important considerations, as follows:

- the methodologies used to obtain LST also produce reliable temperature estimates in the very particular case of geothermal anomalies and are usable for near ground air temperature trends' analysis;
- the PCA allows us to extract the dominant patterns within the time series to detect thermal anomalies, offering a good and easy way to produce very detailed maps of the thermal anomalies in both geothermal areas and in urban areas (UHI);
- the PCA allows for the differentiation of the surface cover without using other re-mote sensing images in VIS/NIR or auxiliary classification products. This differentiation improves the analysis of the thermal behavior of the surfaces.

The two studied cases represent two more demonstrations of the potential of satellite observations in TIR for environmental applications.

Author Contributions: Conceptualization, data processing, investigation and writing—original draft preparation, M.S.; data processing and GIS visualization, F.R.; supervision, S.T. and M.F.B.; visualization, M.M. All authors have read and agreed to the published version of the manuscript.

Funding: This research received no external funding.

Institutional Review Board Statement: Not applicable.

Informed Consent Statement: Not applicable.

Data Availability Statement: The data that support the findings of this study are available from the corresponding author (M.S.) upon reasonable request.

Acknowledgments: Thanks to Enrica Marotta, Gara Avvisati, Pasquale Belviso, Eliana Bellucci Sessa, Teresa Caputo and Vito Romaniello for their constructive comments and helping with the data analysis.

Conflicts of Interest: The authors declare no conflict of interest.

References

1. Buongiorno, M.F.; Pieri, D.; Silvestri, M. Thermal analysis of volcanoes based on 10 years of ASTER data on Mt. Etna. In *Thermal Infrared Remote Sensing*; Springer: Dordrecht, The Netherlands, 2013; pp. 409–428.
2. Pieri, D.; Abrams, M. ASTER watches the world's volcanoes: A new paradigm for volcanological observations from orbit. *J. Volcanol. Geotherm. Res.* **2004**, *135*, 13–28. [CrossRef]
3. Silvestri, M.; Romaniello, V.; Hook, S.; Musacchio, M.; Teggi, S.; Buongiorno, M.F. First Comparisons of Surface Temperature Estimations between ECOSTRESS, ASTER and Landsat 8 over Italian Volcanic and Geothermal Areas. *Remote Sens.* **2020**, *12*, 184. [CrossRef]
4. Fridleifsson, I.B.; Bertani, R.; Huenges, E.; Lund, J.W.; Ragnarsson, A.M.; Rybach, L. The possible role and contribution of geothermal energy to the mitigation of climate change. In Proceedings of the IPCC Scoping Meeting on Renewable Energy Sources, Luebeck, Germany, 20 January 2008; Volume 20, pp. 59–80.
5. Howari, F. Prospecting for geothermal energy through satellite based thermal data: Review and the way forward. *Glob. J. Environ. Sci. Manag.* **2015**, *1*, 265–274.
6. Qin, Q.; Zhang, N.; Nan, P.; Chai, L. Geothermal area detection using Landsat ETM+ thermal infrared data and its mechanistic analysis-A case study in Tengchong, China. *Int. J. Appl. Earth Obs. Geoinf.* **2011**, *13*, 552–559. [CrossRef]
7. Voogt, J.A.; Oke, T.R. Thermal remote sensing of urban climates. *Remote Sens. Environ.* **2003**, *86*, 370–384. [CrossRef]
8. Sobrino, J.A.; Oltra-Carrió, R.; Sòria, G.; Bianchi, R.; Paganini, M. Impact of spatial resolution and satellite overpass time on evaluation of the surface urban heat island effects. *Remote Sens. Environ.* **2012**, *117*, 50–56. [CrossRef]
9. Gillespie, A.; Rokugawa, S.; Matsunaga, T.; Cothern, J.S.; Hook, S.; Kahle, A.B. A temperature and emissivity separation algorithm for advanced spaceborne thermal emission and reflection radiometer (ASTER) images. *IEEE Trans. Geosci. Remote Sens.* **1998**, *36*, 1113–1126. [CrossRef]
10. Silvestri, M.; Marotta, E.; Buongiorno, M.F.; Avvisati, G.; Belviso, P.; Bellucci Sessa, E.; Caputo, T.; Longo, V.; De Leo, V.; Teggi, S. Monitoring of Surface Temperature on Parco delle Biancane (Italian Geothermal Area) Using Optical Satellite Data, UAV and Field Campaigns. *Remote Sens.* **2020**, *12*, 2018. [CrossRef]
11. Caputo, T.; Bellucci Sessa, E.; Silvestri, M.; Buongiorno, M.F.; Musacchio, M.; Sansivero, F.; Vilardo, G. Surface temperature multiscale monitoring by thermal infrared satellite and ground images at Campi Flegrei volcanic area (Italy). *Remote Sens.* **2019**, *11*, 1007. [CrossRef]

engineering proceedings

MDPI

Proceedings

Short Term Load Forecasting Using TabNet: A Comparative Study with Traditional State-of-the-Art Regression Models †

Eugenio Borghini * and Cinzia Giannetti

Faculty of Science and Engineering, Swansea University, Swansea SA1 8EN, UK; c.giannetti@swansea.ac.uk
* Correspondence: eugenio.borghini@swansea.ac.uk

† Presented at the 7th International conference on Time Series and Forecasting, Gran Canaria, Spain, 19–21 July 2021.

Abstract: Electric load forecasting is becoming increasingly challenging due to the growing penetration of decentralised energy generation and power-electronics based loads such as heat pumps and electric vehicles, which adds to a transition to more variable work patterns (accentuated by the COVID-19 pandemic in 2020). In this paper, three different Machine Leaning models are analysed to predict the energy load one week ahead for a period of time including the COVID-19 pandemic. It is shown that, by using the recently proposed TabNet model architecture, it is possible to achieve an accuracy comparable to more traditional approaches based on gradient boosting and artificial neural networks without the need of performing complex feature engineering.

Keywords: short-term electricity demand forecasting; neural networks; TabNet

Citation: Borghini, E.; Giannetti, C. Short Term Load Forecasting Using TabNet: A Comparative Study with Traditional State-of-the-Art Regression Models. *Eng. Proc.* **2021**, *5*, 6. https://doi.org/10.3390/engproc2021005006

Published: 25 June 2021

1. Introduction

Electric power load forecasting is widely recognised as a key task for electrical utilities. Accurate predictions in the short time horizon allow to minimise spinning reserve capacity, plan the generation of electric power and configure cost-effective battery charging schedules [1,2]. In the past few years several models based on artificial neural networks have been proposed and shown to be successful for this task [3,4]. Despite this, model selection is not trivial and heavily depends on several aspects of the specific case under study, such as the time resolution of the available data, the type of climate of the location and the required prediction horizon among others. Moreover, the adoption of distributed energy generation, such as wind turbines and solar photovoltaics, the increasing popularity of low carbon technologies (specially, electric vehicles) and even unusual events such as the ongoing COVID-19 pandemic increment the uncertainty and demand levels experienced by distribution networks.

In this context, the recently proposed TabNet model architecture is analysed and compared with two state-of-the-art models such as gradient boosting based on decision trees and deep neural networks (see [3–9]) in the task of predicting the energy load one week ahead at Stentaway primary substation, UK (the choice of forecast horizon is motivated by a Data Science Challenge recently hosted by Energy Systems Catapult). It was found that the performance achieved by TabNet is comparable with the one exhibited by the more established models, with the advantages of learning directly from the raw data (i.e., no pre-processing is needed) and requiring minimal feature engineering. In addition, given the different nature of TabNet's inductive bias in comparison to more traditional regression algorithms, a further improvement in accuracy was obtained by combining it with the traditional models via ensemble methods.

The article is structured as follows. In Section 2, the description and pre-processing of the employed datasets is given. In Section 3, the three models used for load forecasting are presented. Section 4 is devoted to the analysis of the obtained results. Section 5 contains the summary of the work and some future research lines.

2. Data Description

The historical demand data were collected from the Stentaway Primary substation. They contained average demand power values measured in Megawatts (MW) spanning around 2 1/2 years (between November 2017 and July 2020) and totalling slightly more than 47,000 samples.

Since it is well-known that the weather plays a major role in the energy load, this dataset was complemented with what is known as reanalysis weather data from six sites surrounding the substation extracted using MERRA-2 (the data extraction was based on code available at https://github.com/emilylaiken/merradownload, last accessed on 23 June 2021). Reanalysis is a data processing technique that provides a consistent and complete estimation of weather variables over a period of interest. The process consisted of applying modern forecasting techniques to a blend of actual observations with past short-range weather forecasts, thus imitating for historical data the way in which the day-to-day forecasts are generated. In this way, estimations for the averaged hourly irradiance (W/m^2) and instantaneous surface temperature (^oC) were obtained for six locations that could be interpreted as weather forecasts. The sites corresponded to grid points on the numerical weather prediction grid for dates between January 2015 and July 2020.

Both datasets are publicly available at the Western Power Distribution Open Data Hub site upon login [10].

2.1. Data Pre-Processing

The datasets contained very few erroneous values and gaps (far less than 1% of the samples) which were meaningfully filled. More concretely, the demand dataset presented values that were obviously out of range (both too close to zero and too high) for two weeks in May 2018 and a couple of days in November 2018. All these outliers were replaced by the demand values of the corresponding days from the previous weeks. Regarding the weather data, a few missing values were detected for the temperature at location 4 which were simply filled using the temperature at location 3 since these variables were highly correlated (the correlation coefficient was >0.98).

Finally, the cleaned datasets were merged after linearly interpolating the weather variables to 30 min frequency.

2.2. Feature Extraction

An exploratory data analysis was conducted to unveil patterns and factors that could enhance the predictive value of the original dataset, consisting only of historical demand data and weather reanalysis data.

The most important group of extracted features was derived by studying the auto-correlation of the demand (see Figure 1). As the plot reveals, there were strong daily and weekly patterns in the demand. To account for them, the following features were added to the dataset:

- Hour of the day, day of the week, day of the month, month and year.
- Demand values at the same hour for the whole past week.
- Cyclic versions of hour of the day, day of the month and month, which made explicit the similarity between the end of a period and the beginning of the following one (for instance, the demand around 12:00 PM of a given day tended to be strongly related to the demand around 1:00 AM of the next day) by encoding these features as points in a 2D circle (see [5]).

It was also found that the weather variables produced lagged effects on the demand. After experimenting with different time scales, it was decided to enrich the dataset with the averages of temperature and solar irradiance over periods of 2, 12 and 24 h to capture short-term fluctuations, cyclic day and night patterns and daily trends respectively.

Finally, an ad-hoc strategy was adopted to treat bank holidays and the lockdown period. Specifically, the bank holidays were labelled as a Sunday due to the resemblance of demand patterns between both kind of days, and the lagged demand values were correspondingly shifted to coincide with that of previous Sunday. Since the behaviour of the demand during lockdown was clearly different from that of regular periods (see Figure 2), it was decided to distinguish lockdown days with a flag.

The resulting dataset contained approximately 100 features.

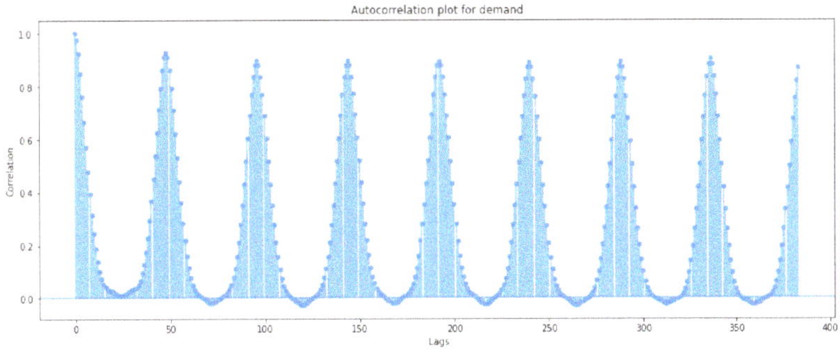

Figure 1. Autocorrelation plot for the demand (the lags are measured at half-hour intervals). There are peaks every 24 h and a slightly higher peak for the same day of the past week.

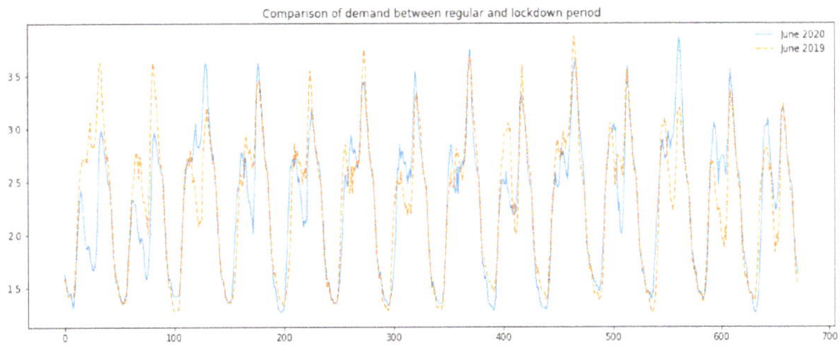

Figure 2. Comparison of demand values between the first two weeks of June 2019 and June 2020 (aligned so that the days of the week coincide).

3. Methodology

The main goal was to forecast one week ahead values of demand (load forecast in MW) using, as model input, its past values in combination with historical and current weather forecast data. As previously stated, the prediction of energy load during the outbreak of the COVID-19 pandemic was one of the main challenges in this study. As it could be expected, the significant change in the energy consumption pattern caused by the various restrictions imposed by the government made it harder to forecast the load for this period. In addition, there is no technique for the short-time load forecasting problem that is known to be superior to all others (see [11]); rather, the best techniques depend heavily on the particular characteristic of the dataset (including factors such as the type of climate and the economic activities at the analysed location, the forecast horizon, etc). For these reasons, three different approaches were contrasted in the present study: gradient boosting tree ensemble model, artificial neural networks and TabNet. The first two techniques are known to achieve state-of-the-art results in several practical tasks and were shown to be successful at short-time load forecasting (see for instance [3–9]). On the other hand, TabNet is a novel deep neural network architecture specially designed to handle tabular data that reportedly outperforms or is on pair with standard neural networks and decision trees based variants [12].

All models were trained to minimise the mean squared difference between the predicted and the actual values of demand one week ahead. Roughly 1 year of data was used (corresponding to the period November 2017–December 2018) as training set, while the remaining weeks (up to July 2020) were used to validate and asses the models' performance using the walk-forward method [2]. Below follows a brief description of each model, together with the specific features and hyperparameters used in each one of them.

CatBoost: CatBoost [13] is an implementation of gradient boosting on decision trees developed by Yandex, which quickly positioned itself as one of the standard methods for learning problems with tabular data, heterogeneous features and complex, non-linear interactions. Gradient boosting is an ensemble method that iteratively improves weak predictors (in the case of CatBoost, decision trees) by performing gradient descent greedily in a certain functional space [14].

All features, both original and extracted, were employed for the CatBoost model. Except for a few relevant hyperparameters that controlled the complexity and regularised the model, the default values were used. These hyperparameters were `n_estimators` (maximum number of trees), `depth` (maximum depth of each decision tree), `max_bin` (number of splits for numerical features) and `rsm` (the proportion of the features considered for each split). Their values were determined by a grid search around initial good values obtained by heuristics and manual experimentation.

Artificial neural network: Artificial neural networks are inspired by a simplified model of how biological neural networks work, and are known to have the capability of learning hidden non-linear and complex pattern in the data. An artificial neural network consists of a directed graph, organised in layers whose nodes are known as neurons. Each neuron applies a non-linear transformation to its input based on learnable parameters and passes the resulting value to neurons in the next layer. These parameters are trained iteratively using stochastic gradient descent with the aim of generating the desired output.

In contrast to the CatBoost model, it was decided to remove several features to reduce multicollinearity issues. Among the time-related features, only the cyclic versions were included and all weather variables were discarded but for the ones corresponding to the two most uncorrelated locations. The total number of neurons was estimated heuristically (proportional to the degrees of freedom of the problem) and it was decided to reduce by a factor of 2 the number of neurons in each hidden layer with the aim of forcing the network to progressively learn more relevant features. The number of neurons in the first hidden layer and the number of layers were determined by a grid search. This resulted in an architecture consisting of four hidden fully connected layers with 64 neurons in the first layer. The non-linear activation ReLU was applied for all layers, while the Adam optimiser was used with the default learning rate 0.001.

TabNet: The new architecture proposed by TabNet learns directly from the raw numerical (not normalised) features of tabular data. The normalisation and feature extraction is somehow embedded in the architecture, since the raw data is filtered by a Batch Normalisation layer and several transformers blocks designed to learn relevant features. One of the salient characteristics of TabNet is the use of a single deep learning block to perform instance-wise feature selection, consisting of a sequential attention mechanism and learnable masks. As a consequence, the accumulated learned weights in this block can be used to interpret the outputs of the model.

For the TabNet model only the cyclic time-related features, the lagged information of the demand and the weather variables of the two most uncorrelated location were employed. The total size of the model was decided by a grid search, following ([12], Guidelines for hyperparameters), to set the values of the hyperparameters width and steps, which are respectively, the number of hidden neurons in each block and the number of hidden blocks.

4. Discussion and Results

The three models considerably beat naive baselines and achieve a steady accuracy across very dissimilar weeks (see Table 1 below). This is consistent with the existing literature and the common consensus that models based in ensemble of regression trees and neural networks are the strongest predictors for generic regression tasks. Although in our tests TabNet did not in general outperform the best traditional model, its accuracy was usually close to it. In addition, since TabNet had an inductive bias of different nature to the traditional regression algorithms it allowed us to obtain a further improvement in accuracy by combining it with the traditional models via ensemble methods. Indeed, it was verified that the simple average of the three models achieved an appreciable higher performance than any single model (see Table 2).

Table 1. R^2 scores and root squared errors for the proposed methods. Here the naive baseline consists of predicting the same as the previous week.

Method	R^2 Score	RMSE	R^2 Score (Lockdown)	RMSE (Lockdown)
CatBoost	**0.9369**	**0.2156**	**0.8562**	**0.2332**
Neural Network	0.9311	0.2254	0.8396	0.2463
TabNet	0.9286	0.2295	0.8424	0.2442
Naive Baseline	0.8740	0.3048	0.7198	0.3256

Table 2. R^2 scores and root mean squared errors for the different averages of the proposed models.

Average	R^2 Score	RMSE
CatBoost+TabNet	0.9477	0.1964
CatBoost+Neural Network	0.9492	0.1936
TabNet+Neural Network	0.9423	0.2062
CatBoost+TabNet+Neural Network	**0.9511**	**0.1898**

Regarding the prediction for the lockdown weeks, it was found that reducing the amount of regular samples in the training sets was beneficial for the performance of the predictive models. Concretely, to generate the predictions on lockdown weeks, only samples starting from 2019 were considered for the training set. The rationale behind this decision is that the reduction allows to give more weight to samples corresponding to the lockdown period. The accuracy attained in this way is comparable to the one obtained for normal times (see Figure 3 and Table 1).

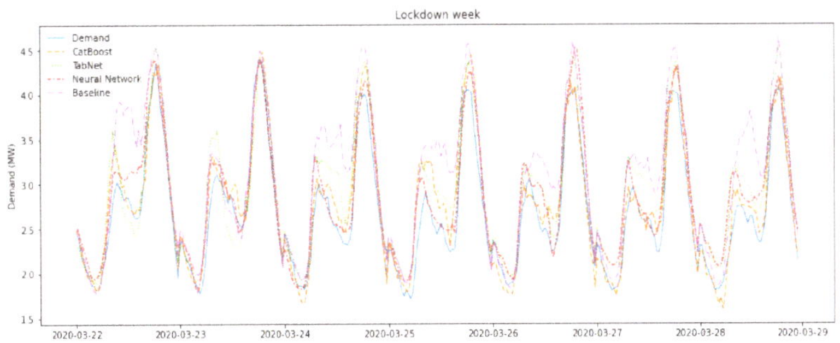

Figure 3. Predictions for the first week of lockdown (from 22 March to 28 March). The consumption pattern is quite different to the one from the previous week.

5. Conclusions

In this study, the performance of the novel TabNet network is compared with two well-established regression models on a short term load forecasting task. It is shown that it is possible to obtain comparable performance to these traditional methods but with little to none feature engineering and data preparation. Moreover, the use of TabNet provides a further boost in the overall accuracy on this task via ensemble methods.

As a future step, it would be interesting to refine the strategy to predict the energy load during the lockdown. As some preliminary evidence suggests, training a strong model on a regular period and then fine-tuning it using data collected during the lockdown (which can be seen as an application of the transfer learning technique) could lead to further improvements in accuracy.

Author Contributions: Conceptualization, E.B. and C.G.; methodology, E.B. and C,G,; software, E.B.; validation, E.B. and G.C.; formal analysis, E.B. and C.G.; investigation, E.B. and C.G.; resources, C.G.; data curation, E.B.; writing—original draft preparation, E.B.; writing—review and editing, C.G.; visualization, E.B.; supervision, C.G.; project administration, C.G.; funding acquisition, C.G. All authors have read and agreed to the published version of the manuscript.

Funding: This research was funded by the UK Engineering and Physical Sciences Research Council (EPSRC) project EP/S001387/1 and the European Regional Development Funds projects IMPACT.

Institutional Review Board Statement: Not applicable.

Informed Consent Statement: Not applicable.

Data Availability Statement: The datasets are publicly available at the Western Power Distribution Open Data Hub site [10] upon login. The code required to generate the results referred in the article will be shared upon request.

Conflicts of Interest: The authors declare no conflict of interest.

References

1. Gross, G.; Galiana, F. Short-term load forecasting. *Proc. IEEE* **1987**, *75*, 1558–1573. [CrossRef]
2. Kaastra, I.; Boyd, M.S. Designing a neural network for forecasting financial and economic time series. *Neurocomputing* **1996**, *10*, 215–236. [CrossRef]
3. Hu, R.; Wen, S.; Zeng, Z.; Huang, T. A short-term power load forecasting model based on the generalized regression neural network with decreasing step fruit fly optimization algorithm. *Neurocomputing* **2017**, *221*, 24–31. [CrossRef]
4. Singh, S.; Hussain, S.; Bazaz, M.A. Short term load forecasting using artificial neural network. In Proceedings of the 2017 Fourth International Conference on Image Information Processing (ICIIP), Shimla, India, 21–23 December 2017; pp. 1–5. [CrossRef]
5. Moon, J.; Park, S.; Rho, S.; Hwang, E. A comparative analysis of artificial neural network architectures for building energy consumption forecasting. *Int. J. Distrib. Sens. Netw.* **2019**, *15*. [CrossRef]
6. Park, D.; El-Sharkawi, M.; Marks, R.; Atlas, L.; Damborg, M. Electric load forecasting using an artificial neural network. *IEEE Trans. Power Syst.* **1991**, *6*, 442–449. [CrossRef]
7. Din, G.M.U.; Marnerides, A.K. Short term power load forecasting using Deep Neural Networks. In Proceedings of the 2017 International Conference on Computing, Networking and Communications (ICNC), Silicon Valley, CA, USA, 26–29 January 2017; pp. 594–598. [CrossRef]
8. Lloyd, J.R. GEFCom2012 hierarchical load forecasting: Gradient boosting machines and Gaussian processes. *Int. J. Forecast.* **2014**, *30*, 369–374. [CrossRef]
9. Hong, T.; Pinson, P.; Fan, S. Global Energy Forecasting Competition 2012. *Int. J. Forecast.* **2014**, *30*, 357–363. 2013.07.001 [CrossRef]
10. Western Power Distribution Open Data Hub Homepage. Available online: https://www.westernpower.co.uk/innovation/pod (accessed on 22 April 2021).
11. Hong, T.; Fan, S. Probabilistic electric load forecasting: A tutorial review. *Int. J. Forecast.* **2016**, *32*, 914–938. 2015.11.011. [CrossRef]
12. Arik, S.Ö.; Pfister, T. TabNet: Attentive Interpretable Tabular Learning. *arXiv* **2019**, arXiv:1908.07442.
13. Prokhorenkova, L.; Gusev, G.; Vorobev, A.; Dorogush, A.V.; Gulin, A. CatBoost: unbiased boosting with categorical features. In *Advances in Neural Information Processing Systems*; Bengio, S., Wallach, H., Larochelle, H., Grauman, K., Cesa-Bianchi, N., Garnett, R., Eds.; Curran Associates, Inc.: New York, NY, USA, 2018; Volume 31.
14. Friedman, J.H. Greedy function approximation: A gradient boosting machine. *Ann. Stat.* **2001**, *29*, 1189–1232. [CrossRef]

engineering proceedings MDPI

Proceedings

Using Learned Health Indicators and Deep Sequence Models to Predict Industrial Machine Health †

Ido Amihai [1,*], Arzam Kotriwala [1], Diego Pareschi [2], Moncef Chioua [3] and Ralf Gitzel [1]

[1] ABB Corporate Research Center, 68526 Ladenburg, Germany; arzam.kotriwala@de.abb.com (A.K.); ralf.gitzel@de.abb.com (R.G.)
[2] ABB, 2629 JD Delft, The Netherlands; diego.pareschi@nl.abb.com
[3] Polytechnique Montréal, Montréal, QC H3T 1J4 QC, Canada; moncef.chioua@polymtl.ca
* Correspondence: ido.amihai@de.abb.com; Tel.: +49-6203-716041
† Presented at the 7th International conference on Time Series and Forecasting, Gran Canaria, Spain, 19–21 July 2021.

Abstract: In this paper, we describe a machine learning approach for predicting machine health indicators with a large time horizon into the future. The approach uses state-of-the-art neural network architectures for sequence modelling and can incorporate numerical-sensor and categorical data using entity embeddings. Moreover, we describe an unsupervised labelling approach where classes are generated using continuous sensor values in the training data and a clustering algorithm. To validate our approach, we performed an ablation study to verify the effectiveness of each of our model's components. In this context, we show that entity embeddings can be used to generate effective features from categorical inputs, that state-of-the-art models, while originally developed for a different set of problems, can nonetheless be transferred to perform industrial asset health classification and provide a performance boost over simpler networks that have been traditionally used, such as relatively shallow recurrent or convolutional networks. Taken together, we present a machine health monitoring system that can accurately generate asset health predictions. This system can incorporate both numerical and categorical information, the current state-of-the-art for sequence modelling, and generate labels in an unsupervised fashion when explicit labels are unavailable.

Keywords: neural networks; time series; sequence modelling; machine health monitoring; predictive maintenance

Citation: Amihai, I.; Kotriwala, A.; Pareschi, D.; Chioua, M.; Gitzel, R. Using Learned Health Indicators and Deep Sequence Models to Predict Industrial Machine Health . *Eng. Proc.* **2021**, 5, 7. https://doi.org/10.3390/engproc2021005007

Published: 25 June 2021

1. Introduction

Modern machine health monitoring systems (MHMS) owe much of their recent success to advances in machine learning algorithms, sensing technologies, and computational power [1–5]. Such systems make use of historical data collected from the monitored equipment, which are used to train machine learning (ML) models for evaluating their health and performance [1], in either a diagnostic or prognostic way, e.g., by remaining useful life estimation (RUL; e.g., [4,6]).

Historically, MHMS were based on ML algorithms that require hand crafted features. However, the utility of such models was limited due to the required domain expertise and inability to cover all spectrum effects, especially nonlinear dependencies in time and domain-specific effects [1]. A mitigation strategy for this problem is to use neural networks (NN), which do not require handcrafted features and can be trained using only the input data (e.g., [1,7–11]).

In the context of sequential data, several NN architecture-types have typically been applied based on their proficiency in learning the temporal dynamic behaviours of systems. In this respect, recurrent neural networks (RNNs) have been extensively used to model sequential data [12]. Although different variants exist, an RNN is normally constructed as an NN with a feedback loop from the previous hidden layer of the network to the next:

$$h(t) = f(h(t-1), X(t); \theta), \quad (1)$$

where h(t) and X(t) are the hidden states and inputs to the network at time t, and θ is the network parameters.

Although RNNs are typically difficult to train due to issues with vanishing and exploding gradients [13], this can be mitigated by using gate functions that regulate the information that passes through the network. This is usually done through long short-term memory (LSTM) or gated recurrent units (GRU) [12], which, instead of the ordinary RNN transition function, involve more complex functions that incorporate gate structures that help regulate the information that passes through the network [14,15]. Other NNs used to model sequential data that are based on RNNs are echo state networks (ESN) [16,17]. ESNs mitigate the vanishing gradient problem by eliminating the need to compute the gradient for the hidden layers of the NN using a sparsely connected RNN called a "reservoir", where the weights are not learned via gradient descent [18].

In addition to RNN based architectures, convolutional NNs (CNNs) have also been used for sequence modelling. CNNs utilize convolutional operations, which are sliding filters that are applied over the data and enable the NN to extract time-invariant nonlinear features [19]. Recently it was demonstrated that CNNs coupled with residual connections, which are connections between an NN layer and a layer it is not directly connected to, can result in highly accurate models for sequential data [19]. An example of this type of architecture is the inception-time network [19], which is one of the architectures we implemented in this research and was inspired by the Inception-v4 architecture [20]. Crucially, it contains "Inception Modules", where the core idea is to simultaneously apply multiple convolutional filters of varying dimensions to the input [21].

Finally, the relatively new transformer architecture-type has also been successfully utilized for sequence modelling (e.g., [22]). These models rely on self-attention mechanisms to model temporal dynamics [23], the most common being the "scaled dot-product attention", "dot-product attention", and "additive attention" [23]. The scaled dot-product attention is computed via the following equation:

$$\text{Attention}(Q,K,V) = \text{softmax}((QK^T)/\sqrt{d_k})\cdot V, \quad (2)$$

where matrices Q, K, and V are generated for each input, and where dk is the dimension of Q, and K. Dot-product attention is identical except that the scaling factor $\sqrt{d_k}$ is not used, and additive attention is computed using a feed-forward NN with a single hidden layer [23]. Although transformers were developed for natural language processing (NLP) applications (e.g., German-English translations), they can be adapted for sequential numerical data, in the simplest case by replacing the embedding layers with fully-connected layers or other layer types that can transform numerical data (e.g., time delay embeddings [22]). Other approaches used for sequence modelling include large memory storage retrieval NNs [9], stacked denoising autoencoders [11], and deep belief networks [8].

Another important issue that arises when developing MHMS stems from the fact that they are typically developed using supervised learning, where ML models are trained to classify the health status of assets based on labelled training examples with a known health status. However, often the relationship between available data and asset health is not known in advance (i.e., the data is unlabelled) and must be determined using statistical, ML, or other methods. To address this issue, we developed an unsupervised approach, where sensor data from the training set was used to generate clusters that represent the asset health status [24].

Currently, the state of the art (SOTA) for processing sensor data are architectures for sequential data modelling such as Res-CNN [25], LSTM fully-convolutional NN [26], inception-time [19], and ResNet [18]. The models were shown to work well on many sequence learning tasks (e.g., [19,23], see [18] for a review). Additionally, these new methods have already been applied in the field of predictive maintenance. For example, ResNet has been used on wind turbine data [27] and bearing data [28] to predict faults. Res-CNN has been applied to motor data [29], and fully-convolutional LSTM used on aircraft engine data [30]. However, to our knowledge, no paper has compared all of the above methods on a single dataset.

In this paper, we describe an ML approach that was used to predict machine health with a large time horizon. Due to the nature of our application, we used a two-week horizon, but the approach can be generalized to other horizons as well. To process the sensor data, we compare all the SOTA architectures named above. Moreover, we also describe the results obtained using a simpler NN baseline model based on bidirectional GRU cells (BiGRU) [24]. Finally, we compared these NN approaches to a random-forest (RF) model, which is a very popular ML approach not based on NNs that performs well on a variety of tasks and does not require special processing for categorical variables [31,32]. Additionally, the inputs to the model are both continuous sensor data and categorical metadata, and we use K-Means clustering to incorporate prior knowledge of the distribution of the predicted variable into our model and generate the predicted variable, as we first described in [24].

We first show that this approach can provide superior predictions of machine health in comparison to a similar model that only incorporates sensor data, similarly to what we previously reported [24]. Moreover, we demonstrate the superiority of SOTA networks over the simpler BiGRU architecture as well as a non-NN approach (RF) for classifying industrial asset health.

2. Methods

2.1. Data

For a more detailed account, see [24]. Briefly, the data consisted of both sensor data collected approximately every 6 h and categorical metadata, over a period of approximately 2.5 years from 51 vibration sensors. The data were divided into training, validation, and test sets, so that approximately the first 2 years of data were used for training and the final 0.5 years of data was split between the validation and test datasets through stratified random shuffling based on the distribution of the predicted variable (defined below). Note that due to important data privacy concerns specified by the owner of the data, some aspects of the data were transformed to maintain data privacy.

2.2. The Predicted Variable

The predicted variable was determined based on the distribution of the sensor data of the training set, as well as practical specifications provided by the owner of the data and only very basic domain knowledge. Specifically, the data owner requested predictions of the systems' health status two weeks into the future. The full method is described in [24], but in brief, we integrated prior knowledge of the predicted values into the architecture of our model so that instead of predicting its value directly, we computed a set of clusters based on its distribution in the training set. We then labelled all our predicted variables based on the nearest cluster centroid calculated through the K-Means algorithm. Since our training data distribution resembled a bimodal distribution, suggesting 2 distinct types of behaviour (see Figure 1), we used the nearest cluster centroid of two possible clusters as the predicted variable.

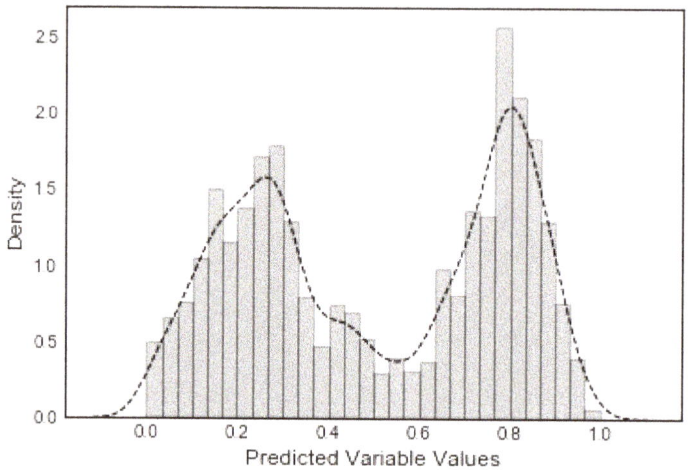

Figure 1. Distribution of the predicted variable in the training set. The dashed line represents a Gaussian kernel density estimation of the distribution (reproduced from [24]).

2.3. Modelling

In the current research, we tested several deep NN architectures for modelling the sensor data (i.e., sequence models). The first was a BiGRU, which we used as a baseline for comparison to different model architectures, and which we also used in a previous study [24]. We compared this relatively simple but popular architecture to several SOTA algorithms as well as a non-NN based approach (RF). First, we trained a transformer model that was slightly modified from [23], where it was used for English to German translation tasks so as to be suitable for sequential numerical data, mainly by replacing its embedding layers with fully connected layers. This stresses the notion that deep learning models that are developed to solve a certain task can often be rather straightforwardly adapted to solve a different task, even when the similarity between the tasks is not apparent. Additional SOTA algorithms that were used were Res-CNN [25], LSTM fully-convolutional NN [26], inception-time [19] and ResNet [18]. The hyperparameters of the models were selected by examining the loss function value on the validation set, and the models were tuned using the logistic loss-function, which is the most commonly used loss-function for binary-classification problems and is almost universally applied [33]:

$$L = -\frac{1}{N}\sum_{i}^{N}\sum_{j}^{M} y_{ij}\log\left(p_{ij}\right), \tag{3}$$

where p is the predicted class and y is the true class label.

In addition, we were provided with metadata in the form of categorical variables that identify important aspects in the equipment, such as its specific type. To incorporate categorical variables in ML models, they are often transformed using one-hot encoding (OHE), where k new binary features are created for k different categories. However, as we stated in [24] when the cardinality of the features is high, OHE requires a large number of computational resources. Additionally, OHE treats the values of categorical variables as independent of each other and often ignores information about the relationships between them [34]. In order to circumvent these issues, we used the categorical metadata to learn entity embeddings, where each categorical variable is mapped to a fixed-size vector space, with parameters that are learned by the model (see [24,34,35]).

The overall modelling approach is presented in Figure 2. The embeddings were concatenated to the outputs of the sequence model component and fed to an FC layer with a rectified linear unit (ReLU) activation function. The outputs of this layer can then be fed to an additional FC layer with a Sigmoid activation function (i.e., the logistic function).

A constant learning rate of 0.001 was used with the Adam optimizer, and models were trained with early stopping, i.e., until we observed an error increase on the validation set [36].

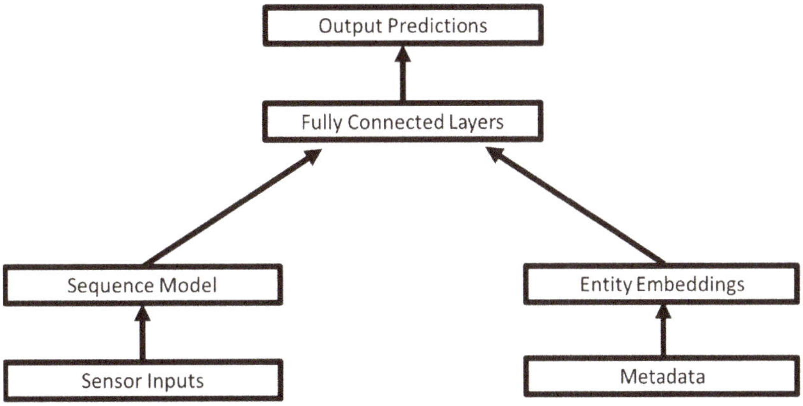

Figure 2. Overall model architecture.

The models were compared using two very popular classification metrics: the F1-score and the Matthews correlation coefficient (MCC) [37].

3. Results

All of the analyses were done using the Python programming language [38]. To assess the importance of the various model components, we performed an ablation study where we systematically removed the main components of our model and observed how it affected performance. In this respect, we compared our approach of using entity embeddings with the BiGRU model to the same model without the embedding inputs. Moreover, we tested a model where the penultimate FC layer was also removed (the first layer of the "fully connected layers" component in Figure 2). Finally, we compared the performance of various sequence models (sequence model component in Figure 2), including SOTA sequence models, as well as an RF model.

The performance of the experimental conditions is summarized in Table 1. The baseline BiGRU model generated an F1 score of 0.876 and an MCC score of 0.747. When entity embeddings were not included in the model, both F1 and MCC scores dropped. Similar results were obtained when the penultimate FC layer was removed, and the concatenated inputs from the BiGRU and embeddings were fed directly into the output layer of the model. Moreover, a model consisting only of the BiGRU component of the model achieved a similar performance, suggesting that the additional FC layer might not be needed when the additional metadata inputs are not included. When SOTA models were used instead of the BiGRU baseline, the model demonstrated an increased performance on both F1, $t(4) = 4.18$, $p < 0.01$, and MCC, $t(4) = 5.43$, $p < 0.01$. RF performed similarly to the BiGRU baseline on the F1 and MCC metrics. However, it also showed a strong bias towards predicting Class 1 (98.59% vs. 81.29% accuracy rates for Class 1 and Class 2, respectively). The F1 differences between SOTA algorithms and RF were marginally significant, $t(4) = 2.03$, $p = 0.056$, and statistically significant when considering only CNN based SOTA algorithms (e.g., Res-CNN, FCN, inception-time and ResNet), which performed best on our task, $t(3) = 4.93$, $p < 0.01$. MCC differences between CNN based SOTA algorithms and RF were marginally significant after correcting for multiple comparisons, $t(3) = 2.55$, $p = 0.04$.

Table 1. Model Classification Performance.

Model	Class 1 Accuracy	Class 2 Accuracy	Overall Accuracy	F1	MCC
BiGRU	85.05	89.6	87.33	0.876	0.747
BiGRU, no entity embed-dings	78.06	92.7	85.4	0.864	0.715
BiGRU, no penultimate FC	78.2	91.8	85.0	0.860	0.707
Only BiGRU	78.36	91.1	84.7	0.856	0.7
Transformer	90.90	85.78	90.26	0.880	0.768
Res-CNN	94.10	87.38	93.26	0.904	0.817
FCN	93.87	90.24	93.42	0.919	0.842
Inception-time	94.63	87.76	93.77	0.909	0.826
ResNet	95.68	85.7	94.43	0.902	0.818
Random-forests	98.59	81.29	89.47	0.890	0.811

4. Discussion

Although fully connected deep learning models have been used in MHMS for many years [39–42], the use of NN approaches that are specialized for sequence models is a relatively recent research trend [43,44]. This is somewhat surprising considering that most industrial data are sensor data, which is by nature sequential. Notably, several studies used recurrent NNs to estimate RUL [45–49] or performance degradation [7,50–52]. Other studies have applied CNN models after transforming sensor data to 2-dimensional, similar to image data that are typically used by CNNs, in order to classify machine faults [53,54] or RUL [55,56]. Yet another research direction has been to transform the sensor signals to the frequency domain before applying CNNs for machine fault diagnosis [21,57–59], while other studies straightforwardly applied CNNs for monitoring the health status of industrial assets using the raw sensor data as the input [60–64]. Importantly, none of the previous studies compared several SOTA sequence models for MHMS on a single dataset [44,65–68], and the current study was the first to apply them in this context. Such models are significantly deeper and computationally more complex than those that were used in most previous studies and were originally developed for applications unrelated to machine health monitoring (e.g., NLP [23]).

The MHMS described in this paper can incorporate SOTA models as well as combine sequential and non-sequential inputs to obtain more accurate predictions, as when using each input type in isolation. Its effectiveness was verified through an ablation study where the main components of the model were systematically removed or altered. Moreover, the proposed MHMS makes use of the predicted variable distribution to derive classes for prediction using unsupervised clustering (see [24]). Such class derivation is especially important in applications where the theoretical variable, e.g., asset health distribution, is not known directly. Our proposed algorithm can be used to derive a proxy of the theoretical variable using a different variable for which the distribution in the training data can be estimated. Moreover, we tested the various SOTA algorithms on our data. While such models, e.g., with a single or few LSTM or GRU layer(s), can work relatively well on industrial tasks, we found that using SOTA models resulted in increased performance on the metrics that we measured, especially CNN based models. What this suggests is that while industrial data might contain important unique features, e.g., features that are representative of industrial asset health might only be discoverable in these data, the SOTA models developed for seemingly unrelated data and tasks are nonetheless also transferrable to these data. This is likely because SOTA sequential models are highly proficient at learning general temporal dynamic behaviour and hence can also be applied here.

In conclusion, we have proposed an MHMS that can handle both numeric and categorical data, can be used in conjunction with SOTA NNs and can be used to predict the health status of industrial assets even when a health status variable is not explicitly provided. Such a system can serve as an integral component of full-fledged predictive maintenance software systems to provide increased automation for asset health inspection.

Author Contributions: Conceptualization, I.A., A.K., M.C., and R.G.; methodology, I.A.; software, I.A.; validation, I.A., A.K., M.C., and R.G.; formal analysis, I.A., A.K., M.C., and R.G.; investigation, I.A., A.K., M.C., and R.G.; data curation, I.A., A.K., M.C., and R.G.; writing—original draft preparation, I.A.; writing—review and editing, I.A., A.K., D.P., M.C., and R.G.; project administration, I.A. and D.P. All authors have read and agreed to the published version of the manuscript.

Funding: This research was funded by ABB.

Institutional Review Board Statement: Not applicable.

Informed Consent Statement: Not applicable.

Data Availability Statement: Restrictions apply to the availability of these data due to privacy concerns pertaining to the data source.

Conflicts of Interest: The authors declare no conflict of interest. The funders had no role in the design of the study; in the collection, analyses, or interpretation of data; in the writing of the manuscript, or in the decision to publish the results.

References

1. Zhao, R.; Wang, D.; Yan, R.; Mao, K.; Shen, F.; Wang, J. Machine Health Monitoring Using Local Feature-Based Gated Recurrent Unit Networks. *IEEE Trans. Ind. Electron.* **2017**, *65*, 1539–1548. [CrossRef]
2. Lund, D.; MacGillivray, C.; Turner, V.; Morales, M. *Worldwide and Regional Internet of Things (IoT) 2014–2020 Forecast: A Virtuous Circle of Proven Value and Demand*; International Data Corporation: Framingham, MA, USA, 2014.
3. Lei, Y.; Jia, F.; Lin, J.; Xing, S.; Ding, S.X. An intelligent fault diagnosis method using unsupervised feature learning towards mechanical big data. *IEEE Trans. Ind. Electron.* **2016**, *63*, 3137–3147. [CrossRef]
4. Yin, S.; Li, X.; Gao, H.; Kaynak, O. Data-Based Techniques Focused on Modern Industry: An Overview. *IEEE Trans. Ind. Electron.* **2015**, *62*, 657–667. [CrossRef]
5. Chen, Z.; Fang, H.; Chang, Y. Weighted data-driven fault detection and isolation: A subspace-based approach and algorithms. *IEEE Trans. Ind. Electron.* **2016**, *63*, 3290–3298. [CrossRef]
6. Jardine, A.K.; Lin, D.; Banjevic, D. A review on machinery diagnostics and prognostics implementing condition-based maintenance. *Mech. Syst. Signal Process.* **2006**, *20*, 1483–1510. [CrossRef]
7. Zhao, R.; Wang, J.; Yan, R.; Mao, K. Machine health monitoring with LSTM networks. In Proceedings of the 10th International Conference on Sensing Technology (ICST), Nanjing, China, 11–13 November 2016; pp. 1–6.
8. Liu, Z.; Jia, Z.; Vong, C.M.; Bu, S.; Han, J.; Tang, X. Capturing high-discriminative fault features for electronics-rich analog system via deep learning. *IEEE Trans. Ind. Inform.* **2017**, *13*, 1213–1226. [CrossRef]
9. He, M.; He, D. Deep Learning Based Approach for Bearing Fault Diagnosis. *IEEE Trans. Ind. Appl.* **2017**, *53*, 3057–3065. [CrossRef]
10. Janssens, O.; Van De Walle, R.; Loccufier, M.; Van Hoecke, S. Deep Learning for Infrared Thermal Image Based Machine Health Monitoring. *IEEE/ASME Trans. Mechatronics* **2017**, *23*, 151–159. [CrossRef]
11. Jiang, G.; He, H.; Xie, P.; Tang, Y. Stacked Multilevel-Denoising Autoencoders: A New Representation Learning Approach for Wind Turbine Gearbox Fault Diagnosis. *IEEE Trans. Instrum. Meas.* **2017**, *66*, 2391–2402. [CrossRef]
12. Hochreiter, S.; Schmidhuber, J. Long short-term memory. *Neural Comput.* **1997**, *9*, 1735–1780. [CrossRef]
13. Bengio, Y.; Simard, P.; Frasconi, P. Learning long-term dependencies with gradient descent is difficult. *IEEE Trans. Neural Networks* **1994**, *5*, 157–166. [CrossRef]
14. Chung, J.; Gulcehre, C.; Cho, K.; Bengio, Y. Empirical evaluation of gated recurrent neural networks on sequence modeling. In Proceedings of the NIPS Workshop on Deep Learning, Montreal, QC, Canada, 8–13 December 2014.
15. Cho, E.; Van Merriënboer, B.; Gulcehre, C.; Bahdanau, D.; Bougares, F.; Schwenk, H.; Bengio, Y. Learning phase representations using RNN encoder-decoder for statistical machine translation. In Proceedings of the Conference for Empirical Methods in Natural Language Processing (EMNLP), Doha, Qatar, 25–29 October 2014; pp. 1724–1734.
16. Gallicchio, C.; Micheli, A. Deep echo state network (DeepESN): A brief survey. *arXiv* **2017**, arXiv:1712.04323.
17. Jaeger, H.; Haas, H. Harnessing nonlinearity: Predicting chaotic systems and saving energy in wireless communication. *Science* **2004**, *304*, 78–80. [CrossRef]
18. Fawaz, H.I.; Forestier, G.; Weber, J.; Idoumghar, L.; Muller, P.-A. Deep learning for time series classification: A review. *Data Min. Knowl. Discov.* **2019**, *33*, 917–963. [CrossRef]
19. Fawaz, H.I.; Lucas, B.; Forestier, G.; Pelletier, C.; Schmidt, D.F.; Weber, J.; Webb, G.I.; Idoumghar, L.; Muller, P.-A.; Petitjean, F. InceptionTime: Finding AlexNet for time series classification. *Data Min. Knowl. Discov.* **2020**, *34*, 1–27. [CrossRef]

20. Szegedy, C.; Ioffe, S.; Vanhoucke, V.; Alemi, A.A. Inception-v4, inception-ResNet and the impact of residual connections on learning. In Proceedings of the Thirty-First AAAI Conference on Artificial Intelligence, San Francisco, CA, USA, 4–9 February 2017; pp. 4278–4284.
21. Szegedy, C.; Liu, W.; Jia, Y.; Sermanet, P.; Reed, S.; Anguelov, D.; Erhan, D.; Vanhoucke, V.; Rabinovich, A. Going deeper with convolutions. In Proceedings of the IEEE Conference on Computer Vision and Pattern Recognition, Boston, MA, USA, 7–12 June 2015; pp. 1–9.
22. Wu, N.; Green, B.; Ben, X.; O'Banion, S. Deep transformer models for time series forecasting: The influenza prevalence case. *arXiv* **2020**, arXiv:2001.08317.
23. Vaswani, A.; Shazeer, N.; Parmar, N.; Uszkoreit, J.; Jones, L.; Gomez, A.N.; Kaiser, L.; Polosukhin, I. Attention is all you need. *Adv. Neural Inf. Process. Syst.* **2017**, *30*, 5998–6008.
24. Amihai, I.; Chioua, M.; Gitzel, R.; Kotriwala, A.M.; Pareschi, D.; Sosale, G.; Subbiah, S. Modeling Machine Health Using Gated Recurrent Units with Entity Embeddings and K-Means Clustering. In Proceedings of the IEEE 16th International Conference on Industrial Informatics, Porto, Portugal, 18–20 July 2018; pp. 212–217.
25. Liu, L.; Chen, S.; Zhang, F.; Wu, F.X.; Pan, Y.; Wang, J. Deep convolutional neural network for automatically segmenting. *Neural Comput. Appl.* **2020**, *32*, 6545–6558. [CrossRef]
26. Karim, F.; Majumdar, S.; Darabi, H.; Chen, S. LSTM Fully Convolutional Networks for Time Series Classification. *IEEE Access* **2018**, *6*, 1662–1669. [CrossRef]
27. Stetco, A.A. Wind Turbine operational state prediction: Towards featureless, end-to-end predictive maintenance. In Proceedings of the International Conference on Big Data (Big Data), Los Angeles, CA, USA, 9–12 December 2019; pp. 4422–4430.
28. Duan, J.S. A novel ResNet-based model structure and its applications in machine health monitoring. *J. Vib. Control* **2021**, *27*, 1036–1050. [CrossRef]
29. Liu, R.; Wang, F.; Yang, B.; Qin, S.J. Multiscale Kernel Based Residual Convolutional Neural Network for Motor Fault Diagnosis Under Nonstationary Conditions. *IEEE Trans. Ind. Informatics* **2020**, *16*, 3797–3806. [CrossRef]
30. Zhang, W. Aero-engine remaining useful life estimation based on 1-dimensional FCN-LSTM neural networks. In Proceedings of the 2019 IEEE Chinese Control Conference (CCC), Guangzhou, China, 27–30 July 2019.
31. Breiman, L. Random forests. *Mach. Learn.* **2001**, *45*, 5–32. [CrossRef]
32. Cutler, A.; Cutler, D.R.; Stevens, J.R. Random forests. In *Ensemble Machine Learning*; Zhang, C., Ma, Y., Eds.; Springer: Boston, MA, USA, 2012; pp. 157–175.
33. Painsky, A.; Wornell, G. on the universality of the logistic loss function. In Proceedings of the IEEE International Symposium on Information Theory (ISIT), Vail, CO, USA, 17–22 June 2018; pp. 936–940.
34. Cheng, G.; Berkhahn, F. Entity embeddings of categorical variables. *arXiv* **2016**, arXiv:1604.06737.
35. de Brébisson, A.; Simon, É.; Auvolat, A.; Vincent, P.; Bengio, Y. Artificial neural networks applied to taxi destination prediction. *arXiv* **2015**, arXiv:1508.00021.
36. Prechelt, L. Early stopping-but when? In *Neural Networks: Tricks of the Trade*; Montavon, G., Orr, G.B., Müller, K.R., Eds.; Springer: Berlin/Heidelberg, Germany, 1998; pp. 55–69.
37. Chicco, D.; Jurman, G. The advantages of the Matthews correlation coefficient (MCC) over F1 score and accuracy in binary classification evaluation. *BMC Genom.* **2020**, *21*, 6. [CrossRef]
38. Oliphant, T.E. Python for scientific computing. *Comput. Sci. Eng.* **2007**, *9*, 10–20. [CrossRef]
39. Li, B.; Chow, M.-Y.; Tipsuwan, Y.; Hung, J. Neural-network-based motor rolling bearing fault diagnosis. *IEEE Trans. Ind. Electron.* **2000**, *47*, 1060–1069. [CrossRef]
40. Samanta, B.; Al-Balushi, K. Artificial neural network based fault diagnostics of rolling element bearings using time-domain features. *Mech. Syst. Signal Process.* **2003**, *17*, 317–328. [CrossRef]
41. Aminian, M.; Aminian, F. Neural-network based analog-circuit fault diagnosis using wavelet transform as preprocessor. *IEEE Trans. Circuits Syst. II Analog. Digit. Signal Process.* **2000**, *47*, 151–156. [CrossRef]
42. Su, H.; Chong, K.T. Induction machine condition monitoring using neural network modeling. *IEEE Trans. Ind. Electron.* **2007**, *54*, 241–249. [CrossRef]
43. Samir, K.; Takehisa, Y. A review on the application of deep learning in system health management. *Mech. Syst. Signal Process.* **2018**, *107*, 241–265.
44. Toh, G.; Park, J. Review of vibration-based structural health monitoring using deep learning. *Appl. Sci.* **2020**, *10*, 1680. [CrossRef]
45. Zheng, S.; Ristovski, K.; Farahat, A.; Gupta, C. Long short-term memory network for remaining useful life estimation. In Proceedings of the IEEE International Conference on Prognostics and Health Management (ICPHM), Dallas, TX, USA, 19–21 June 2017; pp. 88–95.
46. Yuan, M.; Wu, Y.; Li, L. Fault diagnosis and remaining useful life estimation of aero engine using LSTM neural network. In Proceedings of the IEEE International Conference on Aircraft Utility Systems (AUS), Beijing, China, 10–12 October 2016; pp. 135–140.
47. Malhotra, P.; Ramakrishnan, A.; Anand, G.; Vig, L.; Agarwal, P.; Shroff, G. Multi-sensor prognostics using unsupervised health index based on LSTM Encoder-Decoder. *arXiv* **2016**, arXiv:1608.06154.
48. Chen, Y.; Peng, G.; Zhu, Z.; Li, S. A novel deep learning method based on attention mechanism for bearing remaining useful life prediction. *Appl. Soft Comput.* **2020**, *86*, 105919. [CrossRef]

49. Xia, T.; Song, Y.; Zheng, Y.; Pan, E.; Xi, L. An ensemble framework based on convolutional bi-directional LSTM with multiple time windows for remaining useful life estimation. *Comput. Ind.* **2020**, *115*, 103182. [CrossRef]
50. He, M.; Zhou, Y.; Li, Y.; Wu, G.; Tang, G. Long short-term memory network with multi-resolution singular value decomposition for prediction of bearing performance degradation. *Measurement* **2020**, *156*, 107582. [CrossRef]
51. Zhao, R.; Yan, R.; Wang, J.; Mao, K. Learning to monitor machine health with convolutional bi-directional LSTM networks. *Sensors* **2017**, *17*, 273. [CrossRef] [PubMed]
52. Tao, Y.; Wang, X.; Sanches, R.V.; Yang, S.; Bai, Y. Spur gear fault diagnosis using a multilayer gated recurrent unit approach with vibration signal. *IEEE Access* **2019**, *7*, 56880–56889. [CrossRef]
53. Guo, L.; Gao, H.; Huang, H.; He, X.; Li, S. Multifeatures fusion and nonlinear dimension reduction for intelligent bearing condition monitoring. *Shock Vib.* **2016**, *2016*, 1–10. [CrossRef]
54. Janssens, O.; Slavkovikj, V.; Vervisch, B.; Stockman, K.; Loccufier, M.; Verstockt, S.; Van de Walle, R.; Van Hoecke, S. Convolutional Neural Network Based Fault Detection for Rotating Machinery. *J. Sound Vib.* **2016**, *377*, 331–345. [CrossRef]
55. Babu, G.S.; Zhao, P.; Li, X.L. Deep convolutional neural network based regression approach for estimation of remaining useful life. In Proceedings of the International Conference on Database Systems for Advanced Applications, Dallas, TX, USA, 16–19 April 2016; pp. 214–228.
56. Chen, Z.; Shang, L.; Zhou, M. A FP-CNN method for aircraft fault prognostics. In Proceedings of the 3rd International Conference on Automation, Mechanical Control and Computational Engineering (AMCCE), Dalian, China, 12–13 May 2018; pp. 571–579.
57. Wang, J.; Zhuang, J.; Duan, L.; Cheng, W. A multi-scale convolutional neural network for featureless fault diagnosis. In Proceedings of the 2016 International Symposium of Flexible Automation (ISFA), Cleveland, OH, USA, 1–3 August 2016; pp. 1–6.
58. Guennemann, N.; Pfeffer, J. Predicting defective engines using convolutional neural networks on temporal vibration signals. In Proceedings of the First International Workshop on Learning with Imbalanced Domains: Theory and Applications, Skopje, Macedonia, 22 September 2017; pp. 92–102.
59. de Oliveira, M.; Monteiro, A.; Vieira, F.J. A new structural health monitoring strategy based on PZT sensors and convolutional neural networks. *Sensors* **2018**, *18*, 2955. [CrossRef]
60. Wen, L.; Li, X.; Gao, L.; Zhang, Y. A New Convolutional Neural Network-Based Data-Driven Fault Diagnosis Method. *IEEE Trans. Ind. Electron.* **2017**, *65*, 5990–5998. [CrossRef]
61. Abdeljaber, O.; Avci, O.; Kiranyaz, S.; Gabbouj, M.; Inman, D.J. Real-time vibration-based structural damage detection using one-dimensional convolutional neural networks. *J. Sound Vib.* **2017**, *388*, 154–170. [CrossRef]
62. Han, T.; Liu, C.; Yang, W.; Jiang, D. A novel adversarial learning framework in deep convolutional neural network for intelligent diagnosis of mechanical faults. *Knowl. Based Syst.* **2019**, *165*, 474–487. [CrossRef]
63. Dong, H.; Yang, L.; Li, H. Small fault diagnosis of front-end speed controlled wind generator based on deep learning. *WSEAS Trans. Circuits Syst.* **2016**, *15*, 64–72.
64. Lin, Y.; Nie, Z.-H.; Ma, H.-W. Structural Damage Detection with Automatic Feature-Extraction through Deep Learning. *Comput. Civ. Infrastruct. Eng.* **2017**, *32*, 1025–1046. [CrossRef]
65. Zhao, R.; Yan, R.; Chen, Z.; Mao, K.; Wang, P.; Gao, R.X. deep learning and its applications to machine health monitoring. *Mech. Syst. Signal Process.* **2019**, *115*, 213–237. [CrossRef]
66. Baur, M.; Albertelli, P.; Monno, M. a review of prognostics and health management of machine tools. *the international J. Adv. Manuf. Technol.* **2020**, *107*, 2843–2863. [CrossRef]
67. Alshorman, O.; Irfan, M.; Saad, N.; Zhen, D.; Haider, N.; Glowacz, A.; Alshorman, A. A Review of Artificial Intelligence Methods for Condition Monitoring and Fault Diagnosis of Rolling Element Bearings for Induction Motor. *Shock. Vib.* **2020**, *2020*, 1–20. [CrossRef]
68. Thoppil, N.M.; Vasu, V.; Rao, C.S.P. Deep learning algorithms for machinery health prognostics using time-series data: A review. *J. Vib. Eng. Technol.* **2021**. [CrossRef]

engineering proceedings

Proceedings

Analyzing the Effectiveness of COVID-19 Lockdown Policies Using the Time-Dependent Reproduction Number and the Regression Discontinuity Framework: Comparison between Countries †

Shangjun Liu [1,2], Tatiana Ermolieva [2], Guiying Cao [2,*], Gong Chen [1] and Xiaoying Zheng [1,*]

[1] Institute of Population Research, Peking University, Beijing 100871, China; liushangjun@pku.edu.cn (S.L.); chengong@pku.edu.cn (G.C.)
[2] International Institution for Applied Systems Analysis, 2361 Laxenburg, Austria; ermol@iiasa.ac.at
* Correspondence: cao@iiasa.ac.at (G.C.); xzheng@pku.edu.cn (X.Z.)
† Presented at the 7th International conference on Time Series and Forecasting, Gran Canaria, Spain, 19–21 July 2021.

Abstract: This study compares the effectiveness of COVID-19 control policies on the virus's spread and on the change of the infection dynamics in China, Germany, Austria, and the USA relying on a regression discontinuity in time and 'earlyR' epidemic models. The effectiveness of policies is measured by real-time reproduction number and cases counts. Comparison between the two lockdowns within each country showed the importance of people's risk perception for the effectiveness of the measures. Results suggest that restrictions applied for a long period or reintroduced later may cause at-tenuated effect on the circulation of the virus and the number of casualties.

Keywords: COVID-19; regression discontinuity in time and 'earlyR' epidemic models; real-time reproduction number; risk perception; effectiveness of intervention measures

Citation: Liu, S.; Ermolieva, T.; Cao, G.; Chen, G.; Zheng, X. Analyzing the Effectiveness of COVID-19 Lockdown Policies Using the Time-Dependent Reproduction Number and the Regression Discontinuity Framework: Comparison between Countries . *Eng. Proc.* **2021**, *5*, 8. https://doi.org/10.3390/engproc2021005008

Published: 25 June 2021

1. Introduction

COVID-19 is an infectious disease caused by SARS-CoV-2, which has been declared a global public health emergency [1]. As of 29 May 2021, it has affected more than 100million people and resulted in more than 3.5 million deaths globally (WHO). Governments worldwide have implemented similarly strict containment and closure policies to mitigate the pandemic in order to limit the spread of the virus. These restrictive community measures that limit activities or access to resources, facilities, or institutions have been often referred to as "lockdown" measures in Asia, Europe, and America [2,3]. Countries exhibited 'herd behavior' in response to COVID-19 [2] meaning they applied similar restrictive measures. However, the effectiveness of these measures has been different between countries. Previous studies showed that containment measures implemented in countries like China and South Korea have reduced new cases by more than 90%, which has not been the case in many other countries such as Italy, Spain, and the United States [4]. The effectiveness of the social distancing measures was evident in the data of Italy, Germany, and Turkey, but not clearly in the data of the USA and the U.K. [5]. Thus, the public administration community needs to embrace international and comparative perspectives on COVID-19 to inform how governments respond to the crisis, to learn the lessons from more successful governments, and to advance pandemic crisis management [6]. Up to now and currently, the situation is still uncertain, even though the COVID-19 vaccine is being used at full throttle in vaccination campaigns.

Related research shows that policy effectiveness is associated with income groups [2], regional political trust, and compliance [7], as well as country preparedness, socioeconomic factors [4], and a country's values [3]. More and more research has pooled coronavirus data and control measures from countries and regions to compare the effectiveness of public health measures. These studies revealed the requirements needed to enhance scientific analysis and epidemic modeling, and the social and institutional challenges of operating in a global crisis [8]. The USA and Germany both are highly affected countries, with 34.1 million and 3.6 million confirmed cases, ranking 1st and 10th worldwide, respectively, as of 1 June 2021. Austria shares aspects of culture with Germany, and had 0.64 million confirmed cases as of the same date, ranking 38th worldwide. Nevertheless, the pandemic spread patterns of Austria and Germany have been different, especially in the second wave [9]. As for China, The Lancet recognized the quick containment of COVID-19 in China, which sets an encouraging example for other countries [10]. Moreover, these four countries have experienced the whole COVID-19 period, with at least two waves of outbreak, which could help to indicate the long-term effects.

Abundant time series data have been collected, and time-dependent statistical analysis has been widely applied in the public health policy research. Study on Africa used the interrupted time series analysis (ITSA) to analyze the effect of border closure on COVID-19 incidence rates (IRs), which revealed that the implementation of border closures within African countries had minimal effect on the IRs of COVID-19 [11]. The research conducted in England shows that mental health service delivery underwent sizable changes during the first national lockdown by using regression discontinuity in time design (RDiT) [12]. Furthermore, the regression discontinuity design (RDD) has been used by Chinese researchers for examining the lockdown policy effects on air quality, which explores the relationship between anti-epidemic measures and air quality based on the daily data from 326 prefecture-level cities in China [13] and an early assessment with cross-national evidence on the causal impacts of COVID-19 on air pollution by using a RDD approach [14].

This study intends to assess the effectiveness of lockdown COVID-19 control policies on the virus's spread and on the change of the infection dynamics over a year with the event of a resurgence of cases, which have been implemented in China, Germany, Austria, and the USA based on real-time monitoring data and government responses. In this analysis, the different pandemic waves and the characteristics between countries are addressed. This comparative analysis aims to provide important lessons to be learnt from the experiences of these countries. Although the future of the virus is unknown at present, countries should continue to share their experiences, shield populations, and suppress transmission to save lives. In the assessment, the Oxford COVID-19 Government Response Tracker (OxCGRT) was used, which has been also used widely during the pandemic to measure the policies. We focus on the part of containment and closure, including school closing, workplace closing, cancelling public events, restrictions on gathering size, closing public transport, stay-at-home requirements, restrictions on internal movement, and restrictions on international travel. More details can be seen in Reference [15]. Since the vaccination program has been well underway since early 2021, there is hope for a gradual return to normal interaction. However, the virus in different forms poses an ongoing threat. Therefore, we should learn from the knowledge and lessons generated in the lockdown period in order to leverage better public policy to enable more resilient and effective public health services.

2. Data and Methods

2.1. Data Sources

Data used in this analysis are from 1 January to 31 December 2020. We obtained data on policy interventions from the Oxford COVID-19 Government Response Tracker (OxCGRT), which has tracked national government policy measures in response to the COVID-19 pandemic globally for 186 countries, starting from 1 January 2020 (Version 7.0). The database details are described in the working paper [16]. Our main interest is lockdown at the city/country-level, such as stay at home orders and restrictions on movement. Data on COVID-19 daily reported cases were obtained from various official sources, including the European Centre for Disease Prevention Control (ECDC), the Johns Hopkins University

Centre for Systems Science and Engineering (JHU-CSSE) and the Center for Disease Control and Prevention (CDC) [17].

2.2. Epidemics and Regression Discontinuity in Time (RDiT) Model

We used a regression discontinuity in time (RDiT) design to estimate the effectiveness of lockdown policy interventions. RDiT is extended by the regression discontinuity (RD) framework that has applications in several fields. Compared to the standard RD framework, RDiT has been adapted to applications where time is the running variable and treatment begins at a particular threshold in time. In other words, it uses time as the running variable, with a treatment date as the threshold. This approach is close to quasi experimental framework (pre-intervention compared to post-intervention). Papers using RDiT span fields that include public economics, industrial organization, environmental economics, marketing, and international trade [18].

The effectiveness of intervention measures is measured by two ways: real-time reproduction number (R_t) and counts of cases. R_t was estimated by the 'earlyR' epidemic model, which is a simplified version of the model introduced by Anne Cori et al. [19]. Parameter estimates were obtained from the early transmission dynamics in Wuhan, China of COVID-19 project by the China Center for Disease Control and Prevention (CCDC), and the serial interval distribution had a mean ($\pm$SD) of 7.5 $\pm$ 3.4 days (95% CI, 5.3 to 19) [20]. Since the policy interventions may not have immediate effects, we hypothesized a 14-days lag time for counts of cases to coincide with the approximate incubation period of COVID-19.

We took advantage of the pandemic-induced lockdowns as an exogenous policy shock and attempted to retrieve the impact of policy interventions using RDiT approaches. In this approach, we assume the lockdown's start date is when the first "stay at home requirements" become equal to "2", which means to mandate not leaving the house with exceptions for daily exercise, grocery shopping, and 'essential' trips. Alternatively, this was also evaluated as when "restrictions on internal movement" become greater than zero, which means it is recommend not to travel between regions/cities. The usual RDiT regressions were run, both using a polynomial approach and a local linear approach. The equation is as follows:

$$Y_{it} = \alpha_i + \beta_i L_{it} + \gamma X_{it} + f(d_{it}) + \varepsilon_{it} \tag{1}$$

where the outcome variables $Y(R_t$ or *counts of cases*$)$ in country i on date t, Y_{it}, is regressed by treatment variable L_{it}, a dummy variable for pre/post-intervention, a vector of covariates X_{it}, and a flexible nth-order polynomial in $f(d_{it})$, and d_{it} denotes the number of days from lockdown date. The coefficient of interest, β_i, is the treatment effect of the lockdown interventions on outcome variables in country i. In other words, this is the expected difference between the outcome variable before and after the lockdown. Additionally, α_i denotes the country fixed effects and ε_{it} denotes the error term.

As countries implemented more than one lockdown because of the secondary COVID-19 waves, we define the first lockdown as the timing of "stay at home requirements" policy adoption (score becomes equal to "2") and the second lockdown as the timing of re-imposition after subsequent policy easing. Table 1 presents two consequent lockdowns and summarizes the information about the lockdowns in the case study countries during the research period from 1 January to 31 December 2020, i.e., the date on the lockdown, number of COVID-cases on that day, and the policy stringency index. Policy stringency index is one of the composite measures, which combine different indicators into a general index. The details can be found in the codebook [16]. Although there is a lack of information on policy implication and demographic or cultural characteristics, the value and purpose of the indices is to allow for cross-national comparisons of government interventions.

Table 1. Lockdown time and conditions.

Country	Lockdowns	Date	R_t	COVID-19 Cases	Policy Stringency Index
China	1st	1 February	1.27	2089	77.31
	2nd	10 May	0.93	20	81.94
Germany	1st	21 March	1.09	2365	68.06
	2nd	22 October	1.11	5952	60.65
Austria	1st	16 March	0.92	158	81.48
	2nd	17 October	1.49	1747	58.8
USA	1st	15 March	1.63	234	41.2
	2nd	13 October	1.22	52879	66.2

3. Results

3.1. Estimates of R_t

The estimated R_t for all included countries (China, Germany, Austria, and the United States) from 1 January to 1 December 2020 are shown in Figure 1. It is clear that all countries were affected by the pandemic after March 2020, and the changing dynamics of the impacts in the four selected countries was different. China had the highest reproduction number, while the maximum reproduction number in the other three countries seemed to be similar. In the early stages, all countries were exposed to extremely high pandemic risk spread rate, which is indicated in Figure 1, with the highest position of the parameter R_t in all countries. In the period from March to April, R_t gradually declined because of governmental intervention policies to reduce the pandemic spread.

Among the intervention measures, the lockdowns are perhaps the most stringent. Table 1 shows that lockdowns in different case study countries were introduced differently. In China, the first lockdown was implemented on 1 February, i.e., the earliest date of the four studied countries. In Austria and the USA, the first lockdown started almost simultaneously, then followed by the lockdown on 21 March in Germany. From March to May, the curve of R_t was flattened; however, it still fluctuated around $R_t = 1$. It is visible from Figure 1 that after improving the situation as a result of the first lockdown, all four countries experienced repetitive rush increases of the parameter R_t in different subperiods during March and December 2020. This can be explained by the fact that in each country the removal of the lockdown led to the return of the highly epidemic situation because of the insufficient natural immunity among the population.

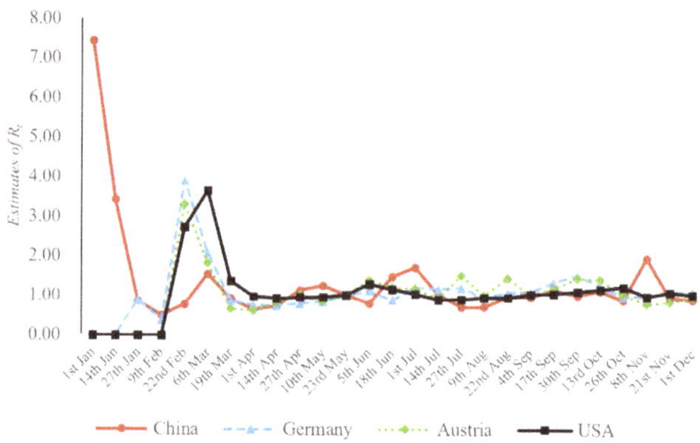

Figure 1. Estimated R_t of four countries from 1 January to 1 December (14-days smoothed).

Figure 2 shows the trend of 14-days average daily cases during this period. For China, the epidemic peak passed with the number of new cases steadily declining and the epidemic under control. The estimated R_t shows fluctuations because China is likely to see sporadic outbreaks of scattered infections or to experience regional outbreaks. For the USA, there was an initial infection peak in April, and the rate of new cases dropped somewhat after the containment interventions. However, it is more of a plateau, and the next peak came in July. Experiencing the temporarily declining, the second wave bounced, increasing exponentially after September. For Germany and Austria, the second wave also came after July, but Austria seemed to control it better.

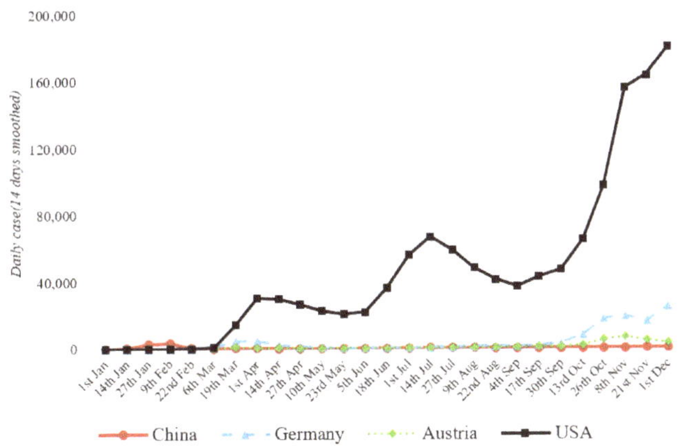

Figure 2. Daily case of four countries from 1 January to 1 December (14-days smoothed).

3.2. Overall Impact of Lockdown Interventions

Figures 3 and 4 show the regression discontinuity in time estimates, including R_t and daily cases. The horizontal axis displays days before and after the complete lockdown at d = 0, the vertical axis defines the value of R_t (or daily cases) in the respective day on the horizontal axis. Both show the prima facie evidence of impacts of the first lockdowns. Especially the R_t shows significant discontinuity for these four countries, and it implies the lockdowns have effects on the spread. Every country had a tendency to flatten R_t after the lockdown. However, the decreasing trend of R_t before d = 0 indicates that some of governmental intervention measures were already implemented before the complete lockdown, such as "keep distance" or "wear masks". As for the daily cases, Germany, Austria, and the USA show a closer discontinuity gap. This can be attributed to the limited cases before the first lockdown time.

Most RDiT models were of good fit. The country-specific linear interaction and quadratic interaction regression results are presented in Table 2 (dependent variable is R_t) and Table 3 (dependent variable is daily cases). For China, the R_t could decline by 0.988 before the first lockdown, and the lockdown brought a 4.457 decrease, which is strongly statistically significant. The quadratic interaction regression results were similar, with a 4.432 decrease. For Germany, compared to the slightly increase, the R_t declined by nearly 2 after the lockdown. Austria showed a 1.201 increase before the first lockdown, while there was a 3.831 decrease after the first lockdown. For the U.S., the R_t could increase by 1.879, while there was a 5.566 decrease after the first lockdown.

As for the analysis of daily cases, the results looked different. For China, the daily cases would increase 2782.4 if there was no intervention of lockdown. The first lockdown decrease of 3274.5 daily cases was strongly statistically significant. However, for Germany, Austria, and the USA, the situation was different. Even though the R_t flattened, the daily cases increased after the first lockdown, with 7162.5, 1473.1, and 37,561.1, respectively. It seems strange but is in line with the facts.

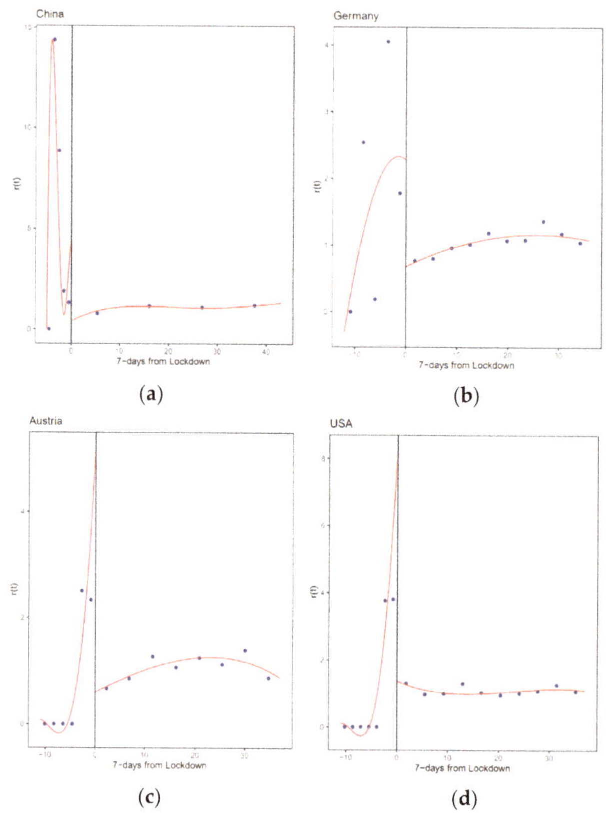

Figure 3. RDiT graphs, (**a**) China (**b**) Germany (**c**) Austria (**d**) USA. 7 days from the first lockdown and R_t (These figures plot regression discontinuity in time estimates. All show a line fitted to those observations by using a local linear approach).

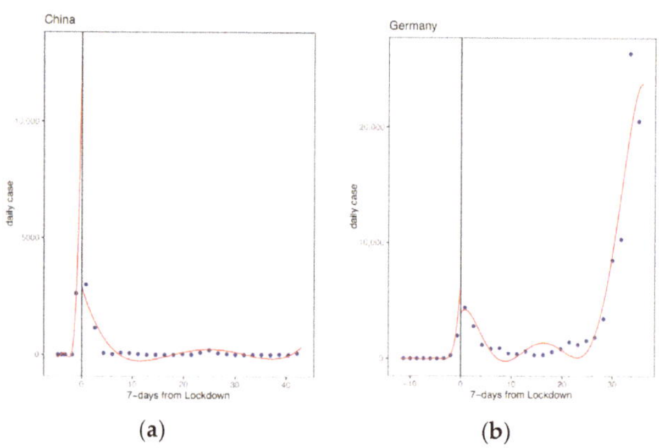

Figure 4. *Cont.*

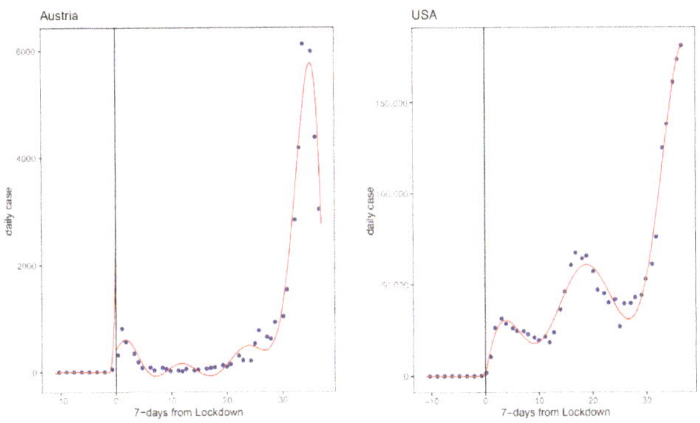

(**c**) (**d**)

Figure 4. RDiT graphs, (**a**) China (**b**) Germany (**c**) Austria (**d**) USA. 7 days from the first lockdown and R_t (These figures plot regression discontinuity in time estimates. All shows a line fitted to those observations by using a local linear approach).

Table 2. RDiT of the effects of first lockdown on COVID-19 R_t across China, Germany, Austria, and the USA.

	Dependent Variable: R_t							
	China		Germany		Austria		USA	
	(1)	(2)	(1)	(2)	(1)	(2)	(1)	(2)
X	−0.988 *	0.007	0.240 ***	0.108 ***	0.289 ***	1.201 ***	0.447 ***	1.879 ***
I(X_2)		0.0001		−0.003 **		0.076 ***		0.119 ***
treatment	−4.457 ***	−4.432 ***	−2.167 ***	−2.059 ***	−1.534 ***	−3.831 ***	−2.605 ***	−5.566 ***
X_treatment	0.999 *	14.458 ***	−0.228 ***	0.107	−0.279 ***	−1.137 ***	−0.448 ***	−1.904 ***
I(X_2):treatment		2.240 ***		0.023		−0.077 ***		−0.119 ***
Constant	2.322	−13.358 ***	3.000 ***	2.280 **	2.403 ***	4.379 ***	3.716 ***	6.819 ***
Adjusted R^2	0.329	0.619	0.178	0.164	0.319	0.504	0.319	0.565
F Statistic	8.842 ***	16.567 ***	4.463 ***	2.880 **	8.493 ***	10.750 ***	8.499 ***	11.171 ***

Note: * p ** p *** $p < 0.01$.

Table 3. RDiT of the effects of lockdowns on COVID-19 daily cases across China, Germany, Austria and the USA.

	Dependent Variable: Daily Cases							
	China		Germany		Austria		USA	
	(1)	(2)	(1)	(2)	(1)	(2)	(1)	(2)
X	526.400 *	2782.400 ***	402.040 ***	550.004 ***	95.700 ***	13.405	2930.453 ***	50.155
I(X_2)		376.000 **		27.681 ***		0.910		121.134 ***
treatment	−1362.27 *	−3274.50 ***	5537.66 **	7162.482 **	1388.330 **	1473.104 **	18,375.940	37,561.110 **
X_treatment	−550.325 **	−2909.15 ***	326.664	1701.041	95.463	237.527		2846.793
I(X_2):treatment		−373.61 **		16.871		7.794		153.278
Constant	2105.600 ***	4737.600 ***	2805.675 *	1631.250	19.982	43.648	74.382	163.018
Adjusted R^2	0.216	0.475	0.382	0.781	0.447	0.768	0.677	0.789
F Statistic	5.41 ***	9.676 ***	10.901 ***	35.251 ***	13.946 ***	32.729 ***	31.495 ***	32.115 ***

Note: * p ** p *** $p < 0.01$.

3.3. Comparative Effectiveness of First and second Lockdown

As discussed earlier, many countries implemented more than one lockdown because of the secondary COVID-19 waves. Therefore, we took the second wave into the consideration in our research. We compared the R_t and daily cases of 25 days before and after each lockdown. The country-specific quadratic interaction regression results are presented in Table 4 (dependent variable is R_t) and Table 5 (dependent variable is daily cases).

For China, the effectiveness of the second lockdown was weaker compared to the first lockdown, with an estimated R_t decrease by 1.556 in the first lockdown and increase by

0.585 in the second lockdown. Meanwhile, the daily cases did not decrease as fast as before (−24.5 vs. −762.3). In Germany, we found out the effect on the R_t was slightly stronger; R_t decreased by 0.715 (from 0.827 to 0.112), although it still was positive. Therefore, there was a significant increase in the daily cases after the second lockdown (4811.651 increase). For Austria, the effect on R_t after the first lockdown and the second lockdown was a 0.564 increase and 0.128 decrease, respectively. It means that the second lockdown contributed to flattening the R_t curve. Meanwhile, we also saw a decrease of daily cases in Austria, with 108.633 and 254.206, respectively. For the USA, there was a significant increase of R_t after the second lockdown, which had a 4.75 increase. Compared to the first lockdown, the situation became worse, with a higher R_t (1.334 vs. 4.750) and daily cases (−2058.49 vs. 2100.23).

It is noted that the results were related to the baseline, namely, the total confirmed and infected cases. For China, the first outbreak was the most serious wave during the COVID-19 period, which affected the country nationwide. Therefore, the first lockdown quickly smoothed the curve and reduced a large number of cases. The second lockdown was introduced at regional level to smooth provincial outbreaks. On the other hand, for Europe and the USA, the pattern was different. The first lockdown in Europe and the USA was introduced when the cases were growing and the epidemics' epicenters were detected in neighboring countries. The second lockdown was implemented to deal with the domestic outbreak.

Table 4. Compared RDiT of the effects of the first and second lockdowns on COVID-19 R_t across China, Germany, Austria, and the USA.

	Dependent Variable: R_t							
	China		Germany		Austria		USA	
	1st	**2nd**	**1st**	**2nd**	**1st**	**2nd**	**1st**	**2nd**
X	0.371 ***	−0.117 ***	−0.282 ***	−0.065 ***	−0.359 ***	0.002	−0.874 ***	−2.352 ***
I(X_2)	0.033 ***	−0.005 ***	−0.005 **	−0.002 ***	−0.006	−0.001 *	−0.041 ***	−0.180 ***
treatment	−1.556 ***	0.585 *	0.827 *	0.112 ***	0.564	−0.128 **	1.334	4.750 ***
X_treatment	−0.359 ***	0.141 **	0.272 ***	0.046 ***	0.336 ***	−0.037 ***	0.752 ***	2.282 ***
I(X_2):treatment	−0.034 ***	0.004 *	0.005 *	0.003 ***	0.007	0.001	0.044 ***	0.178 ***
Constant	1.986 ***	0.529 **	0.027	1.014 ***	0.174	1.566 ***	0.771	−2.742 **
Adjusted R^2	0.986	0.173	0.854	0.964	0.910	0.903	0.649	0.752
F Statistic	685.674 ***	3.098 **	59.560 ***	272.104 ***	91.812 ***	94.318 ***	19.486 ***	22.653 ***

Note: * p ** p *** $p < 0.01$.

Table 5. Compared RDiT of the effects of the first and second lockdowns on COVID-19 daily cases across China, Germany, Austria, and the USA.

	Dependent Variable: Daily Cases							
	China		Germany		Austria		USA	
	1st	**2nd**	**1st**	**2nd**	**1st**	**2nd**	**1st**	**2nd**
X	353.007 ***	7.889 ***	426.831 ***	646.062 ***	36.538 **	67.566 *	735.098 ***	735.098 ***
I(X_2)	9.801 **	0.423 ***	10.875 ***	12.018	1.191 *	4.351 ***	24.305 ***	24.305 ***
treatment	−762.321	−24.508 **	−865.244 *	4811.651 ***	−108.633 *	−254.206	−2058.490 **	2100.234 ***
X_treatment	−243.693	−8.009 ***	−192.626 **	441.737	110.960 ***	208.117 ***	2100.234 **	−35.416 ***
I(X_2):treatment	−20.954 ***	−0.416 ***	−24.520 ***	−46.101 ***	−17.168 ***	−1.549	−35.416 ***	986.184
Constant	2981.537 ***	31.207 ***	4104.298 ***	9243.376 ***	303.808 ***	1558.123 ***	938.363	938.363
Adjusted R^2	0.699	0.768	0.942	0.879	0.972	0.981	0.991	0.991
F Statistic	24.255 ***	34.156 ***	162.698 ***	122.417 ***	226.549 ***	521.901 ***	1091.190 ***	1091.190 ***

Note: * p ** p *** $p < 0.01$.

4. Discussion

We offer a retrospective study that provides cross-national evidence on the causal impacts of policy intervention on COVID-19 spread. A rich database was assembled from various sources, which were analyzed with EarlyR and RDiT models. Overall, the results show that COVID-19-induced lockdowns resulted in a decrease in R_t and daily cases, which varied across different countries. We expected the lockdown could mitigate the spread of COVID-19, but the results were not satisfactory and should be further explained. Comparing different countries, China had the most effective lockdown, which could lower the R_t and decrease the daily cases, while the USA, Germany, and Austria had strongly decreased R_t but presented large daily case enhancement. Comparison between the two lockdowns within each country showed that people's risk perception was relaxed during the second lockdown, especially in Germany (the increased daily cases were the highest in the studied countries) and the USA (the increased reproduction number was the highest in the studied countries).

Our results were similar to the relevant research, which suggested that the stringent lockdown policies adopted in China, Italy, and Spain were among the most effective national-scale policies [21]. China also showed the most effective results in our study. For Germany and Austria, they showed different patterns in our study, although they share common borders in Central Europe and have substantial cultural, historical, and economic ties [22]. The differences may be explained in terms of the fact that the power was consolidated in central governments in Austria, while in Germany, states retain their autonomy [22]. The USA presented the most unexpected results, and some research gives further explanations: namely, except for the timing and strictness of implementing measures [23,24], national culture, economic, and health and social issues also influence the results [25,26]. Most importantly, our results suggest that restrictions applied for a long period or reintroduced late in the pandemic would exert, at best, a weaker, attenuated effect on the circulation of the virus and the number of casualties. Our results support the conclusion of Haug et al. (2020) that lockdowns should be strict and brief [27].

Institutional Review Board Statement: Not applicable.

Informed Consent Statement: Not applicable.

Data Availability Statement: Data on policy interventions from the OxCGRT (http://bsg.ox.ac.uk/covidtracker, accessed on 24 June 2021). COVID-19 daily reported cases were obtained from official sources, including the ECDC (https://www.ecdc.europa.eu/en/covid-19/situation-updates, accessed on 24 June 2021); JHU-CSSE (https://github.com/CSSEGISandData/COVID-19, accessed on 24 June 2021) and the CDC (https://covid.cdc.gov/covid-data-tracker/#datatracker-home, accessed on 24 June 2021).

References

1. Hui, D.S.; Azhar, E.I.; Memish, Z.A.; Zumla, A. Human Coronavirus Infections—Severe Acute Respiratory Syndrome (SARS), Middle East Respiratory Syndrome (MERS), and SARS-CoV-2. In *Reference Module in Biomedical Sciences*; Elsevier: Amsterdam, The Netherlands, 2020.
2. Pincombe, M.; Reese, V.; Dolan, C.B. The effectiveness of national-level containment and closure policies across income levels during the COVID-19 pandemic: An analysis of 113 countries. *Health Policy Plan.* **2021**. [CrossRef]
3. Chen, S.X.; Lam, B.C.; Liu, J.H.; Choi, H.S.; Kashima, E.; Bernardo, A.B. Effects of containment and closure policies on controlling the COVID-19 pandemic in East Asia. *Asian J. Soc. Psychol.* **2021**, *24*, 42–47. [CrossRef] [PubMed]
4. Chaudhry, R.; Dranitsaris, G.; Mubashir, T.; Bartoszko, J.; Riazi, S. A country level analysis measuring the impact of government actions, country preparedness and socioeconomic factors on COVID-19 mortality and related health outcomes. *EClinicalMedicine* **2020**, *25*, 100464. [CrossRef]
5. Thu, T.P.B.; Ngoc, P.N.H.; Hai, N.M. Effect of the social distancing measures on the spread of COVID-19 in 10 highly infected countries. *Sci. Total Environ.* **2020**, *742*, 140430. [CrossRef]
6. You, J. Lessons From South Korea's Covid-19 Policy Response. *Am. Rev. Public Adm.* **2020**, *50*, 801–808. [CrossRef]
7. Bargain, O.; Aminjonov, U. Trust and Compliance to Public Health Policies in Times of Covid-19. *J. Public Econ.* **2020**, *192*, 104316. [CrossRef]
8. Trump, B.D.; Keenan, J.M.; Linkov, I. Multi-Disciplinary Perspectives on Systemic Risk and Resilience in the Time of COVID-19. In *COVID-19: Systemic Risk and Resilience*; Springer: Berlin/Heidelberg, Germany, 2021.
9. Post, L.; Culler, K.; Moss, C.B.; Murphy, R.L.; Achenbach, C.J.; Ison, M.G.; Resnick, D.; Singh, L.N.; White, J.; Boctor, M.J.; et al. Surveillance of the Second Wave of COVID-19 in Europe: Longitudinal Trend Analyses. *JMIR Public Health Surveill.* **2021**, *7*, e25695. [CrossRef] [PubMed]

10. Lancet, T. Sustaining containment of COVID-19 in China. *Lancet* **2020**, *395*, 1230. [CrossRef]
11. Emeto, T.I.; Alele, F.O.; Ilesanmi, O.S. Evaluation of the effect of border closure on COVID-19 incidence rates across nine African countries: An interrupted time series study. *Trans. R. Soc. Trop. Med. Hyg.* **2021**. [CrossRef]
12. Bakolis, I.; Stewart, R.; Baldwin, D.; Beenstock, J.; Bibby, P.; Broadbent, M.; Cardinal, R.; Chen, S.; Chinnasamy, K.; Cipriani, A.; et al. Changes in daily mental health service use and mortality at the commencement and lifting of COVID-19 'lockdown' policy in 10 UK sites: A regression discontinuity in time design. *BMJ Open* **2021**, *11*, e049721. [CrossRef]
13. Song, Y.; Li, Z.; Liu, J.; Yang, T.; Zhang, M.; Pang, J. The effect of environmental regulation on air quality in China: A natural experiment during the COVID-19 pandemic. *Atmos. Pollut. Res.* **2021**, *12*, 21–30. [CrossRef]
14. Dang, H.A.H.; Trinh, T.A. Does the COVID-19 lockdown improve global air quality? New cross-national evidence on its unintended consequences. *J. Environ. Econ. Manag.* **2021**, *105*, 25. [CrossRef]
15. Hale, T.; Angrist, N.; Goldszmidt, R.; Kira, B.; Petherick, A.; Phillips, T.; Webster, S.; Cameron-Blake, E.; Hallas, L.; Majumdar, S.; et al. A global panel database of pandemic policies (Oxford COVID-19 Government Response Tracker). *Nat. Hum. Behav.* **2021**, *5*, 529–538. [CrossRef]
16. Hale, T.; Webster, S. *Oxford COVID-19 Government Response Tracker*; Blavatnik School of Government: Oxford, UK, 2020.
17. Hasell, J.; Mathieu, E.; Beltekian, D.; Macdonald, B.; Giattino, C.; Ortiz-Ospina, E.; Roser, M.; Ritchie, H. A cross-country database of COVID-19 testing. *Sci. Data* **2020**, *7*, 345. [CrossRef]
18. Hausman, C.; Rapson, D. Regression Discontinuity in Time: Considerations for Empirical Applications. *Annu. Rev. Resour. Econ.* **2018**, *10*, 533–552. [CrossRef]
19. Cori, A.; Ferguson, N.M.; Fraser, C.; Cauchemez, S. A New Framework and Software to Estimate Time-Varying Reproduction Numbers During Epidemics. *Am. J. Epidemiol.* **2013**, *178*, 1505–1512. [CrossRef]
20. Li, Q.; Guan, X.; Wu, P.; Wang, X.; Zhou, L.; Tong, Y.; Ren, R.; Leung, K.S.; Lau, E.H.; Wong, J.Y.; et al. Early Transmission Dynamics in Wuhan, China, of Novel Coronavirus–Infected Pneumonia. *N. Engl. J. Med.* **2020**, *382*, 1199–1207. [CrossRef]
21. Dehkordi, A.H.; Alizadeh, M.; Derakhshan, P.; Babazadeh, P.; Jahandideh, A. Understanding epidemic data and statistics: A case study of COVID-19. *J. Med. Virol.* **2020**, *92*, 868–882. [CrossRef]
22. Desson, Z.; Lambertz, L.; Peters, J.W.; Falkenbach, M.; Kauer, L. Europe's Covid-19 Outliers: German, Austrian and Swiss policy responses during the early stages of the 2020 pandemic. *Health Policy Technol.* **2020**, *9*, 405–418. [CrossRef] [PubMed]
23. Coughlin, S.S.; Yiğiter, A.; Xu, H.; Berman, A.E.; Chen, J. Early detection of change patterns in COVID-19 incidence and the implementation of public health policies: A multi-national study. *Public Health Pract.* **2020**, *2*, 100064. [CrossRef] [PubMed]
24. Fouda, A.; Mahmoudi, N.; Moy, N.; Paolucci, F. Comparing the COVID-19 Pandemic in Greece, Iceland, New Zealand, and Singapore. *Icel. N. Z. Singap.* **2020**. [CrossRef]
25. Giamberardino, P.D.; Iacoviello, D. Evaluation of the effect of different policies in the containment of epidemic spreads for the COVID-19 case. *Biomed. Signal Process. Control* **2021**, *65*, 102325. [CrossRef] [PubMed]
26. Wang, Y. Government Policies, National Culture and Social Distancing during the First Wave of the COVID-19 Pandemic: International Evidence. *Saf. Sci.* **2020**, *135*, 105138. [CrossRef]
27. Haug, N.; Geyrhofer, L.; Londei, A.; Dervic, E.; Desvars-Larrive, A.; Loreto, V.; Pinior, B.; Thurner, S.; Klimek, P. Ranking the effectiveness of worldwide COVID-19 government interventions. *Nat. Hum. Behav.* **2020**, *4*, 1303–1312. [CrossRef] [PubMed]

engineering proceedings

MDPI

Proceedings

From Permutations to Horizontal Visibility Patterns of Periodic Series †

Francisco J. Muñoz * and **Juan Carlos Nuño**

Department of Applied Mathematics, Universidad Politécnica de Madrid, 28040 Madrid, Spain; juancarlos.nuno@upm.es

* Correspondence: f.j.munoz.ortega@gmail.com

† Presented at the 7th International conference on Time Series and Forecasting, Gran Canaria, Spain, 19–21 July 2021.

Abstract: Periodic series of period T can be mapped into the set of permutations of $[T-1] = \{1, 2, 3, \ldots, T-1\}$. These permutations of period T can be classified according to the relative ordering of their elements by the horizontal visibility map. We prove that the number of horizontal visibility classes for each period T coincides with the number of triangulations of the polygon of $T+1$ vertices that, as is well known, is the Catalan number C_{T-1}. We also study the robustness against Gaussian noise of the permutation patterns for each period and show that there are periodic permutations that better resist the increase of the variance of the noise.

Keywords: periodic time series; patterns; visibility; triangulation; Gaussian noise

Citation: Muñoz, F.J.; Nuño, J.C. From Permutations to Horizontal Visibility Patterns of Periodic Series. *Eng. Proc.* **2021**, *5*, 9. https://doi.org/10.3390/engproc2021005009

Published: 25 June 2021

1. Introduction

Periodic or noisy periodic time series appear in many natural phenomena. Strictly speaking, real signals are approximately periodic and are considered to incorporate a certain seasonality [1,2]. They can also be solutions of dynamical systems, either discrete or continuous [3,4]. In practice, periodicity is finite, that is, it appears in a finite set of points. However, theoretically, these periodic series extend to infinity and, thus, they allow the consideration of the limit of infinite periods. This paper is focused on the study of the complete set of periodic natural series for each period $T \in \mathbb{N}$. Indeed, there are other classical approaches, such as, for instance, the Fourier analysis, that provide a complete set of solutions to this problem. However, our approach does not pretend to surpass them but, on the contrary, to offer an alternative viewpoint for studying this kind of time series.

A discrete series $\{X_n\}_{n \in \mathbb{N}}$, infinite or not, is said to be periodic if there is a natural number T such that $X_{n+T} = X_n$ for all $n > n_t$, that is, after a transient period n_t. For real valued series X_n, there are infinite periodic series for each period T (see Figure 1 for an example). However, this infinite number of cases can be reduced to a finite number by means of the application of discrete mappings, such as the horizontal visibility map [5]. It is worth pointing out that, for the horizontal visibility map, the time scale is not relevant, since only the values of any pair of points are compared. This means that the same horizontal visibility pattern is obtained if the time units are either seconds or years or, in the case of spatial series, either millimeters or kilometers.

2. Permutations of Periodic Series and Their Horizontal Visibility Patterns

A real valued periodic time series can be mapped into a positive integer series in which the elements within the period are ranked according to their value. For instance, as shown in Figure 1, a period 4 series with real values:
$\{\ldots, 0.5008842, 0.8749973, 0.3828197, 0.8269407, 0.5008842, \ldots\}$
obtained from the logistic equation, $X_n = r\,X_{n-1}\,(1 - X_{n-1})$, $n = 1, 2 \ldots$, is mapped into the positive integer series: $\{\ldots, 4, 1, 3, 2, 4, \ldots\}$. Thus, any real valued time series can be

transformed using this method. Let us assume that the largest value is fixed as the first value in each period. Then, if no equal values occur within the period, there are $(T-1)!$ possible permutations for the period T. Formally, given a periodic series of period T, we define its permutation pattern as the natural numbers that rank the values within the period, starting from the largest, that takes the value T. It is worth noting that all of these permutation patterns are effectively generated by applying the generic function *Rank*, a command that is defined in most programming languages.

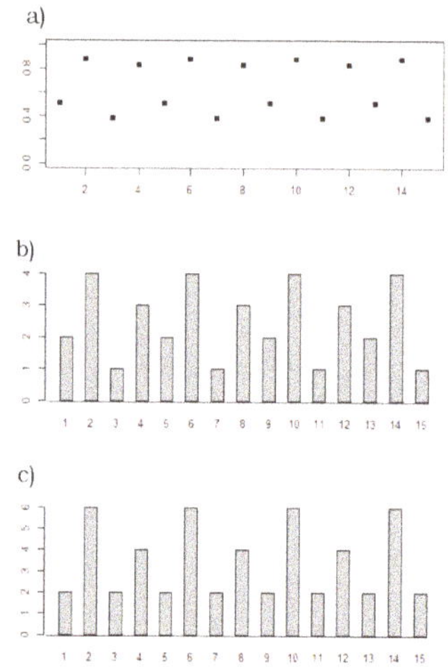

Figure 1. (**a**) $\{\ldots, 0.5008842, 0.8749973, 0.3828197, 0.8269407, 0.5008842, \ldots\}$ is a real-valued time series obtained from the logistic map $X_{n+1} = r\, X_n\,(1 - X_n)$ with initial condition $X_0 = 0.5$, after a transient period of 10^4 time steps. The growth rate value is: $r = 3.5$, which corresponds to a period 4 solution. (**b**) $\{\ldots, 4, 1, 3, 2, (4)\ldots\}$ is the corresponding permutation set obtained by ranking the values of the real-valued series. (**c**) The associated horizontal visibility pattern is: $\{\ldots, 6, 2, 4, 2, (6)\ldots\}$.

All periodic patterns can be obtained as permutations of the set $\{T-1, T-2, \ldots, 2, 1\}$ and studied by applying combinatorial techniques [6,7]. In this context, the question is how these permutation patterns can be classified with regards to a definite order, for example, that turns out from the horizontal visibility map [5].

Two points of the series, X_i and X_j with $i < j$, are said to see each other horizontally if

$$X_i, X_j > X_k \qquad \forall\, i < k < j. \tag{1}$$

Since the horizontal visibility algorithm is only dependent on the relative values between the points of the integer series, it turns out that different permutation patterns could be classified within the same category. Indeed, the application of the horizontal visibility map enables a substantial reduction of the set of all permutation patterns.

In practice, for any permutation pattern, we can obtain the associated horizontal visibility pattern: for each value of the series, we calculate the ordinal of its horizontal visibility basin, that is, the number of points that are horizontally seen from the said value, for example, as shown in Figure 1, the horizontal visibility pattern of the permutation $\{4, 1, 3, 2, (4)\}$ is $\{6, 2, 4, 2, (6)\}$. This means that the largest value has six points in its visibility basin and the remaining points, 1,3,2 have two, four and two points, respectively, in their visibility basins.

In order to count the number of horizontal visibility patterns that exist for each permutation pattern of period T, it is convenient to represent the values of the period as a convex polygon of $T+1$ vertices. To each vertex of this polygon, we assign the corresponding value of the element of the series and read counterclockwise (see Figure 2).

If we link the vertices forming the corresponding horizontal visibility graph, the projection of the edges forms a triangulation of the polygon. If we map each of the permutations into a polygon and compute their triangulation, we obtain all of the possible triangulations of these polygons of $T+1$ vertices. For example, Figure 3 depicts the six polygons that appear for period $T=4$. Please note that the label 4 appears twice, in order to close the period.

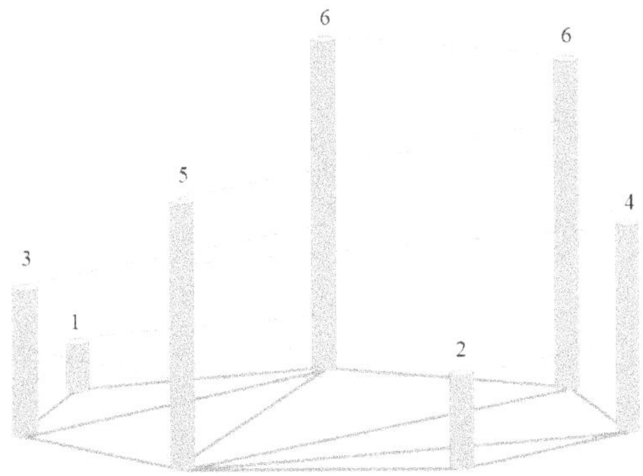

Figure 2. Tridimensional representation of a labeled heptagon that corresponds to the permutation pattern of period $T=6$: $\{6,1,3,5,2,4,(6)\}$. Its horizontal visibility pattern $\{7,2,3,5,2,3,(7)\}$ is obtained summing all the edges of the vertices, counting both 6-vertices. The projection of the horizontal visibility links on the plane yields the triangulation.

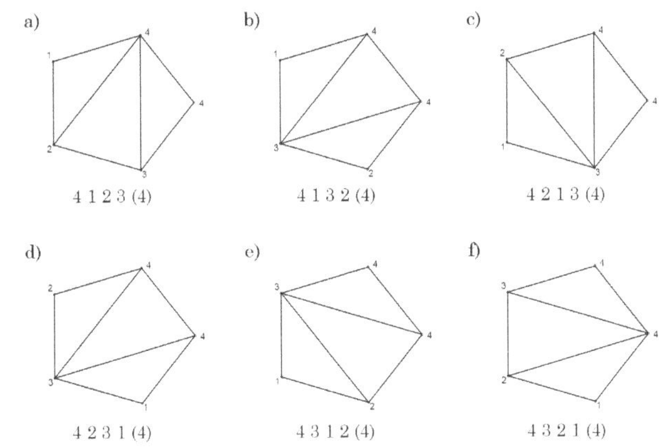

Figure 3. The six polygon triangulations related to the six possible permutation patterns for the period $T=4$ time series. As can be observed, triangulations (**b**,**d**) are equal and correspond to the same horizontal visibility pattern. On the other hand, the other four triangulations (**a**,**c**,**e**,**f**) have a unique correspondence (see Table 1).

The triangulation of convex polygons is a classical problem and it is well known that, for a polygon of $T+1$ vertices, the number of possible triangulations is given by the Catalan number C_{T-1}:

$$C_{T-1} = \frac{(2T-2)!}{T!(T-1)!}.$$

This means that the infinite real valued series of period T can be reduced to C_{T-1} horizontal visibility patterns. Nonetheless, even after this reduction, the number of horizontal visibility patterns increase exponentially with the period:

$$C_{T-1} \approx 0.021\, e^{1.269\, T}.$$

For example, for a period $T = 20$, the number of possible horizontal visibility patterns rises to $1767263190 \approx 1.8\, 10^9$.

Table 1. The 6 and 24 permutation patterns for period $T = 4$ and $T = 5$. These patterns correspond to 5 and 14 different horizontal visibility patterns. The number of interior pinnacles [6] and their values are shown in the fourth and the fifth columns.

Period	Permutation Pattern	H. Visibility Pattern	# Pinnacle	Max.Pinnacle
4	4 1 2 3 (4)	6 2 3 3 (6)	0	
4	4 1 3 2 (4)	6 2 4 2 (6)	1	3
4	4 2 1 3 (4)	5 3 2 4 (5)	0	
4	4 2 3 1 (4)	6 2 4 2 (6)	1	3
4	4 3 1 2 (4)	5 4 2 3 (5)	0	
4	4 3 2 1 (4)	6 3 3 2 (6)	0	
5	5 1 2 3 4 (5)	7 2 3 3 3 (7)	0	
5	5 1 2 4 3 (5)	7 2 3 4 2 (7)	1	4
5	5 1 4 2 3 (5)	6 2 5 2 3 (6)	1	4
5	5 4 1 2 3 (5)	5 5 2 3 3 (5)	0	
5	5 4 1 3 2 (5)	6 4 2 4 2 (6)	1	3
5	5 1 4 3 2 (5)	7 2 4 3 2 (7)	1	4
5	5 1 3 4 2 (5)	7 2 3 4 2 (7)	1	4
5	5 1 3 2 4 (5)	6 2 4 2 4 (6)	1	3
5	5 3 1 2 4 (5)	5 4 2 3 4 (5)	0	
5	5 3 1 4 2 (5)	6 3 2 5 2 (6)	1	4
5	5 3 4 1 2 (5)	6 2 5 2 3 (6)	1	4
5	5 4 3 1 2 (5)	6 3 4 2 3 (6)	0	
5	5 4 3 2 1 (5)	7 3 3 3 2 (7)	0	
5	5 3 4 2 1 (5)	7 2 4 3 2 (7)	1	4
5	5 3 2 4 1 (5)	6 3 2 5 2 (6)	1	4
5	5 3 2 1 4 (5)	5 3 3 2 5 (5)	0	
5	5 2 3 1 4 (5)	6 2 4 2 4 (6)	1	3
5	5 2 3 4 1 (5)	7 2 3 4 2 (7)	1	4
5	5 2 4 3 1 (5)	7 2 4 3 2 (7)	1	4
5	5 4 2 3 1 (5)	6 4 2 4 2 (6)	1	3
5	5 4 2 1 3 (5)	5 4 3 2 4 (5)	0	
5	5 2 4 1 3 (5)	6 2 5 2 3 (6)	1	4
5	5 2 1 4 3 (5)	6 3 2 5 2 (6)	1	4
5	5 2 1 3 4 (5)	6 3 2 4 3 (6)	0	

A property that can be immediately deduced from the polygon triangulations is that the total visibility, $V_{tot}(T)$, of any horizontal visibility pattern is the same for each period [8]: it is the sum of the edges that appear in the triangulated polygon multiplied by 2:

$$V_{tot}(T) = 2\,(\#edges) = 4\,T - 2.$$

It can also be proven that, for each period T, the maximum visibility that any permutation pattern can attain is $V_{max}(T) = T + 2$. In addition, the other $T - 3$ permutation patterns reach other maxima: $T + 1, T, \ldots$ (see Table 2). For instance, for $T = 4$, there are four permutations with the largest visibility $V_{max}(4) = 6$ and two permutation patterns with a maximum visibility of 5. Note that the maximum visibility might not occur for the largest value. For instance, the permutation pattern $\{6, 5, 1, 2, 3, 4\}$ corresponds to the horizontal visibility pattern: $\{6, 7, 3, 4, 4, 4\}$.

Another property that is worth mentioning is the average visibility of a point for each period, $\bar{V}$ [9]. It is obtained from the total visibility of each period divided by the period:

$$\bar{V} = \frac{V_{tot}(T)}{T}.$$

Evidently, it tends to four as T tends to infinity.

Table 2. Number of maximal visibilities for low period series.

T	# Max Visib	Max. Visibilities
2	1	4
3	1	5
4	2	(5,6)
5	3	(5,6,7)
6	4	(5,6,7,8)
7	5	(5,6,7,8,9)
8	6	(5,6,7,8,9,10)

The correspondence between permutation patterns and horizontal visibility patterns is not evident, as shown in Table 3. Columns provide the relation between permutations and horizontal visibility pattens. The first column indicates the number of permutations that are related univocally to a visibility pattern for each period T (rows). Similarly, the entries of the second column provide the number of permutations that are related to two visibility patterns for each period T, and so on. This table forms a reduced schelon matrix with some internal patterns that deserve to be commented on briefly. The entries of the first column grow as 2^{T-2}, whereas for the second column, they grow as 2^{T-4}. For the first three rows, the number of columns with non null entries are 2, 3 and 6. Surprisingly, there are some columns, for example, 7 and 9, which appear for the first time at period $T = 9$ and $T = 11$, respectively. Unfortunately, the table is not complete, so no rigorous conclusions can be drawn for a larger order of columns and rows.

Table 1 also shows the number of maxima (pinnacles) in the permutation patterns [6,10]. For these two periods $T = \{4, 5\}$, there are permutations that have no maximum, while others exhibit only one. The former permutations are related one-to-one to a horizontal visibility pattern. On the other hand, those permutations with one pinnacle can share the same horizontal visibility pattern. As a matter of fact, if the pinnacle takes the value of 3, for each horizontal visibility pattern there are two permutations, whereas if this value is 4, this correspondence is 3 to 1. Table 3 provides this equivalence for each period precisely.

Table 3. Each entry indicates the number of permutations for period T that corresponds to the number of visibility patterns. For instance, for period $T = 5$, there are eight permutation patterns related one-to-one to one visibility pattern. The other two appear each from two permutations and four come from three different permutations (see Table 1). The sum of each row yields the total number of visibility patterns for each period. The total permutation patterns for each period are obtained from each row, multiplying the entry by the value of each column. For example, the sum of the entries of the second row, for $T = 5$, gives the Catalan number $C_4 = 14$ and the weighted sum $(8 \cdot 1 + 2 \cdot 2 + 4 \cdot 3)$ equals the number of permutations $(T-1)! = 4! = 24$.

$T \backslash \#$	1	2	3	4	5	6	7	8	9	10	12	14	15	16	18	20
4	4	1														
5	8	2	4													
6	16	4	8	8		4		2								
7	32	8	16	16	16	8		4		20			8			4
8	64	16	32	32	32	48		8		40	8		48	16		24
9	128	32	64	64	64	96	64	16		80	16	16	96	32		48
10	256	64	128	128	128	192	128	160		160	32	32	192	32	64	96
11	512	128	256	256	256	384	256	320	256	320	64	64	384	64	192	192

3. Patterns of Noisy Periodic Series

As has been described in the previous section, for each period, more than one permutation is reduced to the same horizontal visibility pattern. The question is whether this equivalence remains when the signal is affected by any kind of noise, in particular Gaussian noise. It is expected that, when the intensity of the noise is small, the permutations of the noisy signal fall in the same visibility class as the non-perturbed time series and, consequently, all have the same horizontal visibility pattern. A similar problem has also been studied in [9]. Here, we focus on the problem of robustness against noise, for example, how Gaussian noise affects the permutation patterns as a function of the variance [11].

The way noisy series have been considered is detailed as follows:

(i) For each period T, we generate the $(T-1)!$ synthetic permutations.

(ii) From each of these patterns, a $1000 * T$ series is generated.

(iii) To each of these series we add a random variable according to a normal distribution of null mean and standard deviation (specifically, the R-function *rnorm* [12]).

(iv) We vary the standard deviation from 0 to 6, with increments of 0.1. Consequently, 61 noisy series are generated from the initial permutation.

(v) The visibility algorithm is applied for all of the 61 series, including the periodic synthetic series.

(vi) To compare the noisy series with the periodic one, we count the number of coincidences between each pair of noisy-periodic series.

Figure 4 depicts the proportion of digit coincidences between the horizontal visibility patterns obtained from the noisy permutations as a function of the variance of the Gaussian noise for low periods. The same plot for each of the visibility patterns for period $T = 4$ is presented in Figure 4. The proportion of coincidences is greater in both permutations, corresponding to the same visibility pattern: $\{6, 2, 4, 2\}$.

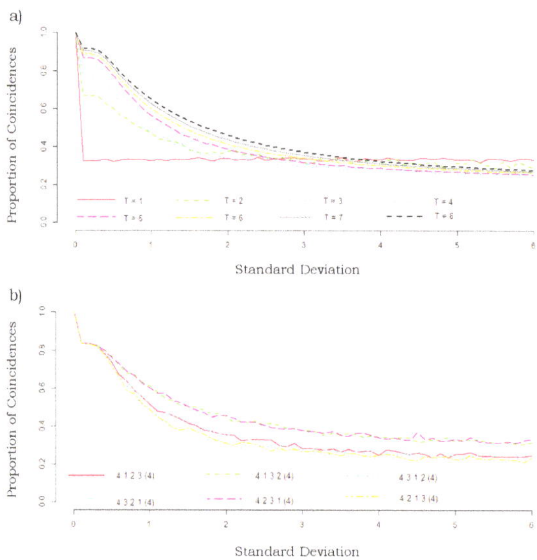

Figure 4. (**a**) Proportion of coincidences (Y-axis) between the digits of the original series formed by repeated permutations and the noisy series that result after applying a Gaussian noise of standard deviation referred to in the X-axis. Please note that the level of coincidences achieved for large values of the standard deviation is compatible with a loss of memory, that is, the loss of any relationship with the original permutation as it is evident for period $T = 1$. (**b**) For $T = 4$, six visibility permutation patterns exist. When a time series formed by the repetition of each pattern is perturbed by a Gaussian noise with a standard deviation given in the X-axis, the proportion of coincidences with the original series decreases as shown in this figure. Note that the two series formed from the permutation patterns that correspond to the same horizontal visibility pattern: $\{6, 2, 4, 2\}$ are more robust against white noise.

4. Concluding Remarks

Periodic or noisy periodic patterns appear in data sets from multiple fields of science [2]. In particular, a huge amount of data is formed by time series in which a unique variable, either discrete or continuous, is presented as a function of time that, indifferently, can also be considered discrete or continuous [3,4]. Many mathematical models also exhibit this oscillatory behavior and have been applied extensively to study its properties. Visibility algorithms are useful tools for the analysis of univariate series, for example, time series [13]. In particular, the horizontal visibility map provides analytical results about different types of series, namely periodic, random, fractional or chaotic [5]. Contrary to the natural visibility algorithm, the properties derived from the horizontal visibility map only depend on the ratio between the values of the points of the series, not on the distance between them. As shown in this paper, this enables a complete reduction of the infinite number of real valued periodic series to a finite set of visibility patterns. We prove that the number of horizontal visibility patterns for any period T is given by the Catalan number C_{T-1}. Despite this huge reduction, the number of horizontal visibility patterns still grows exponentially as a function of T.

This exponential growth contrasts with the low number of visibility patterns that are found in the logistic map [14]. This is a consequence of the form of the field, $f(x;r) = r\,x\,(1-x)$, which sets the following rules of period doubling bifurcations:

1. If previous values are such that $x_1 < x_2$, then the new duplicated values verify $x_{11}, x_{12} < x_{21}, x_{22}$.
2. The new points coming from x_1 and x_2 must be intercalated in time.

For instance, if these rules are applied to each period in the Feigenbaum cascade, a sequence of horizontal visibility patterns appears that, at the limit of the infinite period, converge to the ruler sequence [14,15]. Other unknown integer patterns are to be discovered in each of the infinite period doubling cascades that occur in the bifurcation diagram of the logistic and, in general, in unimodal maps.

Lastly, we would like to point out that it is also possible to obtain an elementary periodic pattern that corresponds to a given horizontal visibility pattern. The algorithm seeks to find a periodic pattern with the minimum positive integers that are compatible with the horizontal visibility map. Starting from the initial pattern $\{1,1,1,1,1,1\}$, the program recurrently increases these values until the given horizontal visibility pattern is obtained. For example, for the horizontal visibility pattern $\{7,2,3,5,2,3,(7)\}$, associated with permutation: $\{6,1,3,5,2,4,(6)\}$, the elementary periodic pattern would be $\{4,1,2,3,1,2,(4)\}$. It is important to remark that this relationship is one-to-one, that is, it is the unique elementary pattern that yields the given horizontal visibility pattern.

Institutional Review Board Statement: Not applicable.

Informed Consent Statement: Not applicable.

Data Availability Statement: Not applicable.

Conflicts of Interest: The authors declare no conflict of interest.

References

1. Barnett, A.G.; Dobson, A.J. *Analysing Seasonal Health Data*; Springer: Berlin/Heidelberg, Germany, 2010.
2. Franses, P.H.; Paap, R. *Periodic Time Series Models*; Oxford University Press: Oxford, UK, 2004.
3. Galor, O. *Discrete Dynamical Systems*; Springer: Berlin/Heidelberg, Germany, 2007.
4. Guckenheimer, J.; Holmes, P. *Nonlinear Oscillations, Dynamical Systems, and Bifurcations of Vector Fields*; Springer Science & Business Media: Berlin/Heidelberg, Germany, 2013.
5. Luque, B.; Lacasa, L.; Ballesteros, F.; Luque, J. Horizontal visibility graphs: Exact results for random time series. *Phys. Rev. E* **2009**, *80*, 046103. [CrossRef] [PubMed]
6. Davis, R.; Nelson, S.A.; Petersen, T.K.; Tenner, B.E. The pinnacle set of a permutation. *Discret. Math.* **2018**, *341*, 3249–3270. [CrossRef]

7. Elizalde, S. A survey of consecutive patterns in permutations. In *Recent Trends in Combinatorics*; Beveridge, A., Griggs, J.R., Hogben, L., Musiker, G., Tetali, P., Eds.; The IMA Volumes in Mathematics and Its Applications 159; Springer International Publishing: Cham, Switzerland, 2016. [CrossRef]
8. Nuño, J.C.; Muñoz, F.J. The partial visibility curve of the Feigenbaum cascade to chaos. *Chaos Solitons Fractals* **2020**. [CrossRef]
9. Núñez, A.; Lacasa, L.; Valero, E.; Gómez, J.; Luque, B. Detecting series periodicity with horizontal visibility graphs. *Int. J. Bifurc. Chaos* **2012**, *22*. [CrossRef]
10. André, D. Étude sur les maxima, minima et séquences des permutations. In *Annales Scientifiques de l'École Normale Supérieure. 3e série, Tome 1, pp. 121–134*; Gauthier-Villars (Éditions scientifiques et médicales Elsevier): Paris, France, 1884.
11. Amigó, J.M. *Permutation Complexity in Dynamical Systems*; Springer Series in Synergetics; Springer: Berlin/Heidelberg, Germany, 2010.
12. Team RC. *R: A Language and Environment for Statistical Computing*; R Foundation for Statistical Computing: Vienna, Austria, 2018; Available online: https://www.R-project.org/ (accessed on 4 June 2021).
13. Lacasa, L.; Luque, B.; Ballesteros, F.; Luque, J.; Nuño, J.C. From time series to complex networks: The visibility graph. *Proc. Natl. Acad. Sci. USA* **2008**, *105*, 4972–4975. [CrossRef] [PubMed]
14. Nuño, J.C.; Muñoz, F.J. Universal visibility patterns of unimodal maps. *Chaos* **2020**. [CrossRef] [PubMed]
15. Nuño, J.C.; Muñoz, F.J. On the ubiquity of the ruler sequence. *arXiv* **2020**, arXiv:2009.14629.

engineering proceedings

MDPI

Proceedings

A Hypothesis Test for the Goodness-of-Fit of the Marginal Distribution of a Time Series with Application to Stablecoin Data †

Mark Levene

Department of Computer Science and Information Systems, Birkbeck, University of London, London WC1E 7HX, UK; mlevene@dcs.bbk.ac.uk

† Presented at the 7th International conference on Time Series and Forecasting, Gran Canaria, Spain, 19–21 July 2021.

Abstract: A bootstrap-based hypothesis test of the goodness-of-fit for the marginal distribution of a time series is presented. Two metrics, the empirical survival Jensen–Shannon divergence ($\mathcal{E}SJS$) and the Kolmogorov–Smirnov two-sample test statistic ($KS2$), are compared on four data sets—three stablecoin time series and a Bitcoin time series. We demonstrate that, after applying first-order differencing, all the data sets fit heavy-tailed α-stable distributions with $1 < \alpha < 2$ at the 95% confidence level. Moreover, $\mathcal{E}SJS$ is more powerful than $KS2$ on these data sets, since the widths of the derived confidence intervals for $KS2$ are, proportionately, much larger than those of $\mathcal{E}SJS$.

Keywords: cryptocurrency; Bitcoin; stablecoin; marginal distribution; heavy-tails; stationary process; stable distribution; goodness-of-fit; survival Jensen–Shannon divergence

Citation: Levene, M. A Hypothesis Test for the Goodness-of-Fit of the Marginal Distribution of a Time Series with Application to Stablecoin Data. *Eng. Proc.* **2021**, *5*, 10. https://doi.org/10.3390/engproc2021005010

Published: 25 June 2021

1. Introduction

The *empirical survival Jensen–Shannon divergence* ($\mathcal{E}SJS$) has recently been proposed as a goodness-of-fit measure of a fitted parametric continuous distribution [1]. However, the important issue of hypothesis testing whether the output $\mathcal{E}SJS$ value is significant was left open.

To alleviate this problem, we propose a hypothesis test based on the parametric bootstrap [2,3], and evaluate the method on time series data [4,5]. As a proof of concept, we chose four cryptocurrency time series, three stablecoin [6] data sets, and, for reference, we employ a fourth, Bitcoin [7], data set. The stablecoins we chose maintain their "stability" by being pegged to the dollar, and thus one would expect their volatility to be low. Apart from the general interest in cryptocurrency time series, it has already been shown that Bitcoin data are heavy-tailed [8]; thus, demonstrating that stablecoins also exhibit heavy tails is interesting in its own right. One reason to experiment with heavy-tailed distributions, such as the α-stable distribution [9] (or simply the stable distribution) employed herein, is that they pose additional problems compared to, say, the normal distribution (in the special case when $\alpha = 2$) due to their variance being infinite (in the more general case when $\alpha < 2$).

The rest of the paper is organised as follows: In Section 2, we introduce the $\mathcal{E}SJS$ and, for comparison purposes, also bring in the well-known *Kolmogorov–Smirnov* two-sample test statistic ($KS2$) [10] Section 6.3. In Section 3, we present a parametric bootstrap-based goodness-of-fit hypothesis test. Time series do not necessarily comprise independent and identically distributed (iid) random variables (as is assumed in, say, [11]), so utilising more general models, such as autoregressive models (as is assumed in, say, [12]), is more appropriate when generating time series bootstrap samples. Here, we assume an autoregressive process of order one [4,5], abbreviated to AR(1), with α-stable innovations, as in [13,14]. In Section 4, we introduce the cryptocurrency time series we experiment with, and fit them to a stable distribution after applying first-order differencing to the raw data, to obtain stationary processes. In particular, we demonstrate that in this case $\alpha < 2$, that is, they are not normally distributed. In Section 5, we apply the goodness-of-fit hypothesis test of Section 3 to the cryptocurrency time series described in Section 4 and discuss the results. Finally, in Section 6, we provide our concluding remarks. We note that all computations were carried out using the Matlab software package.

2. Empirical Survival Jensen–Shannon Divergence

To set the scene, we assume a time series, $\mathbf{x} = \{x_1, x_2, \ldots, x_n\}$, where x_t, for $t = 1, 2, \ldots, n$ is a value indexed by time, t, for example, modelling the movement of a stock price. More specifically, a time series of n values is a random sample generated by a stochastic process that forms a sequence of random variables $\mathbf{X} = X_1, X_2, \ldots, X_n$, where each value x_i is a realisation of the random variable X_i. The stochastic process $\mathbf{X}$ may be a sequence of iids, but, more often than not, a time series exhibits temporal dependencies between its values, which is more realistic. We will also assume that the time series is stationary [4,5]. This makes sense in our context, since we are particularly interested in the marginal distribution of $\mathbf{x}$, which we suppose comes from an underlying parametric continuous distribution D.

The *empirical survival function* of a value z for the time series $\mathbf{x}$, denoted by $\widehat{S}(\mathbf{x})[z]$, is given by

$$\widehat{S}(\mathbf{x})[z] = \frac{1}{n} \sum_{i=1}^{n} I_{\{x_i > z\}}, \tag{1}$$

where I is the indicator function. In the following, we will let $\widehat{P}(z) = \widehat{S}(\mathbf{x})[z]$ stand for the empirical survival function $\widehat{S}(\mathbf{x})[z]$, where the time series $\mathbf{x}$ is assumed to be understood from the context; we will generally be interested in the empirical survival function $\widehat{P}$, which we suppose arises from the survival function P of the parametric continuous distribution D, mentioned above.

The *empirical survival Jensen–Shannon divergence* ($\mathcal{E}SJS$) [1] between two empirical survival functions, $\widehat{Q}_1$ and $\widehat{Q}_2$, arising from the survival functions Q_1 and Q_2, is given by

$$\mathcal{E}SJS(\widehat{Q}_1, \widehat{Q}_2) = \frac{1}{2} \int_0^\infty \widehat{Q}_1(z) \log\left(\frac{\widehat{Q}_1(z)}{\widehat{M}(z}\right) + \widehat{Q}_2(z) \log\left(\frac{\widehat{Q}_2(z)}{\widehat{M}(z)}\right) dz, \tag{2}$$

where

$$\widehat{M}(z) = \frac{1}{2}\left(\widehat{Q}_1(z) + \widehat{Q}_2(z)\right).$$

We note that the $\mathcal{E}SJS$ is bounded and can thus be normalised, so it is natural to assume its values are between 0 and 1; in particular, when $\widehat{Q}_1 = \widehat{Q}_2$ its value is zero. Moreover, its square root is a metric (cf. [1]).

For completeness, we provide the definition of the *Kolmogorov–Smirnov* two-sample test statistic ([10] Section 6.3) between $\widehat{Q}_1$ and $\widehat{Q}_2$ as above, which is given by

$$KS2(\widehat{Q}_1, \widehat{Q}_2) = \max_z |\widehat{Q}_1(z) - \widehat{Q}_2(z)|, \tag{3}$$

where max is the maximum function, and $|v|$ is the absolute value of a number v. We note that $KS2$ is bounded between 0 and 1, and is also a metric.

Now, for a parametric continuous distribution D, we let $\phi = \phi(D, \widehat{P})$ be the parameters that are obtained from fitting D to the empirical survival function, $\widehat{P}$. The distribution D may, in principle, be any continuous distribution, although here we concentrate on the α-stable distribution, since it allows for the modelling of heavy-tailed data, which poses additional problems to those of light-tailed data, due to the variance (and possibly the mean) being infinite. In particular, we have an interest in cryptocurrency data, which is likely to be heavy-tailed [8].

We now let $P_\phi = S_\phi(\mathbf{x})$ be the survival function of $\mathbf{x}$, for D with parameters ϕ. Thus, the empirical survival Jensen–Shannon divergence and the Kolmogorov–Smirnov two-sample test statistic, between $\widehat{P}$ and P_ϕ, are given by $\mathcal{E}SJS(\widehat{P}, P_\phi)$ and $KS2(\widehat{P}, P_\phi)$, respectively. These values provide us with two measures of goodness-of-fit for how well D, with parameters ϕ, is fitted to $\mathbf{x}$ (cf. [1]).

3. A Bootstrap-Based Goodness-of-Fit Hypothesis Test

Our hypothesis test makes use of the parametric bootstrap [2,3]; the pseudocode for the parametric bootstrap in our context is given in Algorithm 1. It takes as input a time series $\mathbf{x}$, the distribution D we hypothesise $\mathbf{x}$ comes from, and the number of bootstrap samples m; in the simulations we use the typical value of $m = 1000$ samples [15]. The algorithm outputs two vectors, $BV\text{-}\mathcal{E}SJS$ and $BV\text{-}KS2$. The first contains m $\mathcal{E}SJS$ values, for $i = 1,2,\ldots,m$, between the empirical survival function $\widehat{P}_i = \widehat{S}(\mathcal{B}_i)$ for the ith bootstrap sample, $\mathcal{B}_i$, and the survival function $P_\phi = S_\phi(\mathbf{x})$ of $\mathbf{x}$, for D with parameters ϕ. Correspondingly, the second contains m $KS2$ values, for $i = 1,2,\ldots,m$, between $\widehat{P}_i$ and P_ϕ. The bootstrap samples are generated by an AR(1) process with α-stable distribution innovations [14] (see also [13]), which is more realistic than assuming that the samples are generated from an iid process, as in [11].

Algorithm 1: Parametric-Boostrap($\mathbf{x}$,D,m).

begin
  Initialise $BV\text{-}\mathcal{E}SJS$ and $BV\text{-}KS2$ as the vector, $\langle 0,0,\cdots,0\rangle$, of m zeros;
  Let n be the number of values in $\mathbf{x}$;
  Let $\phi = \phi(D,\widehat{P})$;
  Let $P_\phi = S_\phi(\mathbf{x})$;
  for $i = 1$ **to** m **do**
    Generate a bootstrap sample $\mathcal{B}_i = x^*_{i1}, x^*_{i2}, \ldots, x^*_{in}$,
    where $\mathcal{B}_i$ is generated from an AR(1) process with innovations derived from D with parameters ϕ;
    Let $\widehat{P}_i = \widehat{S}(\mathcal{B}_i)$;
    Let $BV\text{-}\mathcal{E}SJS(i) = \mathcal{E}SJS(\widehat{P}_i, P_\phi)$;
    Let $BV\text{-}KS2(i) = KS2(\widehat{P}_i, P_\phi)$;
  end for
  return $BV\text{-}\mathcal{E}SJS$ and $BV\text{-}KS2$ sorted in ascending order.
end

As we have assumed that the time series is stationary, the absolute value $|\rho|$ of the parameter ρ of the AR(1) process generating $\mathbf{x}$ should be less than one. For the generation process, we use an estimate $\widehat{\rho}$ of ρ, and, as we will see in Section 4, $|\widehat{\rho}| < 1$ is satisfied for the data sets we employ, as required. We also add a burn-in period of 100 steps to the AR(1) process generated, which we found to be sufficient for the data sets we used.

Given the bootstrap vectors, $BV\text{-}\mathcal{E}SJS$ and $BV\text{-}KS2$, and the output from Algorithm 1, we can form confidence intervals for $\mathcal{E}SJS(\widehat{P}, P_\phi)$ and $KS2(\widehat{P}, P_\phi)$, according to the bootstrap *percentile method* ([16] Section 3.1.2), which is the simplest way to construct a bootstrap confidence interval; see [16] for improvements on the percentile method. We assume that the significance level we are interested in for a hypothesis test is a percentage, and set the significance level to 5%, which is the value we will use in Section 5.

Subsequently, for a one-sided test, we would exclude the highest 5% values from the parametric bootstrap vector, say BV, returned from Algorithm 1, and for a two-sided test we would exclude from BV the lowest 2.5% values and the highest 2.5% values. For both $\mathcal{E}SJS$ and $KS2$ only a one-sided test makes sense, since both metrics are bounded below by zero. Therefore, the null hypothesis is that the distribution of $\widehat{P}$ is D, and so we reject the null hypothesis at the 5% confidence level, if $\mathcal{E}SJS(\widehat{P}, P_\phi)$ or, correspondingly, $KS2(\widehat{P}, P_\phi)$ is greater than the upper bound of the constructed confidence interval, depending on which goodness-of-fit measure we are employing.

4. Cryptocurrencies and Heavy Tails

As a proof of concept, we analysed four time series data sets. These include the prices of three stablecoins [6]: Tether (https://tether.to, accessed on 1 June 2021), DAI (https://makerdao.com, accessed on 1 June 2021) and USDC (https://www.centre.io/usdc, accessed on 1 June 2021), which are all pegged to the dollar. In addition, for comparison purposes, we make use of a fourth time series data set, the price of the archetypal decentralised cryptocurrency, Bitcoin [7], the price of which has previously been hypothesised to follow the heavy-tailed stable distribution [8].

In Table 1, we describe the details of the time series data we used for the empirical validation of the proposed goodness-of-fit method; the data were obtained from Coin Metrics (https://coinmetrics.io, accessed on 1 June 2021). For the stablecoins, 1 is subtracted from the daily closing rate, so that its value is positive if above 1, zero if exactly 1, and negative if below 1. For analysis purposes we applied first-order differencing [4,5] to all the time series, that is, we computed the difference between consecutive observations, which is useful for removing trends, transforming the price time series into a return series (in future work we will also consider analysing the raw data set without differencing; however, since our main aim is to introduce the hypothesis test, for brevity and clarity of exposition we will not consider this further analysis here). The time series, after differencing was applied to the raw data sets, are shown in Figure 1.

Table 1. Description of time series data used for experimentation; #Values is the number of values in the time series.

Currency	#Values	From	Until	Closing Rate
Tether	1264	06 January 2017	15 November 2020	daily
DAI	362	20 November 2019	15 November 2020	daily
USDC	772	28 September 2018	15 November 2020	daily
Bitcoin	8929	01 January 2020	07 January 2021	hourly

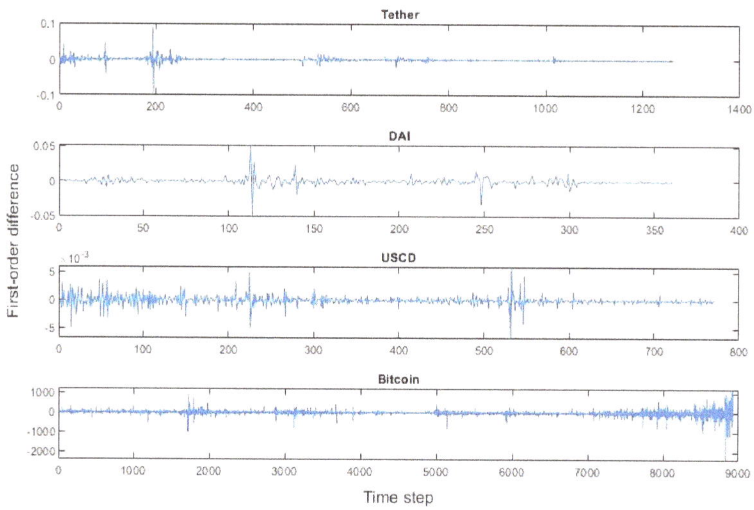

Figure 1. The time series of the four cryptocurrencies after differencing was applied to the raw data sets.

The *α-stable distribution* (or simply the stable distribution) [9] has four parameters: (i) the characteristic exponent $\alpha \in (0,2]$; (ii) the skewness parameter $\beta \in [-1,1]$ (when $\beta = 0$, the distribution is symmetric); (iii) the scale parameter γ; and (iv) the location parameter δ. It is heavy-tailed unless $\alpha = 2$, when the stable distribution reduces to the light-tailed normal distribution with $\beta = 0$. When $\alpha < 2$, the stable distribution is heavy-tailed, its variance as well as all its other higher moments are infinite; in the case of $\alpha \leq 1$, its mean is also infinite. In the following we will refer to a distribution as *stable* when $\alpha < 2$, and *normal* when $\alpha = 2$.

In Figure 2, we show the histograms of the marginal distributions of the four cryptocurrencies overlaid with the curve of the maximum likelihood fit of the normal distribution to the data. It is visually evident that the normal distribution is not a good fit for these data sets. Kurtosis of a distribution, in this case the marginal distribution of a time series, indicates peakedness and tailedness of the data relative to the normal distribution [17] (for ease of comparison with the kurtosis of the normal distribution, which is 3, we will subtract 3 from the kurtosis, giving the *excess kurtosis*). In Table 2, we show the excess kurtosis of the four cryptocurrencies, which provides further evidence that none of them follow a normal distribution, and are in fact heavy-tailed.

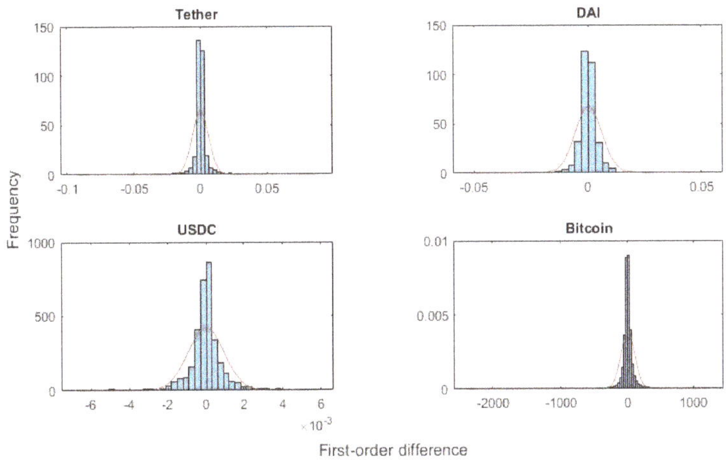

Figure 2. Histograms of the marginal distributions of the four cryptocurrencies, each overlaid with the curve of the maximum likelihood fit of the normal distribution to the data.

Table 2. Excess kurtosis of the four cryptocurrencies.

Currency	Excess Kurtosis
Tether	86.0207
DAI	34.6573
USDC	10.1905
Bitcoin	59.7350

Next, we fitted the stable distribution to the four data sets using the Matlab implementation provided by [18], which is based on the empirical characteristic function method [19]. The fitted parameters are shown in Table 3, noting that in all cases $1 < \alpha < 2$, implying that the means of the marginal distributions are finite but their variances are infinite.

Table 3. Parameters from fits of the stable distribution to the data of the four cryptocurrencies.

Fitted Parameters for Stable Distribution				
Currency	α	β	γ	δ
Tether	1.0111	0.0019	0.0011	0.0001
DAI	1.1953	0.0821	0.0016	0.0003
USDC	1.2259	0.0125	0.0003	0.0000
Bitcoin	1.2261	0.0909	27.9685	7.3644

5. Application of the Goodness-Of-Fit Hypothesis Test to Cryptocurrencies

We apply the bootstrap goodness-of-fit test presented in Section 3, based on the empirical survival Jensen–Shannon divergence ($\mathcal{E}SJS$) and Kolmogorov–Smirnov two-sample test statistic ($KS2$) metrics, to construct 95% confidence intervals for $\mathcal{E}SJS(\widehat{P},P_\phi)$ and $KS2(\widehat{P},P_\phi)$, where $\widehat{P}$ is the empirical survival function of the input time series and P_ϕ is the survival function of time series $\mathbf{x}$, for D with parameters ϕ. When running Algorithm 1, we computed 1000 bootstrap samples, that is, we set $m = 1000$. Moreover, it can be seen in Table 4 that, for all data sets, the estimate $\widehat{\rho}$ of the AR(1) parameter is less than one in absolute value, implying that the generated bootstrap time series, $\mathcal{B}_i$, are stationary as required.

Table 4. Estimates $\widehat{\rho}$ of the parameter ρ of the AR(1) process for the four cryptocurrencies, noting that, when $|\rho| < 1$, the process is stationary.

Currency	$\widehat{\rho}$
Tether	−0.3604
DAI	−0.4045
USDC	−0.4948
Bitcoin	−0.0504

In Tables 5 and 6, we show the results of the bootstrap hypothesis test when employing the $\mathcal{E}SJS$ and $KS2$ metrics, respectively. In particular, *for all* data sets, both metrics are within the 95% confidence interval, and thus with 95% confidence we *cannot* reject the null hypothesis that the marginal distribution of the input time series comes from an α-stable distribution.

The bar chart in Figure 3 shows that *for all* four cryptocurrencies the width of the confidence interval for the $KS2$ goodness-of-fit measure is, proportionately, much larger than that of the $\mathcal{E}SJS$ goodness-of-fit measure. Statistical tests using measures resulting in smaller confidence intervals are normally considered to be more powerful as this implies, with high confidence, that a smaller sample size may be deployed [20].

Finally, to provide contrast to the stable distribution result, we now hypothesise that the marginal distribution of the time series is actually normal (i.e., $\alpha = 2$). We see in Table 7 that, *for all* four cryptocurrencies, we reject the null hypothesis that the marginal distribution is normal, as both the $\mathcal{E}SJS$ and $KS2$ are outside their respective 95% confidence intervals.

Table 5. Parametric bootstrap results for the $\mathcal{E}SJS$ hypothesis test assuming the marginal distribution is stable; LB, UB, CI, Mean and STD stand for lower bound, upper bound, confidence interval, mean of samples and standard deviation of samples, respectively.

Parametric Bootstrap for $\mathcal{E}SJS$ Assuming a Stable Distribution						
Currency	LB of CI	UB of CI	Width of CI	$\mathcal{E}SJS$	Mean	STD
Tether	0.0006	0.0232	0.0226	0.0090	0.0198	0.0741
DAI	0.0030	0.0345	0.0315	0.0156	0.0188	0.0096
USDC	0.0013	0.0247	0.0234	0.0119	0.0133	0.0063
Bitcoin	0.0004	0.0066	0.0062	0.0061	0.0036	0.0016

Table 6. Parametric bootstrap results for the $KS2$ hypothesis test assuming the marginal distribution is stable; LB, UB, CI, Mean and STD stand for lower bound, upper bound, confidence interval, mean of samples and standard deviation of samples, respectively.

Parametric Bootstrap for $KS2$ Assuming a Stable Distribution						
Currency	LB of CI	UB of CI	Width of CI	$KS2$	Mean	STD
Tether	0.0014	0.0308	0.0294	0.0139	0.0289	0.0996
DAI	0.0029	0.0532	0.0503	0.0358	0.0299	0.0136
USDC	0.0035	0.0374	0.0339	0.0219	0.0210	0.0093
Bitcoin	0.0008	0.0103	0.0095	0.0088	0.0057	0.0025

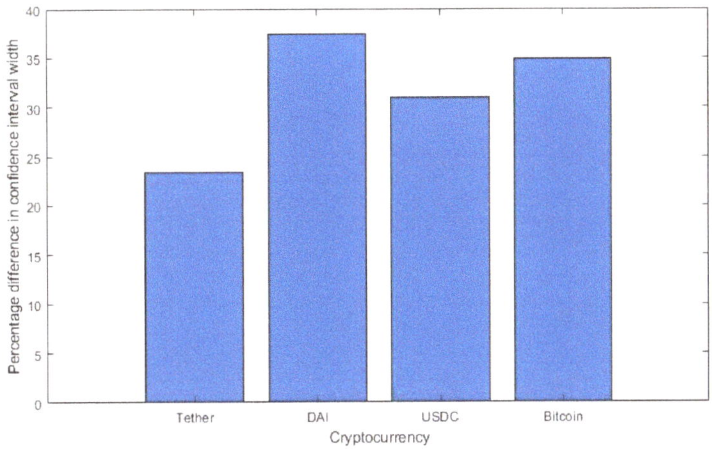

Figure 3. How much larger, proportionately, is the width of the $KS2$ confidence interval compared to that of the $\mathcal{E}SJS$?

Table 7. Parametric bootstrap results for the $\mathcal{E}SJS$ and $KS2$ hypothesis tests assuming the marginal distribution of the time series for the four cryptocurrencies is normal; LB and UB stand for lower and upper bounds of the confidence intervals, respectively, and we abbreviate $\mathcal{E}SJS$ to $\mathcal{E}$ and $KS2$ to K.

Parametric Bootstrap Results Assuming a Normal Distribution						
Currency	LB-$\mathcal{E}$	UB-$\mathcal{E}$	$\mathcal{E}SJS$	LB-K	UB-K	$KS2$
Tether	0.0001	0.0132	0.1440	0.0004	0.0182	0.2162
DAI	0.0003	0.0240	0.1160	0.0006	0.0330	0.1665
USDC	0.0002	0.0147	0.0830	0.0002	0.0227	0.1330
Bitcoin	0.0001	0.0067	0.1218	0.0000	0.0085	0.1708

6. Concluding Remarks

We presented a proof of concept of the bootstrap-based goodness-of-fit test on four cryptocurrency time series, concentrating on the α-stable distribution, which allows for the modelling of heavy-tailed data. Our results demonstrate that, when first-order differenced, the marginal distributions of all four time series are all α-stable with $\alpha < 2$. Moreover, for both $\mathcal{E}SJS$ and $KS2$, the confidence level of the bootstrap-based test is at the 95% level. Furthermore, $\mathcal{E}SJS$ is more powerful than $KS2$ on these data sets, since the widths of the derived confidence intervals for the $KS2$ measure are, proportionately, much larger than those for the $\mathcal{E}SJS$ measure.

We emphasise that the proposed goodness-of-fit test may be applied to any marginal distribution, not just to the heavy-tailed stable distributions. Thus, there is a need to further establish the validity of the proposed hypothesis test on more data sets and on a variety of distributions, which may or may not be heavy-tailed. In addition, it would be useful to look at the assumptions regarding the process underlying the generation of the time series, and to ascertain how this affects the hypothesis test.

References

1. Levene, M.; Kononovicius, A. Empirical survival Jensen–Shannon divergence as a goodness-of-fit measure for maximum likelihood estimation and curve fitting. *Commun. Stat.-Simul. Comput.* **2019**. [CrossRef]
2. Ventura, V. Bootstrap Tests of Hypotheses. In *Analysis of Parallel Spike Trains*; Springer Series in Computational Neuroscience; Grün, S., Rotter, S., Eds.; Springer: Boston, MA, USA, 2010; Volume 7, Chapter 18, pp. 383–398.
3. Pewsey, A. Parametric bootstrap edf-based goodness-of-fit testing for sinh–arcsinh distributions. *TEST* **2018**, *27*, 147–172. [CrossRef]
4. Enders, W. *Applied Econometric Time Series*, 4th ed.; Wiley Series in Probability and Statistics; John Wiley & Sons: Hoboken, NJ, USA, 2014.
5. Chatfield, C.; Xing, H. *The Analysis of Time Series: An Introduction with R*, 7th ed.; Text in Statistical Science; Chapman & Hall: London, UK, 2019.
6. Sidorenko, E. Stablecoin as a new financial instrument. In *Proceedings of International Scientific Conference on Digital Transformation of the Economy: Challenges, Trends, New Opportunities*; Springer Nature: Cham, Switzerland, 2020.
7. Judmayer, A.; Stifter, N.; Krombholz, K.; Weippl, E.; Bertino, E.; Sandhu, R. *Blocks and Chains: Introduction to Bitcoin, Cryptocurrencies, and Their Consensus Mechanisms*; Synthesis Lectures on Information Security; Privacy, and Trust, Morgan & Claypool Publishers: San Francisco, CA, USA, 2017.
8. Kakinaka, S.; Umeno, K. Characterizing cryptocurrency market with Lévy's stable distributions. *J. Phys. Soc. Jpn.* **2020**, *89*, 024802-1–024802-13. [CrossRef]
9. Nolan, J. *Univariate Stable Distributions: Models for Heavy Tailed Data*; Springer Series in Operations Research and Financial Engineering; Springer Nature: Cham, Switzerland, 2020.
10. Gibbons, J.; Chakraborti, S. *Nonparametric Statistical Inference*, 6th ed.; Marcel Dekker: New York, NY, USA, 2021.
11. Cornea-Madeira, A.; Davidson, R. A parametric bootstrap for heavy-tailed distributions. *Econom. Theory* **2015**, *31*, 449–470. [CrossRef]
12. Lin, J.; McLeod, A. Improved Peňa–Rodriguez portmanteau test. *Comput. Stat. Data Anal.* **2006**, *51*, 1731–1738. [CrossRef]
13. Gallagher, C. A method for fitting stable autoregressive models using the autocovariation function. *Stat. Probab. Lett.* **2001**, *53*, 381–390. [CrossRef]
14. Ouadjed, H.; Mami, T. Estimating the tail conditional expectation of Walmart stock data. *Croat. Oper. Res. Rev.* **2020**, *11*, 95–106. [CrossRef]
15. Hesterberg, T. what teachers should know about the bootstrap: Resampling in the undergraduate statistics curriculum. *Am. Stat.* **2015**, *69*, 371–386. [CrossRef] [PubMed]
16. Chernick, M. *Bootstrap Methods: A Guide for Practitioners and Researchers*, 2nd ed.; Wiley Series in Probability and Statistics; John Wiley & Sons: Hoboken, NJ, USA, 2008.
17. DeCarlo, L. On the meaning and use of kurtosis. *Psychol. Methods* **1997**, *2*, 292–307. [CrossRef]
18. Veillette, M. Alpha-Stable Distributions in MATLAB. 2015. Available online: http://math.bu.edu/people/mveillet/html/alphastablepub.html (accessed on 1 June 2021).
19. Koutrouvelis, I. An iterative procedure for the estimation of the parameters of stable laws. *Commun. Stat.-Simul. Comput.* **1981**, *10*, 17–28. [CrossRef]
20. Liu, X. Comparing sample size requirements for significance tests and confidence intervals. *Couns. Outcome Res. Eval.* **2013**, *4*, 3–12. [CrossRef]

engineering proceedings

Proceedings

Improving the Accuracy and Time Interval of Predicting Ambient Parameters Applied to Dynamic Line Rating †

Milenko Kabović , Anka Kabović , Slavica Boštjančič Rakas * and Valentina Timčenko

Mihailo Pupin Institute, University of Belgrade, 11000 Belgrade, Serbia; milenko.kabovic@pupin.rs (M.K.); anka.kabovic@pupin.rs (A.K.); valentina.timcenko@pupin.rs (V.T.)
* Correspondence: slavica.bostjancic@pupin.rs
† Presented at the 7th International conference on Time Series and Forecasting, Gran Canaria, Spain, 19–21 July 2021.

Abstract: This paper addresses wind speed prediction in the dynamic line rating (DLR) environment. We have described architecture of the DLR system as well as the main characteristics of nonlinear forecasting models, such as neural and fuzzy logic networks. Described models were tested and compared using real data (time series with data on wind speed, wind direction, air temperature, and solar radiation). The goal was to increase the accuracy and time of short-term prediction. The results show that neural networks outperform fuzzy logic and that the prediction time interval can be extended up to several hours, with no major compromise of the accuracy.

Keywords: dynamic line rating; fuzzy logic; neural networks; prediction

Citation: Kabović, M.; Kabović, A.; Rakas, S.B.; Timčenko, V. Improving the Accuracy and Time Interval of Predicting Ambient Parameters Applied to Dynamic Line Rating . *Eng. Proc.* **2021**, 5, 11. https://doi.org/10.3390/engproc2021005011

Published: 25 June 2021

1. Introduction

Dynamic Line Rating (DLR) is used to dynamically increase the transmission capacity of overhead lines (OHL), taking into account their thermal state and ambient conditions. The DLR system provides more efficient OHL's load utilization and improves existing assets utilization, which leads to overall cost decrease, and also reduces greenhouse gas emissions (with the integration of renewable energy resources). It is usually integrated with the power utility's supervisory control and data acquisition (SCADA) system, but a stand-alone solution is also possible [1,2]. Information obtained from the DLR system can further be processed by the utility's Energy Management System (EMS).

Besides the main benefit, the increase of the OHL's transmission capacity, DLR system optimizes the transfer of energy from renewable sources by predicting the production of energy that comes from these sources. Therefore, DLR system represents an important part of the energy management system.

Forecasting methods enable the full utilization of DLR systems, allowing operators to respond in a timely manner in the case of unexpected situations, as well as operational planning, particularly in the case of renewable energy sources (especially wind turbines) connected to the grid, and they can also help with the energy trade.

Forecasting is one of the smart grid's key functionalities that can help with the network load balancing, optimization of electricity distribution and failure management. Wind forecasting is very important for managing the production of electricity in wind farms, as well as for the forecasting of the allowed transmission line's current load [3]. Due to its spatial and temporal variability, it is difficult to accurately predict the wind parameters (speed and direction). Different methods are used in practice, like numerical weather prediction, statistical methods that include linear, nonlinear models and hybrid methods, whereas the statistical linear regression models cannot be used for more accurate forecasting, especially in the case of rapid and significant changes of wind parameters. Therefore, nonlinear models such as artificial neural networks and models with fuzzy logic are used. These models are particularly interesting for short-term forecasting (1–4 h). It should be emphasized that the result of the allowed current load forecasting is not supposed to be the value nearest to the one that is obtained by measuring in real time. It should rather provide for a lower limit of the permissible load, such that the transmission system operators (TSO) can ensure that they have the lowest value of the allowed load in real time.

This paper describes basic characteristics of the nonlinear prediction models, as well as the architecture of the DLR system used for the analysis of the described prediction models. Test results of wind speed prediction, using the real data, i.e., time series with data on wind speed, wind direction, air temperature, and solar radiation were presented. The emphasis is on increasing the accuracy and time of short-term prediction, so several models based on the neural networks and fuzzy logic were tested and compared.

2. Tested DLR System

The architecture of the DLR system used for the testing consists of three main parts: (1) a measuring unit (a temperature sensor unit and three weather stations); (2) a DLR server, and (3) work stations; as depicted in Figure 1.

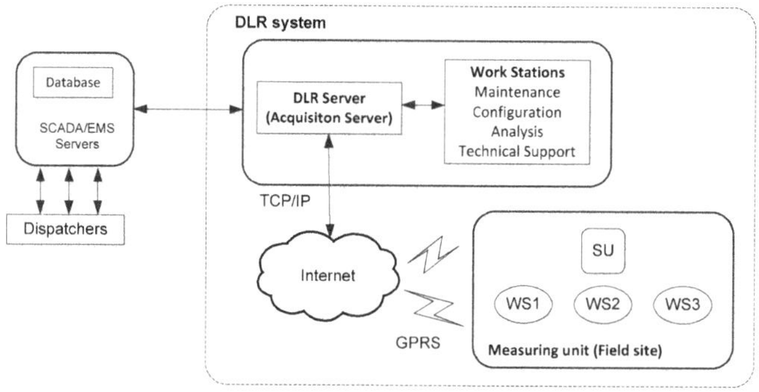

Figure 1. Tested DLR architecture.

A measuring unit consists of sensor unit (SU) and three weather stations (WSs). The sensor unit is mounted on the transmission line and it measures line current, conductor temperature, tension, and/or sag. The weather station is located near the sensor unit (usually mounted on the tower of the overhead line), and it measures ambient parameters (air temperature, solar radiation, and wind speed and direction).

The communication between the sensor unit and acquisition server is provided with GPRS (General Packet Radio Service). DLR server is connected to a SCADA system via secure TCP/IP (Transmission Control Protocol/Internet Protocol) connection. The sensors unit's locations are determined by the minimal wind speed and minimal ground clearance (critical spans). Data collected from measuring units are sent to the DLR server, which processes the data, and determines the conductor ampacity, based on actual conditions of the OHL and ambient parameters. Processed data, which may include alarms, are sent to the control system (SCADA) and work stations. Real-time monitoring of particular OHL's temperature and ampacity, maintenance and configuration of measuring units are performed by work stations.

With the structure shown, the DLR system brings different benefits such as more efficient utilization of transmission line's load and operational flexibility of the transmission system. It improves utilization of the existing assets, reduces greenhouse gas emissions, through optimal integration of renewable energy resources, and improves the security of the power grid's operation in normal operating conditions.

3. Wind Speed Prediction Modeling

3.1. Artificial Neural Networks

Artificial neural networks (ANNs) are widely applied to real-life issues in different areas such as economics, education, engineering, etc. They can be also used for optimization, intrusion detection, and data classification [4]. Artificial neural networks have already been utilized for wind speed and wind power prediction, since they are great identifiers of trends in data and patterns [5]. Several types of neural networks are usually applied for wind speed prediction such as: feed-forward backpropagation (FFBP), multilayer perceptron (MLP), recurrent neural networks (RNN), and radial basis function neural networks (RBFN).

ANNs, by definition, represent a massively parallel distributed processor with the natural ability to memorize experimental knowledge and to use it later. They can learn from examples (past data), recognize a hidden pattern in historical observations, and use them to forecast future data values. They consist of several layers of simple process elements (neurons) that are interconnected. Signals travel from the input layer to the output layer, usually after traversing one or multiple hidden layers. Connections are usually characterized with weights that adapt (increase or decrease) as learning process proceeds. Neurons are the basic elements, and represent the independent computational units [6]. They process received inputs and calculate the outputs by non-linear functions of the sum of the inputs (Figure 2). The threshold is used in such a way that a signal is sent only if the resulting sum of the signals crosses that threshold.

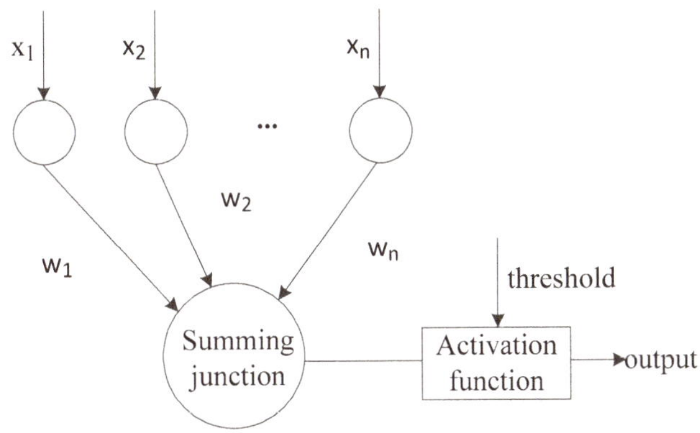

Figure 2. Neuron—the basic element of the neural network.

Mathematical representation of the neuron function is:

$$y_k = f\left(\sum_{j=1}^{n} w_{jk} * x_j + b_k\right), \quad (1)$$

where x_j are the input values, w_{jk} are connection weights, b_k is the bias value, y_k is the output of the neuron, and f is the neuron transfer function.

Neural networks used in this research are FFBP and MLP. FFBP network, presented in Figure 3, is one of the most frequently used types of neural networks for short term predictions of the wind parameters. The advantage of this network is the simple process of parameters setting. Training is performed with a set of input patterns to be learned and the desired outputs for each pattern. Once trained, this type of network can recognize similar patterns very quickly, or the patterns obscured with noise. The back propagation training algorithm is designed to minimize the mean square error across all training patterns [7].

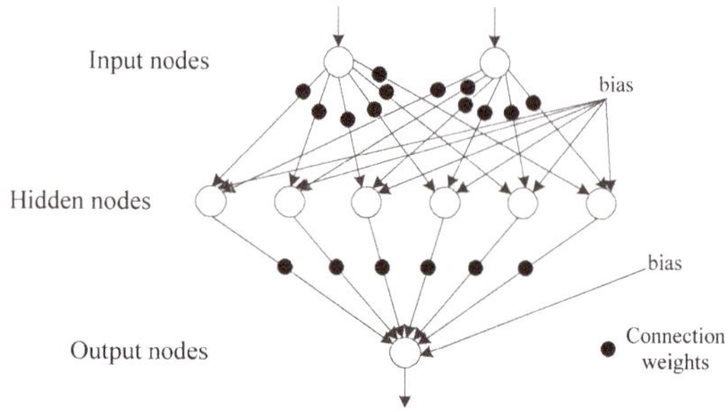

Figure 3. Feed forward neural network.

MLP represents a subset of feed-forward ANN and the most widely-used ANN. It consists of a minimum of three layers (input, output, and hidden layers) [8]. MLP is based on the concept of a feed-forward-flow of information (i.e., the network is organized in an ordered way) and can perform static mapping between the input and the output. It uses a backpropagation for training, which is a supervised learning technique. MLP is fully-connected and each connection between neurons from different layers is associated with a certain weight. MLP can learn non-linear models and models in real-time (on-line learning). The main disadvantages of this network are that the MLP networks with hidden layers produce a non-convex loss function, and there is a possibility of obtaining multiple local minimums. Consequently, in the case of different random weight initializations there is a possibility of obtaining different validation accuracies. Some additional complexity brings the MLPs sensitivity to the feature scaling, and the need to adequately tune a range of hyperparameters, namely the number of hidden neurons, number of layers, and iterations.

3.2. Fuzzy Logic Networks

Fuzzy logic models use sets of data in which the affiliation of the set is not denoted by 0 and 1, but instead it uses values from the interval between 0 and 1. Member functions are used to calculate the degree of data belonging to a given set. In addition, logical rules are used to define the relationship between input and output variables. Depending on the structure of the rules, there are two types of these models: Mamdani and Takagi Sugeno. The output is calculated as the weighted average contribution of each rule. According to the learning method, these models can be classified into five groups: neuro-phase models, genetic algorithm models, clustering algorithms, gradient descent algorithm models, and models with space separation algorithm [9,10].

ANFIS (Adaptive Network-based Fuzzy Inference System) model is one of the fuzzy models frequently used for wind speed and wind power prediction [11]. ANFIS is an ANN based on Takagi–Sugeno fuzzy inference system. It consists of five layers. The first layer is called the fuzzification layer and it takes the input values and determines the membership functions belonging to them. The second layer, denoted as a rule layer, generates the firing strengths for the rules. The third layer normalizes the computed firing strengths. The fourth layer takes the normalized values and the consequence parameters and returns defuzzificated values, which are then passed to the fifth layer that returns the final output [10,12]. In this model, in addition to the number of variables, the number and parameters of member functions, the number of rules and parameters of linear functions, are also determined. For wind speed prediction, some authors used ANFIS model with artificial neural network [13,14].

4. Testing, Results and Discussion

Testing was performed based on the meteorological database that contains meteorological observations from a weather station located on the transmission line tower, which is the part of the installed DLR system. The collected data cover ten days in April of 2018.

The training sequence includes all the data from midnight on 10 April to 6 p.m. (4 p.m.) at 19 April, depending on the length of the test sequence. The test sequence covers the period from 19 April at 6 p.m. (4 p.m.), depending on the desired length, to 19 April at 10 p.m. The time resolution of the data is 5-min. All testing was performed in the R-Studio development environment.

We have tested the following types of models: FFBP, MLP and ANFIS. The "neuralnet" library was used to create a FFBP neural network relying on the "back propagation" or "resilient back propagation" methods, with or without weight backtracking, as well as the modified globally convergent version. MLP model relies on the use of the "nnet" library, while the testing of ANFIS model was performed using the "frbs" library of functions, that includes several subtypes that differ in the methods for adjusting model parameters. The testing was performed with the data taken at 5-min intervals, for different numbers of prediction points (8, 16, 32, 40 and 64) and with two types of input data: (1) different meteorological data as ambient temperature, solar radiation and wind speed and wind direction, and (2) delayed series of wind speed measurements for one, two and three 5-min measurement intervals. In the case of tested fuzzy-based model, there is also the possibility of normalizing the input data, changing of the number of prediction points, as well as choosing the model type.

The comparison is performed by the means of mean absolute error (MAE) value, depending on the model type, number of prediction points (8, 16, 32, 40 and 64), and the length of the test sequence (4 and 6 h), and by matching the actual data values of the wind speed with the predicted ones.

The examples of generated neural networks for eight prediction points for some tested models are presented in Figure 4.

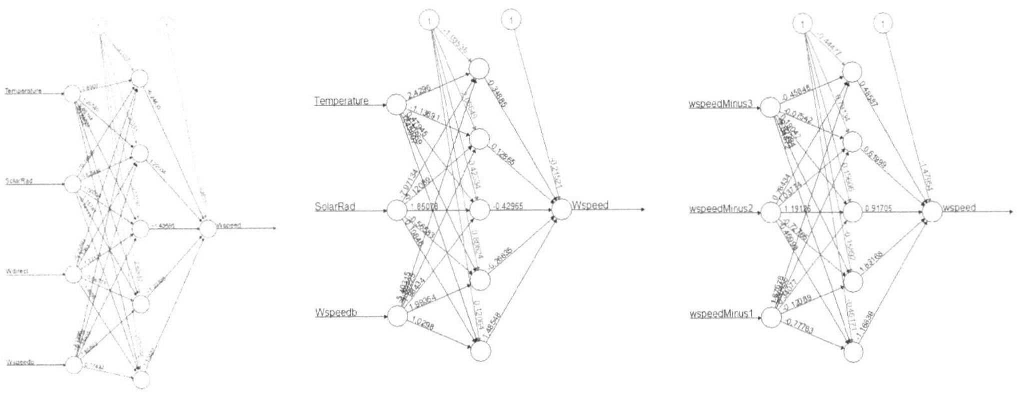

(a) (b) (c)

Figure 4. Neural network with 8 prediction points: (**a**) FFBP with 4 different inputs; (**b**) FFBP with 3 different inputs types; (**c**) FFBP with 3 same delayed inputs types.

The test results of all tested models are presented in Tables 1 and 2, showing the results obtained for different number of prediction points, as well as for different lengths of the test sequence (4 h and 6 h, respectively). Table 1 shows that for the first four models (FFBP and MLP), the MAE does not change very much as the number of prediction points increases. Testing also showed that the change of parameters values of "neuralnet" functions for FFBP model does not affect much the change of the MAE. When increasing the number of neurons in the hidden layer, it reduces the MAE value, but only to the second decimal place, while at the same time it prolongs the modeling time. In the case of 6-h test sequence, with the number of prediction points increased to 64 points and sudden wind speed jumps, in the test sequence; the results show the increase of the MAE value, as well as stronger dependence on the number of prediction points (Table 2). The dependence of the MAE value on the model type is not very pronounced, while there is a noticeable dependence on the number of prediction points. The value of the MAE for the ANN models does not depend much on the type of the model and the number of prediction points, only if there are no sudden changes of the wind speed value in the test sequence that are greater than 2 m/s. This can be clearly seen from the results shown in Table 1.

Table 1. MAE values for different tested models and different number of prediction points with a 5-min resolution, and a 4-h test sequence length.

Tested Models	Prediction Points			
	8	16	32	40
FFBP with four different input types	0.2796	0.2969	0.2755	0.2875
FFBP with three different input types	0.292	0.2586	0.2557	0.2816
FFBP with three same inputs types	0.269	0.25	0.219	0.263
MLP with three same inputs types	0.2786	0.2718	0.2475	0.2879
ANFIS with three different inputs types and with normalization	0.316	0.5069	0.70244	0.6921
ANFIS with three same inputs types and with normalization	0.317	0.309	0.679	0.345

Table 2. MAE values for different tested models and different number of prediction points with a 5-min resolution, and a 6-h test sequence length.

Tested Models	Prediction Points				
	8	16	32	40	64
FFBP with four different input types	0.9267	0.937	0.6755	0.566	0.4615
FFBP with three different input types	0.895	0.947	0.653	0.568	0.4662
FFBP with three same inputs types	0.8353	0.8956	0.6679	0.5827	0.4713
MLP with three same inputs types	0.886	0.879	0.6779	0.594	0.4764
ANFIS with three different inputs types and with normalization	1.03369	1.6294	0.644	0.6035	1.226
ANFIS with three same inputs types and with normalization	0.9919	0.9498	0.961	0.6268	0.6586

For the ANFIS fuzzy logic model, for both test sequence lengths (4 and 6 h), test results have showed that the MAE depends not only on the way the model is generated, but also on the number of points at which the prediction is made. In comparison with the neural networks models, this dependence is here noticeable even with a shorter length of the test sequence. In the case of the large number of prediction points, the tested model does not provide good results, as it gives a constant output value. The best results were obtained with the ANFIS model generated with delayed inputs of the same type, and applied normalization of inputs. The test results have shown that this type of fuzzy model is not suitable for prediction intervals longer than 1.5 h.

Figures 5 and 6 show the prediction results for different testing models and different number of prediction points, which are grouped based on the test sequence length. Figure 5 presents the 4-h test sequence (a total of 48 prediction points with a resolution of 5 min), where graphs present the actual wind speed values as well as the predicted values. Figure 6 shows the results of the extended 6-h long test sequence (with a total of 72 prediction points and 5-min resolution). Figure 6 shows the results of the extended, 6-h long test sequence (with a total of 72 prediction points and 5-min resolution).

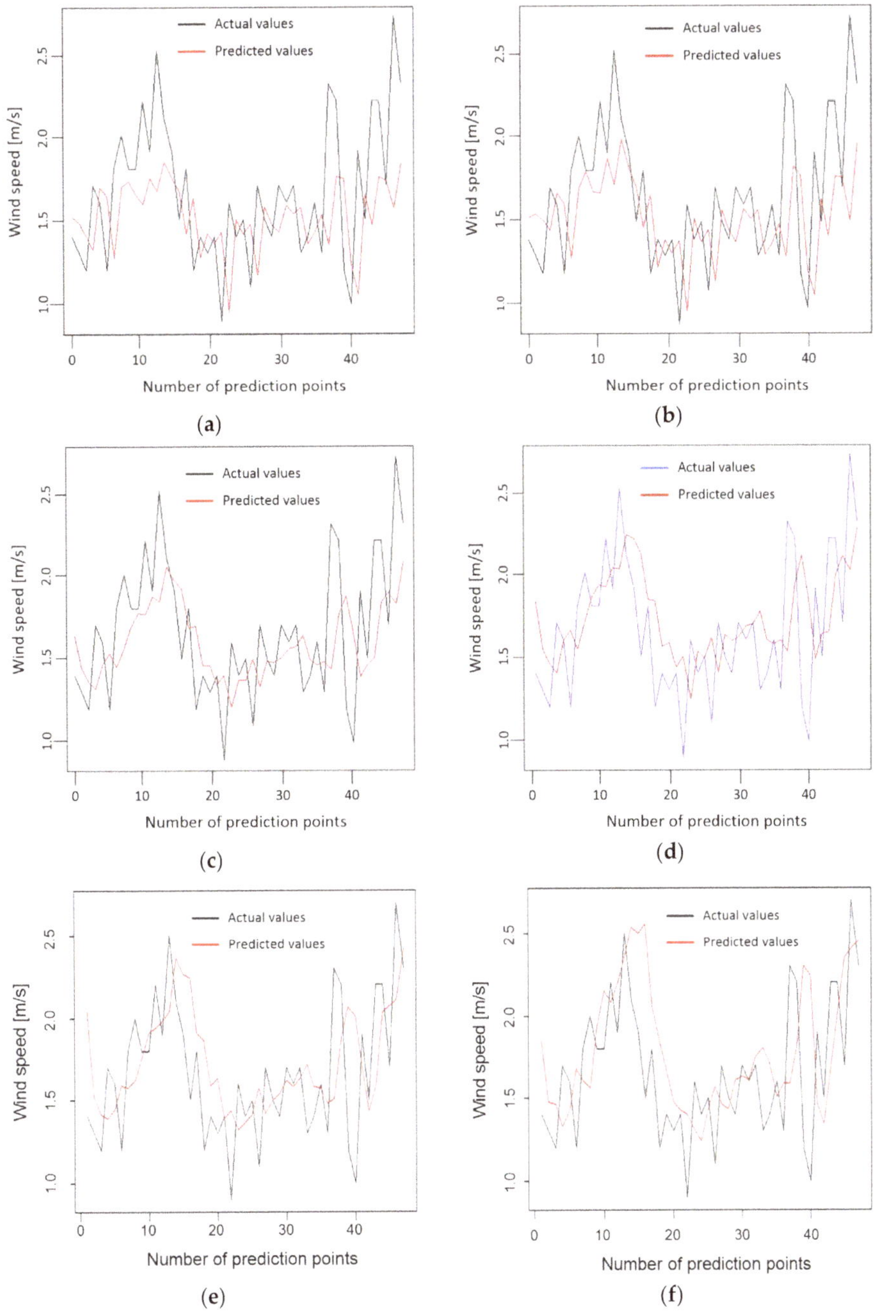

Figure 5. Prediction with the 4-h test sequence: (**a**) FFBP with 4 different input types; (**b**) FFBP with 3 different input types; (**c**) FFBP with 3 same input types; (**d**) MLP; (**e**) ANFIS with 3 different input types; (**f**) ANFIS with 3 same input types.

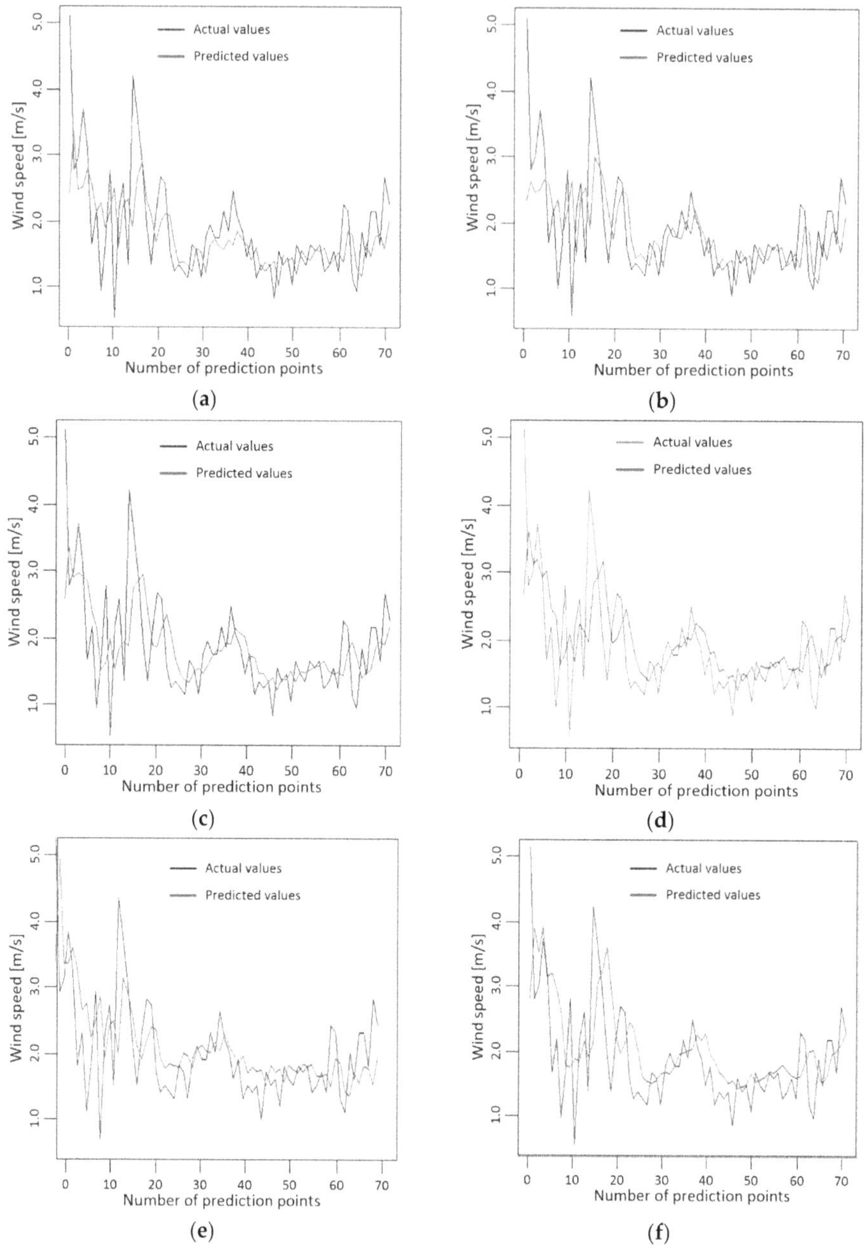

Figure 6. Prediction with the 6-h test sequence: (**a**) FFBP with 4 different input types; (**b**) FFBP with 3 different input types; (**c**) FFBP with 3 same input types; (**d**) MLP; (**e**) ANFIS with 3 different input types; (**f**) ANFIS with 3 same input types.

We can say that when there are no sudden jumps in the wind speed values (greater than 2 m/s) in the test sequence, the number of prediction points for neural network models does not affect much the prediction accuracy, so the prediction can be safely prolonged to 5 h. For the mentioned condition, the prediction accuracy is also not significantly affected by the parameters of the model function, i.e., the number of neurons in the hidden layer of the neural network.

5. Conclusions

In this paper, we have shown the analysis and testing results of the FFBP, and MLP types of neural networks, as well as the ANFIS type of fuzzy logic network, in order to investigate which type has best performance regarding the minimal absolute error and prediction duration of maximum five hours. Test results have shown that both FFBP and MLP types have similar and good performance especially when the test sequence doesn't contain changes of wind speed larger than 2 m/s. Neural networks also outperformed the tested ANFIS fuzzy logic model.

Future work will be focused on the analysis of additional neural network models such as the generalized feed-forward neural network (GFNN) and the recursive radial basis function neural network (RRBFNN), as well as some hybrid models.

Institutional Review Board Statement: Not applicable.

Informed Consent Statement: Not applicable.

Data Availability Statement: Data available on request from the authors.

Acknowledgments: This work was supported in part by the Serbian Ministry of Education, Science and Technological development.

References

1. Bostjancic Rakas, S.; Timcenko, V.; Kabovic, A.; Kabovic, M. Cyber Security Issues in Conductor Temperature and Meteorological Measurement Based DLR System. In Proceedings of the Mediterranean Conference on Power Generation, Transmission, Distribution and Energy Conversion, Belgrade, Serbia, 6–9 November 2016; pp. 1–7. [CrossRef]
2. Uski-Joutsenvuo, S.; Pasonen, R. Maximising Power Line Transmission Capability by Employing Dynamic Line Ratings—Technical Survey and Applicability in Finland; Research Report VTT-R-01604-1; VTT Technical Research Centre of Finland: Espoo, Finland. Available online: http://sgemfinalreport.fi/files/D5.1.55%20-%20Dynamic%20line%20rating.pdf (accessed on 22 May 2021).
3. Nazir, T.M.S.; Alturise, F.; Alshmrany, S.; Nazir, H.M.J.; Bilal, M.; Abdalla, A.N.; Sanjeevikumar, P.; Ali, Z.M. Wind generation forecasting methods and proliferation of artificial neural network: A review of five years research trend. *Sustainability* **2020**, *12*, 3778. [CrossRef]
4. Abiodun, O.I.; Jantan, A.; Omolara, A.E.; Dada, K.V.; Mohamed, N.A.; Arshad, H. State-of-the-art in artificial neural network applications: A survey. *Heliyon* **2018**, *4*, e00938. [CrossRef] [PubMed]
5. Fazelpour, F.; Tarashkar, N.; Rosen, M.A. Short-term wind speed forecasting using artificial neural networks for Tehran, Iran. *Int. J. Energy Environ. Eng.* **2016**, *7*, 377–390. [CrossRef]
6. Hagan, M.T.; Demuth, H.B.; Beale, M.H.; De Jesus, O. *Neural Network Design*, 2nd ed.; Oklahoma State University: Stillwater, OK, USA, 2014.
7. Sreelakshmi, K.; Ramakanthkumar, P. Neural networks for short term wind speed prediction. *Int. J. Comput. Inf. Sci.* **2008**, *2*. [CrossRef]
8. Lawana, S.M.; Abidinb, W.A.W.Z.; Chaic, W.Y.; Baharund, A.; Masri, T. Some methodologies of wind speed prediction: A critical review. *Int. J. Renew. Energy* **2014**, *9*, 41–55. [CrossRef]
9. Babuška, R. Neuro-fuzzy methods for modeling and identification. In *Recent Advances in Intelligent Paradigms and Applications. Studies in Fuzziness and Soft Computing*; Abraham, A., Jain, L.C., Kacprzyk, J., Eds.; Physica: Heidelberg, Germany, 2003; Volume 113, pp. 161–186. [CrossRef]
10. Ross, T.J. *Fuzzy Logic With Engineering Applications*, 4th ed.; John Wiley & Sons: Hoboken, NJ, USA, 2016.
11. Ismail, F.H.; Aziz, M.A.; Hassanien, A.E. Optimizing the parameters of Sugeno based adaptive neuro fuzzy using artificial bee colony: A Case study on predicting the wind speed. In Proceedings of the Federated Conference on Computer Science and Information Systems, Gdansk, Poland, 11–14 September 2016; pp. 645–651.
12. Karaboga, D.; Kaya, E. Adaptive network based fuzzy inference system (ANFIS) training approaches: A comprehensive survey. *Artif. Intell. Rev.* **2019**, *52*, 2263–2293. [CrossRef]
13. Saleh, A.; Moustafa, M.S.; Abdullah, A.A. A hybrid neuro-fuzzy power prediction system for wind energy generation. *Int. J. Electr. Power* **2016**, *74*, 384–395. [CrossRef]
14. Ujjwal, S.; Kukrety, L.; Kakinada, S. Wind power forecasting using artificial neural networks (ANN) and artificial neuro-fuzzy inference system (ANFIS). In *Proceedings of 6th International Conference on Recent Trends in Computing. Lecture Notes in Networks and Systems*; Mahapatra, R.P., Panigrahi, B.K., Kaushik, B.K., Roy, S., Eds.; Springer: Singapore, 2021; Volume 177, pp. 535–541. [CrossRef]

engineering proceedings

Proceedings

Time-Series of Distributions Forecasting in Agricultural Applications: An Intervals' Numbers Approach †

Christos Bazinas, Eleni Vrochidou, Chris Lytridis and Vassilis G. Kaburlasos *

HUMAIN-Lab, Department of Computer Science, International Hellenic University (IHU), 65404 Kavala, Greece; chrbazi@cs.ihu.gr (C.B.); evrochid@cs.ihu.gr (E.V.); lytridic@cs.ihu.gr (C.L.)

* Correspondence: vgkabs@cs.ihu.gr

† Presented at the 7th International conference on Time Series and Forecasting, Gran Canaria, Spain, 19–21 July 2021.

Abstract: This work represents any distribution of data by an Intervals' Number (IN), hence it represents all-order data statistics, using a "small" number of L intervals. The INs considered are induced from images of grapes that ripen. The objective is the accurate prediction of grape maturity. Based on an established algebra of INs, an optimizable IN-regressor is proposed, implementable on a neural architecture, toward predicting future INs from past INs. A recursive scheme tests the capacity of the IN-regressor to learn the physical "law" that generates the non-stationary time-series of INs. Computational experiments demonstrate comparatively the effectiveness of the proposed techniques.

Keywords: agriculture 4.0; big data; computational intelligence; Intervals' Number (IN); non-stationary; prediction regressor model; time-series

Citation: Bazinas, C.; Vrochidou, E.; Lytridis, C.; Kaburlasos, V.G. Time-Series of Distributions Forecasting in Agricultural Applications: An Intervals' Numbers Approach. *Eng. Proc.* **2021**, *5*, 12. https://doi.org/10.3390/engproc2021005012

Published: 25 June 2021

1. Introduction

There is a long-term interest in extending the fourth industrial revolution (Industry 4.0) to agricultural production [1]. Regarding viticulture, in particular, the interest is in minimizing the human presence in the vineyard during production. In the aforementioned context, the accurate prediction of the grape maturity is critical in order to timely engage both human labor and equipment for harvest. Newly introduced technologies from associated areas such as the Internet of Things (IoT), Big Data, and Artificial Intelligence (AI) can be combined with autonomous robotic systems in order to collect and interpret data, monitor and evaluate crop status, and automatically plan effective and timely interventions. An early in-field assessment of fruit maturity level and therefore an estimation on harvest time has the potential to enable sustainable farming by balancing between economy, ecology, and optimal crop quality [2]. However, the development of autonomous robots for agricultural applications faces the daunting scale of the data involved [3]. Our special interest here is in the prediction of grapes maturity level, intended to be integrated into an autonomous grape-harvester robot [4].

Fruit maturity can be studied as a time-series, where the sequence of maturity data, $m_1, m_2, \ldots, m_D$ is indexed in time. The objective is to predict future fruit's maturity level. A number of different models have been used including linear ones such as classic autoregressive (AR), moving average (MA), autoregressive moving average (ARMA), and autoregressive integrated moving average (ARIMA) models; however, those models typically assume stationary time-series [5,6], and they may fail with non-stationary time-series regarding maturity, unless a very high order linear model is used to gain an insight into the system and its underlying laws. The latter needs extensive learning which calls for long processing time as well as computational resources. Therefore, nonlinear models, such as Neural Networks (NN), have been proposed to counter forecasting problems [7–9]. The most straightforward approach for an NN model to learn a time-series is to provide the samples in time to the input of the NN. However, if the time-series are complex, then more past samples are needed; the latter usually results in a complex system of multiple inputs and weights.

This paper uses Intervals' Numbers (INs) for predicting the maturity of grapes based on real-world measurements by a parametric IN-regressor model implementable by a neural network architecture. Note that INs have originally been introduced, under the name Fuzzy Intervals' Numbers (FINs), in the context of fuzzy set theory [10]. The interpretation of INs was later extended, in the context of the Lattice Computing (LC) information processing paradigm [11]. In particular, an IN has been defined as a mathematical object that can be interpreted either as a fuzzy interval or as a distribution of samples; the latter interpretation implies that an IN can potentially represent all-order statistics [11]. The mathematical properties of the set of INs have been studied for a long time. An algebra of INs has been established [10,12,13]. INs have been employed in logic and reasoning applications [14,15]. Furthermore, they have been employed in interpolation/extrapolation applications [16].

INs have already been applied to time-series classification applications regarding electroencephalography (EEG) signals [17]. Recently, INs have been employed toward predicting the maturity of grapes [18]. This work is a follow-up of [18] with the following novelties: first, additional computational experiments are carried out using (1) the previous three and four INs to predict a future IN, (2) fewer data for training; second, a recursive scheme here demonstrates an IN-regressor's capacity to learn the physical "law" that generates the non-stationary time-series of INs regarding grape maturity; third, the problem of time-series forecasting in agriculture is described, in mathematical terms, as a non-stationary time-series forecasting problem thus bringing INs to the foreground as an object for time-series processing in other domains with the advantage that an IN represents a distribution of data including all-order data statistics.

The layout of the paper is as follows: Section 2 presents two IN-regressor models for prediction. Section 3 details experimental application results. Finally, Section 4 summarizes our contribution, and it discusses potential future work extensions.

2. An IN-Regressor Parametric Models for Prediction

Figure 1 displays an IN in its two equivalent representations, namely the membership-function–representation in Figure 1a and the interval–representation in Figure 1b. More specifically, the membership-function–representation is identical to a probability distribution function; therefore, it is amenable to interpretations, whereas its equivalent interval-representation lends itself to useful algebraic operations in the context of mathematical lattice theory.

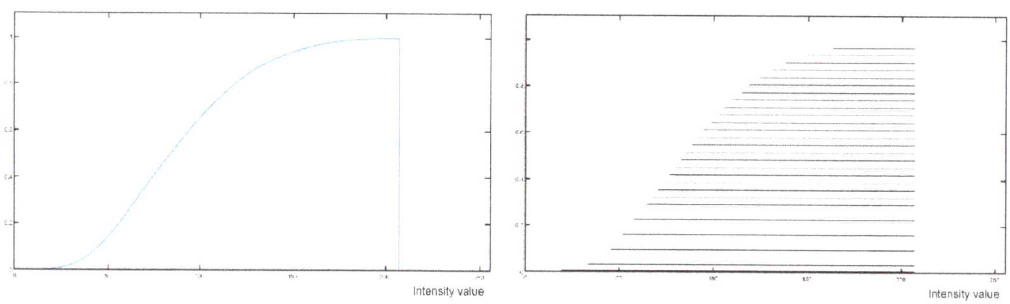

(a) (b)

Figure 1. Two equivalent representations of an IN: (**a**) The membership–function–representation, which is identical to a probability distribution function, and (**b**) the interval–representation which lends itself to useful algebraic operations in the context of mathematical lattice theory.

The space of INs is known to be a cone in a linear space. Therefore, linear models such as ARMA models can be developed. The work in [18] has proposed a nonlinear model in the space of INs implemented by three-layer feed-forward neural network, namely IN-based Neural Network or INNN for short, with $N = 2$ inputs. In this work, the number of inputs of the INNN is increased to N as shown in Figure 2. More specifically, the input to INNN is an N-tuple $(F_{k+1}, \ldots, F_{k+N})$ of INs, where $k \in \{0, \ldots, n-N\}$ and n is the total number of INs in a time-series $F_1, \ldots, F_n$. The INNN is trained to learn mapping an N-tuple $(F_{k+1}, \ldots, F_{k+N})$ to the true output IN F_{k+N+1}. In other words, the nonlinear regressor model implemented by the INNN is trained to learn the physical "law" that generates the non-stationary time-series of INs regarding grape maturity. More specifically, a sliding window of size N INs is used at times $k \in \{0, \ldots, n-N\}$ to generate an N-tuple $(F_{k+1}, \ldots, F_{k+N})$ of INs. We point out that an IN in this work represents a grape image data regarding the maturity of a grape bunch as described in [18].

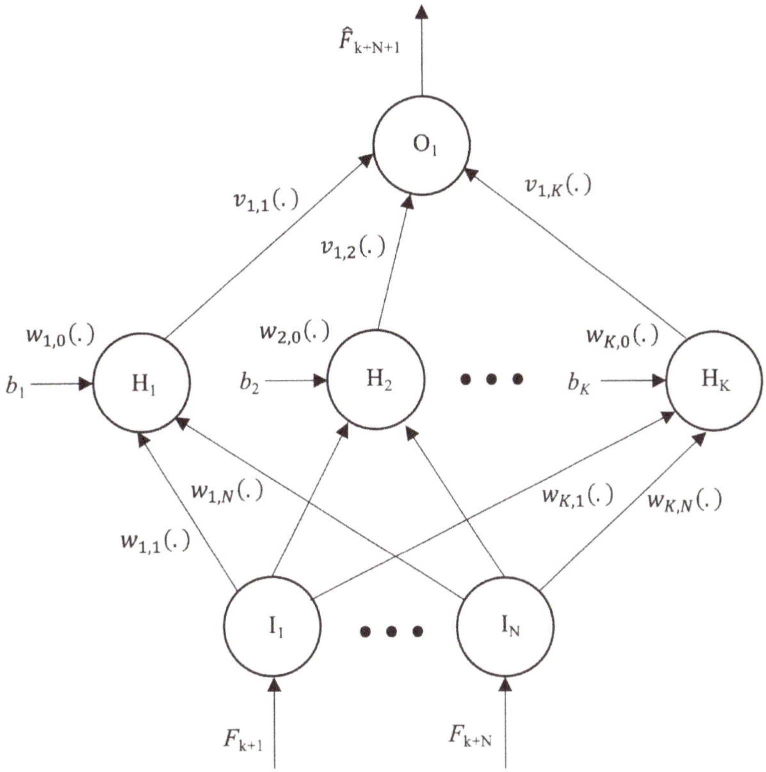

Figure 2. Given a time-series of n INs samples, an $N \times K \times 1$ forward neural network architecture, which operates on INs, can implement the proposed IN-regressor. The output $\hat{F}_{k+N+1}$ at sampling time $k + N$, where $k \in \{0, \ldots, n-N\}$, is an estimate/prediction of the true IN F_{k+N+1} at the next sampling time.

The architecture in Figure 2 is trained to learn a difference equation that calculates an estimate $\hat{F}_{k+N+1}$ of the true future IN F_{k+N+1} based on N past INs $F_{k+1}, \ldots, F_{k+N}$. Learning involves the calculation of a set of parameters that minimize the error between the estimate $\hat{F}_{k+N+1}$ and the true output IN F_{k+N+1} induced from a training image. Algorithm 1 describes the training of the INNN based on a genetic algorithm (GA) [18]. In particular, error minimization is pursued by a GA whose cost function is the metric distance between the estimate $\hat{F}_{k+N+1}$ and the true output IN F_{k+N+1}. The population of chromosomes is a number of parameters including: (a) the set of weights of the neural network, (b) the parameters of the activation function of each neuron and (c) the set of biases for all neurons. In this case, the activation function is a sigmoid.

Algorithm 1. IN-Regressor Training by a Genetic Algorithm (GA)

1. Consider the training data set.
2. Generate an initial population of parameter sets.
3. **for** g generations **do**
4. Evaluate individuals using the distance between the IN-regressor computed output IN (prediction) estimate and the true output IN.
5. Apply the genetic operators.
6. **end for**

In contrast to the INNN model presented in [18], which used only measured INs as inputs, the IN-regressor model in this paper uses, in addition, previous predictions as inputs to calculate predictions for future days. Figure 3 delineates the operation of the recursive IN-regressor with N inputs. In other words, for the calculation of a maturity prediction for a particular day, one or more of the input INs is actually a previous prediction. In this manner, the recursive scheme in Figure 3 tests the capacity of the proposed IN-regressor to learn the physical "law" that generates a time-series of INs regarding grape maturity.

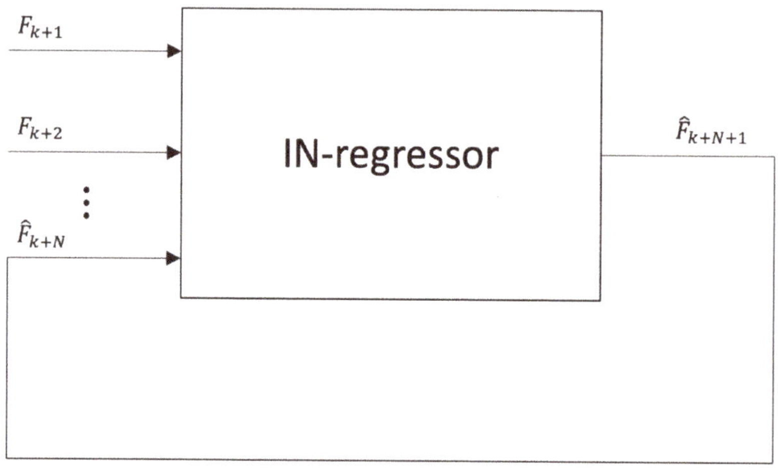

Figure 3. Given a time-series of n INs samples, a recursive IN-regressor estimates/predicts IN $\hat{F}_{k+N+1}$ at time $k + N$, where $k \in \{0, \ldots, n - N\}$, based on past IN predictions such as $\hat{F}_{k+N}$ etc.

In terms of Computational Intelligence, the proposed IN-regressor model can be interpreted as a multilayer Fuzzy Inference System (FIS) for deep learning. In conclusion, knowledge is induced from the data in the form of rules; furthermore, fuzzy lattice reasoning (FLR) explanations of the IN-regressor answers can be given as demonstrated below.

3. Experimental Results

Grape maturity at harvest time is based on the composition balance of several maturity-related chemical compounds and sensory attributes such as color and taste [19]. In order to exploit composition changes and decide on optimal harvest time, it is necessary to perform sensory assessments, i.e., ripeness evaluation, optimally by using non-destructive methods.

Toward this end, Red–Green–Blue (RGB) color imaging has been used to calculate the color intensity distribution on grape images while ripening. More specifically, the green channel histogram was represented by an IN [18].

This work extends the IN-regressor in [18] whose results are partly presented below for comparison reasons. More specifically, in [18], only two inputs were used, i.e., $N = 2$ in Figure 2; whereas, in this work, additional experiments were carried out using both $N = 3$ and $N = 4$ in order to study the robustness of the prediction when incrementally more past data were used for prediction. Moreover, in [18], three different training modes were used, namely (a) only One (the first) data sample, (b) Every Other data sample, and (c) nearly the First Half the data samples; whereas, in this work, an additional training mode was used, that is, (d) nearly the First Third of the data samples were also employed for training in order to study the robustness of the prediction when incrementally more data were used to train the IN-regressor. An IN-regressor was trained by the GA in Algorithm 1.

A trained IN-regressor was tested in two different modes, namely "forward" and "recursive" using all the remaining (non-training) data. During "forward" testing, the N-tuple of INs inputs to the IN-regressor included exclusively real (true) INs induced from images, whereas, during "recursive" testing, the N-tuple of INs input to the IN-regressor progressively included ever more of its previous IN predictions as shown in Figure 3. Both "forward" and "recursive" testing were preceded by the same training.

Each different training/testing mode has a particular experimental value as explained in the following. More specifically, One, First Third, and First Half modes use progressively ever more training data; the Every Other mode indicates the effects of reducing the sampling rate by 2, by sampling every other day. Finally, the recursive scheme, in particular, demonstrates an IN-regressor's capacity to learn the physical "law" that generates the time-series of INs regarding grape maturity toward achieving long-term predictions. In the experiments below, an interval-representation of an IN included $L = 32$ levels; moreover, the time-series of $n = 13$ INs in [18] was used.

The results for $N = 2$ have been detailed in [18]. Tables 1–4 detail the results for $N = 3$. Table 5 summarizes the results for all training/testing modes for all $N = 2$, $N = 3$, and $N = 4$. Table 5 clearly demonstrates that, as N increases, the training error decreases as well as the corresponding standard deviation due to more accurate predictions. A similar observation holds for the testing error, for the same reason, even though the testing error is significantly larger with significantly larger standard deviation. For constant N, as the number of training data increases, so does the error, due to "curve-fitting" problems; nevertheless, often the corresponding standard deviation appears to decrease. However, an IN-regressor demonstrates a good capacity for generalization on the testing data because, for constant N, as the number of training data increases, the error decreases even though it is clearly larger than the corresponding training error. Especially promising is the performance of the IN-regressor in the recursive mode for $N = 4$ when the First Half of the data were used for training. Then, an average of 6.65 was recorded with a standard deviation of 2.32 recorded compared to 4.30 and 0.96, respectively, recorded in the forward mode. The significance of the latter is that the proposed IN-regressor could potentially make accurate long-term predictions, thus providing time to engage both human labor and equipment for grape harvest.

Table 1. Training/Testing distance error Average and Standard Deviation using N = 3 inputs. The training set included only one sample, i.e., a triplet of INs as input and one output.

Training		Testing			
		Forward		Recursive	
Data	Error	Data	Error	Data	Error
$(F_1, F_2, F_3) \rightarrow F_4$	0.20				
		$(F_2, F_3, F_4) \rightarrow F_5$	8.53	$(F_2, F_3, \hat{F}_4) \rightarrow F_5$	13.05
		$(F_3, F_4, F_5) \rightarrow F_6$	10.68	$(F_3, \hat{F}_4, \hat{F}_5) \rightarrow F_6$	14.97
		$(F_4, F_5, F_6) \rightarrow F_7$	12.40	$(\hat{F}_4, \hat{F}_5, \hat{F}_6) \rightarrow F_7$	10.61
		$(F_5, F_6, F_7) \rightarrow F_8$	9.66	$(\hat{F}_5, \hat{F}_6, \hat{F}_7) \rightarrow F_8$	12.02
		$(F_6, F_7, F_8) \rightarrow F_9$	13.54	$(\hat{F}_6, \hat{F}_7, \hat{F}_8) \rightarrow F_9$	19.87
		$(F_7, F_8, F_9) \rightarrow F_{10}$	15.41	$(\hat{F}_7, \hat{F}_8, \hat{F}_9) \rightarrow F_{10}$	27.01
		$(F_8, F_9, F_{10}) \rightarrow F_{11}$	15.00	$(\hat{F}_8, \hat{F}_9, \hat{F}_{10}) \rightarrow F_{11}$	30.48
		$(F_9, F_{10}, F_{11}) \rightarrow F_{12}$	2.60	$(\hat{F}_9, \hat{F}_{10}, \hat{F}_{11}) \rightarrow F_{12}$	19.27
		$(F_{10}, F_{11}, F_{12}) \rightarrow F_{13}$	3.27	$(\hat{F}_{10}, \hat{F}_{11}, \hat{F}_{12}) \rightarrow F_{13}$	21.36
Average	0.20		10.12		18.74
Standard Deviation	0		4.68		6.82

Table 2. Training/Testing distance error Average and Standard Deviation using N = 3 inputs. Every Other data sample, i.e., a triplet of INs as input and one output, was used for training.

Training		Testing			
		Forward		Recursive	
Data	Error	Data	Error	Data	Error
$(F_1, F_2, F_3) \rightarrow F_4$	2.29				
		$(F_2, F_3, F_4) \rightarrow F_5$	4.84	$(F_2, F_3, \hat{F}_4) \rightarrow F_5$	5.76
$(F_3, F_4, F_5) \rightarrow F_6$	4.56				
		$(F_4, F_5, F_6) \rightarrow F_7$	4.67	$(F_4, F_5, \hat{F}_6) \rightarrow F_7$	6.22
$(F_5, F_6, F_7) \rightarrow F_8$	1.77				
		$(F_6, F_7, F_8) \rightarrow F_9$	3.53	$(F_6, F_7, \hat{F}_8) \rightarrow F_9$	23.36
$(F_7, F_8, F_9) \rightarrow F_{10}$	2.87				
		$(F_8, F_9, F_{10}) \rightarrow F_{11}$	5.26	$(F_8, F_9, \hat{F}_{10}) \rightarrow F_{11}$	10.02
$(F_9, F_{10}, F_{11}) \rightarrow F_{12}$	2.93				
		$(F_{10}, F_{11}, F_{12}) \rightarrow F_{13}$	5.30	$(F_{10}, F_{11}, \hat{F}_{12}) \rightarrow F_{13}$	3.19
Average	2.89		4.72		9.71
Standard Deviation	1.05		0.71		8.01

Table 3. Training/Testing distance error Average and Standard Deviation using N = 3 inputs. The training set included approximately the First Third of the total number of data samples.

Training		Testing			
		Forward		Recursive	
Data	Error	Data	Error	Data	Error
$(F_1, F_2, F_3) \rightarrow F_4$	2.69				
$(F_2, F_3, F_4) \rightarrow F_5$	0.77				
$(F_3, F_4, F_5) \rightarrow F_6$	3.62				
$(F_4, F_5, F_6) \rightarrow F_7$	1.10				
		$(F_5, F_6, F_7) \rightarrow F_8$	4.23	$(F_5, F_6, \hat{F}_7) \rightarrow F_8$	8.12
		$(F_6, F_7, F_8) \rightarrow F_9$	3.97	$(F_6, \hat{F}_7, \hat{F}_8) \rightarrow F_9$	14.48
		$(F_7, F_8, F_9) \rightarrow F_{10}$	5.74	$(\hat{F}_7, \hat{F}_8, \hat{F}_9) \rightarrow F_{10}$	20.75
		$(F_8, F_9, F_{10}) \rightarrow F_{11}$	5.73	$(\hat{F}_8, \hat{F}_9, \hat{F}_{10}) \rightarrow F_{11}$	24.01
		$(F_9, F_{10}, F_{11}) \rightarrow F_{12}$	6.37	$(\hat{F}_9, \hat{F}_{10}, \hat{F}_{11}) \rightarrow F_{12}$	22.20
		$(F_{10}, F_{11}, F_{12}) \rightarrow F_{13}$	8.67	$(\hat{F}_{10}, \hat{F}_{11}, \hat{F}_{12}) \rightarrow F_{13}$	22.34
Average	2.05		5.78		18.65
Standard Deviation	1.34		1.69		6.12

Table 4. Training/Testing distance error Average and Standard Deviation using N = 3 inputs. The training set included approximately the First Half of the total number of data samples.

Training		Testing			
		Forward		Recursive	
Data	Error	Data	Error	Data	Error
$(F_1, F_2, F_3) \rightarrow F_4$	3.15				
$(F_2, F_3, F_4) \rightarrow F_5$	0.61				
$(F_3, F_4, F_5) \rightarrow F_6$	4.71				
$(F_4, F_5, F_6) \rightarrow F_7$	1.20				
$(F_5, F_6, F_7) \rightarrow F_8$	1.59				
		$(F_6, F_7, F_8) \rightarrow F_9$	6.56	$(F_6, F_7, \hat{F}_8) \rightarrow F_9$	6.93
		$(F_7, F_8, F_9) \rightarrow F_{10}$	7.07	$(F_7, \hat{F}_8, \hat{F}_9) \rightarrow F_{10}$	10.15
		$(F_8, F_9, F_{10}) \rightarrow F_{11}$	6.81	$(\hat{F}_8, \hat{F}_9, \hat{F}_{10}) \rightarrow F_{11}$	13.90
		$(F_9, F_{10}, F_{11}) \rightarrow F_{12}$	8.12	$(\hat{F}_9, \hat{F}_{10}, \hat{F}_{11}) \rightarrow F_{12}$	11.09
		$(F_{10}, F_{11}, F_{12}) \rightarrow F_{13}$	3.64	$(\hat{F}_{10}, \hat{F}_{11}, \hat{F}_{12}) \rightarrow F_{13}$	13.06
Average	2.25		6.44		11.03
Standard Deviation	1.66		1.67		2.73

Table 5. Training/Testing distance error Average and Standard Deviation (Std) using N ∈ {2,3,4} inputs for four training modes: (1) One sample, (2) Every Other sample, (3) First Third of the samples, and (4) First Half of the samples.

N	Training Mode	Training Error (Average/Std)	Testing Error (Average/Std) Forward	Recursive
2	(1) One sample	0.10/0	11.45/9.17	37.87/9.96
	(2) Every Other sample	3.21/1.26	10.54/8.74	17.85/11.93
	(3) First Third of the samples	5.97/3.28	7.92/3.43	18.54/4.53
	(4) First Half of the samples	5.15/3.79	6.84/4.12	7.57/4.87
3	(1) One sample	0.20/0	10.12/4.68	18.74/6.82
	(2) Every Other sample	2.89/1.05	4.72/0.71	9.71/8.01
	(3) First Third of the samples	2.05/1.34	5.78/1.69	18.65/6.12
	(4) First Half of the samples	2.25/1.66	6.44/1.67	11.03/2.73
4	(1) One sample	0.14/0	14.09/3.68	20.21/8.92
	(2) Every Other sample	1.95/0.48	4.99/1.95	9.64/4.66
	(3) First Third of the samples	0.93/0.32	10.85/4.25	25.89/2.75
	(4) First Half of the samples	1.64/0.99	4.30/0.96	6.65/2.32

An *N*-tuple of INs input to the IN-regressor followed by its corresponding output IN can be interpreted as a fuzzy rule (i.e., knowledge), of a "Mamdani type" FIS, induced from the training data as indicated in Figure 4, where Figure 4a shows the rule's antecedent and Figure 4b shows the rule's consequent.

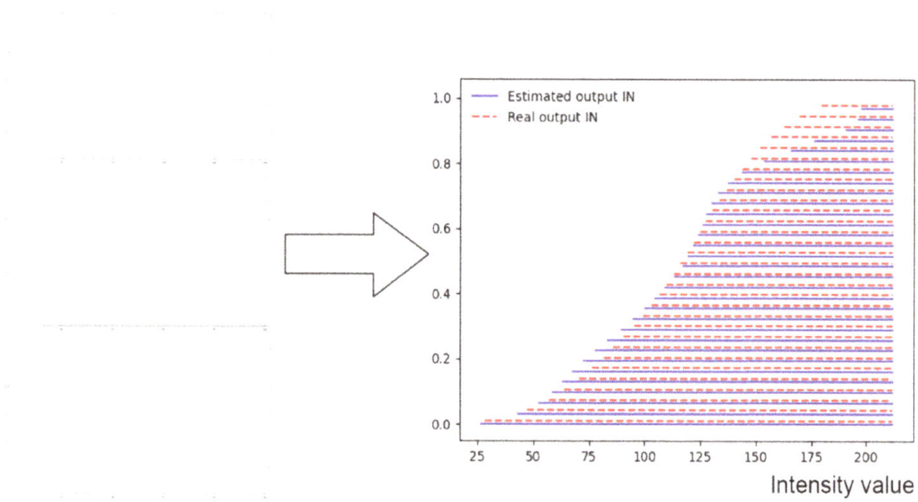

(**a**) (**b**)

Figure 4. The first three INs (**a**) the first three INs F_1, F_2 and F_3, from top to bottom, of the considered time-series of Ins; (**b**) estimated out IN $\hat{F}_4$ (shown in solid intervals) computed in the first line of Table 4 versus the real (true) output IN F_4 (shown in dashed intervals).

4. Discussion and Conclusions

Agriculture 4.0 [1], including viticulture, calls for intelligent decision-making. Of special interest is the accurate prediction of the grape maturity in order to timely engage both human labor and equipment for harvest. This work has proposed a parametric regressor, namely IN-regressor, model for grape maturity prediction.

The IN-regressor processes Intervals' Numbers (INs) with the advantage that an IN represents a distribution of data including all-order data statistics. Hence, instead of representing the maturity status of grapes by few numbers, e.g., the mean and standard deviation of a number of measurements, the maturity status of grapes is represented by a distribution of measurements, i.e., all-order data statistics, toward better decision-making. A neural network architecture, namely INNN, with N inputs (INs) and one output (IN) was shown to implement the proposed IN-regressor.

Extensive computational experiment here has demonstrated that an IN-regressor can accurately predict the grape maturity status, especially for larger N as well for more training data. Therefore, the IN-regressor be used for predicting the grape harvest time. Furthermore, especially promising is a recursive IN-regressor scheme for long-term prediction.

The proposed IN-regressor has been interpreted as a deep learning FIS with a capacity to suggest explanations for its answers by "Mamdani type" fuzzy rules.

Technical future work will pursue one (or more) neural network layer(s) in the input as a filter that normalizes the effects of taken a grape image at different azimuth /altitude /distance /lighting conditions, etc. Furthermore, a faster algorithm for optimization will be pursued instead of a GA. An extension of this work can also demonstrate far more experimental results using data already acquired on-the-field.

As grapes mature, their image statistics change with time. Therefore, since an IN represents a distribution of image statistics regarding grape maturity, it follows that a time-series of INs by definition represents a non-stationary time-series process. Hence, the proposed IN-regressor can be used for predicting a future probability distribution function from past probability distribution functions in a non-stationary time-series. Therefore, apart from agriculture, this work has presented potentially useful instruments for other application domains including the environment [20], medicine [21], econometrics [22], stock-market data [23], and other.

Author Contributions: Conceptualization, V.G.K.; methodology, V.G.K.; software, C.L and C.B.; investigation, E.V.; writing—original draft preparation, V.G.K., E.V., C.B., and C.L.; writing—review and editing, V.G.K. and E.V.; visualization, V.G.K.; supervision, V.G.K.; project administration, V.G.K.; funding acquisition, V.G.K. All authors have read and agreed to the published version of the manuscript.

Funding: We acknowledge support of this work by the project "Technology for Skillful Viniculture (SVtech)" (MIS 5046047) which is implemented under the Action "Reinforcement of the Research and Innovation Infrastructure", funded by the Operational Programme "Competitiveness, Entrepreneurship and Innovation" (NSRF 2014-2020) and co-financed by Greece and the European Union (European Regional Development Fund).

Institutional Review Board Statement: Not applicable.

Informed Consent Statement: Not applicable.

Data Availability Statement: Publicly available datasets were analyzed in this study. This data can be found here: https://github.com/humain-lab/ripeness-estimation-videoframes.

Conflicts of Interest: The authors declare no conflict of interest.

References

1. Rose, D.C.; Wheeler, R.; Winter, M.; Lobley, M.; Chivers, C.-A. Agriculture 4.0: Making it work for people, production, and the planet. *Land Use Policy* **2021**, *100*, 104933. [CrossRef]
2. Sohaib Ali Shah, S.; Zeb, A.; Qureshi, W.S.; Arslan, M.; Ullah Malik, A.; Alasmary, W.; Alanazi, E. Towards fruit maturity estimation using NIR spectroscopy. *Infrared Phys. Technol.* **2020**, *111*, 103479. [CrossRef]
3. Perry, T.S. Want a Really Hard Machine Learning Problem? Try Agriculture, Says John Deere Labs. Available online: https://spectrum.ieee.org/view-from-the-valley/robotics/artificial-intelligence/want-a-really-hard-machine-learning-problem-try-agriculture-say-john-deere-labs-leaders (accessed on 19 May 2021).
4. Vrochidou, E.; Tziridis, K.; Nikolaou, A.; Kalampokas, T.; Papakostas, G.A.; Pachidis, T.P.; Mamalis, S.; Koundouras, S.; Kaburlasos, V.G. An Autonomous Grape-Harvester Robot: Integrated System Architecture. *Electronics* **2021**, *10*, 1056. [CrossRef]
5. Mehdizadeh, S. Using AR, MA, and ARMA Time Series Models to Improve the Performance of MARS and KNN Approaches in Monthly Precipitation Modeling under Limited Climatic Data. *Water Resour. Manag.* **2020**, *34*, 263–282. [CrossRef]

6. Wang, Q.; Li, S.; Li, R.; Ma, M. Forecasting U.S. shale gas monthly production using a hybrid ARIMA and metabolic nonlinear grey model. *Energy* **2018**, *160*, 378–387. [CrossRef]
7. Tealab, A.; Hefny, H.; Badr, A. Forecasting of nonlinear time series using ANN. *Futur. Comput. Inform. J.* **2017**, *2*, 39–47. [CrossRef]
8. Raj, J.S.; Ananthi, J.V. Reccurent Neural Networks and Nonlinear Prediction in Support Vector Machines. *J. Soft Comput. Paradig.* **2019**, *2019*, 33–40. [CrossRef]
9. Tealab, A. Time series forecasting using artificial neural networks methodologies: A systematic review. *Futur. Comput. Inform. J.* **2018**, *3*, 334–340. [CrossRef]
10. Kaburlasos, V.G. FINs: Lattice Theoretic Tools for Improving Prediction of Sugar Production From Populations of Measurements. *IEEE Trans. Syst. Man Cybern. Part B* **2004**, *34*, 1017–1030. [CrossRef] [PubMed]
11. Kaburlasos, V.G. The Lattice Computing (LC) Paradigm. In Proceedings of the the 15th International Conference on Concept Lattices and Their Applications CLA, Tallinn, Estonia, 29 June–1 July 2020; pp. 1–8.
12. Kaburlasos, V.G.; Papadakis, S.E. Granular self-organizing map (grSOM) for structure identification. *Neural Netw.* **2006**, *19*, 623–643. [CrossRef] [PubMed]
13. Kaburlasos, V.G.; Kehagias, A. Fuzzy Inference System (FIS) Extensions Based on the Lattice Theory. *IEEE Trans. Fuzzy Syst.* **2014**, *22*, 531–546. [CrossRef]
14. Kaburlasos, V.G.; Athanasiadis, I.N.; Mitkas, P.A. Fuzzy lattice reasoning (FLR) classifier and its application for ambient ozone estimation. *Int. J. Approx. Reason.* **2007**, *45*, 152–188. [CrossRef]
15. Kaburlasos, V.G.; Papakostas, G.A. Learning Distributions of Image Features by Interactive Fuzzy Lattice Reasoning in Pattern Recognition Applications. *IEEE Comput. Intell. Mag.* **2015**, *10*, 42–51. [CrossRef]
16. Kaburlasos, V.G.; Papakostas, G.A.; Pachidis, T.; Athinellis, A. Intervals' Numbers (INs) interpolation/extrapolation. In Proceedings of the 2013 IEEE International Conference on Fuzzy Systems (FUZZ-IEEE), Hyderabad, India, 7–10 July 2013; pp. 1–8.
17. Vrochidou, E.; Lytridis, C.; Bazinas, C.; Papakostas, G.A.; Wagatsuma, H.; Kaburlasos, V.G. Brain Signals Classification Based on Fuzzy Lattice Reasoning. *Mathematics* **2021**, *9*, 1063. [CrossRef]
18. Kaburlasos, V.G.; Vrochidou, E.; Lytridis, C.; Papakostas, G.A.; Pachidis, T.; Manios, M.; Mamalis, S.; Merou, T.; Koundouras, S.; Theocharis, S.; et al. Toward Big Data Manipulation for Grape Harvest Time Prediction by Intervals' Numbers Techniques. In Proceedings of the 2020 International Joint Conference on Neural Networks (IJCNN), Glasgow, UK, 19–24 July 2020; pp. 1–6.
19. Kangune, K.; Kulkarni, V.; Kosamkar, P. Automated estimation of grape ripeness. *Asian J. Converg. Technol. (AJCT)* **2019**. Available online: https://asianssr.org/index.php/ajct/article/view/792 (accessed on 20 June 2021).
20. Garnot, V.S.F.; Landrieu, L.; Giordano, S.; Chehata, N. Satellite Image Time Series Classification with Pixel-Set Encoders and Temporal Self-Attention. In Proceedings of the IEEE/CVF Conference on Computer Vision and Patern Recognition (CVPR), Seattle, WA, USA, 13–19 June 2020.
21. Dupont, G.; Kalinicheva, E.; Sublime, J.; Rossant, F.; Pâques, M. Analyzing Age-Related Macular Degeneration Progression in Patients with Geographic Atrophy Using Joint Autoencoders for Unsupervised Change Detection. *J. Imaging* **2020**, *6*, 57. [CrossRef]
22. Greene, W.W.H. *Econometric Analysis*, 7th ed.; Prentice-Hall: Hoboken, NJ, USA, 2012; ISBN 978-0-273-75356-8.
23. Barra, S.; Carta, S.M.; Corriga, A.; Podda, A.S.; Recupero, D.R. Deep learning and time series-to-image encoding for financial forecasting. *IEEE/CAA J. Autom. Sin.* **2020**, *7*, 683–692. [CrossRef]

engineering proceedings

MDPI

Proceedings

On the Introduction of Diffusion Uncertainty in Telecommunications' Market Forecasting †

Nikolaos Kanellos *, Dimitrios Katsianis and Dimitrios Varoutas

Department of Informatics and Telecommunications, National and Kapodistrian University of Athens (NKUA), 157 72 Athens, Greece; dkats@di.uoa.gr (D.K.); D.Varoutas@di.uoa.gr (D.V.)
* Correspondence: kanetza@di.uoa.gr

† Presented at the 7th International Conference on Time Series Forecasting, Gran Canaria, Spain, 19–21 July 2021.

Abstract: Long-run forecasts of telecommunication services' diffusion play an important role in policy, regulation, planning and portfolio decisions. Forecasting diffusion of telecommunication technologies is usually based on S-shaped models, mainly due to their accurate long-term predictions. Yet, the use of these models does not allow the introduction of risk in the forecast. In this paper, a methodology for the introduction of uncertainty in the underlying calculations is presented. It is based on the calibration of an Ito stochastic process and the generation of possible forecast paths via Monte Carlo Simulation. Results consist of a probabilistic distribution of future demand, which constitutes a risk assessment of the diffusion process under study. The proposed methodology can find applications in all high-technology markets, where a diffusion model is usually applied for obtaining future forecasts.

Keywords: diffusion modelling; time-series forecasting; forecast uncertainty; Monte Carlo Simulation; risk estimation

Citation: Kanellos, N.; Katsianis, D.; Varoutas, D. On the Introduction of Diffusion Uncertainty in Telecommunications' Market Forecasting. *Eng. Proc.* **2021**, *5*, 13. https://doi.org/10.3390/engproc2021005013

Published: 25 June 2021

1. Introduction

The study of the diffusion process of telecommunication services is of paramount importance in understanding the factors influencing the development of telecommunication networks. For telecommunication service operators, it provides the basis for strategic decisions, such as technology selection and capacity expansion. Moreover, the derived knowledge can be used by policy makers and regulators for shaping market competition.

Based on the findings in [1], telecommunications' demand modelling and forecasting usually involves the use of traditional diffusion theory. Most commonly used diffusion models include the Bass model, the Fisher–Pry model, the Gompertz models and some representatives of the logistic variants. With respect to studies not mentioned, examples of this literature include the work of [2–5] and more recently [6,7].

These S-shaped diffusion models accurately capture the telecommunications' market expectations, but do not provide measures for the inherent uncertainty in their forecasts [1]. Consequently, the decision maker is deprived of the ability to estimate the risk (systematic and/or idiosyncratic based on the diffusion process under study) inherent to the market under study, as well as to investigate the link between this risk and the market's competitive environment.

To cope with this shortcoming, the literature suggests the use of stochastic models, e.g., Geometric Brownian Motion (GBM). In [8], an error factor with normal distribution was used to model uncertainty. In [9], GBM is described as a mathematical tool with the capability of calibrating demand volatility very reasonably and accurately. In [10], GBM was indicated as a good first approximation for uncertainties. In [11], a GBM process with a linear expected growth rate was used to model the stochastic nature of the diffusion process. In [12], GBM modelling was applied to generate sample paths of demand in the semiconductor manufacturing industry. In the telecommunications' market, the relevant literature is quite limited; in [13], four datasets in the energy, transportation, and telecommunication sectors were analyzed using GBM.

Despite their ability to capture and communicate forecast uncertainties better to stakeholders, the above stochastic models have also received some criticism; in [14], it was pointed out that stochastic models cannot capture demand trends as good as S-shaped models. For GBM applications, this can be accounted for by the constant drift rate, which varies significantly from the dynamic growth exhibited in new product demand [11]. To tackle this issue, in [15], a calibrated GBM model with spline interpolation was proposed to address the problem of stochasticity in forecasting diffusion of a new product with scarce historical data; the drift parameter is calculated from the forecasted data provided by a best-fit polynomial model and the volatility parameter is considered equal to the root mean square error (RSME) of the best-fit function found for the drift. This approach enables the stochastic model to capture the dynamics of the product's life cycle; yet, it should not be used in forecasting because of the polynomial model's prowess to overfitting.

To consider both dynamic growth and possible stochasticity in the future demand for telecommunication services, this study suggests the use of an S-curve calibrated generalized Brownian motion—Ito stochastic process. Dynamic growth is captured by a variable drift parameter following the diffusion rate provided by the best-fitting S-shaped model. Furthermore, the diffusion uncertainty is modelled through the volatility parameter, which is defined as the standard deviation of the percentage error of the best-fitting S-shaped model. Since an Ito process is used, the proposed model is valid provided the actual diffusion log changes follow a normal distribution.

The proposed approach offers significant advantages in telecommunications' demand modelling and forecasting over the existing literature. S-curve diffusion modelling has proven its ability to accurately capture growth trends in telecommunication services. This ability is incorporated in the proposed stochastic model by the variable growth rate of the best-fitting S-curve model, which serves as the drift parameter of the model. The main advantage, though, lies with the accurate estimation of the volatility incorporated in the diffusion process under study; the better determination of the data drift highlights the changes in data due to uncertainty, thus allowing for a better determination of diffusion uncertainty. If the diffusion process of an entire market is examined, the calculated volatility reflects the overall market uncertainty, whereas, if the diffusion process of a specific technology on a provider basis is examined, the calculated volatility corresponds to the overall technology uncertainty the provider experiences.

The proposed stochastic process can be used in telecommunications' demand forecasting. Monte Carlo Simulation is deployed to provide the diffusion forecast. Depending on the diffusion process under study, through this analysis, the estimation of both the systematic and the idiosyncratic risk inherent in the telecommunications' services market may also be provided. Moreover, when a specific technology for both the overall market and a provider are examined, through a standard cointegration analysis, the effect of the overall market uncertainty to the provider uncertainty may also be determined.

To indicate the dynamics of the proposed method, a real-world example, based on the diffusion of the mobile market in Greece, is provided. Monte Carlo Simulation outputs of the calibrated stochastic process are compared with the equivalent results from a standard GBM model. Results validate the enhanced uncertainty measurement and diffusion forecast hypothesis.

To conclude, this paper addresses the uncertainty determination problem in the telecommunications' market diffusion processes and the introduction of this uncertainty in diffusion forecasting. The latter allows the estimation of the idiosyncratic and/or the systematic risk inherent in the diffusion process under study. In addition, it provides a way to estimate the effect of the overall market uncertainty to the diffusion of a specific firm/technology.

The rest of the paper is organized as follows. Section 2 provides an overview of the proposed model. Section 3 presents the results, after its application in a telecommunication market paradigm. Finally, Section 4 concludes.

2. Forecasting Telecommunications' Services Diffusion under Uncertainty

The aim of this study is to propose a statistical and simulation-based methodology for forecasting the demand of a telecommunication service in an uncertain and dynamic environment. This methodology builds upon the use of a calibrated generalized Brownian motion—Ito stochastic process. The steps taken in performing the proposed forecasting methodology are represented in Figure 1.

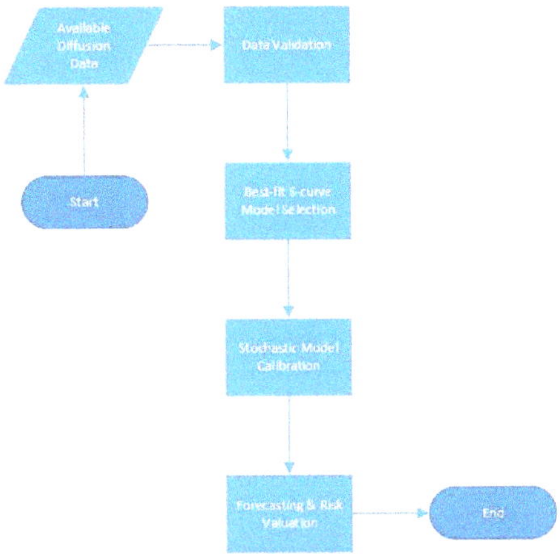

Figure 1. Proposed methodology.

As can be seen in Figure 1, the proposed methodology follows a four-step procedure, comprising Data Gathering and Validation, the Best-fit S-curve Model Selection, the Stochastic Model Calibration and the Forecasting and Risk Valuation. Details of each step are provided below.

2.1. Data Gathering and Validation

The first step of the methodology includes the respective data collection about the telecommunication service diffusion and the validation of this data for use.

Since an Ito process is used, the proposed model is valid provided the actual diffusion log changes follow a normal distribution. Consequently, a normality test has to be deployed to determine if the data set is well-modeled by a normal distribution.

2.2. Best-Fit S-Curve Model Selection

The second step of the methodology includes the selection of the S-curve model that best describes the demand evolution of the diffusion process under study.

The S-shaped diffusion models can be derived from the differential equation represented in (1).

$$\frac{dN(t)}{dt} = \delta \times f(N(t)) \times [K - N(t)] \tag{1}$$

where $N(t)$ represents the penetration estimation, K is the saturation level and δ is the coefficient of diffusion.

From (1), it can be seen that for a diffusion model to produce an estimation, the saturation level K and diffusion coefficient δ have to be determined. While the determination of the saturation level K is most of the times a more straightforward procedure, the diffusion coefficient involves the estimation of model-specific parameters through data-fitting procedures, e.g., in the Bass model, described in (2), in which the diffusion coefficient involves the determination of parameter p—the coefficient of innovation—and parameter q—the coefficient of imitation.

$$A(t) = \frac{m \times (p+q)^2}{p} \times \frac{e^{-(p+q)t}}{\left[\left(\frac{q}{p}\right)e^{-(p+q)t} + 1\right]} \quad (2)$$

where m is the market potential, p is the coefficient of innovation and q is the coefficient of imitation.

The estimation of the parameters of the models under evaluation may be achieved through data fitting, with the use of dedicated software. It should be considered that for an S-curve model to produce valid results, a considerable amount of data is required.

Following the estimation of the required parameters, the selection of the best-fitting model is accomplished with the use of forecast accuracy measures, such as the Mean Absolute Percentage Error (MAPE) and Root Mean Square Error (RMSE). Failure to select the S-curve model that best captures the diffusion process dynamics will have a strong impact on the validity of the results of the proposed methodology.

2.3. *Stochastic Model Calibration*

The proposed model is based on a generalized Brownian motion—Ito stochastic process. The latter was chosen because, unlike GBM, it incorporates drift and volatility coefficients that are functions of the current state and time.

The Ito process is represented by (3):

$$dx = a(x,t)dt + b(x,t)dz \quad (3)$$

where dz is the increment of a Wiener process and $a(x,t)$ and $b(x,t)$ are known (nonrandom) functions serving as the drift and volatility parameters, respectively.

Equation (3) defines two terms that affect the calculated estimation. The first term defines that at each time period, the estimated value will drift up by the expected market growth rate. The second term indicates that a random value, scaled from the volatility coefficient, will be added to or subtracted from the drift value. Hence, estimations follow a series of steps, which result from the interactions of the above two terms and are independent of past estimations (a Markov process property).

When stochastic models are used for diffusion modelling, their parameters are estimated based on historical data, e.g., [16]. In few cases, these parameters are considered variable and are calibrated based on existing data, e.g., [15]. Under the proposed methodology, the drift coefficient of the Ito process is calibrated based on the diffusion rate provided by the best-fitting S-shaped model of Step 1, whereas the volatility coefficient is calculated after the extraction of the drift trend of the data.

2.3.1. Drift Coefficient Calibration

To incorporate the market dynamic growth into the Ito stochastic process, the drift coefficient $a(x,t)$ is set equal to the variable market growth rate provided by the best-fitting S-shaped diffusion model. For S-shaped models providing cumulative penetration estimation, like the logistic family of models, this growth rate may be calculated at any given time period t following (4),

$$\mu(t) = \frac{N(t) - N(t-1)}{N(t-1)} \quad (4)$$

whereas for models providing spot penetration growth, like the Bass model, the growth rate may be calculated using (5).

$$\mu(t) = \frac{\sum_0^t N(t) - \sum_0^{t-1} N(t-1)}{\sum_0^{t-1} N(t-1)} = \frac{N(t)}{\sum_0^{t-1} N(t-1)} \quad (5)$$

Hence, the proposed Ito process, after calibration, is represented by (6):

$$dx = \mu(t)dt + b(x,t)dz \quad (6)$$

It should be noted that in the absence of volatility ($b(x,t) = 0$), the results of (6) converge to the results provided by (1), the S-shaped diffusion model that was used for the forecast.

2.3.2. Volatility Coefficient Estimation

Similar to the work proposed in [15], following the Ito process drift coefficient calibration, the volatility coefficient has to be estimated. Under the proposed methodology, volatility is defined as the standard deviation of the percentage error of the best-fitting S-shaped model for $\mu(t)$. In this way, the volatility coefficient $b(x,t)$ will remain constant throughout the evaluation period and depends on the residuals of the S-curve fitting process. These residuals are considered to be a direct result of the inherent uncertainty in the diffusion process under study. Therefore, the better determination of the data drift highlights the changes in data due to uncertainty, thus allowing for a better determination of diffusion uncertainty.

Moreover, based on this view of the residuals of the S-curve fitting process, when a specific technology for both the overall market and a provider are examined, the effect of the overall market uncertainty to the provider uncertainty may also be determined. This may be achieved through a standard cointegration analysis, provided that both residual data series are integrated of the same order.

2.4. Forecasting and Risk Valuation

Given the best-fitted function to the demand growth as well as the value obtained for the volatility coefficient, the targeted stochastic differential equation is made based on (7).

$$dx = \mu(t)dt + bdz \quad (7)$$

To generate possible demand forecasts, Monte Carlo Simulation is deployed. Outputs include the probabilistic distribution of the future demand for the telecommunication service under evaluation, at a specific time t. It is noted that even though there is no constrain for the forecast period, the larger this period, the higher the data deviations due to the underlying uncertainty.

Besides the generation of future diffusion forecasts, Monte Carlo Simulation may be used to estimate the risk inherent to the diffusion process. The calculated probabilistic distribution constitutes a risk assessment of the forecasted diffusion of the telecommunication service under study. If the diffusion process of an entire market is examined, the calculated volatility reflects the overall market uncertainty, thus enabling the estimation of the market's systematic risk. On the contrary, if the diffusion process of a specific technology on a provider basis is examined, the calculated volatility corresponds to the overall technology uncertainty the provider experiences. This enables the estimation of the total technology risk for the provider, which includes both the systematic and the idiosyncratic technology risk.

Following the risk estimation, the results may be compared to various levels of risk tolerance. This can help telecommunication providers to adjust their strategy regarding technology selection and capacity expansion. Moreover, the derived knowledge can be used by policy makers and regulators for shaping market competition. This concludes the proposed method.

3. Insights from the Greek Mobile Telecommunications Market

To indicate the dynamics of the proposed methodology, a real-world example, based on the diffusion of the mobile market in Greece, is provided.

In its current state, the Greek mobile telecommunications market offers a subscriber the ability to choose between four competing technologies, 2–2.5G, 3G, 4G and 5G. After the introduction of VoLTE, all four technologies may be used for both telephony and data services. These services are included in either prepaid or postpaid packages and are provided by three companies, namely, CosmOTE, Vodafone and Wind, with an expected new entrant, Forthnet. The market is regulated by the National Telecommunications and Post Commission (EETT, https://www.eett.gr (accessed on 28 May 2021)). The responsibility for drafting legislation is retained by the Greek Ministry for Transport and Communications (YME, www.yme.gr (accessed on 28 May 2021)).

The data used in the analysis were published by EETT. These involve the number of active subscriptions per mobile telecommunications service provider from 1998, when the first mobile telecommunications networks were deployed in Greece, to 2019. These data for the incumbent operator CosmOTE and the total market are presented in Figure 2.

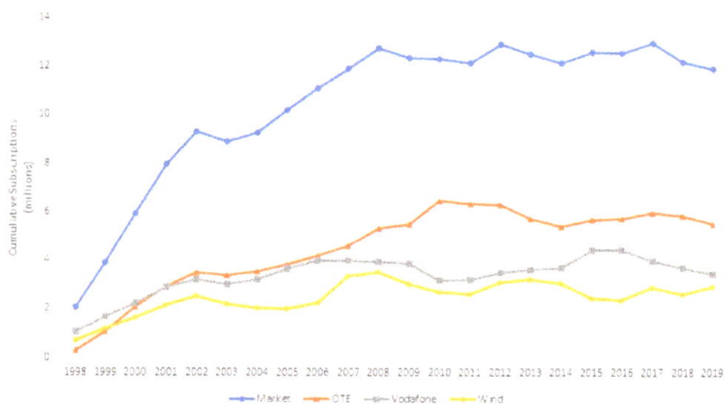

Figure 2. Diffusion of mobile services in Greece.

It can be seen that the incumbent operator CosmOTE captures about 50% of the entire market. The other 50% is split between the other operators, namely, Vodafone and Wind.

3.1. Methodology Application

3.1.1. Data Validation

To be able to apply the proposed methodology, the annual log changes of active subscriptions must be normally distributed. The Anderson–Darling normality test was used for this purpose. Results are presented in Figure 3.

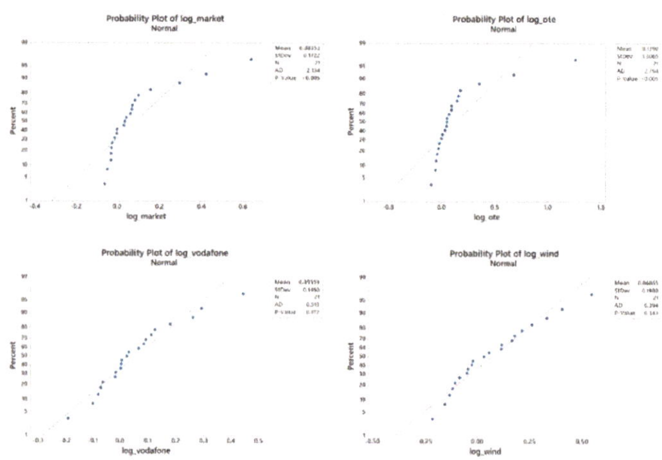

Figure 3. Normality test results.

As can be seen in Figure 3, both the total market and the incumbent operator CosmOTE annual log changes of their active subscriptions are not normally distributed. Subsequently, they cannot be used with the proposed methodology. This is not the case though with Vodafone and Wind, whose data may be used for future demand forecasting with the proposed methodology.

3.1.2. Best-Fitting S-Curve Model Selection

For the purposes of this study, four S-curve diffusion models were evaluated: The Logistic, Fisher–Pry, Gompertz and TONIC models. Moreover, the Mean Absolute Percentage Error (MAPE) was selected as a forecast accuracy measure and calculated in each case. MAPE was calculated for all sets of data and the model for which the smallest statistical error was calculated is consequently considered to be the most appropriate to be used for forecasting future diffusion of mobile services. The results are presented in Table 1.

Table 1. MAPE estimation.

S-Curve Model	Vodafone	Wind
Logistic	6.764817	12.0521
Fisher–Pry	6.764812	12.05204
Gompertz	6.863766	11.57264
TONIC	6.77584	11.57277

Based on the data of Table 1, the best-fitting S-curve model for Vodafone is the Fisher–Pry model, whereas for Wind, the best-fitting S-curve model is Gompertz. Parameter estimation for the best-fitting models are given in Table 2.

Table 2. Best-fitting S-curve model parameter estimation.

Vodafone—Fisher–Pry		Wind—Gompertz	
S	3,674,635	S	2,693,731
a	2.339	a	−0.706
b	0.647	b	0.432

3.1.3. Stochastic Model Calibration

Following the proposed methodology, the Ito process was calibrated based on the data provided by the best-fitting S-curve model. The calculated volatility coefficients are provided in Table 3. For comparison purposes, the equivalent GBM volatility coefficients are also included in Table 3.

Table 3. Calculated volatility coefficients.

Provider	Ito	GBM
Vodafone	8.58%	14.59%
Wind	13.53%	19.87%

It can be seen that the calibration of the Ito process provides results in the smaller volatility coefficient calculation. This is due to the better capturing of the diffusion trend, provided by the S-curve model.

3.1.4. Forecasting and Risk Valuation

To complete the analysis, Monte Carlo Simulation was deployed to forecast diffusion for a period of 6 years (up to 2025). Results were compared with the ones provided by traditional GBM forecasting.

As can be seen in Figure 4, all possible paths provided by the calibrated Ito process are below the saturation point of the total market. On the contrary, for both operators, traditional GBM forecasting provides a significant number of paths that greatly exceed the total market saturation point. This is due to the constant drift rate and the higher volatility coefficient assumed by GBM. Consequently, application results validate the enhanced uncertainty measurement and diffusion forecast hypothesis; the proposed calibrated Ito process outperforms the traditional GBM forecasting.

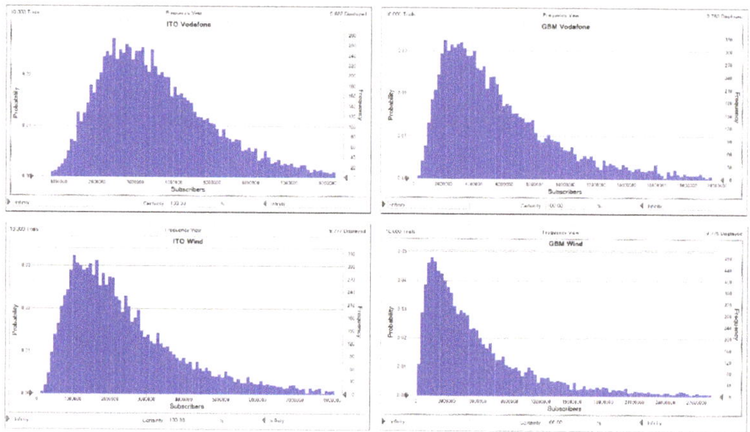

Figure 4. Monte Carlo Simulation results.

Moreover, the calculated probabilistic distribution constitutes a risk assessment of the forecasted Vodafone and Wind mobile services diffusion. Therefore, it corresponds to the overall risk experienced by both providers, which includes both the systematic and the idiosyncratic risk. Results may help telecommunication operators to adjust their strategy. Furthermore, provided that the proposed methodology could be applied to total market diffusion data, the market's systematic risk could be extracted, thus enabling the estimation of the operators' idiosyncratic risk.

4. Conclusions

In this paper, a forecast methodology was suggested for capturing both the dynamism and stochasticity of future demand for telecommunication services. The proposed methodology is based on the calibration of a generalized Brownian motion—Ito stochastic process for use in telecommunications' demand modelling.

Under the proposed methodology, the drift coefficient follows the variable diffusion rate provided by the best-fitting S-shaped diffusion model. Moreover, the volatility parameter is defined as the standard deviation of the percentage error. The calibrated Ito forecast model permits involving possible uncertainty in predicting future demand. The outputs of the proposed forecast model consist of a probabilistic distribution of future demand that constitutes a risk assessment of the forecasted diffusion of the telecommunication services under study.

The performance of the proposed methodology was tested against traditional GBM forecasting. A result comparison confirmed the enhanced uncertainty measurement and the capability of the proposed methodology in demand forecasting in the telecommunications sector.

The proposed methodology contributes well to developing strategic plans in dynamic and uncertain markets when a robust scenario analysis is required. In addition, it is compatible with all S-shaped diffusion models. Therefore, it can be applied over all cases of the high-technology market, where a diffusion model is commonly used for diffusion modelling and obtaining future demand forecasts.

Author Contributions: Conceptualization, N.K.; methodology, N.K.; formal analysis, N.K.; writing—original draft preparation, N.K.; writing—review and editing, D.K.; visualization, N.K.; supervision, D.V.; project administration, D.K. All authors have read and agreed to the published version of the manuscript.

Funding: This research received no external funding.

Institutional Review Board Statement: Not Applicable.

Informed Consent Statement: Not Applicable.

Data Availability Statement: Publicly available datasets were analyzed in this study. This data can be found here: https://www.eett.gr/opencms/opencms/EETT_EN/Journalists/MarketAnalysis/MarketReview/ (accessed on 28 May 2021).

Conflicts of Interest: The views and opinions expressed in this article are those of the authors and do not reflect the official positions of the organizations that the authors are affiliated with. The official positions of the organizations are presented via their public announcements and documents such as decisions, etc. Data used in this study are those of public sources as stated in the document and they do not constitute reproduction of officially adopted documents. The user of opinions and results of this study acknowledges and agrees that authors and their organizations are not liable to any conduct for any user.

References

1. Meade, N.; Islam, T. Forecasting in telecommunications and ICT—A review. *Int. J. Forecast.* **2015**, *31*, 1105–1126. [CrossRef]
2. Fildes, R.; Kumar, V. Telecommunications demand forecasting—A review. *Int. J. Forecast.* **2002**, *18*, 489–522. [CrossRef]
3. Michalakelis, C.; Varoutas, D.; Sphicopoulos, T. Diffusion models of mobile telephony in Greece. *Telecommun. Policy* **2008**, *32*, 234–245. [CrossRef]
4. Christodoulos, C.; Michalakelis, C.; Varoutas, D. Forecasting with limited data: Combining ARIMA and diffusion models. *Technol. Forecast. Soc. Chang.* **2010**, *77*, 558–565. [CrossRef]
5. Dergiades, T.; Dasilas, A. Modelling and forecasting mobile telecommunication services: The case of Greece. *Appl. Econ. Lett.* **2010**, *17*, 1823–1828. [CrossRef]
6. Smail, G.; Weijia, J. Techno-economic analysis and prediction for the deployment of 5G mobile network. In Proceedings of the 2017 20th Conference on Innovations in Clouds, Internet and Networks (ICIN), Paris, France, 7–9 March 2017; pp. 9–16.
7. Jha, A.; Saha, D. Diffusion and forecast of mobile service generations in Germany, UK, France and Italy—A comparative analysis based on bass, Gompertz and simple logistic growth models. In Proceedings of the 26th European Conference on Information Systems: Beyond Digitization—Facets of Socio-Technical Change, ECIS 2018, Portsmouth, UK, 23–28 June 2018.
8. Scitovski, R.; Meler, M. Solving parameter estimation problem in new product diffusion models. *Appl. Math. Comput.* **2002**, *127*, 45–63. [CrossRef]
9. Chou, Y.-C.; Cheng, C.-T.; Yang, F.-C.; Liang, Y.-Y. Evaluating alternative capacity strategies in semiconductor manufacturing under uncertain demand and price scenarios. *Int. J. Prod. Econ.* **2007**, *105*, 591–606. [CrossRef]
10. Yao, T.; Jiang, B.; Young, S.T.; Talluri, S. Outsourcing timing, contract selection, and negotiation. *Int. J. Prod. Res.* **2009**, *48*, 305–326. [CrossRef]
11. Qin, R.; Nembhard, D.A. Demand modeling of stochastic product diffusion over the life cycle. *Int. J. Prod. Econ.* **2012**, *137*, 201–210. [CrossRef]
12. Chou, Y.-C.; Sung, W.-C.; Lin, G.; Jahn, J. A comparative study on the performance of timing and sizing models of capacity expansion under volatile demand growth and finite equipment lifetime. *Comput. Ind. Eng.* **2014**, *76*, 98–108. [CrossRef]
13. Marathe, R.; Ryan, S. On The Validity of The Geometric Brownian Motion Assumption. *Eng. Econ.* **2005**, *50*, 159–192. [CrossRef]
14. Valle, A.D.; Furlan, C. Forecasting accuracy of wind power technology diffusion models across countries. *Int. J. Forecast.* **2011**, *27*, 592–601. [CrossRef]
15. Madadi, N.; Ma'Aram, A.; Wong, K.Y. A simulation-based product diffusion forecasting method using geometric Brownian motion and spline interpolation. *Cogent Bus. Manag.* **2017**, *4*, 1300992. [CrossRef]
16. Huang, M.-G. Real options approach-based demand forecasting method for a range of products with highly volatile and correlated demand. *Eur. J. Oper. Res.* **2009**, *198*, 867–877. [CrossRef]

engineering proceedings

MDPI

Proceedings

Tourism and Big Data: Forecasting with Hierarchical and Sequential Cluster Analysis †

Miguel Ángel Ruiz Reina

Department of Theory and Economic History (Staff of Fundamentals), PhD Program in Economics and Business, University of Malaga, s/n, Plaza del Ejido, 29013 Málaga, Spain; ruizreina@uma.es

† Presented at the 7th International conference on Time Series and Forecasting, Gran Canaria, Spain, 19–21 July 2021.

Abstract: A new Big Data cluster method was developed to forecast the hotel accommodation market. The simulation and training of time series data are from January 2008 to December 2019 for the Spanish case. Applying the Hierarchical and Sequential Clustering Analysis method represents an improvement in forecasting modelling of the Big Data literature. The model is presented to obtain better explanatory and forecasting capacity than models used by Google data sources. Furthermore, the model allows knowledge of the tourists' search on the internet profiles before their hotel reservation. With the information obtained, stakeholders can make decisions efficiently. The Matrix U1 Theil was used to establish a dynamic forecasting comparison.

Keywords: Big Data; forecasting; Google Trends; cluster

check for updates

Citation: Ruiz Reina, M.Á. Tourism and Big Data: Forecasting with Hierarchical and Sequential Cluster Analysis . *Eng. Proc.* **2021**, *5*, 14. https://doi.org/10.3390/engproc2021005014

Published: 28 June 2021

1. Introduction

Big Data is a keyword in digitised markets. Technological development and the incorporation of analysis tools have meant a structural change for organisations, firms and institutions. The interpretation and visualisation of complex data are the core of Data Science [1,2]. Technology companies have the most precious asset in a digitised economic environment: information as a competitive advantage [3].

This new digital economy involves reducing information barriers in markets where intermediaries traditionally existed [4]. Consumers, through their searches on the internet, reveal their intentions. These intentions can be used as a predictive modelling tool for future demands of certain products. Hotel demand in a globalised market can be described through searches for potential consumers [5]. Researchers have paid attention to the selective secondary data sources of the internet network. This means a contribution to traditional analysis [6,7].

Methodologies currently applied have attempted to examine regularities in consumer behaviour data [8–10]. The difficulty lies in trying to explain quantitative and qualitative aspects in the modelling. In the field of time series with high dimensions and complex Big Data problems, attention has been paid to concepts such as "The Freedman's Paradox using an Info-Metrics perspective" [11] or "the power of Text in multidimensional contexts with high frequency" [12].

This article is interested in constructing a Hierarchical and Sequential Cluster Analysis (HSCA) for discrete time series. The analysis carried out focused on the decision-making mechanisms of economic agents for the demand for Hotel Accommodation in Spain (HADS). In particular, there are several generic words that consumers search for on the internet that reveal their intention of HADS. Google Trends (GT) provides an amount of information, which is used in this paper. A better understanding of previous searches can be translated into modelling inputs for structuring the forecasting of HADS.

The contribution of this paper is an improvement to current articles in the literature. The previous methodology has been proven to be an adequate input as a predictive tool, but it lacks classification and hierarchy by topics. The inclusion of a cluster of keywords (124) will allow identifying and segmenting potential consumers. The GT search indexes are for keywords related to tourist interest to visit Spain, and "broad matching" has been used [13,14]. This modelling could be used on internet forecasting for the tourism industry and hospitality, among other fields. Once a volume of temporary searches is known, companies will adjust the offers to their consumers, and there will be a gain in efficiency in decision-making. This fact allows us to model consumer behaviours and to project the regularities of the online tourism market.

Periodicity is essential to reveal systematic behaviours. As we previously cited, a Big Data analysis's difficulty lies in combining qualitative and quantitative research while maintaining traditional modelling standards. We will build the predictions on discrete-time-series variables and seasonal variable dummies (sampling January 2008 to December 2019).

The HSCA method is compared with SARIMA models [15], ADRL + SEASONALITY model [5], Hierarchical Neural Networks (HNN) [16] and Singular Spectrum Analysis (SSA) [5,8]. As a model selection criterion for forecasting, we will use the Matrix U1 Theil decision matrix [5]. The results obtained from the HSCA methodology reveal improvements in predictive capacity about the other models.

The remainder of this investigation is as follows: Section 1.1 provides a review of the existing literature on the forecasting of Big Data applied to Tourism; in Section 2, the theoretical methodology is performed; in Section 3, data analysis of primary and secondary data sources is done; Section 4 is dedicated to discussing the empirical results obtained after applying the methods proposed. Finally, Section 5 is for the mains conclusions obtained and bibliographic references.

1.1. Literature Review

The grouping in time series occurs when we are interested in the collection into categories or clusters. Nowadays, the application is interesting for finance, economics, medicine, engineering, or computing [17–19]. Clustering approaches for time series are time series clustering by features [20–22], clustering models in time series [23–25], or dependency clustering models [26,27].

Regarding predictive modelling of the use of GT, it should be noted that it is relatively recent. The new datasets from Google resources are a disruptive change in the traditional analysis of HDAS worldwide. The model's predictive capacity evolution was determined by techniques previously developed by mathematicians and statisticians. The conventional scientific research was joined by technology development, meaning a breakthrough summarised in Big Data Technologies.

In the scientific literature published using GT in tourism, we would highlight studies with an extensive literature review [9,10], or new modelling and forecasting developments. These studies have found standard results in the forecasting techniques concerning other fields such as parametric and non-parametric techniques [8].

In recent years, authors have published papers with secondary databases from Google. In addition, Neural Networks, Machine Learning, Statistical Methods, and traditional Econometrics have been used as forecasting methods in the tourism sector. Recently, attention has been paid to the spurious relationship between GT Searches and tourism demand [14].

Hierarchical algorithm approaches for clusters have been applied to tourism but have always been used to cross-section data. In particular, secondary data obtained from the travel and tourism competitiveness index are analysed to create clusters. Subsequently, multidimensional scaling techniques are applied to detect the most and most minor influential determinants in tourist destinations' competitiveness [28].

Moreover, a causality method called Granger Causality and seasonality testing has recently been developed, supposing an improvement to Granger's traditional process of causality [5,29,30]. Furthermore, a new dimensionless model selection criterion has recently emerged called the Matrix U1 Theil. This new criterion is a comparative advantage compared to usual forecasting criteria such as Root of the Mean Square Error, Mean Absolute Error, Theil inequality index, and Diebold–Mariano criterion [5].

2. Methods

This methodological section will develop a new cluster criterion named Hierarchical and Sequential Clustering Analysis (HSCA). This grouping methodology was designed to classify the amount of information existing on the internet network. HSCA will improve and overcome the limitations of keywords previously used in econometric modelling [5]. For this, some properties are cited for modelling with large volumes of data. The first property is Effectiveness and Replicability criteria; the use of HSCA can be replicated in other fields related to Big Data. A second property, identifying clusters with correlation and testing criteria, reveals the importance and causality in our explanatory variables' modelling. A third property, Noise Tolerance and Outliers Values working with large volumes of data, makes the usual theoretical assumptions to be relaxed in favour of accessible interpretation and usability of the model. Finally, a property, Parsimony Criterion, will determine the best model with the least number of explanatory variables.

In real Big Data applications, it is not easy to find a single algorithm that meets the properties described above. The diagram (Figure 1) represents the sequence from a universe of words related to a variable of interest to predict. The graph shows how the keywords initially relate to each cluster and the predicted variable.

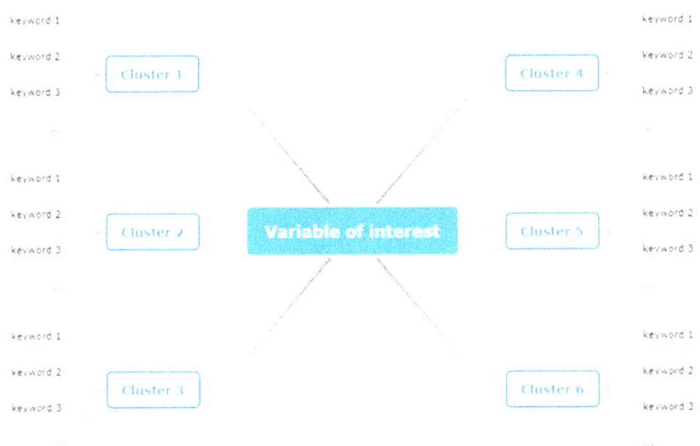

Figure 1. Clustering scheme for a predictive variable (Variable of interest). Own Elaboration.

2.1. Hierarchical and Sequential Clustering Analysis (HSCA)

In this subsection, we will describe the HSCA method. We could divide the methodology into the following sequential steps:

First step: Relevant explanatory variables ($keywords_t$) are selected for forecasting $\{keywords_{mt} \in \mathbb{R}^+;\ m = 1,2,3\ldots;\ t \in T = 1,2,3\ldots T\}$.

In our model, $keywords_t$ are words that future consumers search on the internet before their tourist demand, for instance, Google searches and "broad matching" such as "visit Spain", "rent a car in Spain", or "Weather in Spain" among others. The search words and clusters obtained from GT will be presented in the data section.

Second step: the words of the first step are organised by clusters (topics). $\{keywords_{mlt} \subset cluster_{lt};\ \forall\,(cluster_{1t}, cluster_{2t}\ldots, cluster_{lt});\ l = 1,2,3\ldots\}$.

Third step: auxiliary regressions (y_t and $(keyword_{m1t}, keyword_{m2t}, \ldots, keyword_{mlt})$ are expressed in natural logarithms) are performed for the same forecasting variable (y_t) classified by the cluster. The hierarchy of each group is determined by its R^2. The models present the same dependent variable, and the explanatory variables are different in each grouping.

$$y_t = f(cluster_{1t}) + \sum_{i=1}^{12} \alpha_i w_i + u_{1t} = \sum_{m=1}^{j} \beta_m keyword_{m1t} + \sum_{i=1}^{12} \alpha_i w_i + u_{1t} \tag{1}$$

$$y_t = f(cluster_{2t}) + \sum_{i=1}^{12} \lambda_i w_i + u_{2t} = \sum_{m=1}^{k} \delta_m keyword_{m2t} + \sum_{i=1}^{12} \lambda_i w_i + u_{2t} \tag{2}$$

$$y_t = f(cluster_{lt}) + \sum_{i=1}^{12} \tau_i w_i + u_{lt} = \sum_{m=1}^{o} \psi_m keyword_{mlt} + \sum_{i=1}^{12} \tau_i w_i + u_{lt} \tag{3}$$

where w_i $(for\ monthly\ data\ i = 1, 2, \ldots, 12)$ is a deterministic seasonal dummy and uses the HAC covariance method [31].

$$\begin{aligned} w_1 &= -1,\ for\ others\ w_i = 0 \\ w_1 &= -1,\ w_2 = 1\ for\ others\ w_i = 0; \\ w_1 &= -1,\ w_3 = 1\ for\ others\ w_i = 0; \\ &\vdots \\ w_1 &= -1,\ w_{12} = 1\ for\ others\ w_i = 0 \end{aligned} \tag{4}$$

Once the regressions and tests of individual significance of the parameters were made, we determine the most relevant keywords within each cluster. The model selection criteria that verify the clustering procedure developed in this article are the usual ones from Akaike (AIC) and Hannan–Quinn [32]. For instance, to contrast any keyword, we define the null hypothesis as the statement that narrows the model and the alternative hypothesis as the broader one [32].

$$\begin{aligned} &y_t = f(cluster_{1t}) + \sum_{i=1}^{12} \alpha_i w_i + u_{1t} = \sum_{m=1}^{j} \beta_m keyword_{m1t} + \sum_{i=1}^{12} \alpha_i w_i + u_{1t} \\ &H_o : \beta_m = 0 \\ &H_1 : \beta_m \neq 0 \end{aligned} \tag{5}$$

Fourth step: after the most relevant words of each cluster were selected, a final preliminary auxiliary regression is performed with the most pertinent explanatory variables of each group.

$$\begin{aligned} y_t &= f(\widehat{cluster}_{1t}) + f(\widehat{cluster}_{2t}) + \cdots + f(\widehat{cluster}_{lt}) + \sum_{i=1}^{12} \vartheta_i w_i + \varepsilon_t = \\ &= \sum_{m=1}^{j} \gamma_1 \widehat{keyword}_{m1t} + \sum_{m=1}^{k} \phi_1 \widehat{keyword}_{m2t} + \cdots + \sum_{m=1}^{l} \omega_1 \widehat{keyword}_{mlt} + \sum_{i=1}^{12} \vartheta_i w_i + \varepsilon_t \end{aligned} \tag{6}$$

The model is simplified under the parsimony criterion, seeking the fewest number of significant explanatory variables with explanatory capacity.

$$\begin{aligned} y_t &= f(\widehat{\widehat{cluster}}_{1t}) + f(\widehat{\widehat{cluster}}_{2t}) + \cdots + f(\widehat{\widehat{cluster}}_{lt}) + \sum_{i=1}^{12} \vartheta_i w_i + \widehat{\widehat{\varepsilon}}_t = \\ &= \sum_{m=1}^{j} \gamma_1 \widehat{\widehat{keyword}}_{m1t} + \sum_{m=1}^{k} \phi_1 \widehat{\widehat{keyword}}_{m2t} + \cdots + \sum_{m=1}^{l} \omega_1 \widehat{\widehat{keyword}}_{mlt} + \sum_{i=1}^{12} \vartheta_i w_i + \widehat{\widehat{\varepsilon}}_t \end{aligned} \tag{7}$$

The interpretation of coefficients are elasticities, and the dummy variables are semi-elasticities [33].

2.2. Comparison of Forecasting and Evaluation

Forecasting and control problems are closely linked. To forecast, we will define the following expression for our modelling as follows:

$$\begin{aligned} E(y_{t+h} | x_t, w_i) &= E\Big(\sum_{m=1}^{j} \gamma_1 \widehat{\widehat{keyword}}_{m1t+h} + \sum_{m=1}^{k} \phi_1 \widehat{\widehat{keyword}}_{m2t+h} + \cdots \\ &+ \sum_{m=1}^{l} \omega_1 \widehat{\widehat{keyword}}_{mlt+h} + \sum_{i=1}^{12} \vartheta_i w_i \Big) \end{aligned} \tag{8}$$

where h represents the time horizon, and the residuals of the forecasting are white noise $\left\{E(\widehat{\varepsilon}_{t+h}|x_{t+h}, w_i) = 0;\ \text{var}(\widehat{\varepsilon}_{t+h}|x_{t+h}, w_i) = \sigma^2_{\widehat{\varepsilon}};\ \text{cov}(\widehat{\varepsilon}_{t+h}|x_{t+h}, w_i) = 0\right\}$.

As a model selection criterion, we will base ourselves on the Matrix U1 Theil decision matrix. A dimensionless matrix is designed for the decision to select predictive models [5].

3. Data

Data were collected from Jan. 2008 to Dec. 2019. Therefore, we can differentiate two data sources, on the one hand, the official data sources from the INE (Spanish National Statistics Institute (Instituto Nacional de Estadística) https://ine.es/ (accessed on 24 June 2021).) for the predicted variable (HDAS), and the explanatory variables are obtained from Big Data secondary sources, in particular, from the GT tool.

HDAS presents some relevant characteristics in the time series analysis; it is worth noting the high seasonality and a growing trend throughout the period analysed (Figure 2).

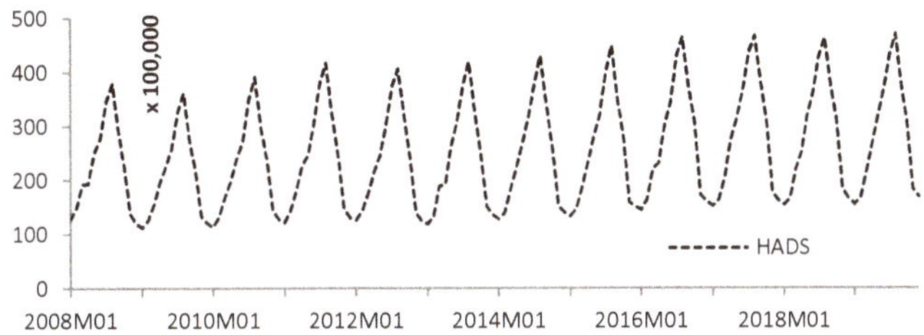

Figure 2. Number of HADS (January 2008 to December 2019). Data source: INE. Own Elaboration.

From a statistical point of view, it should be noted that the maximum values for each year occur in the summer season, the highest value in August 2019 with 46,998,612 hotel overnights in Spain, and the lowest value in January 2009 with 11,203,819. For the 144 observations analysed (Table 1), the existence of unit roots (ADF (p-value) = 0.85) and stationarity variance (KPSS (p-value) = 0.56) should be highlighted [34,35]. The KPSS (stationary variance) results allow us in our modelling to adjust dummy variables for the repetitive behaviours of the series (seasonality).

Table 1. Descriptive Statistics and Stationary Analysis of HADS (Jan. 2008 to Dec. 2019). Own Elaboration.

Mean	Maximum	Minimum	ADF (p-Value)	KPSS (p-Value)	Observations
24,989,874	46,998,612	11,203,819	0.85	0.56	144

The sample period includes 18,000 contemporary observations. From INE data, there are 144 for the variable to be predicted (HADS). The search terms related to planning a visit to Spain were collected from GT and are presented in Table A1 (see Appendix A). In this document, we worked with 17,856 observations of search variables contemporary to the HADS variable. The information is summarised in nine clusters with 124 search terms related to hotel tourism demand from January 2008 to December 2019 for tourists worldwide. All the keywords were searched using "broad match" and combination with other terms. e.g., entering "Spain Hotel", "Spain culture", and so on [13].

4. Results

In the following section of empirical results, we describe a training period between January 2008 and December 2018, with a testing sample to forecast 12 months in 2019. The applied methodology is previously mentioned in Section 3—Table 2 shows the most relevant keywords within each tourist interest cluster. Regarding the hierarchy, we can indicate that all the keywords finally selected in each group are the most descriptive capacity. For example, finding all values between 0.95 and 0.99, highlighting the terms related to the "social" cluster, which shows that these search engines have a high explanatory capacity, highlighting "Airbnb", "Youtube", "English", "Tripadvisor", "Twiter". However, the differences between the clusters and their hierarchy are minimal. An aspect to highlight is that the dummy variables described for systematic seasonality were relevant for all models in all the sets.

Table 2. Summary of clusters and keywords (broad matching) relevant for HADS. Sample January 2008–December 2018. Own Elaboration.

Cluster	Relevant Keywords	R-Squared
Sports	sport	0.95
Laws	visa	0.97
Transport	car, flight	0.98
Seasonality	summer, winter	0.95
Social	Airbnb, Youtube, English, Tripadvisor, Twiter	0.99
Welfare	Android, Xiaomi	0.98
Searches	low-cost, Spain Tourism, visit Spain	0.98
Culture	alcohol, city breaks, monuments, architecture	0.97
Places	Beach, Canary Island, Alhambra, Plaza de España, Sagrada Familia	0.98

Once the main information clusters were selected to predict the variable of interest, we carried out final modelling for the set of variables in the groups to choose the best regressors to evaluate their predictive capacity. In our modelling, we expressed all the variables in natural logarithms, except the seasonal dummy variables, with the p-values in parentheses. We obtain the following result as follows:

$$\begin{aligned}\widehat{y}_t =\ & \underset{(0.00)}{15.90} + \underset{(0.04)}{0.08}\,Airbnb_t + \underset{(0.00)}{0.06}\,Apple_t - \underset{(0.00)}{0.12}\,car_t + \underset{(0.00)}{0.03}\,city\_breaks_t + \underset{(0.00)}{0.07}\,flight_t \\ & - \underset{(0.00)}{0.08}\,Samsung_t + \underset{(0.01)}{0.03}\,sport_t + \underset{(0.00)}{0.13}\,visa_t + \underset{(0.06)}{0.07}\,visit\_Spain_t + \sum_{i=1}^{12} \vartheta_i w_i\end{aligned} \quad (9)$$

$$\begin{aligned}\sum_{i=1}^{12} \vartheta_i w_i =\ & \underset{(0.00)}{0.10}\,w_2 + \underset{(0.00)}{0.34}\,w_3 + \underset{(0.00)}{0.50}\,w_4 + \underset{(0.00)}{0.69}\,w_5 + \underset{(0.00)}{0.82}\,w_6 \\ & + \underset{(0.00)}{1.05}\,w_7 + \underset{(0.00)}{1.16}\,w_8 + \underset{(0.00)}{0.90}\,w_9 + \underset{(0.00)}{0.69}\,w_{10} + \underset{(0.00)}{0.18}\,w_{11} + \underset{(0.00)}{0.10}\,w_{12}\end{aligned} \quad (10)$$

$$\begin{aligned}& T = 132;\ Sample:\ 2008M01\ 2018M12; \\ & Method:\ Least\ Squares\ HAC\ standard\ errors\ \&\ covariance \\ & (Bartlett\ kernel,\ Newey-West\ fixed = 5.0000); \\ & R^2 = 0.99;\ Adjusted\ R^2 = 0.99; \\ & AIC = -3.04;\ P-Value\ (Wald\ F-Statistic) = 0\end{aligned} \quad (11)$$

The final model selected presents a high explanatory capacity $R^2 = 0.99$. All the parameter interpretations are studied as the percentage increases of the regressors (1%). For instance, the variable "Airbnb" implies an increase of HADS of 8%; in the explanatory variables, the variables "flight" and "visit Spain" are interpreted as a 7% increase in HADS. It is interesting to mention that the variables "Car" (−0.12) have a negative sign and "flight" (0.07) represents a positive sign. The technological variables (Samsung, Apple), "sports", and "City Breaks" are relevant.

The prediction of the final HSCA model is compared to other models cited in the Introduction section. The comparative graph of the forecasting time series can be seen in Figure 3.

Table 3 below shows the comparison between the HSCA model and the other predictive models (ADRL + SEASONALITY, SARIMA, HNN, SSA) using Matrix U1 Theil (values more significant than one will indicate better predictive capacity than HSCA; otherwise, we find values less than 1). The HSCA model shows the best predictive power in test h = 3 and h = 6. For a time horizon of h = 12, it would be below ADRL + SEASONALITY and HNN.

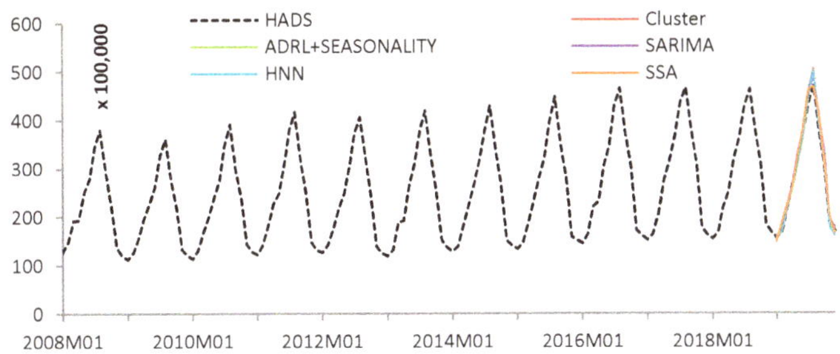

Figure 3. Out-sample forecast HADS h = 12 (January 2018 to December 2019). Own Elaboration.

Table 3. Summary of forecasting accuracy. Out-Sample training Jan. 2019–Dec. 2019. Own Elaboration.

	HSCA	ADRL + SEASONALITY	SARIMA	HNN	SSA
HSCA (h = 3)	1.00	0.39	0.36	0.43	0.15
HSCA (h = 6)	1.00	0.50	0.69	0.92	0.39
HSCA (h = 12)	1.00	1.14	0.86	1.13	0.79

5. Conclusions

In the present investigation, a grouping model was developed for hotel accommodation forecasting (HADS). The properties described in the methodological section were central to the research (Section 3). Databases from primary (INE) and secondary (GT) sources were studied. The HSCA model shows a forecasting and causality capacity. A total of 124 Keywords were analysed in a time series from January 2008 to December 2019 (18,000 observations, including HADS). We determined the primary search keywords by topic (Table 2). The hierarchy of each cluster was also fixed.

Furthermore, this research was compared with other models with high predictive capacity, such as ADRL + SEASONALITY: SARIMA, HNN and SSA. Analysing the Matrix U1 Theil results for time horizons $h = 3$, we found HSCA (coefficients less than 1) as the best model. For an annual time horizon, we discovered that ADRL + SEASONALITY (1.14) and HNN (1.13) performed better results than HSCA. Let us compare the causal explanatory capacity ($R^2 = 0.99$). We can say that HSCA is the best since it includes many more explanatory variables (search topics) than the rest of the models studied. With the information obtained from the HSCA model, it is possible to adjust tourist profiles based on their searches. Primary and secondary tourism industries can benefit from this knowledge of the global market.

We can deduce that previous studies' explanatory capacity was improved from this work, providing relevant and novel information to the scientific literature. Furthermore, this research is the basis for future empirical work related to stakeholders' Big Data field and decision-making. Currently, the most developed economies are focused on a digital environment. Both firms and consumers are expanding their activities on digital platforms, which makes it possible to measure market actions. Furthermore, the engineering of search engines such as Google comes from valuable information to improve the predictive capacity of the models. The results presented in this study refer to consumers' active search, but the data generated can generate predictive information for future tourism consumers. The impact on this type of study's economy supposes a paradigm shift in traditional tourism analysis studies.

The study was applied to the tourism field. However, this methodology can be applied to the finance, insurance or airline field, where decision-making is critical in competitive markets.

Funding: This research received no external funding.

Institutional Review Board Statement: Not applicable.

Informed Consent Statement: Not applicable.

Data Availability Statement: The datasets generated during and analysed during the current study are available in the repository (https://drive.google.com/file/d/1T5_dQGxCLIG7IKMOPvgFUD9lsCe344fn/view?usp=sharing) (accessed on 23 June 2021).

Acknowledgments: The author wishes to acknowledge the support given by the University of Malaga. PhD. Program in Economics and Business, effective from 16 July 2013. Especially to Associate Professor Antonio Caparrós Ruiz from the University of Malaga for reviewing this work. Group of research: "SEJ157-INIDICADORES SOCIALES".

Conflicts of Interest: The authors declare no conflict of interest.

Appendix A

Table A1. Keywords and clusters correlated with HADS (broad matching). January 2008 to December 2019. Own Elaboration.

Sports	Laws	Transport	Seasonality	Social	Welfare	Searches	Culture	Places
Sport	Taxes	Transport	Weather	Spanish People	Hospitality	Trip Spain	Monuments	Beach
Football	Tax free	Flight	Winter	Mind	Environment	Visit Spain	Musueums	Mountain
Basketball	Laws	Train	Summer	vegan Spain	Relax	Spain Tourism	Congress	Island
Athletics Spain	Schengen	Roads	Autumn	English	Stress	Hotel Spain	Study	nature
Swimming Spain	Spain passport	Cruise ships	Spring	French	Life style	Apartment Spain	Disco	Mediterranean area
Volleyball Spain	Visa Spain	Helicopter Spain	Climate Change	Italian	Hospital	Best travel	Concert	Canary Island
Tennis Spain	Spain travel insurance	Bus Spain	Easter week Spain	German	Apple Spain	Resort	Food	Zoo Spain
Boxing Spain	Medical certificate Spain	Car Spain	Christmas Spain	Facebook	Android Spain	Ecotourism	Wine	Andalusia Spain
Soccer Spain	Spain driving license	Tolls Spain	-	Twitter	Samsung Spain	Family Trip	theme parks Spain	Catalonia Spain
Hockey on ice Spain	-	Motorhomes Spain	-	Tripadvisor	Xiaomi Spain	low cost	nightlife Spain	Alcázar de Toledo
Baseball Spain	-		-	Hotels.com Spain	Huawei Spain	Rural Spain	Spain architecture	Monasterio del Escorial
-	-	-	-	Booking.com Spain	-	Agriculture Spain	alcohol	Palacio Real
-	-	-	-	Wimdu	-	Fishing Spain	City Breaks	Muralla de Ávila
-	-	-	-	Kayak Spain	-	Livestock Spain	-	Alcázar de Segovia
-	-	-	-	Airbnb	-	Blitz	-	Valencian Community Spain
-	-	-	-	Instagram	-	-	-	Plaza de España
-	-	-	-	Youtube	-	-	-	Teatro Romano de Mérida
-	-	-	-	Terrorism	-	-	-	Acueducto de Segovia
-	-	-	-	Overtourism	-	-	-	Mezquita de Córdoba
-	-	-	-	Tourism Phobia	-	-	-	Sagrada Familia
-	-	-	-	Wifi Spain	-	-	-	La Giralda
-	-	-	-	3G, 4G and 5G Spain	-	-	-	La Alhambra and Tours

References

1. Kleinberg, J.; Lakkaraju, H.; Leskovec, J.; Ludwig, J.; Mullainathan, S. Human decisions and machine predictions. *Q. J. Econ.* **2018**, *133*, 237–283. [CrossRef]
2. Carrizosa, E.; Guerrero, V.; Morales, D.R. Visualising data as objects by DC (difference of convex) optimisation. *Math. Program.* **2018**, *169*, 119–140. [CrossRef]
3. Mikalef, P.; Pappas, I.O.; Krogstie, J.; Giannakos, M. Big data analytics capabilities: A systematic literature review and research agenda. *Inf. Syst. e-Bus. Manag.* **2018**. [CrossRef]
4. Palos-Sanchez, P.R.; Correia, M.B. The collaborative economy based analysis of demand: Study of airbnb case in Spain and Portugal. *J. Theor. Appl. Electron. Commer. Res.* **2018**. [CrossRef]
5. Ruiz-Reina, M.Á. Big Data: Does it really improve Forecasting techniques for Tourism Demand in Spain? In *International Conference on Time Series and Forecasting*; Godel Impresiones Digitales S.L.: Granada, Spain, 2019; pp. 694–706.
6. Song, H.; Li, G. Tourism demand modelling and forecasting—A review of recent research. *Tour. Manag.* **2008**, *29*, 203–220. [CrossRef]
7. Pan, B.; Wu, D.C.; Song, H. Forecasting hotel room demand using search engine data. *J. Hosp. Tour. Technol.* **2012**, *3*, 196–210. [CrossRef]
8. Wu, D.C.; Song, H.; Shen, S. New developments in tourism and hotel demand modeling and forecasting. *Int. J. Contemp. Hosp. Manag.* **2017**, *29*, 507–529. [CrossRef]
9. Mariani, M.; Baggio, R.; Fuchs, M.; Höepken, W. Business intelligence and big data in hospitality and tourism: A systematic literature review. *Int. J. Contemp. Hosp. Manag.* **2018**. [CrossRef]
10. Li, J.; Xu, L.; Tang, L.; Wang, S.; Li, L. Big data in tourism research: A literature review. *Tour. Manag.* **2018**, *68*, 301–323. [CrossRef]
11. Macedo, P. Freedman's Paradox: An Info-Metrics Perspective. In *International Conference on Time Series and Forecasting*; Godel Impresiones Digitales S.L.: Granada, Spain, 2019; pp. 665–676.
12. Gabrielyan, D.; Masso, J.; Uuskula, L. Powers of Text. In *International Conference on Time Series and Forecasting*; Godel Impresiones Digitales S.L.: Granada, Spain, 2019; pp. 677–693.
13. Choi, H.; Varian, H. Predicting the Present with Google Trends. *Econ. Rec.* **2012**, *88*, 2–9. [CrossRef]
14. Bokelmann, B.; Lessmann, S. Spurious patterns in Google Trends data—An analysis of the effects on tourism demand forecasting in Germany. *Tour. Manag.* **2019**, *75*, 1–12. [CrossRef]
15. Box, G.E.P.; Jenkins, G.M.; Reinsel, G.C. *Time Series Analysis: Forecasting and Control*, 4th ed.; John Wiley & Sons: Hoboken, NJ, USA, 2013.
16. Athanasopoulos, G.; Hyndman, R.J.; Kourentzes, N.; Petropoulos, F. Forecasting with temporal hierarchies. *Eur. J. Oper. Res.* **2017**, *262*, 60–74. [CrossRef]
17. Liao, T.W. Clustering of time series data—A survey. *Pattern Recognit.* **2005**, *38*, 1857–1874. [CrossRef]
18. Aghabozorgi, S.; Shirkhorshidi, A.S.; Wah, T.Y. Time-series clustering—A decade review. *Inf. Syst.* **2015**, *53*, 16–38. [CrossRef]
19. Caiado, J.; Maharaj, E.A.; D'Urso, P. Time-series clustering. In *Handbook of Cluster Analysis*; Chapman & Hall/CRC: Boca Raton, FL, USA, 2015; pp. 241–264.
20. Kakizawa, Y.; Shumway, R.H.; Taniguchi, M. Discrimination and clustering for multivariate time series. *J. Am. Stat. Assoc.* **1998**, *93*, 328–340. [CrossRef]
21. Scotto, M.G.; Alonso, A.M.; Barbosa, S.M. Clustering time series of sea levels: Extreme value approach. *J. Waterw. Port Coast. Ocean Eng.* **2010**, *136*, 215–225. [CrossRef]
22. D'Urso, P.; Maharaj, E.A.; Alonso, A.M. Fuzzy clustering of time series using extremes. *Fuzzy Sets Syst.* **2017**, *318*, 56–79. [CrossRef]
23. Alonso, A.M.; Berrendero, J.R.; Hernández, A.; Justel, A. Time series clustering based on forecast densities. *Comput. Stat. Data Anal.* **2006**, *51*, 762–766. [CrossRef]
24. Scotto, M.G.; Barbosa, S.M.; Alonso, A.M. Model-based clustering of Baltic sea-level. *Appl. Ocean Res.* **2009**, *31*, 4–11. [CrossRef]
25. Vilar, J.A.; Alonso, A.M.; Vilar, J.M. Non-linear time series clustering based on non-parametric forecast densities. *Comput. Stat. Data Anal.* **2010**, *54*, 2850–2865. [CrossRef]
26. Alonso, A.M.; Peña, D. Clustering time series by linear dependency. *Stat. Comput.* **2019**, *29*, 655–676. [CrossRef]
27. Alonso, A.M.; Galeano, P.; Peña, D. A robust procedure to build dynamic factor models with cluster structure. *J. Econom.* **2020**, *216*, 3552. [CrossRef]
28. Chávez, J.C.N.; Torres, A.I.Z.; Torres, M.C. Hierarchical Cluster Analysis of Tourism for Mexico and the Asia-Pacific Economic Cooperation (APEC) Countries. *Rev. Tur. Anál.* **2016**, *27*, 235–255. [CrossRef]
29. Granger, C.W.J. Investigating Causal Relations by Econometric Models and Cross-spectral Methods. *Econometrica* **1969**, *37*, 424–438. [CrossRef]
30. Ruiz-Reina, M.Á. Forecasting using Big Data: The case of Spanish Tourism Demand. In *International Conference on Time Series and Forecasting*; Godel Impresiones Digitales S.L.: Granada, Spain, 2019; pp. 782–789.
31. Newey, W.K.; West, K.D. A Simple, Positive Semi-Definite, Heteroskedasticity and Autocorrelation Consistent Covariance Matrix. *Econometrica* **1987**, *55*, 703–708. [CrossRef]
32. Greene, W.W.H. *Econometric Analysis*, 7th ed.; Prentice Hall: Hoboken, NJ, USA, 2012.

33. Peng, B.; Song, H.; Crouch, G.I.; Witt, S.F. A Meta-Analysis of International Tourism Demand Elasticities. *J. Travel Res.* **2015**, *54*, 611–633. [CrossRef]
34. Dickey, D.A.; Fuller, W.A. Likelihood Ratio Statistics for Autoregressive Time Series with a Unit Root. *Econometrica* **1981**, *49*, 1067–1072. [CrossRef]
35. Kwiatkowski, D.; Phillips, P.C.B.; Schmidt, P.; Shin, Y. Testing the null hypothesis of stationarity against the alternative of a unit root. How sure are we that economic time series have a unit root? *J. Econom.* **1992**, *54*, 159–178. [CrossRef]

engineering proceedings

MDPI

Proceedings

Analyzing Seasonality in Hydropower Plants Energy Production and External Variables †

Eralda Gjika [1,*], Lule Basha [1], Aurora Ferrja [1] and Arbesa Kamberi [2]

1 Faculty of Natural Science, University of Tirana, 1001 Tirana, Albania; lule.hallaci@fshn.edu.al (L.B.); aurora.simoni@fshn.edu.al (A.F.)
2 Albanian Power Corporation, 1001 Tirana, Albania; kamberia@kesh.al
* Correspondence: eralda.dhamo@fshn.edu
† Presented at the 7th International Conference on Time Series and Forecasting, Gran Canaria, Spain, 19–21 July 2021.

Abstract: This study is focused on energy production in Albania which involves different types of infrastructure at the various points of the energy production and distribution chain, as well as monitoring and early warning systems. At a time of rapid climate change, estimating the appropriate dimensions and design of such infrastructure and systems becomes crucial. The main objective is to analyze the seasonality pattern and main external climacteric factors, such as precipitation, average temperature, and water inflow. This work deals with the seasonality patterns of climacteric factors affecting energy production and considers different statistical learning methods for prediction.

Keywords: time series; prediction; energy; seasonality; climacteric

check for updates

Citation: Gjika, E.; Basha, L.; Ferrja, A.; Kamberi, A. Analyzing Seasonality in Hydropower Plants Energy Production and External Variables. *Eng. Proc.* **2021**, *5*, 15. https://doi.org/10.3390/engproc2021005015

Published: 25 June 2021

1. Introduction

About 20% of the total installed capacity for electricity generation in Europe is from hydropower [1]. In Albania, the country's needs for electricity are met mainly by the hydropower plants and less by the thermo power plants. The hydropower plants provide about 94% of the produced electricity, while the rest is produced by thermo power plants that use residual fuel oil as fuel and, in special cases, steam coal. Substantial drought in recent years has significantly reduced water levels for energy production in the Drin River cascade, generating by this way the lowest levels for the last 30 years. The cascade built in the Drin River basin is the largest not only for Albania but also in the Balkans for its installed capacity and the size of hydro technical works. The Albanian Electric Power Corporation (KESH), having in operation 79% of the production capacity in the country from the Drin cascade, manages to supply about 70–75% of the demand for electricity. KESH is not only one of the producers of electricity from important hydropower sources in the region, but it is also considered a factor with regional impact on the safety of hydro cycles [2].

Albania has established ALPEX as its energy exchange and electricity market operator which marks a further step towards the country's integration as part of the European energy market and contributes to improving the investment climate in the country and attracting foreign investment in the energy sector [3,4]. The Albanian Energy Corporation (KESH) is the main public producer of electricity in the country and Drin is the longest river in the Albanian territories, with a length of 160 km. Figures 1 and 2 show the position of the three HPP situated in Drin river cascade.

The approximate distance between the three HPPs is the same. Therefore, it is very important to take into account the fact that water flows through the Drin River and external sources (snowmelt, precipitation, etc.) are exploited through the cascade for energy production. Fierza, being the first HPP, uses natural inputs to produce energy and serves as a flow regulator for two ongoing HPPs. The electricity system in Albania is divided into three main sectors: the generation, transmission, and distribution sector. In Albania, the manufacture sector (KESH) produces energy based on the demands of the distribution operator (OSHE). So, why is it important to forecast power generation from HPPs in Albania? From this [5] 2018 report, Albania has a great potential for "hydroelectric energy" with eight main rivers crossing a river basin with over 57% of the current management extent, with an average altitude of 700 m above the level of sea and a perennial flow of 1245 m^3/s, for a combined water supply of 40 billion cubic meters. An overview of the situation: 2100 MW the total installed capacity; Up to 615 MW of further potential capacity; above 1785 MW concession warded, eligible for partnerships. The water reserves valuing per capita second in the whole Europe makes the country offer an average cost of hydro production starting around 35 Euro/MWh.

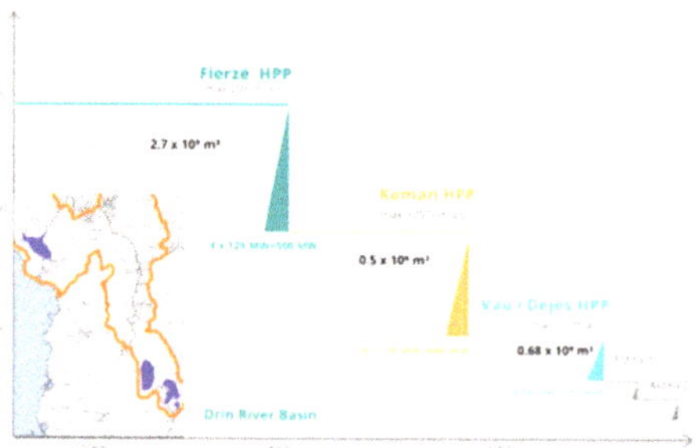

Figure 1. Drin cascade and the three main HPP (Source: http://www.kesh.al/en/asset/drini-cascade/ (accessed on 28 November 2020)).

Figure 2. HPP in Drin cascade: Fierza, Koman and Vau-Dejes (Source: Google Earth).

2. Objective of the Study

Electricity demand and supply depend on many factors, the most important of which are climacteric indicators. In the production of electricity through hydropower plants, water resources play an important role. Among the factors that are likely to affect both variability in the supply and absolute availability of water are decreasing snow cover, increases in rainfall in hilly areas, drier conditions in the lowlands, as well as reduction in the capacity of soils to retain water due to land degradation and impacts of multiple stressors on vegetation and forests. Soil saturation can lead to sudden peaks in water inflow, even with mild precipitation. Global weather systems are also destabilized, leading to longer consecutive periods of precipitation or dry weather, as well as changes in how overall dynamics play out at very local levels due to factors such as topography [6,7]. Thus, increased weather variability will mean that reducing risk becomes more important than optimizing infrastructure for typical conditions. Increasing energy production by using half-empty reservoirs it may not be a problem if this can reduce potential disasters. Although increased investment costs are mostly the result of measures to decrease vulnerability to future climate shifts, this may affect infrastructure and supply chains for other options as well, so that the cost of HPP relative to other energy sources with low Greenhouse Gases (GHG) emissions is not likely to change noticeably. However, other factors may change the relative cost and outcomes of investments, such as absolute reduction in water availability for a region, increasing opportunities for wind power, considering that weather systems will contain more energy, and reduced cost of solar technologies, as a result of large investments in improved technologies globally [8].The increased anticipated incidence of extreme events is an argument for choosing numerous small-scale power plants, rather than investing in large-scale power plants, to reduce the impacts of disasters. Smaller plants are easier to retrofit and adapt as climate conditions change over the coming years. Also, it becomes important to ensure energy supply and access with a wider mix and range of options, both to compensate for seasonal variability and reduced predictability, and to mitigate impacts of disasters.

One of the main objectives of this study is to analyze the seasonality pattern and the correlation among some climacteric external factors which may affect the energy production in the Drin cascade and further use these variables as explanatory variables in energy production. The prediction of the capacities of energy produced will help the stakeholders and decision makers (such as the government) to better take precautions on demand and supply of the energy for the country needs and region. Because Albania is heavily reliant on hydropower electricity production some vulnerability in the future may be the reduction of power generation due to severe drought which will result in less electricity produced by the hydropower plants. The heavy reliance on hydropower sources may be appropriate for reducing greenhouse gas emissions and improving air quality in Albania but can increase vulnerability to climate change. During last few years, a decrease in precipitation was observed and increased temperature in summer season as well. These changes could reduce annual average electricity output from Albania's large hydropower plants (LHPPs) by about 15% and from small hydropower plants (SHPPs) by around 20% by 2050. Global climate change may affect the provision of energy from solar and wind generation. A likely increase in the global solar radiation and the hours of sunshine duration will lead to an increase in the use of solar energy for different energy services, but at the moment the main interest is focused on energy produced by HPP and the capacity of production. In their study, ref [9] point out that spring shifting to earlier in the season may leave reservoirs half-empty if managers expect later floods that never arrive, with adverse consequences for hydropower production and later winter floods is some coastal areas of the Mediterranean may encounter reservoirs that have already been filled which may increase downstream flood risk [9,10]. Also, unpredictable reservoir storage could affect hydroelectric power production and the energy market [11]. There are many research works focused on seasonality pattern of external factors affecting energy produced by HPP. A review of these works is presented in [12–14]. The relationship between energy production season and climacteric variables is also discussed in [15–17].

3. Time Series Analysis

Electricity produced by hydropower plants is likely to be influenced by climatic factors and their seasonal patterns. It is expected that underlying causal dynamics affecting water inflow will follow a seasonal pattern (related to snow smelting, precipitation, upstream water use, capacity of vegetation and soils to retain water), so that correlations to any single factor will vary over the year. The spatial distribution of precipitation, topography, and time-lag between the time of precipitation and time of water inflow need to be considered [18]. Crucially, for the future, domino effects are likely to arise connected to aspects such as the fact that existing water management systems, reservoirs, and natural bodies of water that retain water upstream will not be sufficient to handle extended periods of high precipitation or indeed, peaks connected to extreme weather. This leads to non-linear situations, and an asymmetry in the impacts of variability in terms of aspects of systems affected and the time scale. Wet periods (or rapid snow smelting) may also lead to short term flooding, infrastructure damage, and possibly dam collapse. Also, extended dry periods will lead to forest fires or collapse of forests due to drought and increase in diseases. Forests are also exposed to the increased force in wind, avalanches, landslides, and erosion in mountain areas connected to more intense precipitation. Impacts on forests have long term and sometimes irreversible consequences, which thus may affect future dynamics of hydropower energy production.

In this study, we have considered four time series with monthly observation and duration from January 1991 to December 2016, in total 312 observations. We considered the monthly average temperature (Celsius degree. Source World Bank); monthly average rainfall (in millimeters. Source: World Bank), Water inflow in Fierza (in m3/second, Source: KESH); and total energy produced by three HPP of Drin cascade (Fierza-Koman-VD measured in GWh, Source: KESH).

Previous studies were based on the analyses of these variables and their importance in energy production showing the effect of these variables in energy demand and production, and how seasonality patterns affect these components of energy sectoring Albania [19].

Observing the four time series in Figure 3, we can agree on the fact that no clear linear trend is observed and that perhaps a seasonal pattern is present in each of the time series. The time series have no missing data and the presence of outliers is not significant.

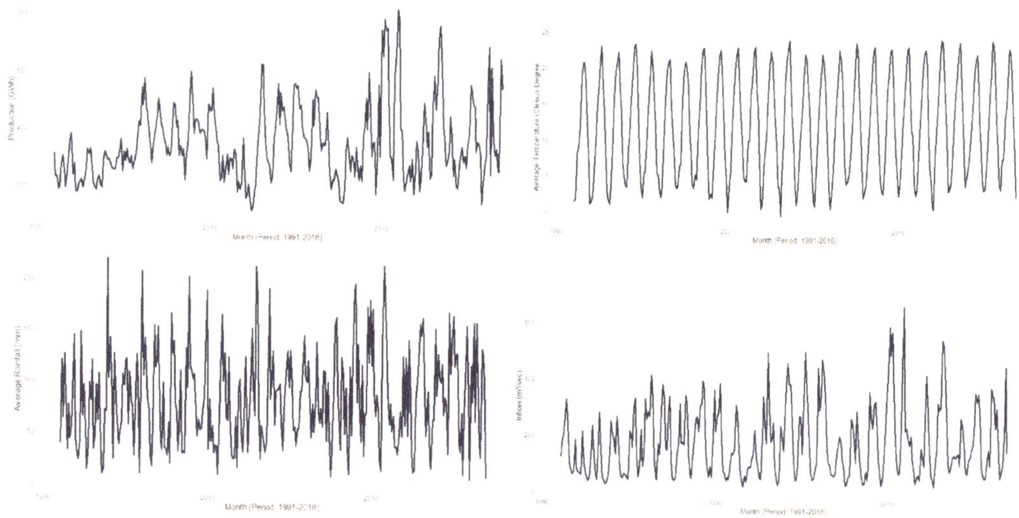

Figure 3. Time series of average temperature, average rainfall, water inflow, and energy produced by three HPP in Drin cascade.

Figure 4 shows the correlation plot among the variables by season. We may notice that there is a clear positive correlation between inflows in Fierza and production of the cascade which is most evident and strong during spring season (correlation = 0.664) when precipitation and snowmelt flows are higher. Inflows in Fierza and rainfalls have a positive correlation during autumn and winter season (correlation = 0.542). For the water inflow time series and energy produced in the cascade (from three HPP), we observe a significant change during 2010. This change will be also analyzed in the seasonal plot below.

Given that Albania is a Mediterranean country where seasons are clearly observed, we expect seasonality in the pattern of the time series taken into consideration for the study. For a better view of the correlation among the time series chosen we can also use the correlation plot. The correlation plot shows a positive correlation between the energy produced and water inflow (correlation = 0.64). However, because of the fact that water inflow is not mainly affected by the precipitations we observe a low value of the correlation between production and rainfall. A negative correlation, also with a low value (correlation = −0.38) is observed between the production and the average temperature.

Given the monthly frequency of our data and the efficiency of such predictions in long term, we decided to use as training set 80% of the observations and as testing set

20% of the observations. Another issue was to take the observation for year 2010 in our train dataset. So, we decide to have this representation 80% (250 observations) and 20% (62 observations).

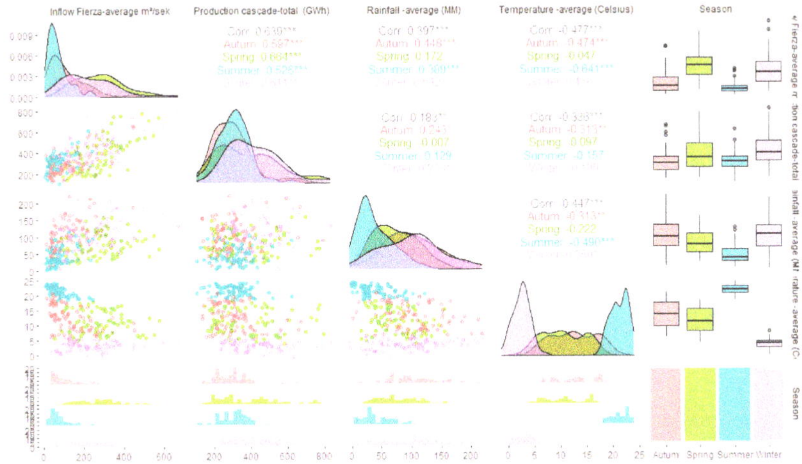

Figure 4. Correlation plot, Boxplot and density plot of the time series (Significance codes: * low correlation; ** average correlation; *** considerable correlation) (Source: Authors).

The seasonality of the monthly average temperature was confirmed by the seasonal graphs in Figure 5. The minimum average temperature is observed during winter and spring and the maximum average temperature is observed during the summer (July and August). The monthly average rainfall time series also expose presence of seasonality with high levels of rainfall during the wet months of autumn, winter and spring and with low levels during the summer. Carefully observing the seasonal plot for the time series of the water inflow, we can see the pattern of the time series for year 2010 which is significantly seen (in the two first seasonal plots) with high levels in almost all the months. We also observe high levels of inflows during the first months and the last months of the year. This phenomenon may be due to the increase in the level of inflows from natural causes such as precipitation and temperature which affects the snow melting and increase by this way the water level of the river Drin. We mention here again that the river cascade is positioned in the north side of Albania (the Alps). In the energy production time series, we also observe the same behavior as in the water inflow with high levels of production during the first months and the last months of the year. This is because the energy produced by HPP is positively correlated with the levels in the basin and water inflow.

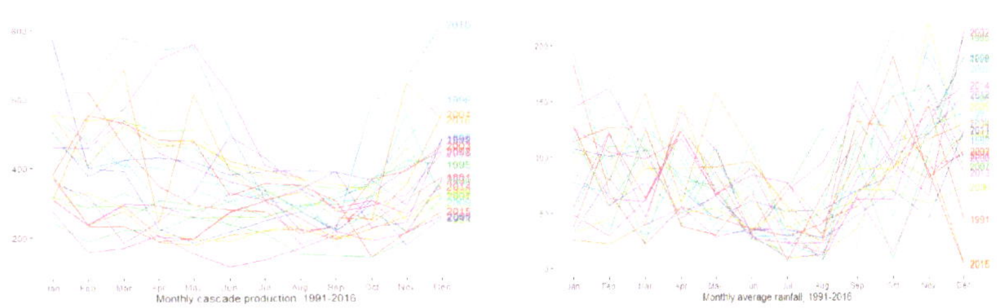

(a) (b)

Figure 5. Seasonal plot of the time series: (**a**) monthly total production; (**b**) monthly average rainfall) (period: 1991–2016) (Source: Authors).

4. Results

4.1. Models

The statistical time series models such as Naïve, autoregressive integrated moving average (ARIMA) [20] and exponential smoothing (Holt-Winters, ETS [21]) are most commonly used when it comes to monthly prediction especially with seasonality patterns. They have also become as popular as they are suitable for non-professionals, and they offer high accuracy and efficiency when it comes to non-complex time series data. Many competitions have shown that these methods outperform machine learning methods in many situations [22–24]. The advantage of classical univariate prediction methods is that they perform well when the volume of the data is considerable [25]. Neural networks are becoming more and more popular due to their ability to consider complexity and historical patterns in time series, being used as alternatives in different situations [26]. In this study our focus was to understand the relation between the external climacteric factors affecting the energy production by HPP. Below, we provide some results when using some statistical methods and neural networks with external variables which is the challenge for the future of our study.

We started by modeling our time series as a univariate time series and we also considered some statistical models using external factors among those presented above. Because the energy production time series show no linear trend, we decided not to go for the standard models such as naïve or drift because of the non-satisfactory visual results in prediction. ARIMA models consider in particular the linear behavior of the time series and stable seasonality; the ETS model takes into consideration the main components and in particularly the seasonality nature of the time series. Artificial Neural Networks (ANNs) are special mathematical models used also in prediction. They allow complex nonlinear relationships between the dependent variable and the independent(s) variable(s) used as explanatory variable(s). Neural networks are not based on an explicit stochastic model, so in most of the time we obtain prediction intervals by simulating future sample paths. The training process ofan ANN will depend on the activation function and the method used for finding the opportune weights recursively. Occasionally, we begin the training process of an ANN by choosing randomly the input values and then apply weights to each observation that will pass on information to a hidden layer where the information will be handled by an activation function. There are many studies on the performance of ANN in different types of data [27,28].

During the time series models progressions a considerable research is also made in the hybridization between ANN and classical time series models, in order to consolidate and benefit from the advantages of both models [29,30]. The automotive process is very easy in R, so we used forecast libraries which offer many facilities of these models [21,31].

4.2. Model Performance Measures

The accuracy of the models is evaluated based on accuracy measures such as error measures and information criteria. Bias and accuracy are then analyzed for every model and based on a critical judgment we have given our proposals for future work. The selection of the "best" model between all proposed was affected also on subjective indicators observed in the behavior of the time series such as seasonality [32,33]. Here, we used Mean Error (ME), Mean Absolute Error (MAE), Mean Absolute Percentage Error (MAPE), symmetric MAPE (sMAPE), and Root Mean Square Error (RMSE), calculated from the equations below:

$$MAE = \frac{1}{n}\sum_{i=1}^{n}|\hat{X}_t - X_t| \tag{1}$$

$$ME = \frac{1}{n}\sum_{i=1}^{n}(\hat{X}_t - X_t) \tag{2}$$

$$MAPE = \left(\frac{1}{n}\sum_{i=1}^{n}\frac{|\hat{X}_t - X_t|}{|X_t|}\right)\cdot 100\% \tag{3}$$

$$sMAPE = \left(\frac{1}{n} \sum_{i=1}^{n} \frac{2|\hat{X}_t - X_t|}{|X_t| + |\hat{X}_t|} \right) \quad (4)$$

$$RMSE = \sqrt{\frac{1}{n} \sum_{i=1}^{n} (\hat{X}_t - X_t)^2} \quad (5)$$

where X_t denote the observation at time t and $\hat{X}_t$ denote the estimated time series. Also, we graphed the fitted values (January 1991 up to January 2013) and the predicted values and compared them graphically with the testing time series (starting from February 2011 up to December 2016). For forecast of neural net models with external regressors, we need to have future values of the external regressor to be fed in the forecast function. We can use the test values for the inflow regressor. More than one external regressors can be used in the forecast procedure of neural network models. Error measurements should be small if the predicted values are close to the true values and will be large otherwise. The error measurements expressed in Figure 6 are computed using the training set used to fit the model and are referred to as the training errors. In general, we are focused on the accuracy of the predictions that we obtain when we apply a method to previously unseen test set, so we also calculate and evaluate the performance of the model based on out of sample set. From Figure 6, we observe that the Neural Network model (NNETAR) has the lowest values of the errors for the in-sample set compared to all the other models considered. We know that there is no guarantee that the method with the lowest training error will also have the lowest test error so we should evaluate the accuracy of the model based also in out-of-sample performance. Over fitting happens commonly when our statistical learning procedure is trying hard to discover patterns in the training set and we notice in most of the time low values for training set which are accompanied by large values for the testing set. Nonetheless, because many statistical learning methods seek to minimize the training error, we almost always expect these values to be smaller than the testing errors. Bias is also important when it comes to improving forecasting accuracy [34]. Bias is calculated as the average of the difference between the real values (y_i) and the predicted values $(\tilde{y}_i)$ by the model: $bias = mean(y_i - \tilde{y}_i)$.

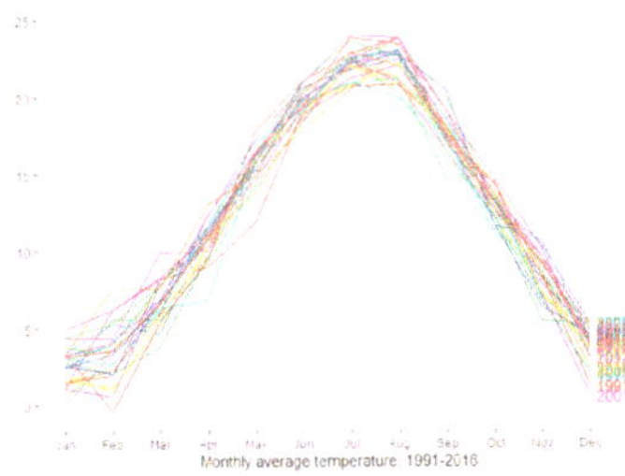

(a)

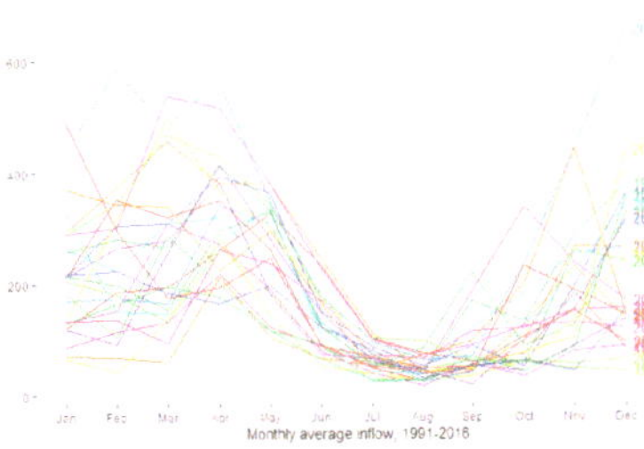

(b)

Figure 6. Seasonal plot of the time series: (**a**) monthly average temperature; (**b**) monthly average inflow) (period: 1991–2016) (Source: Authors).

Figure 7 shows the situation of the in-sample and out-of-sample bias and accuracy (RMSE) for the proposed models. We notice that among the models, NN in both cases (in-sample and out-of-sample) has the lowest values of accuracy (RMSE) and bias (very close to 0). The fact that this model offers good performance indicators, in both sets, ranks it among the best models to be used for forecasting purposes. The artificial neural network learns using the patterns of the time series seasonal cycles.

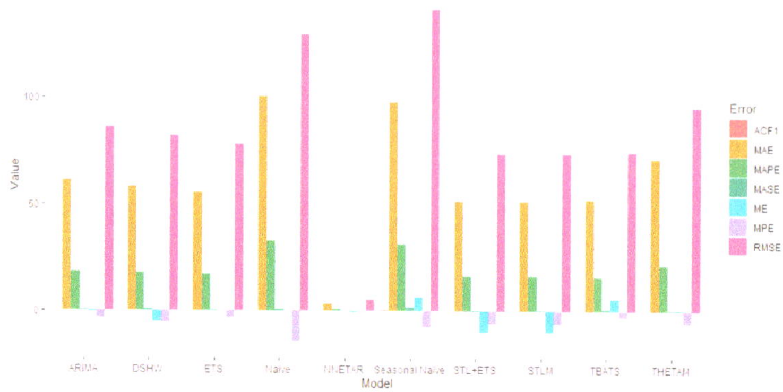

Figure 7. Error values for different models (in sample errors) (Source: Authors).

Figure 8 shows the comparison of accuracy versus bias of the proposed models for in-sample and out-of-sample data. We may observe that NNETAR has better accuracy and the lower bias for both in-sample and out-of-sample data compared to other models. Figure 9 shows the test data and forecasts according to each method reviewed in this study. Here, too, we graphically note the goodness of the NNETAR method in the prediction of monthly seasonal time series.

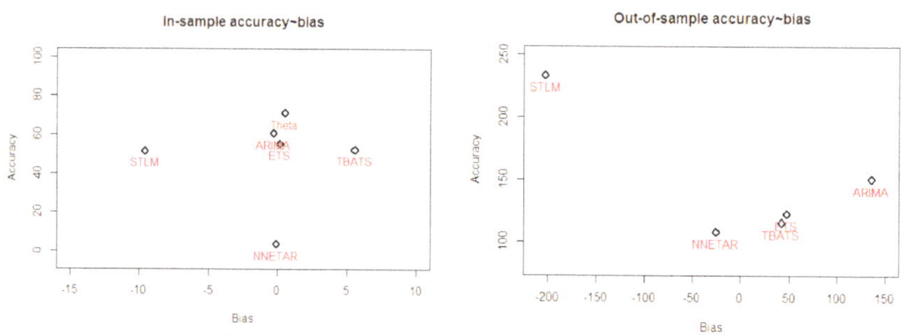

(**a**) (**b**)

Figure 8. (**a**) In-Sample; (**b**) Out-of-sample: accuracy versus bias of the proposed models.

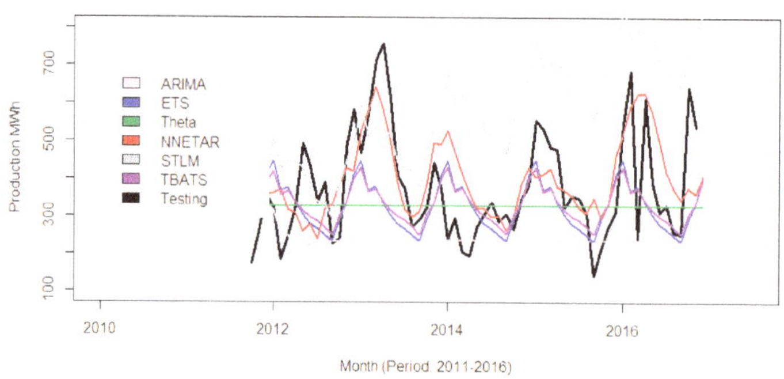

Figure 9. Predictions and testing data (Source: Authors).

5. Conclusions

Electricity produced by hydropower plants is likely to be influenced by climatic factors and their seasonal patterns. As such, analyzing these patterns and the correlation among climacteric external factors is becoming one of the challenges to obtain accurate predictions of the amount of energy a hydropower plant could produce during a given season. Many statistical learning techniques are used to obtain accurate predictions such as ARIMA, ETS, NN, TBATS, STLM, etc. In this study, we analyzed the seasonality of patterns of the monthly average temperature, monthly average precipitations, monthly average inflow in the first HPP, and the total monthly amount of energy produced in the cascade of Drin River positioned in the Alps of Albania.

We are aware of the enormous work to be done with the data presented here, especially for the energy production by HPP which is highly affected by climacteric factors. We have considered many models which are chosen based on the seasonality cycles of the data. Among the models we tested, we observed and confirmed that neural networks have managed to capture these seasonal cycles and providing good forecasts for monthly energy produced in the cascade. At the end of this work, we must admit that there is always uncertainty that will affect our predictions and there is always a challenge to obtain better predictions through hybrid machine learning models.

Institutional Review Board Statement: Not applicable.

Informed Consent Statement: Not applicable.

Data Availability Statement: Not applicable.

References

1. European Environment Agency (EEA). Report 2005. Available online: https://www.eea.europa.eu/publications/report_2005_0802_115659/at_download/file (accessed on 28 November 2020).
2. KESH Albania. Available online: http://www.kesh.al/en/asset/drini-cascade/ (accessed on 28 November 2020).
3. Available online: https://pressroom.ifc.org/all/pages/PressDetail.aspx?ID=26083 (accessed on 5 December 2020).
4. Available online: https://crossbowproject.eu (accessed on 15 January 2021).
5. Available online: https://adviser.albaniaenergy.org/wp-content/uploads/2018/11/Albania-hydroelectric-sector-towards-large-and-sustainable-developments-kept-by-Dr-Lorenc-Gordani.pdf (accessed on 15 January 2021).
6. Martz, F.; Vuosku, J.; Ovaskainen, A.; Stark, S.; Rautio, P. The Snow Must Goon: Ground Ice Encasement, Snow Compaction and Absence of Snow Differently Cause Soil Hypoxia, CO_2 Accumulation and Tree Seedling Damagein Boreal Forest. *PLoS ONE* **2016**, *11*, e0156620. [CrossRef]
7. Confortola, G.; Soncini, A.; Bocchiola, D. Climate change will affect hydrological regimes in the Alps. *J. Alp. Res. Rev. Géogr. Alp.* **2013**, 101–103. [CrossRef]
8. Saberian, A.; Hizam, H.; Radzi, M.A.M.; AbKadir, M.Z.A.; Mirzaei, M. Modelling and Prediction of Photovoltaic Power Output Using Artificial Neural Networks. *Int. J. Photoenergy* **2014**, *2014*, 469701. [CrossRef]
9. Blöschl, G.; Hall, J.; Parajka, J.; Perdigão, R.A.; Merz, B.; Arheimer, B.; Aronica, G.T.; Bilibashi, A.; Bonacci, O.; Borga, M.; et al. Climate Change Shifts the Timing of European Floods. *Sci. News Science* **2017**, *357*, 588–590. [CrossRef]
10. Blöschl, G.; Hall, J.; Viglione, A.; Perdigão, R.A.; Parajka, J.; Merz, B.; Lun, D.; Arheimer, B.; Aronica, G.T.; Bilibashi, A.; et al. Changing climate both increases and decreases European river floods. *Nature* **2019**, *573*, 108–111. [CrossRef]
11. Wagner, T.; Themeßl, M.; Schüppel, A.; Gobiet, A.; Stigler, H.; Birk, S. Impacts of climate change on stream flow and hydropower generation in the Alpineregion. *Env. Earth Sci.* **2017**, *76*, 1–22. [CrossRef]
12. Engeland, K.; Borga, M.; Creutin, J.D.; François, B.; Ramos, M.H.; Vidal, J.P. Space-time variability of climate variables and intermittent renewable electricity production—A review. *Renew. Sustain. Energy Rev.* **2017**, *79*, 600–617. [CrossRef]
13. Schaeffer, R.; Szklo, A.S.; de Lucena, A.F.; Borba, B.S.; Nogueira, L.P.; Fleming, F.P.; Troccoli, A.; Harrison, M.; Boulahya, M.S. Energy sector vulnerability to climate change: A review. *Energy* **2012**, *38*, 1–12. [CrossRef]
14. Chandramowli, S.N.; Felder, F.A. Impact of climate change on electricity systems and markets—A review of models and forecasts. *Sustain. Energy Technol. Assess* **2014**, *5*, 62–74. [CrossRef]
15. Sun, T.; Zhang, T.; Teng, Y.; Chen, Z.; Fang, J. Monthly Electricity Consumption Forecasting Method Based on X12 and STL Decomposition Model in an Integrated Energy System. *Math. Prob. Eng.* **2019**, *2019*, 9012543. [CrossRef]
16. Usha, T.; Balamurugan, S. Seasonal Based Electricity Demand Forecasting Using Time Series Analysis. *Circuits Syst.* **2016**, *7*, 3320–3328. [CrossRef]
17. Kalimoldayev, M.; Drozdenko, A.; Koplyk, I.; Marinich, T.; Ab-dildayeva, A.; Zhukabayeva, T. Analysis of modern approaches for the prediction of electric energy consumption. *Open Eng.* **2020**, *10*, 350–361. [CrossRef]

18. Crochet, P.; Jóhannesson, T.; Jónsson, T.; Sigurðsson, O.; Björnsson, H.; Pálsson, F.; Barstad, I. Estimating the Spatial Distribution of Precipitation in Iceland Usinga Linear Model of Orographic Precipitation. *J. Hydrometeorol.* **2007**, *8*, 1285–1306. [CrossRef]
19. Gjika, E.; Ferrja, A.; Kamberi, A. A Study on the Efficiency of Hybrid Models in Forecasting Precipitations and Water Inflow Albania Case Study. *Adv. Sci. Technol. Eng. Syst. J.* **2019**, *4*, 302–310. [CrossRef]
20. Box, G.E.; Hunter, W.H.; Hunter, S. *Statistics for Experimenters*; John Wiley and Sons: New York, NY, USA, 1978; Volume 664.
21. Hyndman, R.; Koehler, A.B.; Ord, J.K.; Snyder, R.D. *Forecasting with Exponential Smoothing: The State Space Approach*; Springer Science & Business Media: Secaucus, NJ, USA, 2008.
22. Makridakis, S.; Hibon, M. The M3-Competition: Results, conclusions and implications. *Int. J. Forecast.* **2000**, *16*, 451–476. [CrossRef]
23. Makridakis, S.; Spiliotis, E.; Assimakopoulos, V. The M4 competition: Results, findings, conclusion and way forward. *Int. J. Forecast.* **2018**, *34*, 802–808. [CrossRef]
24. Makridakis, S.; Spiliotis, E.; Assimakopoulos, V. Statistical and machine learning forecasting methods: Concern and ways forward. *PLoS ONE* **2018**, *13*, e0194889. [CrossRef] [PubMed]
25. Bandara, K.; Bergmeir, C.; Smyl, S. Forecasting across time series databases using recurrent neural networks on groups of similar series: A clustering approach. *Expert Syst. Appl.* **2020**, *140*, 112896. [CrossRef]
26. Hewamalage, H.; Bergmeir, C.; Bandara, K. Recurrent Neural Networks for Time Series Forecasting: Current status and future directions. *Int. J. Forecast.* **2021**, *37*, 388–427. [CrossRef]
27. Del Angel, R.G. Financial time series forecasting using Artificial Neural Networks. *Rev. Mex. Econ. Finanz. Nueva Época* **2020**, *15*, 105–122.
28. Tealab, A. Time series forecasting using artificial neural networks methodologies: Asystematic review. *Future Comput. Inform. J.* **2018**, *3*, 334–340. [CrossRef]
29. Wang, L.; Zou, H.; Su, J.; Li, L.; Chaudhry, S. An ARIMA-ANN hybrid model for time series forecasting. *Syst. Res. Behav. Sci.* **2014**, *30*, 244–259. [CrossRef]
30. Young, C.C.; Liu, W.C.; Hsieh, W.L. Predicting the water level fluctuation in an alpine lake using physically based, artificial neural network, and time series forecasting models. *Math. Probl. Eng.* **2015**, *2015*, 708204. [CrossRef]
31. Hyndman, R.J.; Khandakar, Y. Automatic Time Series Forecasting: The Forecast Package for R. *J. Stat. Softw.* **2008**, *27*, 1–22. [CrossRef]
32. Hyndman, R.J.; Koehler, A.B. Another look at measures of forecast accuracy. *Int. J. Forecast.* **2006**, *22*, 679–688. [CrossRef]
33. Hyndman, R.J. Measuring Forecast Accuracy. 2014. Available online: https://pdfs.semanticscholar.org/af71/3d815a7caba8dff7248ecea05a5956b2a487.pdf (accessed on 28 November 2020).
34. Spiliotis, E.; Petropoulos, F.; Assimakopoulos, V. Improving the forecasting performance of temporal hierarchies. *PLoS ONE* **2019**, *14*, e0223422. [CrossRef]

engineering proceedings

MDPI

Proceedings

Enhanced Day-Ahead PV Power Forecast: Dataset Clustering for an Effective Artificial Neural Network Training †

Andrea Matteri *, Emanuele Ogliari and Alfredo Nespoli

Politecnico di Milano, Dipartimento di Energia, Via La Masa, 34, 20156 Milan, Italy; emanuelegiovanni.ogliari@polimi.it (E.O.); alfredo.nespoli@polimi.it (A.N.)
* Correspondence: andrea.matteri@polimi.it
† Presented at the 7th International Conference on Time Series and Forecasting, Gran Canaria, Spain, 19–21 July 2021.

Abstract: The increasing integration of renewable energy sources into the existing energy supply structure is challenging due to the intermittency typical of these energy sources, which implies problems of reliability and scheduling of grid operation. Concerning solar energy, the solar forecast tool predicts the photovoltaic (PV) power production and therefore permits a more efficient grid management. In this paper, the combination of clustering techniques and ANNs (Artificial Neural Networks) for day-ahead PV power forecast is analyzed. Clustering techniques are exploited to divide a dataset into different classes of days with similar weather conditions. Then, a dedicated ANN is developed for every group. The main goal is to assess the forecast improvement determined by the combination of ANNs and dataset clustering methods. Different combinations are compared on a real case study: a PV facility in SolarTech^{LAB}, in Politecnico di Milano.

Keywords: power forecast; photovoltaic; artificial neural network; clustering; clearness index; k-means; classification; random forest

check for updates

Citation: Matteri, A.; Ogliari, E.; Nespoli, A. Enhanced Day-Ahead PV Power Forecast: Dataset Clustering for an Effective Artificial Neural Network Training. *Eng. Proc.* **2021**, *5*, 16. https://doi.org/10.3390/engproc2021005016

Published: 28 June 2021

1. Introduction

The ongoing energy transition is progressively redefining structure and arrangement of the current energy system. A crucial challenge is represented by the large penetration of RES (Renewable Energy Sources) into the existing power supply structure. A grid operator should be able to ensure the balance between the electricity production and consumption any moment, accommodating expected and unexpected changes on both sides. RES have dynamic nature and large variability depending on geographical locations and weather conditions. For instance, concerning PV (photovoltaic) plants, the power output depends on several meteorological variables such as solar irradiance, air temperature, cloud variation, wind speed and so on, intrinsically intermittent and non-controllable: these aspects imply problems of reliability, stability, and scheduling of the power supply structure [1].

Reliable forecast tools allow the prediction of the expected power production and its fluctuations, leading to a more efficient grid management [2]: for this reason, power forecast research field is presently receiving unprecedent attention from the scientific community. The current work is focused specifically on day-ahead PV power forecast.

According to literature, solar forecast methods can be categorized in: statistical methods, physical methods, Machine Learning (ML) methods and hybrid methods [2–4]. Statistical methods are capable, given a time series of historical data, to reconstruct the relationship between solar irradiance or PV power output and meteorological parameters. Moreover, they do not require physical knowledge about a system to model it [3]. Physical methods, mainly consisting of Numerical Weather Predictions (NWP), model the interactions between solar radiation and atmospheric components by means of differential equations and do not require historical data [5,6]. ML methods mimics the capability of human brain to learn from experience and can solve even problems which cannot be represented explicitly. As with statistical methods, to perform a prediction, they require historical data but not physical knowledge of the modeled system [2]. Artificial Neural Networks (ANNs) are a ML method commonly involved in PV power forecast. Finally, hybrid methods consist of combinations between other forecast methods, with the purpose of solving the weaknesses of individual ones and benefiting from their advantages [7,8].

In the current work, as forecast models, several combinations between ANNs and clustering techniques are proposed. Clustering is an unsupervised machine learning technique that allows the partitioning of a dataset into groups of samples presenting similarities [9,10]. In the following, different clustering criteria are applied to divide the days in a dataset into different classes according to their weather conditions. Once a partition is defined, a specific ANN is developed for every cluster: each ANN is trained using only samples belonging to a certain cluster and is used to forecast PV power production only in the weather conditions typical of that cluster. The similarity between PV power curves registered in similar weather conditions is therefore exploited to construct optimized forecast models.

The aim of this paper is to assess whether it possible to improve the training of artificial neural networks for day-ahead PV power forecast by dividing a dataset through clustering techniques and, in the case of a positive answer, to identify the best-performing dataset partition in terms of forecast accuracy between the proposed ones.

2. Case Study and Procedure

Different combinations between clustering techniques and artificial neural networks are tested, validated, and compared on a real case study: a PV facility in SolarTech^{LAB}, at Politecnico di Milano [11]. However, the proposed procedure is valid for PV plants of all sizes. The available dataset contains historical data about measured power and predicted weather parameters, namely temperature, global horizontal irradiance, global plane-of-array irradiance and wind speed. The predicted weather parameters are provided as input to the proposed prediction models, whose output is compared with the measured power data to assess the forecast performance. Data are recorded on an hourly basis for a total amount of 840 days in the time span comprised between January 2017 and September 2020.

The overall PV power forecast process can be summarized as the iterative multi-step procedure represented in Figure 1. In the following, details about every single step are provided.

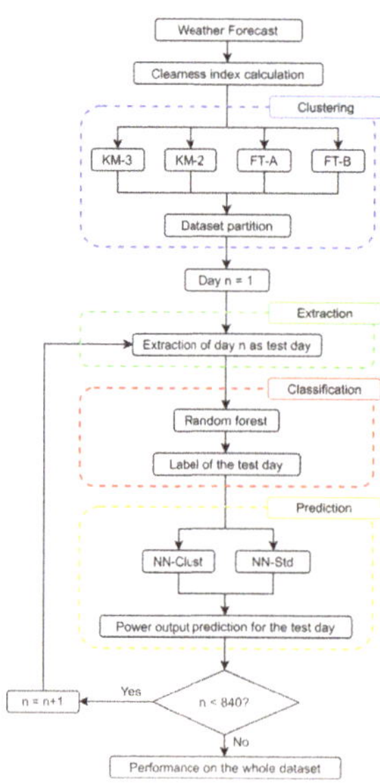

Figure 1. Proposed forecast procedure.

2.1. Clustering Phase

In the clustering phase, to obtain a proper partition, the daily clearness index (K_t) is employed in clustering as meaningful parameter for day type estimation [12–14]. It is defined as:

$$K_t = G/G_0 \tag{1}$$

In the equation, G is the daily global horizontal irradiation, while G_0 represents the corresponding daily extraterrestrial horizontal irradiation. Hence, K_t is a dimensionless quantity employed in day type clustering thanks to its capability to remove the seasonal dependence from solar irradiation, isolating the information content about weather conditions [9]. Large values of clearness index indicate clear sky conditions, while low values represent overcast sky conditions. Starting from the previously described dataset, the K_t value for each day is computed by means of the same procedure applied by ESRA (European Solar Radiation Atlas) [15].

Then, four different dataset division criteria are proposed, namely: FT-A, FT-B, KM-3, and KM-2. All the approaches are based on the clearness index K_t and, as previously mentioned, aim to divide the dataset in classes according to weather conditions of single days.

FT-A (Fixed Threshold set A) and FT-B (Fixed Threshold set B) are not properly clustering algorithms, but they perform a partition relying of fixed threshold values of clearness index defined in scientific literature [16,17]. In detail, they divide the dataset in three different weather classes based on the thresholds summarized Table 1.

Table 1. Clearness index partition.

Weather Conditions	FT-A	FT-B
Sunny	$K_t > 0.45$	$K_t > 0.65$
Partially cloudy	$0.25 < K_t < 0.45$	$0.35 < K_t < 0.65$
Cloudy	$K_t < 0.25$	$K_t < 0.35$

Both KM-3 and KM-2 are based on the k-means clustering algorithm. The choice of k-means instead of other possible clustering algorithms is related to its simplicity in implementation and its efficiency. It is worth noticing that the application of k-means algorithm based on a single parameter (i.e., the clearness index) corresponds to a fixed-thresholds-based partition where the thresholds are set automatically by the algorithm instead of by an external intervention (as in FT-A and FT-B). The difference between KM-3 and KM-2 consists of the choice of the number of clusters (K). KM-3 adopts $K = 3$ for a homogeneous comparison with the fixed-thresholds-based approaches (i.e., FT-A and FT-B). KM-2 exploits some proper indexes to select the best possible dataset partition in terms of clustering quality, namely: Silhouette index, Davies-Bouldin index and Calinski–Harabasz index.

Given a generic dataset $X = \{X_1, X_2, \ldots, X_N\}$, containing N elements and partitioned into K clusters $C = \{C_1, C_2, \ldots, C_K\}$, these indexes can be computed as follows.

The Silhouette index [18] (computed as global value) is defined as:

$$S(C) = \frac{1}{N} \cdot \sum_{j=1}^{K} \frac{1}{m_j} \cdot \sum_{i=1}^{m_j} \frac{b_i^j - a_i^j}{\max\{a_i^j, b_i^j\}} \tag{2}$$

In the equation: m_j is the number of elements in the generic cluster C_j; a_i^j is the average distance between the i_{th} element in the cluster C_j and the other elements in the same cluster; b_i^j is the minimum average distance between the i_{th} element in the cluster C_j and all the elements belonging to clusters C_k, with $k = \{1, 2, \ldots, K\}$ and $k \neq j$. The optimal number of clusters is the one that maximizes the value of Silhouette index.

The Davies-Bouldin index [18] is defined as:

$$DB(C) = \frac{1}{K} \cdot \sum_{i=1}^{K} \max_{i \neq j} \frac{\Delta(C_i) - \Delta(C_j)}{\delta(C_i, C_j)} \quad (3)$$

In the equation: $\Delta(C_i)$ is the within-cluster distance; $\delta(C_i, C_j)$ is the between cluster distance. The optimal clustering solution is the one that minimizes the Davies-Bouldin index value.

The Calinski–Harabasz index [19] is defined as:

$$CH(C) = \frac{\sum_{i=1}^{K} m_i \cdot ||G_i - G||^2}{\sum_{i=1}^{K} \sum_{X \in C_i} ||X - G_i||^2} \cdot \frac{N - K}{K - 1} \quad (4)$$

In the equation: m_i is the number of elements in the cluster C_i; G_i is the barycentre of the cluster C_i (in the case of k-means clustering, it corresponds to the centroid); and G is the barycentre of the entire dataset (the overall mean of the data). The optimal clustering solution is the one that maximizes the value of Calinski–Harabasz index.

These indexes are computed in function of different numbers of clusters and, applying a majority voting procedure, $K = 2$ is selected as the optimal dataset partition.

2.2. Extraction Phase

The extraction phase corresponds to the extraction of a test day, consisting of 24 consecutive hourly samples, from the initial dataset. This day constitutes the test set on which the prediction performance is computed. The cluster of origin of the extracted day is assumed to be unknown, as it would be in a real day-ahead power forecast. For a complete and reliable prediction performance assessment, all days available in the dataset are extracted one by one in different iterations.

2.3. Classification Phase

In the classification phase, the most suitable cluster for the test day is identified. Once labeled, the test day is assigned to the proper ANN, which perform the power prediction in the test day weather conditions. Therefore, this phase represents an additional step with respect to the single-network-based forecast, where the inputs are directly provided to the unique ANN available. As classifier, the random forest model is chosen, among all the possible algorithms, thanks to its flexibility, fast implementation, and easy tuning [20]. The classifier optimization consists of a proper selection of number of trees and input features based on out-of-bag classification error. The optimal configuration consists of a structure with 60 trees that takes global horizontal irradiance and global plane-of-array irradiance as input features.

2.4. Prediction Phase

Lastly, in the prediction step, different neural networks are developed to predict the PV power output in the extracted test days. Two different approaches are adopted, namely NN-Clust and NN-Std.

NN-Clust represents the clustering-based approach. In this approach, only days belonging to the same cluster of the test day are used or the training of each ANN. Then, the trained ANN predicts PV power output for the test day, characterized by weather conditions similar to those of samples involved in training. For the training of each ANN, 10% of samples contained in a given cluster is randomly extracted as validation set, while the remaining 90% constitutes the training set. Moreover, an ensemble of 10 independent trials is implemented to enhance the generalization capability of the model. To optimize the hidden layer size, a sensitivity analysis is carried out for every ANN corresponding to a different cluster. In practical terms, the sensitivity analysis studies the trade-off between performance and computational cost, analyzing the value of the Mean Square Error in function of a variable number of hidden neurons. The predicted weather parameters available in the dataset, namely temperature, global horizontal irradiance, and global plane-of-array irradiance and wind speed, are provided as input features to all the networks.

On the other hand, NN-Std represents the most common forecast approach in scientific literature, involving a single neural network, and it is developed for comparison with the previously described clustering-based approach. For the sake of a fair comparison, NN-Std must present several similarities with NN-Clust: same number of neurons in the hidden layer, same input features and same days predicted as test. The crucial difference between NN-Clust and NN-Std is that the latter is trained with days extracted from all clusters.

3. Error Metrics

Given a forecast output P and an observed output $\hat{P}$ several error metrics are defined and adopted in this work for performance evaluation.

The Normalized Mean Absolute Error (NMAE) estimates the average magnitude of the errors for a set of N predictions divided by the plant net capacity C:

$$NMAE_{\%} = \frac{1}{N} \cdot \sum_{h=1}^{N} \frac{P_h - \hat{P}_h}{C} \cdot 100 \tag{5}$$

The Root Mean Square Error (RMSE) is computed using the square of the difference between observed and predicted values, and therefore penalizes large gaps:

$$RMSE = \sqrt{\frac{1}{N} \cdot \sum_{h=1}^{N} (P_h - \hat{P}_h)^2} \tag{6}$$

The normalized Root Mean Square Error (nRMSE) corresponds to the ratio between RMSE and the maximum observed power output in the considered time frame:

$$nRMSE_{\%} = \frac{RMSE}{\max(P_h)} \cdot 100 \tag{7}$$

The Weighted Mean Absolute Error (WMAE) is based on the total energy production:

$$WMAE_{\%} = \frac{\sum_{h=1}^{N} |P_h - \hat{P}_h|}{\sum_{h=1}^{N} P_h} \tag{8}$$

Finally, the Envelope-weighted Mean Absolute Error (EMAE), introduced in [21], aims to provide a measure of forecast accuracy in the interval between 0% and 100%:

$$EMAE_{\%} = \frac{\sum_{h=1}^{N} |P_h - \hat{P}_h|}{\sum_{h=1}^{N} \max(P_h, \hat{P}_h))} \tag{9}$$

4. Results and Discussion

The groups identified by the different dataset partitioning methods proposed are different and quite unbalanced in terms of numerosity, as reported in Table 2. In general, the cluster corresponding to sunny conditions is the largest while the others, in comparison, contain much less elements. The only exception is represented by FT-B, providing a more homogeneous grouping where sunny days and partially cloudy days clusters have comparable size. The numerosity of a cluster is relevant by the point of view of the forecast: ANNs trained using too few elements could present poor generalization performance.

Concerning the forecast accuracy, several comparisons are performed. First, the NN-Clust models developed are compared to the corresponding NN-Std to evaluate the performance enhancement allowed by the proposed methodology. The performance improvements computed according to all the evaluation metrics are reported in Table 3.

Table 2. Clusters numerosity with different partitions.

Weather Conditions	FT-A	FT-B	KM-3	KM-2
Sunny days	641	339	511	618
Partially cloudy days	125	369	193	-
Cloudy days	74	132	136	222
Total	840	840	840	840

Table 3. Performance improvement given by NN-Clust with respect to NN-Std.

Method	Cluster	ΔNMAE	ΔRMSE	ΔnRMSE	ΔWMAE	ΔEMAE
KM-3	3	6.0%	3.9%	7.9%	7.0%	4.2%
KM-2	2	4.6%	1.9%	6.5%	6.0%	3.4%
FT-A	3	3.9%	2.0%	5.4%	5.2%	3.2%
FT-B	3	5.8%	3.3%	7.1%	6.1%	3.9%

Independently from the error metric and the dataset partition considered, the approach involving clustering (i.e., NN-Clust) outperforms the one based on a single-network prediction (i.e., NN-Std). The largest improvement recorded consists of an error reduction of 7.9% in nRMSE with KM-3, while smallest one consists of an error reduction of 1.9% in RMSE with KM-2. Therefore, weather type clustering is demonstrated to be effective and beneficial when combined to ANN with the goal to optimize their training.

Then, a comparison between different dataset partitioning criteria, always in terms of prediction performance, is carried out and visually represented in Figure 2. The spider-web chart is represented normalizing all the error metrics, i.e., dividing them by the corresponding maximum recorded value.

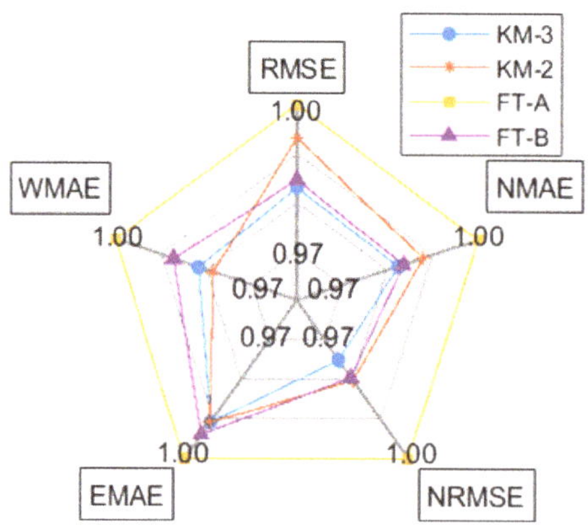

Figure 2. Comparisons between different partitioning methods.

Comparing all the approaches that divide the dataset in 3 clusters, it is observed that the clustering-based approach, i.e., KM-3, outperform both FT-B and FT-A, based on fixed threshold values of clearness index. Therefore, at equal number of clusters identified, the clustering-based methods exhibit better performance.

On the other hand, comparing the clustering-based approaches, i.e., KM-3 and KM-2, four error metrics out of five highlight the superiority of clustering method KM-3, even if the error reduction allowed with respect to KM-2 is always limited. This means that the optimal dataset partition in terms of clustering quality does not necessarily imply the best prediction performance of the forecast model.

Among all the dataset partitioning methods considered, KM-3 reveals to be the best-performing one by the point of view of forecast accuracy.

Lastly, the "best" and "worst" days in terms of forecast performance, corresponding respectively to minimum and maximum recorded values of EMAE, are extracted and analyzed for each cluster identified by KM-3, i.e., the best-performing partitioning criterion. For these days, the actual power curve (P_m) and the ones forecast by NN-Clust and NN-Std approaches are depicted and compared in Figure 3.

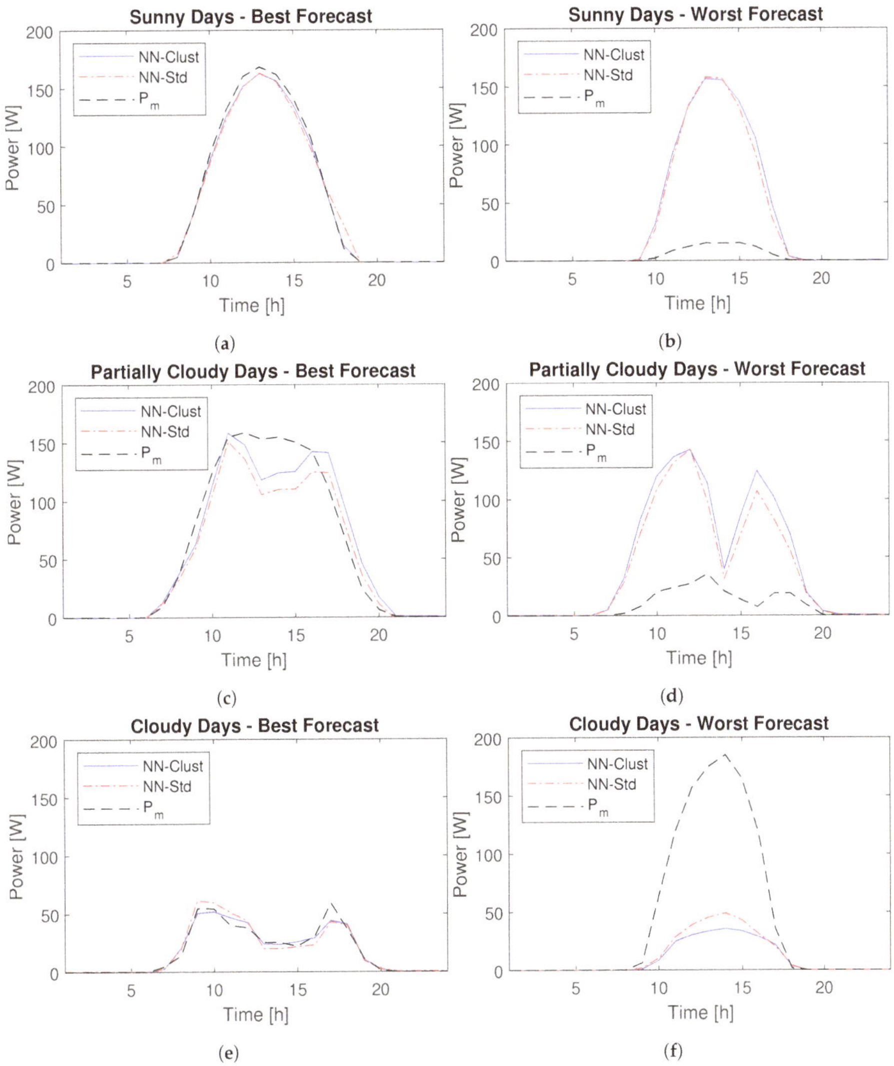

Figure 3. Forecast and actual power curves in: "best" (**a**) and "worst" (**b**) sunny days, "best" (**c**) and "worst" (**d**) partially cloudy days, and "best" (**e**) and "worst" (**f**) cloudy days.

In the "best" case for sunny days, both NN-Clust and NN-Std approaches accurately approximate the smooth power curve typical of sunny days. The "best" partially cloudy day presents an actual power trend not as smooth as a typical sunny day, but not even much irregular. Indeed, this day shows one of the highest clearness index value (0.53) among the partially cloudy days cluster. The forecast curves accurately approximate the actual one except for a small region around the central hours of the day, where NN-Clust clearly outperforms NN-Std. The "best" cloudy day presents the irregular PV power trend typical of overcast sky conditions. NN-Clust outperforms NN-Std in terms of forecast error, but both models are capable of accurately approximating the actual trend.

The "worst" days always correspond to errors in weather forecast, when the real weather characteristics of a given day turned out to be completely different from the expected ones. In this condition, the forecast power either strongly overestimate or underestimate the measured one. It is, therefore, observed that with inaccurately predicted weather parameters in input to an ANN, the forecast performance exhibits a heavy deterioration.

5. Conclusions

With the increasing RES penetration in the energy mix, reliable forecast tools allow the prediction of the expected power production and its fluctuations, leading to a more efficient grid management. The current work focuses on PV power output prediction and proposes several combinations of ANNs and clustering techniques for an enhanced day-ahead forecast. The aim of this work is to assess whether it possible to improve the training of artificial neural networks for day-ahead PV power forecast by dividing a dataset through clustering techniques and, in the case of a positive answer, to identify the best-performing dataset partition in terms of forecast accuracy between the proposed ones. The methodologies proposed are tested and validated on a real case study, a PV facility located in Politecnico di Milano, the SolarTech^{LAB}.

The conclusions drawn from the analysis of the results are summarized in the following:

- The proposed procedure, based on a day type clustering according to weather conditions, is beneficial for ANNs training. Indeed, the performance obtained with clustering-based approaches always outperform those of their non-clustering-based counterpart. The NN-Clust (clustering-based) approach based on KM-3, i.e., the best-performing combination, presents an improvement of 4.2% in EMAE with respect to the corresponding NN-Std (non-clustering-based) approach.
- Comparing all the approaches identifying a constant number of clusters (i.e., FT-A, FT-B and KM-3, identifying $K = 3$ clusters), it is observed that the clustering-based partition is more effective than clearness-index-fixed-threshold-based ones in terms of forecast performance.
- Comparing the clustering-based approaches (i.e., KM-3 and KM-2), it is observed that the optimal dataset partition in terms of cluster quality do not necessarily lead to the best forecast result. Therefore, a partition showing good scores according to a quality evaluation criterion do not necessarily imply a good effectiveness in an application.
- The forecast performance is strongly influenced by the inaccuracies in weather parameters prediction, which can heavily affect the final result.

Author Contributions: Conceptualization, A.M., E.O. and A.N.; data curation, A.N.; formal analysis, A.M.; methodology, A.M. and E.O.; software, A.M.; supervision, E.O. and A.N.; visualization, A.M.; writing—original draft, A.M.; writing—review and editing, A.M., E.O. and A.N. All authors have read and agreed to the published version of the manuscript.

Funding: This research received no external funding.

Institutional Review Board Statement: Not applicable.

Informed Consent Statement: Not applicable.

Data Availability Statement: Data is contained within the article.

Conflicts of Interest: The authors declare no conflict of interest.

References

1. Voyant, C.; Notton, G.; Kalogirou, S.; Nivet, M.L.; Paoli, C.; Motte, F.; Fouilloy, A. Machine learning methods for solar radiation forecasting: A review. *Renew. Energy* **2017**, *105*, 569–582. [CrossRef]
2. Mellit, A.; Pavan, A.M.; Ogliari, E.; Leva, S.; Lughi, V. Advanced methods for photovoltaic output power forecasting: A review. *Appl. Sci.* **2020**, *10*, 487. [CrossRef]
3. Sobri, S.; Koohi-Kamali, S.; Rahim, N.A. Solar photovoltaic generation forecasting methods: A review. *Energy Convers. Manag.* **2018**, *156*, 459–497. [CrossRef]
4. Ahmed, R.; Sreeram, V.; Mishra, Y.; Arif, M.D. A review and evaluation of the state-of-the-art in PV solar power forecasting: Techniques and optimization. *Renew. Sustain. Energy Rev.* **2020**, *124*, 109792. [CrossRef]
5. Yang, D.; Kleissl, J.; Gueymard, C.A.; Pedro, H.T.; Coimbra, C.F. History and trends in solar irradiance and PV power forecasting: A preliminary assessment and review using text mining. *Sol. Energy* **2018**, *168*, 60–101. [CrossRef]
6. Larson, V.E. Forecasting Solar Irradiance with Numerical Weather Prediction Models. *Sol. Energy Forecast. Resour. Assess.* **2013**, 299–318.
7. Guermoui, M.; Melgani, F.; Gairaa, K.; Mekhalfi, M.L. A comprehensive review of hybrid models for solar radiation forecasting. *J. Clean. Prod.* **2020**, *258*, 120357. [CrossRef]
8. Ogliari, E.; Dolara, A.; Manzolini, G.; Leva, S. Physical and hybrid methods comparison for the day ahead PV output power forecast. *Renew. Energy* **2017**, *113*, 11–21. [CrossRef]
9. Jimenez-Perez, P.F.; Mora-Lopez, L. Modeling and forecasting hourly global solar radiation using clustering and classification techniques. *Sol. Energy* **2016**, *135*, 682–691. [CrossRef]
10. Xu, R.; Wunsch, D. Survey of clustering algorithms. *IEEE Trans. Neural Netw.* **2005**, *16*, 645–678. [CrossRef] [PubMed]
11. SolarTech Lab. Available online: http://www.solartech.polimi.it (accessed on 16 April 2021).
12. Udo, S.O. Sky conditions at Ilorin as characterized by clearness index and relative sunshine. *Sol. Energy* **2000**, *69*, 45–53 [CrossRef]
13. Page, J. The Role of Solar-Radiation Climatology in the Design of Photovoltaic Systems. In *Practical Handbook of Photovoltaics*; McEvoy, A., Markvart, T., Castañer, L., Eds.; Academic Press: Cambridge, MA, USA, 2012; pp. 573–643.
14. Brownson, J.R. Measure and Estimation of the Solar Resource. In *Solar Energy Conversion Systems*; Brownson, J.R., Ed.; Academic Press: Cambridge, MA, USA, 2014; pp.199–235.
15. Scharmer, K.; Greif, J. *The European Solar Radiation Atlas: Fundamentals and Maps*; Presses des Mines: Paris, France, 2000; Volume 1.
16. Leva, S.; Dolara, A.; Grimaccia, F.; Mussetta, M.; Ogliari, E. Analysis and validation of 24 hours ahead neural network forecasting of photovoltaic output power. *Math. Comput. Simul.* **2017**, *131*, 88–100. [CrossRef]
17. Kudish, A.I.; Ianetz, A. Analysis of daily clearness index, global and beam radiation for Beer Sheva, Israel: Partition according to day type and statistical analysis. *Energy Convers. Manag.* **1996**, *37*, 405–416. [CrossRef]
18. Petrovic, S. A Comparison Between the Silhouette Index and the Davies-Bouldin Index in Labelling IDS Clusters. In Proceedings of the 11th Nordic Workshop on Secure IT-Systems, Linköping, Sweden, 19–20 October 2006; pp. 53–64.
19. Chowdhury, S.A.; Riccardi, G.; Alam, F. Unsupervised Recognition and Clustering of Speech Overlaps in Spoken Conversations. In Proceedings of the Workshop on Speech, Language and Audio in Multimedia (SLAM2014), Penang, Malaysia, 11–12 September 2014; pp. 62–66.
20. Cutler, A.; Cutler, D.R.; Stevens, J.R. Random Forests. In *Ensemble Machine Learning*; Zhang, C., Ma, Y., Eds.; Springer: Berlin, Germany, 2012; pp. 157–175.
21. Dolara, A.; Grimaccia, F.; Leva, S.; Mussetta, M.; Ogliari, E. Comparison of training approaches for photovoltaic forecasts by means of machine learning. *Appl. Sci.* **2018**, *8*, 228. [CrossRef]

engineering proceedings

MDPI

Proceedings

Bernoulli Time Series Modelling with Application to Accommodation Tourism Demand †

Miguel Ángel Ruiz Reina

Department of Theory and Economic History (Staff of Fundamentals), University of Malaga, PhD Program in Economics and Business, s/n, Plaza del Ejido, 29013 Málaga, Spain; ruizreina@uma.es

† Presented at the 7th International Conference on Time Series and Forecasting, Gran Canaria, Spain, 19–21 July 2021.

Abstract: In this research, a new uncertainty method has been developed and applied to forecasting the hotel accommodation market. The simulation and training of Time Series data are from January 2001 to December 2018 in the Spanish case. The Log-log BeTSUF method estimated by GMM-HAC-Newey-West is considered as a contribution for measuring uncertainty vs. other prognostic models in the literature. The results of our model present better indicators of the RMSE and Ratio Theil's for the predictive evaluation period of twelve months. Furthermore, the straightforward interpretation of the model and the high descriptive capacity of the model allow economic agents to make efficient decisions.

Keywords: Time Series; forecasting; bernoulli; ratio theil; generalised method of moments

check for updates

Citation: Ruiz Reina, M.Á. Bernoulli Time Series Modelling with Application to Accommodation Tourism Demand . *Eng. Proc.* **2021**, *5*, 17. https://doi.org/10.3390/engproc2021005017

Published: 28 June 2021

Publisher's Note: MDPI stays neutral with regard to jurisdictional claims in published maps and institutional affiliations.

1. Introduction

Statistical Learning is a branch of science that is based on learning patterns and identifying structures in data collection. Researchers develop theories using algorithms from Statistics, Mathematics, Machine Learning, Artificial Intelligence, Deep Learning, or mixed models. The applications of these methodologies are fundamental tasks of the description of the study and forecasting. The main difference, according to the statistical analysis, is the lack of a prior assumption of information and that knowledge is obtained from the data. In this paper, we will focus on the connection between Statistical Learning Theory and Econometrics [1].

The methodologies based on the use in the measurement of uncertainty can be classified into three broad categories: survey-based, model-based, and using economics and financial indexes as proxies [2]. Historically, the development of uncertainty measures has been based on the study of variance and its main distribution moments. The use of Entropy as an information measure has led to contributions in the field of uncertainty study [3]. Information Theory supposes the ordering of the results and the derivation in new conclusions. Maximum Entropy expresses the greatest uncertainty concerning the set of information analyzed [4]. Entropy, based on the Shannon Entropy concept [5], is a powerful tool for approximating exponential distributions and groups of families [6]. Despite the versatility and flexibility of the uncertainty study, it has not been widely used in empirical economic studies. The reason for this may be that the Statistical Learning approach has not been developed following an orthodox approach. In this study, we propose a sequential method for identifying uncertainty patterns in dynamic decision-making by agents based on an uncertainty function that we will call Bernoulli Time Series Modeling (BeTSM). The empirical application of this work is in the prediction of tourist hotel accommodation in Spain for decription (January 2001 to December 2018) and the out-sample period January to December 2019.

A theoretical framework is developed, and the dataset used to obtain the result was the case of Tourism markets for accommodation in Spain. In particular, we will model the decision of tourist accommodation in choosing apartments versus hotels. On a monthly database, the National Statistical Institute (INE) of Spain offers statistics related to tourism based on the National Survey of Tourist Accommodation. This database has been important in the study of the tourism market, assuming that this study is a methodological contribution on the disaggregated behaviour of uncertainty and the study of unobserved components of the Time Series [7]. The empirical results reveal an interesting cyclical movement of seasonality in decision-making. In the training period between January 2001 and December 2018, we have observed repetitive patterns: in the low season months for Spanish tourism there was less uncertainty and in the months of high demand, there was greater uncertainty in tourist decision-making. The study and description with forecasting tasks imply contributions for researchers or policy-makers.

In this article, we extend the use of BeTSM to a causality model with the use of the uncertainty factor based on the variance of BeTSM. We will call the uncertainty factor Bernoulli Time Series Uncertainty Function (BeTSUF), and it will be inserted in the predictive model called log-log BeTSUF. The easy interpretation with elasticities and the predictive capacity characterizes this method. Due to the simultaneous causality that occurs in the causal model, we will work with the Generalised Method of Moments corrected by the weighting of the Heteroskedasticity and Autocorrelation Consistent matrix (GMM + HAC-Newey-West). This method of estimation, based on a matrix of instruments, allows obtaining consistency properties of the estimated parameters. The results of our forecasting model improve the data of models contrasted in the tourism forecasting literature such as the Entropy model [8], Seasonal Autoregressive Integrated Moving Average (SARIMA) [9] and Autoregressive Distributed Lags extended to Seasonality (ARDL + Seasonality) [10]. The results of Ratio Theil's ($RT's\ U_1$) verify these empirical results. In this paper, we work with models with Seasonality mainly because previous studies of uncertainty analysis have demonstrated their existence [8].

The remainder of this investigation is as follows: Section 1.1 provides a review of the existing literature on the forecasting Tourism; in Section 2, the theoretical methodology is developed; in Section 3, data analysis of Open Data sources is done, as is the application of the modeling; Section 4 is dedicated to the main conclusions obtained after applying the methods proposed. Finally, bibliographic references are shown.

1.1. Literature Review

This subsection cites and reviews the relevant literature and the most-used models in hotel accommodation forecasting from the last 50 years. There are great reviews of the literature that highlight the predictive capacity of the models in Time Series, Econometric Modeling, Neural Networks or other relevant frameworks in the tourism field [11–14]. In the big data field, there are also several reviews applied to tourism forecasting [15–17]. As we referred to before, three forecasting models are extracted from the review that we will use as a comparison for our contribution to the literature: Entropy Model, Seasonal Autoregressive Integrated Moving Average (SARIMA), and Auto-regressive Distributed Lags extended to Seasonality (ARDL + Seasonality).

In our work, we performed measurements for binary choices of tourist accommodation. The use of a binary choice series can occur in many areas where the temporary problem to solve could be used in chemical, industrial, or socio-economic processes.

Some discrete Time Series methods, such as the Poisson distribution approach (model for counts), or continuous methods with a constant coefficient of variation (e.g., gamma) have been developed [18] for the use of clinical trial comparing the evaluation of logistic regression and Cox Regression with binary results in a fixed period [19]. For a better reading of these processes, deeper readings are recommended [20].

In the area where we will develop our empirical study, tourist accommodation markets and their decisions are unexplored using Bernoulli distribution in Time Series as far as our knowledge reaches. The development of the BeTSM is a contribution to the literature applied to Social Sciences. The crossover study of tourist accommodation in Hotels and the appearance of a competitor, such as a tourist apartment, has not been widely addressed in the measurement of final accommodation decisions. Researchers on tourism accommodation markets have focused their attention on the appearance since the global crisis of 2008 in studies of applications of apartment tourist offers such as Airbnb [21], apartment prices [22], the quality of accommodation services [23], and the effect of images on the final accommodation decision [24]. For more detail on forecasting and tourist accommodation, the reader has bibliographic reviews of papers at the beginning of this section.

In the field of Statistical Learning applied to the measurement of uncertainty in tourist accommodation, the introduction of Entropy in decision-making stands out. Of particular interest is the use of the Shannon Entropy dynamic to quantify the randomness in the decision between tourist accommodation in apartments and hotels. The authors highlight the descriptive and predictive goodness compared to the SARIMA predictive models, the measurement of the improvement in forecasting capacity is carried out with $RT's$ U_1 [8]. This relative ratio can be classified in the measurements of goodness of the ex-post prediction by its interpretability [25]. It should be noted that previous studies of Entropy applied to tourism reveal a cyclical behaviour compatible with seasonal flows [8].

From the reviewed literature, we observe in the empirical results section that our model shows improvements in the forecasts made for the Spanish hotel market. In the next sections, we will detail the theoretical modeling that we will apply in later sections.

2. Methods

In this methodological section, we will focus on the theoretical development of BeTSM. This modeling allows the descriptive, control and forecasting tasks to be carried out on events with two possible Time Series results. Once we have described the modeling of the temporal choice options, we will work with a log-log model of Time Series to perform forecasting. For this, we will introduce an uncertainty factor described by Bernoulli Time Series Uncertainty Function (BeTSUF). The inclusion of this factor implies simultaneous causality for the log-log model, violating the usual assumption of exogeneity in econometric models. We propose the GMM + HAC-Newey-West matrix. Forecasting tasks will be compared with automatic TRAMO-SEATS for SARIMA models [26] and causality models such as Autoregressive Distributed Lags Extended to Seasonality, in addition to the causality model with Entropy factor [27]. For the evaluation of the prediction, we propose the Root Mean Squared Error (RMSE) criterion and the relative dimensionless criterion of $RT's$ U_1 [10]. In the following paragraphs, we will describe the application methodology in the empirical section.

2.1. Bernoulli Time Series Modeling

In this subsection, we will define a binary decision mathematically over time. Suppose we are in a mutually exclusive and binary random situation in a time $t = 1, \ldots, T$. For the application of our model in real cases, we will assume that in each monthly period, the tourist market decides between staying in a hotel or in a tourist apartment.

Let us consider that the temporary binary realization takes a value of zero or one, assuming that each temporary decision is individual and independent of the previous one. In a period the Bernoulli density function $X_t \sim Be\,(p_t)$ could be expressed as follows:

$$f(x_t) = {p_t}^{x_t}(1 - p_t)^{1-x_t};\, x_t = \{0,1\} \tag{1}$$

Given the values of x_t in each time t, the formulation would be:

$$f(x_t, p_t) = \begin{cases} 1 - p_t & for \quad x_t = 0 \\ p_t & for \quad x_t = 1 \\ 0 & in\ other\ cases \end{cases} \tag{2}$$

Probability of a successful event is defined (It would be expressed by the number of times "n_t" or "m_t" that an event occurs $\forall\, t$ and the number of possible cases $(n_t + m_t)$ in that period. Alternatively, we can define the opposite event $1 - p_t = m_t/(n_t + m_t)$. In our work "$n_t$" represents numbers of overnight hotels and "m_t" represents te numbers of overnight apartments $p_t = n_t/(n_t + m_t)\ \forall\ t$. For our work, we propose a chronologically ordered distribution of independent random variables called Bernoulli Time Series Modeling (BeTSM), proportioning information from each period, t, on the probability of an event.

For this work, we are interested in measuring the uncertainty in each period t; for this, we will use the contemporary variance of each event defined by the expression $\text{var}[x_t] = p_t(1 - p_t)$. Situations of uncertainty can be summarized with a minimal variance $\text{var}[x_t] = 0$ if $p_t = 0$ and a maximum variance $\text{var}[x_t] = 0.25$ if $p_t = 0.5$.

The chronologically ordered distribution of $\text{var}[x_1], \text{var}[x_2], \ldots, \text{var}[x_t]$ is named the Bernoulli Time Series Uncertainty Function (BeTSUF).

In the example that we will develop in the empirical section, we will define the probability of success equal to the one for accommodation in hotels. Otherwise, we will consider that accommodation is produced in a tourist apartment. The sequence of all data collected chronologically will assume that the variance of this series will determine our Time Series (BeTSM) and the measurement of uncertainty. The ordering of the sequences of variances will represent what we have theoretically called BeTSUF.

2.2. Log-Log Modeling BeTSUF: Estimated by Generalized Method Moments HAC-Newey-West (GMM + HAC-Newey-West)

In a random context, we propose the introduction of the BeTSUF to carry out forecasting and control tasks. A statistical problem generated by the use of the uncertainty factor is the endogeneity of the regressors due to simultaneous causality of variables, not fulfilling the exogeneity and relevance conditions usually required for the estimation by the Instrumental Variables method. Furthermore, in the case of the existence of heteroscedasticity, we find a problem of efficiency in the parameters estimated. For this, we propose the estimation method Generalized Method of the Moments with the efficient residual matrix of Heteroskedasticity and Autocorrelation Consistent (HAC-Newey-West). HAC-Newey-West estimators of the variance-covariance matrix circumvent this issue [28]. For the theoretical development, we will rely on matrix expressions. Our modeling would be as follows:

$$Y = X\beta + BeTSUF\delta + \varepsilon \tag{3}$$

where X and BeTSUF are the matrices of explanatory variables (endogenous and exogenous expressed in logarithms with base 10). β and δ are the vector of parameters to be estimated consistently through GMM + HAC-Newey-West. For this estimation, it is necessary to use a list of instruments. Z is the matrix of instruments that must satisfy the relevance condition ($\text{cov}(X, Z) \neq 0$) and exogeneity of the instruments ($\text{cov}(Z, \varepsilon) \neq 0$). The range condition must be fulfilled for the model to be at least identifiable [29]. The estimated parameters are obtained from the following expression:

$$\widehat{\beta}^{GMM} = \left(X'Z\widehat{\Omega}^{-1}Z'X\right)^{-1}\left(X'Z\widehat{\Omega}^{-1}Z'Y\right) \tag{4}$$

The matrix of the residuals is defined $\widehat{\Omega} = \sum_{t=1}^{T} Z_t Z_t' \widehat{\varepsilon}_t^2$, which allows us to obtain the estimators $\widehat{\beta}^{GMM}$ efficiently. In testing the orthogonality of the instruments (null hypothesis $E(Z_t\varepsilon_t) = 0$), we use the test statistic $J^{GMM} = \left(Z'\widehat{\varepsilon}^{GMM}\right)'\widehat{\Omega}^{-1}\left(Z'\widehat{\varepsilon}^{GMM}\right)/t$.

The residuals of the statistic are obtained from the estimation by GMM + HAC-Newey-West $\widehat{\varepsilon}^{GMM} = Y - X\widehat{\beta}^{GMM} - BeTSUF\,\widehat{\delta}^{GMM}$. Finally, the asymptotic distribution of the statistic is $J^{GMM} \underset{a}{\sim} \chi^2_{m-k}$ where m is the number of instruments and k is the number of endogenous regressors. Once the estimation of the β and δ parameter vectors has been carried out by the GMM + HAC-Newey-West, we can guarantee the asymptotic consistency property of the estimators [30,31].

For the empirical application, our dependent variable will be the number of hotel overnight stays. The explanatory variable will be the number of accommodations in tourist apartments and also the uncertainty factor BeTSUF. In subsequent sections, we will work with the model expressed in logarithms called log-log BeTSUF.

2.3. Accuracy of the Predictive Capacity of the Models

We propose an evaluation for the time horizon $h = 12$ with the value predicted $\hat{y}_{t+h}$ and real value y_{t+h}. Specifically, we use two model selection criteria based on the prediction; on the one hand, we will use the Root Mean Squared Error (RMSE) [25]:

$$RMSE = \sqrt{\frac{\sum_{h=1}^{H} (\hat{y}_{t+h} - y_{t+h})^2}{h}} \quad (5)$$

On the other hand, we will propose the relative criterion $RT's\ U_1$, which is designed to perform model comparisons for prediction periods with the time horizon h. The benchmark is based on the concept of inequality of Theil U_1 [32] and developed for forecastings comparisons between the results of modeling [10]:

$$U_1 = \frac{\left[\frac{1}{h}\sum_{i=1}^{N}(y_{T+h} - \hat{y}_{T+h})^2\right]^{1/2}}{\left[\frac{1}{h}\sum_{i=1}^{N}(y_{T+h})^2\right]^{1/2} + \left[\frac{1}{h}\sum_{i=1}^{N}(\hat{y}_{T+h})^2\right]^{1/2}} \quad (6)$$

$$RT's_{y_{it},y_{jt}} = \frac{U_1^{y_{it}}}{U_1^{y_{jt}}} \quad (7)$$

The values of the $RT's\ U_1$ will determine which model has the most significant predictive capacity if it is equal to one; the models i and j will present the same predictive power. For values greater than one, the numerator model will show a worse predictive capacity than the denominator. For values between zero and one, the numerator model will present a better predictive capacity. In the next empirical section, we will show a comparative table with the most outstanding results in an annual forecast.

3. A Case Study in the Social Sciences: The Dichotomy of Choice between Hotels and Tourist Apartments

In this section, we divide two subsections: on the one hand, data and correlations, and on the other hand, empirical results. The first part presents the variables under study and the instruments for the estimation of elasticities; in the second subsection, we apply modeling to these data. For our analysis, we have modeled the contemporary choice in the field of empirical application in the tourism sector. The objective is to analyze how the probabilities of accommodation in one place or another are distributed through the market for hotel demand and tourist apartments in Spain. With the temporal analysis, we will observe how tourists in the Spanish market reflect their housing interests in terms of probabilities.

3.1. Data and Correlations

In this subsection, we will carry out an analysis of the modeling presented in the previous sections. In particular, Open Data resources available in the INE for the application of the statistical model in the field of social sciences. We will consider a monthly training analysis period from January 2001 to December 2018, the evaluation of the predictive capacity will be carried out for all the months of 2019. The descriptive statistics are those shown in Table 1.

Table 1. Descriptive Statistics (January 2001 to December 2018). Own Elaboration.

	y_t	x_t	$BeTSUF_t$
Mean	22,934,393	5,891,936	0.163905
Median	21,721.214	4,858,973	0.162965
Maximum	46,657,187	12,520,497	0.221206
Minimum	9,797,644	3,302,242	0.118740
Std. Dev.	9,463,750	2,433,807	0.023312
Skewness	0.565925	1,234,384	0.312166
Kurtosis	2.330	3.408	2.441
Observations	216	216	216

In Table 1, we identify the variable y_t as the number of hotel accommodations, the variable x_t as the number of accommodations in tourist apartments and $BeTSUF_t$ is the uncertainty factor described in the methodological section.

The descriptive data with 216 observations without missing values are for the training period and the aggregate data from the Spanish Tourism market. The descriptive statistics in the table indicate the maximum values of the accommodations found at their maximum values in August 2017. On the other hand, the maximum value of the uncertainty factor is given in January 2001. The minimum amounts are different chronologically for the different variables, for the hotel demand Time Series it occurs in December 2001, for the tourist apartments demand it happened in January 2010 and for the uncertainty factor in May 2012. As for the Skewness and Kurtosis, we can highlight that the three series present positive asymmetry and this could be determined by the strong seasonality where the predominant months are June, July and August compared to the remaining nine that generally show lower values.

Referring to the methodological section, with the modeling that we present, we must use instrumental variables for the variables of our model due to simultaneous causality. In the following Table 2, the correlations between the explanatory variables of our model are presented $(x_t, BTSUF_t)$ and the list of instruments: rural apartments (z_{1t}, z_{1t-1}) and accommodation in campsites (z_{2t}, z_{2t-1}).

Table 2. Cross correlations for explanatory and instruments variables. Sample January 2001–December 2018. Own Elaboration.

	x_t	$BTSUF_t$	z_{1t}	z_{2t}	z_{1t-1}	z_{2t-1}
x_t	1.00 (----)					
$BTSUF_t$	0.15 (0.03)	1.00 (----)				
z_{1t}	0.74 (0.00)	−0.24 (0.00)	1.00 (----)			
z_{2t}	0.94 (0.00)	0.03 (0.61)	0.84 (0.00)	1.00 (----)		
z_{1t-1}	0.38 (0.00)	−0.39 (0.00)	0.49 (0.00)	0.33 (0.00)	1.00 (----)	
z_{2t-1}	0.66 (0.00)	−0.13 (0.06)	0.53 (0.00)	0.56 (0.00)	0.84 (0.00)	1.00 (----)

In Table 2 of cross-correlations, we can observe that all the instruments meet the relevance conditions since all the instrumental variables are correlated with the regressors. In Table 2, we find the value of the statistic and in parentheses the p-value under the hypothesis of no correlation between explanatory variables and the list of instruments. The results present all p-values less than 0.05, the correlation between contemporary camping sites and the uncertainty factor (0.61) being the only one greater. For the rest of the lagged variables, the relevance assumption is fulfilled. Taking this matrix into account, we can consider that a priori the list of instruments is valid for estimating through the GMM + HAC-Newey-West method.

3.2. Empirical Results

In this subsection, we work with the application of the data collected to the modeling log-log BeTSUF described in the methodological section. Given the results of the correlation matrix in Table 2, we proceed to carry out the estimation through the GMM + HAC-Newey-West. The estimation and training period of the model is for 216 months (from 2001 to 2018). The purpose of establishing an analysis period with sample prediction is to obtain a robust model to face a prediction scenario with the most significant guarantees.

In our model (8) the parameters estimated should be interpreted as elasticities. From the results obtained, we can verify the signature of the estimated parameters, the contrast z-statistic obtained in the consistent HAC matrix is shown in parentheses. Values greater than ± 2 imply that parameters are significant in the modeling. The modeling of resids present a white noise structure with a Seasonal Autoregressive structure $SAR(1,12)$. Given this modeling log-log BeTSUF, we can highlight the high explanatory capacity $R^2 = 0.998$.

In this case, the model presents overidentification; the contrast J allows us to contrast the exogeneity of the instruments. Taking into account that the empirical value shows a probability of 0.1517, we cannot reject the hypothesis of exogeneity of the instruments with a 95% confident.

$$\begin{aligned}
&\log y_t = \underset{(-7.1837)}{-1.2077} + \underset{(138.7423)}{0.9869} \log x_t - \underset{(-40.7203)}{1.5317} \log BeTSUF + \widehat{\varepsilon}_t \\
&R^2 = 0.9998;\ \mathrm{Prob}\ J - Statistic = 0.1517 \\
&\widehat{\varepsilon}_t = \underset{(7.0338)}{0.5316}\, \widehat{\varepsilon}_{t-1} + \underset{(7.0338)}{0.8149}\, \widehat{\varepsilon}_{t-12} + \hat{e}_t \\
&Sample\ of\ Estimation:\ 2001.1\ to\ 2018.12 \\
&Instruments\ list: \log(z_{1t}), \log(z_{1t-1}), \log(z_{2t}), \log(z_{2t-1})
\end{aligned} \tag{8}$$

According to the validation of the model log-log BeTSUF, we proceed to interpret the estimated parameters of the model. The first aspect to highlight is that the signs obtained are as expected; the relationship between hotel accommodation and tourist apartments is positive. The elasticity is 0.9869, which implies a direct relationship between both variables analyzed. Second, the inverse relationship between BeTSUF and the hotel accommodation variable should be highlighted. According to our model, we can interpret that when there is more significant uncertainty, hotel accommodations lose demand in favour of tourist accommodation. In particular, when there is an increase of one per cent in uncertainty, hotel accommodations decrease their demand by 1.5317 ceteris paribus.

After analyzing the descriptive capacity of the model and its validation, we will focus on the forecasting capacity and its comparison with other forecasting models through RMSE and $RT's\ U_1$. Regarding our causal model estimated through GMM weighted by HAC-Newey-West, it is worth highlighting the goodness of the predictive capacity of the model based on the scale of the data.

The forecasting period is between January 2019 to December 2019. According to Table 3, our model log-log BeTSUF presents the minimum values (78,507.36) of the widespread RMSE criterion, giving a better predictive capacity compared to forecasting models. The Entropy model (107,581) showing proximity in terms of the predictive power of a complete cycle of twelve months. The rest of the models present very high values, which are considered worse than the model exposed to our work.

Table 3. Summary of forecasting accuracy (RMSE). Out-Sample training January 2019–December 2019. Own Elaboration.

	log-log *BeTSUF*	***Entropy***	***ADRL+Seasonality***	***SARIMA***
RMSE ($h = 12$)	78,507.36	107,581	1,524,295	1,528,357

As a relative measure of prediction calculation, we observe the $RT's\ U_1$ of the estimated models in Table 4. It should be noted that all are greater than 1, and our benchmark method is the model exposed in our methodological development. The scores obtained for the ADRL + Seasonality and SARIMA methodology are widely worse (19 times worse) than our estimated model with the uncertainty factor. The closest model compared to our benchmark method is that of Entropy, taking a value higher than 1.37 times.

Table 4. Summary of forecasting accuracy ($RT's\ U_1$). Out-Sample training January 2019–December 2019. Own Elaboration.

	log-log *BeTSUF*	*Entropy*	*ADRL+Seasonality*	*SARIMA*
$RT's\ U_1\ (h = 12)$	1	1.3696	19.0210	19.0699

The following Figure 1 shows the predictions with a time horizon of twelve months. From a graphic point of view, it is difficult to differentiate between models, but the use of ratios allows us to quantify the benefits of our model proposed in the methodological section. In Table 4, we can verify that the best model is the log-log BeTSUF. Finally, in the conclusions section, we will specify the advantages, advances and limitations of the use of this proposed methodology.

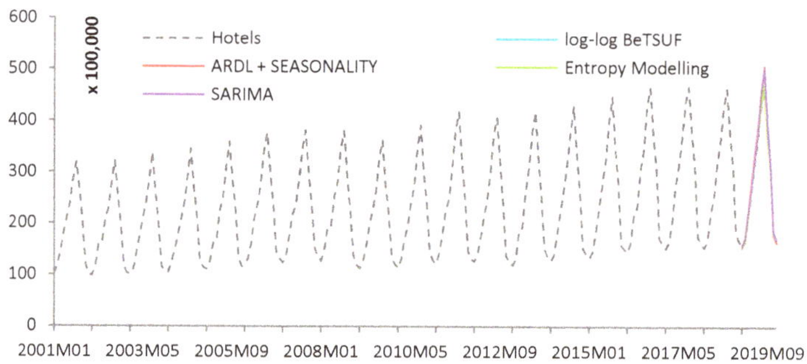

Figure 1. Out-sample forecast hotel accommodation h = 12 (January 2019 to December 2019). Own Elaboration.

4. Conclusions

In the scientific article, we have developed modeling under the assumption of BeTSM with Application to Accommodation Tourism Demand. The objective covered is to create a Statistical Learning approach through the analysis of behavioural patterns of the Hotel accommodation market, in particular, we have modeled the market decision between hotel accommodation and tourist apartments with data from INE Spain. The use of the uncertainty factor described in the methodological section allows us to analyze how unobservable information BeTSUF is transmitted from one variable to another in the sense of causality. In theoretical terms, it assumes that an applied quantity function is used instead of a state counter such as Entropy [8].

Log-log BeTSUF is worth highlighting robust theoretical and empirical properties. The property of consistency of the estimators of the explanatory variables of the model and the efficient use of the residuals to carry out inference tasks with GMM + HAC-Newey-West, giving a solution to the problem of causal simultaneity found in the theoretical modeling.

According to our results, we have found a high explanatory capacity of the model with a high $R^2 = 0.998$. The easy interpretability measured in elasticities, from which we can deduce that the variables of hotel accommodation and tourist apartments present a unitary elasticity (0.9869); the uncertainty factor provides added value in the modeling, knowing that the unitary elasticity of this uncertainty factor allows us to see the transfer of information that occurs from one variable to another. Meaning that an increase of 1% in uncertainty presents a decrease of 1.5317 in demand for hotel accommodation in favour of the apartments; regarding the predictive capacity, our modeling log-log BeTSUF gave the lowest RMSE and the best relative criterion of $RT's\ U_1$ for the models presented in this paper for the same period of forecasting.

This study provides knowledge about the uncertainty that has to be measured. As it was introduced in this article, it is possible to consider this modeling to explain situations in Computer Science, Engineering, Physics, Mathematics and many other applications.

As can be seen, taking into account the possible limitations that the researcher could find with the application of this technique, this article contributes to the scientific literature and adds forecasting tools. The study exposed continues to open the field of forecasting and control for the advancement of these techniques from a theoretical and empirical point of view. This debate should always be based on robustness criteria, which implies sensitivity to changes in specific factors to be tested and insensitive to changes in outliers in practice. The work developed is in the context of uncertainty, and this work has been a contribution to a real forecasting problem [33].

Funding: This research received no external funding.

Institutional Review Board Statement: Not applicable.

Informed Consent Statement: Not applicable.

Data Availability Statement: The datasets generated during and analyzed during the current study are available in the repository (https://drive.google.com/file/d/1T5_dQGxCLlG7IKMOPvgFUD9lsCe344fn/view?usp=sharing (accessed on 22 June 2021)).

Acknowledgments: The author wishes to acknowledge the support given by the University of Malaga. PhD. Program in Economics and Business, effective from 16 July 2013. Especially to Associate Professor Antonio Caparrós Ruiz from the University of Malaga for reviewing this work. Group of research: "SEJ157-INIDICADORES SOCIALES".

Conflicts of Interest: The author declares no conflict of interest.

References

1. Maasoumi, E.; Medeiros, M.C. The link between statistical learning theory and econometrics: Applications in economics, finance, and marketing. *Econ. Rev.* **2010**, *29*, 470–475. [CrossRef]
2. Shoja, M.; Soofi, E.S. Uncertainty, information, and disagreement of economic forecasters. *Econ. Rev.* **2017**, *36*, 796–817. [CrossRef]
3. Maasoumi, E. A compendium to information theory in economics and econometrics. *Econ. Rev.* **1993**, *12*, 137–181. [CrossRef]
4. Jaynes, E.T. Information theory and statistical mechanics. *Phys. Rev.* **1957**, *106*, 620–630. [CrossRef]
5. Shannon, C.E. A mathematical theory of communication. *Bell Syst. Tech. J.* **1948**, *27*, 379–423. [CrossRef]
6. Lehmann, E.L.; Casella, G. *Theory of Point Estimation*, 2nd ed.; Springer: Berlin, Germany, 1998.
7. Harvey, A. Chapter 7 forecasting with unobserved components time series models. *Handb. Econ. Forecast.* **2006**. [CrossRef]
8. Ruiz-Reina, M.Á. Entropy of tourism: The unseen side of tourism accommodation. In Proceedings of the International Conference on Applied Research in Business, Management and Economics, Barcelona, Spain, 12–14 December 2019.
9. Box, G.E.P.; Jenkins, G.M.; Reinsel, G.C. *Time Series Analysis: Forecasting and Control*, 4th ed.; John Wiley & Sons: Hoboken, NJ, USA, 2013.
10. Ruiz-Reina, M.Á. Big data: Does it really improve forecasting techniques for tourism demand in spain? In *International Conference on Time Series and Forecasting*; Godel Impresiones Digitales S.L.: Granada, Spain, 2019; pp. 694–706.
11. Li, G.; Song, H.; Witt, S.F. Recent developments in econometric modeling and forecasting. *J. Travel Res.* **2005**. [CrossRef]
12. Song, H.; Li, G. Tourism demand modelling and forecasting-A review of recent research. *Tour. Manag.* **2008**, *29*, 203–220. [CrossRef]
13. Peng, B.; Song, H.; Crouch, G.I. A meta-analysis of international tourism demand forecasting and implications for practice. *Tour. Manag.* **2014**. [CrossRef]
14. Jiao, E.X.; Chen, J.L. Tourism forecasting: A review of methodological developments over the last decade. *Tour. Econ.* **2019**. [CrossRef]
15. Wu, D.C.; Song, H.; Shen, S. New developments in tourism and hotel demand modeling and forecasting. *Int. J. Contemp. Hosp. Manag.* **2017**, *29*, 507–529. [CrossRef]
16. Mariani, M.; Baggio, R.; Fuchs, M.; Höepken, W. Business intelligence and big data in hospitality and tourism: A systematic literature review. *Int. J. Contemp. Hosp. Manag.* **2018**. [CrossRef]
17. Li, J.; Xu, L.; Tang, L.; Wang, S.; Li, L. Big data in tourism research: A literature review. *Tour. Manag.* **2018**, *68*, 301–323. [CrossRef]

18. Zeger, S.L.; Qaqish, B. Markov regression models for time series: A quasi-likelihood approach. *Biometrics* **1988**, *44*, 1019–1031. [CrossRef] [PubMed]
19. Peduzzi, P.; Holford, T.; Detre, K.; Chan, Y.K. Comparison of the logistic and Cox regression models when outcome is determined in all patients after a fixed period of time. *J. Chronic Dis.* **1987**, *40*, 761–767. [CrossRef]
20. Hung, Y.; Zarnitsyna, V.; Zhang, Y.; Zhu, C.; Wu, C.F.J. Binary time series modeling with application to adhesion frequency experiments. *J. Am. Stat. Assoc.* **2008**, *103*, 1248–1259. [CrossRef] [PubMed]
21. Bakker, M.; Twining-Ward, L. *Tourism and the Sharing Economy: Policy and Potential of Sustainable Peer-to-Peer Accommodation*; The World Bank: Washington, DC, USA, 2018.
22. Portolan, A. The impacts of private accomodation attributes on tourism demand. In *DIEM: Dubrovnik International Economic Meeting*; Sveučilište u Dubrovniku: Dubrovnik, Croatia, 2013.
23. Juaneda, C.; Raya, J.M.; Sastre, F. Pricing the time and location of a stay at a hotel or apartment. *Tour. Econ.* **2011**, *17*, 321–338. [CrossRef]
24. Ert, E.; Fleischer, A.; Magen, N. Trust and reputation in the sharing economy: The role of personal photos in Airbnb. *Tour. Manag.* **2016**, 62–73. [CrossRef]
25. Hyndman, R.J.; Koehler, A.B. Another look at measures of forecast accuracy. *Int. J. Forecast.* **2006**, *22*, 679–688. [CrossRef]
26. Hyndman, R.J.; Khandakar, Y. Automatic time series forecasting: The forecast package for R. *J. Stat. Softw.* **2008**, *27*. [CrossRef]
27. Ruiz-Reina, M.Á. Entropy of Tourism: The Unseen Side of Tourism Accommodation. Available online: https://www.dpublication.com/wp-content/uploads/2019/12/424.pdf (accessed on 22 June 2021).
28. Hayashi, F. *Econometrics*; Princeton University Press: Princeton, NJ, USA, 2000.
29. Chow, G. *Econometrics*; McGraw-Hill Book Company: New York, NY, USA, 1983.
30. Sargan, J.D. The estimation of relationships with autocorrelated residuals by the use of instrumental variables. *J. R. Stat. Soc. Ser. B* **1959**, *21*, 91–105. [CrossRef]
31. Hall, A.R. *Advanced Texts in Econometrics: Generalized Method of Moments*; Oxford University Press: Oxford, UK, 2005.
32. Christ, C.F.; Theil, H. Economic forecasts and policy. *Econometrica* **1962**. [CrossRef]
33. Gill, P.E.; Murray, W.; Saunders, M.A.; Tomlin, J.A.; Wright, M.H.; George, B. Dantzig and systems optimization. *Discret. Optim.* **2008**, 151–158. [CrossRef]

engineering proceedings

MDPI

Proceedings

A Mathematical Investigation of a Continuous Covariance Function Fitting with Discrete Covariances of an AR Process †

Johannes Korte *, Till Schubert, Jan Martin Brockmann and Wolf-Dieter Schuh

Institute of Geodesy and Geoinformation, University of Bonn, 53115 Bonn, Germany; schubert@geod.uni-bonn.de (T.S.); brockmann@geod.uni-bonn.de (J.M.B.); schuh@uni-bonn.de (W.-D.S.)
* Correspondence: korte@geod.uni-bonn.de; Tel.: +49-228-73-3576
† Presented at the 7th International Conference on Time Series and Forecasting, Gran Canaria, Spain, 19–21 July 2021.

Abstract: In this paper, we want to find a continuous function fitting through the discrete covariance sequence generated by a stationary AR process. This function can be determined as soon as the Yule–Walker equations are found. The procedure consists of two steps. At first the inverse zeros of the characteristic polynomial of the AR process must be fixed. The second step is based on the fact that an AR process can also be seen as a difference equation. By solving this difference equation, it is possible to determine a class of functions from which a candidate for a continuous covariance function can be determined. To analyze if this function is applicable as a positive definite covariance function, it is analyzed mathematically in view of the power spectral density compared to the characteristics of the power spectral density for the discrete covariances. Then it is shown that this function is positive semi-definite. At the end, a simulation of a stationary AR(3) process is elaborated to illustrate the derived properties.

Keywords: AR process; continuous covariance function; Fourier transform; power spectral density; positive definiteness; signal prediction

check for updates

Citation: Korte, J.; Schubert, T.; Brockman, J.M.; Schuh, W.-D. A Mathematical Investigation of a Continuous Covariance Function Fitting with Discrete Covariances of an AR Process. *Eng. Proc.* **2021**, 5, 18. https://doi.org/10.3390/engproc2021005018

Published: 28 June 2021

1. Introduction

In geodesy the observations or analyzed signals are often discrete measurements which repeat at regular distances. For example, deformation observations (repeating in time) or data from satellite missions such as GOCE (repeating in time and space). It is a common way to describe regular and equidistant signals by auto regressive moving average (ARMA) processes [1–5].

Within this contribution we focus on the analysis of the AR part. This AR part defines the causal link between an observation and its predecessors. Additionally, Least Squares Collocation (LSC) (see e.g., [6–8]) and kriging [9,10] are benefiting from the use of AR processes. For example, the inverse of a covariance matrix, based on AR process, is a band matrix witch bandwidth equals the order of the AR process (see e.g., [11]).

However, to activate the full potential of LSC a continuous covariance function is indispensable. With this function it will be possible to predict a pseudo signal between the observations. Furthermore, it is possible to predict the signal outside the observation field and not only for a multiple of the sampling rate. Moreover, refs. [12,13] used a continuous covariance function to switch from one functional to another (Like using sea level heights to calculate sea level changes which are proportional to the stream velocity).

In this paper, we want to find an analytical description of a continuous function fitting through the discrete sequence of covariances generated by any stationary AR process. This function is derived from the coefficients of the AR process and the discrete covariances using a system of equations with a unique solution. The resulting function should be positive definite, and its spectrum is expected to correspond to the spectrum of the discrete AR process. It will turn out that this function is the continuous solution of the difference equation and correctly interpolates the discrete covariance sequence with appropriate basis functions, indeed following sampling theory/convolution theorem. In the end an example is given by a simulated AR process and the accompanying continuous covariance function as well as the two spectra are estimated.

2. Continuous Covariance Function

To find a suitable covariance function for any stationary AR process the definition of AR processes is a good starting point. Especially the transfer from the AR process to the difference equation approach will lead us to the continuous representation we looked for (The following definitions could be found in [14–18]. Here the notation of [17] is used).

2.1. Construction of a Continuous Covariance Function

The process $\mathcal{S}_t$ is called one-dimensional *AR process of order p* (AR(p) process) if it is described by the recursive equation

$$\mathcal{S}_t = \alpha_1 \mathcal{S}_{t-1} + \alpha_2 \mathcal{S}_{t-2} + \ldots + \alpha_p \mathcal{S}_{t-p} + \mathcal{E}_t \tag{1}$$

where $\alpha_1, \alpha_2, \ldots, \alpha_p$ are the coefficients of the AR process and $\mathcal{E}_t$ is a i.i.d. sequence with variance $\sigma_{\mathcal{E}}^2$ [17] (p. 58, Equation (3.4.31)). We assume that $\alpha_p \neq 0$, as otherwise the AR(p) process is also an AR($p-1$) process so that AR(p) is not well-defined (In addition, if $\alpha_p = 0$ some formulas in this paper cannot be used).

An important quantity is the *zeros* (ζ_l) *of an AR(p) process* defined by the zeros of the characteristic polynomial

$$\begin{aligned} \chi(x) &= 1 - \alpha_1 x^1 - \alpha_2 x^2 - \ldots - \alpha_p x^p \\ &= (x - \zeta_1)(x - \zeta_2)\ldots(x - \zeta_p), \end{aligned} \tag{2}$$

see e.g., [17] (p. 58, Equation (3.4.32)).

An alternative definition is given by the *auxiliary equation* if we interpret the AR process as a difference equation which has the general solution (see [19], p. 134, Equation (3.33))

$$\begin{aligned} b(x) &= x^p - \alpha_1 x^{p-1} - \alpha_2 x^{p-2} - \ldots - \alpha_p \\ &= (x - p_1)(x - p_2)\ldots(x - p_p). \end{aligned} \tag{3}$$

These zeros p_l only occur as real values or in pairs of complex conjugated zeros. Bear in mind that the zeros of the characteristic polynomial from Equation (2) are linked to the zeros of the auxiliary equation (cf. Equation (3)) by $\zeta_l = 1/p_l$. AR processes are stationary if and only if the ζ_l are outside the unit circle, such that $|p_l| < 1$. In the following we will restrict to p_l for simplicity.

With this definitions in mind, the discrete covariances Σ_j of an AR(p) process, are linked with each other by the *Yule–Walker (YW) equations* (see e.g., [17], p. 59, Equation (3.4.36)), i.e.,

$$\Sigma_0 = \alpha_1 \Sigma_1 + \alpha_2 \Sigma_2 + \ldots + \alpha_p \Sigma_p \qquad + \sigma_{\mathcal{E}}^2 \tag{4}$$

$$\Sigma_j = \alpha_1 \Sigma_{|j-1|} + \alpha_2 \Sigma_{|j-2|} + \ldots + \alpha_p \Sigma_{|j-p|} \qquad \text{if } j \neq 0. \tag{5}$$

The YW equation of higher order than 0 (Equation (5)) are basically homogeneous difference equations of order p,

$$\Sigma_j - \alpha_1 \Sigma_{|j-1|} - \alpha_2 \Sigma_{|j-2|} - \ldots - \alpha_p \Sigma_{|j-p|} = 0. \tag{6}$$

The general solution to the difference equation can be expressed by the powers of the zeros p_l of the auxiliary Equation (3). The particular solution is fixed by the boundary conditions using the discrete covariances determined from the YW equations (cf. Equation (5)),

$$\Sigma_j = \sum_{l=1}^{p} A_l p_l^{|j|}. \tag{7}$$

Here A_l are coefficients which are complex if and only if the corresponding p_l is complex. Furthermore, if there is a pair of complex conjugated p_l then A_l occur also as complex conjugated pairs (see e.g., [18], p. 134, Equation (3.5.44) or [19], p.163, f.).

At this point a new but now continuous function is defined, which can be seen as the continuous covariance function $\gamma(h) : \mathbb{R} \to \mathbb{R}$ for any AR(p) process,

$$\gamma(h) := \sum_{l=1}^{p} A_l p_l^{|h|}. \tag{8}$$

Here A_l and p_l are the same as in Equation (7), but the domain changed. $j \in \mathbb{N}_0$ is replaced by $h \in \mathbb{R}$.

Attentive readers will have noticed that $\gamma(h)$ is complex if any $p_l \in \mathbb{R}^-$. Then $\gamma(h) \in \mathbb{C}$ for all $h \notin \mathbb{N}_0$. One important convention that will help with this inconsistency is the use of the real part $\mathrm{Re}(\gamma(h))$ of the complex function (see Figure 1). This condition will not have any impact if $\gamma(h)$ is real (what is mostly the case), and furthermore a covariance function for real valued signals is defined to be a function in $\mathbb{R}$ not in $\mathbb{C}$.

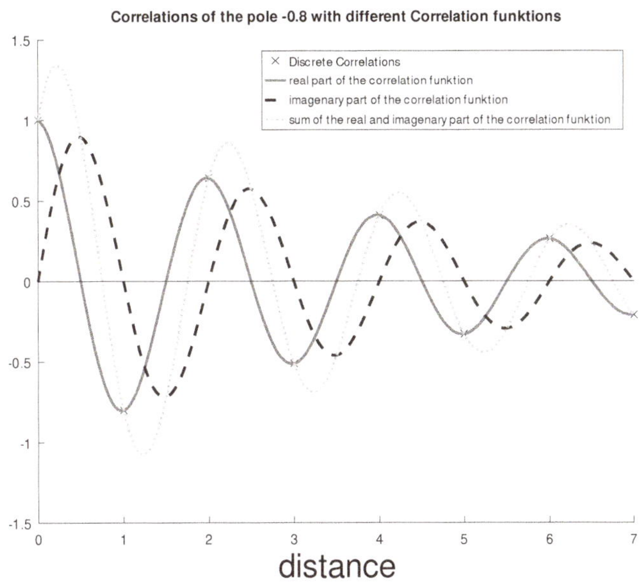

Figure 1. Real part, imaginary part and sum of both parts of a complex covariance function of an AR process with pole -0.8.

2.2. Properties of the Continuous Covariance Function

Since with $\gamma(h)$ from Equation (8) a suitable function is found to fit through the discrete covariances from Equation (7), we want to analyze the power spectral densities of the continuous and the discrete functions. On this basis we can demonstrate that the Fourier transform of the continuous covariance function is positive semi-definite.

Initially the problem restricted to AR processes of order 1 and order 2 with two complex conjugated zeros. On the one hand any AR(p) process can be dissected into a product of AR(1) and AR(2) processes. This is a linear function, so the power spectral function is the product of the corresponding AR(1) processes and AR(2) processes. On the other hand, Equation (8) shows that the covariance function is a weighted sum of the real valued zeros, or pairs of complex conjugated zeros. So, the zeros are also in a linear relation, and so is the Fourier transform. So, it is only necessary to examine the spectrum and the Fourier transform for the first order AR process and second order AR process with two complex conjugated zeros.

For these specific types of AR processes there is an analytical solution to switch from AR coefficients α_l to the zeros p_l (see [20]),

$$\text{for order } p = 1 \qquad \alpha_1 = p_1 \tag{9}$$

$$\text{and for order } p = 2 \qquad \begin{cases} \alpha_1 = p_1 + p_2 \\ \alpha_2 = -p_1 p_2 \end{cases} \quad \text{with } p_1 = p_2^*. \tag{10}$$

2.2.1. Power Spectral Density

The power spectral density for an AR(p) process is well known (see e.g., [16], p. 244, Equation (11.20)) and is described by the transfer function

$$\mathcal{H}^2(\nu) = \frac{\sigma_{\mathcal{E}}^2}{|1 - \sum_{l=1}^{p} \alpha_l e^{-i2\pi\nu l}|^2}. \tag{11}$$

In consideration of Equations (9) and (10) the power spectral density for any AR(1) process and AR(2) process with complex conjugated zero can be calculated explicitly. So, the power spectral density AR(1) process is generated via

$$\mathcal{H}^2(\nu) = \frac{\sigma_{\mathcal{E}}^2}{1 - 2p_1 \cos(2\pi\nu) + p_1^2}. \tag{12}$$

For the AR(2) process with zeros $p_1 = p_2^*$ the power spectral density is a little more complicated and turns out to be

$$\mathcal{H}^2(\nu) = \frac{\sigma_{\mathcal{E}}^2}{1 - 2(p_1+p_2)\cos(2\pi\nu) + p_1p_2(2 + 2\cos(4\pi\nu) - (p_1+p_2)\cos(2\pi\nu) + p_1p_2) + p_1^2 + p_2^2}. \tag{13}$$

Using the Fourier transform of the covariance function $\gamma(h)$

$$\Gamma(\nu) := \mathcal{F}\{\gamma(h)\}(\nu) = \sum_{l=1}^{p} A_l \frac{-2\ln(p_l)}{(\ln(p_l))^2 + (2\pi\nu)^2} \tag{14}$$

is an alternative way to derive the power spectral density (For further derivations of the Fourier transform see Appendix A). However, $\mathcal{H}^2(\nu) \neq \Gamma(\nu)$. Especially Equation (11) shows $\mathcal{H}^2(\nu)$ as a function whose only parameter ν arise as power of the complex function $e^{-i2\pi l}$. Therefore $\mathcal{H}^2(\nu)$ is a repetitive function with period 1. In contrast, Equation (14) shows $\Gamma(\nu)$ is aperiodic function with $\lim_{\nu\to\infty} \Gamma(\nu) = 0$. To understand this circumstance two theorems are of importance:

1. The discrete covariances of an AR(p) process (Σ_j) are equivalent to the product of the Dirac comb with the continuous covariance function $\gamma(h)$.
2. The convolution theorem shows that multiplication in time domain results in convolution in frequency domain.

Combining these two pieces of information shows indeed that $\mathcal{H}^2(\nu) \neq \Gamma(\nu)$ but

$$\mathcal{H}^2(\nu) = \Gamma(\nu) * \left(\sum_{k=-\infty}^{\infty} \delta(x-k) \right) \tag{15}$$

where $\sum_{k=-\infty}^{\infty} \delta(x-k)$ is the Dirac comb of distance 1 and $\Gamma(\nu) * \left(\sum_{k=-\infty}^{\infty} \delta(x-k)\right)$ is the convolution of $\Gamma(\nu)$ and the Dirac comb. The transitions from continuous functions to discrete sequences as well as the resulting Fourier transforms are shown in Figure 2 (For a more detailed method of calculation for AR(1) and AR(2) processes see Appendix C).

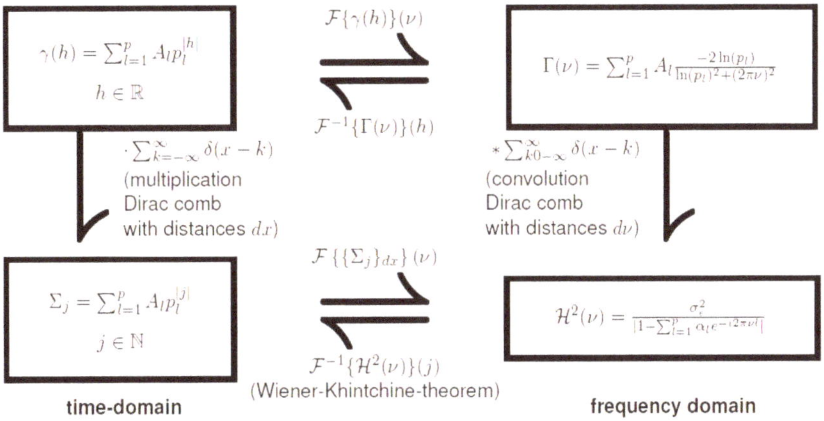

Figure 2. Magic square for the convolution of a continuous covariance function with a Dirac comb.

2.2.2. Positive Semi-Definite Function

Equation (15) shows that knowing if $\mathcal{H}^2(\nu)$ is positive semi-definite is not sufficient to guarantee that the Fourier transform of the continuous function $\gamma(h)$ is positive semi-definite too. Therefore, the explicit Fourier transforms of the AR(1) and AR(2) process are derived here. For the case of the AR(1) process it is easy to see that for the Fourier transform of the covariance function $\gamma(h)$

$$\Gamma(\nu) = \frac{\sigma_{\mathcal{E}}^2}{1-p^2} \frac{-2\ln(p)}{(\ln(p))^2 + (2\pi\nu)^2} > 0 \quad \forall \nu \in \mathbb{R}. \tag{16}$$

holds (For the derivation of this formula see Appendix B.1). Neither the squared terms could be less than 0 nor $1-p^2$ or $-\ln(p)$ due to the fact that p lies within the unit circle.

For the AR(2) process things are not that obvious. In Appendix B.2 it is demonstrated that the Fourier transform of $\gamma(h)$ is

$$\Gamma(\nu) = 2\mathrm{Re}\left(\frac{-\sigma_{\mathcal{E}}^2 p_1}{(p_2-p_1)(1-p_1^2)(1-p_1p_2)} \frac{-2\ln(p_1)}{(\ln(p_1))^2 + (2\pi\nu)^2}\right). \tag{17}$$

To work with complex valued fractions, it is necessary to eliminate the imaginary part in the denominator. This is done by multiplying each term of the sum with the complex conjugated denominator divided by itself. Afterwards it is simple to pick the real part. To simplify the formula, we use the polar coordinates $p_1 = re^{i\phi}$, and $p_2 = re^{-i\phi}$ with $0 < r < 1$ and $0 \le \phi \le \pi$. So, it can be shown that the numerator is positive ($\Gamma(\nu) \ge 0$) if and only if

$$\underbrace{-\ln(r)\coth(-\ln(r))}_{:=f(r)} \ge \underbrace{\phi\cot(\phi)}_{:=g(\phi)} \tag{18}$$

The function $f(r)$ and $g(\phi)$ are displayed in Figures 3 and 4 (Since $g(\phi) = g(-\phi)$, the negative values of ϕ are not needed). On the one hand $f(r)$ is a declining function with infimum 1. On the other hand, the function $g(\phi)$ is also declining, with a maximum of 1. Thus, infimum $f(r) \ge \max g(\phi)$, which evaluate that Equation (18) is always satisfied.

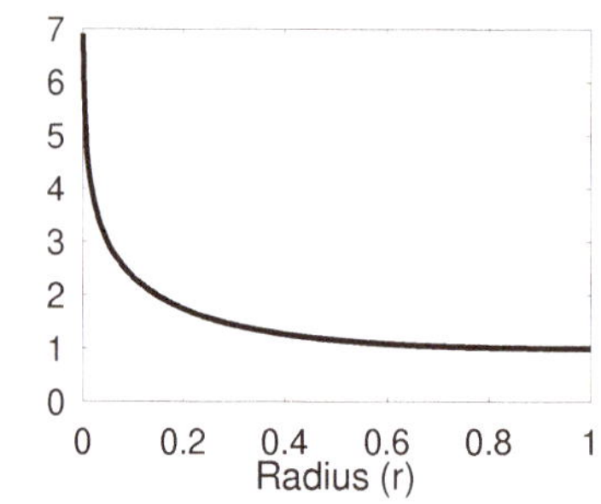

Figure 3. The function $f(r) = -\ln(r)\coth(-\ln(r))$ for $0 < r < 1$.

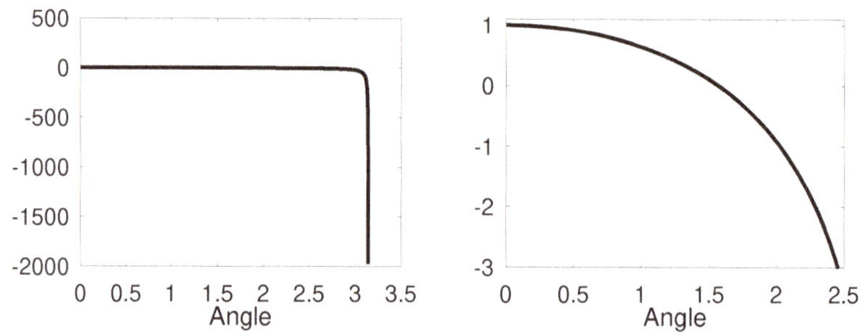

Figure 4. The function $g(\phi) = \phi\cot(\phi)$ for $0 \leq \phi \leq \pi$, and an enlarged section of the beginning.

3. Simulation

To visualize the results from Section 2 of an AR(3) process with two complex conjugated zeros was generated as an example. First to guarantee stationarity the roots are chosen as

$$\begin{bmatrix} p_1 \\ p_2 \\ p_3 \end{bmatrix} = \begin{bmatrix} -0.252 - 0.126i \\ -0.252 + 0.126i \\ 0.306 \end{bmatrix}.$$

Let the variance of the white noise be $\sigma_\epsilon^2 = 1$. After deriving the coefficients α_l, $l \in \{1, 2, 3\}$, using Equation (3), we estimate the discrete covariances (Σ_j) of the AR(3) process by the reorganized YW equations (see [21], p. 32, Equation (183)). With Equation (8) a continuous function $\gamma(h)$ is fitted through the discrete covariances (see the left upper corner of Figure 5). Subsequently the power spectral density is set first by Equation (11) using the coefficients α_l and secondly by Equation (14). For the second step the coefficients A_l are estimated by solving the system of equations

$$\begin{bmatrix} \Sigma_0 \\ \Sigma_1 \\ \Sigma_2 \end{bmatrix} = \begin{bmatrix} 1 & 1 & 1 \\ p_1 & p_2 & p_3 \\ p_1^2 & p_2^2 & p_3^2 \end{bmatrix} \begin{bmatrix} A_1 \\ A_2 \\ A_3 \end{bmatrix}. \quad (19)$$

Here the zeros p_l and covariances Σ_l are known. Each row represents Equation (7) for $l \in \{1, 2, 3\}$. The coefficients A_l are used in Equation (8) to estimate the power spectral density of the continuous covariance function. It must be mentioned that this example is an extreme one where the Fourier transform of the continuous covariance function has high values for frequencies higher the Nyquist frequency $\nu_n = 0.5$. The Fourier transform of the continuous covariance function is not periodic at all (compare right upper corner of Figure 5). However, periodicity is the characteristic of the spectral density of a discrete AR process. Therefore, the periodicity is a result of the convolution of the Dirac comb and $\Gamma(\nu)$.

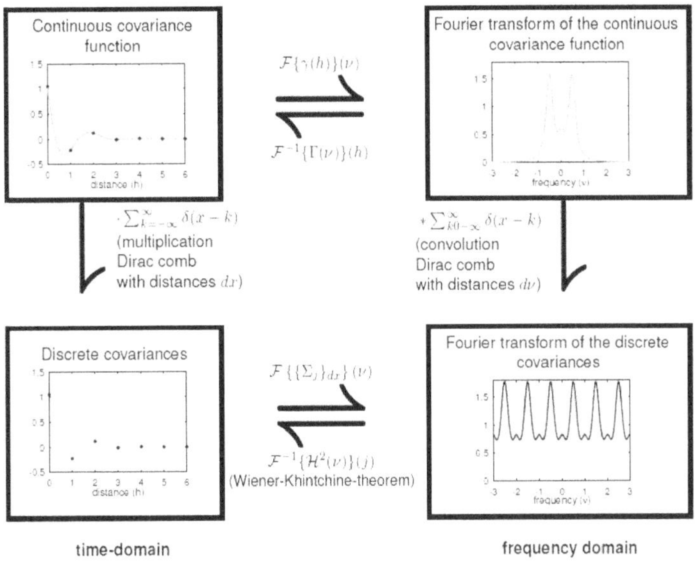

Figure 5. Magic square for a convolution of a continuous covariance function of an AR(3) process with a Dirac comb.

4. Conclusions and Outlook

In this paper, it was shown that the choice of a valid continuous covariance function for AR(p) processes is given by the function

$$\gamma(h) = \sum_{l=1}^{p} A_l p_l^{|h|}.$$

Here $h \in \mathbb{R}$ is the lag, p_l are the roots of the characteristic polynomial, and A_l follows from the unique solution of Equation (14) for p arbitrarily chosen discrete covariances Σ_{j_1}, Σ_{j_2}, ..., Σ_{j_p} (with $j_i \in \mathbb{N}_0$, and $j_i \neq j_k \Leftrightarrow i \neq k$):

$$\begin{bmatrix} A_1 \\ A_2 \\ \dots \\ A_p \end{bmatrix} = \begin{bmatrix} p_1^{j1} & p_2^{j1} & \dots & p_p^{j1} \\ p_1^{j2} & p_2^{j2} & \dots & p_p^{j2} \\ \dots & \dots & \dots & \dots \\ p_1^{jp} & p_2^{jp} & \dots & p_p^{jp} \end{bmatrix}^{-1} \begin{bmatrix} \Sigma_{j1} \\ \Sigma_{j2} \\ \dots \\ \Sigma_{jp} \end{bmatrix}.$$

Due to the convolution theorem, the power spectral density of $\gamma(h)$ might be different to the power spectral density of the discrete AR(p) process. Nevertheless, the proof was given that $\gamma(h)$ is still positive semi-definite (cf. Section 2.2.2), and consequently meets all conditions for a suitable covariance function. The Fourier transforms $\Gamma(\nu)$ and $\mathcal{H}^2(\nu)$ may not vary much for $\nu \in [-1, 1]$ and the simulation (see Section 3) is an extreme example. Anyway $\gamma(h)$ is an exponential function, so it is easy to use it as functional for covariance function propagation or LSC.

In further works the continuous covariance function $\gamma(h)$ could be extended to a function for an autoregressive moving average (ARMA) process to examine its properties. It is not yet demonstrated that the oscillation of $\text{Re}(\gamma(h))$ leads towards the minimal frequency if there is a real negative zero ($p_l = 0$ for any l).

Author Contributions: Conceptualization, methodology, formal analysis, investigation, software, validation, visualization, data curation, writing—Original draft preparation: J.K., writing—Review and editing: J.K., T.S., J.M.B. and W.-D.S., funding acquisition, project administration, resources, supervision: W.-D.S. All authors have read and agreed to the published version of the manuscript.

Funding: This research is funded by the Deutsche Forschungsgemeinschaft (DFG, German Research Foundation)—Grant No. 435703911 (SCHU 2305/7-1 'Nonstationary stochastic processes in least squares collocation—NonStopLSC').

Conflicts of Interest: The authors declare no conflict of interest.

Appendix A. General Fourier Transform of an AR(p) Process

In this part the Fourier transform of a function $\gamma(h) = \sum_{l=1}^{p} A_l p_l^{|h|}$ with $h \in \mathbb{R}$ is computed,

$$\Gamma(\nu) = \int_{-\infty}^{\infty} \sum_{l=1}^{p} A_l p_l^{|h|} e^{i2\pi\nu h} dh \qquad = \sum_{l=1}^{p} A_l \underbrace{\int_{-\infty}^{0} p_l^{-h} e^{i2\pi\nu h} dh}_{\frac{1}{-\ln(p_l)-i2\pi\nu}} + \underbrace{\int_{-\infty}^{0} p_l^{h} e^{i2\pi\nu h} dh}_{\frac{1}{-\ln(p)+i2\pi\nu}}$$

$$= \sum_{l=1}^{p} A_l \frac{-2\ln(p_l)}{(\ln(p_l))^2 + (2\pi\nu)^2}$$

Please note that A_l and p_l might be complex, but this will not have any influence on the equations.

Appendix B. Explicit Fourier Transform of the AR(1) Process and AR(2) Process with Two Complex Conjugated Zeros

In this section, the explicit Fourier transform $\Gamma(\nu)$ of the continuous covariance function $\gamma(h)$ for the orders $p = 1$ and $p = 2$ are given as function of h, and the zeros p_l.

Appendix B.1. Fourier Transform of the Continuous Covariance Function of AR(1) Processes

First the discrete covariance Σ_0 must be computed as function of the zero p_1. This is done by the reorganized YW equations (see [21], p. 32, Equation (183)):

$$\left(\begin{bmatrix} -1 & 0 \\ \alpha_1 & -1 \end{bmatrix} + \begin{bmatrix} 0 & \alpha_1 \\ 0 & 0 \end{bmatrix} \right)^{-1} \begin{bmatrix} -\sigma_\epsilon^2 \\ 0 \end{bmatrix} = \begin{bmatrix} \Sigma_0 \\ \Sigma_1 \end{bmatrix}$$

$$\Leftrightarrow \frac{1}{1-\alpha_1^2} \begin{bmatrix} -1 & -\alpha_1 \\ -\alpha_1 & -1 \end{bmatrix} \begin{bmatrix} -\sigma_\epsilon^2 \\ 0 \end{bmatrix} = \begin{bmatrix} \Sigma_0 \\ \Sigma_1 \end{bmatrix}$$

$$\Rightarrow \Sigma_0 = \frac{\sigma_\epsilon^2}{1-\alpha_1^2}$$

Further Equation (7) gives $\Sigma_0 = A_1$ and in combination with Equation (9) the deduction is

$$A_1 = \frac{\sigma_\epsilon^2}{1-p_1^2}.$$

Insert A_1 in Equation (14) for order $p = 1$ to obtain

$$\Gamma(\nu) = \frac{\sigma_\epsilon^2}{1-p_1^2} \frac{-2\ln(p_1)}{(\ln(p_1))^2 + (2\pi\nu)^2}.$$

Appendix B.2. Fourier Transform of the Continuous Covariance Function of AR(2) Processes with Two Complex Conjugated Zeros

Like in the last subsection the discrete covariances Σ_0 and Σ_1 are computed by the reorganized YW equations:

$$\left(\begin{bmatrix} -1 & 0 & 0 \\ \alpha_1 & -1 & 0 \\ \alpha_2 & \alpha_1 & -1 \end{bmatrix} + \begin{bmatrix} 0 & \alpha_1 & \alpha_2 \\ 0 & \alpha_2 & 0 \\ 0 & 0 & 0 \end{bmatrix} \right)^{-1} \begin{bmatrix} -\sigma_\epsilon^2 \\ 0 \\ 0 \end{bmatrix} = \begin{bmatrix} \Sigma_0 \\ \Sigma_1 \\ \Sigma_2 \end{bmatrix}$$

$$\Leftrightarrow \frac{\sigma_\epsilon^2}{-\alpha_2^2 + \alpha_2^2 + \alpha_2(1+\alpha_1^2) + \alpha_1^2 - 1} \begin{bmatrix} 1-\alpha_2 \\ \alpha_1 \\ \alpha_1^2 - \alpha_2^2 + \alpha_2 \end{bmatrix} = \begin{bmatrix} \Sigma_0 \\ \Sigma_1 \\ \Sigma_2 \end{bmatrix}.$$

With the transformation from α_1, α_2 to p_1, p_2 (cf. Equation (10)) the discrete covariances are set by

$$\Sigma_0 = \frac{-(1+p_1p_2)\sigma_\epsilon^2}{(p_1^2-1)(1-p_2^2)(1-p_1p_2)} \qquad \Sigma_1 = \frac{-(p_1+p_2)\sigma_\epsilon^2}{(p_1^2-1)(1-p_2^2)(1-p_1p_2)}.$$

This time Equation (7) is an equation system in the two variables A_1 and A_2. Here the first and second discrete covariances (Σ_0, Σ_1) are used:

$$\begin{bmatrix} \Sigma_0 \\ \Sigma_1 \end{bmatrix} = \begin{bmatrix} 1 & 1 \\ p_1 & p_2 \end{bmatrix} \begin{bmatrix} A_1 \\ A_2 \end{bmatrix} \qquad \Leftrightarrow \begin{bmatrix} A_1 \\ A_2 \end{bmatrix} = \begin{bmatrix} 1 & 1 \\ p_1 & p_2 \end{bmatrix}^{-1} \begin{bmatrix} \Sigma_0 \\ \Sigma_1 \end{bmatrix}$$

$$\Leftrightarrow \begin{bmatrix} A_1 \\ A_2 \end{bmatrix} = \frac{1}{p_2 - p_1} \begin{bmatrix} p_2 & -1 \\ -p_1 & 1 \end{bmatrix} \begin{bmatrix} \Sigma_0 \\ \Sigma_1 \end{bmatrix}$$

$$\Rightarrow A_1 = \frac{p_2\Sigma_0 - \Sigma_1}{p_2 - p_1}; \qquad \Rightarrow A_2 = \frac{p_1\Sigma_0 - \Sigma_1}{p_1 - p_2}$$

Including the solution for Σ_0 and Σ_1 to obtain A_1 and A_2 as function of p_1 and p_2 leads to

$$A_1 = \frac{-p_1\sigma_\epsilon^2}{(p_2-p_1)(1-p_1^2)(1-p_1p_2)}; \qquad A_2 = \frac{-p_2\sigma_\epsilon^2}{(p_1-p_2)(1-p_2^2)(1-p_1p_2)}$$

Due two $p_1 = p_2^*$, inserting A_1 and A_2 in Equation (14) leads to the sum of two complex conjugated values. This is equally to two times the real part of the complex value:

$$\begin{aligned} \Gamma(\nu) &= A_1 \frac{-2\ln(p_1)}{(\ln(p_1))^2 + (2\pi\nu)^2} + A_2 \frac{-2\ln(p_2)}{(\ln(p_2))^2 + (2\pi\nu)^2} \\ &= 2\mathrm{Re}\left(A_1 \frac{-2\ln(p_1)}{(\ln(p_1))^2 + (2\pi\nu)^2} \right) \\ &= 2\mathrm{Re}\left(\frac{-\sigma_\mathcal{E}^2 p_1}{(p_2-p_1)(1-p_1^2)(1-p_1p_2)} \frac{-2\ln(p_1)}{(\ln(p_1))^2 + (2\pi\nu)^2} \right) \end{aligned}$$

Appendix C. Convolution of the Fourier Transform of a Continuous Covariance Function of an AR Process with a Dirac Comb

The convolution theorem means if $F(\nu)$ and $G(\nu)$ are the Fourier transforms of the function $f(h)$ and $g(h)$, then

$$\mathcal{F}\{f(h)g(h)\}(\nu) = F(\nu) * G(\nu).$$

In this context, $f(x) = \gamma(h)$ (see Equation (8)) and $F(\nu) = \Gamma(\nu)$ (see Equation (14)). For $g(x) = \sum_{k=-\infty}^{\infty} \delta(x-k)$ (the Dirac comb of distance $dx = 1$) the Fourier transform is again a Dirac comb of distance $d\nu = 1/dx = 1$. So, in the time domain is the same function as in the frequency domain (for $\nu = x$: $G(\nu) = g(x)$). Using these results leads to

$$\begin{aligned}
\mathcal{F}\left\{\gamma(h)\sum_{k=-\infty}^{\infty}\delta(x-k)\right\}(\nu) &= \Gamma(\nu) * \left(\sum_{k=-\infty}^{\infty}\delta(\nu-k)\right) \\
&= \int_{-\infty}^{\infty}\Gamma(u)\left(\sum_{k=-\infty}^{\infty}\delta(u-\nu-k)\right)du \\
&\overset{(i)}{=} \sum_{k=-\infty}^{\infty}\underbrace{\int_{-\infty}^{\infty}\Gamma(u)\delta(u-(\nu+k))du} \\
&\overset{(ii)}{=} \sum_{k=-\infty}^{\infty}\qquad \Gamma(\nu+k) \\
&\overset{(iii)}{=} \sum_{k=-\infty}^{\infty}\overbrace{\int_{-\infty}^{\infty}\gamma(h)e^{-i2\pi\nu h}e^{-i2\pi kh}dh} \\
&= \sum_{l=1}^{p}A_l\sum_{k=-\infty}^{\infty}\int_{-\infty}^{\infty}p_l^{|h|}e^{-i2\pi\nu h}e^{-i2\pi kh}dh.
\end{aligned}$$

In step (i) the sum and the integral are exchanged. Step (ii) represents the ability of the Dirac impulse that $\int_{-\infty}^{\infty} f(u)\delta(u-x)du = f(x)$. Finally, (iii) uses the frequency shift of the inverse Fourier transform (see e.g., [16], p. 26, Table 2.2). Using this function to compute the power spectral density for an AR(1) process will result in Equation (12) or in the case of an AR(2) process with two complex conjugated zeros in Equation (13).

References

1. Förstner, W. Determination of the additive noise variance in observed autoregressive processes using variance component estimation technique. *Stat. Decis.* **1985**, *2*, 263–274.
2. Förstner, W.; Wrobel, B.P. *Photogrammetric Computer Vision–Statistics*; Springer: Berlin/Heidelberg, Germany, 2016; Volume 6.
3. Koch, K. Rekursive Numerische Filter. *Z. Für Vermess.* **1975**, *100*, 281–292.
4. Schuh, W.D. *Tailored Numerical Solution Strategies for the Global Determination of the Earth's Gravity Field*; Mitteilungen Der Geodätischen Institute, Technische Universität Graz (TUG): Graz, Austria, 1996; Volume 81.
5. Zeng, W.; Fang, X.; Lin, Y.; Huang, X.; Zhou, Y. On the total least-squares estimation for autoregressive model. *Taylor Fr.* **2018**, *50*, 186–190. [CrossRef]
6. Krarup, T. *A Contribution to the Mathematical Foundation of Physical Geodesy*; Number 44 in Meddelelse; Danish Geodetic Institute: Copenhagen, Denmark, 1969.
7. Moritz, H. *Advanced Least-Squares Methods*; Number 175 in Reports of the Department of Geodetic Science, Ohio State University Research Foundation: Columbus, OH, USA, 1972.
8. Moritz, H. *Least-Squares Collocation*; Number 75 in Reihe A; Deutsche Geodätische Kommission: München, Germany, 1973.
9. Dermanis, A. Kriging and collocation: A comparison. *Manuscr. Geod.* **1984**, *9*, 159–167.
10. Schuh, W.D. Signalverarbeitung in Der Physikalischen Geodäsie. In *Handbuch Der Geodäsie, Erdmessung Und Satellitengeodäsie*; Freeden, W., Rummel, R., Eds.; Springer Reference Naturwissenschaften; Springer: Berlin/Heidelberg, Germany, 2016; pp. 73–121. [CrossRef]
11. Schuh, W.D.; Brockmann, J. The Numerical Treatment of Covariance Stationary Processes in Least Squares Collocation. In *Handbuch Der Geodäsie*; Freeden, W., Ed.; Springer: Berlin/Heidelberg, Germany, 2018; [CrossRef]
12. Moritz, H. *Advanced Physical Geodesy*; Wichmann: Karlsruhe, Germany, 1980.
13. Reguzzoni, M.; Sansó, F.; Venuti, G. The Theory of General Kriging, with Applications to the Determination of a Local Geoid. *Geophys. J. Int.* **2005**, *162*, 303–314. [CrossRef]
14. Box, G.E.P.; Jenkins, G.M.; Reinsel, G.C. *Time Series Analysis: Forecasting and Control, Fourth Edition*; Wiley Series in Probability and Statistics; John Wiley & Sons: Hoboken, NJ, USA, 2008; [CrossRef]
15. Brockwell, P.J.; Davis, R.A. *Time Series Theory and Methods*, 2nd ed.; Springer Series in Statistics; Springer: New York, NY, USA, 1991; [CrossRef]
16. Buttkus, B. *Spectral Analysis and Filter Theory in Applied Geophysics*; Springer: Berlin/Heidelberg, Germany, 2000; [CrossRef]
17. Hamilton, J.D. *Time Series Analysis*; Princeton University Press: Princeton, NJ, USA, 1994.

18. Priestley, M.B. *Spectral Analysis and Time Series*; Academic Press: London, UK; New York, NY, USA, 1981.
19. Goldberg, S. *Introduction to Difference Equations*, Reprint ed.; Dover Publications: Mineola, NY, USA, 1986.
20. Viète, F. *Opera mathematica*. 1579. Reprinted in Leiden, Netherlands, 1646. [CrossRef]
21. Schuh, W.D.; Krasbutter, I.; Kargoll, B. Korrelierte Messung—Was Nun? In *Zeitabhängige Messgrößen—Ihre Daten Haben (Mehr-) Wert*; Neuner, H., Ed.; DVW-Schriftenreihe; Wißner: Augsburg, Germany, 2014; Volume 74, pp. 85–101.

engineering proceedings

Proceedings

Asymptotic Distributions of M-Estimates for Parameters of Multivariate Time Series with Strong Mixing Property †

Alexander Kushnir * and Alexander Varypaev

Institute of Earthquake Prediction Theory and Mathematical Geophysics of Rassian Academy of Sciences, 113556 Moscow, Russia; avalex89@gmail.com
* Correspondence: afkushnir@gmail.com
† Presented at the 7th International conference on Time Series and Forecasting, Gran Canaria, Spain, 19–21 July 2021.

Abstract: The publication is devoted to studying asymptotic properties of statistical estimates of the distribution parameters $\boldsymbol{u} \in R^q$ of a multidimensional random stationary time series $\boldsymbol{z}_t \in R^m$, $t \in \mathbb{Z}$ satisfying the strong mixing conditions. We consider estimates $\overset{\wedge}{\boldsymbol{u}}{}_n^{\delta}\left(\overline{\boldsymbol{z}}_n\right)$, $\overline{\boldsymbol{z}}_n = \left(\boldsymbol{z}_1^{\mathrm{T}}, \ldots, \boldsymbol{z}_n^{\mathrm{T}}\right)^{\mathrm{T}} \in R^{mn}$ that provide in asymptotic $n \to \infty$ the maximum values for some objective functions $Q_n\left(\overline{\boldsymbol{z}}_n; \boldsymbol{u}\right)$, which have properties similar to the well-known property of local asymptotic normality. These estimates are constructed by solving the equations $\boldsymbol{\delta}_n\left(\overline{\boldsymbol{z}}_n; \boldsymbol{u}\right) = \mathbf{0}$, where $\boldsymbol{\delta}_n\left(\overline{\boldsymbol{z}}_n; \boldsymbol{u}\right)$ are arbitrary functions for which $\delta_n\left(\overline{\boldsymbol{z}}_n; \boldsymbol{u}\right) - \underset{\boldsymbol{h}}{\mathrm{grad}} Q_n\left(\overline{\boldsymbol{z}}_n; \boldsymbol{u} + n^{-1/2}\boldsymbol{h}\right) \to 0$ $(n \to \infty)$ in $P_{n,\boldsymbol{u}}\left(\overline{\boldsymbol{z}}_n\right)$-probability uniformly on $\boldsymbol{u} \in U$, were U is compact in R^q. In many cases, the estimates $\overset{\wedge}{\boldsymbol{u}}{}_n^{\delta}\left(\overline{\boldsymbol{z}}_n\right)$ have the same asymptotic properties as well-known M-estimates defined by equations $\overset{\wedge}{\boldsymbol{u}}{}_n^{Q}\left(\overline{\boldsymbol{z}}_n\right) = \underset{\boldsymbol{u} \in U}{arg\, max} Q_n\left(\overline{\boldsymbol{z}}_n; \boldsymbol{u}\right)$ but often can be much simpler computationally. We consider an algorithmic method for constructing estimates $\overset{\wedge}{\boldsymbol{u}}{}_n^{\delta}\left(\overline{\boldsymbol{z}}_n\right)$, which is similar to the accumulation method first proposed by R. Fischer and rigorously developed by L. Le Cam. The main theoretical result of the article is the proof of the theorem, in which conditions of the asymptotic normality of estimates $\overset{\wedge}{\boldsymbol{u}}{}_n^{\delta}\left(\overline{\boldsymbol{z}}_n\right)$ are formulated, and the expression is proposed for their matrix of asymptotic mean-square deviations $\lim\limits_{n \to \infty} n \mathrm{E}_{n,\boldsymbol{u}}\left\{\left(\overset{\wedge}{\boldsymbol{u}}{}^{\delta}\left(\overline{\boldsymbol{z}}_n\right) - \boldsymbol{u}\right)\left(\overset{\wedge}{\boldsymbol{u}}{}^{\delta}\left(\overline{\boldsymbol{z}}_n\right) - \boldsymbol{u}\right)^{\mathrm{T}}\right\}$.

Keywords: random time series; estimation of distribution parameters; local asymptotical normality; function of estimation quality; asymptotically efficient estimates

Citation: Kushnir, A.; Varypaev, A. Asymptotic Distributions of M-Estimates for Parameters of Multivariate Time Series with Strong Mixing Property . *Eng. Proc.* **2021**, *5*, 19. https://doi.org/10.3390/engproc2021005019

Published: 28 June 2021

1. Introduction. Methods of Construction Asymptotically Efficient Estimates for Parameters of Stationary Time Series

In applications of mathematical statistics to modern problems of data analysis in natural science and technology, it is often impossible to use the classical observation models in the form of a sequence of independent identically distributed random variables (i.i.d. model). As a rule, the i.i.d. model does not provide sufficient accuracy of statistical inferences about the unknown parameters of the investigated physical processes, distorted by noise, if both of them are stationary random processes.

Thus, it is important to generalize the classical results of the statistical theory of parameter estimation, developed for the i.i.d. model, in order to apply them to actual practical problems in the analysis of real physical processes.

In modern systems for analyzing physical wave fields, a large number of parameters are simultaneously measured, and many sensors are used to improve the accuracy of the analysis. That is, *multidimensional* time series $z_t \in R^m$, $t \in \mathbb{Z}$ are subjected to statistical processing, and *vector* parameters are estimated as a result of this processing.

For many statistical models of multivariate time series, it is impossible to synthesize statistically efficient estimates $\hat{u}_n^{ef}\left(\bar{z}_n\right)$ of vector parameters u for which the standard deviation matrices are minimal for any finite size n of observations and are equal to the inverse Fisher information matrix:

$$K_n^{ef}(u) = \mathrm{E}_{n,u}\left\{\left(\hat{u}_n^{ef}\left(\bar{z}_n\right) - u\right)\left(\hat{u}_n^{ef}\left(\bar{z}_n\right) - u\right)^{\mathrm{T}}\right\} = J_n^{-1}(u), \tag{1}$$

where $J_n(u) = \int\limits_{R^{mn}} \left(\nabla_u p_{z,n}\left(\bar{x}_n;\, u\right)\right)\left(\nabla_u p_{z,n}\left(\bar{x}_n;\, u\right)\right)^{\mathrm{T}} p_{z,n}^{-1}\left(\bar{x}_n;\, u\right) d\,\bar{x}_n;$

$$\bar{x}_n = \left(x_1^{\mathrm{T}}, \ldots, x_n^{\mathrm{T}}\right)^{\mathrm{T}} \in R^{mn};\ \nabla_u p_z\left(\bar{x}_n; u\right) = \left(\frac{\partial}{\partial u_k} p_z\left(\bar{x}_n; u\right),\ k \in \overline{1,q}\right)^{\mathrm{T}};$$

$p_z\left(\bar{x}_n; u\right)$ is the probability density of the observations $\bar{z}_n$.

At the same time, asymptotically efficient (AE) estimates $\hat{u}_n^{ae}\left(\bar{z}_n\right)$ can be constructed for a wide class of multivariate time series with interdependent elements z_t possessing a strong mixing property [1]. For AE-estimates, equality (1) is attained asymptotically for $n \to \infty$:

$$K^{ae}(u) = \lim_{n\to\infty} n\mathrm{E}_{n,u}\left\{\left(\hat{u}_n^{ae}\left(\bar{z}_n\right) - u\right)\left(\hat{u}_n^{ae}\left(\bar{z}_n\right) - u\right)^{T}\right\} = \lim_{n\to\infty} nJ_n^{-1}(u).$$

They can be found in the class $\mathcal{R}$ of *regular* estimates $\hat{u}\left(\bar{z}_n\right)$ for which the random quantities $\sqrt{n}\left(\hat{u}\left(\bar{z}_n\right) - u\right)$, $u \in U$ have limit distributions with finite second moments. This statement is one of the results of the extensive asymptotic theory of statistical inference for random time series, which is most fully presented in [2]. Fundamental results in this theory were obtained in the known publications [3–6]. In these books, sufficient conditions were established under which AE-estimates exist for many probabilistic models of random time series and continuous processes.

The main condition under which the AE-estimates can be constructed is the local asymptotic normality (LAN) of the likelihood ratio $\mathrm{L}_n\left(\bar{z}_n\right)$ of observations $\bar{z}_n$ [3]. It means that the likelihood ratio of the observations $\bar{z}_n$ admits the following asymptotic expansion:

$$L_n\left(\bar{z}_n\right) = ln\frac{p_{z,n}\left(\bar{z}_n;\, u + n^{-1/2}h\right)}{p_{z,n}\left(\bar{z}_n;\, u\right)} = h^{\mathrm{T}}\Delta_n\left(\bar{z}_n;\, u\right) - \frac{1}{2}h^{\mathrm{T}}\Gamma_n(u)h + \alpha_n\left(\bar{z}_n;\, u, h\right), \tag{2}$$

where $\lim\limits_{n\to\infty}\Gamma_n(u) = \Gamma(u) = \lim\limits_{n\to\infty} n^{-1}J_n(u)$; $\Delta_n\left(\bar{z}_n;\, u\right) \in R^q$ is a family of statistics for which probability distributions tend as $n \to \infty$ to the q-dimensional Gaussian distributions with the parameters $(0,\Gamma(u))$ uniformly in $u \in U$; $\alpha_n\left(\bar{z}_n; u, t\right) \to 0$ $(n \to \infty)$ in $dP_n\left(\bar{z}_n\right)$-probability uniformly in $u \in U$; $|h| < c$ where c is any number.

Many publications, for example, [7–14], have been devoted to proving the LAN property for various probabilistic models of time series other than the i.i.d model. The results of research in this direction, obtained up to the end of the twentieth century, are summarized in the monograph [2]. It was shown that the LAN property is inherent in a wide class of multidimensional time series and continuous random processes.

The formulation of the LAN condition (2) largely determined the further development and practical applications of the asymptotic estimation theory. In the well-known monograph [6], it is shown that under the LAN condition, the maximum likelihood estimate belongs to the class $\mathcal{R}$ of regular statistical estimates and is an AE-estimate.

At the same time, using the decomposition (2) of the likelihood function of observations, new AE-estimates were constructed, which differ from the traditional maximum likelihood estimates and are computationally simpler. An elegant and, in many cases, the most computationally simple method for constructing AE-estimates, was proposed in [3,4]. It is based on R. Fisher's [15] idea of "improving" the quality of some "simple" estimate to the quality of an AE-estimate. In mentioned publications, L. Le Cam showed that the AE-estimate can be obtained using the equation:

$$\hat{u}_n^{ae}\left(\bar{z}_n\right) = u_n^*\left(\bar{z}_n\right) - n^{-1/2}\Gamma_n^{-1}\left(u_n^*\left(\bar{z}_n\right)\right)\Delta_n\left(\bar{z}_n; u_n^*\left(\bar{z}_n\right)\right), \quad (3)$$

where $u_n^*\left(\bar{z}_n\right)$ is an arbitrary $\sqrt{n}$-consistent estimate of the parameter u for which the quantities $\sqrt{n}\left(u^*\left(\bar{z}_n\right) - u\right)$, $u \in U$, $n \in \mathbb{Z}$ have the property: for any $\varepsilon > 0$ there is $C_\varepsilon > 0$, such that $\sup_{u \in U, n \in \mathbb{Z}^+}\left[P_{n,u}\left\{\left|\sqrt{n}\left(u^*\left(\bar{z}_n\right) - u\right)\right| > C_\varepsilon\right\}\right] < \varepsilon$.

Note that Equation (3) defines a whole class of AE-estimates, the quality of which is asymptotically equivalent to the quality of the ML-estimate, since $\Delta_n\left(\bar{z}_n; u\right)$, $\Gamma_n(u)$ in the LAN expansion (2) and the $\sqrt{n}$-consistent estimate $u_n^*\left(\bar{z}_n\right)$ are not unique functions. For this reason, in many practically important cases, formula (3) allows one to obtain AE-estimates, which are computationally much simpler than ML-estimates.

2. Construction of M-Estimates for Parameters of Stationary Time Series with Suitable Asymptotical Properties

The AE-estimates have some disadvantages from the point of view of practical applications. First, they can be synthesized only if the probability density $p_{n,z}\left(\bar{x}_n; u\right)$ of the observations $\bar{z}_n$ is fully known. In practice, some important details of this density are often not fully defined. Only a certain class $\mathcal{K}$ is known to which this density belongs. Second, the quality of AE-estimates is often unstable to deviations of the actual density $p_{n,z}\left(\bar{x}_n; u\right)$ from the assumed one for which they were synthesized. Even a small deviation from the expected density can lead to a significant loss in the accuracy of the AE-estimate.

In the publications [16,17], methods were developed for constructing estimates that are *robust* to changes in the distribution of observations, and in many applications, such robust estimates are preferable to AE-estimates. A robust estimate $\hat{u}\left(\bar{z}_n\right)$ is constructed by finding the global maximum of a certain objective function $Q_n\left(\overline{z}_n;\boldsymbol{u}\right)$(a criterion of estimation quality), which differs from likelihood function:

$$\hat{\boldsymbol{u}}\left(\overline{z}_n\right) = \underset{\boldsymbol{u}\in U}{arg\,max} Q_n\left(\overline{z}_n;\boldsymbol{u}\right). \quad (4)$$

In addition to robust estimates, estimates synthesized using Equation (4) arise in other problems of mathematical statistics. The examples include Bayesian estimation problems, estimation problems with interfering (nuisance) parameters, problems arising in the analysis of natural and economic dynamical systems.

The estimates obtained as the maxima of some objective functions $Q_n\left(\overline{z}_n;\boldsymbol{u}\right)$ were called "M-estimates". Apart from books [16,17], they were considered in many other publications, for example, in [18,19]. In most of these publications, the M-estimates were constructed and analyzed for the i.i.d. model of random observations.

The authors are not aware of publications in which the asymptotic properties of M-estimates were studied with a sufficient level of mathematical rigor for multidimensional stationary random time series that have a strong mixing property. The authors are also unaware of publications devoted to the construction of computationally simple estimates that are asymptotically equivalent in quality to M-estimates.

In this paper, we consider an approach to solving these problems from the standpoint of view of the asymptotic theory of statistical inference [2], which is based on Le Cam's concept of local asymptotically normality.

We suppose that random objective function $Q_n\left(\overline{z}_n;\boldsymbol{u}\right)$ is twice differentiable in $P_{n,\boldsymbol{u}}$-probability with respect to components of the vector $\boldsymbol{u} \in U$; that is, there exist the following family of vector statistics $\boldsymbol{d}_n\left(\overline{z}_n;\boldsymbol{u}\right)$ and matrix function $\boldsymbol{F}_n\left(\overline{z}_n;\boldsymbol{u}\right)$:

$$\begin{gathered}\boldsymbol{d}_n\left(\overline{z}_n;\boldsymbol{u}\right) = \left(d_{n,k}\left(\overline{z}_n;\boldsymbol{u}\right) = \tfrac{\partial}{\partial u_k} Q_n\left(\overline{z}_n;\boldsymbol{u}\right),\, k \in \overline{1,q}\right)^{\mathrm{T}} = \nabla_{\boldsymbol{u}} Q_n\left(\overline{z}_n;\boldsymbol{u}\right) \in R^q,\\ \boldsymbol{F}_n\left(\overline{z}_n;\boldsymbol{u}\right) = \left[\tfrac{\partial}{\partial u_l} d_{n,k}\left(\overline{z}_n;\boldsymbol{u}\right),\, k,l \in \overline{1,q}\right] = \Delta_{\boldsymbol{u}} Q_n\left(\overline{z}_n;\boldsymbol{u}\right) \in R^{q\times q}.\end{gathered} \quad (5)$$

In this case, the M-estimate (4) is *one* of the roots $\tilde{\boldsymbol{u}}_n\left(\overline{z}_n\right)$ of the following equation system with respect to the parameter $\boldsymbol{u}$:

$$\boldsymbol{d}_n\left(\overline{z}_n;\boldsymbol{u}\right) = 0. \quad (6)$$

In this paper, we show how to find the estimate $\hat{\boldsymbol{u}}_n^{\delta}\left(\overline{z}_n\right)$, which is a root of the equation system (6), and, at the same time, it is an $\sqrt{n}$-consistent estimate of the parameter $\boldsymbol{u}$. It is proved in Theorem 1 that under certain restrictions, such an estimate $\hat{\boldsymbol{u}}_n^{\delta}\left(\overline{z}_n\right)$ can be found using the algorithm

$$\hat{\boldsymbol{u}}_n^{\delta}\left(\overline{z}_n\right) = \boldsymbol{u}_n^*\left(\overline{z}_n\right) - n^{-1/2}\boldsymbol{\Phi}_n\left(\boldsymbol{u}_n^*\left(\overline{z}_n\right)\right)\delta_n\left(\overline{z}_n;\boldsymbol{u}_n^*\left(\overline{z}_n\right)\right), \quad (7)$$

where $\delta_n\left(\overline{z}_n;\boldsymbol{u}\right) = n^{-1/2}\boldsymbol{d}_n\left(\overline{z}_n;\boldsymbol{u}\right)$; $\Phi_n(\boldsymbol{u}) = n^{-1}\mathrm{E}_{\boldsymbol{u}}\left\{F_n\left(\overline{z}_n;\boldsymbol{u}\right)\right\}$; $\boldsymbol{u}_n^*\left(\overline{z}_n\right)$ is any $\sqrt{n}$-consis- tent estimate of the parameter $\boldsymbol{u}$.

Conditions are formulated in Theorem 1 on the family of statistics $\delta_n\left(\overline{z}_n;\boldsymbol{u}\right)$ and the sequence of the matrix functions $\Phi_n(\boldsymbol{u})$ that are sufficient for the asymptotic normality of the estimate (7): $\mathfrak{L}\left\{\sqrt{n}\left(\overset{\wedge\delta}{u}_n\left(\bar{z}_n\right)-u\right)\right\}\to \mathbb{N}(0,D(u))\ (n\to\infty)$, where the asymptotic covariance matrix $D(u)=\lim_{n\to\infty} n\mathrm{E}_u\left\{\left(\overset{\wedge\delta}{u}_n\left(\bar{z}_n\right)-u\right)\left(\overset{\wedge\delta}{u}_n\left(\bar{z}_n\right)-u\right)^{\mathrm{T}}\right\}$ is equal to

$$D(u)=\Phi^{-1}(u)\Psi(u)\Phi^{-1}(u)\Psi(u)=\lim_{n\to\infty}\mathrm{E}_u\left\{\delta_n\left(\bar{z}_n;u\right)\delta_n^{\mathrm{T}}\left(\bar{z}_n;u\right)\right\}\Phi(u)=\lim_{n\to\infty}\Phi_n(u).$$

The corollary of Theorem 1 describes a method for constructing another estimate $\tilde{u}_n^{\delta}\left(\bar{z}_n\right)$ that has the same asymptotical distribution as the estimate (7) but does not require an auxiliary $\sqrt{n}$-consistent estimate $u_n^*\left(\bar{z}_n\right)$.

Note that the statements of Theorem 1 and the corollary were formulated earlier in [20]. In our paper, the above statements are proved under more general assumptions, and simpler proofs are given.

Theorem 1. A. *There exists a $\sqrt{n}$-consistent estimate $u_n^*\left(\bar{z}_n\right)$ of the parameter u.*

B. *Let the family of statistics $\delta_n\left(\bar{z}_n,u\right)\in R^m$, $u\in U$, and the sequence of positive definite symmetric $q\times q$-matrix functions $\Phi_n(u)$ satisfy the following constraints:*

B1. *For each value of the parameter $u\in U$, the sequence of statistics $\delta_n\left(\bar{z}_n,u\right)$ is asymptotically normal with zero mean and the covariance matrix $\Psi(u)$:*

$$\mathfrak{L}\left\{\delta_n\left(\bar{z}_n,u\right)\right\}\to\mathcal{N}(0,\Psi(u))(n\to\infty)$$

where $\Psi(u)=\lim_{n\to\infty}\mathrm{E}_u\left\{\delta_n\left(\bar{z}_n,u\right)\delta_n^{\mathrm{T}}\left(\bar{z}_n,u\right)\right\}$.

B2. *For each value of the parameter $u\in U$, the following asymptotic expansion of the statistic $\delta_n\left(\bar{z}_n,u\right)$ holds:*

$$\delta_n\left(\bar{z}_n;u+n^{-1/2}h\right)=\delta_n\left(\bar{z}_n;u\right)+\Phi_n(u)h+\beta_n\left(\bar{z}_n;u,h\right),\ |h|<c\ for\ \forall c;$$

where $\sup_{u\in U,|h|<c} P_{n,u}\left\{\left|\beta_n\left(\bar{z}_n;u,h\right)\right|>\varepsilon\right\}\to 0\ (n\to\infty)$ *for any* $\varepsilon>0$;

$$\inf_{n\in\mathbb{Z}^+,u\in U}\det\Phi_n(u)>d;\ \lim_{n\to\infty}\sup_{u\in U}\|\Phi_n^{-1}(u)-\Phi^{-1}(u)\|=0;\ \sup_{u\in U}\|\Phi^{-1}(u)\|<C;$$

$\Phi^{-1}(u)$ is a continuous function of $u\in U$.

Then the following statement is true:

For any $\sqrt{n}$-consistent estimate $u_n^*\left(\bar{z}_n\right)$ of the parameter $u\in U$, the statistic

$$\overset{\wedge\delta}{u}_n\left(\bar{z}_n\right)=u_n^*\left(\bar{z}_n\right)-n^{-1/2}\Phi_n^{-1}\left(u_n^*\left(\bar{z}_n\right)\right)\delta_n\left(\bar{z}_n;u_n^*\left(\bar{z}_n\right)\right)\tag{8}$$

is the $\sqrt{n}$-consistent and asymptotically normal estimate of the parameter $u\in U$ with the moments $(0,D(u))$:

$$\mathfrak{L}\left\{\sqrt{n}\left(\overset{\wedge\delta}{u}_n\left(\bar{z}_n\right)-u\right)\right\}\to\mathbb{N}(0,D(u))\ (n\to\infty)$$

where $\mathbf{D}(\mathbf{u}) = \mathbf{\Phi}^{-1}(\mathbf{u})\mathbf{\Psi}(\mathbf{u})\mathbf{\Phi}^{-1}(\mathbf{u})$.

Corollary 1. **(a)** *Let, for any* $n \in \mathbb{Z}^+$, *a statistic* $\tilde{\mathbf{u}}_n^{\delta}\left(\bar{z}_n\right)$ *be the root of the equation* $\boldsymbol{\delta}_n\left(\bar{z}_n; \mathbf{u}\right) = \mathbf{0}$ *with respect to the parameter* $\mathbf{u} \in U$ *with probability equal to* 1.

(b) *Let the statistic* $\tilde{\mathbf{u}}_n^{\delta}\left(\bar{z}_n\right)$ *also is a* $\sqrt{n}$*-consistent estimate of the parameter* $\mathbf{u} \in U$. *Then the statistic* $\tilde{\mathbf{u}}_n^{\delta}\left(\bar{z}_n\right)$ *is asymptotically normal with the moments* $(0, \mathbf{D}(\mathbf{u}))$.

Remark 1. **(a)** *The statement similar to Statement (T1) of Theorem 1 was proved in [3,4] in the case when the objective function* $Q_n\left(\bar{z}_n; \mathbf{u}\right)$ *is the likelihood function of* $\bar{z}_n$ *having the LAN property (2). In this case* $\boldsymbol{\delta}_n\left(\bar{z}_n; \mathbf{u}\right) \equiv \boldsymbol{\Delta}_n\left(\bar{z}_n; \mathbf{u}\right)$, *the matrix function* $\mathbf{\Phi}_n(\mathbf{u}) \equiv \mathbf{\Gamma}_n(\mathbf{u})$ *and*

$$\mathfrak{L}\left\{\Delta_n\left(\bar{z}_n; \mathbf{u}\right)\right\} \to \mathbb{N}(0, \Gamma(\mathbf{u}))\,(n \to \infty); \Gamma(\mathbf{u}) = \lim_{n \to \infty} n^{-1} J_n(\mathbf{u}),$$

where $J_n(\mathbf{u})$ *is the Fisher matrix. It follows from Theorem 1, that in this case*

$$\mathbf{D}(\mathbf{u}) = \mathbf{\Gamma}^{-1}(\mathbf{u})\mathbf{\Gamma}(\mathbf{u})\mathbf{\Gamma}^{-1}(\mathbf{u}) = \mathbf{\Gamma}^{-1}(\mathbf{u}).$$

Consequently, the statistic

$$\hat{\mathbf{u}}_n^{\Delta}\left(\bar{z}_n\right) = \mathbf{u}_n^*\left(\bar{z}_n\right) - n^{-1/2}\Gamma_n^{-1}\left(\mathbf{u}_n^*\left(\bar{z}_n\right)\right)\Delta_n\left(\bar{z}_n; \mathbf{u}_n^*\left(\bar{z}_n\right)\right)$$

is asymptotically normal with the parameters $(0, \Gamma(\mathbf{u}))$, *and hence, it is the asymptotically efficient estimate of the parameter* $\mathbf{u}$.

(b) *It follows from the corollary of Theorem 1 that a statistic* $\tilde{\mathbf{u}}_n^{\Delta}\left(\bar{z}_n\right)$, *which has the property:* $\boldsymbol{\Delta}_n\left(\bar{z}_n; \tilde{\mathbf{u}}_n^{\Delta}\left(\bar{z}_n\right)\right) = \mathbf{0}$ *with probability equal to one, and at the same time is a* $\sqrt{n}$*-consistent estimate of the parameter* $\mathbf{u} \in U$, *is asymptotically normal with the moments* $(0, \Gamma(\mathbf{u}))$. *Consequently, the statistic* $\tilde{\mathbf{u}}_n^{\Delta}\left(\bar{z}_n\right)$ *is the asymptotically efficient estimate of the parameter* $\mathbf{u} \in U$.

Thus, Theorem 1 is, in some sense, an extension of Le Cam's results to the case of an arbitrary objective function $Q_n\left(\bar{z}_n; \mathbf{u}\right)$ whose gradient satisfies conditions B1, B2 of Theorem 1.

3. Proof of Theorem 1

In the course of proving Theorem 1, we will omit, if it is obvious, the dependence of functional quantities on the observations $\bar{z}_n$ and sometimes denote their dependence on the parameter $\mathbf{u}$ by a subscript.

In these notations, the definition of the estimate $\hat{\mathbf{u}}_n\left(\bar{z}_n\right)$ can be written as

$$\hat{\mathbf{u}}_n = \mathbf{u}_n^* - n^{-1/2}\Phi_n^{-1}(\mathbf{u}_n^*)\delta_n(\mathbf{u}_n^*).$$

Then we can write the following chain of equalities:

$$\sqrt{n}\left(\hat{u}_n - u\right) = \sqrt{n}(u_n^* - u) - \Phi_n^{-1}(u_n^*)\delta_n(u_n^*) = \\ = -\Phi_n^{-1}(u)\delta_n(u) + \left[\sqrt{n}(u_n^* - u) - \Phi_n^{-1}(u_n^*)\delta_n(u_n^*) + \Phi_n^{-1}(u)\delta_n(u)\right] = \\ = -\Phi_n^{-1}(u)\delta_n(u) + \xi_{n,u}(u_n^*), \tag{9}$$

where $\xi_{n,u}(u_n^*) = \sqrt{n}(u_n^* - u) + \boldsymbol{\Phi}_n^{-1}(u_n^*)[-\delta_n(u_n^*) + \delta_n(u)]$. It follows from (9):

$$\Phi_n(u_n^*)\xi_{n,u}(u_n^*) = -\delta_n(u_n^*) + \delta_n(u) + \Phi_n(u_n^*)\sqrt{n}(u_n^* - u) = \rho_{n,u}(u_n^*). \tag{10}$$

By denoting $\tau_{n,u}^* = \sqrt{n}(u_n^* - u)$, we obtain from (10):

$$\delta_n(u + \tau_{n,u}^*/\sqrt{n}) - \delta_n(u) = \boldsymbol{\Phi}_n(u + \tau_{n,u}^*/\sqrt{n})\tau_{n,u}^* - \rho_{n,u}(\tau_{n,u}^*), \tag{11}$$

where the random quantities $\tau_{n,u}^*$, $n \in \mathbb{Z}^+$, $u \in U$ have the property: for any $\varepsilon > 0$ there is $S_\varepsilon > 0$ such that $\sup\limits_{u \in U, n \in \mathbb{Z}^+} \left[P_{n,u}\{|\tau_{n,u}^*| > S_\varepsilon\}\right] < \varepsilon$.

At the same time, from condition B2 of Theorem 1, we obtain:

$$\delta_n\left(\bar{z}_n; u + n^{-1/2}h\right) - \delta_n\left(\bar{z}_n; u\right) = \boldsymbol{\Phi}_n(u)h + \beta_{n,u}\left(\bar{z}_n; h\right), \tag{12}$$

where $\sup\limits_{u \in U, |h| < c} P_{n,u}\left\{\left|\beta_{n,u}\left(\bar{z}_n; h\right)\right| > \varepsilon\right\} \to 0\,(n \to \infty)$.

The comparison Equations (11) and (12) allow us to prove the following Lemma.

Lemma 1. *Under the conditions of Theorem 1, the following convergences take place for any* $\varepsilon > 0$*:*
(a) $\lim\limits_{n\to\infty}\sup\limits_{u \in U} P_{n,u}\{|\rho_{n,u}(u_n^*)| > \varepsilon\} = 0$, **(b)** $\lim\limits_{n\to\infty}\sup\limits_{u \in U} P_{n,u}\{|\xi_{n,u}(u_n^*)| > \varepsilon\} = 0$.

The proof of Lemma 1 is given in Section 5.

The following statement will be needed below.

Lemma 2. *Let some random variables* φ_n *and* η_n *have the properties:*
(a) $\lim\limits_{n\to\infty} \mathfrak{L}_n\{\varphi_n\} = \lim\limits_{n\to\infty} P_n\{\varphi_n < x\} = F(x)$; **(b)** *for any* $\varepsilon > 0$ $\lim\limits_{n\to\infty} P_n\{|\eta_n| > \varepsilon\} = 0$.
Then $\lim\limits_{n\to\infty} \mathfrak{L}_n\{\varphi_n + \eta_n\} = \lim\limits_{n\to\infty} P_n\{\varphi_n + \eta_n < x\} = F(x)$.

The proof of Lemma 2 is quite simple, and we omit it.

Taking into account Equations (9)–(12) and statements of Lemmas 1 and 2, we can write the following equalities:

$$\mathfrak{L}\left\{\sqrt{n}\left(\hat{u}_n - u\right)\right\} = \lim_{n\to\infty} \mathfrak{L}\left\{\Phi_n^{-1}(u)\delta_n(u) + \xi_{u,n}\right\} = \lim_{n\to\infty} \mathfrak{L}\left\{\Phi_n^{-1}(u)\delta_n(u)\right\},$$

where the existence of the limits follows from conditions B1, B2 of Theorem 1. According to conditions B1 of Theorem 1, we have:

$\lim\limits_{n\to\infty} \mathfrak{L}\{\delta_n(u)\} = \mathbb{N}(0; \Psi(u))$ where $\boldsymbol{\Psi}(u) = \lim\limits_{n\to\infty} \mathrm{E}_n\{\delta_n(u)\delta_n^{\mathrm{T}}(u)\}$

Therefore:

$\lim\limits_{n\to\infty} \mathfrak{L}\{\Phi_n^{-1}(u)\delta_n(u)\} = \mathbb{N}(0; D(u))$, where $D(u) = \boldsymbol{\Phi}^{-1}(u)\boldsymbol{\Psi}(u)\boldsymbol{\Phi}^{-1}(u)$. □

4. Proof of Corollary

Under the conditions B1, B2 of Theorem 1, the statistic $\overset{\wedge\delta}{u}_n\left(\bar{z}_n\right)$ in Equation (8) is asymptotically normal with the moments $(0, \boldsymbol{D}(\boldsymbol{u}))$ for *any* $\sqrt{n}$-consistent estimate $\boldsymbol{u}_n^*\left(\bar{z}_n\right)$. Consequently, due to condition (**b**) of the corollary, the statistic

$$\overset{\wedge\delta}{u}_n\left(\bar{z}_n\right) = \tilde{u}_n^\delta\left(\bar{z}_n\right) + n^{-1/2}\boldsymbol{\Phi}_n^{-1}\left(\tilde{u}_n^\delta\left(\bar{z}_n\right)\right)\delta_n\left(\bar{z}_n; \tilde{u}_n^\delta\left(\bar{z}_n\right)\right)$$

is asymptotically normal with the moments $(0, \boldsymbol{D}(\boldsymbol{u}))$.

But by virtue of condition (**a**) of the corollary, we have that $\overset{\wedge\delta}{u}_n\left(\bar{z}_n\right) = \tilde{u}_n^\delta\left(\bar{z}_n\right)$ with probability equal to one. Hence, the statistic $\tilde{u}_n^\delta\left(\bar{z}_n\right)$ is asymptotically normal with the moments $(0, \boldsymbol{D}(\boldsymbol{u}))$. □

5. Proof of Lemma 1

(**a**) For any $\varepsilon > 0$, $q > 0$ and $\boldsymbol{u} \in U$, we can write the following equation:

$$\begin{gathered} P_{n,u}\{|\rho_{n,u}(\tau_n^*)| > \varepsilon\} = \\ = P_{n,u}\{|\rho_{n,u}(\tau_n^*)| > \varepsilon \cap |\tau_{n,u}^*| \le q\} + P_{n,u}\{|\rho_{n,u}(\tau_n^*)| > \varepsilon \cap |\tau_{n,u}^*| > q\}. \end{gathered} \tag{13}$$

Let denote $P_{n,u}(\{|\rho_{n,u}(\tau_n^*)| > \varepsilon\} \mid \{|\tau_{n,u}^*| < q\})$ the conditional probability of the event $\{|\rho_{n,u}(\tau_n^*)| > \varepsilon\}$ under the condition of the event $\{|\tau_{n,u}^*| < q\}$. Then (13) can be rewritten as:

$$\begin{gathered} P_{n,u}\{|\rho_{n,u}(\tau_n^*)| > \varepsilon\} = P_{n,u}(\{|\rho_{n,u}(\tau_n^*)| > \varepsilon\} \mid \{|\tau_{n,u}^*| \le q\})P_{n,u}\{|\tau_{n,u}^*| \le q\} + \\ + P_{n,u}(\{|\rho_{n,u}(\tau_n^*)| > \varepsilon\} \mid \{|\tau_{n,u}^*| > q\})P_{n,u}\{|\tau_{n,u}^*| > q\}. \end{gathered} \tag{14}$$

According to (11), there is $C_\varepsilon > 0$ such that $\sup\limits_{u \in U, n \in \mathbb{Z}^+}\left[P_{n,u}\{|\tau_{n,u}^*| > C_\varepsilon\}\right] < \varepsilon$ for any $\varepsilon > 0$. It follows then from (14) that for any $\varepsilon > 0$ and $\boldsymbol{u} \in U$

$$P_{n,u}\{|\rho_{n,u}(\tau_n^*)| > \varepsilon\} < P_{n,u}(\{|\rho_{n,u}(\tau_n^*)| > \varepsilon\} \mid \{|\tau_{n,u}^*| < C_\varepsilon\}), \tag{15}$$

where $\boldsymbol{\rho}_{n,u}(\boldsymbol{\tau}_{n,u}^*) = \boldsymbol{\delta}_n(\boldsymbol{u} + \boldsymbol{\tau}_{n,u}^*/\sqrt{n}) - \boldsymbol{\delta}_n(\boldsymbol{u}) - \boldsymbol{\Phi}_n(\boldsymbol{u} + \boldsymbol{\tau}_{n,u}^*/\sqrt{n})\boldsymbol{\tau}_{n,u}^*$.

According to (12), for any $\varepsilon > 0$, $\boldsymbol{u} \in U$ and $|\boldsymbol{h}| < C_\varepsilon$

$$\sup_{u \in U} P_{n,u}\left\{\left|\beta_{n,u}\left(\bar{z}_n; \boldsymbol{h}\right)\right| > \varepsilon\right\} \to 0\,(n \to \infty), \tag{16}$$

where $\boldsymbol{\beta}_{n,u}\left(\bar{z}_n; \boldsymbol{h}\right) = \boldsymbol{\delta}_n\left(\bar{z}_n; \boldsymbol{u} + n^{-1/2}\boldsymbol{h}\right) - \boldsymbol{\delta}_n\left(\bar{z}_n; \boldsymbol{u}\right) - \boldsymbol{\Phi}_n(\boldsymbol{u})\boldsymbol{h}$,

It follows from (15), (16) that for any $\varepsilon > 0$ $\lim\limits_{n \to \infty}\sup\limits_{u \in U} P_{n,u}\{|\rho_{n,u}(\boldsymbol{u}_n^*)| > \varepsilon\} = 0$.

(**b**) Since $|\xi_{n,u}(\boldsymbol{u}_n^*)| \le \|\Phi_{n,u}^{-1}(\boldsymbol{u}_n^*)\||\rho_{n,u}(\boldsymbol{u}_n^*)|$, to prove statement (**b**) of Lemma 1, it suffices to check that $\|\Phi_n^{-1}(\boldsymbol{u}_n^*)\|$ is bounded in probability. Since $\boldsymbol{\Phi}_n^{-1}(\boldsymbol{u})$ satisfies conditions B2 of Theorem 1, for any $\varepsilon > 0$ there exists $C_\varepsilon > 0$ that for all n the following inequality holds: $P_{n,u}\{\|\Phi_{n,u}^{-1}(\boldsymbol{u}_n^*)\| \ge C_\varepsilon\} < \varepsilon$. So, we can write:

$$\begin{gathered} P_{n,u}\{|\xi_{n,u}(\boldsymbol{u}_n^*)| > \varepsilon\} = P_{n,u}(\{|\rho_{n,u}(\boldsymbol{u}_n^*)| > \varepsilon\} \cap (\|\Phi_{n,u}^{-1}(\boldsymbol{u}_n^*)\| < C_\varepsilon)) + \\ + P_{n,u}(\{|\rho_{n,u}(\boldsymbol{u}_n^*)| > \varepsilon\} \cap \{\|\Phi_{n,u}^{-1}(\boldsymbol{u}_n^*)\| \ge C_\varepsilon\}) \le \\ \le P_{n,u}(\{|\rho_{n,u}(\boldsymbol{u}_n^*)| > \varepsilon\} \cap (\|\Phi_{n,u}^{-1}(\boldsymbol{u}_n^*)\| < C_\varepsilon)) + \varepsilon. \end{gathered}$$

Since $|\rho_{u,n}(\boldsymbol{u}_n^*)|$ satisfies statement (**a**) of Lemma 1, one can find a number N_ε such that $\sup\limits_{u \in U, n > N_\varepsilon} P_{u,n}\{|\xi_{u,n}(\boldsymbol{u}_n^*)| > \varepsilon\} < 2\varepsilon$. □

6. Conclusions

The paper investigates the asymptotic properties of statistical estimates for the vector parameter $\boldsymbol{u} \in R^q$ of a stationary multidimensional random time series $z_t \in R^m$, $t \in \mathbb{Z}$ satisfying the strong mixing conditions. We have considered estimates $\tilde{\boldsymbol{u}}\left(\bar{z}_n\right)$ that are solutions of the equations $\nabla_{\boldsymbol{u}} Q_n\left(\bar{z}_n; \boldsymbol{u}\right) = 0$, $\bar{z}_n = \left(z_1^{\mathrm{T}}, \ldots, z_n^{\mathrm{T}}\right)^{\mathrm{T}}$, where $Q_n\left(\bar{z}_n; \boldsymbol{u}\right)$ is some objective function for which $\nabla_{\boldsymbol{u}} Q_n\left(\bar{z}_n; \boldsymbol{u}\right)$ satisfies the constraints of Theorem 1. We have proved that under these constraints, the estimates $\tilde{\boldsymbol{u}}\left(\bar{z}_n\right)$ are $\sqrt{n}$-consistent and asymptotically normal with a limit covariance matrix uniquely determined by the objective function $Q_n\left(\bar{z}_n; \boldsymbol{u}\right)$.

The results of this paper are a generalization of the methods for constructing and analyzing the asymptotic properties of M-estimates, which were previously studied for the case of independent identically distributed observations.

Institutional Review Board Statement: The management of the institute does not object to the publication of the article materials.

Informed Consent Statement: Both authors agreed with content of this paper to be published.

Data Availability Statement: It is theoretical paper, there is no data.

Conflicts of Interest: The authors declare no conflict of interest.

References

1. Billingsley, P. Convergence of Probability Measures. In *Wiley Series in Probability and Statistics*; John Wiley and Sons, Inc.: New York, NY, USA, 1999.
2. Taniguchi, M.; Kakizawa, Y. *Asymptotic Theory of Statistical Inferences for Time Series*; Springer Series in Statistics; Springer: New York, NY, USA, 2000.
3. Le Cam, L. *Locally Asymptotically Normal Families of Distributions*; University California Publication Statistics: San Diego, CA, USA, 1960; Volume 3, pp. 37–99.
4. Le Cam, L. *Asymptotic Methods in Statistical Decision Theory*; Springer: New York, NY, USA; Berlin, Germany, 1986.
5. Le Cam, L.; Lo Yang, G. *Asymptotics in Statistics*; Springer: New York, NY, USA, 1990.
6. Ibragimov, I.A.; Has'minskii, R.Z. *Statistical Estimation. Asymptotic Theory. Applications of Mathematics*; Springer: Berlin/Heidelberg, Germany, 1981; Volume 16.
7. Kushnir, A.F. Asymptotically optimal tests for a regression problem of testing hypotheses. *Teor. Veroyatnost. Primenen.* **1968**, *13*, 682–700. (In Russian) [CrossRef]
8. Kushnir, A.F.; Pinskii, A.I. Asymptotically optimal tests of testing hypothesis for an interdependent sample. *Teor. Veroyatnost. Primenen.* **1971**, *16*, 280–291. (In Russian) [CrossRef]
9. Rousas, G.G. *Contiguity of Probability Measures. Some Applications in Statistics*; Cambridge University Press: Cambridge, UK, 1972.
10. Devies, R.B. Asymptotic Inference in Stationary Gaussian Time Series. Advances in applied probability. *Appl. Probab. Trust.* **1973**, *5*, 469–497. [CrossRef]
11. Dzhaparidze, K.O.; Yaglom, A.M. Spectrum parameter estimation in time series analysis. In *Developments in Statistics*; Krishnaiah, P.R., Ed.; Academic Press: New York, NY, USA, 1983; Volume 4, pp. 1–181.
12. Dzhaparidze, K.O. *Parameter Estimations and Hypothesis Testing in Spectral Analysis of Stationary Time Series*; Springer: New York, NY, USA, 1986.
13. Liptser, R.S.; Shiryayev, A.N. Statistics of Random Processes. In *Applications of Mathematics*; Springer: Berlin/Heidelberg, Germany; New York, NY, USA, 1978; Volume 5, p. 6.
14. Kutoyants, Y.A. *Parameter Estimation for Stochastic Processes*; Heldermann: Berlin, Germany, 1984.
15. Fisher, R.A. *Theory of Statistical Estimations, Proceedings of Cambridge Philosophical Society*; Cambridge University Press: Cambridge, UK, 1925; Volume 22, pp. 700–725.
16. Huber, P.J. *Robust Statistics*; Wiley Series in Probability and Statistics; John Wiley and Sons: New York, NY, USA, 1981.
17. Huber, P.J.; Ronchetti, E.M. *Robust Statistics*; Wiley Series in Probability and Statistics; John Wiley and Sons: New York, NY, USA, 2009.
18. Newey, W.K.; McFadden, D. Large Sample Estimation and Hypothesis Testing. In *Handbook of Econometrics*; Engle, R.F., McFadden, D.L., Eds.; Elsevier: Amsterdam, The Netherlands, 1986; Volume 4, Chapter 36.

19. Borovkov, A.A. *Mathematical Statistics*; Gordon and Breach Science Publishers: Amsterdam, The Netherlands, 1998.
20. Kushnir, A.F. Identification algorithms for linear systems with correlated input and output noise. *Probl. Inf. Transm.* **1987**, *23*, 139–150. (In Russian)

engineering proceedings

Proceedings

Bayesian Robust Multivariate Time Series Analysis in Nonlinear Models with Autoregressive and t-Distributed Errors †

Alexander Dorndorf [1], Boris Kargoll [2], Jens-André Paffenholz [3] and Hamza Alkhatib [1,*]

1 Geodätisches Institut, Leibniz Universität Hannover, Nienburger Str. 1, D-30167 Hannover, Germany; dorndorf@gih.uni-hannover.de
2 Institute of Geoinformation and Surveying, Anhalt University of Applied Sciences, Seminarplatz 2a, D-06846 Dessau-Rosslau, Germany; boris.kargoll@hs-anhalt.de
3 Institute of Geo-Engineering, Clausthal University of Technology, Erzstraße 18, D-38678 Clausthal-Zellerfeld, Germany; jens-andre.paffenholz@tu-clausthal.de
* Correspondence: alkhatib@gih.uni-hannover.de; Tel.: +49-5117622464

† Presented at the 7th International conference on Time Series and Forecasting, Gran Canaria, Spain, 19–21 July 2021.

Abstract: Many geodetic measurement data can be modelled as a multivariate time series consisting of a deterministic ("functional") model describing the trend, and a stochastic model of the correlated noise. These data are also often affected by outliers and their stochastic properties can vary significantly. The functional model of the time series is usually nonlinear regarding the trend parameters. To deal with these characteristics, a time series model, which can generally be explained as the additive combination of a multivariate, nonlinear regression model with multiple univariate, covariance-stationary autoregressive (AR) processes the white noise components of which obey independent, scaled t-distributions, was proposed by the authors in previous research papers. In this paper, we extend the aforementioned model to include prior knowledge regarding various model parameters, the information about which is often available in practical situations. We develop an algorithm based on Bayesian inference that provides a robust and reliable estimation of the functional parameters, the coefficients of the AR process and the parameters of the underlying t-distribution. We approximate the resulting posterior density using Markov chain Monte Carlo (MCMC) techniques consisting of a Metropolis-within-Gibbs algorithm.

Keywords: multivariate time series; nonlinear Bayesian regression model; AR process; scaled t-distribution; partially adaptive estimation; robust parameter estimation; GNSS time series

Citation: Dorndorf, A.; Kargoll, B.; Paffenholz, J.-A.;Alkhatib, H. Bayesian Robust Multivariate Time Series Analysis in Nonlinear Models with Autoregressive and t-Distributed Errors. *Eng. Proc.* **2021**, 5, 20. https://doi.org/10.3390/engproc2021005020

Published: 28 June 2021

1. Introduction

Adjustment calculus offers a rich toolbox of statistical models and procedures for parameter estimation and hypothesis testing based on given numerical observations (cf. [1]). Such models usually consist of a deterministic functional model (e.g., a linear model describing some trend function), a correlation model (e.g., in the form of a variance-covariance matrix or an autoregressive (AR) error process), and a stochastic model (i.e., a probability distribution of the observation errors or the innovations of the AR error process). The stochastic model is often taken to be some multivariate normal distribution, which, however, easily leads to erroneous estimation results if the observations are afflicted by outliers. To take outliers into account, the normal distribution can be replaced by some outlier distribution, for example, a heavy-tailed t-distribution (cf. [2]). A multivariate time series model, including a nonlinear functional model and an autoregressive observation error model with t-distributed innovations, was suggested and investigated in [3] and [4]. A shortcoming of that model is that it does not include prior knowledge about the parameters of the functional, correlation or stochastic model, the information about which may readily be available. Therefore, the current paper describes a Bayesian extension of that time series model, which can be expected to result in more robust and more accurate parameter estimates (cf. [5]).

A general Bayesian estimation approach in the specific context of models based on the t-distribution was introduced by [6]. Due to the complexity of such a model, the posterior density function must be approximated numerically or analytically. For numerical approximation, Monte-Carlo (MC) simulation and, in particular, Markov-Chain Monte-Carlo (MCMC) methods, which are suitable also for multivariate distributions, have been applied routinely (cf. [7]). In particular, the Gibbs sampler and the Metropolis–Hastings algorithm have been employed for (non-robust) Bayesian estimation of the parameters of a linear functional model with autoregressive moving-average (ARMA) and normally distributed errors [8]. MCMC methods have also been applied in the context of the robust Bayesian estimation of ARMA models [9] and AR models [10], with one additional (directly observed) mean parameter in the functional model. In both studies, outliers within the auto-correlated errors and within the uncorrelated innovations were modeled as normally distributed random variables with variances inflated by unknown multipliers. Thus, the stochastic error model was based on a discrete mixture of normal distributions, not on the t-distribution. To incorporate an automatic model selection procedure regarding the AR/ARMA model into the adjustment, the preceding studies also included unknown index parameters, taking the value 0 in case the corresponding AR (or MA) coefficient is 0 (or not significant) and taking the value 1 otherwise. Prior distributions for all of the parameter groups and the likelihood function for the data were fixed, and sampling distributions were then derived in order to obtain a numerical approximation of the posterior distribution for all the unknowns. In [11], an MCMC-based computational algorithm was proposed, to facilitate Bayesian analysis of real data when the error structure can be expressed as a p-order AR model.

The paper is organized as follows: First, the Bayesian multivariate time series model with AR and t-distributed errors is described in detail in Section 2. It is shown how the generic deterministic functional model, the AR process and the t-distribution model are first combined to a likelihood function and how prior information about the model parameters to be estimated is taken into account by means of a specified prior density. Here, we denote unknown parameters with Greek letters, random variables with calligraphic letters, and constants with Roman letters. Furthermore, we distinguish between a random variable (e.g., $\mathcal{L}_t$) and its realization (l_t). Matrices are shown, as usual, as bold capital letters and vectors as bold small letters. Section 3 outlines an MCMC algorithm for determining the posterior density of the unknown parameters of the functional model, the coefficients of the AR process and the scale parameter, as well as the degrees of freedom of the t-distribution. In Section 4, a time series model for GNSS observations of a circle in 3D is proposed, and the results of a Monte Carlo simulation are discussed. These findings are used to evaluate the performance of the implemented Metropolis–Hastings-within Gibbs algorithm in this scenario.

2. The Bayesian Time Series Model

We assume that an N-dimensional time series $(\boldsymbol{\mathcal{L}}_t) = \left(\begin{bmatrix}\mathcal{L}_{1,t} & \cdots & \mathcal{L}_{N,t}\end{bmatrix}^T\right)$ is observed at equi-spaced time instances t without data gaps. The observation model consists of the three interconnected model components,

$$\mathcal{L}_{k,t} = h_{k,t}(\beta_1, \ldots, \beta_m) + \mathcal{E}_{k,t}, \tag{1}$$

$$\mathcal{E}_{k,t} = \alpha_{k,1}\mathcal{E}_{k,t-1} + \ldots + \alpha_{k,p_k}\mathcal{E}_{k,t-p_k} + \mathcal{U}_{k,t}, \tag{2}$$

$$\mathcal{U}_{k,t} \mid \psi_k^2, \nu_k \overset{\text{ind.}}{\sim} t_{\nu_k}(0, \psi_k^2, \nu_k), \tag{3}$$

where (1) defines the "observation equations", (2) the "error equations" and (3) defines the probability distribution of the innovations. The parameters of this observation model are combined within the vector,

$$\boldsymbol{\theta} = \begin{bmatrix} \boldsymbol{\beta}^T & \boldsymbol{\alpha}^T & \boldsymbol{\psi}^T & \boldsymbol{\nu}^T \end{bmatrix}^T, \quad (4)$$

with

$$\boldsymbol{\beta} = \begin{bmatrix} \beta_1 & \cdots & \beta_m \end{bmatrix}^T \quad (5)$$

$$\boldsymbol{\alpha} = \begin{bmatrix} \boldsymbol{\alpha}_1^T & \cdots & \boldsymbol{\alpha}_N^T \end{bmatrix}^T = \begin{bmatrix} \alpha_{1,1} & \cdots & \alpha_{1,p_1} & \cdots & \alpha_{N,1} & \cdots & \alpha_{N,p_N} \end{bmatrix}^T \quad (6)$$

$$\boldsymbol{\psi} = \begin{bmatrix} \psi_1 & \cdots & \psi_N \end{bmatrix}^T \quad (7)$$

$$\boldsymbol{\nu} = \begin{bmatrix} \nu_1 & \cdots & \nu_N \end{bmatrix}^T. \quad (8)$$

On the one hand, the parameters $\boldsymbol{\theta}$ are treated as variables of the likelihood function $f_{\mathcal{L}|\boldsymbol{\Theta}}(\boldsymbol{L}|\boldsymbol{\theta})$, defined by the observation model (1)–(3). On the other hand, the parameters $\boldsymbol{\theta}$ are viewed as a realization of a random vector $\boldsymbol{\Theta}$, having a specified pdf independent of the observables. According to the Bayes theorem, this prior density $f_{\boldsymbol{\Theta}}(\boldsymbol{\theta})$ and the likelihood function $f_{\mathcal{L}|\boldsymbol{\Theta}}(\boldsymbol{L}|\boldsymbol{\theta})$ are connected to the posterior density $f_{\boldsymbol{\Theta}|\mathcal{L}}(\boldsymbol{\theta}|\boldsymbol{L})$ via proportionality relationship

$$f_{\boldsymbol{\Theta}|\mathcal{L}}(\boldsymbol{\theta}|\boldsymbol{L}) \propto f_{\boldsymbol{\Theta}}(\boldsymbol{\theta}) \cdot f_{\mathcal{L}|\boldsymbol{\Theta}}(\boldsymbol{L}|\boldsymbol{\theta}), \quad (9)$$

which serves as the foundation of the inference of the parameters and the adjustment of the observations. The details of this model are described in the following.

The observation equations: Equation (1) reflects the idea that geodetic measurements $\mathcal{L}_{k,t}$ are approximated by a "deterministic" model using mathematical functions $h_{k,t}(\boldsymbol{\beta})$, which are assumed to be partially differentiable. The index k refers to the time series surveyed by the kth sensor or sensor component, and the time instances $t = 1, \ldots, n$ are the same for all sensors. In some applications, the functional model $h_{k,t}(\boldsymbol{\beta})$ takes the form

$$h_{k,t}(\boldsymbol{\beta}) = \boldsymbol{X}_{k,t}\boldsymbol{\beta}, \quad (10)$$

of a "linear model", where $\boldsymbol{X}$ denotes the design matrix and has a full rank. Since geodetic observables can generally not be modeled using a deterministic model alone, random deviations $\mathcal{E}_{k,t}$ are added to absorb the remaining effects. It is assumed that the instruments used to survey the observables are calibrated, so that no systematic errors occur. Thus, the expected values of the random deviations are assumed to be 0.

The error equations: Equation (2) is included to take account of auto-correlations within each of the N time series. Since the different sensors or sensor components may have different noise characteristics, AR processes, with individual orders $p_1, \ldots, p_N$ and sets of coefficients $\boldsymbol{\alpha}_1, \ldots, \boldsymbol{\alpha}_N$, are selected. The noise characteristics are assumed to be constant throughout the measurement period. For practical purposes, the AR processes considered are therefore required to be (asymptotically) covariance-stationary. The random variables $\mathcal{U}_{k,t}$ are referred to as "innovations". Since the observation window is finite, ranging from $t = 1, ..., n$, the error equations involve errors at times $t = 0, -1, ...$. To ensure asymptotic covariance-stationarity and the computability of the recursive equations, these quantities are set as equal to 0 (cf. [12]). This initial distortion of the AR process fades out as the process advances in time.

The stochastic model: The innovations of an AR process are usually assumed to be Gaussian white noise. Since the assumption of normal distributions is unrealistic in some geodetic applications, for example, due to outliers, the heavy-tailed t-distributions are employed here. These are defined by the probability density function (pdf),

$$f(x) = \frac{\Gamma\left(\frac{\nu+1}{2}\right)}{\Gamma\left(\frac{\nu}{2}\right)\sqrt{(\nu\pi)}\psi}\left[1 + \frac{(x-\mu)^2}{\nu\psi^2}\right]^{-\frac{\nu+1}{2}}, \tag{11}$$

where Γ is the gamma function. Since the expected values of the random deviations $\mathcal{E}_{k,t}$ should be 0, we can also restrict the location parameter μ to 0. Since the noise of different sensors or sensor components may exhibit different levels of variance and outliers, each time series involves a t-distribution with individual scale parameter ψ_k^2 and degree of freedom (df) ν_k. It should be mentioned that the alternative usage of a multivariate t-distribution (as defined in [2]) involves a single df and would therefore not allow for the modeling of distinct outlier characteristics within the different time series.

The likelihood function: A likelihood function $f_{\mathcal{L}|\Theta}(\boldsymbol{L}|\boldsymbol{\theta})$ can be obtained by combining the observation Equation (1), the error Equation (2) and the stochastic model of the innovations (3). To do so, the well-known method of conditional likelihoods in connection with AR processes with various forms of non-Gaussian innovations is applied (cf. [13–15]). Assuming the AR processes to be invertible, the error Equation (2), in terms of their numerical realizations, can be rewritten as "innovation equations,"

$$u_{k,t} = e_{k,t} - \alpha_{k,1}e_{k,t-1} - \ldots - \alpha_{k,p_k}e_{k,t-p_k}. \tag{12}$$

As the errors $e_{k,t}$ contained in the observation Equation (1) can be expressed as

$$e_{k,t} = \ell_{k,t} - h_{k,t}(\boldsymbol{\beta}), \tag{13}$$

the innovation Equation (12) become

$$u_{k,t} = e_{k,t} - \alpha_{k,1}(\ell_{k,t-1} - h_{k,t-1}(\beta)) - \ldots - \alpha_{k,p_k}(\ell_{k,t-p_k} - h_{k,t-p_k}(\beta)). \tag{14}$$

The conditional likelihood function is then obtained as the product of the univariate pdf (11), evaluated at all the stochastically independent innovations $u_{k,t}$ with location $\mu = 0$, associated scale factor ψ_k^2 and df ν_k, that is,

$$\mathrm{L}(\boldsymbol{\theta}|\boldsymbol{L}) = f_{\mathcal{L}|\Theta}(\boldsymbol{L}|\boldsymbol{\theta}) = \prod_{k=1}^{N}\prod_{t=1}^{n}\frac{\Gamma\left(\frac{\nu_k+1}{2}\right)}{\Gamma\left(\frac{\nu_k}{2}\right)\sqrt{\nu_k\pi}\psi_k}\left[1 + \frac{u_{k,t}^2}{\nu_k\psi_k^2}\right]^{-\frac{\nu_k+1}{2}}. \tag{15}$$

For the purpose of maximum likelihood (ML) estimation, the logarithm of this likelihood function is easier to handle (see [3]). In that contribution, a computationally convenient ML estimation of the model parameters was achieved by rewriting the t-distributions as conditional normal distributions with latent variables; these variables play the role of weights in an iteratively reweighted least-squares algorithm. As the main innovation of the current contribution, the likelihood function (15) is incorporated into a Bayesian model instead, which is described in the following.

The Bayesian model: In this contribution, both informative and non-informative prior information is considered. The result of using a fully non-informative prior is that the posterior density follows directly from the likelihood function. In the case of an informative prior, a joint pdf must be specified for the random vector $\boldsymbol{\Theta}$. This task is simplified by the assumption of stochastic independence of the parameter groups $\boldsymbol{\beta}$, $\boldsymbol{\alpha}$, $\boldsymbol{\psi}$ and $\boldsymbol{\nu}$, so that the factorization property,

$$f_{\Theta}(\boldsymbol{\theta}) = f_{\Theta}(\boldsymbol{\beta},\boldsymbol{\alpha},\boldsymbol{\psi},\boldsymbol{\nu}) = f_{\Theta}(\boldsymbol{\beta})f_{\Theta}(\boldsymbol{\alpha})f_{\Theta}(\boldsymbol{\psi})f_{\Theta}(\boldsymbol{\nu}), \tag{16}$$

holds. Consequently, individual prior densities can be specified for these parameter groups. As the prior density of $\boldsymbol{\beta}$ depends on the selected functional model $h(\boldsymbol{\beta})$, its specification is fixed after the introduction of the application example in Section 4. In contrast, the prior densities for $\boldsymbol{\alpha}$, $\boldsymbol{\psi}$ and $\boldsymbol{\nu}$ do not depend on the choice of the functional model but mainly on the precision of the sensors or instruments employed. In the case that no prior information is available for the employed sensors or instruments, it is still possible to define prior densities for these three groups of parameters. Due to the assumption, in connection with the error Equation (2) and the stochastic model (3), that the N time series is stochastically independent, the prior density can be further simplified to

$$f_\Theta(\boldsymbol{\alpha}) = f_\Theta(\boldsymbol{\alpha}_1) f_\Theta(\boldsymbol{\alpha}_2) \cdots f_\Theta(\boldsymbol{\alpha}_N), \tag{17}$$

$$f_\Theta(\boldsymbol{\psi}) = f_\Theta(\psi_1) f_\Theta(\psi_2) \cdots f_\Theta(\psi_N), \tag{18}$$

$$f_\Theta(\boldsymbol{\nu}) = f_\Theta(\nu_1) f_\Theta(\nu_2) \cdots f_\Theta(\nu_N). \tag{19}$$

Consequently, as far as the parameters $\boldsymbol{\psi}$ and $\boldsymbol{\nu}$ are concerned, only univariate prior densities $f_\Theta(\psi_k)$ and $f_\Theta(\nu_k)$ need to be specified. When it is known that the scale factor ψ is between $\psi_{\min}$ and $\psi_{\max}$, the prior density defining the continuous uniform distribution $U(\psi_{\min}, \psi_{\max})$ can be used as a weak form of prior information. The specification of the prior for the df ν_k can be based, on the one hand, on the requirement $\nu_k > 2$, so that the variance of the t-distributed random variables is defined. On the other hand, the t-distribution is practically indistinguishable from a normal distribution for dfs greater than 120 (cf. [16]), so that the upper limit $\nu_k \leq 120$ can be fixed. In the absence of further information about the dfs, the prior density defining the continuous uniform distribution $U(2, 120)$ is reasonable. The auto-correlations of a time series may, for instance, be induced by calibration corrections within the measurement device, by movements of the measured object, or by a combination of the two effects. Therefore, a general definition of the prior density of the AR coefficients is not trivial. For this reason, a non-informative prior density is specified under the additional assumption that all of the AR coefficients are stochastically independent.

3. The Developed MCMC Algorithm

Because of the use of the Student distribution for the white measurement noise (Equation (15)), an analytical solution of the posterior density based on the Bayes theorem (Equation (9)) is not possible, so it can only be solved numerically. The general solution approach is based on generating a so-called Markov Chain for the unknown posterior density using the MCMC method. MCMC algorithms are commonly used in all fields of statistics because of their versatility and generality. When an MCMC method is applied to solve the posterior density function given in (9), it is usually realized with a Gibbs sampler. An implementation of a Gibbs sampler relies on the availability of the complete conditional pdfs of all parameters of interest in our particular problem (cf. Equation (4)). However, the complete conditional pdfs of all parameters of interest are not readily available. In such cases, a Metropolis–Hastings (MH) method can be incorporated within a Gibbs sampler to draw samples from the parameters, the full conditional pdf of which cannot be analytically determined. In this paper, we demonstrate the development of such an algorithm, known as Metropolis–Hastings-within Gibbs. In this algorithm, the Gibbs sampler is used to generate the Markov Chain for $\boldsymbol{\theta}|\boldsymbol{L}$, and within the Gibbs sampler the MH algorithm is used to generate random numbers. For a clearer presentation, the solution algorithm is encapsulated by two separate functions. For the Gibbs sampler, the outer function is given by,

$$\begin{aligned}
&\text{for: } j = 1, \ldots, \mathbb{M} \\
&\text{Draw } \boldsymbol{\psi}^j | \boldsymbol{\beta}^{j-1}, \boldsymbol{\nu}^{j-1}, \boldsymbol{\alpha}^{j-1}, \boldsymbol{L} && \sim f_{\boldsymbol{\Theta}|\mathcal{L}}(\boldsymbol{\psi}|\boldsymbol{\beta}, \boldsymbol{\nu}, \boldsymbol{\alpha}, \boldsymbol{L}) \\
&\text{Draw } \boldsymbol{\beta}^j | \boldsymbol{\psi}^{j}, \boldsymbol{\nu}^{j-1}, \boldsymbol{\alpha}^{j-1}, \boldsymbol{L} && \sim f_{\boldsymbol{\Theta}|\mathcal{L}}(\boldsymbol{\beta}|\boldsymbol{\psi}, \boldsymbol{\nu}, \boldsymbol{\alpha}, \boldsymbol{L}) \\
&\text{Draw } \boldsymbol{\nu}^j | \boldsymbol{\beta}^{j}, \boldsymbol{\psi}^{j}, \boldsymbol{\alpha}^{j-1}, \boldsymbol{L} && \sim f_{\boldsymbol{\Theta}|\mathcal{L}}(\boldsymbol{\nu}|\boldsymbol{\beta}, \boldsymbol{\psi}, \boldsymbol{\alpha}, \boldsymbol{L}) \\
&\text{Draw } \boldsymbol{\alpha}^j | \boldsymbol{\beta}^{j}, \boldsymbol{\psi}^{j}, \boldsymbol{\nu}^{j}, \boldsymbol{L} && \sim f_{\boldsymbol{\Theta}|\mathcal{L}}(\boldsymbol{\alpha}|\boldsymbol{\beta}, \boldsymbol{\psi}, \boldsymbol{\nu}, \boldsymbol{L})
\end{aligned} \tag{20}$$

where $\mathbb{M}$ is the length of the MCMC. Such a Markov chain ensures the convergence of the distribution of the samples to the target distribution after a few burn-in periods $\mathbb{V}$ [17]. We observe that the full conditional pdf shown in (20) does not fit to any known pdf and, therefore, we cannot directly draw samples from it.

However, there are two challenges to calculating the conditional posterior densities. The first challenge is that it results from the likelihood function and the prior density. Consequently, changing the distributional assumption for the prior density results in a new conditional posterior density. While this challenge can be overcome with a small amount of effort, the second challenge is much more fundamental. For the calculation of the conditional posterior densities, several integrals have to be solved and this may not be analytically possible. To overcome these challenges, the MH algorithm is used to draw the required random numbers. The general algorithm for drawing a random number θ_i^j from the posterior density $f_{\boldsymbol{\Theta}|\mathcal{L}}(\boldsymbol{\theta}|\boldsymbol{L})$ follows from the following steps:

$$\begin{aligned}
&\text{1. Generate: } \theta_i^{\text{new}} \sim N\left(\theta_i^{j-1}, \lambda_{\theta_i}\right) \\
&\text{2. Set: } \boldsymbol{\theta}^{\text{new}} = \left[\theta_1^j, \theta_2^j, \ldots, \theta_i^{\text{new}}, \ldots, \theta_{m-1}^{j-1}, \theta_m^{j-1}\right]^T \\
&\boldsymbol{\theta}^{\text{old}} = \left[\theta_1^j, \theta_2^j, \ldots, \theta_i^{j-1}, \ldots, \theta_{m-1}^{j-1}, \theta_m^{j-1}\right]^T \\
&\text{3. Calculate: } \xi = \min\left[1, \frac{f_{\boldsymbol{\Theta}}(\boldsymbol{\theta}^{\text{new}})\mathrm{L}(\boldsymbol{\theta}^{\text{new}}|\boldsymbol{L})}{f_{\boldsymbol{\Theta}}\left(\boldsymbol{\theta}^{\text{old}}\right)\mathrm{L}\left(\boldsymbol{\theta}^{\text{old}}|\boldsymbol{L}\right)}\right] \\
&\text{4. Accept or Reject: } \tau \sim U(0,1) \\
&\text{if } \tau \leq \xi: \quad \theta_i^j = \theta_i^{\text{new}} \\
&\text{else}: \quad \theta_i^j = \theta_i^{j-1},
\end{aligned} \tag{21}$$

where m in Equation (22) denotes the length of the parameter vector. The results of the Metropolis–Hastings-within Gibbs are $\mathbb{M}$ random realizations of the unknown parameters $\boldsymbol{\beta}, \boldsymbol{\alpha}, \boldsymbol{\psi}$ and $\boldsymbol{\nu}$ from the posterior density $f_{\boldsymbol{\Theta}|\mathcal{L}}(\boldsymbol{\theta}|\boldsymbol{L})$. The estimated values $\hat{\boldsymbol{\theta}}$ for the unknown parameters with their variance–covariance matrix (VCM) result from (cf. [5]):

$$\hat{\theta}_i = \frac{1}{\mathbb{M} - \mathbb{V}} \sum_{j=\mathbb{V}+1}^{\mathbb{M}} \theta_i^j \quad ; \quad \hat{\boldsymbol{\Sigma}}_{\theta_{i,s}} = \frac{1}{\mathbb{M} - \mathbb{V}} \sum_{j=\mathbb{V}+1}^{\mathbb{M}} \left(\theta_i^j - \hat{\theta}_i\right)\left(\theta_s^j - \hat{\theta}_s\right). \tag{22}$$

4. Closed Loop Monte Carlo Simulation

4.1. The Framework of the Simulation

In our Closed Loop Monte Carlo simulation (CLS), we rely on the real-world application demonstrated in [3], in which we used a multi-sensor-system (MSS) composed of a laser scanner and two firmly attached pieces of GNSS equipment (see [18] for details). We consider a multivariate, non-linear regression model in terms of a circle in $N = 3$ dimensions that has the following six parameters: two for the orientation (φ and ω) of its unit normal vector; one for the radius (r); and three for the circle center (c_x, c_y, c_z). The observable 3D circle points are described by

$$l_{x,t} := l_{1,t} = h_{1,t}(\boldsymbol{\beta}) = r\cos(\kappa_t)\cos(\varphi) + c_x \tag{23}$$
$$l_{y,t} := l_{2,t} = h_{2,t}(\boldsymbol{\beta}) = r\cos(\kappa_t)\sin(\varphi)\sin(\omega) + r\sin(\kappa_t)\cos(\omega) + c_y \tag{24}$$
$$l_{z,t} := l_{3,t} = h_{3,t}(\boldsymbol{\beta}) = -r\cos(\kappa_t)\sin(\varphi)\cos(\omega) + r\sin(\kappa_t)\sin(\omega) + c_z, \tag{25}$$

with $t = 1, \ldots, 2000$ and where $\kappa_t = \frac{2\pi}{200} \cdot t$ represents fixed rotation angles around the z-axis. In this simulation, the functional parameters are the circle parameters $\boldsymbol{\beta}$, which were assumed to take the true values

$$\boldsymbol{\beta} = \begin{bmatrix} c_x & c_y & c_z & r & \omega & \varphi \end{bmatrix}^T = \begin{bmatrix} 1716.0 & 3012.0 & 1064.0 & 30.0 & 0.0019\,\text{rad} & -0.0013\,\text{rad} \end{bmatrix}^T. \tag{26}$$

The random deviations $\mathcal{E}_t$ were generated by the AR(2) processes

$$e_{k,t} = \sum_{j=1}^{2} \alpha_{k,j} e_{k,t-1} + u_{k,t}, \quad \text{for } k = 1, 2, 3, \tag{27}$$

with true coefficients

$$\alpha_{1,1} = 0.57, \alpha_{1,2} = 0.11, \alpha_{2,1} = 0.67, \alpha_{2,2} = 0.22, \alpha_{3,1} = 0.35, \alpha_{3,2} = 0.55.$$

The innovations of these processes are sampled from the scaled t-distributions

$$u_{k,t} \overset{\text{ind.}}{\sim} t_{\nu_k}(0, \psi_k^2), \tag{28}$$

with true scale parameters

$$\psi_x := \psi_1 = 0.2, \quad \psi_y := \psi_2 = 0.2, \quad \psi_z := \psi_3 = 0.4$$

and true dfs

$$\nu_x := \nu_1 = 8, \quad \nu_y := \nu_2 = 10, \quad \nu_z := \nu_3 = 4.$$

In Equation (16), the prior density has been introduced for the general Bayes model. In this simulation, only prior densities for the functional parameters $\boldsymbol{\Theta}$ (see Equation (26)) are assumed to be known:

$$f_{\boldsymbol{\Theta}}(\boldsymbol{\beta}) = f_{\boldsymbol{\Theta}}(c_x, c_y, c_z) f_{\boldsymbol{\Theta}}(r) f_{\boldsymbol{\Theta}}(\omega) f_{\boldsymbol{\Theta}}(\varphi). \tag{29}$$

In Equation (29), we assume that the prior information for the center of the circle, the radius and the angles are independent of each other. The reason for this assumption is that this information is obtained from data sheets or from calibrations. These are specified as follows.

The prior density of the circle center: The prior for the center of the circle is the knowledge that it must be located approximately in the middle between the observations constituting a circle. The location parameter μ_c for the definition of the prior density of the circle center is thus dependent on the observations. Therefore, we consider the prior of the circle center as weak prior information. Hence, we use the identity matrix as a VCM for $\boldsymbol{\Sigma}_c$. The uncertainty of a coordinate component of the prior information of the circle center is thus significantly larger than the assumed measurement precision of a single observation. As a pdf for the prior $f_{\boldsymbol{\Theta}}(c_x, c_y, c_z)$, the multivariate normal distribution is assumed with the expected value μ_c and VCM $\boldsymbol{\Sigma}_c$.

The prior density for the radius: The prior information for the radius of the circle is the result of a calibration measurement using a laser tracker. The manufacturer specifies the accuracy of a single point measurement with this instrument as the maximum permissible error of $MPE_{x,y,z} = \pm 15\mu m + 6\frac{\mu m}{m}$. During the measurement, the diameter of the wing, on which the GNSS antenna was mounted, was determined. For this purpose, the 3D coordinates were measured on the left and right hand side and the Euclidean distance was calculated, which corresponds to the diameter. In total, the radius was estimated four times in this way. By averaging these results, $\mu_r = 30.0026$ was obtained as the location parameter for the prior density of the radius. The scale parameter for the prior is the result of the standard deviation $\sigma_r = 0.0043$ of the four determined radii of the laser tracker measurement. The measurement results of the laser tracker are assumed to be normally distributed. Hence, the normal distribution is assumed for the prior density of the radius of the circle.

The prior density for the rotation angles: The prior information for the rotation angles is derived from the instrument's levelling. The manufacturer specifies the accuracy of the bubble level as ±0.0047 rad for a 99.9% confidence interval. It is assumed that this information refers separately to one vial axis, so that $f_{\Theta}(\omega)$ and $f_{\Theta}(\varphi)$ are independent of each other. Due to the specification of the accuracy by means of the confidence interval, the uniform distribution is used as the prior family for the angles. The limit of the confidence interval is used as the limit for the uniform distribution, from which follows $a_\omega = -0.0047$ rad and $b_\omega = 0.0047$ rad. This results in the prior density:

$$f_{\Theta}(\omega) = \begin{cases} \frac{1}{b_\omega - a_\omega} & a_\omega \leq \omega \leq b_\omega \\ 0 & \text{else} \end{cases}. \tag{30}$$

For φ, the density of the prior is identical to that of the previous formula.

4.2. Results of the Simulation

The results described in this section were achieved with $\mathbb{M} = 10,000$ iterations and a burn-in period with $\mathbb{V} = 3000$ for the MCMC algorithm. We compare in the following three estimation procedures: Bayesian informative (Bayes Inf) using prior knowledge about the functional parameters $\boldsymbol{\beta}$, Bayesian non-informative (Bayes Non-Inf) without using prior knowledge about $\boldsymbol{\beta}$ and the EM (expectation–maximization) algorithm developed in [3]. Before the results of the three estimation methods from the entire CLS are compared, the result of the Bayes Non-Inf solution is considered in more detail. For this purpose, the result of the Markov chains from a single simulation solution is considered in Figure 1. In total, Markov chains were generated for 18 unknown parameters. Figure 1 only shows a representative selection of the results and is limited to the results of the z-component and only one rotation angle. For the other components, the generated chains correspond to the behavior of the z-component. The green dashed line in Figure 1 shows the true value of the parameters used to generate the simulated observations. The red dashed line, or the red cross on the secondary diagonal, show the estimated parameters resulting from the Markov chains. On the diagonal, the distribution of the generated random numbers of the chain is shown as a histogram. The histograms show how the generated random numbers scatter around the estimated parameter and that the true value is close to the estimated parameter. The secondary diagonal of the figure represents the scatter of the generated random numbers depending on two parameters. From this distribution, the correlation of the generated chain between two parameters can be calculated. A dependency can be seen, especially for the scaling parameter and the df of the z-components. This also holds for the coefficients of the AR(2) process of the z-component. The same can be noticed for the x- and y-components. For all other parameters, the major axes of the ellipse are rather parallel to the axes of the expected values, which corresponds to a correlation around zero. The fact that, for example, the scale parameter and the df show stronger correlation behavior is also to be expected. The reason for this is that, with a smaller scale factor, more observations lie in the tails of the distribution, which leads to a smaller df. In contrast, if the scaling factor is large, there are fewer observations in the tails of the distribution, so a larger df can be selected.

To compare the results of the three approaches, the adjusted observations $\hat{l}$ are used instead of the estimated parameters $\hat{\beta}$. The reason for this is that, in all estimation procedures, the estimated parameters are closer to the nominal value of the simulation parameters and, therefore, it is difficult to judge which approach produces the best results. On the other hand, the predicted observations $\hat{l}$ include the cumulative estimation uncertainties of all parameters $\hat{\beta}$, allowing for an easier comparison. Furthermore, we restrict ourselves here to the representation of the z-component because $\hat{l}_z$ is most sensitive to inaccuracies in the estimated angles $\hat{\omega}$ and $\hat{\varphi}$. The results for the x- and y-components show a result similar to that of the model shown in Figure 2 for the z-component.

Before discussing the results of the three approaches, Figure 2 is first explained. The $\hat{l}$ were calculated with the estimated parameters $\hat{\beta}$ using Equation (25). Subsequently, the $\hat{l}_z$ were reduced by the true value for the observations $E(l_z)$, which is why the predicted observations scatter around 0. The dashed line is the mean value of observation $l_{z,t}$, which results from the 10,000 results of the CLS for one observation t. The dashed blue line of the mean value of the EM algorithm cannot be seen in the figure because it is overlaid by the dashed red line. The colored area shows, for observation t, the 95% confidence interval that results from the 10,000 predicted observations $\hat{l}_{z,t}$. The colored lines >0 show the maximum deviation from the true value for the observation $\hat{l}_{z,t}$ that appeared in the 10,000 simulations. The colored lines <0 show the minimum deviation from the true value for the corresponding approaches.

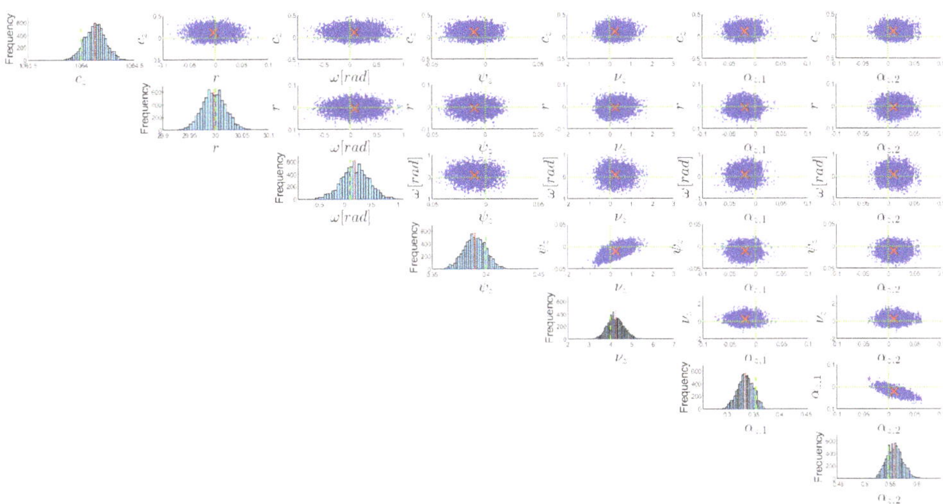

Figure 1. Result of the generated Markov chain after burn–in for Bayes Non–Inf: The main diagonal shows the distribution of the generated random numbers of a parameter. The secondary diagonal shows the correlation behavior of the generated random numbers at time j between two parameters.

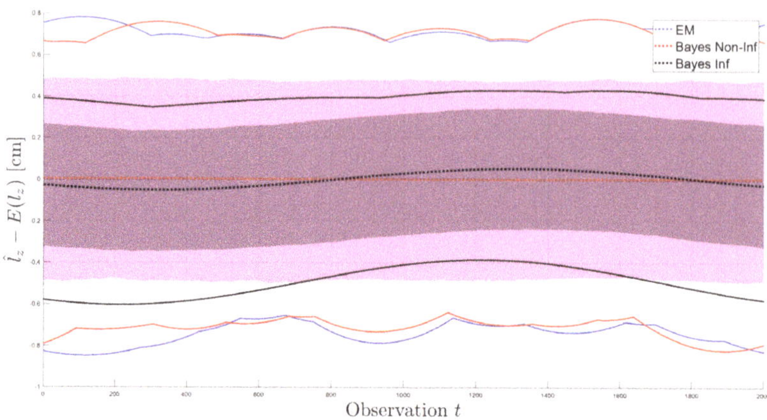

Figure 2. Result of the 10,000 CLS for the 2000 predicted observations of the z–component $\hat{l}_z$ of the EM algorithm, Bayes non–informative (Bayes Non–Inf) and Bayes informative (Bayes Inf).

It is expected that the result of the EM algorithm is identical to the Bayes Non-Inf solution. This can be seen clearly in the result of the mean, where the two dashed lines (blue and red) overlap almost completely and only deviate from each other by a maximum of $\approx$0.002 cm. Furthermore, the mean values of the two estimation procedures are identical to the nominal value of 0 for all predicted observations within 10^{-2} cm. This deviation can be explained by the fact that the CLS was only performed 10,000 times and not infinitely often. For the confidence interval of the EM and Bayes Non-Inf solution, a similar behavior can be seen as for the mean value. For most observations t, the two confidence intervals overlap almost perfectly and, only for a few observations, a difference of at most $\approx$0.05 cm can be identified. However, this identical behavior is not seen in the maximum and minimum deviations of the blue and red lines, where the Bayes Non-Inf solution more often has smaller minimum deviations than the EM results. The reason for this has not yet been analysed in more detail and will be addressed in future work.

In the result of the mean value of Bayes Inf, a deviation from the mean value of EM and Bayes Non-Inf can be seen. The solution of the mean value of Bayes Inf oscillates cyclically around the nominal value of 0 with a maximum deviation of $\pm$0.05 cm, whereby this deviation is smaller by a factor of 10 than the scale factor ψ_z of the measurement noise. The mean values of the EM and Bayes also oscillate around 0, but this is smaller by a factor of 100 and therefore cannot be seen visually in Figure 2. The influence of the prior information $f_{\Theta}(\beta)$ on the observations $\hat{l}_{z,t}$ can be seen well in the confidence interval and the lines marked in black. These are clearly closer to the true value than in the EM algorithm and Bayes Non-Inf. It is of particular interest that the maximum deviation lies in the 95% confidence interval of EM and Bayes Non-Inf. This is not always the case for the minimum deviation of Bayes Inf, but the black line is still closer to zero than the blue and red lines. The reason for this is mainly due to the prior information for the angles $f_{\Theta}(\omega)$ and $f_{\Theta}(\varphi)$, which improves the estimation of ω and φ.

To compare the results $\hat{l}_x$, $\hat{l}_y$ and $\hat{l}_z$ of the three approaches, the RMSE of the true observations is calculated for each CLS result. From these RMSE values, the statistical measures in Table 1 were calculated for each approach. All results in the table show the same behavior as the results previously displayed in Figure 2. The EM and Bayes Non-Inf results are almost identical, with the small deviation explained by the different methods used to estimate the parameters. In the Bayes Inf solution, however, all statistical measures are smaller. For the mean, the RMSE of Bayes Inf is about 37% smaller than the RMSE of the EM algorithm, and for the median the difference is slightly larger, at about 41%. Only for the maximum RMSE value from the 10,000 simulations is the difference between EM and

Bayes Informative significantly smaller at about 13%, but the maximum RMSE of Bayes Inf is still significantly smaller than the result from EM.

Table 1. Root-Mean-Square-Error (RMSE) for the predicted observations to the true observations.

Methode	Mean [cm]	Median [cm]	Min [cm]	Max [cm]	σ [cm]
EM	0.107	0.103	0.013	0.337	0.041
Bayes Non-Inf	0.104	0.100	0.012	0.303	0.040
Bayes Inf	0.067	0.060	0.004	0.290	0.032

5. Conclusions

To achieve an Bayesian adaptive robust adjustment of a multivariate regression time series with outlier-afflicted/heavy-tailed and autocorrelated errors, we described the theory and implementation of an MCMC based approach consisting of a Metropolis-within-Gibbs algorithm. In particular, the Gibbs sampler and the Metropolis–Hastings algorithm have been employed for robust Bayesian estimation of the parameters of a non-linear functional model with AR and t-distributed errors. An advantage of this procedure compared to the EM algorithm, besides the capability to process additional prior knowledge, is that the approximation of the posterior model parameters is feasible without linearization of the functional model. Furthermore, the approximation of the VCM $\hat{\Sigma}_\theta$ of the estimated parameters can be derived directly from the generated chains. CLS showed that the bias of the parameter estimates and the adjusted observations, as well as the RMSE, are significantly reduced when an informative Bayesian approach is used, but only under the condition that good prior information, that is, information that contains the true value, is assumed to be known. Otherwise, the prior can also have the opposite effect and may cause the estimated parameters to deteriorate.

Author Contributions: Conceptualization, H.A., A.D., B.K., J.-A.P.; methodology, A.D., H.A. and B.K.; software, A.D., B.K. and H.A.; validation, A.D. and H.A.; formal analysis, A.D. and H.A.; investigation, A.D., B.K., and H.A.; writing—original draft preparation, A.D., H.A. and B.K.; writing—review and editing, H.A. and B.K.; visualization, H.A. and A.D.; supervision, H.A., B.K. and J.-A.P.; project administration, H.A., B.K. and J.-A.P.; funding acquisition, H.A., B.K. and J.-A.P. All authors have read and agreed to the published version of the manuscript.

Funding: This research was supported by the Deutsche Forschungsgemeinschaft (DFG, German Research Foundation)—project number 386369985.

Institutional Review Board Statement: Not applicable.

Informed Consent Statement: Not applicable.

Conflicts of Interest: The authors declare no conflict of interest.

References

1. Koch, K.R. *Parameter Estimation and Hypothesis Testing in Linear Models*; Springer: Berlin, Germany, 1999.
2. Lange, K.L.; Little, R.J.A.; Taylor, J.M.G. Robust Statistical modeling using the t-distribution. *J. Am. Stat. Assoc.* **1989**, *84*, 881–896. [CrossRef]
3. Alkhatib, H.; Kargoll, B.; Paffenholz, J.-A. Further results on robust multivariate time series analysis in nonlinear models with autoregressive and t-distributed errors. In *Time Series Analysis and Forecasting*; Valenzuela, O., Rojas, F., Pomares, H., Rojas, I., Eds.; ITISE 2017; Contributions to Statistics; Springer: Cham, Switzerland, 2018; pp. 25–38. [CrossRef]
4. Kargoll, B.; Omidalizarandi, M.; Loth, I.; Paffenholz, J.A.; Alkhatib, H. An iteratively reweighted least-squares approach to adaptive robust adjustment of parameters in linear regression models with autoregressive and t-distributed deviations. *J. Geod.* **2018**, *92*, 271–297. [CrossRef]
5. Koch, K.R. *Introduction to Bayesian Statistics*, 2nd ed.; Springer: Berlin/Heidelberg, Germany, 2007. [CrossRef]
6. Geweke, J. Bayesian treatment of the independent Student-t linear model. *J. Appl. Econom.* **1993**, *8*, 19–40. [CrossRef]
7. Koch, K.R. Monte Carlo methods. In *Mathematische Geodäsie/Mathematical Geodesy*; Freeden, W., Ed.; Springer Spektrum: Berlin/Heidelberg, Germany, 2020; pp. 445–475. [CrossRef]

8. Chib, S.; Greenberg, E. Bayes inference in regression models with ARMA(p,q) errors. *J. Econom.* **1994**, *64*, 183–206. [CrossRef]
9. Barnett, G.; Kohn, R.; Sheather, S. Robust Bayesian estimation of autoregressive-moving-average models. *J. Time Ser. Anal.* **1997**, *18*, 11–28. [CrossRef]
10. Barnett, G.; Kohn, R.; Sheather, S. Bayesian estimation of an autoregressive model using Markov chain Monte Carlo. *J. Econom.* **1996**, *74*, 237–254. [CrossRef]
11. Alkhamisi, M.A.; Shukur, G. Bayesian analysis of a linear mixed model with AR(p) errors via MCMC. *J. Appl. Stat.* **2011**, *32*, 741–755. [CrossRef]
12. Priestley, M.B. *Spectral Analysis and Time Series*; Academic Press: Cambridge, UK, 1981.
13. Bera, A.K.; Jarque, C.M. Model specification tests: A simultaneous approach. *J. Econom.* **1982**, *20*, 59–82. [CrossRef]
14. McDonald, J.B. Partially adaptive estimation of ARMA time series models. *Int. J. Forecast.* **1989**, *5*, 217–230. [CrossRef]
15. Tiku, M.L.; Wong, W.-K.; Vaughan, D.C.; Bian, G. Time series models in non-normal situations: symmetric innovations. *J. Time Ser. Anal.* **2000**, *21*, 571–596. [CrossRef]
16. Koch, K.R. Expectation maximization algorithm and its minimal detectable outliers. *Stud. Geophys. Geod.* **2017**, *61*, 1–18. [CrossRef]
17. Robert, C.P.; Casella, G. *Monte Carlo Statistical Methods*; Springer: New York, NY, USA, 2005.
18. Paffenholz, J.-A. Direct Geo-Referencing of 3D Point Clouds with 3D Positioning Sensors. Ph.D. Thesis, Deutsche Geodätische Kommission, Munich, Germany, 2012; Series C, No. 689.

engineering proceedings

MDPI

Proceedings

Comparative Analysis of Statistical and Analytical Techniques for the Study of GNSS Geodetic Time Series †

Paola Barba, Belén Rosado *, Javier Ramírez-Zelaya and Manuel Berrocoso

Laboratorio de Astronomía, Geodesia y Cartografía, Departamento de Matemáticas, Facultad de Ciencias, Campus de Puerto Real, Universidad de Cádiz, 11510 Puerto Real, Spain; paola.barbaceballos@alum.uca.es (P.B.); javierantonio.ramirez@uca.es (J.R.-Z.); manuel.berrocoso@uca.es (M.B.)

* Correspondence: belen.rosado@uca.es

† Presented at the 7th International Conference on Time Series and Forecasting, Gran Canaria, Spain, 19–21 July 2021.

Abstract: GNSS systems allow precise resolution of the geodetic positioning problem through advanced techniques of GNSS observation processing (PPP or relative positioning). Current instrumentation and communications capabilities allow obtaining geocentric and topocentric geodetic high frequencies time series, whose analysis provides knowledge of the tectonic or volcanic geodynamic activity of a region. In this work, the GNSS time series study was carried out through the use and adaptation of R packets to determine their behavior, obtaining displacement velocities, noise levels, precursors in the time series, anomalous episodes, and their temporal forecast. Statistical and analytical methods were studied, for example, ARMA, ARIMA models, least-squares methods, wavelet functions, and Kalman techniques. To carry out a comparative analysis of these techniques and methods, significant GNSS time series obtained in geodynamically active regions (tectonic and/or volcanic) were considered.

Keywords: GNSS time series analysis; statistical methods; R software

check for updates

Citation: Barba, P.; Rosado, B.; Ramírez-Zelaya, J.; Berrocoso, M. Comparative Analysis of Statistical and Analytical Techniques for the Study of GNSS Geodetic Time Series. *Eng. Proc.* **2021**, *5*, 21. https://doi.org/10.3390/engproc2021005021

Published: 28 June 2021

1. Introduction

GNSS systems (GPS—USA, Glonass—Russia, Galileo—European Union, and Beidou—China), initially designed for land, sea, and air navigation, possess the ability to use specific and advanced techniques and methods to provide precisions sufficient to evaluate the movement of tectonic plates, the volcanic activity, or possible hillside landslides from the velocities obtained in positioning successive subcentimeter precision.

Geodynamic GNSS geodetic networks are based on permanent stations that operate continuously, even for high sample rates. The positions obtained make up time series, initially in geocentric Cartesian coordinates (X, Y, Z). To facilitate the notion of displacement on the surface of the Earth, these coordinates are transformed into a topocentric system (e, n, u). The analysis of these series provides the displacement velocity vector as well as the anomalies that may have occurred in the time period defined by the series. For this, and according to the objective of the study, different analytical or statistical methods of time series analysis were used.

In this work, a review of geodetic time series analysis methods and techniques is presented, and the GNSS positioning of some stations of the SPINA network (South of the Antarctic Peninsula and North of Africa) that present very significant particularities, are evaluated, e.g., SEVI (Seville) and CAAL (Calar Alto, Almería). The R language was used to design and develop new applications and/or adapt existing packages to the case of topocentric time series. Finally, a comparative analysis of the techniques and methods used was carried out, and the optimal procedure was proposed for the cases studied, taking into account the results obtained.

2. Time Series GNSS Geodetics

GNSS data were analyzed by using scientific software Bernese v5.0 [1]. Along with the parameter estimation process, carrier phase double difference data were used in ionospheric delay–free mode. Tropospheric errors were handled by using a combination of the a priori Saastamoinen model [2] and Neill mapping functions [3]. Tropospheric parameters were estimated hourly, and ambiguities were fixed for the baseline by using the ionosphere-free observable with an a priori ionospheric model for determining the wide lane ambiguity [4]. Ocean tide loading displacement corrections from the Onsala Observatory were also introduced. Normal equations were computed for each daily solution.

It was considered as a VILL reference station (Villafranca) because it belongs to the IGS network and, therefore, has geocentric coordinates and high precision ITRF2008 velocities [5]. The result of this treatment was a geocentric Cartesian time series (X, Y, Z) of subcentimeter accuracy. For their geodynamic interpretation, they were transformed into topocentric coordinates (east, north, elevation). In Rosado et al. 2019 [6], the algorithm used for this coordinate transformation is described in detail.

3. Statistical and Analytical Methods and Techniques

Topocentric GNSS time series are affected by various sources of error from the spatial constellation, the GNSS signal propagation medium, and the tracking station. Therefore, the precision of the ephemeris, of the corrections of the satellite oscillators, of the parameters of the Earth's rotation; the influence of the ionosphere and the troposphere; station stability, multipath effect, electromagnetic signal interference, etc., decisively influence the quality of the calculated time series. The existence of anomalous observations, the loss of observations due to obstacles, the noise introduced by other signals, etc., make a prior descriptive analysis of the series obtained necessary. Through this analysis of the raw series, outliers, gross errors, and, especially, the noise level of the series were detected. These parameters recommended an a priori methodological procedure to be followed.

To eliminate or reduce the noise level of the series, various time series filtering techniques were considered, methodologically grouped into initial filters (1–σ, 2–σ, Outlier R), analytical filters (Kalman, wavelets), and statistical filters (ARMA/ARIMA). Once this process was carried out, adjustment techniques were applied in order to extract the information on the geodynamic behavior of the GNSS series considered. In this process, it was essential to clearly define the objective pursued and the series to be analyzed. A distinction was made between the horizontal components (east, north) and, on the other hand, the vertical component (elevation) between linear and non-linear behaviors, between series that present anomalies due to events of a tectonic or volcanic nature, etc. All this made it impossible to establish a single procedure for each and every one of the GNSS geodetic series. Rather an adaptation of techniques and methods was carried out according to the geodynamic process under study. Figure 1 shows the adjustment techniques used in this work: linear adjustment, Create and Analyze Time Series (CATS), Seasonal-Trend Loess Decomposition (STL), Kalman Adjustment, and ARMA/ARIMA. These procedures developed were all carried out in R software.

These procedures are succinct and conceptually described below, resulting, however, in a greater depth in those that, due to their specificity, are practically exclusive for the GNSS series.

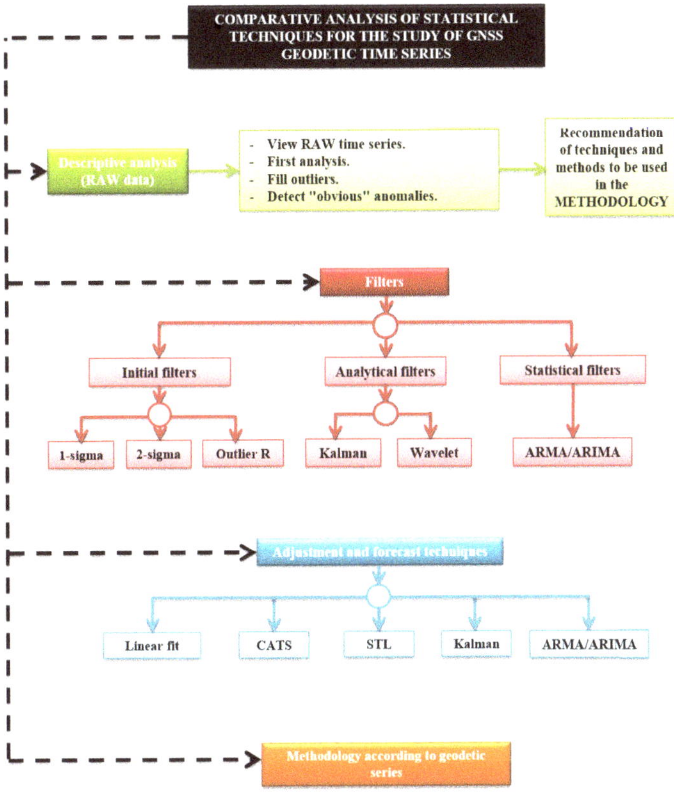

Figure 1. Scheme of statistical and analytical techniques and methods for the treatment of GNSS time series.

3.1. Initial Filters of the Series

The objective of any initial filtering, which was applied to the GNSS series, consists of the elimination of data with very different values, outliers, from the rest of the series. The 1σ and 2σ filters are based on the distance of the series points from a simple linear regression line. Depending on the chosen filtering, a greater (1σ) or less (2σ) number of data is eliminated from the series. In the case of non-linear series, this process is carried out by linear sections within the series. On the other hand, R contains a package, *forecast*, to filter time series data that are based on the Box–Cox transform [7,8], which is done by the *tsoutliers()* function. It was used to achieve greater linearity, homoscedasticity, and a tendency to a normal distribution of the values of the series.

3.2. Predictive Filtering: Kalman, ARIMA, ARMA

3.2.1. Kalman

For this filtering, it is necessary to know what the dynamic linear models are like. Assuming they are known, we proceed to define the Kalman filtering. The Kalman filter is of a predictive–corrective type; as the parameter θ_t that determines the state of the model at time t is calculated, and the estimation of the observations of the series is calculated [9]. Assuming $\theta_0 \sim N(m_0, C_0)$:

$$\theta_t = G_t\theta_{t-1} + c_i + R_i W_t$$

To calculate the estimate of the data of the series, the following is used:

$$y_t = F_t\theta_t + d_t + v_t$$

3.2.2. ARIMA Model

ARIMA (integrated moving average autoregressive) models are given by the $ARIMA(p,d,q)$, deal with stationary time series, and are made up of three models: the autoregressive (AR), the integrated (I), and the mean mobile (MA) model, which are defined, respectively, by p, d, and q. By uniting these three models, we get the ARIMA model, which is given by:

$$\phi_p(B)(1-B)^d Y_t = \phi_0 + \theta_q(B)e_t$$

where e_t represents the errors produced at time t and Y_t of the data of the series. Additionally:

$$\phi_p(B) = 1 - \phi_1 B - \phi_2 B^2 - \ldots - \phi_p B^p$$

$$\theta_q(B) = 1 - \theta_1 B - \theta_2 B^2 - \ldots - \theta_q B^q$$

where B is the delay operator.

3.2.3. ARMA Model

ARMA models, defined by $ARMA(p,q)$, deal with non-stationary series and are given by the union of autoregressive models (AR (p)) and moving average models (MA (q)). Therefore, by joining the expressions of both models, we obtained the expression of the ARMA model:

$$\phi_p(B)Y_t = \phi_0 + \theta_q(B)e_t$$

where $\phi_p(B)$ and $\theta_q(B)$ are defined in the same way as in the ARIMA model.

3.3. *Wavelet Analysis*

The wavelet transform decomposes a signal using functions (wavelets) well localized in both the physical space (time) and spectral space (frequency), generated from each other by translation and dilation [10]. The wavelet continuous transform tries to express a signal $x(t)$, continuous in time, by an expansion of proportional coefficients to the inner product between the signal and different scaled and translated versions of a function prototype ψ. This function, known as the mother wavelet or wavelet function, provides a decomposition of the data in the time-frequency plane, along with successive scales. This time-frequency transformation depends on two parameters, the scale parameter a, which is related to the frequency, and the time parameter b, related to the translation of function ψ in the time domain. The continuous wavelet transform is obtained by:

$$CWT(a,b) = \frac{1}{\sqrt{a}} \int_{-\infty}^{\infty} x(t)\psi(\frac{t-b}{a})$$

where ψ is the mother wavelet.

3.4. *CATS Analysis*

CATS adjustment (Create and Analyze Time Series) is based on stochastic analysis of the GNSS series using Maximum Likelihood Estimation (MLE). This estimate is optimal for the study of noise in a time series. This method makes it possible to simultaneously estimate the noise amplitudes, the linear trend, the periodic signal, and the amplitudes of the existing discontinuities, as well as the uncertainty of these parameters [11]. This setting makes it possible to differentiate between the linear and non-linear parts of the series. The linear part includes the calculation of outliers, the trend, sudden jumps (e.g., earthquakes), and sinusoidal terms. In non-linearity, different types of specific noise models are solved, e.g., white noise and power noise. For the analysis of the GNSS coordinate series, the following functional model is considered:

$$x(t) = a + bt + \sum_{j=1}^{2}(A\sin(\omega_j t) + B\cos(\omega_j t)) + \sum_{j=1}^{n} C_j H(t - T_j)$$

where x is the value of the GNSS coordinate at time t; a is the initial value; b is the velocity; ω_1 and ω_2 are the angular frequencies of the annual and semi-annual harmonic components; and A_j and B_j are the amplitudes of the sine and cosine, respectively. The coefficients C_j are the magnitudes of the discontinuities described by the following Heasivide function:

$$H(\tau) = \begin{cases} 0 & si\ \tau < 0 \\ 1 & si\ \tau \geq 0 \end{cases}$$

and the time instant of the discontinuity T_j. The number of discontinuities in each series is given by n. Therefore, the parameters to be estimated are the initial value a, the velocity b, the sine and cosine amplitudes of the annual and semi-annual harmonic components A_j and B_j, and the coefficients C_j of the magnitudes of the discontinuities considered.

To estimate the noise components using the MLE, the probability function is maximized by fitting the covariance matrix of the data. The resulting expression is given by:

$$lik(\hat{v}, C) = \frac{1}{(2\pi)^{\frac{N}{2}}(detC)^{\frac{1}{2}}} e^{-0.5\hat{v}^T C^{-1}\hat{v}}$$

Taking the natural logarithm, we obtain:

$$MLE = ln[lik(\hat{v}, C)] = -\frac{1}{2}[ln(detC) + \hat{v}^T C^{-1}\hat{v} + Nln(2\pi)]$$

where N is the number of epochs or observations, C is the covariance matrix of the data, and $\hat{v}$ are the post-fit residuals of a model applied to the original series using least squares with the same covariance matrix C.

Therefore, we are going to assume that the matrix C is a combination of two sources of error, a white noise component and a power series noise component, so that:

$$C = a_\omega^2 I + b_k^2 J_k$$

where a_ω and b_ω are the amplitudes of white noise and color noise, respectively. The identity matrix, I, is the covariance matrix of the white noise evoking independence in the time of this type of process. The matrix J_k is the noise covariance matrix of a power series with a spectral index k and is calculated by means of fractional models integrated in such a way that:

$$J_k = TT^T$$

where T is a transformation matrix obtained from:

$$T = \triangle T^{-k/4} = \begin{pmatrix} \psi_0 & 0 & 0 & \cdots & 0 \\ \psi_1 & \psi_0 & 0 & \cdots & 0 \\ \psi_2 & \psi_1 & \psi_0 & \cdots & 0 \\ \vdots & \vdots & \vdots & \ddots & \vdots \\ \psi_{n-1} & \psi_{n-2} & \psi_{n-3} & \cdots & \psi_0 \end{pmatrix}$$

where $\triangle T$ is the sample interval and,

$$\psi_n = \frac{\frac{k}{2}(1-\frac{k}{2})\ldots(n-1-\frac{k}{2})}{n!} = \frac{\Gamma(n-\frac{k}{2})}{n!\Gamma(-\frac{k}{2})}$$

When n tends to inf, ψ_n can be approximated by

$$\psi_n = \frac{n^{-\frac{k}{2}-1}}{\Gamma(-\frac{k}{2})}$$

Therefore, using MLE, we could fit the coordinate time series to a standard model accurate acoustic by estimating the noise amplitudes for a model, assuming that it is a combination of white noise and power series noise (WN + PLN, is say, White Noise + Power-Law Noise). This approach is based on the recent general formula of the covariance matrix for a power series process, allowing us to estimate noise amplitudes and spectral index together with the rest of the parameters of the station's motion model.

The stochastic properties and the linear parameters were adjusted together in one way, iterative through a function, to maximize them. The function to be maximized chose a model noise level and estimated the linear parameters, on which a new set of waste was calculated. Using these residuals and the covariance matrix, the value of likelihood and a new noise model with a higher likelihood value were chosen. This process was repeated until the likelihood function reached its maximum value.

3.5. STL Decomposition

STL decomposition (Seasonal and Trend decomposition procedure based on Loess) additively decomposes a time series into its three components trend, seasonality, and irregularities [12]. The time series can contain gaps due to various factors. These do not have a negative influence on the decomposition of the time series. Local regression (Loess) was used to estimate the three components of the series. STL decomposition consisted of two processes: internal and external. In the internal process, in each position, the values of the trend and seasonality components were estimated and updated with the Loess regression. In the external process, the irregularities component of the series was obtained. The trend and seasonality components were smoothed. However, both components were affected by the variation of the series, which could be solved by applying a filter to the seasonality component. This filter was composed of three models of moving averages and the Loess regression [12].

4. Application of Methodology Developed and/or Adapted R

4.1. Description of Selected Series from the Spina Network

Selected time series came from permanent geodetic stations located in the south of the Iberian Peninsula and North Africa, which constitute the SPINA network. This geodetic network is composed of 7 networks of permanent GPS stations: RAP, MERISTEMUM, IGS, IGN, REGAM, RENEP, and ERVA. Each of these networks is made up of GPS stations located in Andalusia, Murcia, the Valencian Community, the south of Portugal, and the north of Africa [10]. We used the position time series derived from daily observations and processed the positioning with respect to the IGS station located in Villafranca (VILL) to get site displacements. Figure 2 shows the horizontal displacement rates at GPS sites in the south of the Iberian Peninsula and North Africa, estimated from GPS time series data (January 2005 to January 2014) [13]. All GPS solutions were realized in the ITRF2005 global reference frame.

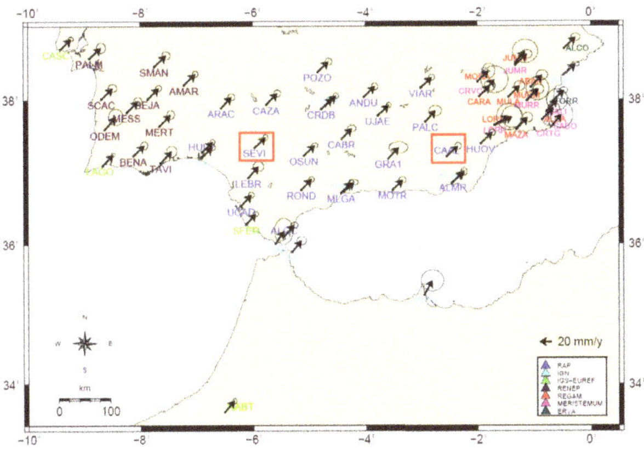

Figure 2. Horizontal displacement rates at GPS sites in the south of the Iberian Peninsula and North Africa, estimated from GPS time series data with 95% confidence level error ellipses. Different colors are used for different networks. The red rectangles indicate the selected stations. Adapted from [13].

4.2. Results

The filters explained in this work were applied to the time series of the SEVI and CAAL stations. Figures 3–6 show the results of the SEVI station and, Figures 7–10 show the results of the CAAL station.

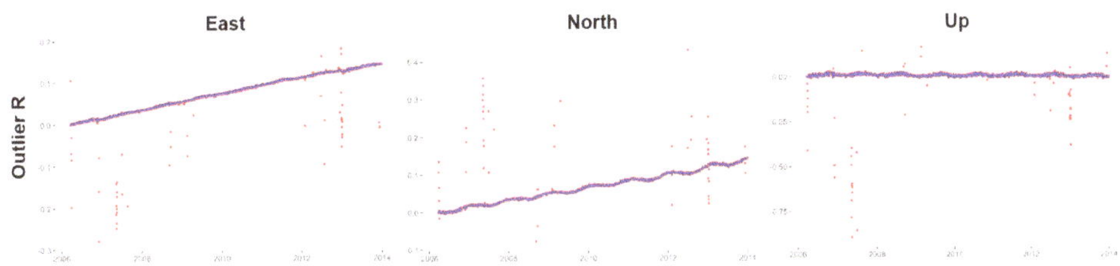

Figure 3. Topocentric time series east, north, and the elevation of the SEVI station with Outlier R filter. Red dots indicate RAW series and blue line indicates Outlier R filter.

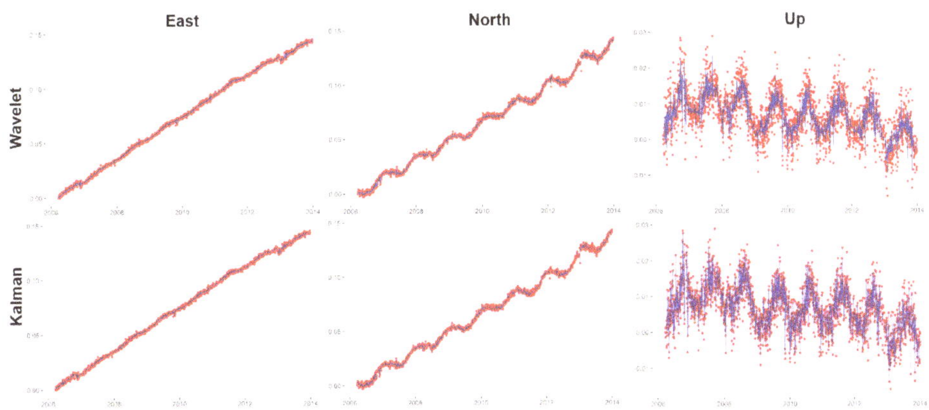

Figure 4. Topocentric time series east, north, and the elevation of the SEVI station with the wavelet filter result (**first line**) and the Kalman filter result (**second line**). Red dots indicate Outlier R series and blue line indicates filter.

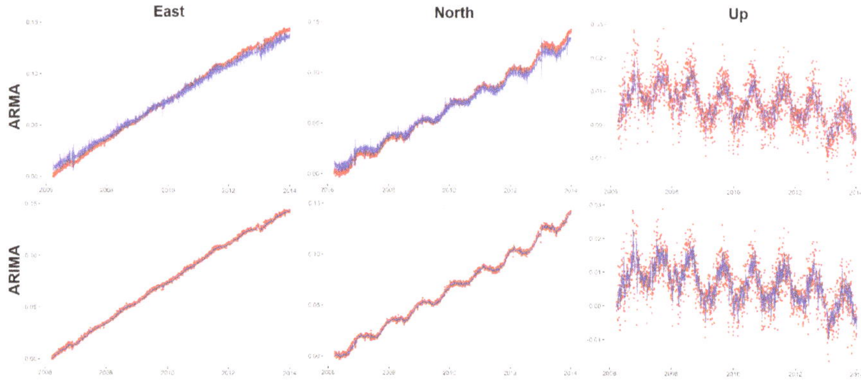

Figure 5. Topocentric time series east, north, and the elevation of the SEVI station with the ARMA filter result (**first line**) and the ARIMA filter result (**second line**). Red dots indicate Outlier R series and blue line indicates filter.

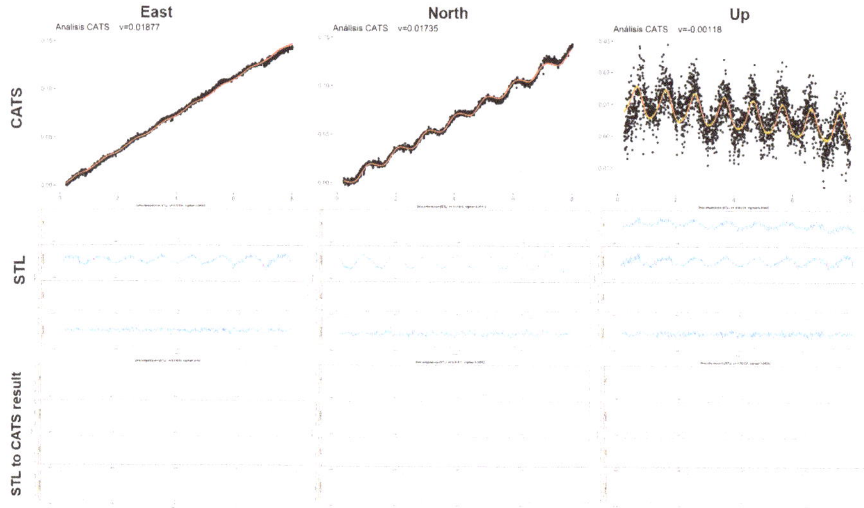

Figure 6. Topocentric time series east, north, and the elevation of the SEVI station with the CATS result (**first line**), the STL decomposition result (**second line**), and the STL decomposition to CATS result (**third line**). In the CATS series, black dots indicate the Outlier R series and red line indicates the CATS result.

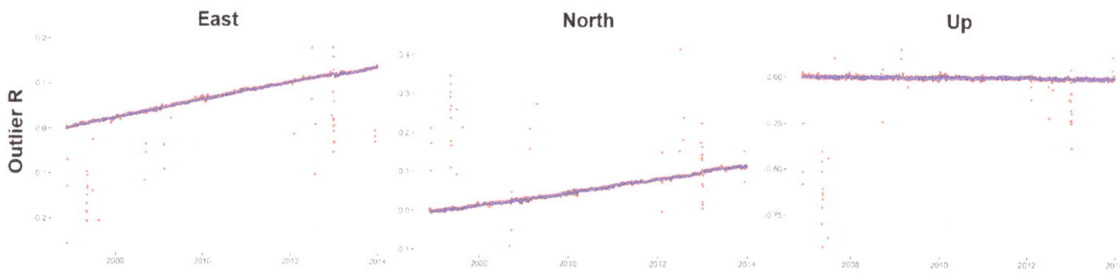

Figure 7. Topocentric time series east, north, and the elevation of the CAAL station with Outlier R filter. Red dots indicate RAW series and blue line indicates Outlier R filter.

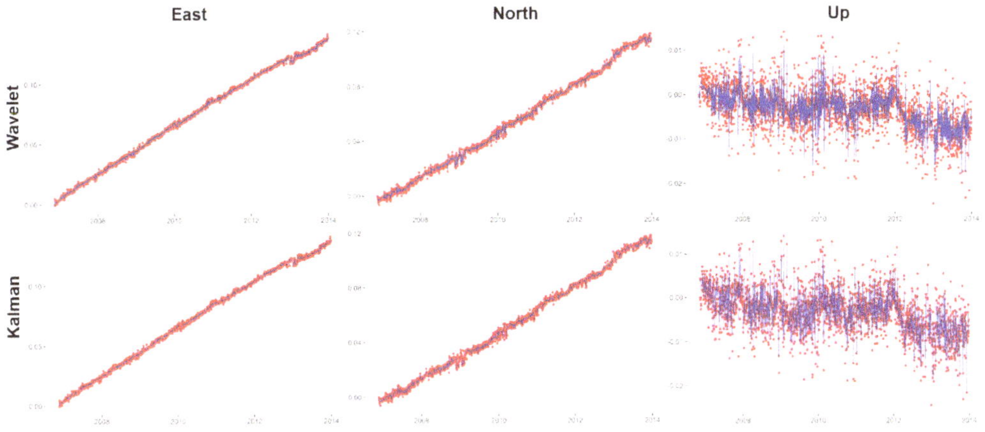

Figure 8. Topocentric time series east, north, and the elevation of the CAAL station with the wavelet filter result (**first line**) and the Kalman filter result (**second line**). Red dots indicate Outlier R series and blue line indicates filter.

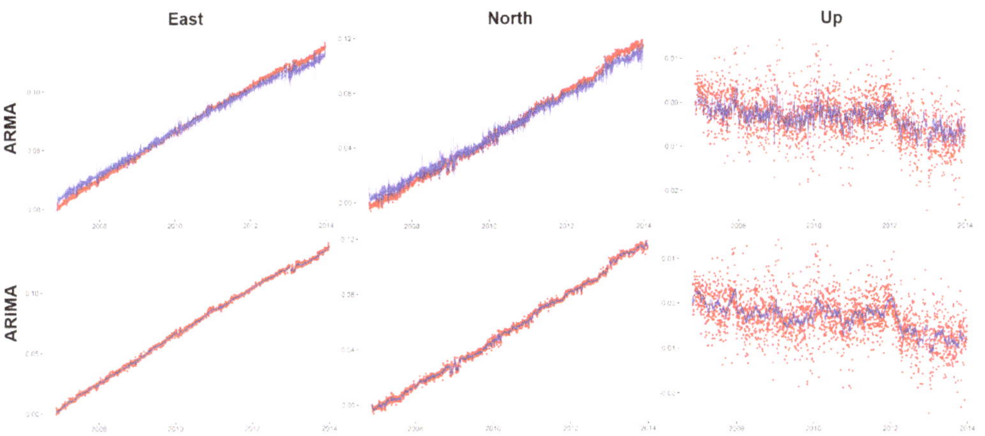

Figure 9. Topocentric time series east, north, and the elevation of the CAAL station with the ARMA filter result (**first line**) and the ARIMA filter result (**second line**). Red dots indicate Outlier R series and blue line indicates filter.

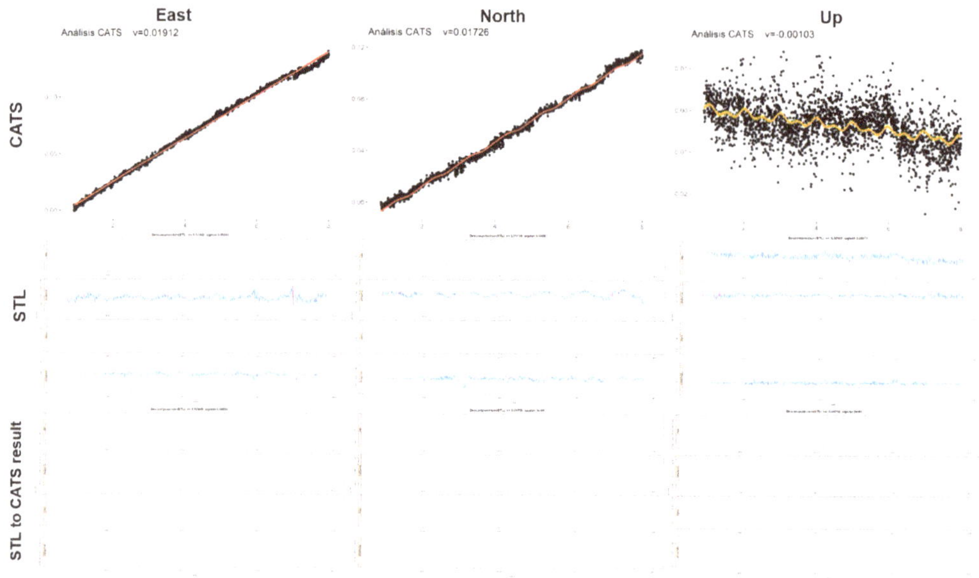

Figure 10. Topocentric time series east, north, and the elevation of the CAAL station with the CATS result (**first line**), the STL decomposition result (**second line**), and the STL decomposition to the CATS result (**third line**). In the CATS series, black dots indicate the Outlier R series and red line indicates the CATS result.

5. Conclusions

GNSS time series analysis seeks to know the behavior and level of existing geodynamic activity. The geodynamic model is obtained from the velocities of the displacements of each station in the region. That is the starting point for the calculation of the stress and strain models. The GNSS experimental process involves multiple factors that can introduce deviations and dispersions in the values of the GNSS series and, consequently, in the models and results obtained.

In this work, a brief review of analysis techniques and methods for GNSS time series was carried out. Filtering, filtering–fitting, and fitting techniques were analyzed. The need for a descriptive analysis of the RAW series was previously established. Anomalous values, gaps, and dispersion of the series were detected. It was also used to detect changes in the trend or seasonality of the GNSS series.

Among the exclusive filtering techniques, outliers R was more effective and adaptable for both linear and non-linear series, whereas the processes 1 *sigma* and 2 *sigma*, especially in non-linear cases, were not applicable to the entire series.

The following were considered as filtering–fitting techniques: Kalman, ARMA/ARIMA, and wavelets. The Kalman and ARMA filters presented more dispersion in the result than ARIMA and wavelets. In series fitting, Kalman and ARIMA obtained smoother curves than ARMA and wavelets, and, therefore, they were more effective in forecasting series. ARIMA and wavelets better adjusted those internal changes in the series providing information on the level of geodynamic activity and the possible detection of seismic events.

CATS-R software provided a series of adjustments, adapted to controlled changes on antenna changes, receivers, firmware, etc. It is a very reliable technique when calculating velocities and, especially, when fitting the elevation component. The STL package that allowed decomposition of the time series into trend, seasonal, and reminder and was analyzed and applied. Its versatility and precision were verified once any of the other techniques had been applied and the series had been purified of adverse effects (outliers, gaps, dispersions, deviations, etc.).

Finally, there is no standardized procedure for any time series. Really, the descriptive analysis informs about the processes to consider in its treatment.

Data Availability Statement: Data supporting reported results can be found at http://www.ign.es, https://www.epncb.oma.be, https://renep.dgterritorio.gov.pt, https://sitmurcia.carm.es, https://www.juntadeandalucia.es and http://icv.gva.es.

Acknowledgments: We are thankful to all individuals and institutions contributing to GPS data collection and dissemination. Thanks are due to RAP (IECA, Instituto de Estadística y Cartografía de Andalucía, Junta de Andalucía), REGAM (Consejería de Obras Públicas y ordenación del Territorio, Región de Murcia) and MERISTEMUM (Consejería de Agua, Agricultura y Medio Ambiente, Región de Murcia) for the Murcia Region CGPS Networks, ERVA (Generalitat Valenciana) for Valencian Reference Stations Network, IGN (Instituto Geográfico Nacional), IGS (International GNSS Service), ROA (Real Instituto y Observatorio de la Armada) and from Portugal permanent stations by RENEP (Direcção Geral do Território). We would also like to thank Onsala Space Observatory for their calculation of the ocean tide loading parameters. The contributions of both editor and of the reviewers are very much appreciated, which greatly helped to improve the manuscript.

References

1. Dach, R.; Hugentobler, U.; Fridez, P.; Meindi, M. *Bernese GPS Software Ver. 5.0 User Manual*; Astronomical Institute, University of Bern: Bern, Switzerland, 2007.
2. Saastamoinen, J. Contribution of the theory of atmospheric refraction. *Geod. Bull.* **1973**, *107*, 13–34. [CrossRef]
3. Niell, A. Global mapping functions for the atmosphere delay at radio wavelengths. *J. Geophys. Res.* **2011**, *101*, 3227–3246. [CrossRef]
4. Mervart, L. Ambiguity Resolution Techniques in Geodetic and Geodynamic Applications of the Global Positioning System. Ph.D. Thesis, University of Bern, Bern, Switzerland, 1995.
5. Altamimi, Z.; Collilieux, X.; Metivier, L. ITRF2008: An improved solution of the International Terrestrial Reference Frame. *J. Geod.* **2011**, *85*, 457–473. [CrossRef]
6. Rosado, B.; Fernández-Ros, A.; Berrocoso, M.; Prates, G.; Gárate, J.; De Gil, A.; Geyer, A. Volcano-tectonic dynamics of Deception Island (Antarctica): 27 years of GPS observations (1991–2018). *J. Volcanol. Geotherm. Res.* **2019**, *381*, 57–82. [CrossRef]
7. Box, G.E.P.; Cox, D.R. An analysis of transformations. *J. R. Stat. Soc.* **1964**, *26*, 211–252. [CrossRef]
8. Peña, D.; Peña, J. A normality test based on the Box-Cox transformation. *Span. Stat.* **1986**, 33–46.
9. Prates, G. GNSS-GPS Geodetic Time Series Treatment for Volcanic Activity Monitoring and Surveillance: Spatial Dilatometer and Inclinometer. Applied to Deception (Antarctica) and El Hierro (Canaries Islands). Ph.D. Thesis, Facultad de Ciencias, Cadiz University, Cádiz, Spain, 2012.
10. Rosado, B. *Modeling of Surface Deformation in Tectonic Areas Using the Walevets Theory*; Application to the SPINA Network. Final Master's Project; Cadiz University: Cádiz, Spain, 2014.
11. Williams, S.D.P. CATS: GPS coordinate time series analysis software. *GPS Solut.* **2008**, *12*, 147–153. [CrossRef]
12. Cleveland, R.B.; Cleveland, W.S.; McRae, J.E.; Terpenning, I.J. STL: A seasonal-trend decomposition procedure based on loess. *J. Off. Stat.* **1990**, *6*, 3–33.
13. Rosado, B.; Fernández-Ros, A.; Jiménez, A.; Berrocoso, M. Modelo de deformación horizontal GPS de la región sur de la Península Ibérica y norte de áfrica (SPINA). *Boletín Geológico y Min.* **2017**, *128*, 141–156. [CrossRef]

engineering proceedings

Proceedings

A Systematic Review of Python Packages for Time Series Analysis †

Julien Siebert *, Janek Groß and Christof Schroth

Fraunhofer Institut for Experimental Software Engineering IESE, Fraunhofer Platz 1, 67663 Kaiserslautern, Germany; janek.gross@iese.fraunhofer.de (J.G.); christof.schroth@iese.fraunhofer.de (C.S.)
* Correspondence: julien.siebert@iese.fraunhofer.de

† Presented at the 7th International Conference on Time Series and Forecasting, Gran Canaria, Spain, 19–21 July 2021.

Abstract: This paper presents a systematic review of Python packages with a focus on time series analysis. The objective is to provide (1) an overview of the different time series analysis tasks and preprocessing methods implemented, and (2) an overview of the development characteristics of the packages (e.g., documentation, dependencies, and community size). This review is based on a search of literature databases as well as GitHub repositories. Following the filtering process, 40 packages were analyzed. We classified the packages according to the analysis tasks implemented, the methods related to data preparation, and the means for evaluating the results produced (methods and access to evaluation data). We also reviewed documentation aspects, the licenses, the size of the packages' community, and the dependencies used. Among other things, our results show that forecasting is by far the most frequently implemented task, that half of the packages provide access to real datasets or allow generating synthetic data, and that many packages depend on a few libraries (the most used ones being numpy, scipy and pandas). We hope that this review can help practitioners and researchers navigate the space of Python packages dedicated to time series analysis. We also provide an updated list of the reviewed packages online.

Keywords: time series analysis; Python; review

Citation: Siebert, J.; Groß, J.; Schroth, C. A Systematic Review of Python Packages for Time Series Analysis. *Eng. Proc.* **2021**, 5, 22. https://doi.org/10.3390/engproc2021005022

Published: 28 June 2021

1. Introduction

A time series is a set of data points generated from successive measurements over time. The analysis of this type of data has found application in many fields, from finance to health, including the monitoring of computer networks or the environment. The current trend of reducing the cost of sensors and data storage, the increasing performance of Big Data and data analysis technologies such as machine learning or data mining, are opening up more and more possibilities to acquire and analyze temporal data. Moreover, as the number of time series analysis application cases rises, more and more data scientists, data engineers, analysts, and software engineers have to use dedicated time series analysis libraries.

In this article, we systematically review Python packages dedicated to time series analysis. Python is one of the programming languages of choice for data scientists (See the different surveys performed by Kaggle from 2017 until 2020: https://www.kaggle.com/kaggle-survey-2020 (accessed on 24 June 2021). Data scientists are not only responsible for analyzing data; their task is also to ensure that services based on these analyses reach a sufficient level of maturity to be deployed and maintained in production. In this context, we review not only the analysis tasks implemented in the packages, but also several factors external to the tasks themselves, such as which dependencies are used or how big the community behind the development of the package in question is. Our goal is not to evaluate the quality of the implementations themselves but to provide a structured overview that is useful for data scientists confronted with time series analysis (and faced with having to choose which packages to rely on), the scientific community, and the community of Python developers working in this field. This paper is structured as follows: Related work is introduced in Section 2; the search methodology and the search results are described in Sections 3 and 4, respectively; threats to validity are discussed in Section 5; and Section 6 concludes the paper.

2. Related Work

Time series analysis is a broad research field covering many application domains. The literature contains many reviews, either focusing on analysis tasks and methods (see, for instance, these reviews on forecasting [1–4], clustering and classification [5–9], anomaly detection [10–12], changepoint analysis [13–15], pattern recognition [16,17], or dimensionality reduction [18]) or focusing on a specific application domain (see, for instance, these surveys on finance [19], IoT and Industry 4.0 [20–22], or health [23]). Over time, several formal definitions and reviews of time series analysis tasks have been published; see, for example [24–26].

However, existing implementations (software packages or libraries) are often listed—usually in a non-systematic way—in textbooks (like [27,28] for R, or [29] for Python) or gray literature (for example, Towards Data Science (https://towardsdatascience.com/), KDnuggets (https://www.kdnuggets.com/) or Machine Learning Mastery (https://machinelearningmastery.com/), and few papers actually systematically review packages or libraries in a specific language. For example, Ref. [30] reviewed packages for analyzing animal movement data in R, and [31] surveyed R packages for hydrology. With respect to Python, we found several reviews of packages for different domains: social media content scrapping [32], topological data analysis [33], or data mining [34]. For time series analysis in Python, the only related work we could find is [35], where the authors review packages focusing on forecasting.

There is, to the best of our knowledge, no systematic review of Python packages for generic time series analysis.

3. Methodology

We conducted a systematic literature review according to [36]. However, these guidelines focus on printed literature, not on software packages. Hence, we adjusted these methods. Our search process is illustrated in Figure 1. We conducted a search in both literature databases and code repositories (GitHub). The following sections provide more details on the different steps of the search itself.

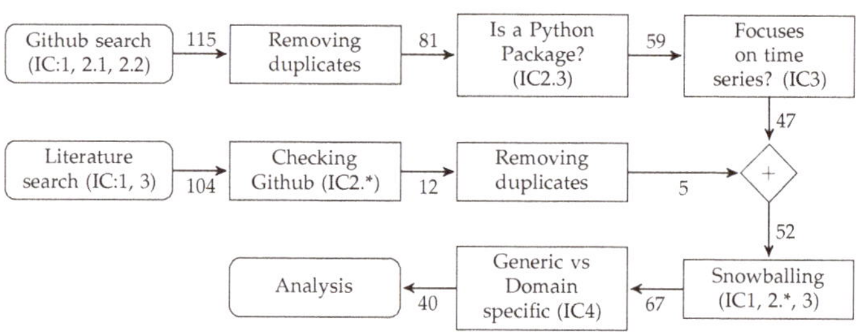

Figure 1. Search and filtering process overview. Edge labels indicate the number of repositories left after each step.

3.1. Research Questions

We already stated our goal and the context we set for this review in the introduction. We formalize this context as follows: We want to **analyze** Python packages dedicated to time series analysis **for the purpose of** structuring the available implementations (we explicitly exclude the purpose of evaluating them) **with respect to** the implemented time series analysis tasks **from the viewpoint of** practitioners **in the context of** building data-driven services on top of these implementations. Our research questions are:

- **RQ1** Which time series analysis tasks exist? And which of these are implemented in maintained Python packages?
- **RQ2** How do the packages support the evaluation of the produced results?
- **RQ3** How do the packages support their usage, and what insights can we gain to estimate the durability of a given package and make an informed choice about its long-term use?

3.2. Inclusion Criteria

To guide our review and filter relevant packages, we defined the following inclusion criteria (IC): The package should be open source, written in Python, and available on GitHub (**IC1**). The package should be actively maintained (last commit within less than 6 months) (**IC2.1**); it should have more than 100 GitHub stars (**IC2.2**); and it should be listed in PyPI (PyPI is the Python Package Index, a repository of software for the Python programming language, see https://pypi.org/) and be installable via pip (pip is the Python Package Installer, see https://pip.pypa.io/en/stable/) or conda (conda is a Python package management system and environment management provided by the Anaconda distribution, see https://docs.conda.io/en/latest/) (**IC2.3**). The package should explicitly target time series analysis (**IC3**). We excluded packages that can be used for time series analysis (as building blocks) but whose main purpose is not time series analysis per se (for example, generic scientific computing packages such as scipy or numpy, packages dedicated to data manipulation or storage such as pandas, or generic machine learning or data mining packages such as scikit-learn). Finally, we focused our search on packages offering methods that tend to be domain-agnostic (**IC4**) and excluded domain-specific packages. Domain-specific packages are packages aiming to solve time series analysis in a specific domain (for example, audio, finance, geoscience, etc.). They usually focus on specific types and formats of time series and domain related analysis tasks.

3.3. Searching Open-Source Repositories in GitHub

In order to filter GitHub repositories, we selected a list of topics (https://github.com/topics (accessed on 1 March 2021)), filtered the results by language (Python, IC1), by number of stars (at least 100, IC2.2), and considered only repositories that were updated after July 2020 (IC2.1).

In order to select a list of relevant topics, we first manually selected a list of eight Python packages known to be used in time series analysis (i.e., a seeds set): pandas, numpy, scipy, statsmodel, ruptures, tsfresh, tslearn, and sktime; as well as a sample of the packages using the topic "time-series". We examined the topics used by these packages and then extended this list of topics with different spellings while manually double-checking their existence in GitHub. We considered a total of 16 different topics (see Table 1). The first search led to a total of 115 repositories.

Table 1. List of topics used to conduct the search on GitHub.

time-series	time-series-regression	signal-processing	time-series-classification
time-series-analysis	time-series-forecast	time-series-visualization	time-series-decomposition
time-series-forecasting	time-series-data-mining	timeseries	timeseries-forecasting
time-series-prediction	time-series-segmentation	timeseries-analysis	time-series-clustering

3.3.1. Removing Duplicates

We found 24 unique repositories that were duplicated (i.e., listed in more than one topic). After duplicate removal, 81 unique repositories remained.

3.3.2. Checking If the Repository Contains the Code of a Python Package

We restricted our search to packages that are referenced by PyPI and can be installed with pip or conda (IC2.3). Note that the repository name might not reflect the package name (if one exists). For example, the repository https://github.com/PyWavelets/pywt

(accessed on 24 June 2021) contains the source code for the package named pywavelets. The repository https://github.com/angus924/rocket (accessed on 24 June 2021) does not contain the source code for the Python package rocket. We therefore checked each of the 81 repositories manually and excluded 22 repositories, which yielded a total of 59 remaining repositories that contain the source code of a Python package.

3.3.3. Including only Packages Focused on Time Series Analysis

Finally, we manually checked whether the focus of the package is time series analysis (IC3). After exclusion, 47 remaining packages were kept for further analysis.

3.4. *Searching Scientific Bibliographic Databases*

The search for packages only in a repository might not be sufficient to cover all existing packages. For example, one of our seed packages (namely tsfresh) was not uncovered by the search. Hence, we extended our search to existing literature and software databases (in march 2021). We used the bibliographic databases IEEE Xplore (https://ieeexplore.ieee.org), ACM Digital Library (https://dl.acm.org/), Web of Science (https://www.webofknowledge.com), and Scopus (https://www.scopus.com/), as well as the Journal of Open Source Software (JOSS) (https://joss.theoj.org/), and Zenodo (https://zenodo.org/). For IEEE Xplore, ACM Digital Library, Web of Science, and Scopus, we limited ourselves to the search string `''Python'' AND ''time series''` in the document title. For the Journal of Open Source Software (JOSS), we first used the key words `''time series''` and then filtered the results by language (the query used is: https://joss.theoj.org/papers/search?q=time+series (accessed on 1 March 2021)). For Zenodo, we also used the search string `''Python'' AND ''time series''`, limited the search to the software category and removed the duplicates (e.g., different versions of the same software). The full query for Zenodo is: https://zenodo.org/search?page=1&size=200&q=%22time%20series%22%20AND%20%22python%22&sort=mostrecent&type=software (accessed on 1 March 2021). We only included references that matched our inclusion criteria IC1, IC2.*, and IC3. Table 2 summarizes our search results.

Table 2. Literature search results.

Data Source	Number of Hits	Number of Included Documents	Included References
IEEE Xplore	1	0	
ACM Digital Library	2	1	[37]
Web of Science	10	4	[37–40]
Scopus	12	4	[37–40]
JOSS	21	1	[41]
Zenodo	68	6	[42–47]

We manually cross-checked the results obtained from GitHub with the results obtained by our literature search. Out of the eleven packages resulting from our literature search, only five repositories were not already in the GitHub search results: tsfresh, neurodsp, EoN, nolds, and pastas.

3.5. *Snowballing*

In order to extend our search, we used a snowballing approach. We first manually reviewed the package documentations in order to find links to other similar packages. Only two packages—tsfresh (https://tsfresh.readthedocs.io/en/latest/text/introduction.html#what-else-is-out-there (accessed on 24 June 2021)) and sktime (https://www.sktime.org/en/latest/related_software.html (accessed on 24 June 2021))—actually document related packages. Second, we manually reviewed the documentation and the GitHub repositories of all packages to find related publications. We then reviewed the papers to find new packages (i.e., we performed a single backward snowballing pass). Out of a total of 79 packages, 15 new packages were included after the snowballing phase, for a total of 67 packages.

3.6. Generic vs. Domain-Specific Packages (IC4)

Finally, we classified the packages in two categories: domain-specific and generic. As previously defined, we consider domain-specific packages to be packages aiming to solve time series analysis in a specific domain (for example, audio, finance, geoscience, etc.) and generic packages as those offering methods that tend to be domain-agnostic. Out of the 67 packages, 27 packages were categorized as domain-specific and 40 packages as generic.

3.7. Data Extraction and Categorization

We manually extracted relevant information about the packages from their documentation pages and code. For the categorization, we used an iterative, bottom-up approach. Two researchers first proposed category definitions and then categorized the packages. A third researcher was responsible for resolving disagreements. Iterations were performed until the categories and results were consolidated.

4. Results

4.1. RQ1: Implementation of the Time Series Analysis Tasks

To answer our research question RQ1, we first reviewed the task definitions present in the literature and then analyzed the 40 packages classified as generic to extract information about which tasks have been implemented in the packages.

4.1.1. Task Definitions

Time series analysis tasks are formally defined in the literature. Reviews like [24–26,48] define the following tasks: **Indexing (query by content)**: given a time series and some similarity measure, find the nearest matching time series [24–26]. **Clustering**: find groups (clusters) of similar time series [24–26,48]. **Classification**: assign a time series to a predefined class [24–26,48]. **Segmentation (Summarization)**: create an accurate approximation of a time series by reducing its dimensionality while retaining its essential features [24–26,48]. **Forecasting (Prediction)**: given a time series dataset up to a given time t_n, forecast the next values [24,25]. **Anomaly Detection**: find abnormal data points or subsequences (also called discords) [24,25]. **Motif Discovery**: find every subsequence (called motif) that appears recurrently in a time series [24,25,48]. **Rules Discovery (Rule Mining)**: find the rules that may govern associations between sets of time series or subsequences [25,48].

Esling and Agon also define implementation components [24]: **preprocessing** (e.g., filtering noise, removing outliers, or imputing missing values), **representation** (e.g., dimensionality reduction, finding fundamental shape characteristics), **similarity measures**, and **indexing schemes**.

4.1.2. Implemented Tasks

While analyzing the packages, we found packages explicitly mentioning the tasks corresponding to our literature review. We found 20 packages explicitly providing forecasting methods (T1), 6 packages providing classification methods (T2), 6 packages providing clustering methods (T3), 6 packages providing anomaly detection methods (T4), and 4 packages providing segmentation methods (T5). We classified four packages under the category pattern recognition(T6), encompassing both indexing and motif discovery tasks. We also classified five packages under the category change point detection (T7), which was not in our literature review. Finally, we could not find any package explicitly mentioning the rules discovery task.

Considering the implementation components, we found 4 packages explicitly providing *dimensionality reduction* methods (DP1), 17 packages explicitly providing *missing values imputation* methods (DP2), 16 packages explicitly providing *decomposition* methods (e.g., decomposing time series into trends, seasonal components, or frequency components) (DP3), 24 packages explicitly providing generic *transformation and features generation* methods (DP4), and 7 packages explicitly providing methods for computing similarity measures (DP5). Table 3 gives an overview of our categorization of the packages.

Table 3. Classification of packages. Tasks: T1 (forecasting), T2 (classification), T3 (clustering), T4 (anomaly detection), T5 (segmentation), T6 (pattern recognition), T7 (change point detection). Data Preparation (also called implementation components): DP1 (dimensionality reduction), DP2 (missing values imputation), DP3 (decomposition), DP4 (preprocessing), DP5 (similarity measures). Evaluation: E1 (model selection, hyperparameter search, feature selection), E2 (metrics and statistical tests), E3 (visualization). Datasets: D1 (synthetic data generation) and D2 (contains datasets). Documentation: Do1 (dedicated documentation), Do2 (notebook: directly executable (+), present (*)), Do3 (API reference), Do4 (install guide), Do5 (user guide).

Package Name	Tasks							Data Preparation					Evaluation			Data		Documentation				
	T1	T2	T3	T4	T5	T6	T7	DP1	DP2	DP3	DP4	DP5	E1	E2	E3	D1	D2	Do1	Do2	Do3	Do4	Do5
arch	+													+		+	+	+	*	+	+	+
atspy	+								+	+	+		+	+	+				+		+	+
banpei				+			+														+	+
cesium											+							+	*	+	+	+
darts	+								+	+	+			+	+	+	+	+		+	+	+
deeptime	+		+					+		+	+			+	+	+		+	+		+	+
deltapy	+		+					+		+	+	+						+	+	+	+	+
dtaidistance			+									+			+			+	+	+	+	+
EMD-signal										+					+			+	+	+	+	+
flood-forecast	+								+		+			+	+			+	+	+	+	+
gluonts	+								+		+			+	+	+	+	+		+	+	+
hcrystalball	+								+		+		+	+	+	+	+	+	*	+	+	+
hmmlearn	+										+					+		+	*		+	+
hypertools			+					+	+		+				+			+	*	+	+	+
linearmodels														+			+		*		+	+
luminaire	+			+			+		+	+	+		+					+			+	
matrixprofile			+	+	+	+			+			+			+		+	+		+	+	+
mcfly		+											+		+			+			+	+
neuralprophet	+								+	+	+				+			+	*	+	+	+
nolds									+					+		+	+	+		+	+	+
pmdarima	+									+	+		+	+	+		+		*		+	+
prophet	+						+			+				+	+			+	*	+	+	+
pyaf	+								+	+	+		+	+	+		+	+		+	+	+
pycwt										+							+	+	+	+	+	+
pydlm	+									+			+	+	+			+	*	+	+	+
pyFTS	+										+		+	+	+	+	+		+		+	+
pyodds				+									+	+	+			+	*	+	+	+
pytorchts	+								+		+				+	+	+	+	+	+	+	+
pyts		+			+	+			+	+	+	+		+	+	+	+	+	*	+	+	
PyWavelets										+	+					+	+	+	*	+	+	+
ruptures							+							+	+	+		+	+		+	+
scikit-multiflow		+		+			+		+		+			+	+	+		+	*	+	+	
seglearn											+						+	+	*	+	+	+
sktime	+	+		+	+				+	+	+	+	+	+	+		+	+	*	+	+	+
sktime-dl	+	+											+						+		+	+
statsmodels	+							+	+	+			+	+	+	+	+	+	*	+	+	+
stumpy					+	+						+						+	*	+	+	+
tftb										+	+			+	+	+		+	+	+	+	+
tsfresh									+		+		+	+			+	+		+	+	+
tslearn		+	+			+					+	+		+		+	+	+			+	+
Total	**20**	**6**	**6**	**6**	**4**	**4**	**5**	**4**	**17**	**16**	**24**	**7**	**13**	**23**	**25**	**16**	**19**	**34**	**30**	**28**	**40**	**37**
	T1	T2	T3	T4	T5	T6	T7	DP1	DP2	DP3	DP4	DP5	E1	E2	E3	D1	D2	Do1	Do2	Do3	Do4	Do5
	Tasks							**Data Preparation**					**Evaluation**			**Data**		**Documentation**				

Forecasting is by far the most frequently implemented task. There is no significant difference, in terms of number of packages, between the other tasks. However, we need to be cautious when interpreting these numbers. First, the tasks as formally defined in the literature might not be explicitly mentioned in the packages documentation or code. Second, the delineation between a task and the methods used to implement it is sometimes blurry and context dependent. For example, one can perform change point detection for the sake of finding time points where some time series properties change and, as a consequence, raising alarms in a production system, or use it as a preprocessing step for segmenting a time series into different phases. Another example are forecasting models, which can also be applied for outlier detection.

4.2. RQ2: Evaluation of the Produced Results

To answer our research question RQ2, we extracted information about the evaluation of the outcomes produced by the packages. We came up with two main clusters: functions that facilitate the evaluation itself (E1, E2, E3) and functions for either generating synthetic data or downloading existing datasets (D1, D2). We found 13 packages explicitly providing methods for model selection, hyperparameter search, or feature selection (E1), 20 packages explicitly providing evaluation metrics and statistical tests (E2), and 25 packages providing visualization methods (E3). Concerning the data, we found 16 packages explicitly providing functions for generating synthetic time series data (D1), and 19 packages providing access to time series datasets (D2). A large majority of the packages provide a way to evaluate the results produced. Only 4 packages have not been classified in any of the E or D classes.

4.3. RQ3: Package Usage and Community

To answer our research question RQ3, we extracted information about the documentation, the dependencies, and the community supporting the packages. For instance, GitHub provides many statistics about a repository (e.g., the number of stars, forks, issues) that can be used to get a first idea of the liveliness of the different packages. We used the number of GitHub stars and forks to estimate the community behind each package. Figure 2a shows the distribution of stars and forks for all 40 packages. Another piece of information that is relevant to practitioners are the licenses under which the implementations are available. Figure 2b shows the distribution of the licenses used among the 40 repositories.

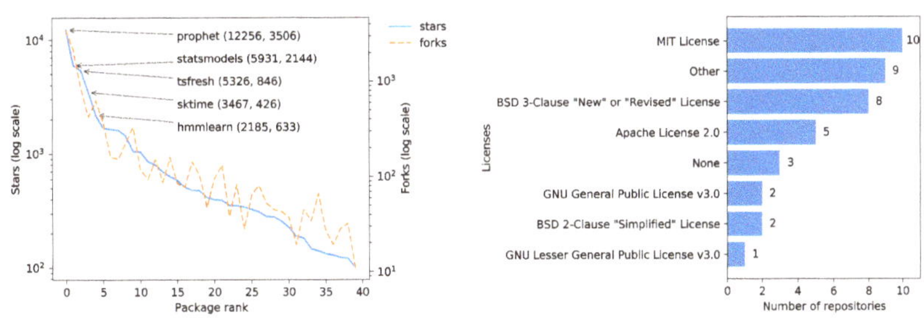

(a) (b)

Figure 2. (a) Distribution of stars and forks for all 40 repositories (log scale). The repositories are ranked by the number of stars, in descending order. (b) Distribution of licenses (number of repositories per license). None means that no license information was available from GitHub directly.

We also investigated the dependencies used by each of the selected 40 packages. We used the Python program johnnydep (https://pypi.org/project/johnnydep/ (accessed on 24 June 2021)) to automatically collect the dependencies without installing the packages directly. We only looked at direct dependencies required for the installation of the package. We did not consider specific installation options such as dev or test. We did not search for all dependencies recursively. Here is an example of how we called the program `johnnydep`

`PACKAGENAME --fields=ALL --no-deps --output-format=json`. The dependencies of two packages could not be retrieved automatically (cesium and deeptime). We also manually cross-checked the dependencies and filled in the missing ones. Table 4 shows which dependencies are used the most by the packages.

Table 4. Ranking of the most frequently used dependencies.

Package Name	Used	Rank	Package Name	Used	Rank
numpy	37	1	torch	6	8
scipy	30	2	numba	6	8
pandas	23	3	cython	6	8
scikit-learn	21	4	tensorflow	5	9
matplotlib	16	5	seaborn	4	10
statsmodels	8	6	future	4	10
tqdm	7	7	joblib	4	10

Almost all packages (37) depends upon numpy. The top 5 dependencies are numpy, scipy (scientific computing), pandas (data manipulation), scikit-learn (machine learning), and matplotlib (visualization).

Finally, we investigated five documentation aspects (Do1-Do5). We found that 30 packages provide a separate documentation page (Do1). The other ten packages use the README of the repository file as documentation. 18 packages provide notebooks directly executable without installation via a link to either mybinder.org (https://mybinder.org/) or Google Colab (https://colab.research.google.com/ (accessed on 24 June 2021)) (Do2 +), 12 packages provide stand-alone notebook files to be downloaded (Do2 *), and 10 packages do not provide any notebook file at all. 28 packages provide an API reference (Do3). All packages provide an installation page (Do4) and almost all packages (38) provide user guides in the form of static examples or tutorials.

5. Discussion and Threats to Validity

In this section, we discuss the choices we made and that may affect the validity of this review.

This review focused on GitHub. Gitlab and Sourceforge were checked manually, but we decided not to include them as sources due to the insufficient number of results.

We limited ourselves to packages with at least 100 stars. This somehow arbitrary limit led us to exclude packages with a number of stars close to 100 (e.g., the stingray package with 93 stars at the time of the search). We excluded packages that were not maintained but might have been relevant for practitioners. An example is the pyflux package (forecasting). We also excluded repositories that are not Python packages. This led us to discard interesting repositories like ad_examples (which provides state-of-the-art anomaly detection methods) and many repositories containing code scripts associated with scientific papers.

Concerning the search process, we used a mix of literature databases and GitHub topics together with a snowballing approach to find relevant packages. The reason forthis was that several known packages could not be found automatically. For example, the package cesium does not list any topic and therefore was not found in our first GitHub search. It was found after snowballing. Another example is tsfresh, which was missing in the first GitHub search and was found in the literature search. The problem may be the language filter (strictly Python), as tsfresh lists some of the topics we searched for ("time-series").

We tried to automate some of the tasks (e.g., filtering repositories that contain Python packages or finding the dependencies), using both PyPI and GitHub API, or the johnnydep tool. There were false positives and false negatives. This led us to manually cross check the results obtained from our automated search.

Whether a package focuses on time series analysis or not can sometimes be fuzzy. For example, we decided to leave the topic of survival analysis out of this review. We initially found two packages: lifelines and scikit-survival. The same applies to the boundary between generic and domain-specific packages. We took a conservative approach to keep our survey sufficiently focused.

As already mentioned above, the definition of what should be regarded as a task vs. an "implementation component" is difficult, as a strict boundary may not even exist. Moreover, it is sometimes not clear what methods the packages provide without actually installing them and testing them. Indeed, the documentation might not be complete or the vocabulary used may differ from one package to another. One solution was to check the code itself. Here again, the search strings used play an important role in avoiding false negatives.

6. Conclusions

This paper presented a systematic review of Python packages dedicated to time series analysis. The search process led to a total of 40 packages that were analyzed further. We proposed a categorization of the packages based on the analysis tasks implemented, the methods related to data preparation, the means for evaluating the results produced, and the kind of documentation present, and also looked at some development aspects (licenses, stars, dependencies). We also discussed the search process with its possible bias and the challenges we encountered while searching for and reviewing the relevant packages. The scope of this survey does, however, not include any evaluation of the implementations or the results they would produce, for example, on benchmark datasets. We hope that this review can help practitioners and researchers navigate the space of Python packages dedicated to time series analysis. Since the packages will evolve, we plan to maintain an updated list of the reviewed packages online at https://siebert-julien.github.io/time-series-analysis-python/.

Author Contributions: J.S. designed the search process. All authors contributed equally to the results evaluation and to the manuscript. All authors have read and agreed to the published version of the manuscript.

Funding: This research received no external funding.

Institutional Review Board Statement: Not applicable.

Informed Consent Statement: Not applicable.

Data Availability Statement: Not applicable.

Acknowledgments: The authors want to thank Anna Maria Vollmer and Markus Löning for their valuable feedback.

Conflicts of Interest: The authors declare no conflict of interest.

References

1. Hendikawati, P.; Subanar; Abdurakhman; Tarno. A survey of time series forecasting from stochastic method to soft computing. *J. Phys. Conf. Ser.* **2020**, *1613*, 012019. [CrossRef]
2. Mahalakshmi, G.; Sridevi, S.; Rajaram, S. A survey on forecasting of time series data. In Proceedings of the 2016 International Conference on Computing Technologies and Intelligent Data Engineering (ICCTIDE'16), Kovilpatti, India, 7–9 January 2016; IEEE: Piscataway, NJ, USA, 2016; pp. 1–8. [CrossRef]
3. Panigrahi, S.; Behera, H.S. Fuzzy Time Series Forecasting: A Survey. In *Computational Intelligence in Data Mining*; Advances in Intelligent Systems and Computing Ser; Behera, H.S., Nayak, J., Naik, B., Pelusi, D., Eds.; Springer: Singapore, 2020; pp. 641–651.
4. Tealab, A. Time series forecasting using artificial neural networks methodologies: A systematic review. *Future Comput. Inform. J.* **2018**, *3*, 334–340. [CrossRef]
5. Abanda, A.; Mori, U.; Lozano, J.A. A review on distance based time series classification. *Data Min. Knowl. Discov.* **2019**, *33*, 378–412. [CrossRef]
6. Aghabozorgi, S.; Seyed Shirkhorshidi, A.; Ying Wah, T. Time-series clustering—A decade review. *Inf. Syst.* **2015**, *53*, 16–38. [CrossRef]
7. Bagnall, A.; Lines, J.; Bostrom, A.; Large, J.; Keogh, E. The great time series classification bake off: A review and experimental evaluation of recent algorithmic advances. *Data Min. Knowl. Discov.* **2017**, *31*, 606–660. [CrossRef] [PubMed]

8. Fawaz, H.I.; Forestier, G.; Weber, J.; Idoumghar, L.; Muller, P.A. Deep learning for time series classification: A review. *Data Min. Knowl. Discov.* **2019**, *33*, 917–963. [CrossRef]
9. Susto, G.A.; Cenedese, A.; Terzi, M. Time-Series Classification Methods: Review and Applications to Power Systems Data. In *Big Data Application in Power Systems*; Elsevier: Amsterdam, The Netherlands, 2018; pp. 179–220. [CrossRef]
10. Ayadi, A.; Ghorbel, O.; Obeid, A.M.; Abid, M. Outlier detection approaches for wireless sensor networks: A survey. *Comput. Netw.* **2017**, *129*, 319–333. [CrossRef]
11. Cook, A.A.; Misirli, G.; Fan, Z. Anomaly Detection for IoT Time-Series Data: A Survey. *IEEE Internet Things J.* **2020**, *7*, 6481–6494. [CrossRef]
12. Wu, H.S. A survey of research on anomaly detection for time series. In Proceedings of the 2016 13th International Computer Conference on Wavelet Active Media Technology and Information Processing (ICCWAMTIP), Chengdu, China, 16–18 December 2016; Institute of Electrical and Electronics Engineers, Inc.: Piscataway, NJ, USA, 2016; pp. 426–431. [CrossRef]
13. Aminikhanghahi, S.; Cook, D.J. A Survey of Methods for Time Series Change Point Detection. *Knowl. Inf. Syst.* **2017**, *51*, 339–367. [CrossRef]
14. Sharma, S.; Swayne, D.A.; Obimbo, C. Trend analysis and change point techniques: A survey. *Energy Ecol. Environ.* **2016**, *1*, 123–130. [CrossRef]
15. Truong, C.; Oudre, L.; Vayatis, N. Selective review of offline change point detection methods. *Signal Process.* **2020**, *167*, 107299. [CrossRef]
16. Torkamani, S.; Lohweg, V. Survey on time series motif discovery. *WIREs Data Min. Knowl. Discov.* **2017**, *7*, e1199. [CrossRef]
17. Wang, J.; Chen, Y.; Hao, S.; Peng, X.; Hu, L. Deep Learning for Sensor-based Activity Recognition: A Survey. *Pattern Recognit. Lett.* **2018**. [CrossRef]
18. Badhiye, S.S. A Review on Time Series Dimensionality Reduction. *HELIX* **2018**, *8*, 3957–3960. [CrossRef]
19. Sezer, O.B.; Gudelek, M.U.; Ozbayoglu, A.M. Financial time series forecasting with deep learning: A systematic literature review: 2005–2019. *Appl. Soft Comput.* **2020**, *90*, 106181. [CrossRef]
20. Lepenioti, K.; Bousdekis, A.; Apostolou, D.; Mentzas, G. Prescriptive analytics: Literature review and research challenges. *Int. J. Inf. Manag.* **2020**, *50*, 57–70. [CrossRef]
21. Mohammadi, M.; Al-Fuqaha, A.; Sorour, S.; Guizani, M. Deep Learning for IoT Big Data and Streaming Analytics: A Survey. *IEEE Commun. Surv. Tutor.* **2018**, *20*, 2923–2960. [CrossRef]
22. Zhao, Y.; Zhang, C.; Zhang, Y.; Wang, Z.; Li, J. A review of data mining technologies in building energy systems: Load prediction, pattern identification, fault detection and diagnosis. *Energy Built Environ.* **2019**. [CrossRef]
23. Zeger, S.L.; Irizarry, R.; Peng, R.D. On time series analysis of public health and biomedical data. *Annu. Rev. Public Health* **2006**, *27*, 57–79. [CrossRef]
24. Esling, P.; Agon, C. Time-series data mining. *ACM Comput. Surv.* **2012**, *45*, 1–34. [CrossRef]
25. Fakhrazari, A.; Vakilzadian, H. A survey on time series data mining. In Proceedings of the 2017 IEEE International Conference on Electro Information Technology (EIT), Lincoln, NE, USA, 14–17 May 2017; pp. 476–481. [CrossRef]
26. Keogh, E.; Kasetty, S. On the Need for Time Series Data Mining Benchmarks: A Survey and Empirical Demonstration. *Data Min. Knowl. Discov.* **2003**, *7*, 349–371. [CrossRef]
27. Cowpertwait, P.S.P.; Metcalfe, A.V. *Introductory Time Series with R*; Use R!; Springer: Dordrecht, The Netherlands; New York, NY, USA, 2009.
28. Shumway, R.H.; Stoffer, D.S. *Time Series Analysis and Its Applications: With R Examples*, 4th ed.; Springer: Cham, Switzerland, 2017.
29. Nielsen, A. *Practical Time Series Analysis: Prediction with Statistics and Machine Learning*, 1st ed.; O'Reilly: Beijing, China, 2019.
30. Joo, R.; Boone, M.E.; Clay, T.A.; Patrick, S.C.; Clusella-Trullas, S.; Basille, M. Navigating through the r packages for movement. *J. Anim. Ecol.* **2020**, *89*, 248–267. [CrossRef]
31. Slater, L.J.; Thirel, G.; Harrigan, S.; Delaigue, O.; Hurley, A.; Khouakhi, A.; Prosdocimi, I.; Vitolo, C.; Smith, K. Using R in hydrology: A review of recent developments and future directions. *Hydrol. Earth Syst. Sci.* **2019**, *23*, 2939–2963. [CrossRef]
32. Thivaharan, S.; Srivatsun, G.; Sarathambekai, S. A Survey on Python Libraries Used for Social Media Content Scraping. In Proceedings of the International Conference on Smart Electronics and Communication (ICOSEC 2020), Trichy, India, 10–12 September 2020. [CrossRef]
33. Ray, J.; Trovati, M. A survey of topological data analysis (TDA) methods implemented in python. *Lect. Notes Data Eng. Commun. Technol.* **2018**, *8*, 594–600.
34. Stancin, I.; Jovic, A. An overview and comparison of free Python libraries for data mining and big data analysis. In Proceedings of the 2019 42st International Convention on Information and Communication Technology, Electronics and Microelectronics (MIPRO), Opatija, Croatia, 20–24 May 2019; pp. 977–982. [CrossRef]
35. Januschowski, T.; Gasthaus, J.; Wang, Y. Open-Source Forecasting Tools in Python. *Foresight Int. J. Appl. Forecast.* **2019**, *55*, 20–26.
36. Kitchenham, B.; Brereton, P. A systematic review of systematic review process research in software engineering. *Inf. Softw. Technol.* **2013**, *55*, 2049–2075. [CrossRef]
37. Burns, D.M.; Whyne, C.M. Seglearn: A Python Package for Learning Sequences and Time Series. *J. Mach. Learn. Res.* **2018**, *19*, 3238–3244.
38. Christ, M.; Braun, N.; Neuffer, J.; Kempa-Liehr, A.W. Time Series FeatuRe Extraction on basis of Scalable Hypothesis tests (tsfresh—A Python package). *Neurocomputing* **2018**, *307*, 72–77. [CrossRef]

39. Alexandrov, A.; Benidis, B.; Bohlke-Schneider, M.; Flunkert, V.; Gasthaus, J.; Januschowski, T.; Maddix, D.C.; Rangapuram, S.; Salinas, D.; Schulz, J.; et al. GluonTS: Probabilistic and Neural Time Series Modeling in Python. *J. Mach. Learn. Res.* **2020**, *21*, 1–6.
40. Faouzi, J.; Janati, H. Pyts: A python package for time series classification. *J. Mach. Learn. Res.* **2020**, *21*, 1–6. Available online: http://jmlr.org/papers/v21/19-763.html (accessed on 24 June 2021).
41. Law, S. STUMPY: A Powerful and Scalable Python Library for Time Series Data Mining. *J. Open Source Softw.* **2019**, *4*, 1504. [CrossRef]
42. Collenteur, R.; Bakker, M.; Caljé, R.; Schaars, F. Pastas: Open-Source Software for the Analysis of Hydrogeological Time Series. Available online: https://zenodo.org/record/4277358 (accessed on 24 June 2021)
43. Miller, J.C.; Ting, T. EoN (Epidemics on Networks): A Fast, Flexible Python Package for Simulation, Analytic Approximation, and Analysis of Epidemics on Networks. 2020. Available online: https://zenodo.org/record/3572756 (accessed on 24 June 2021).
44. Schölzel, C. Nonlinear Measures for Dynamical Systems. Available online: https://zenodo.org/record/3814723 (accessed on 24 June 2021).
45. Silva, P.C.D.L.E.; Júnior, C.A.S.; Alves, M.A.; Silva, R.C.P.; Vieira, G.L.; Lucas, P.D.O.E.; Sadaei, H.J.; Guimarães, F.G. PYFTS/pyFTS: Stable Version 1.6. 2019. Available online: https://zenodo.org/record/2669398 (accessed on 24 June 2021).
46. Snow, D.; Baltacı, F. firmai/atspy: Zenodo. 2020. Available online: https://zenodo.org/record/4270168 (accessed on 24 June 2021).
47. Team, T.O.D. Obspy 1.0.0. 2016. Available online: https://zenodo.org/record/46151 (accessed on 24 June 2021).
48. Fu, T.C. A review on time series data mining. *Eng. Appl. Artif. Intell.* **2011**, *24*, 164–181. [CrossRef]

engineering proceedings

Proceedings

Comparative Analysis of Non-Linear GNSS Geodetic Time Series Filtering Techniques: El Hierro Volcanic Process (2010–2014) †

Belén Rosado *, Javier Ramírez-Zelaya, Paola Barba, Amós de Gil and Manuel Berrocoso

Laboratorio de Astronomía, Geodesia y Cartografía, Departamento de Matemáticas, Facultad de Ciencias, Campus de Puerto Real, Universidad de Cádiz, 11510 Puerto Real, Spain; javierantonio.ramirez@uca.es (J.R.-Z.); paola.barbaceballos@alum.uca.es (P.B.); amos.degil@uca.es (A.d.G.); manuel.berrocoso@uca.es (M.B.)
* Correspondence: belen.rosado@uca.es
† Presented at the 7th International conference on Time Series and Forecasting, Gran Canaria, Spain, 19–21 July 2021.

Abstract: GNSS geodetic time series analysis allows the study of the geodynamic behavior of a specific terrestrial area. These time series define the temporal evolution of the geocentric or topocentric coordinates obtained from geodetic stations, which are linear or non-linear depending, respectively, on the tectonic or volcanic–tectonic character of a region. Linear series are easily modeled but, for the study of nonlinear series, it is necessary to apply filtering techniques that provide a more detailed analysis of their behavior. In this work, a comparative analysis is carried out between different filtering techniques and non–linear GNSS time series analysis: 1sigma–2sigma filter, outlier filter, wavelet analysis, Kalman filter and CATS analysis (Create and Analyze Time Series). This comparative methodology is applied to the time series that describe the volcanic process of El Hierro island (2010–2014). Among them, the time series of the slope distance variation between FRON (El Hierro island) and LPAL (La Palma island) stations is studied, detecting and analyzing the different phases involved in the process.

Keywords: GNSS time series analysis; wavelet analysis; El Hierro volcanic process

Citation: Rosado, B.; Ramírez-Zelaya, J.; Barba, P.; de Gil, A.; Berrocoso, M. Comparative Analysis of Non-Linear GNSS Geodetic Time Series Filtering Techniques: El Hierro Volcanic Process (2010–2014). *Eng. Proc.* **2021**, 5, 23. https://doi.org/10.3390/engproc2021005023

Published: 28 June 2021

1. Introduction

GNSS–GPS systems are very effective tools in the study of the geodynamic behavior of a region. From the processing of the GPS observations, geodetic time series of sub–centimetric accuracy are obtained. The analysis of these time series is essential to understand the geodynamic behavior, even distinguishing between tectonic and volcanic activities in those places where the geodynamics presents these or other complex situations.

In this work, a mathematical procedure is established for the study of nonlinear geodetic time series. The methodology consists of a pre-treatment of the time series to eliminate anomalous values that disturb the subsequent adjustments. This is performed using the Outlier R filter. To these filtered series, the analytical Kalman and wavelet filters, the statistical ARMA and ARIMA filters, and the CATS and STL techniques of linear fitting are applied.

This methodology is applied to the time series obtained from the FRON station, located on El Hierro island, in the Canary Islands. These series span from 2010 to 2014; therefore, they reflect the volcanic process that began on the island in July 2011. Due to the volcanic–tectonic geodynamic behavior of the region, the time series present non–linear characteristics. As a comparison, the time series of the IZAN station on Tenerife island are shown, which are not affected by volcanic activity. Therefore, these time series are linear.

2. Site Description

El Hierro island is located southwest of the Canary archipelago. The island's morphology has been interpreted as a triple volcanic rift: NE, NW and S rifts, with axes diverging about 120°, Figure 1 [1]. In July 2011, an increase in surface deformation and seismicity on the island was detected. The climax of this unrest was a submarine eruption first detected on 10 October 2011 [2], and located at about 2 km SW of La Restinga, the southernmost village of the El Hierro island. The eruption ceased on 5 March 2012, although deformation and seismic activity did not cease after the eruption [1].

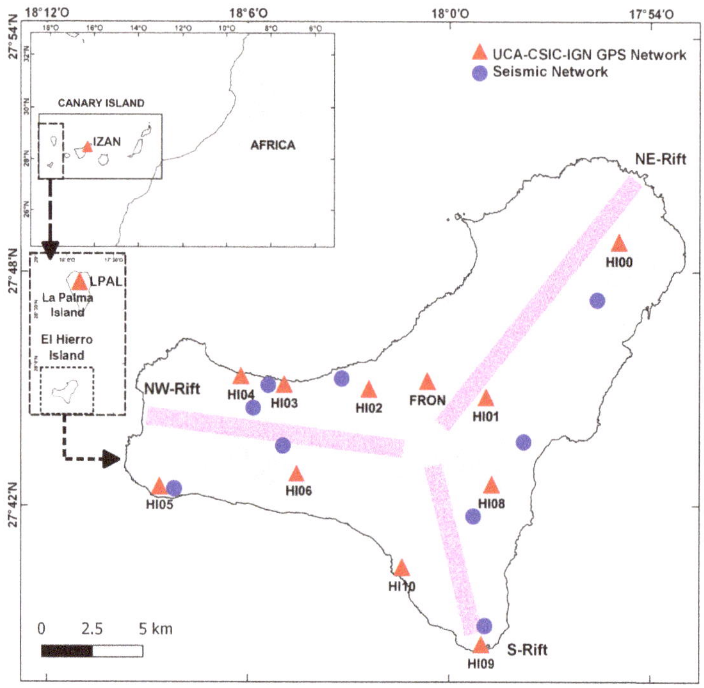

Figure 1. Map of the El Hierro island. The GPS stations of the UCA–CSIC–IGN geodynamic network in El Hierro, the IZAN GPS stations in Tenerife island, the LPAL in La Palma island, and the seismic stations are shown.

On the island, there was only one geodetic benchmark from which global navigation satellite systems (GNSS) provided continuous and publicly accessible data from the beginning of the volcanic unrest. This is the FRON station, located at the Frontera municipality in the El Golfo valley, and maintained by the Canarian Regional Government. Because of the ground deformation detected there through geodetic processing of global positioning system (GPS) data, other GNSS–GPS receivers were deployed by the Spanish National Geographic Institute (IGN) throughout a geodetic benchmarks' network designed by the Laboratory of Astronomy, Geodesy and Cartography of Cadiz University (Figure 1). A first set of four benchmarks in the El Golfo valley was continuously observed near the end of July, forming an almost straight line (HI01, HI02, HI03, and HI04), but it was only later near the eruption's start that four other benchmarks were continuously observed, forming a three–tipped star covering all of the island spatially (HI00, HI05, HI08, and HI09). In addition, HI10 was continuously observed, while three others were only periodically observed (HI06, HI07 and HI11) [2,3].

In the rest of the islands of the Canary archipelago, there are other permanent GNSS–GPS stations managed by the IGN, among them IZAN on Tenerife island and LPAL on La Palma island. The LPAL station also belongs to the international IGS network (Figure 1).

3. Methodology

The evolution of the eruptive process has been studied from the analysis of the geodetic time series of the GNSS–GPS stations located on the islands. The GPS observations have been processed using the Bernese scientific software v5.0 [4]. The IGS LPAL station (La Palma island) has been used as a reference station, in sessions of 24 h and 30 s of sampling frequency, using the ITRF2008 reference frame [5]. For each observation session, geocentric coordinates (X, Y, Z) have been obtained with sub–centimeter accuracy. From a given initial time, the time series of the topocentric coordinates (east, north, up) and the time series of the distance variation between the reference station, LPAL, and the corresponding station have been built.

These time series can be linear or non-linear depending on the tectonic or volcanic-tectonic character of a region. Figure 2 shows the topocentric time series and distance variation time series between the LPAL–IZAN and LPAL–FRON stations from 2010 to 2014. The IZAN station, located on Tenerife island, is not affected by the volcanic process of El Hierro; therefore, it presents a linear behavior in all its components (Figure 2a). On the contrary, the FRON station is located on El Hierro island, so its time series are non-linear (Figure 2b). Linear series are easily modeled but, for the study of nonlinear series, it is necessary to apply filtering techniques that provide a more detailed analysis of their behavior.

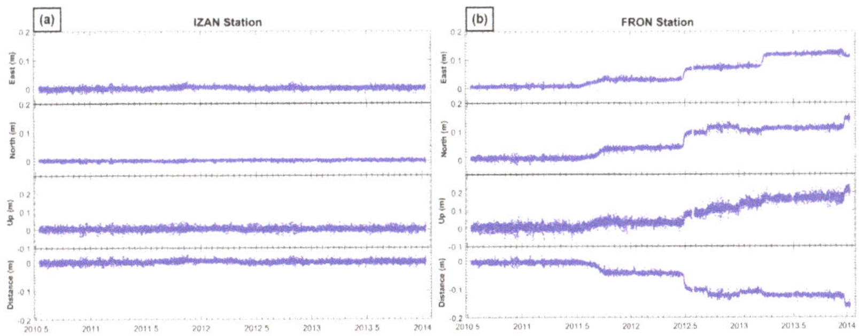

Figure 2. Topocentric time series and distance variation time series between LPAL stations and (**a**) IZAN, Tenerife island, and (**b**) FRON, El Hierro, from 2010 to 2014.

The methodology is summarized in Figure 3. For a better understanding, the distance variation time series between FRON and LPAL is shown with the result of each technique, but the next section shows the result for all the time series of FRON station.

First, a descriptive analysis of the raw data is carried out (Figure 4a). Thus, an initial visualization of the data is carried out, detecting errors due to different causes both physical and instrumental. Due to these anomalous data, it is necessary to treat the time series to eliminate outliers and noise that distort the subsequent analysis. For this reason, an initial filtering of the data is carried out using the Outlier R filter (Figure 4b). The objective of this filter is to ensure that the filtered series has linearity, homoscedasticity and follows a normal distribution. To achieve this objective, this filter uses the Box–Cox transform [6].

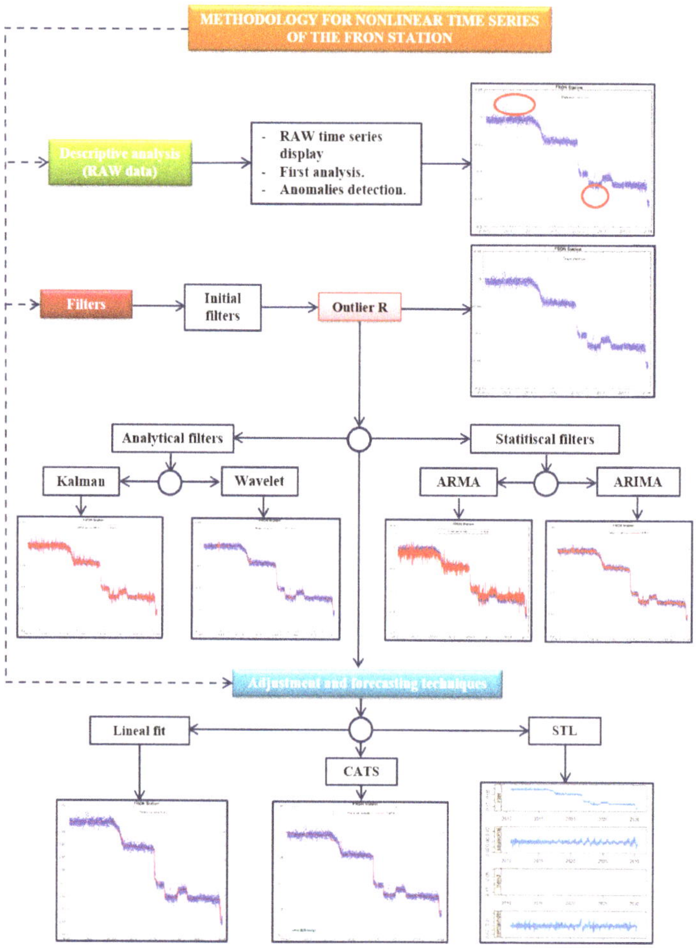

Figure 3. Scheme of the methodology.

Subsequently, different analytical filtering techniques are applied to the filtered time series: Kalman and wavelet. The Kalman filter is an algorithm for updating, observation by observation, the linear projection of a system of variables on the set of available information, as new information becomes available. The Kalman filter makes it possible to easily calculate the likelihood of a linear, uni-equation or multi-equation dynamic model, estimating the parameters of the model, as well as obtaining predictions from these types of models [2,7]. The application of this filter to the time series is shown in Figure 4c.

Wavelet techniques allow to divide a complex function into simpler ones and study them separately. To apply the wavelet transform to a series of numerical data, it is necessary to implement the discrete wavelet transform (DWT) [8]. The objective of applying the DWT to a vector is to obtain a transformed vector that has in the middle, known as the high part (details), the same high–frequency information as the original vector and, in another half, known as the low part (approximations), the low–frequency information. Wavelet transforms comprise a large set of shapes. Over time, different versions of wavelets have been developed, which have given rise to families of wavelets [9]. In this work, the Coiflets family has been used, and specifically, the Coiflets of order 5. The result of this technique is shown in Figure 4d.

On the other hand, statistical filters are applied: ARMA and ARIMA. The ARMA model is given by the composition of autoregressive models (AR) and moving average models (MA). On the other hand, the ARIMA model results from the union of the autoregressive (AR), integrated and moving average (MA) models [10]. The results of both filters are shown in Figure 4e,f, respectively.

Finally, as adjustment and forecasting techniques, a linear adjustment, the CATS adjustment and the STL decomposition are performed. In order to carry out a linear fit and due to the non–linear characteristics of the time series, it is necessary to carry out this fit in parts (Figure 4g). The CATS adjustment (Create and Analyze Time Series) [11] consists of decomposing the time series in order to calculate the trend, the amplitudes of the sinusoidal terms and the magnitudes of the discontinuities that the series presents [8] (Figure 4f). On the other hand, the STL decomposition (Seasonal and Trend decomposition procedure based on Loess) decomposes a time series into its three components: trend, seasonality and irregularities using local regression (loess) [12].

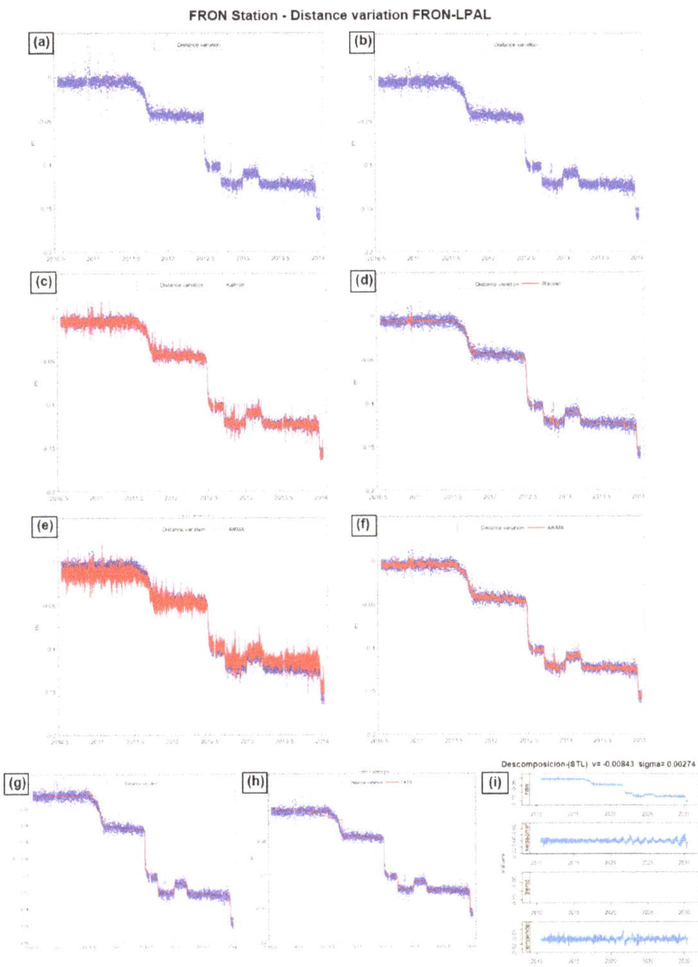

Figure 4. Distance variation time series between FRON and LPAL (**a**), with the results of the filters: (**b**) Outlier R, (**c**) Kalman, (**d**) wavelet, (**e**) ARMA, (**f**) ARIMA, (**g**) linear fit, (**h**) CATS fit and (**i**) STL decomposition.

4. Results

Figure 5 shows the results of applying the filters exposed in the methodology to the topocentric time series east (a), north (b) and height (c) of the FRON station, and to the distance variation between FRON–LPAL (d). The figures show: outlier R series (blue), wavelet series (red), Kalman series (pink), ARMA series (green), ARIMA series (orange), CATS series (light blue). The earthquakes of magnitude greater than 4 (light green color) that occurred in the region between 2011 and 2014 are also represented, obtained from the seismic catalog provided by the IGN (www.ign.es, accessed on 1 June 2021). The black line represents the earthquake of magnitude 5.1 that occurred on 27 December 2013.

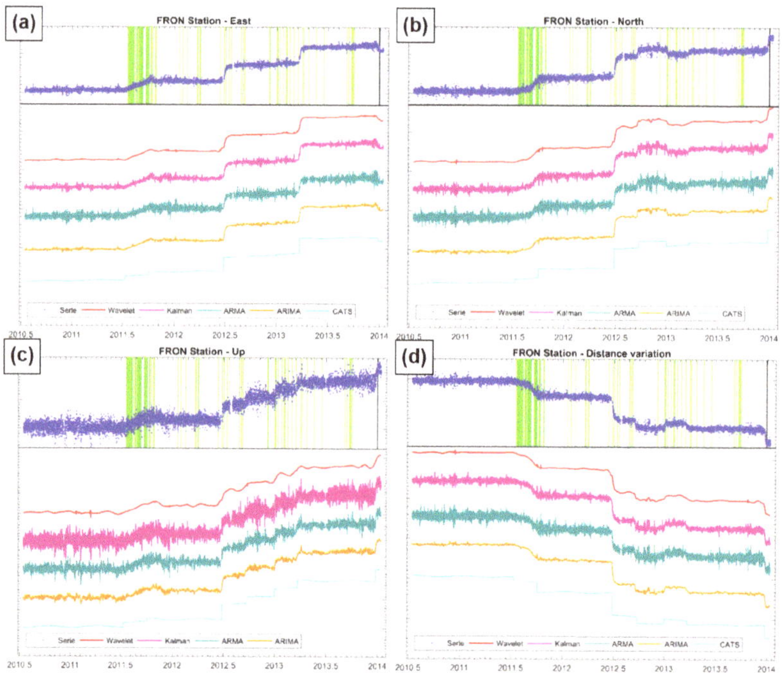

Figure 5. Topocentric time series east (**a**), north (**b**) and height (**c**) of the FRON station, and the distance variation between FRON–LPAL (**d**). They show: Outlier R series (blue), wavelet series (red), Kalman series (pink), ARMA series (green), ARIMA series (orange), CATS series (light blue). The earthquakes of magnitude greater than 4 (light green color) that occurred in the region between 2011 and 2014 are also represented, and the earthquake of magnitude 5.1 that occurred on 27 December 2013 is represented in black.

Data Availability Statement: Data supporting reported results can be found at http://www.ign.es, https://www.epncb.oma.be and http://www.grafcan.es.

Acknowledgments: We are thankful to all individuals and institutions contributing to GPS data collection and dissemination. Thanks are due to GRAFCAN (Gobierno de Canarias), IGN (Instituto Geográfico Nacional) and IGS (International GNSS Service). We would also like to thank Onsala Space Observatory for their calculation of the ocean tide loading parameters. The contributions of both editor and of the reviewers are very much appreciated, which greatly helped to improve the manuscript.

References

1. García, A.; Fernández-Ros, A.; Berrocoso, M.; Marrero, J.M.; Prates, G.; De la Cruz-Reyna, S.; Ortiz, R. Magma displacements under insular volcanic fields, applications to eruption forecasting: El Hierro, Canary Islands, 2011–2013. *Geophys. J. Int.* **2014**, *197*, 322–334. [CrossRef]
2. Prates, G.; García, A.; Fernández-Ros, A.; Marrero, J.M.; Ortiz, R.; Berrocoso, M. Enhancement of sub-daily positioning solutions for surface deformation surveillance at El Hierro volcano (Canary Islands, Spain). *Bull. Volcanol.* **2013**, *75*, 724. [CrossRef]
3. Martín, M.; Rosado, B.; Berrocoso, M. Description of El Hierro volcanic process (2011–2014) from variability analysis of topocentric coordinates obtained by GNSS observations. In Proceedings of the ITISE International work-conference On Time Series, Granada, Spain, 25–27 June 2014; Volume 2, pp. 1293–1298, ISBN 978-84-15814-97-4.
4. Dach, R.; Hugentobler, U.; Fridez, P.; Meindi, M. *Bernese GPS Software ver. 5.0 User Manual*; Astronomical Institute, University of Bern: Bern, Switzerland, 2007.
5. Altamimi, Z.; Collilieux, X.; Metivier, L. ITRF2008: An improved solution of the International Terrestrial Reference Frame. *J. Geod.* **2011**, *85*, 457–473. [CrossRef]
6. Box, G.E.P.; Cox, D.R. An analysis of transformations. *J. R. Stat. Soc.* **1964**, *26*, 211–252. [CrossRef]
7. Larson, K.; Poland, M.; Miklius, A. Volcano monitoring using GPS: Developing data analysis strategies based on the june 2007 Kilauea Volcano intrusion and eruption. *J. Geophys. Res.* **2010**, *115*, B07406. [CrossRef]
8. Rosado, B.; Fernández-Ros, A.; Jiménez, A.; Berrocoso, M. Modelo de deformación horizontal GPS de la región sur de la Península Ibérica y norte de áfrica (SPINA). *Boletín Geológico Minero* **2017**, *128*, 141–156. [CrossRef]
9. Rosado, B.; Fernández-Ros, A.; Berrocoso, M.; Prates, G.; Gárate, J.; De Gil, A.; Geyer, A. Volcano-tectonic dynamics of Deception Island (Antarctica): 27 years of GPS observations (1991–2018). *J. Volcanol. Geotherm. Res.* **2019**, *381*, 57–82. [CrossRef]
10. Box, G.E.P.; Jenkins, F.M. *Time Series Analysis: Forecasting and Control*, 2nd ed.; Holden-Day: Oakland, CA, USA, 1976.
11. Williams, S.D.P. CATS: GPS coordinate time series analysis software. *GPS Solut.* **2008**, *12*, 147–153. [CrossRef]
12. Cleveland, R.B.; Cleveland, W.S.; McRae, J.E.; Terpenning, I.J. STL: A seasonal-trend decomposition procedure based on loess. *J. Off. Stat.* **1990**, *6*, 3–33.

engineering proceedings

MDPI

Proceedings

Predicting the Window Opening State in an Office to Improve Indoor Air Quality †

Thi Hao Nguyen [1,*], Anda Ionescu [1], Olivier Ramalho [2] and Evelyne Géhin [1]

1 Univ Paris-Est Creteil, CERTES, F-94010 Creteil, France; ionescu@u-pec.fr (A.I.); gehin@u-pec.fr (E.G.)
2 Scientific and Technical Center for Building, 77447 Champs-sur-Marne, France; olivier.RAMALHO@cstb.fr
* Correspondence: thi-hao.nguyen@u-pec.fr

† Presented at the 7th International Conference on Time Series and Forecasting, Gran Canaria, Spain, 19–21 July 2021.

Abstract: Window operation is among one of the most influential factors on indoor air quality (IAQ). In this paper, we focus on the modeling of the windows' opening state in a real open-plan office with five windows. The IAQ of this open-plan office was monitored over a whole year along with the opening state of the windows. A k-Nearest Neighbor (k-NN) classification model was implemented, based on a long time series of both indoor and outdoor monitored environmental factors such as temperature and relative humidity, and CO_2 indoor concentration. In addition, the month, the day of the week and the time of the day were included. The obtained model for the window state prediction performs well with an accuracy of 92% for the training set and 86% for the testing set.

Keywords: k-nearest neighbor classification; time series; autocorrelation function; indoor environment; windows state prediction

Citation: Nguyen, T.H.; Ionescu, A.; Ramalho, O.; Géhin, E. Predicting the Window Opening State in an Office to Improve Indoor Air Quality. *Eng. Proc.* **2021**, *5*, 24. https://doi.org/10.3390/engproc2021005024

Published: 28 June 2021

1. Introduction

Indoor air quality (IAQ) is, nowadays, an essential research topic, as we spend more than 90% of our time indoors [1]. The opening state of windows has an important influence on IAQ; therefore, it is necessary to understand and model the relationship between them [2].

Previous studies mostly used logistic regression to compute the correlation between the probability of a window opening and environmental stimuli to predict the probability of a window opening/closing event [3,4]. For this approach, all the observations need to be independent, and the outcomes of the model are usually complex equations which may not be easily understandable and interpreted.

In the last decades, many studies have used Machine Learning (ML) and their research application to the environment is not an exception. In 2014, D'Oca et al. tried to apply ML by using a data-mining approach to discover patterns of window opening and closing behavior in offices [5]. In this study, a huge amount of detailed data was needed and the authors mainly focused on obtaining distinct behavioral patterns of the window tilting angle, instead of for its opening state for a group of windows as was the case in our study. Many ML algorithms, such as Decision Trees, Support Vector Machines, k-Nearest Neighbor and Ensemble classification, can be applied for our study case. The k-NN classification is recommended as 'a theoretically optimal method of classification' [6]. Indeed, the best results were obtained on our case by using k-NN classification. To the best of our knowledge, this method has not yet been applied to predicting the state of window opening, but it has recently been used in a related topic of IAQ, which is occupancy detection [7]. This paper presents the ability of a k-NN classifier to predict the state of window opening in an open-plan office, as presented hereafter.

2. Methodology

2.1. Study Case and Parameters Selection

The studied open-plan office is located in the suburban town of Champs-sur-Marne, France. The surface and the volume of the office are 132 m^2 and 364 m^3, respectively; it is used by 6 to 15 people, from 8:00 a.m. to 6:00 p.m. from Monday to Friday.

Measurement devices were installed inside and outside the office. The monitoring was performed over a full year, in 2014. Temperature (T), relative humidity (RH), carbon dioxide (CO_2) and particulate matter were monitored every minute, during the whole year. The five windows of the office were equipped with sensors that detected each opening or closing event [8].

According to some previous studies, the outdoor temperature and indoor CO_2 concentration were the two most important variables in determining the probability of opening/closing windows, followed by indoor air temperature, and outdoor and indoor relative humidity [3,4,9]. In addition, non-environmental factors, that is, seasonal change, time of the day and personal preference, also affect the window-opening probability [10]. Thus, in our model, the following variables were used: month, day of the week, time of the day, indoor CO_2 concentration, and both indoor and outdoor temperature (T) and relative humidity (RH). The main statistics of these environmental parameters are displayed in Table 1.

Table 1. The statistics for the environmental parameters.

Features	Indoor CO_2 (ppm)	Indoor T (°C)	Outdoor T (°C)	Indoor RH (%)	Outdoor RH (%)
Max value	1144	31.3	35.6	74.6	100.0
Min value	416.8	15	−4.3	18.3	26.9
Mean value	501.1	23	13.5	44.2	82.2
Median value	480.5	22.4	13.5	42.9	86.7
Std value	64.3	2.3	6	9.3	16.2

In order to obtain more information about the monitored time series, the autocorrelation function (ACF) was calculated (using hourly averaged data). The ACF of a time series $Y(t)$ provides a measure of the correlation between y_t and y_{t+k}, where $k = 0, \ldots, K$ ($k \in \mathbb{Z}$, K is not larger than $T/4$) and y_t is assumed to be the realization of a stochastic process. According to [11], the autocorrelation r_k for lag k is:

$$r_k = \frac{c_k}{c_0}, \quad (1)$$

where:

$$c_k = \frac{1}{T} \sum_{t=1}^{T-k} (y_t - \overline{y})(y_{t+k} - \overline{y}) \quad (2)$$

and c_0 is the sample variance, $\overline{y}$ is the sample mean of the time series; T is the number of observations.

Figure 1 presents the ACFs for all the quantitative variables used in this study.

From the results presented in Figure 1, one can notice that the state of the environment at one sample (hour) has the highest correlation with the next sample. In other words, the previous hour of environmental data also has an important impact on the current information. Therefore, this implies that the previous hour of environmental data also has an important impact on the current state of the window. Hence, we decided to use the information on both the previous and current samples for the input to the predicting model.

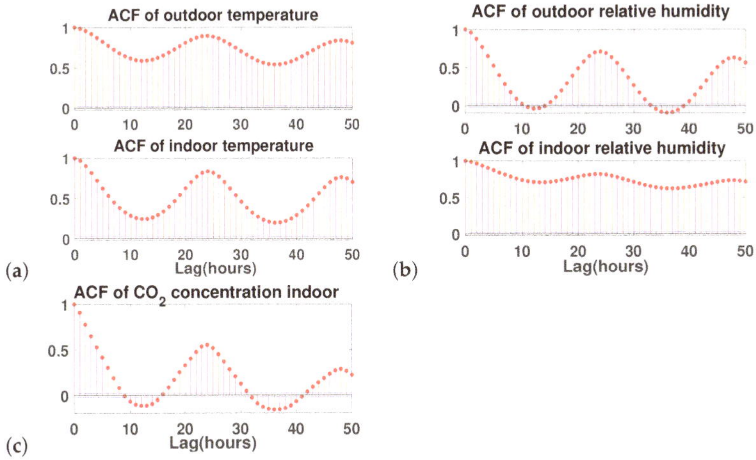

Figure 1. Autocorrelation values of environmental variables: (**a**) temperature, (**b**) relative humidity, and (**c**) Carbon dioxide concentration.

We notice that the autocorrelation becomes zero after around 8 h for indoor CO_2 and outdoor RH. By contrast, indoor RH decreases very slowly. The same pattern can be found for outdoor, and also indoor, air temperature. This reveals the persistence of T and RH indoors, which means that a value at time t of the temperature or indoor relative humidity can have an impact on a value a long time later. We also note that the ACF of the CO_2 concentrations and RH outdoors becomes negative and remains at low levels, then switches back to positive values after a lag of 17 h. As for T outdoors and RH indoors, the autocorrelations persist in the positive for long delays. In general, temperatures and humidity depict the same structures of spectral variability as CO_2: two fundamental frequency peaks at $(24\ h)^{-1}$ and $(12\ h)^{-1}$. The ACF of CO_2 and outdoor RH alternates sign every 8 h on a lag of 24 h. This implies that, instead of using the information from the 'previous hour', in the real-time system, we could use the values of the environmental data from 'the previous 24 h' as an input for this model, which are much easier to access than the 'previous hour' data for a real-time application.

2.2. Classification Model Implementation

The hourly averaged values of the selected parameters were used. A linear interpolation was applied in order to replace missing values. Then, the responses were categorized into four different groups, labelled as follows:

- ALL CLOSED: less than 1 window is opened ($N < 1$)
- MOSTLY CLOSED: from 1 to less than 2 windows are opened ($1 \leq N < 2$)
- MOSTLY OPENED: from 2 to less than 4 windows are opened ($2 \leq N < 4$)
- ALL OPENED: 4 windows or more are opened ($N \geq 4$)

The non-environmental parameters' distribution profiles and the initial statistics of these four groups during the year 2014 are displayed in Figure 2.

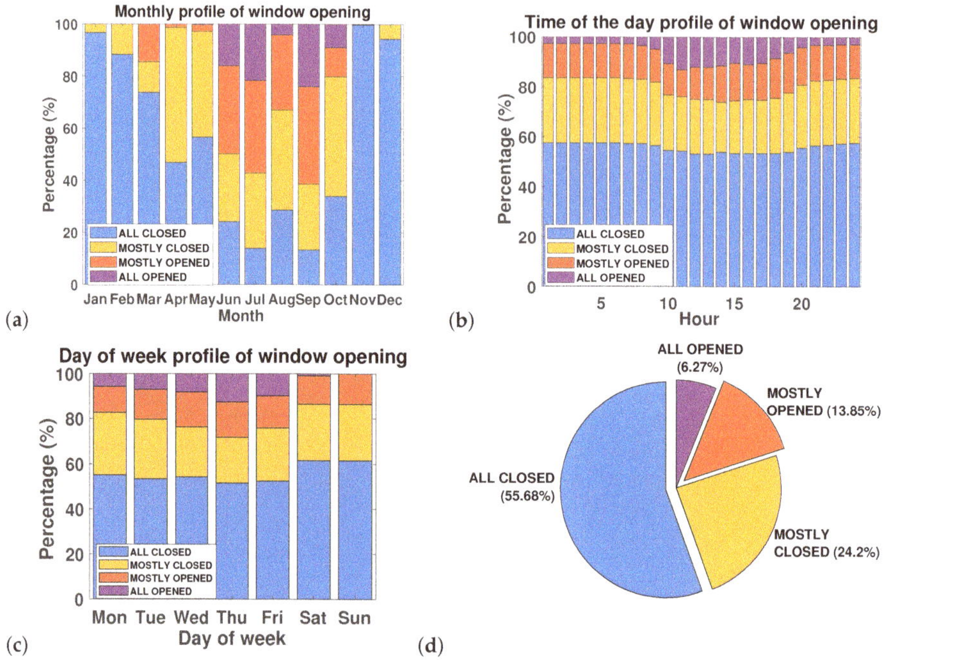

Figure 2. Distribution profile of window opening according to the (**a**) Month, (**b**) Hour of the day and (**c**) Day of the week. (**d**) Statistics for window opening categories.

Firstly, the time series data was divided into sets of consecutive 23 h periods. Next, every 20 first hours of each set were used for training and the other 3 h were used for testing. This results in 7600 h for the training and 1140 h for the testing set (380 sets in total). The reason for choosing a set of 23 h instead of 24 h was that we wanted to achieve an equal distribution of the 'time of the day' in both training and testing sets. This can avoid only training on the same specific hours (1 a.m. to 9 p.m., for example, and always testing on the same 3 h in the evening, starting from 10 p.m.).

A Classification Learner Application provided by Matlab software via the Statistics and Machine Learning Toolbox was used to build the classifier. This application trains models to classify data using supervised machine learning. Based on the amount of data that we have, we applied a 10-fold cross validation for the training step, which helps us to limit the overfitting problem. Regarding the setting parameters of our classification model, the Euclidean distance was adopted. Concerning the number of nearest neighbors, for k = 1, we archived the highest accuracy, so the label of a 'nearest neighbor' is selected.

3. Results and Discussion

The output of the Classification Learner App shows that a fine k-NN model has been obtained with an accuracy of 92.2%. Using this trained k-NN classifier, we predicted the testing set and compared it to the monitored value, obtaining a value of 86.1% for accuracy. A confusion matrix for this test set is displayed in Figure 3. The highest recall value (true positive rate) is obtained when predicting the 'ALL CLOSED' state of the group of windows (93.9%) while the lowest belongs to the 'MOSTLY OPENED' label (only 70.3%). Regarding precision values (positive predictive values), the highest value is still obtained by the 'ALL CLOSED' state; however, the lowest value corresponds to the 'ALL CLOSED' label.

In addition, the statistics for the accuracy of each month, the hour of the day and the day of the week in the testing set are shown in Tables 2–4, respectively, where the lower values mostly belong to the summer season (Jun–Sep, except for April), day-time periods

(10 a.m.–5 p.m., except for 4 p.m.) and the working day (Mon–Fri), which mostly contains the labels 'ALL OPENED' and 'MOSTLY OPENED'.

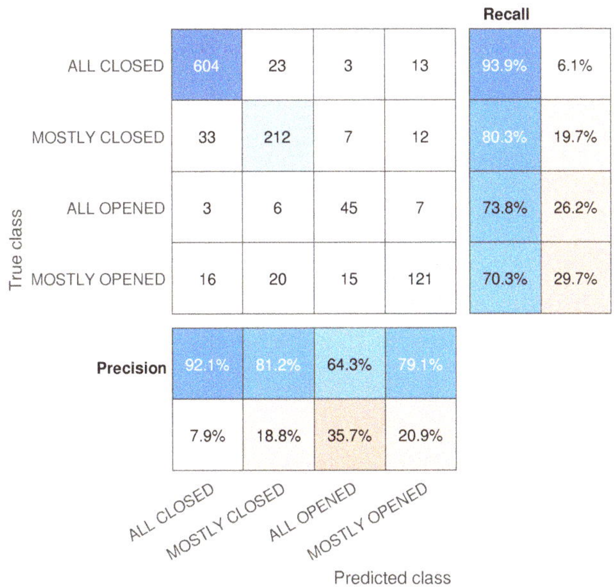

Figure 3. Confusion matrix, precision and recall value (in percentage %) for each label of the test set.

Table 2. The statistics for the accuracy of each month in the testing set.

Month	Jan	Feb	Mar	Apr	May	Jun	Jul	Aug	Sep	Oct	Nov	Dec
No. of samples	96	87	96	96	96	93	99	96	93	99	93	96
Accuracy	0.99	0.91	0.85	0.77	0.92	0.83	0.77	0.76	0.71	0.89	0.97	0.98

Table 3. The statistics for the accuracy of each hour of the day in the testing set.

Hour	1st	2nd	3rd	4th	5th	6th	7th	8th	9th	10th	11th	12th
No. of samples	44	44	46	48	49	49	48	48	48	48	48	48
Accuracy	0.91	0.91	0.96	0.96	0.96	0.98	0.98	0.96	0.90	0.67	0.73	0.81
Hour	**13th**	**14th**	**15th**	**16th**	**17th**	**18th**	**19th**	**20th**	**21st**	**22nd**	**23rd**	**24th**
No. of samples	48	48	48	48	48	48	48	48	48	48	47	45
Accuracy	0.85	0.81	0.85	0.90	0.79	0.88	0.81	0.73	0.81	0.85	0.89	0.78

Table 4. The statistics for the accuracy of each day of the week in the testing set.

Day	Mon	Tue	Wed	Thu	Fri	Sat	Sun
No. of samples	162	161	166	162	164	161	164
Accuracy	0.86	0.84	0.83	0.84	0.76	0.96	0.93

Even though the accuracy of the training set is not so high, this is explained by the unequal proportion in each label group, especially the small amount for the 'ALL OPENED' label (6.3% as in Figure 2b). Therefore, the model tends to 'learn well' with other dominant labels more than with this label. In the future, we can improve this by having an unbiased data set or by providing different weights for each label to penalize misclassification. In addition, the initial set of variables could include the rate of variation of the environmental factors to help improve the performance of the model.

4. Conclusions

In this study, we have obtained a k-NN classification model to predict the opening state for a group of windows in an open-plan office by using both environmental and non-environmental parameters of previous and current samples, including: month, day of the week, time of the day, indoor CO_2 concentration, and both indoor and outdoor temperature and relative humidity. A validation test has been used to compare the outputs of the model and the measured window states observed during the year 2014. We could then use this model by including it in real-time indoor air quality prediction, in order to optimize the action to be taken to reduce the exposure of the occupants.

References

1. Indoor Air Division, Office of Atmostpheric and Indoor Air Programs. *Congress on Indoor Air Quality: Assessment and Control of Indoor Air Pollution*; Technical Report; U.S. Environmental Protection Agency: Washington, DC, USA, 1989.
2. Jian, Y.; Guo, Y.; Liu, J.; Bai, Z.; Li, Q. Case study o fwindow opening behavior using field measurement results. *Build. Simul.* **2011**, *4*, 107–116. [CrossRef]
3. Andersen, R.; Fabi, V.; Toftum, J.; Corgnati, S.P.; Olesen, B.W. Window opening behaviour modelled from measurements in Danish dwellings. *Build. Environ.* **2013**, *69*, 101–113. [CrossRef]
4. Yao, M.; Zhao, B. Window opening behavior of occupants in residential buildings in Beijing. *Build. Environ.* **2017**, *124*, 441–449. [CrossRef]
5. D'Oca, S.; Hong, T. A data-mining approach to discover patterns of window opening and closing behavior in offices. *Build. Environ.* **2014**, *82*, 726–739. [CrossRef]
6. Hastie, T.; Tibshirani, R.; Friedman, J. *The Elements of Statistical Learning*; Springer Series in Statistics; Springer: New York, NY, USA, 2001.
7. Dai, X.; Liu, J.; Zhang, X. A review of studies applying machine learning models to predict occupancy and window-opening behaviours in smart buildings. *Energy Build.* **2020**, *223*, 110–159. [CrossRef]
8. Ramalho, O.; Ouaret R.; Ionescu A.; Le Ponner E.; Candau Y. *TRIBU–Suivi dynamique en Temps Réel de la qualité de l'air Intérieur dans un environnement de BUreaux. Contributions des sources et Modèle prévisionnel rapport, PRIMEQUAL APR EIAI/projet TRIBU*; Technical Report; Scientific and Technical Center for Building (CSTB): Marne-la-Vallée, France, 2016.
9. Fabi, V.; Andersen, R.; Corgnati, S.; Olesen, B. Occupants' window opening behaviour: A literature review of factors influencing occupant behaviour and models. *Build. Environ.* **2012**, *58*, 188–198. [CrossRef]
10. Pan, S.; Xiong, Y.; Han, Y.; Zhang, X.; Xia, L.; Wei, S.; Wu, J.; Han, M. A study on influential factors of occupant window-opening behavior in an office building in China. *Build. Environ.* **2018**, *133*, 41–50. [CrossRef]
11. Box, G.; Jenkins, G.M.; Reinsel, G. *Time Series Analysis: Forecasting and Control*, 3rd ed.; Prentice Hall: Englewood Cliffs, NJ, USA, 1994.

engineering proceedings

MDPI

Proceedings

Factors Affecting Transport Sector CO_2 Emissions in Eastern European Countries: An LMDI Decomposition Analysis †

Souhir Abbes [1,2]

1 Laboratory of Economics and Development, University of Sfax, Sfax 3029, Tunisia; souhir.abbes@hotmail.fr
2 Laboratory of CEARC/OVSQ, University of Versailles Saint-Quentin-en-Yvelines, 78000 Versailles, France
† Presented at the 7th International conference on Time Series and Forecasting, Gran Canaria, Spain, 19–21 July 2021.

Abstract: In this paper, we use the Logarithmic Mean Divisia Index (LMDI) to apply decomposition analysis on Carbon Dioxide (CO_2) emissions from transport systems in seven Eastern European countries over the period between 2005 and 2015. The results show that "economic activity" is the main factor responsible for CO_2 emissions in all the countries in our sample. The second factor causing increase in CO_2 emissions is the "fuel mix" by type and mode of transport. Modal share and energy intensity affect the growth of CO_2 emissions but in a less significant way. Finally, only the "population" and "emission coefficient" variables slowed the growth of these emissions in all the countries, except for Slovenia, where the population variable was found to be responsible for the increase in CO_2 emissions. These results not only contribute to advancing the existing literature but also provide important policy recommendations.

Keywords: CO_2 emissions; transport sector; LMDI; economic activity; modal share; energy intensity; Eastern Europe

Citation: Abbes, S. Factors Affecting Transport Sector CO_2 Emissions in Eastern European Countries: An LMDI Decomposition Analysis . *Eng. Proc.* **2021**, *5*, 25. https://doi.org/10.3390/engproc2021005025

Published: 28 June 2021

1. Introduction and Theoretical Background

Recent studies by the European Environment Agency suggest that transport activities contribute 28.5% of total CO_2 emissions, and around 33.1% of final energy consumption in the European Union. Emissions from this sector have increased from 945.1 million tons in 1990 to 1169.6 million tons in 2015. On the other hand, the share of renewable energy used for transport in the EU rose from 7.4% in 2017 to 8.1% in 2018, which is well below the EU target of 10% set for 2020. Overall, some EU countries have succeeded in reducing their own emissions, while others are still struggling to achieve such objectives, notably Eastern European countries.

Many tools have been developed by economists and mathematicians to study the relationship between transport activities and their environmental effects, and to examine key factors that are thought to contribute to CO_2 emissions in particular.

The first theory in this regard is based on the Granger causality and Co-integration approach. This method examines the effects of a wide range of variables (urbanization, energy consumption energy efficiency, car ownership, economic activity, etc.) on CO_2 emissions from the transport sector. Studies include Gonzales and Marrero [1], Lu et al. [2] and Abbes and Bulteau [3].

Another theory focuses on optimization, either to forecast energy demand and CO_2 emissions, or to analyze energy planning for sustainable development [4–6].

Finally, the most widely used technique is the decomposition methods based on the redefined Laspeyres index method developed by Sun [7] and the Logarithmic Mean Divisia index method (LMDI; Ang and Choi [8]). At the beginning, the decomposition technique has been used to assess the total energy consumption caused by the energy crisis. Later, this technique was generalized for uses and applications in other sectors, particularly the transport sector, in the 1990s and 2000s. This method allows us to quantify the contributions of various factors to CO_2 emissions from the transport sector. The basic idea is that transport CO_2 emissions is the sum of CO_2 emissions from each transportation mode. To extend the analysis, other sub-category levels can be added, such as the decomposition of emissions from the ith transportation mode to emissions coming from fuel type j in year t. Other variables such as population, energy consumption, motorization and economic growth can be introduced into these sub-categories to denote the various "effects" that contribute to transport CO_2 emissions.

One of the first works to use the decomposition method is that of Scholl et al. [9] who studied CO_2 emissions from passenger transport resulting from changes in transport activity, modal structure, CO_2 intensity, energy intensity and fuel mix in nine OECD countries between 1973 and 1992. One year later, Schipper et al. [10] used decomposition analysis to explain the change in energy consumption and carbon emissions from freight transport in 10 industrialized countries from 1973 to 1992, by introducing the following factors: transport activity, modal share and energy intensity. The two studies by Timilsina and Shrestha [11,12] were conducted in 12 countries in Asia, and 20 countries in Latin America and the Caribbean during 1980–2005.

Similarly, Papagiannaki and Diakoulaki [13] studied the variation in CO_2 emissions from passenger cars using decomposition analysis in Greece and Denmark over the period between 1990 and 2005. The variables used are car ownership, type of fuel mixture, annual mileage travelled, engine size or capacity, car engine technology, economic growth and population. The LMDI-I method was applied by Wang et al. [14] in China between 1985 and 2009, in order to obtain a decomposition of CO_2 emissions from transport. For the same country but with a different period from 1995 to 2006, Wang et al. [15] used the full decomposition approach to construct a decomposition model that summarises the impact of road freight transport-related factors on carbon emissions, and to predict its trend. In addition, Andreoni and Galmarini [16] used the decomposition analysis to investigate the main factors influencing CO_2 emissions from transport activities in the maritime and aviation sectors in 14 EU Member States, and in Norway. Similarly, a decomposition model was applied in Sweden by Eng-Larsson et al. [17]. They analysed the relationship between economic growth, freight transport, energy consumption, transport intensity and fuel carbon intensity. Guo et al. [18] presented the characteristics of CO_2 emissions from the transport sector in 30 Chinese provinces and analyzed the driving factors behind these emissions using the LMDI method. More recently, Fan and Lei [19] constructed a generalized multivariate Fisher's index decomposition model to identify potential drivers of carbon emissions in Beijing's transport sector from 1995 to 2012. Given the results, economic growth, energy intensity, and population size are considered to be the main drivers of CO_2 emission increases in the transport sector. Finally, to assess the Moroccan road transport sector from an environmental perspective, Kharbach and Chfadi [20] quantified the contributions of some key factors to CO_2 emissions from the sector using decomposition analysis for the period 2000–2011.

2. Specification of the Model and Results

Understanding the impact of transport activities on the environmental quality is becoming increasingly important as general environmental concerns are making their way into the main public policy agenda in the EU. To this end, time series variables from 2005 to 2015 were used in seven Western European countries (Bulgaria, Estonia, Latvia, Lithuania, Poland, Romania and Slovenia) to investigate the factors affecting CO_2 emissions from the transport sector. The annual data have been extracted from the Eurostat database and European Commission Reports.

We use then the Logarithmic Mean Divisia Index, both in its additive and multiplicative form, to investigate the effect of several factors thought to be responsible for CO_2 emissions in the transport sector.

2.1. The Model and the Variable

The decomposition methods allow us to quantify the contributions of various factors to CO_2 emissions from the transport sector. The basic idea is that transport CO_2 emissions are the sum of CO_2 emissions from each transportation mode. To extend the analysis, other sub-categories levels can be added, such as decomposing emissions from the *i*th transportation mode, to emissions coming from fuel type *j* in year *t*. Other variables such as population, energy consumption, motorization and economic growth can be introduced into these sub-categories to denote the various "effects" that contribute to transport CO_2 emissions.

Mathematically, the application of a Divisia decomposition analysis in transport involves the use of the following equation:

$$CO2_t = \sum_{i,j} CO2_{ijt} \quad (1)$$

where $CO2_t$ are transport sector emissions in a given country in year *t*. *i*, which denotes the mode of transport (road, air, rail, sea and, finally, pipeline transport), and *j*, the type of fuel (i.e., diesel, motor gasoline, biofuels and kerosene).

Equation (1) can further be decomposed to include other sub-categories of variables:

$$CO2_t = \sum_{i,j} \frac{CO2_{ijt}}{CE_{ijt}} \times \frac{CE_{ijt}}{CE_{it}} \times \frac{CE_{it}}{CE_t} \times \frac{CE_t}{GDP_t} \times \frac{GDP_t}{POP_t} \times POP_t \quad (2)$$

CE refers to energy consumption, *GDP* is the gross domestic product and *POP* the population. Finally, Equation (2) is written:

$$CO2_t = \sum_{i,j} EC_{ijt} \times RC_{ijt} \times RM_{it} \times IE_t \times GDP_t \times POP_t \quad (3)$$

where EC_{ijt} is the emission coefficient or CO_2 intensity of a fuel *j* from the *i*th transport mode in year *t*;

RC_{ijt} refers to the fuel mix (i.e., share of consumption of a fuel *j* in the *i*th transportation mode);

RM_{it} is the modal mix given by the energy consumption of the *i*th transport mode to the total energy consumption of the transport sector;

IE_t refers to Energy intensity of transport for year *t* (total energy consumption from transport to GDP);

GDP_t measure the GDP per capita; and finally,

POP_t is the population of the country under study in year *t*.

According to the additive form of the LMDI (Ang, [21,22]), the change in CO_2 emissions can then be calculated using the formula:

$$\Delta CO2 = CO2_t - CO2_{t-1} = \Delta EC + \Delta RC + \Delta RM + \Delta IE + \Delta GDP + \Delta POP \quad (4)$$

The decomposition of each effect between the year *t* and *t-1* is given by the following formulas:

$$\Delta EC = \sum_{i,j} \Delta EC_{ij} = \sum_{i,j} L(CO2_{ijt}, CO2_{ijt-1}) \ln\left(\frac{EC_{ijt}}{EC_{ijt-1}}\right) \quad (5)$$

$$\Delta RC = \sum_{i,j} \Delta RC_{ij} = \sum_{i,j} L(CO2_{ijt}, CO2_{ijt-1}) \ln\left(\frac{RC_{ijt}}{RC_{ijt-1}}\right) \quad (6)$$

$$\Delta RM = \sum_{i,j} \Delta RM_{ij} = \sum_{i,j} L(CO2_{ijt}, CO2_{ijt-1}) \ln\left(\frac{RM_{it}}{RM_{it-1}}\right) \quad (7)$$

$$\Delta IE = \sum_{i,j} \Delta IE_{ij} = \sum_{i,j} L(CO2_{ijt}, CO2_{ijt-1}) \ln\left(\frac{IE_t}{IE_{t-1}}\right) \tag{8}$$

$$\Delta GDP = \sum_{i,j} \Delta GDP_{ij} = \sum_{i,j} L(CO2_{ijt}, CO2_{ijt-1}) \ln\left(\frac{GDP_t}{GDP_{t-1}}\right) \tag{9}$$

$$\Delta POP = \sum_{i,j} \Delta POP_{ij} = \sum_{i,j} L(CO2_{ijt}, CO2_{ijt-1}) \ln\left(\frac{POP_t}{POP_{t-1}}\right) \tag{10}$$

Equation (4) can finally be extended:

$$\begin{aligned} CO2_t - CO2_{t-1} = {} & \sum_{i,j} L(CO2_{ijt}, CO2_{ijt-1}) \ln\left(\frac{EC_{ijt}}{EC_{ijt-1}}\right) + \\ & \sum_{i,j} L(CO2_{ijt}, CO2_{ijt-1}) \ln\left(\frac{RC_{ijt}}{RC_{ijt-1}}\right) + \sum_{i,j} L(CO2_{ijt}, CO2_{ijt-1}) \ln\left(\frac{RM_{it}}{RM_{it-1}}\right) + \\ & \sum_{i,j} L(CO2_{ijt}, CO2_{ijt-1}) \ln\left(\frac{IE_t}{IE_{t-1}}\right) + \sum_{i,j} L(CO2_{ijt}, CO2_{ijt-1}) \ln\left(\frac{GDP_t}{GDP_{t-1}}\right) + \\ & \sum_{i,j} L(CO2_{ijt}, CO2_{ijt-1}) \ln\left(\frac{POP_t}{POP_{t-1}}\right) \end{aligned} \tag{11}$$

Given that:

$$\begin{aligned} L(a,b) &= \frac{(a-b)}{(\ln a - \ln b)} & if\ \ a \neq b \\ &= a & if\ a = b \end{aligned} \tag{12}$$

We have the next condition:

$$\begin{aligned} L(CO2_{ijt}, CO2_{ijt-1}) &= \frac{(CO2_{ijt} - CO2_{ijt-1})}{(\ln CO2_{ijt} - \ln CO2_{ijt-1})} & if\ \ CO2_{ijt} \neq CO2_{ijt-1} \\ &= CO2_{ijt} & if\ CO2_{ijt} = CO2_{ijt-1} \end{aligned} \tag{13}$$

2.2. Empirical Results

In the following, we explain the results obtained by applying the additive form of LMDI (Equation (4)) after the calculation of the net effect of each variable in our model.

The average annual change (Table 1) is based on the calculation of the annual change in CO_2 emissions for the study period. The results show that all the countries in our sample have experienced strong growth in CO_2 emissions from the transport sector. Economic activity (i.e., GDP per capita) is the major factor causing the increase in these emissions, while the population variable was found to be an important factor explaining the decrease in CO_2 emissions, except for Slovenia.

Table 1. Average annual change in CO_2 emissions and underling factors.

Country	Variation of CO_2 Emissions	EC	RC	RM	IE	GDP	POP	Main Factors
Bulgaria	261	−24	−18	82	33	241	−53	RM, IE, GDP
Estonia	39	−8	12	−6	6	41	−6	RC, IE, GDP
Latvia	14	−40	12	−15	7	86	−36	RC, IE, GDP
Lithuania	97	−57	21	−2	39	158	−62	RC, IE, GDP
Poland	1054	−211	−6	0	−345	1637	−21	GDP
Romania	363	−51	12	−53	107	448	−100	RC, IE, GDP
Slovenia	86	−9	16	−15	47	31	16	RC, IE, GDP, POP

Source: Calculation of the author.

As shown in this table, energy intensity (IE) increases the CO_2 emissions in all the countries except Poland. In the latter, the consumed energy per unit of GDP was reduced during the study period. The results show that Poland is also an exception when it comes to the emissions of CO_2 per unit of consumed fuel (variable *EC*).

It is also important to note that the modal mix *RM* contributed directly to the decline of CO_2 emissions in most countries in our sample. However, the impact of this factor is relatively small: 13% (45 mt instead of 39 mt) for Estonia, 2% (99 mt instead of 97 mt) for Lithuania, 12.7% (416 mt instead of 363 mt) for Romania and 15% (101 mt instead of 86 mt) for Slovenia. For Latvia, the impact of this factor is important as it contributes to the deterioration of emissions by a significant value. Similarly, this factor is an important contributor to the increase in CO_2 emissions in Bulgaria due to the national policy of this country consisting of the absence of a rigorous control of vehicle age and emissions. This factor (*RM*) has no impact on the growth of CO_2 emissions from transport in Poland. As mentioned above, the annual improvement of the energy intensity of transport also had a considerable impact on the increase in emissions in our sample; the adjustment of this factor comes from the adjustment of diesel consumption (Table 2).

Table 2. Fuel indicators in the transport sector.

	2005					2015				
Country	Total	Diesel	Motors Gazoline	Bio-Fuels	Kerosene	Total	Diesel	Motors Gazoline	Bio-Fuels	Kerosene
					Fuel Share					
	Mtoe [1]		%			Mtoe			%	
Bulgaria	2.6	65.4	26.9	0	7.7	3.426	64.2	25.7	4.3	5.8
Estonia	0.7	50	42.9	0	7.1	0.854	65.6	28.2	0.4	5.8
Latvia	1.055	58.3	36	0	5.7	1.314	69.2	18.3	1.9	10.6
Lithuania	1.445	68.8	27.7	0	3.5	1.97	76.1	15.2	3.6	5.1
Poland	12.47	55.3	42.3	0	2.4	17.3	59.5	31.9	4.5	4.1
Romania	4.1	58.6	39	0	2.4	5.74	68.1	23.2	3.5	5.2
Slovenia	1.5	52	46	0	2	1.822	72.3	23.9	1.6	2.2
					Emission Coefficient					
	Mt [2]	%				Mt		%		
Bulgaria	7.5	72	26.6	0	1.4	9.41	78.6	18.1	2.2	1.1
Estonia	2.025	56.8	42	0	1.2	2.3745	71.2	27.3	0.2	1.3
Latvia	2.93	64.9	37.4	0	1	3.185	78.5	18.8	1.1	1.6
Lithuania	4.225	74.6	24.8	0	0.6	5.15	83.5	13.6	1.9	1
Poland	35.4	58.5	41.2	0	0.3	46	64.9	32.2	2.7	0.2
Romania	11.75	62.6	37	0	0.4	15.45	72.8	24.7	1.9	0.6
Slovenia	4.377	57.1	42.5	0	0.4	5.41	77.3	21.4	0.9	0.4

Source: Calculation of the author. [1] Mtoe: Million Tons of Oil Equivalent; [2] Mt: Millions of tons.

The emission coefficient has a negative influence on the growth of CO_2 emissions in all the countries in our sample, so this influence is very important. This factor can vary the average increase in emissions, which would have been 8% higher in Bulgaria (285 mt instead of 261 mt), 17% higher in Estonia (47 mt instead of 39 mt), 286% in Latvia (54 mt instead of 14 mt), 37% in Lithuania (154 mt instead of 97 mt), 17% in Poland (1265 mt instead of 1054 mt), 12% in Romania (414 mt instead of 363 mt) and 9% in Slovenia (95 mt instead of 86 mt).

3. Conclusions

In this study, we have carried out a decomposition of transport CO_2 emission elements using the Divisia index in its additive and multiplicative forms and some EU countries as the sample.

According to the results found using the LMDI method, economic activity is the main factor responsible for CO_2 emissions in all countries in our sample. Fuel mix is the second most important CO_2 emitting factor. Modal share and energy intensity also affect CO_2 emissions, but to a lesser extent. On the contrary, the emission factor and population variables reduced the growth of these emissions. Note that all variables have met their respected signs, respectively, except for the population factor in the case of Slovenia.

Since the exchange of goods within and between EU countries is intense, this explains the important impact of the economic activity on CO_2 emissions. Decoupling the increase in CO_2 emissions from economic growth and transport energy demand remains an important issue within the EU economies. On the one hand, implementing intelligent transport systems and encouraging the use of environmentally friendly transport modes and energies are still valid strategies. On the other hand, many other measures (fuel taxation, subsidies and other fiscal instruments, registration tax, etc.) are not yet in place in the majority of the countries in our sample (Bulgaria, Estonia, Lithuania and Poland, for example).

Cleaner fuels and CO_2 efficient cars are also needed in all countries. Unfortunately, according to OECD statistics (2017), the level of investment in transport infrastructure is less than 1% of GDP.

Data Availability Statement: Eurostat; European Union statistical Pocketbooks.

References

1. Gonzalez, R.M.; Marrero, G. The effect of dieselization in passenger cars emissions for Spanish regions: 1998–2006. *Energy Policy* **2012**, *51*, 213–222. [CrossRef]
2. Lu, I.J.; Lewis, C.; Lin, S.J. The forecast of motor vehicle, energy demand and CO_2 emission from Taiwan's road transportation sector. *Energy Policy* **2010**, *38*, 2952–2961. [CrossRef]
3. Abbes, S.; Bulteau, J. Growth in transport sector CO_2 emissions in Tunisia: An analysis using a bounds testing approach. *Int. J. Global Energy Issues* **2018**, *41*, 176–197. [CrossRef]
4. Shakya, S.R.; Shrestha, R.M. Transport sector electrification in a hydropower resource rich developing country: Energy security, environmental and climate change co-benefits. *Energy Sustain. Dev.* **2011**, *15*, 147–159. [CrossRef]
5. Hickman, R.; Banister, D. Looking over the horizon: Transport and reduced CO_2 emissions in the UK by 2030. *Transp. Policy* **2007**, *14*, 377–387. [CrossRef]
6. Almodóvar, M.; Angulo, E.; Espinosa, J.L.; García-Ródenas, R. A modeling framework for the estimation of optimal CO_2 emission taxes for private transport. *Procedia Soc. Behav. Sci.* **2011**, *20*, 693–702. [CrossRef]
7. Sun, J.W. Changes in energy consumption and energy intensity: A complete decomposition model. *Energy Econ.* **1998**, *20*, 85–100. [CrossRef]
8. Ang, B.W.; Choi, K.H. Decomposition of aggregate energy and gas emission intensities for industry: A refined Divisia index method. *Energy J.* **1997**, *18*, 59–73. [CrossRef]
9. Scholl, L.; Schipper, L.; Kiang, N. CO_2 emissions from passenger transport: A comparison of international trends from 1973 to 1992. *Energy Policy* **1996**, *24*, 17–30. [CrossRef]
10. Schipper, L.; Schall, L.; Price, L. Energy use and carbon emissions from freight in 10 industrialized countries: An analysis of trends from 1973 to 1992. *Transp. Res. Part D Transp. Environ.* **1997**, *2*, 57–76. [CrossRef]
11. Timilsina, G.R.; Shrestha, A. Factors affecting transport sector CO_2 emissions growth in Latin American and Caribbean countries: An LMDI decomposition analysis. *Int. J. Energy Res.* **2009**, *33*, 396–414. [CrossRef]
12. Timilsina, G.R.; Shrestha, A. Transport sector CO_2 emissions growth in Asia: Underlying factors and policy options. *Energy Policy* **2009**, *37*, 4523–4539. [CrossRef]
13. Papagiannaki, K.; Diakoulaki, D. Decomposition analysis of CO_2 emissions from passenger cars: The cases of Greece and Denmark. *Energy Policy* **2009**, *37*, 3259–3267. [CrossRef]
14. Wang, W.W.; Zhang, M.; Zhou, M. Using LMDI method to analyze transport sector CO_2 emissions in China. *Energy* **2011**, *36*, 5909–5915. [CrossRef]
15. Wang, T.; Li, H.; Zhang, J.; Lu, Y. Influencing Factors of Carbon Emission in China's Road Freight Transport. *Procedia Soc. Behav. Sci.* **2012**, *43*, 54–64. [CrossRef]
16. Andreoni, V.; Galmarini, S. European CO_2 emission trends: A decomposition analysis for water and aviation transport sectors. *Energy* **2012**, *45*, 595–602. [CrossRef]
17. Eng-Larsson, F.; Lundquist, K.J.; Oloflander, L.; Wandel, S. Explaining the cyclic behavior of freight transport CO_2-emissions in Sweden over time. *Transp. Policy* **2012**, *23*, 79–87. [CrossRef]
18. Guo, B.; Geng, Y.; Franke, B.; Hao, H.; Liu, Y.; Chiu, A. Uncovering China's transport CO_2 emission patterns at the regional level. *Energy Policy* **2014**, *74*, 134–146. [CrossRef]
19. Fan, F.; Lei, Y. Decomposition analysis of energy-related carbon emissions from the transportation sector in Beijing. *Transp. Res. Part D* **2016**, *42*, 135–145. [CrossRef]
20. Kharbach, M.; Chfadi, T. CO_2 Emissions in Moroccan Road Transport sector: Divisia, Cointegration, and EKC analyses. *Sustain. Cities Soc.* **2017**, *35*, 396–401. [CrossRef]
21. Ang, B.W. The LMDI approach to decomposition analysis: A practical guide. *Energy Policy* **2005**, *33*, 867–871. [CrossRef]
22. Ang, B.W. *A Simple Guide to LMDI Decomposition Analysis*; Department of Industrial and Systems Engineering National, University of Singapore: Singapore, 2016.

engineering proceedings

MDPI

Proceedings

Business Days Time Series Weekly Trend and Seasonality †

Karlis Gutans

Department of Computer Sciences, University of Latvia, LV-1586 Riga, Latvia; karlis.gutans@gmail.com

† Presented at the 7th International conference on Time Series and Forecasting, Gran Canaria, Spain, 19–21 July 2021.

Abstract: The world changes at incredible speed. Global warming and enormous money printing are two examples, which do not affect every one of us equally. "Where and when to spend the vacation?"; "In what currency to store the money?" are just a few questions that might get asked more frequently. Knowledge gained from freely available temperature data and currency exchange rates can provide better advice. Classical time series decomposition discovers trend and seasonality patterns in data. I propose to visualize trend and seasonality data in one chart. Furthermore, I developed a calendar adjustment method to obtain weekly trend and seasonality data and display them in the chart.

Keywords: calendar adjustment; business week; seasonal plot

check for updates

Citation: Gutans, K. Business Days Time Series Weekly Trend and Seasonality. *Eng. Proc.* **2021**, *5*, 26. https://doi.org/10.3390/engproc2021005026

Published: 28 June 2021

1. Introduction

Economic digital transformation, and Green Call are current European Commission programs that have a billion-euro budget. There are still sites, where there are published raw data and either no or weak or paid statistics.

One example is meteorological weather data. Latvia pays for data gathering at meteo stations all over the country, but statistics are for money. While tourism has been suffering big losses recently, that could be improved so that people are more informed about local weather conditions.

Another example is currency exchange rates. European Central Bank publishes raw data and their charts (https://www.ecb.europa.eu/stats/policy_and_exchange_rates, accessed on 25 June 2021), but there is no information regarding trends and seasonality patterns. Furthermore, in the charts, weekly data frequency is missing. The UK's favorite currency site has more charts, statistics, and trend information (https://www.exchangerates.org.uk, accessed on 25 June 2021), but there is also missing calculated trend and seasonality patterns and weekly data charts. Figure 1 shows data visualization examples from ECB and UK currency exchange sites.

Trend and seasonality pattern discovery and their visualization is described and summarized in the free online book "Forecasting: Principles and Practice" written by Hyndman and Athanasopoulos [1]. With many solutions for everyday forecasting needs, in chapter 12 there are also mentioned issues that are challenging to tackle. One of them is weekly data processing. I also tried to find satisfactory weekly data analysis on the Internet, but unsuccessfully. To deal with this issue, I thought of the weekly data calendar adjustment method and seasonal plot enrichment with seasonality calculations. This paper reports on my progress so far and provides some calculations of the proposed method.

In this paper, I take formulas from the book's chapter 6, on time series decomposition. Meteorological data are from the Latvia meteo site for the city Liepaja (https://www.meteo.lv/meteorologija-datu-meklesana, accessed on 25 June 2021). Currency exchange rates are from the ECB site. ECB publishes current rates for 32 currency pairs.

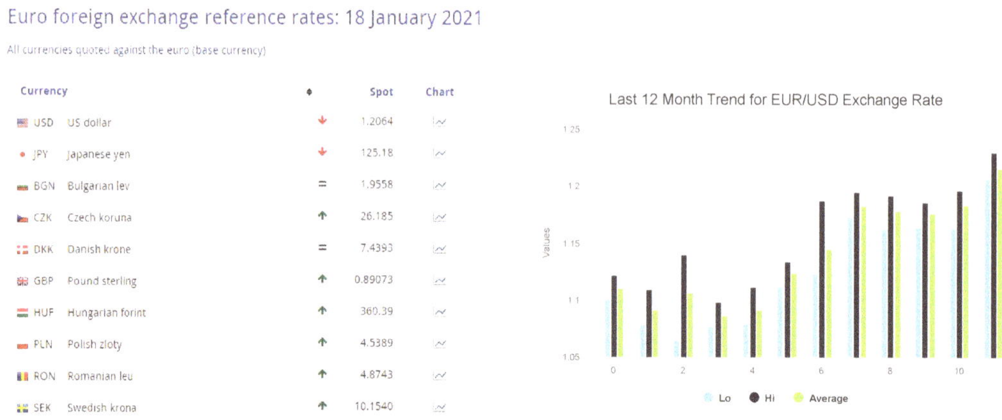

(a) (b)

Figure 1. Examples from currencies exchange rates sites. (**a**) ECB exchange rates. (**b**) UK's site trend statistics.

2. Proposal

Data in seasonal plots provide a lot of information in a small space. Time series highs and lows in different periods of time, when expressed using blocks of plain text or tables, are lengthy and overwhelming. Time series decomposition in trend and seasonal components provide additional quantitative characteristics, which are usually plotted in separate graphics. I suggest adding a seasonality component to seasonal plot.

Time series decomposition can be applied to monthly data. I propose also incorporating the decomposition into more frequent time periods. Therefore, I introduce the time period keews, which are similar to weeks, but with better calendar characteristics. It is much easier to perform the calculations if a year, instead of average number of weeks 52.18, has exactly 48 keews, and each month is 4 keews.

Seasonal plot with seasonal component is described in Section 2.1, calendar adjustment with keew in Section 2.2, more complex case for currency exchange rates in Section 2.3.

2.1. Temperature Seasonal Plot

Time series decomposition equation is $y_t = S_t + T_t + R_t$, where y_t is the data, S_t is the seasonal component, T_t is the trend-cycle component, and R_t is the remainder component, all at period t. Python library Statsmodels has freely available formula implementation (https://www.statsmodels.org/stable/generated/statsmodels.tsa.seasonal.seasonal_decompose.html, accessed on 25 June 2021).

Liepaja is a city in western Latvia, located on the Baltic Sea. It is a popular summer vacation destination due to sandy beaches and music festivals. Figure 2 shows an example of Liepaja monthly temperature seasonal plot with added seasonality estimation. The monthly data consists of the average actual temperature in Liepaja at 12 o'clock each day. The seasonal plot includes last 5 years of data so as not to become too overwhelming. The figure shows that the hottest month is August and trend of the last three years is a temperature increase by approximately a degree.

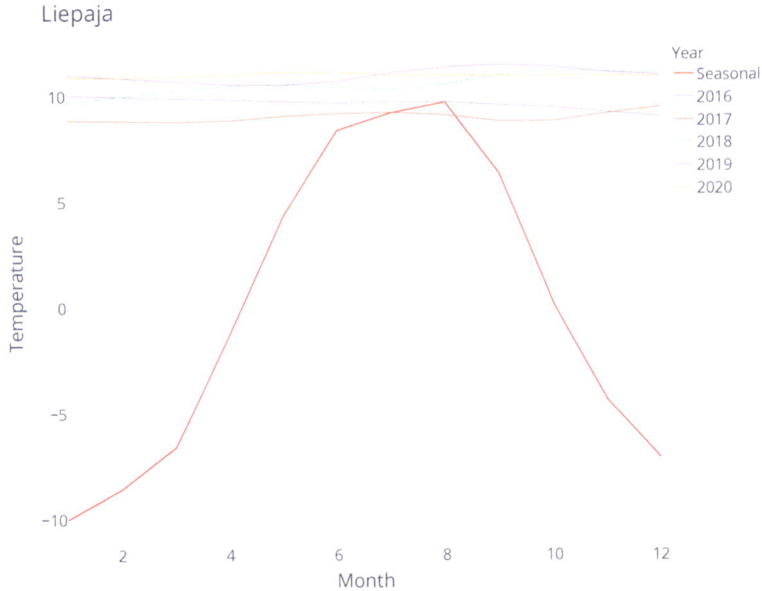

Figure 2. Liepaja monthly temperature seasonal plot.

The used time series decomposition model is a naive approach from 1920s and more sophisticated models are proposed. In this paper, I focus on decomposition possibilities in general, but in each case should be considered usage of more specific decomposition models [2–4], etc.

2.2. Calendar Adjustment with Keew

Current Python decomposition formula implementation has no clear way of doing calculations for weeks. There is parameter period that can be provided, but in a year, the average number of weeks is 52.18. I propose to introduce a concept of time period—keews. In this case, a year will have exactly 48 keews, and each month—4 keews.

Four keews will be in one month boundaries and they will end on the following month days:

1. The 4th keew will end on the month's last day;
2. The 2nd keew will end in the middle of month on day 15;
3. The 3rd keew will end on the following days:
   (a) For months with 31 days, it will be day 23, so that 4th keew and 3rd keew will be equally 8 days long;
   (b) For months with 30 days, it will also be day 23, so that all months but February will have the same 3rd keew end day;
   (c) For February, the 3rd keew will end on day 22.
4. The 1st keew will end on day 7 with no additional consideration.

Table 1 gives a summary of keews.

Table 1. Keews.

Keew 1	**Keew 2**	**Keew 3**	**Keew 4**
Day 1–Day 7	Day 8–Day 15	Day 16–Day 23 February 22	Day 24–End of Month February 23
7 days	8 days	7, 8 days	6, 7, 8 days

Keews consist of all days in their month–day range.

Figure 3 shows an example of corresponding Liepaja keew temperature seasonal plot with seasonality estimation. This is more precise picture for summer tempretures in Liepaja. It shows that the hottest keew is at the end of July and that summers in Liepaja can also have colder keews in June. This should be taken in consideration when planning vacations.

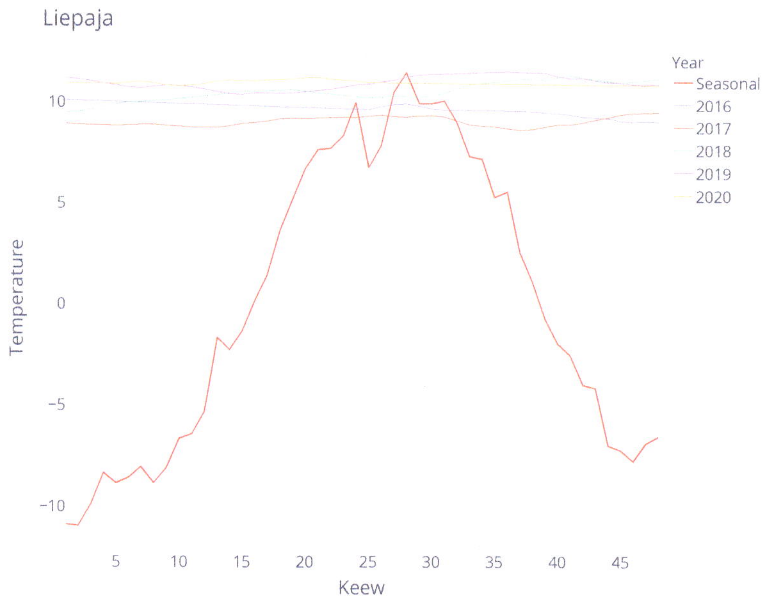

Figure 3. Liepaja keew temperature seasonal plot.

2.3. Exchange Rate Seasonal Plot with Seasonality Estimation

More complex data is currency exchange rates. Exchange rates by ECB are given on business days. Exchange rates can have different strong trends during a year. I propose to also display these data with keew seasonal plot together with seasonality estimation.

Due to the fact that data are only for business days, keews will have less meaningful days. For the last 5 years, there are keews with 3 business days in 1% cases, 4 business days in 8% cases, 5 business days in 48% cases and 6 business days in 43% cases. The good thing is that the majority of keews have 5 and 6 business days.

Figure 4 shows exchange rates for EUR/USD currency pair in a keew seasonal plot with seasonality estimation. In this case, seasonality is calculated with Python decomposition multiplicative model, seasonal mean is the arithmetic mean of the 1st keews of years. To add the seasonal component to seasonal plot, it should be expressed in seasonal plot scale; therefore, in the figure, the seasonality line is given by multiplying the seasonality component with the seasonal mean. Year trend lines show exchange rates on the last business day in the keew.

In general, the keews end dates are suitable for keeping in one month boundaries. It is then easier to compare the displayed results with month estimations. However, different keews end dates can be chosen to better suit further prediction needs. Furthermore, some research on Forex calendar effects show that not all business days are equal one to other [5–8].

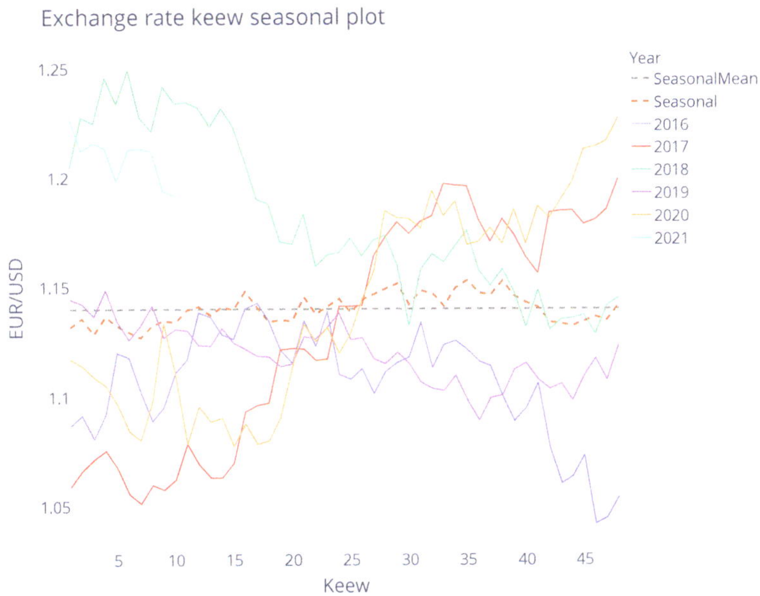

Figure 4. Exchange rate keew seasonal plot.

3. Results

Firstly, the purpose of the statistics calculation is to ascertain that keew trend and seasonality characteristics are similar to monthly estimations. Secondly, it is to find the best seasonality estimations to include in the seasonal plot.

I pick the best model by testing different types of models and data forms. Calculations are based on classical decomposition. It has two forms: an additive decomposition and a multiplicative decomposition. As the purpose is to find and use only seasonal components, then data in models also can be in different forms. I choose to test the usual end of the period data, and also period arithmetic mean and 1st and 3rd quartile arithmetic mean.

Trend and seasonality strength can be measured as described in Hyndman and Athanasopoulos, 2018 Chapter 6.7 [1].

The results are labeled in the following way:

1. F—strength of decomposition component;
2. F_T—strength of trend;
3. F_S—seasonal strength;
4. F_A—additive decomposition model strength;
5. F_M—multiplicative decomposition model strength;
6. F_E—strength calculated on end of period data;
7. F_N—strength calculated on arithmetic mean data;
8. F_D—strength calculated on 1st and 3rd quartile arithmetic mean data.

F_{SME} means Seasonal component strength calculated with multiplicative decomposition model on end-of-period data.

A keew seasonal plot is a suitable way of presenting data for 5 years, so all time series is analysed starting from year 2015. One ECB currency pair exchange rates does not have data for the whole period; therefore, it is omitted. Another one seasonal component data values are all equal to 0, so it is omitted too.

3.1. Monthly Trend and Seasonality

The strength of the trend is bigger than seasonal strength in all data sets.

The strength of trend differs most between additive and multiplicative decompositions. Multiplicative decomposition models has average strength ≈0.97, while additive average is ≈0.81. The best strength of trend average results is for $F_{TMN} \approx 0.974$.

Seasonal strength is approximately the same in all data sets. The best seasonal strength average results is for $F_{SMN} \approx 0.26$.

3.2. *Business Weekly Trend and Seasonality*

The results of business weekly data analysis are similar to monthly data analysis. The strength of the trend is bigger than seasonal strength in all data sets.

The strength of the trend differs most between additive and multiplicative decompositions. Multiplicative decomposition models have an average strength of ≈0.971, while the additive average is ≈0.806. The best strength of trend average results are for $F_{TMN} \approx 0.971$.

Seasonal strength is approximately the same in all data sets. The best seasonal strength average results are for $F_{SMD} \approx 0.266$.

For example, data sets the best seasonal estimation to add in keew seasonal plot is from multiplicative decomposition calculated on 1st and 3rd quartile arithmetic mean data. As the difference between models seasonal estimations are not considerable, all of them can be used in data plotting.

4. Conclusions

Keew seasonal plot with added seasonality estimation provides more detailed view on data, while maintaining at least several characteristics of monthly estimations. Predictions can be based on five year history observations plotted in one chart. In the coming months, I will also work on proposing calendar adjustments for businesses' daily trends and seasonality, and I will search for the best chart to display them.

Institutional Review Board Statement: Not applicable.

Informed Consent Statement: Not applicable.

Data Availability Statement: Meteorological data are from the Latvia meteo site for the city Liepaja (https://www.meteo.lv/meteorologija-datu-meklesana, accessed on 25 June 2021). Currency exchange rates are from the ECB site (https://www.ecb.europa.eu/stats/policy_and_exchange_rates/euro_reference_exchange_rates/html/index.en.html, accessed on 25 June 2021).

References

1. Hyndman, R.J.; Athanasopoulos, G. *Forecasting: Principles and Practice*, 2nd ed.; OTexts: Melbourne, Australia, 2018. Available online: OTexts.com/fpp2 (accessed on 5 March 2020).
2. Quan, J.; Zhan, W.; Chen, Y.; Wang, M.; Wang, J. Time series decomposition of remotely sensed land surface temperature and investigation of trends and seasonal variations in surface urban heat islands. *J. Geophys. Res. Atmos.* **2016**, *121*, 2638–2657. [CrossRef]
3. Clevel, R.B.; Clevel, W.S.; McRae, J.E.; Terpenning, I. STL: A seasonal-trend decomposition. *J. Off. Stat.* **1990**, *6*, 3–73.
4. Chen, D.; Zhang, J.; Jiang, S. Forecasting the Short-Term Metro Ridership with Seasonal and Trend Decomposition Using Loess and LSTM Neural Networks. *IEEE Access* **2020**, *8*, 91181–91187. [CrossRef]
5. Ito, T.; Yamada, M. *Puzzles in the Forex Tokyo "Fixing": Order Imbalances and Biased Pricing by Banks (No. w22820)*; National Bureau of Economic Research: Cambridge, MA, USA, 2016.
6. Ben-David, I.; Birru, J.; Prokopenya, V. Uninformative feedback and risk taking: Evidence from retail forex trading. *Rev. Financ.* **2018**, *22*, 2009–2036. [CrossRef]
7. Popović, S.; Durović, A. Intraweek and intraday trade anomalies: evidence from FOREX market. *Appl. Econ.* **2014**, *46*, 3968–3979. [CrossRef]
8. Dailydytė, I.; Bužienė, I. Black friday and other effects-are they still sustainable in financial markets? *J. Secur. Sustain. Issues* **2020**, *9*, 4.

engineering proceedings

MDPI

Proceedings

Time Series Chlorophyll-A Concentration Data Analysis: A Novel Forecasting Model for Aquaculture Industry †

Elias Eze [1,*], Sam Kirby [2], John Attridge [2] and Tahmina Ajmal [1]

1 Institute for Research in Applicable Computing (IRAC), School of Computer Science and Technology, University of Bedfordshire, Luton LU1 3JU, UK; tahmina.ajmal@beds.ac.uk
2 Chelsea Technology Group, 55 Central Avenue, West Molesey, Surrey KT8 2QZ, UK; skirby@chelsea.co.uk (S.K.); jattridge@chelsea.co.uk (J.A.)
* Correspondence: elias.eze1@beds.ac.uk
† Presented at the 7th International conference on Time Series and Forecasting, Gran Canaria, Spain, 19–21 July 2021.

Abstract: Eutrophication in fresh water has become a critical challenge worldwide and chlorophyll-a content is a key water quality parameter that indicates the extent of eutrophication and algae concentration in a body of water. In this paper, a forecasting model for the high accuracy prediction of chlorophyll-a content is proposed to enable aquafarm managers to take remediation actions against the occurrence of toxic algal blooms in the aquaculture industry. The proposed model combines the ensemble empirical mode decomposition (EEMD) technique and a deep learning (DL) long short-term memory (LSTM) neural network (NN). With this hybrid approach, the time-series data are firstly decomposed with the aid of the EEMD algorithm into manifold intrinsic mode functions (IMFs). Secondly, a multi-attribute selection process is employed to select the group of IMFs with strong correlations with the measured real chlorophyll-a dataset and integrate them as inputs for the DL LSTM NN. The model is built on water quality sensor data collected from the Loch Duart salmon aquafarm in Scotland. The performance of the proposed novel hybrid predictive model is validated by comparing the results against the dataset. To measure the overall accuracy of the proposed novel hybrid predictive model, the Mean Absolute Error (MAE), Mean Square Error (MSE), Root Mean Square Error (RMSE), and Mean Absolute Percentage Error (MAPE) were used.

Keywords: water quality; aquaculture; forecasting; chlorophyll-a time-series data; deep learning LSTM

Citation: Eze, E.; Kirby, S.; Attridge, J.; Ajmal, T. Time Series Chlorophyll-A Concentration Data Analysis: A Novel Forecasting Model for Aquaculture Industry . *Eng. Proc.* **2021**, *5*, 27. https://doi.org/10.3390/engproc2021005027

Published: 29 June 2021

1. Introduction

Eutrophication in freshwater bodies is an organic process usually caused by the increased enrichment of nutrients which can pollute water quality and adversely affect aquatic ecosystems. The extent of eutrophication in fresh water can be estimated through chlorophyll-a concentration monitoring. In the aquaculture industry, this natural process of nutrient enrichment also results in structural changes to the aquatic ecosystem through increased algae production, the depletion of fish species, and the prevalent degradation of overall water quality [1,2]. Chlorophyll-a concentration is representative of the state of freshwater quality and has generally been used as a key indicator for measuring algal blooms [3].

According to Gao and Zhang [4], eutrophication has become a ubiquitous fresh-water-quality pollutant in China. Similarly, a study conducted by Jules et al. [5] estimated the annual damage costs of the eutrophication of fresh water in England and Wales to be $105–160 million (£75.0–114.3 m). Given the link between the adverse effect of eutrophication in freshwater and the stagnation of wild fishery populations, the aquaculture industry has emerged as a crucial means of providing protein to our constantly growing population. Therefore, the monitoring of water quality parameters (for instance, algal biomass and cyanobacteria) through chlorophyll-a concentration is increasingly favoured over laboratory analysis and similar traditional methods because of the high cost and labour-intensive requirements associated with them [6]. The effective monitoring and prediction of chlorophyll-a concentrations is a promising approach for the routine estimation of phytoplankton biomass in the aquaculture ecosystems of the Nile tilapia (*Oreochromis niloticus*) [7]. Sensory monitoring of the chlorophyll-a concentration is an effective approach for reliably assessing the trophic state of freshwater bodies given its strong affinity to the abundance of phytoplankton, cyanobacteria, and biomass, which affect the turbidity and general colouration of fresh water [8].

Several studies have been conducted to establish a means of coping with water quality impairments caused by algal biomass using conventional numerical modelling methods, least squares support vector regression (LSSVR), neural networks methods such as Radial Basis Function neural network (RBFNN), Back Propagation neural network (BPNN) algorithms, and machine learning methods to predict chlorophyll-a concentrations as an indicator for future water quality changes [9–12]. However, the challenge with traditional numerical methods, LSSVR, and neural networks such as RBFNN and BPNN is the inherent weakness of the long-term dependency problem. Research has shown that deep learning long short-term memory (LSTM) neural networks can overcome the above-mentioned weakness and can provide efficient applicability and reliability for water quality parameter prediction [13,14]. Additionally, combining the ensemble empirical mode decomposition (EEMD) method with deep learning LSTM neural network has demonstrated clear advantages over traditional LSTM neural networks in terms of improved water quality parameter prediction accuracy in the aquaculture environment [13]. In this paper, a novel deep learning-based hybrid chlorophyll-a prediction model for the aquaculture industry is proposed.

2. Data Source

2.1. The Study Area Description and Datasets Analysis

Loch Duart is an independent Scottish salmon aquafarm industry, which has its headquarters in Scourie, Sutherland, in north-west Scotland. The salmon farming company owns and operates eight sea-sites and two hatcheries in Sutherland and the Outer Hebrides. In Loch Duart, salmon are hatched and grown in the cold, clear fresh water of north-west Scotland. The salmon farming company annually harvests approximately 5000 tons of fresh salmon. Chlorophyll-a (µg/L) time-series data were collected via a TriLux multi-parameter sensor probe. The sensor deployment took place at one of their sheltered sites along the coast (see Figure 1a). The telemetry unit was secured to the metal walkway around the outside of the net pens and the sensor was situated on the outside of one of the outermost pens, nearest to the feed barge.

A TriLux multi-parameter fluorometer/sensor (see Figure 1b) developed by Chelsea Technology Group was used for measuring and collecting a total of 22,708 sets of a non-linear, non-stationary water-quality parameter time-series dataset at Loch Duart salmon aquafarm between May and October 2020. The water quality parameters include chlorophyll-a (470), turbidity, and chlorophyll-a (530).

Generally, the 470 channel measures chlorophyll fluorescence from direct excitation of chlorophyll-a that usually strongly correlates with phytoplankton biomass in freshwater. Table 1 shows the list of other sensors developed by Chelsea Technology Group and the corresponding parameters that each of them measures.

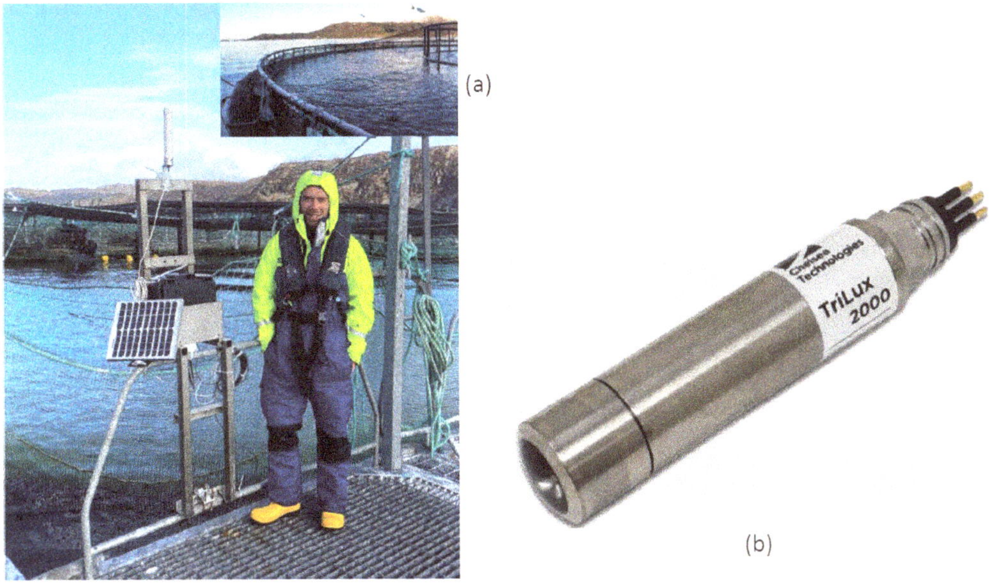

Figure 1. (**a**) Installation site of the TriLux multiparameter fluorometer at the salmon aquafarm, with the inset image depicting the larger part of the salmon cage; (**b**) Chelsea Technologies' TriLux multiparameter fluorometer which monitors three key algal parameters in a single probe [15].

Table 1. Chelsea Technology Group Fluorometers/sensors and the parameters they measure [6].

		Fluorometers							**Active Fluorometers**					**Optical Sensors**		
		UniLux	TriLux	UviLux	VLux AlgaePro	VLux TPro	VLux FuelPro	VLux OilPro	LabSTAF	FastOcean APD	FastOcean	Act2 Lab	FastBallast	PAR Sensor	GlowTracka	UniLux Turbidity
Fluorometers	Chlorophyll-a	X	X		X	X	X	X	X	X	X	X	X			
	Phycobiliproteins	X	X		X	X	X	X	X	X	X	X	X			
	Fluorescein	X														
	Rhodamine	X														
	BTEX			X			X									
	PAH			X				X								
	Tryptophan			X		X										
	CDOM			X		X	X	X								

Table 1. *Cont.*

		Fluorometers							Active Fluorometers					Optical Sensors		
		UniLux	TriLux	UviLux	VLux AlgaePro	VLux TPro	VLux FuelPro	VLux OilPro	LabSTAF	FastOcean APD	FastOcean	Act2 Lab	FastBallast	PAR Sensor	GlowTracka	UniLux Turbidity
Active Fluorometers	Variable Fluorescence								✖	✖	✖	✖	✖			
	Fluorescence Light Curves (FLC)								✖		✖					
	Phytoplankton Primary Productivity								✖	✖	✖					
	Phytoplankton Cell Counting												✖			
Optical Sensors	PAR													✖		
	Bioluminescence														✖	
	Turbidity	✖	✖		✖	✖	✖	✖								✖
	Absorbance					✖	✖	✖								

2.2. Data Pre-Treatment, Filling and Correction

Water-quality parameter time-series dataset defects usually result in excessive deviation between the measured original water-quality parameter values and the prediction results. The basis of accurate time-series analysis and the development of effective and reliable predictive models is high-quality sample data. To provide a concise, accurate dataset for the prediction model and improve prediction accuracy, the measured water-quality parameter data was carefully pre-processed. Generally, the issue of missing data is often inevitable with automatic water quality monitoring systems. The water-quality parameters like turbidity, chlorophyll-a (470), and chlorophyll-a (530) were automatically measured for 10 months at 10 min intervals. To fill in any missing data, a filling-in approach called linear interpolation algorithm [16] is applied to achieve a better estimation effect that can accurately approximate the missing data values. In data analysis, a linear interpolation algorithm assumes the ratio of two separate known data and a single unknown datum to be a linear interrelation. Therefore, to obtain the missing, unknown water quality parameter value, the linear interpolation technique applies the slope of the presumed line to compute the time-series dataset increment.

Definition 1. *Time series nature of the measured parameter (Chlorophyll-a (470)).*

The automated water quality sensory system at Loch Duart salmon aquafarm measures the time series water quality parameters at a constant time interval everyday which can be denoted as β, so that n length time-series of the measured parameters' datasets is defined as (1)

$$S_{i,n} = \{(X_{i,\,1}, T_1), (X_{i,\,2}, T_2), \cdots, (X_{i,\,n}, T_n)\} \tag{1}$$

where $X_{i,\,l}$ represents the values of the measured i^{th} time-series water-quality parameters by the automatic sensory system at time T_l $(1 \le i \le \beta,\ 1 \le l \le n)$, so that for a given T_l, the sampling time interval is constant at $\Delta T = (T_{l+1} - T_l) = 5$ min. Therefore, if the original value $X_{i,\,l}$ is missing, its estimated value $\hat{X}_{i,\,l}$ can be obtained with the problem of minimum, which is given as $|\hat{X}_{i,\,l} - X_{i,\,l}|$, changed into the missing value estimation problem. Based on the measured data $X_{i,\,x}$ and $X_{i,\,y}$ at time $T_{i,\,x}$ and $T_{i,\,y}$, respectively, the linear imputation function $L(t)$ could be formulated for the time-series water-quality parameter monitoring systems as:

$$L(t) = X_{i,\,x} + \left(\frac{X_{i,\,x} - X_{i,\,y}}{T_{i,\,x} - T_{i,\,y}}\right) \cdot \left(t - T_{i,\,x}\right). \tag{2}$$

For any missing time-series water-quality parameter data at any given moment, the linear interpolation algorithm firstly finds the two closest moments $T_{i,\,x}$ and $T_{i,\,y}$ $\left(T_{i,\,x} < t < T_{i,\,y}\right)$, and estimates the lost data value at time t with the help of the known measured data $X_{i,\,x}$ and $X_{i,\,y}$ of $T_{i,\,x}$ and $T_{i,\,y}$ moments based on Equation (2), i.e., $\hat{X}_n = L(t)$.

3. Proposed Model

The EEMD technique and deep learning LSTM NN were merged to form the chlorophyll-a hybrid prediction model. A detailed implementation processes of the applied EEMD technique is shown in full in [13]. The LSTM deep learning NN approach is described in full detail in Section 3.1. The original chlorophyll-a (470) dataset is decomposed effectively by the application of the EEMD technique into n disparate IMFs and a residual item. The IMF components that are contained within individual frequency bands are independently different and usually change with the variation of the chlorophyll-a (470) time-series data $x(t)$. Likewise, the trend of $x(t)$ is generally demonstrated by the corresponding ensemble residual item as the output of the decomposition process implementation.

3.1. Deep Learning LSTM Neural Networks

Deep learning LSTM NNs are a special type of recurrent NN (RNN) with significant improvement in the ability to learn long-term dependencies which gives it an advantage over other artificial neural networks such as BPNN and RBFNN. Figure 2a illustrates a typical schematic diagram of a traditional RNN node with the previous hidden state represented by h_{t-1}, activation tanh function, current input sample by X_t, current output by h_t, and the current hidden state by h_t. As depicted in Figure 2, all RNNs generally have the form of a chain of repeating modules of NNs. These repeating modules generally have a very basic structure in standard RNNs, like a single tanh layer only. However, a deep learning LSTM which stores information with the aid of purpose-built memory cells maintains similar chain-like structure, but with a differently structured repeating module (see Figure 2b).

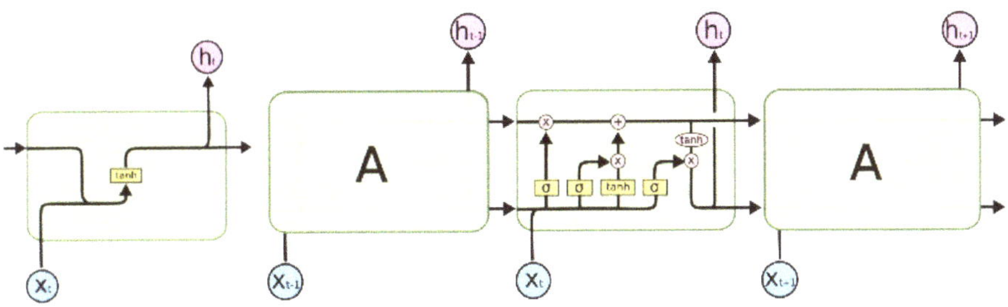

(a) (b)

Figure 2. (**a**,**b**): Typical schematic diagram of (**a**) Traditional RNN node, and (**b**) Chained LSTM blocks.

The equations below illustrate the calculation processes involved in deep learning LSTM NNs.

(a) Forget gate equation:

$$F_t = \sigma\left(W_f \times [h_{t-1}, X_t] + b_f\right) \quad (3)$$

where F_t represents a vector that has a range from 0 to 1 as its values; W_f, σ, and b_f represent the weight matrices, sigmoid function, and the bias of forget gate, respectively. The σ is used to find out whether the new information is unnecessary, in which case the information ignored and discarded, or necessary and used for updating. Finally, the tanh function is used to add weight to individual values that pass and determines their level of relevance, and ranges from -1 to 1. Inside the input gate and the output gate, same operations are repeated, which are shown in (4)–(7).

(b) Input gate equations:

$$I_t = \sigma(W_i \times [h_{t-1}, X_t] + b_i) \quad (4)$$

$$\hat{I}_t = \tan\mathrm{h}(W_i \times [h_{t-1}, X_t] + b_i) \quad (5)$$

(c) Output gate equations:

$$O_t = \sigma(W_o \times [h_{t-1}, X_t] + b_o) \quad (6)$$

$$h_t = O_t \times \tan\mathrm{h}(C_t) \quad (7)$$

(d) Cell state equation:

$$C_t = \left\{(F_t \times C_{t-1}) + \left(I_t \times \hat{I}_t\right)\right\} \quad (8)$$

where W_i and W_o denote the weight matrixes, b_i and b_o denote the bias vectors of the network of both input gate and output gate, and the hyperbolic tangent function is denoted by the tanh function.

3.2. Proposed Water Quality Prediction Model

The proposed hybrid EEMD-LSTM deep learning NN-based water-quality parameter prediction model is depicted in Figure 3. With the proposed novel water quality forecasting model, the measured real water-quality parameter content dataset undergoes decomposition processes into disparate components by applying the EEMD method for the purpose of improving the prediction accuracy of the proposed predictive model. The full procedures demonstrated in Figure 3 show the three important steps which were followed in developing the novel hybrid water quality parameters prediction solution. Firstly, the water quality parameters dataset $x(t)$ generates multiple, distinct IMF components and a corresponding residue $R_N(t)$ from the decomposition processes via the applied EEMD method in the input layer of Figure 3. The decomposition of $x(t)$ is carried out by means of an iterative sifting procedure as given below:

$$x(t) = \sum_{i=1}^{N} IMF_i(t) + R_N(t) \quad (9)$$

Subsequently, the separate IMF components and their corresponding residue undergo a process of normalization in the second step and are then used for prediction by the DL LSTM in the hidden layer of Figure 3. Lastly, in step three, individual prediction results undergo a reverse normalization process before they are efficiently combined together with the aid of a summation operation by the summation function to get the final predicted values as shown in the output layer of Figure 3. As clearly illustrated using the extended forecasting model with multiple hidden DL LSTM layers ($\mathrm{LSTM}_{1,1}$, $\mathrm{LSTM}_{1,2}$, ... , $\mathrm{LSTM}_{m,1}$, up to $LSTM_{m,n}$) in Figure 3, individual hidden layers of the stacked DL LSTM are equipped with multiple memory cells which earn the proposed prediction model the name '*deep learning*' NN [17].

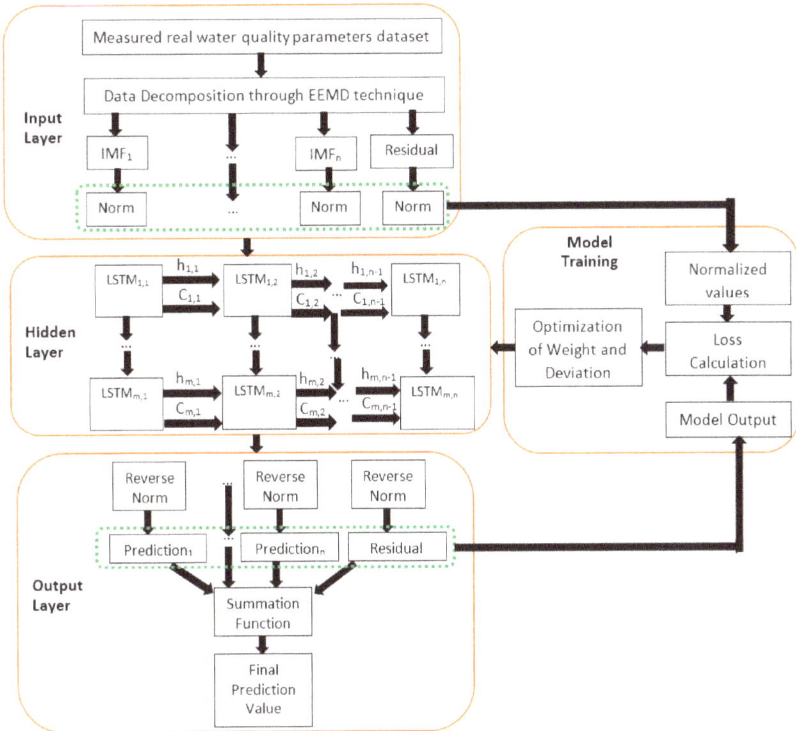

Figure 3. Proposed hybrid EEMD–LSTM deep learning water quality prediction Model.

4. Performance Evaluation

For the evaluation of the proposed hybrid EEMD–LSTM deep learning water-quality prediction model, four performance evaluation metrics were introduced to evaluate its prediction accuracy. These metrics include MAE, MSE, RMSE, and MAPE. The mathematical formulae are expressed as follows:

$$\mathrm{MAE} = \frac{1}{n}\sum_{i=1}^{n}|M_i - F_i| \tag{10}$$

$$\mathrm{MSE} = \frac{1}{n}\sum_{i=1}^{n}(M_i - F_i)^2 \tag{11}$$

$$\mathrm{RMSE} = \sqrt{\frac{1}{n}\sum_{i=1}^{n}(M_i - F_i)^2} \tag{12}$$

$$\mathrm{MAPE} = \frac{1}{n}\sum_{i=1}^{n}\left|\frac{M_i - F_i}{M_i}\right|. \tag{13}$$

In (10)–(13) above, n denotes the number of data points in the dataset, and V_i and F_i represent the measured real chlorophyll-a values and the forecasted values, respectively.

The closer these four performance evaluation metrics tend towards 0, the higher the overall forecasting and fitting accuracy of the proposed solution.

5. Results and Discussions

In this study, decomposing the Chelsea's TriLux multiparameter fluorometer measured chlorophyll-a content dataset is an intrinsic aspect of the novel prediction model for ensuring high short-term prediction accuracy. The EEMD method decomposes the real chlorophyll-a content dataset into seven individually stable IMF components (IMF 1–7) and one residual item as depicted in Figure 4a,b. The obtained IMFs from the original chlorophyll-a (470) dataset decomposition with the EEMD method is shown in Figure 4a,b.

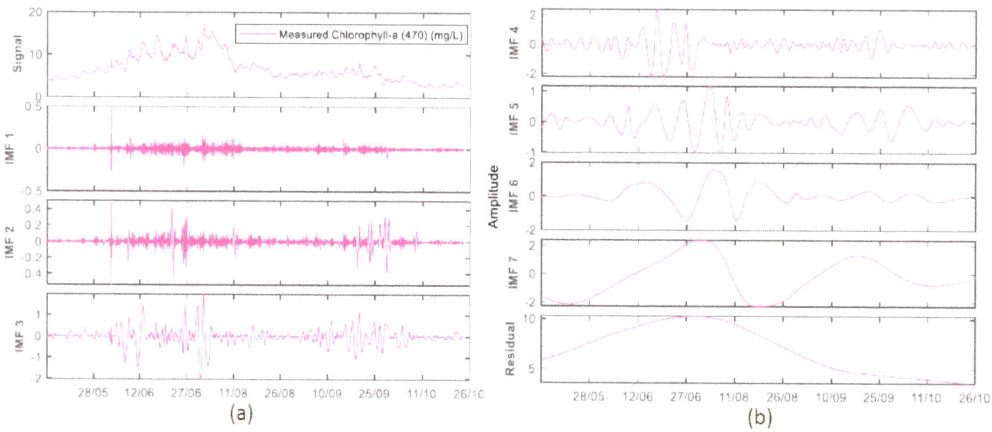

Figure 4. (**a**,**b**). Chlorophyll-a (470) dataset decomposition through the EEMD method showing (**a**) 1 to 3 of the resultant 7 IMFs, and (**b**) 4 to 7 of the resultant 7 IMFs.

The graphs in Figure 5a,b clearly show that the novel hybrid forecasting model provided good results for short-term (6 h) and long-term (24 h) forecast scenarios. With chlorophyll-a (470) concentration data, the matching trends in both Figure 5a,b further show that the model can successfully predict, with a high level of accuracy, the presence of algal bacteria such as cyanobacteria, which is a harmful alga that produces odorous and toxic substances leading to severe problems for different species of fish in the aquaculture industry.

The proposed model improved the prediction accuracy due to the application of the EEMD method, which enabled the predictive model to manifest the temporal features of the chlorophyll-a (470) content time-series data. This was done through the multi-feature selection process of the EEMD method which allowed for the selection of certain groups of IMFs that strongly correlate with the Chelsea's TriLux multi-parameter fluorometer measured chlorophyll-a data and integrate them into inputs for the deep learning LSTM neural network. Table 2 and Figure 6 present the error statistics for both 6 h and 24 h forecast results. Although these are minimal errors, the overall prediction accuracy could be further improved with an increase in data availability because the deep learning LSTM chain structure tends to be more complex and performs better with big data.

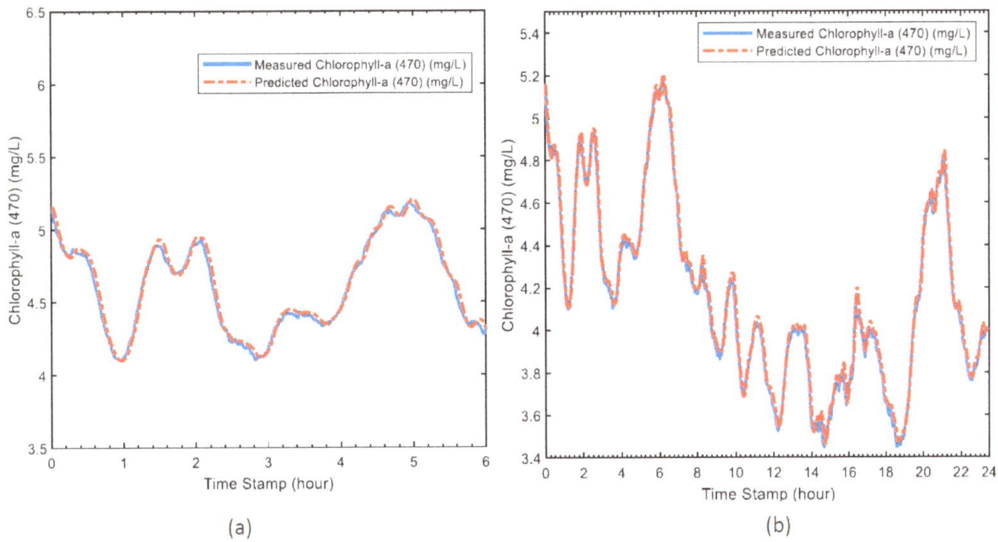

Figure 5. (**a**,**b**). Performance comparison of real Chlorophyll-a (470) parameter values and the predicted values: (**a**) half-day (6 h), and (**b**) one day (24 h) prediction results.

Table 2. Error statistics for 6 h and 24 h chlorophyll-a (470) content prediction.

Error Statistics	6 Hour Prediction	24 Hour Prediction
MSE	0.0013	0.0019
MAE	0.0277	0.0337
RMSE	0.0356	0.0417
MAPE	0.0070	0.0076

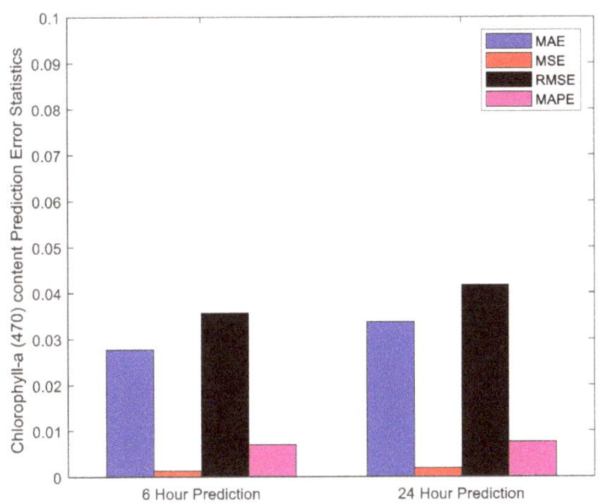

Figure 6. Chlorophyll-a (470) content prediction error statistics for 6 h and 24 h.

6. Conclusions

Timely prediction of toxic algal blooms with the help of real chlorophyll-a (470) sensor time-series data in aquatic ecosystems can allow for the effective operation and management of the aquaculture industry by providing useful information that can facilitate the decision-making process in aquafarming. In this study, we present a novel hybrid model to forecast chlorophyll-a content through the combination of the potential of the EEMD technique and a DL LSTM neural network approach. The actual experimental data from Loch Duart Salmon aquafarm show that the proposed model provides impressive results with high prediction accuracy. For future work, varieties of water quality parameter time-series datasets measured from different aquafarming sites will be considered to broaden the application horizon of the proposed forecasting model.

Funding: This research was funded by Innovate UK/BBSRC (ref: 86204028, BB/S020896/1).

Acknowledgments: The authors wish to thank the anonymous reviewers for their comments and suggestions, which have greatly helped in improving the quality of the paper.

Conflicts of Interest: The authors declare no conflict of interest. The funders had no role in the design of the study; in the collection, analyses, or interpretation of data; in the writing of the manuscript, or in the decision to publish the results.

References

1. Chislock, M.F.; Doster, E.; Zitomer, R.A.; Wilson, A.E. Eutrophication: Causes, Consequences, and Controls in Aquatic Ecosystems. *Nat. Educ. Knowl.* **2013**, *4*, 1–10.
2. Howarth, R.; Chan, F.; Conley, D.J.; Garnier, J.; Doney, S.C.; Marino, R.; Billen, G. Coupled biogeochemical cycles: Eutrophication and hypoxia in temperate estuaries and coastal marine ecosystems. *Front. Ecol. Environ.* **2011**, *9*, 18–26. [CrossRef]
3. Kim, B.C.; Jung, S.M.; Jang, C.W.; Kim, J.K. Comparison of BOD, COD and TOC as the indicator of organic matter pollution in streams and reservoirs of Korea. *J. Korean Soc. Environ. Eng.* **2007**, *29*, 640–643.
4. Gao, C.; Zhang, T. Eutrophication in a Chinese context: Understanding various physical and socio-economic aspects. *Ambio* **2010**, *39*, 385–393. [CrossRef] [PubMed]
5. Pretty, J.N.; Mason, C.F.; Nedwell, D.B.; Hine, R.E.; Leaf, S.; Dils, R. Environmental Costs of Freshwater Eutrophication in England and Wales. *Environ. Sci. Technol.* **2003**, *37*, 201–208. [CrossRef] [PubMed]
6. Chelsea Technologies. Aquaculture. Available online: https://chelsea.co.uk/application-category/aquaculture (accessed on 13 April 2021).
7. El-Otify, A.M. Evaluation of the physicochemical and chlorophyll-a conditions of a subtropical aquaculture in Lake Nasser area, Egypt. *Beni-Suef Univ. J. Basic Appl. Sci.* **2015**, *4*, 327–337. [CrossRef]
8. Ha, N.T.; Koike, K.; Nhuan, M.T. Improved Accuracy of Chlorophyll-a Concentration Estimates from MODIS Imagery Using a Two-Band Ratio Algorithm and Geostatistics: As Applied to the Monitoring of Eutrophication Processes over Tien Yen Bay (Northern Vietnam). *Remote Sens.* **2013**, *6*, 421–442. [CrossRef]
9. Shumway, S.E. A review of the effects of algal blooms on shellfish and aquaculture. *J. World Aquac. Soc.* **1990**, *21*, 65–104. [CrossRef]
10. Shin, Y.; Kim, T.; Hong, S.; Lee, S.; Lee, E.; Hong, S.; Lee, C.; Kim, T.; Park, M.S.; Park, J.; et al. Prediction of chlorophyll-a concentrations in the Nakdong River using machine learning methods. *Water* **2020**, *12*, 1822. [CrossRef]
11. Wang, X.; Wang, G.; Zhang, X. Prediction of Chlorophyll-a content using hybrid model of least squares support vector regression and radial basis function neural networks. In Proceedings of the 2016 Sixth International Conference on Information Science and Technology (ICIST), Dalian, China, 6–8 May 2016; pp. 366–371.
12. Syariz, M.A.; Lin, C.H.; Nguyen, M.V.; Jaelani, L.M.; Blanco, A.C. WaterNet: A convolutional neural network for chlorophyll-a concentration retrieval. *Remote Sens.* **2020**, *12*, 1966. [CrossRef]
13. Eze, E.; Ajmal, T. Dissolved Oxygen Forecasting in Aquaculture: A Hybrid Model Approach. *Appl. Sci.* **2020**, *10*, 7079. [CrossRef]
14. Hu, Z.; Zhang, Y.; Zhao, Y.; Xie, M.; Zhong, J.; Tu, Z.; Liu, J. A water quality prediction method based on the deep LSTM network considering correlation in smart mariculture. *Sensors* **2019**, *19*, 1420. [CrossRef] [PubMed]
15. Chelsea Technologies. TriLux. Available online: https://chelsea.co.uk/products/trilux/ (accessed on 13 April 2021).
16. Pan, L.; Li, J.; Luo, J. A temporal and spatial correction based missing values imputation algorithm in wireless sensor networks. *Chin. J. Comput.* **2010**, *33*, 1–10. [CrossRef]
17. Jason Brownlee, Stacked Long Short-Term Memory Networks Develop Sequence Prediction Models in Keras. 14 August 2019. Available online: https://machinelearningmastery.com/stacked-long-short-term-memorynetworks/ (accessed on 19 February 2021).

engineering proceedings

MDPI

Proceedings

Semiparametric Block Bootstrap Prediction Intervals for Parsimonious Autoregression †

Jing Li

Department of Economics, Miami University, Oxford, OH 45056, USA; lij14@miamioh.edu

† Presented at the 7th International conference on Time Series and Forecasting, Gran Canaria, Spain, 19–21 July 2021.

Abstract: This paper investigates the research question of whether the principle of parsimony carries over into interval forecasting, and proposes new semiparametric prediction intervals that apply the block bootstrap to the first-order autoregression. The AR(1) model is parsimonious in which the error term may be serially correlated. Then, the block bootstrap is utilized to resample blocks of consecutive observations to account for the serial correlation. The Monte Carlo simulations illustrate that, in general, the proposed prediction intervals outperform the traditional bootstrap intervals based on nonparsimonious models.

Keywords: block bootstrap; interval forecasting; principle of parsimony; semiparametric

Citation: Li, J. Semiparametric Block Bootstrap Prediction Intervals for Parsimonious Autoregression. *Eng. Proc.* **2021**, *5*, 28. https://doi.org/10.3390/engproc2021005028

Published: 28 June 2021

1. Introduction

It is well known that a parsimonious model may produce superior out-of-sample *point* forecasts compared to a complex model with overfitting issue, see [1]. One objective of this paper is to examine whether the principle of parsimony (POP) can be extended to *interval* forecasts. Toward that end, this paper proposes a semiparametric block bootstrap prediction intervals (BBPI) based on a parsimonious first-order autoregression AR(1). By contrast, the standard or iid bootstrap prediction intervals developed by Thombs and Schucany [2] (called TS intervals thereafter) are based on a dynamically adequate AR(p), where p can be large.

A possibly overlooked fact is that there is inconsistency between the ways of obtaining point forecasts and interval forecasts, in terms of whether POP is applied. When the goal is the point forecast, the models selected by information criteria of AIC and BIC are typically parsimonious, but not necessarily adequate (see Enders Enders [3] for instance). However, POP is largely forgone by the classical Box–Jenkins prediction intervals and the TS intervals; both require serially uncorrelated error terms, and the chosen models can be very complicated.

This paper attempts to address that inconsistency. The key is to note that the essence of time series forecasting is to utilize the serial correlation, and there are multiple ways to do that. One way is to use a dynamically adequate AR(p) with serially uncorrelated errors, which is fully parametric. This paper instead employs a parsimonious AR(1) with possibly serially correlated errors. Our model is semiparametric since no specific function form is assumed for the error process. Our semiparametric approach of leaving some degree of serial correlation in the error term is similar to the famous Cochrane–Orcutt procedure of Cochrane and Orcutt [4].

We employ the AR(1) model in order to generate the bootstrap replicate. In particular, we are not interested in the autoregressive coefficient, and for our purposes, it becomes irrelevant that the OLS estimate may be inconsistent due to the autocorrelated error. Using the AR(1) has another advantage: the likelihood of multicollinearity is minimized, which can result in a more efficient estimate for the coefficient. On the other hand, we do want to make use of the correlation structure in the error, and that is fulfilled by using the block bootstrap. More explicitly, the block bootstrap redraws with replacement random blocks of consecutive residuals of the AR(1). The blocking is intended to preserve the time dependence structure.

Constructing the BBPI involves three steps. In step one, the AR(1) regression is estimated by ordinary least squares (OLS), and the residual is saved. In step two, a *backward* AR(1) regression is fitted, and random blocks of residuals are used to generate the bootstrap replicate. In step three the bootstrap replicate is used to run the AR(1) regression again and random blocks of residuals are used to compute the bootstrap out-of-sample forecast. After repeating steps two and three many times, the BBPI is determined by the percentiles of the empirical distribution of the bootstrap forecast (In the full-length version of the paper, which is available upon request, we discuss technical issues such as correcting the bias of autoregressive coefficients, selecting the block size, choosing between overlapping and non-overlapping blocks, and using the stationary bootstrap developed by Politis and Romano [5]).

We implement the Monte Carlo experiment that compares the average coverage rate of the BBPI to the TS intervals. There are two main findings. The first is that the BBPI dominates when the error term shows a strong serial correlation. The second is that the BBPI always outperforms the TS intervals for the one-step forecast. For a longer forecast horizon, the TS intervals may perform better. This second finding highlights a tradeoff between preserving correlation and adding variation when obtaining the bootstrap intervals. The block bootstrap achieves the former but sacrifices the latter.

There is a growing body of literature on the bootstrap prediction intervals. Important works include Thombs and Schucany [2], Masarotto [6], Grigoletto [7], Clements and Taylor [8], Kim [9], Kim [10], Staszewska-Bystrova [11], Fresoli et al. [12], and Li [13]. The block bootstrap is developed by Künsch [14]. This work distinguishes itself by applying the block bootstrap to interval forecasts based on univariate AR models. The remainder of the paper is organized as follows. Section 2 specifies the BBPI. Section 3 conducts the Monte Carlo experiment. Section 4 concludes.

2. Semiparametric Block Bootstrap Prediction Intervals

Let $\{y_t\}$ be a strictly stationary and weakly dependent time series with mean of zero. In practice, y_t may represent the demeaned, differenced, detrended or deseasonalized series. At first, it is instructive to emphasize a fact: there are multiple ways to model a time series. For instance, suppose the data generating process (DGP) is an AR(2) with serially uncorrelated errors:

$$y_t = \phi_1 y_{t-1} + \phi_2 y_{t-2} + e_t, \quad (1)$$

where e_t can be white noise or martingale difference. Then, we can always rewrite Equation (1) as an AR(1) with new error v_t, and the new error follows an AR(1) process, so is serially correlated:

$$y_t = \phi y_{t-1} + v_t \quad (2)$$

$$v_t = \rho v_{t-1} + e_t \quad (3)$$

where $\phi_1 = \rho + \phi$ and $\phi_2 = -\rho\phi$ by construction. The point is, the exact form of the DGP does not matter. In this example, it can be AR(1) or AR(2). What matters is the serial correlation of y_t, which can be captured by Equation (1), or Equation (2) along with Equation (3) equally well. This example indicates that it is plausible to obtain forecasts based on the parsimonious AR(1) model, as long as the serial correlation in v_t has been accounted for, even if the "true" DGP is a general AR(p).

There is concern that the estimated coefficient of $\hat{\phi}$ in Equation (2) will be inconsistent due to the autocorrelated error v_t. However, this issue is largely irrelevant here because our focal point is forecasting y, not estimating the coefficient. One may use the generalized least squares method such as Cochrane–Orcutt estimation to mitigate the effect of serial correlation bias. Our Monte Carlo experiment shows that the proposed intervals perform well even without correcting the serial correlation bias.

Using the parsimonious model (2) has two benefits that are overlooked in the forecasting literature. First, notice that y_{t-1} is correlated with y_{t-2}. As a result, there is the issue of multicollinearity (correlated regressors) for Equation (1), but not Equation (2). The absence of multicollinearity can reduce the variance and improve the efficiency of $\hat{\phi}$, which explains why a simple model can outperform a complicated model in terms of out-of-sample forecasting. Second, it is well known that the autoregressive coefficient estimated by OLS can be biased—see Shaman and Stine [15], for instance. As more coefficients need to be estimated in a complex AR model, its forecast can be less accurate than that of a parsimonious model.

2.1. Iterated Block Bootstrap Prediction Intervals

The goal is to find the prediction intervals for future values $(y_{n+1}, y_{n+2}, \ldots, y_{n+h})$, where h is the maximum forecast horizon, after observing $\Omega = (y_1, \ldots, y_n)$. This paper focuses on the bootstrap prediction intervals because (i) they do not assume the distribution of y_{n+i} conditional on Ω is normal, and (ii) the bootstrap intervals can automatically take into account the sampling variability of the estimated coefficients.

The TS intervals of Thombs and Schucany [2] are based on a "long" p-th order autoregression:

$$y_t = \psi_1 y_{t-1} + \psi_2 y_{t-2} + \ldots + \psi_p y_{t-p} + e_t. \tag{4}$$

The TS intervals assume that the error e_t is serially uncorrelated, because the standard or iid bootstrap only works in the independent setting. This assumption of independent errors requires that the model (4) be dynamically adequate, i.e., a sufficient number of lagged values should be included. It is not uncommon that the chosen model can be complicated (e.g., for series with a long memory), which contradicts the principle of parsimony.

Actually, the model (4) is just a finite-order approximation if the true DGP is an ARMA process with AR(∞) representation. In that case, the error e_t is always serially correlated no matter how large p is. This extreme case implies that the assumption of serially uncorrelated errors can be too restrictive in practice.

This paper relaxes that independence assumption, and proposes the block bootstrap prediction intervals (BBPI) based on a "short" autoregression. Consider the AR(1), the simplest one:

$$y_t = \phi_1 y_{t-1} + v_t. \tag{5}$$

Most often, the error v_t is serially correlated, so model (5) is inadequate. Nevertheless, the serial correlation in v_t can be utilized to improve the forecast. Toward that end, the block bootstrap will later be applied to the residual

$$\hat{v}_t = y_t - \hat{\phi}_1 y_{t-1}, \tag{6}$$

where $\hat{\phi}_1$ is the coefficient estimated by OLS.

But first, any bootstrap prediction intervals should account for the sampling variability of $\hat{\phi}_1$. This is accomplished by running repeatedly the regression (5) using the bootstrap replicate, a pseudo time series. Following Thombs and Schucany [2] we generate the bootstrap replicate using the *backward* representation of the AR(1) model

$$y_t = \theta_1 y_{t+1} + u_t. \tag{7}$$

Note that the regressor is lead not lag. Denote the OLS estimate by $\hat{\theta}_1$, and the residual by $\hat{u}_t$:

$$\hat{u}_t = y_t - \hat{\theta}_1 y_{t+1}, \tag{8}$$

then one series of the bootstrap replicate $(y_1^*, \ldots, y_n^*)$ is computed in a backward fashion as (starting with the last observation, then moving backward)

$$y_n^* = y_n, y_t^* = \hat{\theta}_1 y_{t+1}^* + \hat{u}_t^*, \quad (t = n-1, \ldots 1). \tag{9}$$

By using the backward representation we can ensure the conditionality of AR forecasts on the last observed value y_n. Put differently, all the bootstrap replicate series have the same last observation, $y_n^* = y_n$. See Figure 1 of Thombs and Schucany [2] for an illustration of this conditionality.

In Equation (9), the randomness of the bootstrap replicate comes from the pseudo error term $\hat{u}_t^*$, which is obtained by the block bootstrap as follows:

1. Save the residual of the backward regression $\hat{u}_t$ given in Equation (8).
2. Let b denote the block size (length). The first (random) block of residuals is

$$B_1 = (\hat{u}_{i1}, \hat{u}_{i1+1}, \ldots, \hat{u}_{i1+b-1}), \tag{10}$$

where the index number $i1$ is a random draw from the discrete uniform distribution between 1 and $n - b + 1$. For instance, let $b = 3$ and suppose a random draw produces $i1 = 20$, then $B_1 = (\hat{u}_{20}, \hat{u}_{21}, \hat{u}_{22})$. In this example the first block contains three consecutive residuals starting from the 20th observation. By redrawing the index number with replacement we can obtain the second block $B_2 = (\hat{u}_{i2}, \hat{u}_{i2+1}, \ldots, \hat{u}_{i2+b-1})$, the third block $B_3 = (\hat{u}_{i3}, \hat{u}_{i3+1}, \ldots, \hat{u}_{i3+b-1})$, and so on. We stack up these blocks until the length of the stacked series becomes n. $\hat{u}_t^*$ denotes the t-th observation of the stacked series.

Resampling blocks of residuals is intended to preserve the serial correlation of the error term in the parsimonious model. Generally speaking, the block bootstrap can be applied to any weakly dependent stationary series. Here it is applied to the residual of the short autoregression.

After generating the bootstrap replicate series using Equation (9), next, we refit the model (5) using the bootstrap replicate $(y_2^*, \ldots, y_n^*)$. Denote the newly estimated coefficient (called bootstrap coefficient) by $\hat{\phi}_1^*$. Then, we can compute the iterated block bootstrap l-step forecast $\hat{y}_{n+l}^*$ as

$$\hat{y}_n^* = y_n, \hat{y}_{n+l}^* = \hat{\phi}_1^* \hat{y}_{n+l-1}^* + \hat{v}_l^*, \quad (l = 1, \ldots, h) \tag{11}$$

where the pseudo error $\hat{v}_l^*$ is obtained by block bootstrapping the residual (6). For example, let $h = 8, b = 4$. Then two blocks of residuals (6) are randomly drawn, and they are $B_1 = (\hat{v}_{i1}, \hat{v}_{i1+1}, \hat{v}_{i1+2}, \hat{v}_{i1+3})$, $B_2 = (\hat{v}_{i2}, \hat{v}_{i2+1}, \hat{v}_{i2+2}, \hat{v}_{i2+3})$. Notice that $\hat{v}_l^*$ in Equation (11) represents the l-th observation of the stacked series

$$\{\hat{v}_l^*\}_{l=1}^h = \{\hat{v}_{i1}, \hat{v}_{i1+1}, \hat{v}_{i1+2}, \hat{v}_{i1+3}, \hat{v}_{i2}, \hat{v}_{i2+1}, \hat{v}_{i2+2}, \hat{v}_{i2+3}\}. \tag{12}$$

The ordering of B_1 and B_2 in the stacked series (12) does not matter. It is the ordering of the observations within each block that matters. That within-block ordering preserves the temporal structure.

Notice that the block bootstrap has been invoked twice: first it is applied to $\hat{u}_t$ (8), then it is applied to $\hat{v}_t$ (6). The first application adds randomness to the bootstrap replicate y_t^*; whereas the second application randomizes the predicted value $\hat{y}_{n+l}^*$.

To get the BBPI, we need to generate C series of the bootstrap replicate (9), use them to fit the model (5), and use Equation (11) to obtain a series of the iterated block bootstrap l-step forecasts

$$\{\hat{y}_{n+l}^*(i)\}_{i=1}^C \tag{13}$$

where i is the index. The l-step iterated BBPI at the α nominal level are given by

$$l - \texttt{step Iterated BBPI (IBBPI)} = \left[\hat{y}_{n+l}^*\left(\frac{1-\alpha}{2}\right), \hat{y}_{n+l}^*\left(\frac{1+\alpha}{2}\right)\right] \tag{14}$$

where $\hat{y}^*_{n+l}\left(\frac{1-\alpha}{2}\right)$ and $\hat{y}^*_{n+l}\left(\frac{1+\alpha}{2}\right)$ are the $\left(\frac{1-\alpha}{2}\right)$100-th and $\left(\frac{1+\alpha}{2}\right)$100-th percentiles of the empirical distribution of $\{\hat{y}^*_{n+l}(i)\}_{i=1}^C$. Throughout this paper, we let $\alpha = 0.90$. To avoid the discreteness problem, one may let $C = 999$, see Booth and Hall [16]. In this paper we use $C = 1000$ and find no qualitative difference.

Basically, we apply the percentile method of Efron and Tibshirani [17] to construct the BBPI. De Gooijer and Kumar [18] emphasize the percentile method performs well when the conditional distribution of the predicted values is unimodal. In preliminary simulation, we conduct the DIP test of Hartigan and Hartigan [19] and find that the distribution is indeed unimodal.

2.2. Direct Block Bootstrap Prediction Intervals

We call the BBPI (14) iterated because the forecast is computed in an iterative fashion: in Equation (11), the previous step forecast $\hat{y}^*_{n+l-1}$ is used to compute the next step $\hat{y}^*_{n+l}$. Alternatively, we can use the bootstrap replicate $(y^*_1, \ldots, y^*_n)$ to run a set of *direct* regressions using only one regressor. In total there are h direct regressions. More explicitly, the l-th direct regression uses y^*_t as the dependent variable and y^*_{t-l} as the independent variable. Denote the estimated direct coefficient by $\hat{\rho}^*_l$. The residual is computed as

$$\hat{\eta}_{t,l} = y^*_t - \hat{\rho}^*_l y^*_{t-l}. \tag{15}$$

Then, the direct bootstrap forecast is computed as

$$\hat{y}^{d*}_{n+l} = \hat{\rho}^*_l y_n + \hat{\eta}^*_l \tag{16}$$

where $\hat{\eta}^*_l$ is a random draw with replacement from the empirical distribution of $\hat{\eta}_{t,l}$. The l-step direct BBPI at the α nominal level is given by

$$l-\texttt{step Direct BBPI (DBBPI)} = \left[\hat{y}^{d*}_{n+l}\left(\frac{1-\alpha}{2}\right), \hat{y}^{d*}_{n+l}\left(\frac{1+\alpha}{2}\right)\right] \tag{17}$$

where $\hat{y}^{d*}_{n+l}\left(\frac{1-\alpha}{2}\right)$ and $\hat{y}^{d*}_{n+l}\left(\frac{1+\alpha}{2}\right)$ are the $\left(\frac{1-\alpha}{2}\right)$100-th and $\left(\frac{1+\alpha}{2}\right)$100-th percentiles of the empirical distribution of $\{\hat{y}^{d*}_{n+l}(i)\}_{i=1}^C$.

There are other ways to obtain the direct prediction intervals. For example, the bootstrap replicate $(y^*_1, \ldots, y^*_n)$ can be generated based on the backward form of direct regression. Ing [20] compares the mean-squared prediction errors of the iterated and direct point forecasts. In the next section, we will compare the iterated and direct BBPIs.

3. Monte Carlo Experiment

3.1. Error Distributions

This section compares the performances of various bootstrap prediction intervals using the Monte Carlo experiment. First, we investigate the distribution of error terms. Following Thombs and Schucany [2], the data generating process (DGP) is an AR(2):

$$y_t = \phi_1 y_{t-1} + \phi_2 y_{t-2} + u_t \tag{18}$$

where $\phi_1 = 0.75, \phi_2 = -0.5$, $t = 1, \ldots, 55$. The error u_t follows an independently and identically distributed process. Three distributions are considered for u_t: the standard normal distribution, the exponential distribution with mean of 0.5, and mixed normal distribution $0.9N(-1,1) + 0.1N(9,1)$. The exponential distribution is skewed; the mixed normal distribution is bimodal and skewed. All distributions are centered to have zero mean.

We compare three bootstrap prediction intervals. The iterated block bootstrap prediction intervals (IBBPI) are based on the "short" AR(1) regression (5) and its backward form (7). The TS intervals of Thombs and Schucany [2] are based on the "long" AR(2) regression (18) and its backward form. Finally, the direct block bootstrap prediction intervals (DBBPI) are based on a series of first-order direct autoregressions. Each bootstrap prediction intervals are obtained from the empirical distribution of 1000 bootstrap forecasts. That is, we let $C = 1000$ in Equation (13) for the IBBPI, and so on. For the IBBPI and DBBPI, the block size b is 4. The TS intervals use the iid bootstrap, so $b = 1$.

The first 50 observations ($n = 50$) are used to fit the regression. Then, we evaluate whether the last five observations are inside the prediction intervals. That is, we focus on the out-of-sample forecasting. The main criterion for comparison is the average coverage rate (ACR):

$$\texttt{ACR}(h) = m^{-1} \sum_{i=1}^{m} 1(y_{n+h} \in \texttt{Prediction Intervals}) \tag{19}$$

where $1(.)$ denotes the indicator function. The number of iteration is set as m = 20,000. The forecast horizon h ranges from 1 to 5. The nominal coverage α is 0.90. The intervals whose ACR is closest to 0.90 are deemed the best.

Figure 1 plots the ACR against h, in which the ACRs of the IBBPI, TS intervals, and DBBPI are denoted by circle, square and star, respectively. In the leftmost graph, the error follows the standard normal distribution. It is shown that the ACR of the IBBPI is closest to the nominal coverage 0.90, followed by the TS intervals. The DBBPI have the worst performance. For instance, when $h = 5$, the IBBPI has ACR of 0.883, the TS intervals have ACR of 0.854, and the DBBPI has ACR of 0.829.

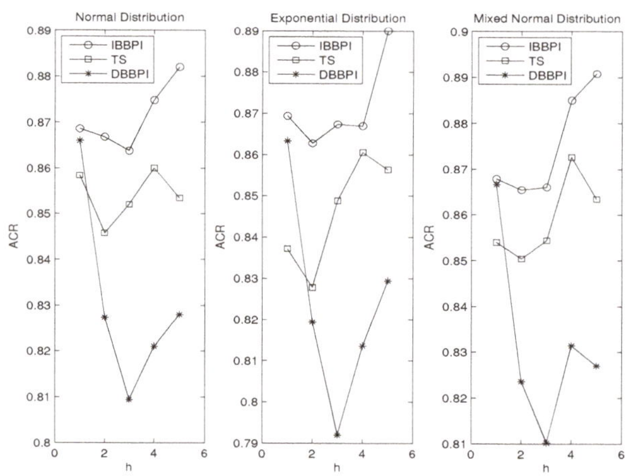

Figure 1. Error Distributions.

The ranking remains largely unchanged for the exponential distribution (in the middle graph) and mixed normal distribution (in the rightmost graph). Overall, Figure 1 indicates that (i) the IBBPI has the best performance, and (ii) the DBBPI has the worst performance. Finding (ii) complements Ing [20], which shows that the iterated *point* forecast outperforms the direct point forecast. Finding (i) is new. By comparing the three graphs in Figure 1, we see no significant change in ACRs as the error distribution varies. This is expected because all intervals are bootstrap intervals that do not assume normality.

3.2. *Autoregressive Coefficients*

Now, we consider varying autoregressive coefficients in Equation (18):

$$\phi_1 = 0.75, \phi_2 = -0.5 \quad (\texttt{stationary AR(2)}) \tag{20}$$

$$\phi_1 = 1.0, \phi_2 = -0.24 \quad (\texttt{stationary AR(2)}) \tag{21}$$

$$\phi_1 = 1.2, \phi_2 = -0.2 \quad (\texttt{non-stationary AR(2)}) \tag{22}$$

where $t = 1, \ldots, 55$ and $u_t \sim \mathtt{iidn}(0, 1)$. The leftmost graph in Figure 2 looks similar to that in Figure 1 because the same DGP is used. In the middle graph we see no change in the ranking. The rightmost graph is interesting, where the sum of autoregressive coefficients is $1.2 - 0.2 = 1$. Therefore, the series becomes nonstationary (having unit root). Obviously, nonstationarity causes distortion in the coverage rate, particularly when h is large. In light of this, we recommend applying the prediction intervals to the differenced data if unit roots are present. It is surprising to see the direct intervals are the best in the presence of nonstationarity, which may be explained by the fact that they are based on the direct regression (More simulations can be found in the full-length version of the paper where we examine the effects of sample sizes and sizes of blocks, and we compare block bootstrap intervals vs stationary bootstrap intervals, and overlapping vs non-overlapping blocks).

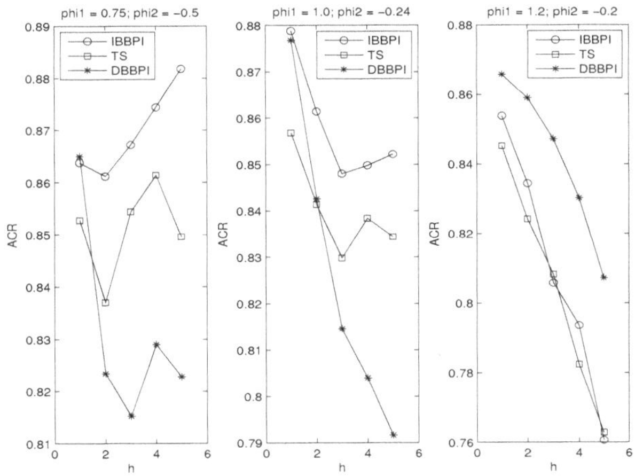

Figure 2. Autoregressive Coefficients.

3.3. *Principle of Parsimony*

So far the DGP has been the AR(2). Next we use an ARMA(1,1) as the new DGP:

$$y_t = \phi y_{t-1} + u_t + \theta u_{t-1} \tag{23}$$

where $t = 1, \ldots, 55$ and $u_t \sim \mathtt{iidn}(0, 1)$. In theory, there is AR(∞) representation for this DGP. Thus, any AR(p) model is a finite-order approximation.

We verify the principle of parsimony (POP) in two ways. Figure 3 compares the iterated block bootstrap prediction intervals based on the AR(1) regression, to the TS intervals based on the AR(2) regression (TS2, denoted by diamond), the AR(3) regression (TS3, denoted by square) and the AR(4) regression (TS4, denoted by star). For the TS intervals, we do not check whether the residual is serially uncorrelated. That job is left to Figure 4.

Figure 3 uses three sets of ϕ and θ. We see the block bootstrap intervals have the best performance in all cases. The performance of the TS deteriorates as the autoregression gets longer. This is the first evidence that POP may be applicable for interval forecasts.

The second evidence is presented in Figure 4, where the TS intervals are based on the autoregression whose order is determined by the Breusch–Godfrey test, which is appropriate since the regressors are lagged dependent variables (so the regressors are not strictly exogenous). This is how the model selection works. We start from the AR(1) regression. If the residual passes the Breusch–Godfrey test, then the AR(1) regression is chosen for constructing the TS intervals. Otherwise, we move to the AR(2) regression, apply the Breusch–Godfrey test again, and so on. In the end, the TS intervals are based on an adequate autoregression with serially uncorrelated errors.

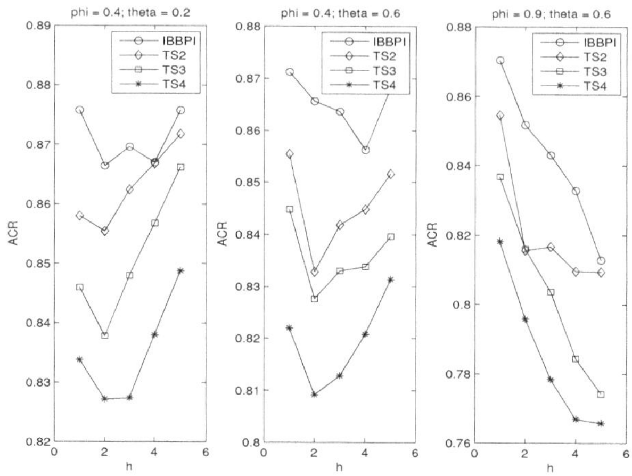

Figure 3. Parsimony I.

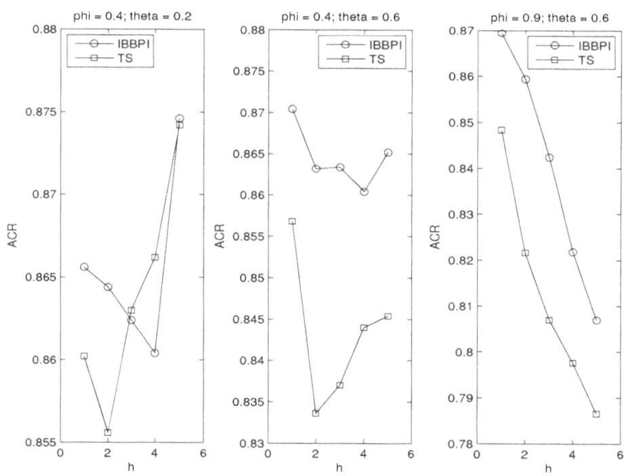

Figure 4. Parsimony II.

In the leftmost graph of Figure 4, $\phi = 0.4, \theta = 0.2$. We see the BBPI outperforms the TS intervals when h equals 1 and 2; for higher h their ranking reverses. In the middle and rightmost graphs, more serial correlation is induced as θ rises from 0.2 to 0.6, and as ϕ rises from 0.4 to 0.9. In those two graphs, the BBPI dominates the TS intervals.

The fact that the BBPI fails to dominate the TS intervals in the leftmost graph indicates a tradeoff between preserving serial correlation and adding variation. Remember that the BBPI uses the block bootstrap that emphasizes preserving serial correlation. By contrast, the TS intervals use the iid bootstrap, which can generate more variation in the bootstrap replicate than the block bootstrap.

Keeping that in mind, then the leftmost graph makes sense. In that graph, θ is 0.2, close to zero. That means the ARMA(1,1) model is essentially an AR(1) model with weakly correlated errors. For such series preserving correlation becomes secondary.

Therefore, the TS intervals may perform better than the BBPI in the presence of weakly correlated errors. It is instructive to consider the limit, when the serial correlation becomes 0 and the error term becomes serially uncorrelated. Then, the block size should reduce to 1, and the block bootstrap degenerates to the iid bootstrap, which works best in the independent setting.

Finally, from Figures 3 and 4 we notice that when $h = 1$, the BBPI always outperforms the TS intervals, no matter the serial correlation is weak or strong. This fact adds value to the BBPI for the short-horizon forecasts.

4. Conclusions

This paper proposes new prediction intervals by applying the block bootstrap to the first-order autoregression. The AR(1) model is parsimonious in which the error term can be serially correlated. Then, the block bootstrap is utilized to resample blocks of consecutive observations in order to maintain the time series structure of the error term. The forecasts can be obtained in an iterated manner, or by running direct regressions. The Monte Carlo experiment shows (1) there is evidence that the principle of parsimony can be extended to interval forecast; (2) there is a trade-off between preserving correlation and adding variation; (3) the proposed intervals have superior performance for one-step forecast.

References

1. James, G.; Witten, D.; Hastie, T.; Tibshirani, R. *An Introduction to Statistical Learning: With Applications in R*; Springer: Berlin/Heidelberg, Germany, 2013.
2. Thombs, L.A.; Schucany, W.R. Bootstrap prediction intervals for autoregression. *J. Am. Stat. Assoc.* **1990**, *85*, 486–492. [CrossRef]
3. Enders, W. *Applied Econometric Times Series*, 3rd ed.; Wiley: Hoboken, NJ, USA, 2009.
4. Cochrane, D.; Orcutt, G.H. Application of least squares regression to relationships containing auto-correlated error terms. *J. Am. Stat. Assoc.* **1949**, *44*, 32–61.
5. Politis, D.N.; Romano, J.P. The Stationary Bootstrap. *J. Am. Stat. Assoc.* **1994**, *89*, 1303–1313. [CrossRef]
6. Masarotto, G. Bootstrap prediction intervals for autoregressions. *Int. J. Forecast.* **1990**, *6*, 229–239. [CrossRef]
7. Grigoletto, M. Bootstrap prediction intervals for autoregressions: Some alternatives. *Int. J. Forecast.* **1998**, *14*, 447–456. [CrossRef]
8. Clements, M.P.; Taylor, N. Boostrapping prediction intervals for autoregressive models. *Int. J. Forecast.* **2001**, *17*, 247–267. [CrossRef]
9. Kim, J. Bootstrap-after-bootstrap prediction intervals for autoregressive models. *J. Bus. Econ. Stat.* **2001**, *19*, 117–128. [CrossRef]
10. Kim, J. Bootstrap prediction intervals for autoregressive models of unknown or infinite lag order. *J. Forecast.* **2002**, *21*, 265–280. [CrossRef]
11. Staszewska-Bystrova, A. Bootstrap prediction bands for forecast paths from vector autoregressive models. *J. Forecast.* **2011**, *30*, 721–735. [CrossRef]
12. Fresoli, D.; Ruiz, E.; Pascual, L. Bootstrap multi-step forecasts of non-Gaussian VAR models. *Int. J. Forecast.* **2015**, *31*, 834–848. [CrossRef]
13. Li, J. Block Bootstrap Prediction Intervals for Parsimonious First-Order Vector Autoregression. *J. Forecast.* **2021**, *40*, 512–527. [CrossRef]
14. Künsch, H.R. The Jackknife and the Bootstrap for General Stationary Observations. *Ann. Stat.* **1989**, *17*, 1217–1241. [CrossRef]
15. Shaman, P.; Stine, R.A. The bias of autoregressive coefficient estimators. *J. Am. Stat. Assoc.* **1988**, *83*, 842–848. [CrossRef]
16. Booth, J.G.; Hall, P. Monte Carlo approximation and the iterated bootstrap. *Biometrika* **1994**, *81*, 331–340. [CrossRef]
17. Efron, B.; Tibshirani, R.J. *An Introduction to the Bootstrap*; Chapman and Hall: London, UK, 1993.
18. De Gooijer, J.G.; Kumar, K. Some recent developments in non-linear time series modeling, testing, and forecasting. *Int. J. Forecast.* **1992**, *8*, 135–156. [CrossRef]
19. Hartigan, J.A.; Hartigan, P.M. The DIP test of unimodality. *Ann. Stat.* **1985**, *13*, 70–84. [CrossRef]
20. Ing, C.K. Multistep Prediction in Autogressive Processes. *Econom. Theory* **2003**, *19*, 254–279. [CrossRef]

engineering proceedings

MDPI

Proceedings

Analysis of Different GNSS Data Filtering Techniques and Comparison of Linear and Non-Linear Times Series Solutions: Application to GNSS Stations in Central America for Regional Geodynamic Model Determination †

Javier Ramírez-Zelaya *, Belén Rosado, Paola Barba, Jorge Gárate and Manuel Berrocoso

Laboratorio de Astronomía, Geodesia y Cartografía, Departamento de Matemáticas, Facultad de Ciencias, Campus de Puerto Real, Universidad de Cádiz, 11510 Puerto Real, Spain; belen.rosado@uca.es (B.R.); paola.barbaceballos@alum.uca.es (P.B.); jorge.garate@uca.es (J.G.); manuel.berrocoso@uca.es (M.B.)
* Correspondence: javierantonio.ramirez@uca.es
† Presented at the 7th International Conference on Time Series and Forecasting, Gran Canaria, Spain, 19–21 July 2021.

Abstract: At present, different methods are used for processing GPS time series data obtained from a network of GNSS stations. Solutions converted to velocity and displacement allow the generation of different geodynamic models in areas influenced by tectonic and volcanic activity. This study focuses on the comparative analysis of the solutions obtained through different processing techniques: Precise Point Positioning (PPP) and Relative Positioning using specialized scientific software (Bernese 5.2). Another important objective of this study is the analysis of the convergence of linear and non-linear time series to determine the accuracy in each component (east, north, up), in addition to the application of statistical techniques and data filtering (1-sigma, 2-sigma, kalman, wavelets, and CATS analysis) to check the behavior of the series. These processing and analysis techniques will be applied to different series obtained from the main stations used for tectonic and volcanic monitoring in the Central America region (Guatemala, El Salvador, Honduras, Nicaragua, and Costa Rica) in order to establish a regional geodynamic model.

Keywords: GNSS data filtering techniques; GNSS time series analysis; CATS analysis; geodynamic model

Citation: Ramírez-Zelaya, J.; Rosado, B.; Barba, P.; Gárate, J.; Berrocoso, M. Analysis of Different GNSS Data Filtering Techniques and Comparison of Linear and Non-Linear Times Series Solutions: Application to GNSS Stations in Central America for Regional Geodynamic Model Determination. *Eng. Proc.* **2021**, *5*, 29. https://doi.org/10.3390/engproc2021005029

Published: 30 June 2021

1. Introduction

GNSS (Global Navigation Satellite Systems) are passive navigation systems based on radio-frequency-emitting satellites providing a space-time reference frame with continuous global coverage that is available to any number of users, regardless of existing atmospheric conditions. GNSS networks are defined as a set of GNSS satellite continuous tracking stations called CORS (Continuous Operating Reference Stations) that are strategically distributed in specific territories, providing real-time or post-processing services to solve the problem of absolute geodetic positioning to any users located in the territory or adjacent areas. There is an international GNSS network known as the IGS (International GNSS Service) which contributes to the International Terrestrial Reference Frame (ITRF). IGS stations are used as reference stations for any geodetic or geodynamic process research anywhere on Earth [1].

This work focuses on the study of the geodynamic behavior of the Central America region (Guatemala, El Salvador, Honduras, Nicaragua, and Costa Rica), a highly active area in terms of natural hazards due to different geological phenomena (earthquakes, volcanic eruptions, tsunamis, landslides, floods, etc. Different permanent and semi-permanent GNSS-GPS station networks, such as COCONet (Continuously Operating Caribbean GPS Observational Network) and IGS, are used as measurement instruments, providing free access to the data. The purpose of this study is to determine an absolute regional geodynamic model obtained from GPS data processing and analysis techniques (PPP and Relative) as well as the application of specific data filtering techniques (1-sigma, 2-sigma, kalman, wavelets, and CATS analysis) [2] to determine the displacement at strategic points in the area, thus obtaining the deformation of the Earth's surface. The GNSS stations used are: GUAT (Guatemala), SSIA (El Salvador), TEG2 (Honduras), MANA (Nicaragua), JAPO (Nicaragua), CN22 (Nicaragua), CN30 (Nicaragua) and VERA (Costa Rica).

2. Site Description

Geodynamic Framework of the Central America Region

The Central America Region is subject to the interaction of the Caribbean, North America, Cocos and Nazca tectonic plates, whose relative velocity is 2 to 9 cm per year. It is also responsible for the active volcanism in the region and the high rate of shallow and intermediate seismicity. Seismic events recorded with magnitudes from 5.5 to 7.9 on the Richter scale have occurred in this region, causing a great deal of destruction; the seismic sources are due to active interplate and intraplate tectonics.

There are important seismotectonic structures that occur intraplate: the Nicaraguan depression, the Polochic–Motagua fault system in Guatemala. The subduction zone is the main tectonic structure and source of seismic and volcanic activity in the countries of the region. It extends along the coasts of Central America on the Pacific. The zone of Wadati–Benioff, the volcanic arc, has a dip towards the northeast with angles of 60° to 80° and seismicity at a depth of up to 200 km.

The Mesoamerican trench includes segments of 100 and 300 km length that are distinguished in their strike and dip angle [3]. The seismic occurrence rate in the Mesoamerican trench [4] establishes that the most important seismic events have occurred in the trench segments off the coast of Guatemala, El Salvador, and Nicaragua. In this environment, there are also volcanic hazards associated with the collapse of large volumes of volcanic buildings, as in Vulcanian, Plinian, and Strombolian eruptions. The frequency of volcanic collapses in Central America is scarcely known due to the absence of precise dating of the deposits [5]. The main risk is due to the movement of large amounts of debris that can move towards nearby populations with the risk of being buried and resulting in loss of human life.

3. Methods, Techniques, and Results

3.1. GNSS-GPS Data Processing Methods and Presentation of the Regional Geodynamic Model Obtained

The geodetic positioning method PPP (Precise Point Positioning) uses GNSS data from a single station and is therefore independent of other reference stations. This technique achieves its maximum precision (centimeter) using auxiliary data provided by IGS: ephemeris, clock corrections, Earth orientation parameters, atmospheric refraction parameters (ionosphere, troposphere), etc. [6]. One of the peculiarities of this technique is that it necessarily requires solving the ambiguities of the satellites in order to guarantee precise coordinates, as it is made by the JPL GIPSY software. In this study, Bernese 5.2 scientific software was used to process daily (24 h) GPS data from 8 permanent and semi-permanent stations with an observation sample rate of 30 s [7].

In stations influenced by the tectonic-volcanic effect, relative data treatment and processing was carried out, making use of simultaneous measurements of the different satellites in order to recognize and cancel orbital errors, satellite clock errors, and means of signal propagation (troposphere and ionosphere) through double satellite–receiver differences. This method makes it possible to calculate the difference between two positions with subcentimeter accuracy, thus requiring that one of the stations be recognized through a reference frame. The best accuracies are obtained with the relative processing technique, which focuses on calculating the distances between the GPS antenna and the satellite through the carrier wave itself by means of interferometric processes [8]. The final calculation is obtained by combining this method with the differential method; that is, one of the receivers must be on a point with known and reliable coordinates [9].

With regard to geodetic control, the variations of the absolute coordinates obtained through a geodetic calculation and adjustment process are analyzed in both Cartesian coordinates (X, Y, Z) and topocentric (east, north, up) using different techniques that guarantee millimeter-level precision to provide optimal results for regional tectonic or volcanic surface deformation parameters [10] of the region. This implies the use of precise auxiliary parameters such as precise ephemeris, corrections of the satellite clocks and the tropospheric models, as well as data processing and filtering methods capable of jointly managing the results. Figure 1 shows the absolute geodynamic model obtained from the analysis of the stations time series: GUAT, SSIA, TEG2, MANA, JAPO, CN22, CN30 and VERA. The Table 1 contains coordinates and velocities of the GNSS stations analyzed.

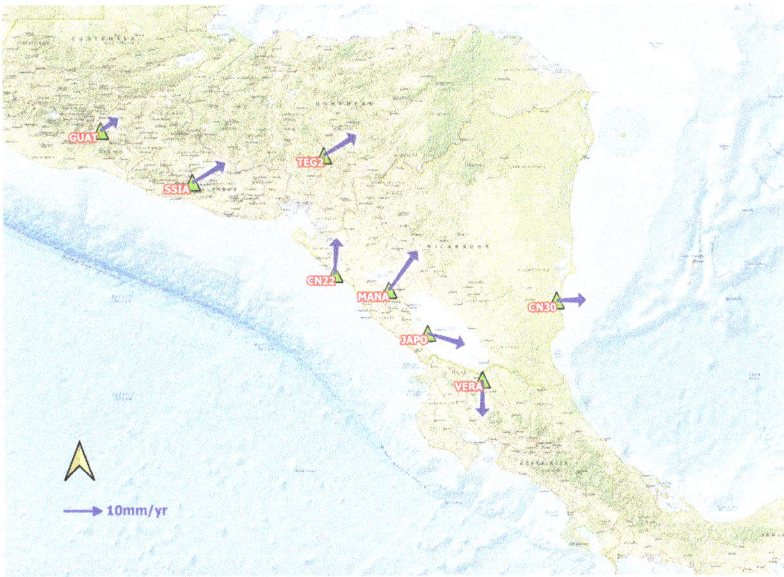

Figure 1. Geodynamic model of the Central America region.

Table 1. Coordinates and velocities of the GNSS stations.

Station	Lon	Lat	Ve (m/yr)	Vn (m/yr)	Vu (m/yr)
MANA	−86.2490012	12.1489402	0.007	0.010	−0.004
JAPO	−85.6784012	11.5259143	0.009	−0.003	0.004
CN30	−83.7720477	11.9935771	0.008	0.000	−0.015
CN22	−87.0446810	12.3841118	0.001	0.009	−0.012
GUAT	−90.5053700	14.4559600	0.005	0.003	0.000
SSIA	−89.1430700	13.7100400	0.008	0.005	−0.002
VERA	−84.8685000	10.8539000	0.000	−0.009	−0.004
TEG2	−87.2056000	14.0901000	0.008	0.005	0.000

3.2. GPS Data Filtering Techniques and Analyzed Linear and Non-Linear Time Series

The previous model was obtained from the data processing (PPP and Relative) and a further analysis of the linear and non-linear time series. It was carried out using different mathematical and statistical techniques for detecting the behavior of the series once the outliers were filtered. Such outliers can introduced by the physical environment or by the instrument itself. The use of the different filtering techniques (1-sigma, 2-sigma, kalman, wavelets, and CATS analysis) applied to each time series minimizes noise and improves the resolution or accuracy of the results, more specifically in the determination of the velocity or displacement parameter in the east, north, and elevation components of each point. The filtering techniques used initially are 1-sigma and 2-sigma to search for anomalous values, considering their deviations through a linear regression method. Then we apply a Kalman predictive-corrective filter and in turn perform a harmonic analysis using the wavelets filtering technique (Figure 2) in order to reduce the noise in these series. Finally, for the calculation of the displacement parameter, correction of offsets, and definition of the series trend, we use the CATS analysis software (Create and Analyze Time Series) (Figure 3), proving to be the one that best fits the geodynamic model.

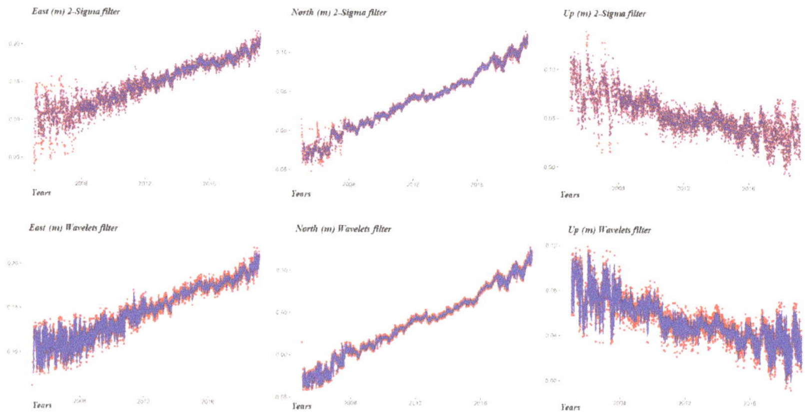

Figure 2. MANA time series solutions, with 2-sigma and wavelets filtering applied.

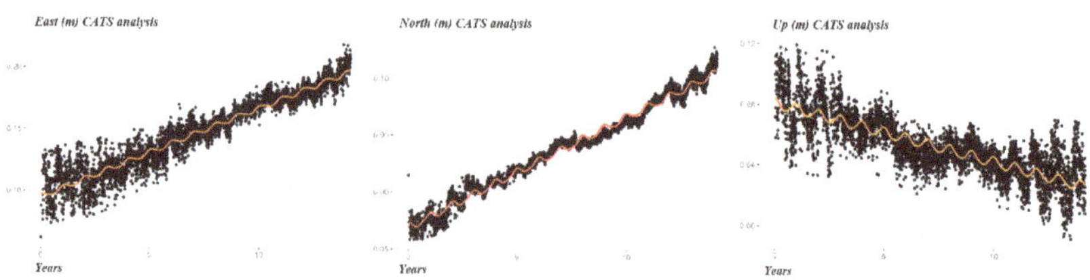

Figure 3. MANA time series solutions, with CATS analysis software applied.

3.3. *Analysis of VERA Time Series Solutions*

We observed an important change in the displacement of the VERA station, specifically in the horizontal component (east, north) produced by the earthquake occurring on 5 September 2012, 8 km northwest of Sámara, Guanacaste province, Costa Rica; the east seismic event had a 7.6 magnitude and 18 km depth (Figure 4). Table 2 shows the final velocities values (east, north, up) from the VERA GNSS station in three different phases (absolute series, pre-seismic phase and post-seismic phase).

Table 2. Velocities of the VERA station in three different phases.

Time	Ve (m/yr)	Vn (m/yr)	Vu (m/yr)
RAW (Black)	0.0001	−0.0091	−0.0041
PRE-S (Blue)	0.0014	0.0134	−0.0028
POST-S (Green)	0.0083	0.0034	−0.0041

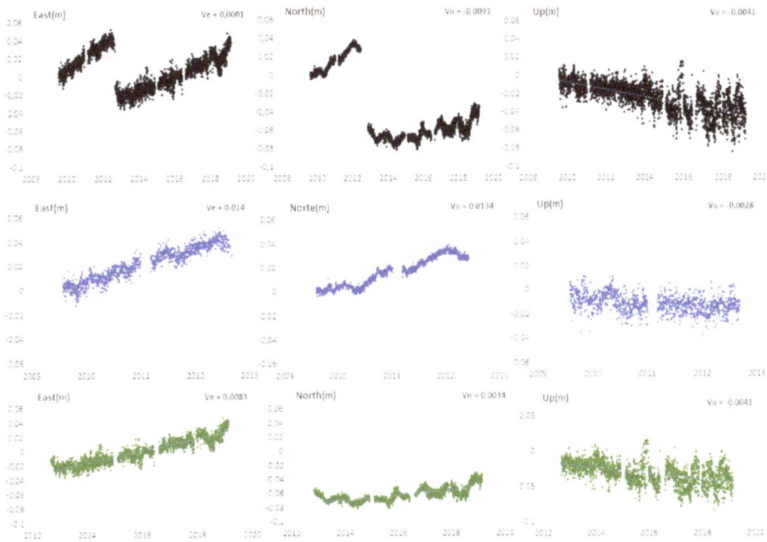

Figure 4. VERA time series in three different phases. This figure shows the changes in series trend (absolute series "black", pre-seismic series "blue" and post-seismic series "green") GNSS data analyzed: 2008–2019.

3.4. Analysis of JAPO Time Series Solutions

The JAPO station, located on the Concepción Volcano of Ometepe Island, Nicaragua, is used to monitor volcanic activity and is part of the Conceptepe GNSS Network. The Concepción volcano has permanent activity, and around it there are many landslides that cause a great deal of destruction. In the JAPO time series (non-linear), changes in displacement can be observed due to the continuous volcano activity (Figure 5). Table 3 shows the final velocities values (east, north, up) from the JAPO GNSS station using CATS analysis technique.

Table 3. Velocities of JAPO station using CATS analysis.

Station	Ve (m/yr)	Vn (m/yr)	Vu (m/yr)
JAPO	0.0094	−0.0027	0.0037

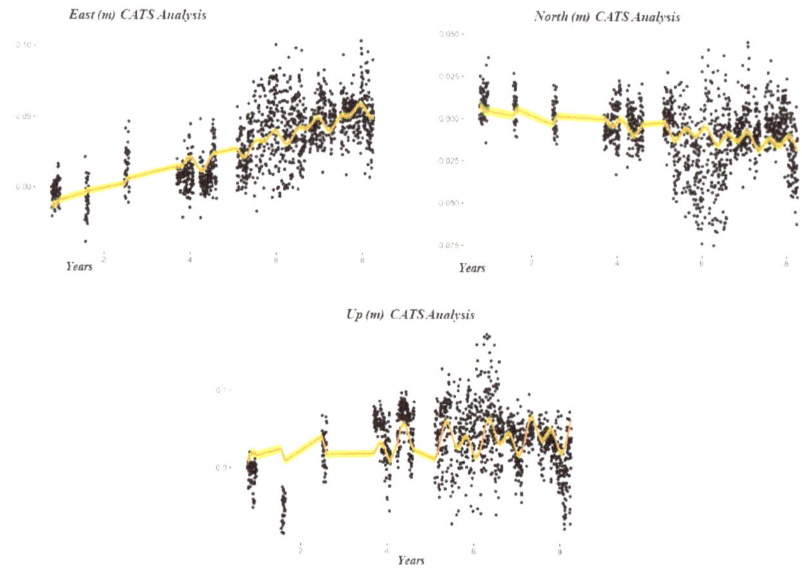

Figure 5. JAPO time series, station located on the Concepción volcano, Ometepe, Nicaragua. GNSS data analyzed: 2010–2017.

4. Conclusions

This work shows the geodynamic behavior of the Central America region and adjacent areas. It also studies the quality of the observations of the reference stations and permanent and semi-permanent GNSS vertices to guarantee maximum precision (subcentimeter) in the time series obtained.

JAPO, VERA, and CN30 stations show changes in their behavior due to being influenced by volcanic activity, earthquake occurrence, or the lack of observations, respectively.

GNSS data filtering and fitting techniques (1-sigma, 2-sigma, wavelets, and CATS analysis) improve the precision of the results by eliminating offset in the components (east, north, up) of a data series; however, its application in nonlinear series can give incorrect solutions. Despite being included in the IGS network, the MANA permanent station, being immersed in the area of greatest seismic activity, means that it does not meet the necessary requirements for geodynamic surveillance in real time.

The VERA station is subject to constant changes in its displacement due to the different earthquakes that have occurred in the area due to local faulting and the Cocos–Caribe plate collision, one of the most important earthquakes in recent years occurring on 5 September 2012, 8 km northwest of Sámara, Guanacaste province, Costa Rica, with a magnitude of 7.6 and a depth of 18 km.

The JAPO time series shows changes in displacement due to the continuous activity of the Concepción volcano and the landslides that occur in this area. The permanent JAPO station is very important for the surveillance and monitoring not only of the Concepción volcano, but also for Maderas volcano located at the southern end of the Ometepe island. However, better solutions could be obtained using other processing techniques and reference stations with precise coordinates.

Data Availability Statement:

– GPS data of the MANA, GUAT, SSIA and TEG2 stations belong to the IGS network and are available at the following link: https://geodesy.noaa.gov/CORS/data.shtml.

– GPS data of the CN22, CN30, and VERA stations belong to the COCONet network and are available at the following link: http://coconetserver.ineter.gob.ni/coconetgsac/.

– GPS data of the JAPO station belong to the Conceptepe network of INETER (Instituto Nicaragüense de Estudios Territoriales), Universidad de Cádiz (manuel.berrocoso@uca.es). Data belong to a research project and are not public.

Acknowledgments: We are thankful to all individuals and institutions contributing to GPS data collection and dissemination. Thanks are due to IGS (International GNSS Service), COCONet–UNAVCO (Continuously Operating Caribbean GPS Observational Network), INETER (Instituto Nicaragüense de Estudios Territoriales). We especially appreciate the contributions and suggestions of the editors and reviewers who helped improve this manuscript.

References

1. Hofmann-Wellenhof, B.; Lichtenegger, H.; Wasle, E. *GNSS: Global Navigation Satellite Systems—GPS, GLONASS, GALILEO and More*; Springer Science and Business Media: Berlin, Germany, 2008.
2. Williams, S.D.P. CATS: GPS coordinate time series analysis software. *GPS Solut.* **2008**, *12*, 147–153. [CrossRef]
3. Burbach, G.; Frohlich, C.; Pennington, W.D.; Matumoto, T. Seismicity and tectonics of the subducted Cocos plate. *J. Geophys. Res. Solid Earth* **1984**, *89*, 7719–7735. [CrossRef]
4. White, R.A.; Ligorria, J.P.; Cifuentes, I.L.; Rose, W.L. Seismic history of the Middle America subduction zone along El Salvador, Guatemala, and Chiapas, Mexico: 1526–2000. In *Natural Hazards in El Salvador*; Rose, W.I.; Bommer, J.J.; López, D.L.; Carr, M.J.; Major, J.J., Eds.; GSA SPECIAL PAPERS; Geological Society of America: McLean, VA, USA, 2004; Volume 375, pp. 379–396; ISBN 9780813723754. [CrossRef]
5. Segall, P. *Earthquake and Volcano Deformation*; Princeton University Press: Princeton, NJ, USA, 2010; 458p.
6. Zumberge, J.F.; Heflin, M.B.; Jefferson, D.C.; Watkins, M.M.; Webb, F.H. Precise point positioning for the efficient and robust analysis of GPS data from large networks. *J. Geophys. Res.* **1997** *102*, 5005–5017. [CrossRef]
7. Dach, R.; Lutz, S.; Walser, P.; Fridez, P. (Eds.) *Bernese GPS Software*, version 5.2; User Manual; Astronomical Institute, University of Bern, Printing Office of the University of Bern: Bern, Switzerland, 2015.
8. Prates, G.; García, A.; Fernández-Ros, A.; Marrero, J.M.; Ortiz, R.; Berrocoso, M. Enhancement of sub-daily positioning solutions for surface deformation surveillance at El Hierro volcano (Canary Islands, Spain). *Bull. Volcanol.* **2013**, *75*, 724. [CrossRef]
9. Rosado, B.; Fernández-Ros, A.; Jiménez, A.; Berrocoso, M. Modelo de deformación horizontal GPS de la región sur de la Península Ibérica y norte de áfrica (SPINA). *Boletín Geológico y Minero* **2017**, *128*, 141–156. [CrossRef]
10. Rosado, B.; Fernández-Ros, A.; Berrocoso, M.; Prates, G.; Gárate, J.; De Gil, A.; Geyer, A. Volcano-tectonic dynamics of Deception Island (Antarctica): 27 years of GPS observations (1991–2018). *J. Volcanol. Geotherm. Res.* **2019**, *381*, 57–82. [CrossRef]

engineering proceedings

Proceedings

Meta-Parameter Selection for Embedding Generation of Latency Spaces in Auto Encoder Analytics †

Maria Walch [1,2,3,*,‡], Peter Schichtel [2,3], Dirk Lehmann [2,4,5] and Amala Paulson [2,3]

1 AG Algebra, Geometrie und Computeralgebra, TU Kaiserslautern, 67663 Kaiserslautern, Germany
2 IAV GmBH, 10587 Berlin, Germany; peter.schichtel@iav.de (P.S.); dirk.lehmann@iav.de (D.L.); amala.paulson@iav.de (A.P.)
3 FLaP Lab, 67663 Kaiserslautern, Germany
4 Fakultaet fuer Informatik, Ostfalia University of Applied Sciences, 38302 Wolfenbuettel, Germany
5 Fakultaet fuer Informatik, Institut fuer Simulation und Graphik, University of Magdeburg, 39106 Magdeburg, Germany
* Correspondence: walch@rhrk.uni-kl.de
† Presented at the 7th International conference on Time Series and Forecasting, Gran Canaria, Spain, 19–21 July 2021.
‡ Current address: FLaP Lab, Deutsches Forschungszentrum fuer kuenstliche Intelligenz, Trippstadterstrasse 122, 67663 Kaiserslautern, Germany.

Abstract: Picking an appropriate parameter setting (*meta-parameters*) for visualization and embedding techniques is a tedious task. However, especially when studying the latent representation generated by an autoencoder for unsupervised data analysis, it is also an indispensable one. Here we present a procedure using a cross-correlative take on the meta-parameters. This ansatz allows us to deduce meaningful meta-parameter limits using OPTICS, DBSCAN, UMAP, t-SNE, and k-MEANS. We can perform first steps of a meaningful visual analysis in the unsupervised case using a vanilla autoencoder on the MNIST and DeepVALVE data sets.

Keywords: dimension reduction techniques; multi-dimensional spaces; big data; time series; autoencoder

Citation: Walch, M.; Schichtel, P.; Paulson, A.; Lehmann, D. Meta-Parameter Selection for Embedding Generation of Latency Spaces in Auto Encoder Analytics. *Eng. Proc.* **2021**, *5*, 30. https://doi.org/10.3390/engproc2021005030

Published: 1 July 2021

1. Introduction

High-dimensional data creates the need for simplification, of which low-dimensional embeddings as well as data visualization constitute two closely related methodologies. Their goal is to preserve the main patterns within the data and obtain a less complex data representation, which for two or three-dimensional embeddings grants also direct visual access on the data. It is well known that finding a low-dimensional data embedding is a meticulous, parameter- and data dependent task for which optimization may be difficult [1]. However, in our approach, we take into account that even the visualization space for an appropriate embedding is related to a set of visualization parameters, which we call meta-parameters.These are not directly optimized over, but introduce bias in the visualization itself when chosen poorly. One example the reader might know is the fact that DBSCAN suffers from the curse of dimensionality, when the minimal number of neighboring points n_{samples} is chosen unfortunately [2,3]. For our investigation, we chose the challenging setting of data (namely MNIST [4] and DeepVALVE [5]) compressed within the latency space of an autoencoder.

1.1. Why Are Autoencoders Interesting?

The idea of autoencoders exists for more than 30 years [6] and the applications are presently widespread. They range from generalization to classification tasks, denoising, anomaly detection, recommender systems, clustering and dimensionality reduction with stunning results [7,9–13]. Within this work, we focus on the latter two use cases, wherein autoencoders perform unsupervised feature extraction and dimensionality reduction [14,15]. Autoencoders consist of an encoder-decoder structure as explained in Figure 1.

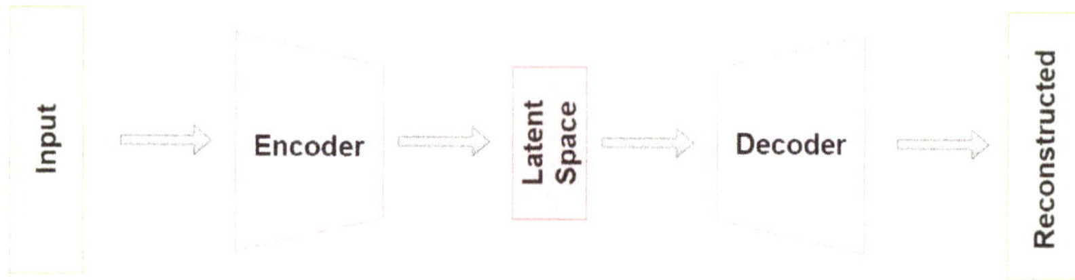

Figure 1. Architecture of an autoencoder. The left side constitutes the **encoder** while the mirror image around the middle is called **decoder**. The exact composition of the layer structure is given in Appendix A.

To achieve their above-mentioned goal, the data is embedded within a **latency space** via the encoder. Usually, the latent dimension is much smaller than the one of the original data set. This kind of setting is also known as bottleneck architecture. From this embedding, the original data representation is reconstructed by the decoder. The system is trained by minimizing the reconstruction error. Conceptually, autoencoders can be seen as a nonlinear generalization of PCA [16]. Under postulation of the manifold hypothesis [17], in some settings, they are supposed to learn the intrinsic low-dimensional data manifold embedded (nonlinearly) into the high-dimensional data observation space. Even more, in this vein they can be interpreted as a nonlinear embedding approach on their own. In the context of unlabeled high-dimensional data sets and especially time series, autoencoders have shown to be powerful tools for unsupervised analysis tasks [15,18]. Yet it has become clear in several applications that the "classical" loss term might not be enough to capture the desired behavior [19]. For this reason, some researchers try to ameliorate the reliability and efficiency of their autoencoder models by introducing additional, task dependent loss-terms (e.g., Ref. [20] introduced a topological loss term to preserve connected components within the data; Ref. [21] introduced a perceptual loss to improve image classification; Ref. [22] introduced a loss term to fix class centroids within a classification task).

1.2. Our Approach

In this work, we approach this problem upside down. We develop methods to investigate the autoencoder's capability to conform to the manifold hypothesis in a visual and qualitative way, which integrates into the general trend of visualization methods gaining more importance over the last while [23–25]. Our goal is to give data scientists a non-mathematical and interpretable tool at hand to monitor and supervise the nonlinear embedding process whose result constitutes the latency space. To do so, we proceed as follows: First, we must formulate our concepts. To make clear what is new to our approach, we must distinguish it from classical parameter and hyperparameter tuning models.

Definition 1 (Parameters)**.** *Parameters are the quantities that determine the actual shape of the data manifold.*

Intuitively, parameters determine the "physics" of our data under consideration. In the case of our autoencoder, they are given by the trainable weights.

Definition 2 (Hyperparameters)**.** *Hyperparameters are the quantities that determine the performance, the setup, and the training of our neural, data driven model in a metrisable way.*

A summary of our autoencoder model and the corresponding hyperparameters can be found in Tables A1 and A2 in Appendix A. The decoder is just a mirror in our case. (Although sometimes a weight tie is implemented too, we adhere from this technique here).

Definition 3 (Meta-parameters). *Meta-parameters are the quantities that determine the performance of our neural, data driven model in a non-metrisable way.*

So, it becomes clear why standard (hyper-)parameter optimization methods cannot be applied to the present purpose: Lacking a metric, there is now quantifiable (stochastic) optimization procedure to find an optimal embedding. For this reason, we took a step back on to a qualitative level and performed a cross-correlative study including t-SNE, UMAP, k-MEANS, DBSCAN and OPTICS.

1.3. Embedding and Visualization Methods

The use of visualization methods to analyze structures of interest for a higher-dimensional space by a visual inspection of a lower-dimensional embedding has become a popular approach in recent years, compare [26–35]. Usually, embedding schemes are classified and distinguished based on their embedding properties, e.g., to discriminate linear and nonlinear embeddings. Thus, to cover an appropriate set of embedding techniques for reasons of comparison, our approach covers a comparative study of different embedding techniques. In the following, a short description of these methods is given. Table A3 in Appendix B states the meta-parameters and their default values.

1.3.1. t-SNE

The t-SNE algorithm assigns mutual "neighborhood"-probabilities based on a distance metric (most commonly the Euclidean one) between points, and successively tries to minimize the Kullback–Leibler divergence. The most important hyperparameter is the *perplexity*, which defines the minimum number of neighborhood points. However, the hyperparameters of the intrinsic optimization algorithm also have crucial impact on the final 2- or 3-dimensional embedding [36,37].

1.3.2. UMAP

This algorithm represents an advancement with respect to t-SNE by constructing a "fuzzy simplicial complex" on the data. However, choosing the appropriate radius for the related Cěch complex is a meticulous task. Additionally, the choice of the metric and the minimum number of neighboring points determine the resulting 2- or 3-dimensional embedding. Like t-SNE, UMAP's dependence on the metrified minimum point distance makes it prone to the curse of dimensionality [38].

1.3.3. k-MEANS

K-Means minimizes the metric distance of data points to predefined cluster centers. This also constitutes its major drawback, aside from not being able to identify noise and imposing complexity on all cluster shapes [39].

1.3.4. DBSCAN

Unlike k-MEANS, DBSCAN is a density-based method able to identify noise and clusters of all shapes. Its main hyperparameters are ϵ, the critical value for which points are seen to belong to the same cluster, and n_{samples}, the minimum number of points that shall belong to one cluster. As ϵ is chosen globally, DBSCAN has its difficulties with clustering heterogeneous data [40].

1.3.5. OPTICS

OPTICS has many commonalities with DBSCAN. The most substantial difference to DBSCAN is that ϵ is chosen from a dendrogrammatic graph called the *reachability plot*. This is based on one of its two main parameters: the *reachability distance*. This expresses the smallest distance for an object p with respect to another object o, such that p is directly density-reachable from o if o is a core object. Intuitively, a core object is one that lies in the ϵ vicinity of n_{samples}. The reachability plot depicts the reachability distances for each object in the cluster ordering. Clusters within the data set are regions where the reachability distance between points are small, so they correspond to "valleys" within the reachability plot. The reachability plot is rather insensitive to ϵ and n_{samples}, but if ϵ is too small, then too many points will have an undefined reachability distance. In contrast to DBSCAN, OPTICS has difficulties when clustering homogeneous data [41].

1.4. Organization and Contribution of the Paper

The main part of our work is given by Section 2, where we elaborate on the nature of our cross-correlative approach before demonstrating how our iterative and interactive cross-study systematically leads to more stable meta-parameter settings on MNIST in Section 2.1. Secondly, we apply our procedure to the DeepVALVE time series data in Section 2.2. In Section 3 we study the visualizations generated by the found meta-parameters. Finally, in Section 4, we conclude on the range of visualization meta-parameters and their connection to unsupervised learning. The contributions of this work are

- autoencoder study on DeepVALVE data set
- cross-correlative study of embedding technologies
- procedure to gain manageable meta-parameter ranges
- visual analysis of autoencoder latency spaces

2. Cross-Correlative Study on Meta-Parameters

For our comparative meta study of dimension reduction algorithms, we define the meta-parameters θ_m to be

$$\theta_m := \bigcup_{i \in I} \theta_{m_i}, \tag{1}$$

where I is the space of values the individual meta-parameters θ_{m_i} may take, see Table A3. A meta-parameter set of a concrete visualization might be a k-dimensional vector embedded into a k-dimensional meta-parameter space. To elucidate this, considering multi-parameter visualization such as the radial visualization method introduced by [42], one faces a (meta-) parameter space k with $2n$ parameters ($k = 2n$), n being the number of data dimensions. Finding a good meta-parameter combination introduces generally an NP-hard issue to optimize the meta-parameters in k-dimensions (within the single algorithm regime). Thus, our working hypothesis states insight can be gained about θ_m by cross-studying θ_m from a multi-algorithmic point of view:

$$\theta_m \approx \theta_{m,\mathcal{A}} \quad := \bigcup_{i \in I;\, \mathcal{A}_j \in \mathcal{A}} \theta_{m_i,\mathcal{A}_j}, \tag{2}$$

where $\mathcal{A}$ denotes the set of algorithms and $\theta_{m_i,\mathcal{A}_j}$ denotes the m_i-th meta-parameter of algorithm $\mathcal{A}_j$. Doing so saves the trouble of solving the (k-dimensional) meta-parameter problem for one specific algorithm. Instead, we iter- and interactively tune $\theta_{m_i,\mathcal{A}_j}$ mutually to approach a valuable embedding and visual representation for the data in touch. Let $\mathcal{R}_{\text{method}}$ be the range for the cardinality of cluster centers with respect to one of the methodologies as quoted above. Then our evaluation results in a cross-correlative range matrix

$$\widehat{\mathcal{R}} := \begin{array}{c} \\ \text{t-SNE} \\ \text{UMAP} \\ \text{k-M} \\ \text{DBS} \\ \text{OPT} \end{array} \begin{array}{c} \begin{array}{ccccc} \text{t-SNE} & \text{UMAP} & \text{k-MEANS} & \text{DBSCAN} & \text{OPTICS} \end{array} \\ \begin{bmatrix} \mathcal{R}_{\text{t-SNE}} & \delta_{\text{t-SNE,UMAP}} & \delta_{\text{t-SNE,k-M}} & \delta_{\text{t-SNE,DBS}} & \delta_{\text{t-SNE,OPT}} \\ \delta_{\text{UMAP,t-SNE}} & \mathcal{R}_{\text{UMAP}} & \delta_{\text{UMAP,k-M}} & \delta_{\text{UMAP,DBS}} & \delta_{\text{UMAP,OPT}} \\ \delta_{\text{k-M,t-SNE}} & \delta_{\text{k-M,UMAP}} & \mathcal{R}_{\text{k-M}} & \delta_{\text{k-M,DBS}} & \delta_{\text{k-M,OPT}} \\ \delta_{\text{DBS,t-SNE}} & \delta_{\text{DBS,UMAP}} & \delta_{\text{DBS,k-M}} & \mathcal{R}_{\text{DBS}} & \delta_{\text{DBS,OPT}} \\ \delta_{\text{OPT,t-SNE}} & \delta_{\text{OPT,UMAP}} & \delta_{\text{OPT,k-M}} & \delta_{\text{OPT,DBS}} & \mathcal{R}_{\text{OPT}} \end{bmatrix} \end{array}. \tag{3}$$

Herein, $\delta_{i,j}$ denotes the intersection of the range of cluster center cardinalities for two methods i, j:

$$\delta_{i,j} = \mathcal{R}_i \bigcap \mathcal{R}_j . \tag{4}$$

By definition, the matrix in Equation (3) is symmetric around the diagonal. The goal is now to find the minimum of the $\delta_{i,j}$ to come as close to the true intrinsic dimension of the data manifold as possible.

2.1. MNIST

The MNIST data set is a well-known image data set containing the digitalization of around 60,000 handwritten digits from zero to nine. Many studies performed with this data set may be found in the literature [43,44]. Therefore, we omit any additional details of this data set except the fact that it is labeled, i.e., for each picture we know which digit is actually depicted. We start our analysis with the reachability plot for the OPTICS algorithm. For computational reasons we fix ϵ to 3.5, see Appendix C.1.

As shown on the right-hand side of Figure 2, no meaningful structures can be found for $n_{\text{samples}} < 15$ as all points are qualified as noise, which refines the order of magnitude mentioned in [41] for meaningful n_{samples}. The general features of the reachability plot itself are known to be stable under some (meaningful) variations of the meta-parameters ϵ and n_{samples} [41]. Valleys in this plot, as shown on the left-hand side of Figure 2, may be connected to clustered structures in the studied latency space as explained in Section 1.3. Tuning $\epsilon = 1.85$, i.e., the red dashed line in Figure 2, we can identify at least six independent structures at the same resolution scale. We also show other, rather poorly tuned values for ϵ, i.e., $\epsilon \in (1.50, 1.85, 2.50)$, indicated by the black dashed lines.

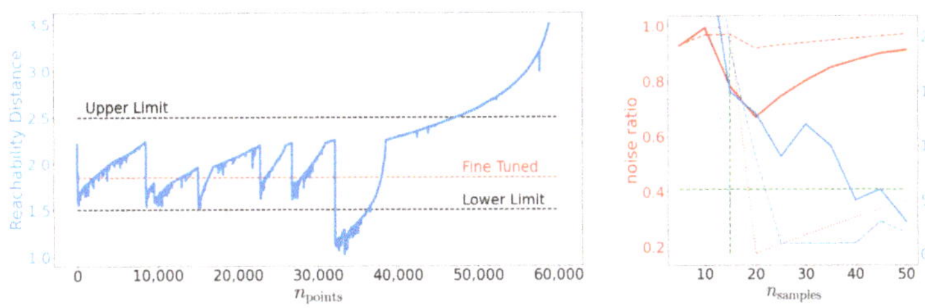

Figure 2. Left: Reachability plot for the OPTICS algorithm on MNIST. This plot is produced using $\epsilon = 3.5$ and $n_{\text{samples}} = 25$. **Right**: The number of identified noise points as well as the number of found clusters as function of n_{samples} for OPTICS. We display different selections with $\epsilon \in (1.50, 1.85, 2.50)$ indicated by the dashed, solid, and dotted line, respectively. The green dashed lines indicate the limits deduced so far.

To bolster this observation, we study the 2D embeddings as computed by t-SNE and UMAP in Figure 3. By eye we can see that both methods give a different perspective on the structure of the latent space. Using t-SNE alone we might identify between six and eleven structurally independent components. On the other hand, UMAP would provide us with six or maybe seven independent structures. Especially the derived *upper* bounds are very subjective. How should the gaps actually look to be counted as independent? At this point we see how the cross-correlative nature of our approach adds value. By now we have clearly established a lower limit of six cluster structures using Figure 2 (left) and Figure 3 (left and middle). In addition, we have limited $n_{\text{samples}} > 15$. At the right-hand side of Figure 2, we show the number of identified clusters as well as the noise ratio for OPTICS as a function of n_{samples} for different values of ϵ. We observe that it actually is the fine-tuned ϵ run which yields the best signal-to-noise ratio while simultaneously respecting the derived lower limits on n_{cluster}. Indicating the so far deduced boundaries by green dashed lines we can set an upper limit on the number of identified clusters. Again, we have settled for rather conservative boundaries by working with $n_{\text{samples}} > 15$. Using the best signal-to-noise ratio, both from Figures 2 and 3, yields $n_{\text{samples}} = 20$ and thus an upper bound of 13 clusters instead of 18. Using this knowledge, let us study the next embedding tool on our list: DBSCAN. As OPTICS and DBSCAN are closely related we can use the already identified values of ϵ and n_{samples} as starting points. This greatly reduces the meta-parameter space to be explored. Indeed, as we can see in Figure 3, DBSCAN favors slightly higher ϵ and lower values of n_{samples} than OPTICS. However, as OPTICS requires values for ϵ and n_{samples} high enough to not fall into the unstable regime, one should also choose n_{samples} for DBSCAN not too low. This "unstable" behavior can be observed also in Figure 3 for values of $n_{\text{samples}} < 15$. Hence we transfer the OPTICS limit to DBSCAN here and arrive at a fine-tuned limit of 11 clusters. So, in total we find

$$\begin{aligned} 11 &< n_{\text{clusters}} &< 18 \\ 1.5 &< \epsilon_{\text{OPTICS}} &< 2 \\ 15 &< n_{\text{samples, OPTICS}} &< 25 \\ 1.9 &< \epsilon_{\text{DBSCAN}} &< 2.2 \\ 15 &< n_{\text{samples, DBSCAN}} &< 20 \end{aligned} \tag{5}$$

Again, we emphasize that wherever necessary we use very conservative heuristics. Therefore, the suggested limits in Equation (5) capture the full structure of the latent representation as produced by our autoencoder.

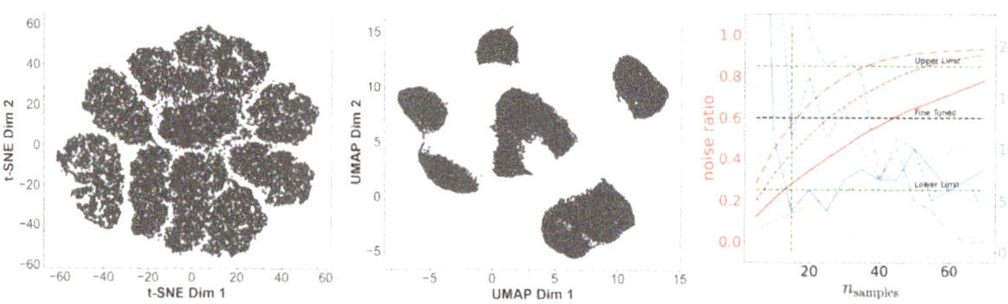

Figure 3. Left and **Middle**: Structure of the latent space distribution of MNIST as identified by the t-SNE respectively UMAP embedding. **Right**: n_{cluster} (blue) and noise ratio (red) as a function of n_{samples} with $\epsilon = 1.85$ (dashdot), $\epsilon = 2.0$ (dashed), $\epsilon = 2.2$ (solid), and $\epsilon = 2.5$ (dotted) for DBSCAN. The green dashed lines indicate the limits deduced so far.

2.2. DeepVALVE

The DeepVALVE data set consists of a series (in total around 25,000) of random opening and closing events of an industrial valve as described in [5]. A part of these events is shown in Figure 4.

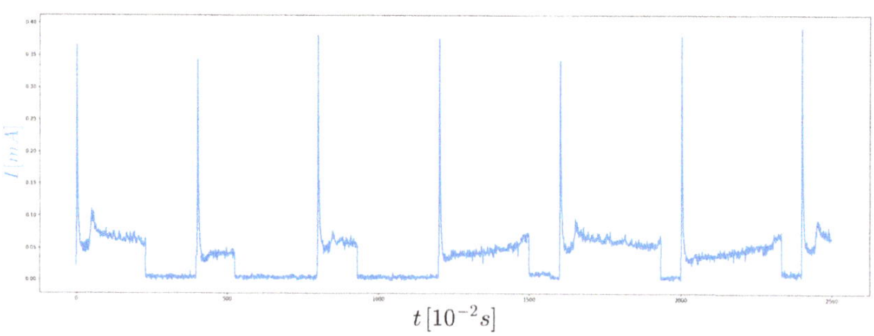

Figure 4. Part of the DeepVALVE time series data set: The blue line represents the measured electrical current driving the engine of the valve.

The allowed labels are: START, LOSE, LINEAR, STUCK, END. Thus, as in the case of MNIST, we have a completely labeled data set where we know the cluster cardinality beforehand, see Appendix D for more examples. As we deal with a time series data set, we must specify the way we feed our data to the neural network. Denoting our time series with $X_{0:T}$, we extract windows at time step t of window size w, i.e., $X_{t-w:t}$. A batch is then created by randomly sampling t. As in this case our latent space is three-dimensional, we are actually able to plot it. The found structure for $w = 10$ is shown in Figure 5. We observe an ellipsoidal structure which is typical for quasi-periodic structures, as indicated in [45]. This is not surprising regarding the recurring opening and closing events of the valve. Now we want to apply the investigative pipeline we developed in Section 2.1. Hence again we start with the OPTICS reachability plot in Figure 6. We can identify several bigger and smaller structures. The reachability graph yields at least three or even four and more structures.

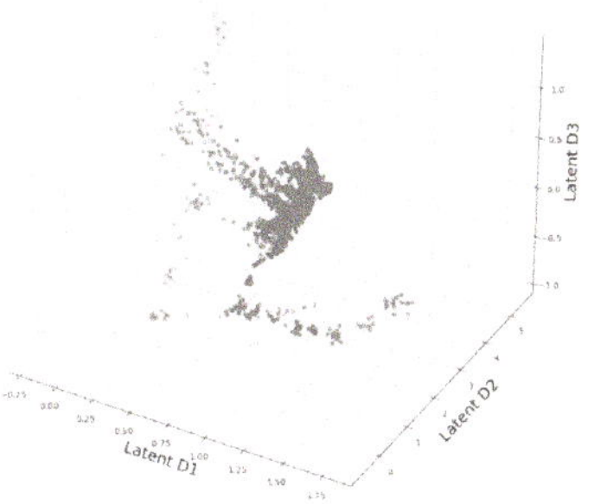

Figure 5. 3D presentation of the latent space of DeepVALVE dataset as computed by our autoencoder.

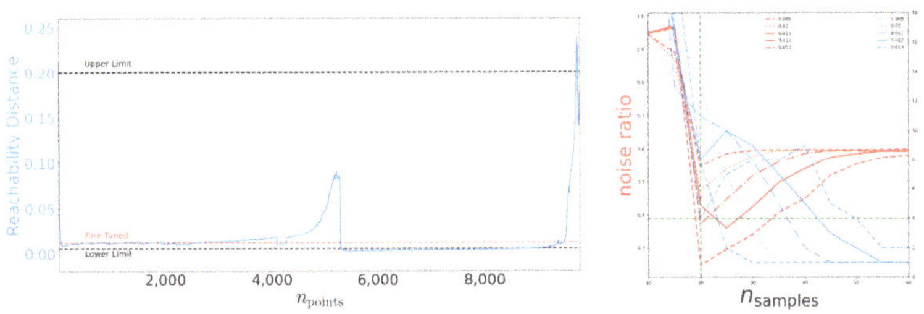

Figure 6. Left: Reachability plot for the OPTICS algorithm on DeepVALVE. This plot is produced using $\epsilon = 0.25$ and $n_{samples} = 25$. **Right**: Noise ratio and number of found cluster as deduced from the OPTICS reachability plot as a function of $n_{samples}$ with $\epsilon \in [0.009, 0.014]$.

Adding the knowledge of Figure 7 we can estimate the lower limit of identified structures as four. Following Section 2.1 one can estimate $n_{samples} > 20$ from the signal-to-noise ratio on the right-hand side of Figure 6. Again, we fine-tune ϵ using the reachability graph. We identify $\epsilon = 0.02$ using this optical procedure. On the right-hand side of Figure 6 we show runs with different fine-tuned ϵ values. Indeed, the visual tuning turns out to be not sensitive enough and the actual range for epsilon is rather in the range of 0.01. We use this figure to estimate the upper limit of identified clusters to be 13.

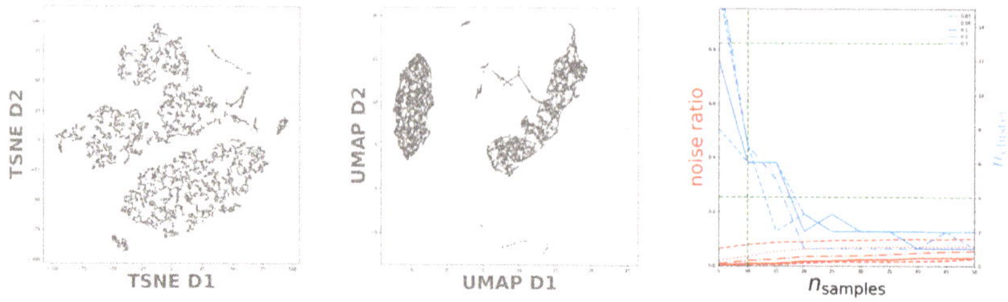

Figure 7. Left and **Middle**: Structure of the latent space distribution of DeepVALVE as identified by the t-SNE respectively UMAP embedding. **Right**: The number of identified noise points as well as the number of found clusters as function of n_{samples} for DBSCAN. We display different selections with $\epsilon \in (0.03, 0.06, 0.1, 0.2, 0.3)$ indicated by the dashed, dotted and solid line, respectively. The green dashed lines indicate the limits deduced so far.

In Figure 7 one can observe the (within the context of temporal data emerging) fact that outliers can be detected with UMAP more easily than with t-SNE [46,47]. In addition to that, UMAP also preserves global structures better than t-SNE, although there are more advanced methods such as dynamic t-SNE including a notion of temporal coherence that allows for better cluster separation [48]. Summing up, from the t-SNE plot, in view of cluster sizes and distances with no specific meaning, one can identify (conservatively estimated) 7 clusters. However, the UMAP plot in the middle of Figure 7 indicates around 5 clusters. Using the limits deduced so far we study DBSCAN on the right-hand side of Figure 7. As with MNIST we observe that DBSCAN prefers slightly different values for ϵ. So, in total we find

$$\begin{aligned} 4 &< n_{\text{clusters}} < 13 \\ 0.009 &< \epsilon_{\text{OPTICS}} < 0.013 \\ 20 &< n_{\text{samples, OPTICS}} < 50 \\ 0.03 &< \epsilon_{\text{DBSCAN}} < 0.3 \\ 10 &< n_{\text{samples, DBSCAN}} < 20 \end{aligned} \quad (6)$$

3. Visualization of Clustered Data

In Section 2 we estimated the meta-parameters of our benchmark data set MNIST and our testing case DeepVALVE within Equations (5) and (6) respectively. However, how does this help us to gain a better *visual* understanding of the data set under investigation? Using our set of meta-parameters, we can now study the t-SNE and UMAP embeddings for our OPTICS, DBSCAN and k-MEANS clustering methods to obtain a first grasp on how well the data are classified and separated within the latent space. From Equation (5) we chose settings as disclosed in Table 1.

Table 1. Meta-parameters used for the visualizations in Figure 8–11.

Method	MNIST	DeepVALVE
OPTICS	$\epsilon = 1.85$, $n_{\text{samples}} = 20$	$\epsilon = 0.012$, $n_{\text{samples}} = 25$
DBSCAN	$\epsilon = 2.0$, $n_{\text{samples}} = 20$	$\epsilon = 0.2$, $n_{\text{samples}} = 15$
K-MEANS	$n_{\text{clusters}} = 11$	$n_{\text{clusters}} = 6$

In Figure 8 we show the clusters found by OPTICS, DBSCAN, and k-MEANS projected onto the t-SNE embedding. We observe that both OPTICS and DBSCAN exhibit oversimplification as has already been visible in Figure 3. Additional structures are only indicated, as few points have been assigned to them. K-MEANS, however, though able to resolve much more substructure, tends also to split certain structures which the other methods clearly identified as belonging together. The reason is that the predefiniton of cluster cardinalities introduces some bias. We observe a similar behavior when using the UMAP embedding in Figure 9 instead. This provides us with the possibility of a direct comparison between t-SNE and UMAP embeddings, which is not possible a priori.

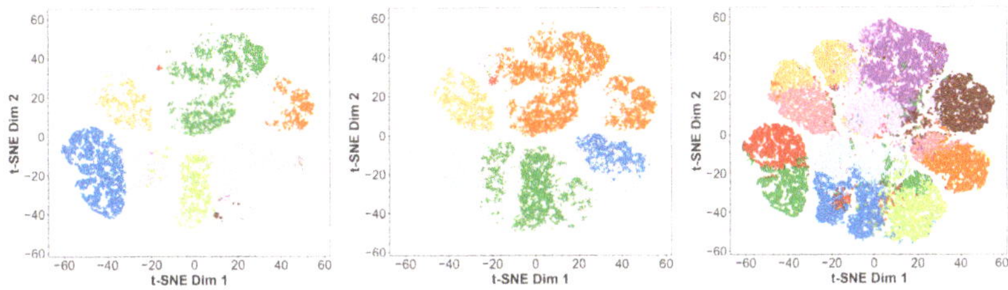

Figure 8. T-SNE embedding of the latent space of our MNIST autoencoder. In color the points belonging to identified cluster structures. From left to right: OPTICS, DBSCAN, k-MEANS.

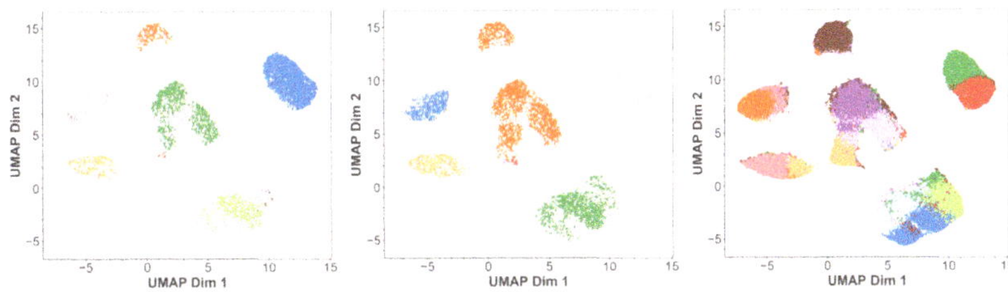

Figure 9. UMAP embedding of the latent space of our MNIST autoencoder. In color the points belonging to identified cluster structures. From left to right: OPTICS, DBSCAN, k-MEANS.

Let us now apply the same procedure to our test data set DeepVALVE. Again, using the values from Table 1 we project the found clusters onto the t-SNE, respectively the UMAP embeddings. In Figures 10 and 11 we can see real structural differences of the DeepVALVE dataset to the MNIST dataset, Figures 8 and 9. Figure 10 (left and middle) clearly reveals that OPTICS is much more sensitive to heterogeneities within the data.

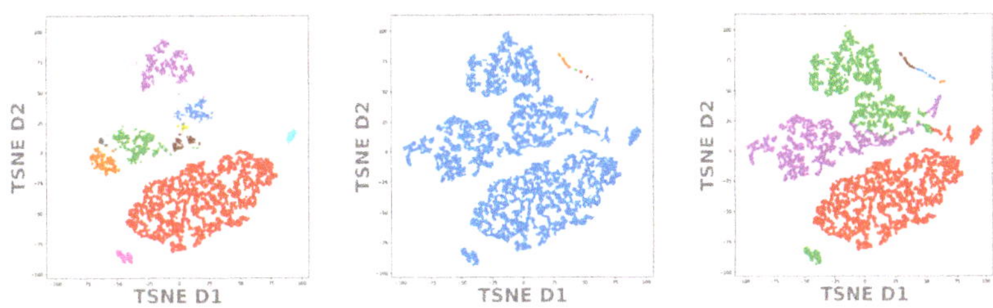

Figure 10. T-SNE embedding of the latent space of our DeepVALVE autoencoder. In color the points belonging to identified cluster structures. From left to right: OPTICS, DBSCAN, k-MEANS.

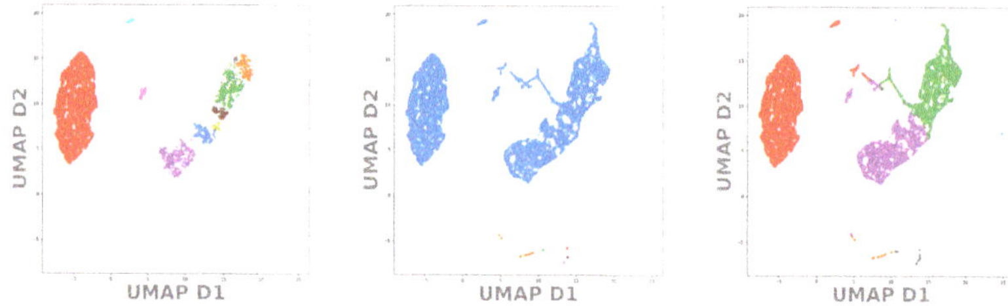

Figure 11. UMAP embedding of the latent space of our DeepVALVE autoencoder. In color the points belonging to identified cluster structures. From left to right: OPTICS, DBSCAN, k-MEANS.

This can be an advantage but also a disadvantage: As DeepVALVE is a huge data set with densely distributed points, density-based clustering methods—and especially OPTICS—find more clusters for smaller training sets. For DeepVALVE, We observed a huge difference between 10,000 and 60,000 points (10,000 depicted in Figure 10). The reason is that larger distributions become "filled in" the more samples are drawn from the true distribution. k-MEANS, on the other hand, constitutes a biased version of clustering, which reveals itself for the MNIST as well as for the DeepVALVE data set within the t-SNE as well as within the UMAP embedding. A comparison of Figures 10 and 11 reveals the main advantage claimed for UMAP in the literature: That it can depict and preserve (global) similarities better [49]. This is even more critical for time series than for image data, as time series segmentation often exhibits not as many labels as classification tasks for image data. Hence the procedural error by choosing wrong cluster cardinalities rises significantly. Thus, our pipeline involving the cross-correlative usage of clusterings and embeddings raises awareness of this fact as well as giving a first hint onto the scale at which cluster center cardinalities can be expected.

4. Conclusions

Summing up what we have done and learned so far, we can identify four main benefits of our approach:

(i) We developed a pipeline to obtain a visual grasp on the generalization capacity of a vanilla autoencoder.
(ii) We use clustering and embedding methods in a cross-correlative way to fine-tune their observational capabilities.
(iii) This cross-correlative ansatz allows better capture of the interrelation between the (transformed) data and the visualizations and embeddings.
(iv) Doing so, structural differences between data sets become apparent, which allows obtaining a first apprehension of an unknown data set without prior knowledge.

4.1. The Generalization Capacity vs. the Manifold Hypothesis

One should keep in mind the reason for investigating the latency space in this detailed fashion: We want to have a grasp on the generalization assumption. This is connected, but not identical to the manifold hypothesis as presented in the introduction. For both of our data sets we know the cluster center cardinalities beforehand and hence we can evaluate the individual performance of our clustering algorithms on the latent space. However, if this is not the case—which it should be for unsupervised learning tasks—our cross-correlative ansatz can give a first hint.

4.2. Meta-Parameter Fine-Tuning

In Equations (5) and (6) we present the results of our (visual) meta-parameter fine-tuning. Especially Figures 2 and 6 reveal how visual investigation ameliorates our results. Although these clustering and embedding methods work well within certain ranges of parameters, as e.g., Ref. [41] points out and investigates in detail for OPTICS, visual methods and their consecutive analysis can really suffer from poorly chosen meta-parameters. So, by working in a cross-correlative way one introduces a level of quantitivity that one would completely loose when restricting to one method.

4.3. Interrelation between Data and Methodology

In Figure 12 the latent space of the DeepVALVE dataset is investigated using our three different clustering methods, and one can clearly see that something goes wrong for OPTICS. So why is this the case? DeepVALVE is a dense temporal data set, and one would expect the clusters corresponding to the temporal labels to lie at the "edges" of the quasi-periodic structure depicted in Figure 12. However, unlike DBSCAN, OPTICS uses not a point value, but a hierarchical scale range for the reachability distance. Thus, if we have a really dense data set and comparatively few samples to estimate its distribution, it might identify large parts of the data set as noise. This can happen neither with DBSCAN nor k-MEANS. Henceforth, we have another demonstration that also visual methods should be taken with a grain of salt at least in the unsupervised case.

4.4. Structural Differences between Data Sets

In Sections 2.1 and 2.2 we studied two structurally different data sets with the same analysis pipeline as developed in Section 2. Although MNIST constitutes a 2*D* image dataset, DeepVALVE consists of temporal measurements of a physically non-trivial process and hence exhibits more structure, as depicted in Figure 4. This is clearly visible from the clustering parameters ϵ and n_{samples}, indicating DeepVALVE is a much denser data set than MNIST, as well as from the respective visualizations. Especially in Figure 8 to Figure 11 this shows itself, as discussed in Section 3.

4.5. Future Outlook and Comparison to Other Work

In [46] a deep convolutional autoencoder was used as a dimensionality reduction method for the subsequent 2*D* visualization using PCA, UMAP and t-SNE. They too developed a pipeline for a quantitative investigation; however, in contrast to our work, they did not use the visualization and embedding methods in a cross-correlative way. As our results indicate, e.g., in Figures 2 and 6, this adds value to the inter-correlated usage of density-based clustering methods. For future investigation, we plan to migrate our visual meta-parameter selection pipeline (partly) to the hyperparameter learning level. Especially the qualitative analyses in Figures 2 and 6 would profit from a deeper, quantitative treatment. Furthermore, we would like to investigate the conjunction between the cardinality of training samples necessary to obtain a "good" estimate on the data distribution and data density in a more sophisticated manner. Especially temporal data sets are prone to heterogeneities that even have physical meaning rather than just being clustering or embedding artefacts. Having performed this comprehensive study, we are keen to walk one step further on this road.

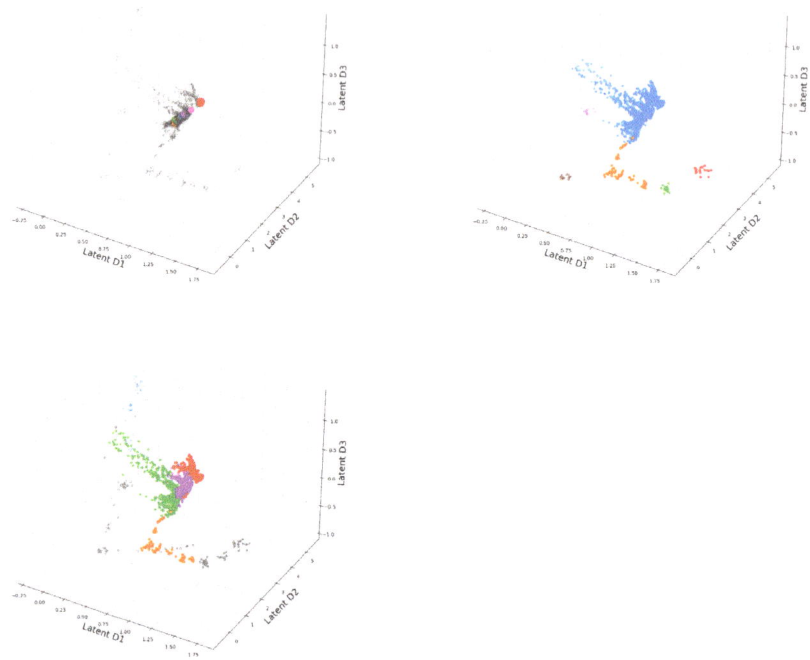

Figure 12. 3D presentation of the latent space of DeepVALVE dataset using OPTICS, DBSCAN, K-Means clustering, respectively.

Data Availability Statement: MNIST is available from http://yann.lecun.com/exdb/mnist/. DeepValve is a company-internal IAV dataset. It will be published in an anonymised fashion following this publication.

Appendix A. Autoencoder Hyperparameters and Architecture for Reproducibility

In Table A1 our choices for the autoencoder hyperparameters are listed. Please note that if not mentioned otherwise, the default values of PyTorch (Version 1.8.1) are used.

Table A1. The hyperparameters used for training our model.

Hyperparameter	Values
Learning Rate	0.001
Optimizer	Adam
Random Seed	0
Activation Function of hidden layers	ReLU
Activation Function of output layer	Sigmoid
Epochs	100
Batch Size	100
Loss	Mean Square Error

Table A2 summarizes the encoder-decoder structure of the autoencoder as well as the final validation loss.

Table A2. Architecture of the encoder chosen for the given data set and achieved validation loss. The architecture numbers represent the number of neurons per layer.

Data Set	Input Size	Architecture	L_{val}
MNIST	784	$400 \rightarrow 8$	2.16×10^{-2}
DeepVALVE	10	$16 \rightarrow 8 \rightarrow 3$	2.1×10^{-5}

Please note that the decoder is a mirror of the encoder. Therefore, we omitted the numbers in Table A2.

Appendix B. Meta-Parameter Default Values

Table A3. List of meta-parameters used in this study.

Embedding Method	Meta-Parameters Used and Their Default Values
t-SNE	$n_{components} = 2$, $random_{state} = 0$
UMAP	$n_{neighbors} = 15$, $min_{dist} = 0.1$
DBSCAN	$\epsilon = 0.5$, $n_{samples} = 5$
OPTICS	$\epsilon = 2.0$, $n_{samples} = 5$
K-Means	$n_{clusters} = 8$, init = 'random', $n_{init} = 20$, $iter_{max} = 300$, tol $= 1 \times 10^{-4}$, $random_{state} = 0$

Appendix C. Additional Material for MNIST

Appendix C.1. Reachability Plots

In Figure A1 we show additional plots using different values for $n_{samples}$ and ϵ.

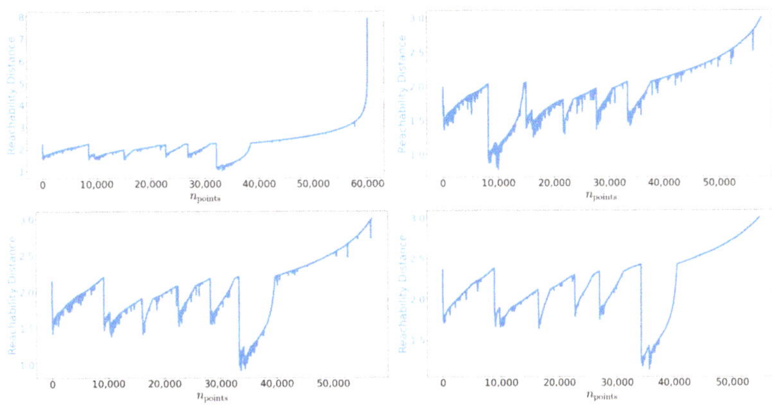

Figure A1. OPTICS reachability plot for MNIST using. Left upper: $\epsilon = \infty$ and $n_{samples} = 25$. Right upper: $\epsilon = 3.0$ and $n_{samples} = 15$. Left lower: $\epsilon = 3.0$ and $n_{samples} = 20$. Right lower: $\epsilon = 3.0$ and $n_{samples} = 35$.

As stated in [41] the key features of this plot are rather stable against different choices of the meta-parameters.

Appendix C.2. Reconstructed Digits

For MNIST we can qualitatively check the identified structures. For all three clustering approaches we construct a cluster center. For k-MEANS this is done automatically by the algorithm. On the other hand, for OPTICS and DBSCAN we just use the center of mass of all points belonging to a given cluster. We then reconstruct the images by sending these points through the decoder.

In Figure A2 we present the reconstructions corresponding to the right-hand side of Figure 8, respectively Figure 9 in the main text. We observe that indeed most of the digits could be identified. However, digit 4 is missing, while digit 1 and 9 are doubled. A behavior we already observed in Section 2.1.

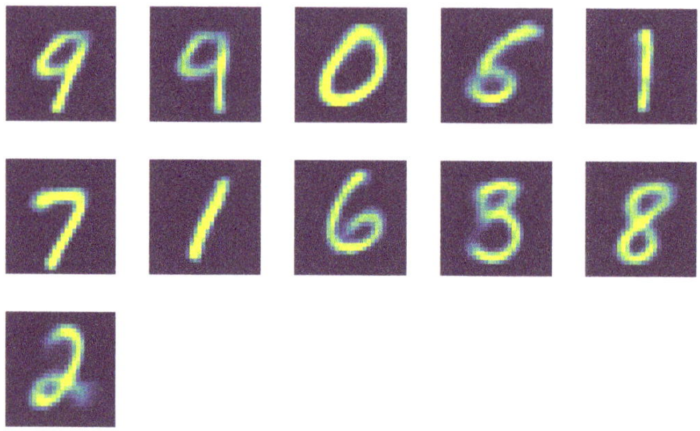

Figure A2. Reconstructed images of the centroids of the cluster using K-Means clustering with $n_{\text{clusters}} = 11$.

Once we increase the allowed number of clusters to $n_{\text{clusters}} = 18$, as shown in Figure A3, we observe that now all digits are present. However, we also have quite some doubling in digits 0 to 4.

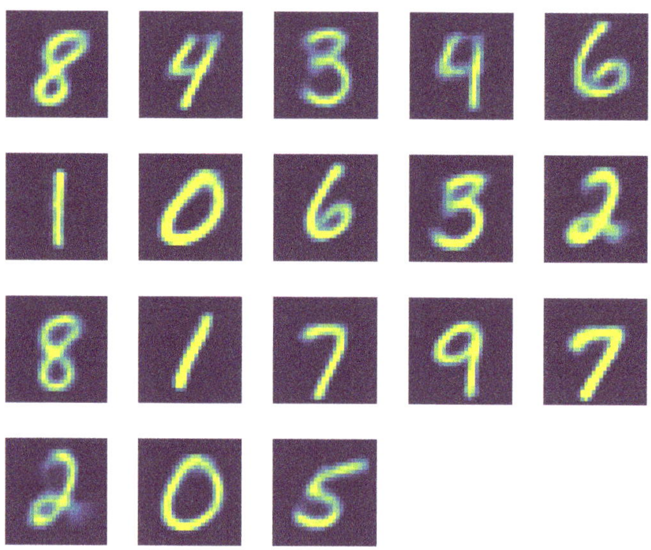

Figure A3. Reconstructed images of the centroids of the cluster using K-Means clustering with $n_{\text{clusters}} = 18$.

As displayed in Figure A4 a similar behavior emerges when we use DBSCAN instead. Using the values from Table 1 we recover most digits except 8 and 9. Again for the other digits we have several clusters they belong to.

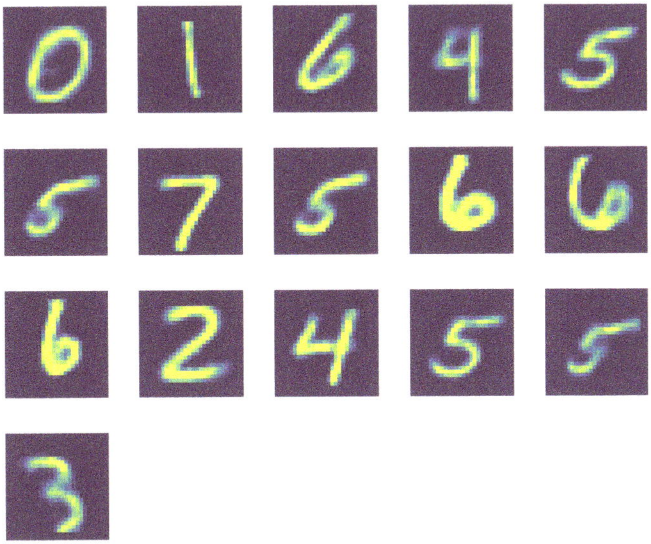

Figure A4. Reconstructed images of the centroids of the cluster using DBSCAN clustering with $\epsilon = 2.0$ and $n_{\text{samples}} = 20$.

Finally in Figure A5 we show the reconstructed digits for OPTICS. Again, we observe missing digits, 3 and 5 this time, as well as two versions of 4.

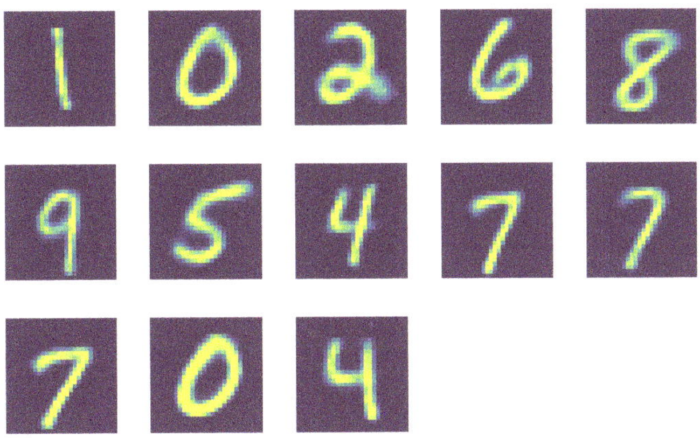

Figure A5. Reconstructed images of the centroids of the cluster using OPITCS clustering with $\epsilon = 1.85$ and $n_{\text{samples}} = 20$.

Interestingly, k-MEANS has trouble locating different digits when compared to OPTICS and DBSCAN. The latter two behave rather similar again.

Appendix D. Additional Material for DeepVALVE

In Figure A6 we show additional labeled data samples from [5].

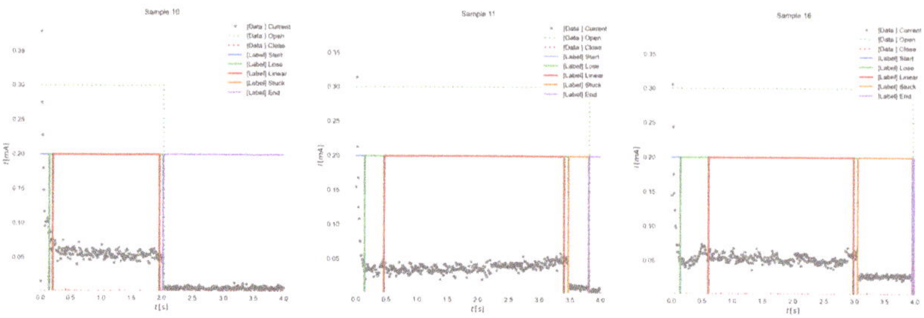

Figure A6. Additional labeled data samples for DeepVALVE.

References

1. Donoho, D.L. High-dimensional data analysis: The curses and blessings of dimensionality. In Proceedings of the AMS Conference on Math Challenges of the 21st Century, Los Angeles, CA, USA, 7–12 August 2000.
2. Sembiring, R.W.; Mohamad Zain, J.; Abdullah, E. Dimension Reduction of Health Data Clustering. *arXiv* **2011**, arXiv:1110.3569.
3. Chen, Y.; Tang, S.; Bouguila, N.; Wang, C.; Du, J.; Li, H. A Fast Clustering Algorithm based on pruning unnecessary distance computations in DBSCAN for High-Dimensional Data. *Pattern Recognit.* **2018**, *83*. [CrossRef]
4. Deng, L. The mnist database of handwritten digit images for machine learning research. *IEEE Signal Process. Mag.* **2012**, *29*, 141–142. [CrossRef]
5. Ahmed, S.; Schichtel, P.; von der Ohe, T. Sensorlose Prozesse mit kuenstlicher Intelligenz erfassen und steuern. *MTZextra* **2018**, *23*, 42–45. [CrossRef]
6. Rumelhart, D.; Hinton, G.; Williams, R. *Parallel Distributed Processing. Volume 1: Foundations*; MIT Press: Cambridge, UK, 1986; Chapter Learning Internal Representations by Error Propagation.
7. Bank, D.; Koenigstein, N.; Giryes, R. Autoencoders. *arXiv* **2021**, arXiv:2003.05991.
8. Rossler, A.; Cozzolino, D.; Verdoliva, L.; Riess, C.; Thies, J.; Niessner, M. FaceForensics++: Learning to Detect Manipulated Facial Images. In Proceedings of the IEEE/CVF International Conference on Computer Vision (ICCV), Seoul, Korea, 27 October–2 November 2019.
9. Zhang, Y.; Lee, K.; Lee, H. Augmenting Supervised Neural Networks with Unsupervised Objectives for Large-scale Image Classification. In Proceedings of the 33rd International Conference on Machine Learning, New York, NY, USA, 19–24 June 2016; Balcan, M.F., Weinberger, K.Q., Eds.; PMLR: New York, NY, USA, 2016; Volume 48, pp. 612–621.
10. Zhang, B.; Qian, J. Autoencoder-based unsupervised clustering and hashing. *Appl. Intell.* **2021**, *51*, 493–505. [CrossRef]
11. Li, X.; Zhang, T.; Zhao, X.; Yi, Z. Guided autoencoder for dimensionality reduction of pedestrian features. *Appl. Intell.* **2020**, *50*, 4557–4567. [CrossRef]
12. Ferreira, D.; Silva, S.; Abelha, A.; Machado, J. Recommendation System Using Autoencoders. *Appl. Sci.* **2020**, *10*, 5510. [CrossRef]
13. Takeishi, N.; Kalousis, A. Physics-Integrated Variational Autoencoders for Robust and Interpretable Generative Modeling. *arXiv* **2021**, arXiv:2102.13156.
14. Goodfellow, I.; Bengio, Y.; Courville, A. *Deep Learning*; MIT Press: Cambridge, MA, USA, 2016. Available online: http://www.deeplearningbook.org (accessed on 7 May 2021).
15. Lee, W.; Ortiz, J.; Ko, B.; Lee, R.B. Time Series Segmentation through Automatic Feature Learning. *arXiv* **2018**, arXiv:1801.05394.
16. Dunteman, G.H. *Principal Component Analysis*; SAGE Publications: Thousand Oaks, CA, USA, 1989.
17. Fefferman, C.; Mitter, S.; Narayanan, H. Testing the Manifold Hypothesis. *arXiv* **2013**, arXiv:1310.0425v2.
18. Ryck, T.D.; Vos, M.D.; Bertrand, A. Change Point Detection in Time Series Data using Autoencoders with a Time-Invariant Representation. *arXiv* **2021**, arXiv:2008.09524.
19. Lapuschkin, S.; Wäldchen, S.; Binder, A.; Montavon, G.; Samek, W.; Müller, K.R. Unmasking Clever Hans Predictors and Assessing What Machines Really Learn. *Nat. Commun.* **2019**, *10*. [CrossRef]
20. Moor, M.; Horn, M.; Rieck, B.; Borgwardt, K.M. Topological Autoencoders. *arXiv* **2019**, arXiv:1906.00722.
21. Pihlgren, G.G.; Sandin, F.; Liwicki, M. Improving Image Autoencoder Embeddings with Perceptual Loss. *arXiv* **2020**, arXiv:2001.03444.

22. Zhu, Q.; Zhang, R. A Classification Supervised Auto-Encoder Based on Predefined Evenly-Distributed Class Centroids. *arXiv* **2020**, arXiv:1902.00220.
23. Chel, S.; Gare, S.; Giri, L. Detection of Specific Templates in Calcium Spiking in HeLa Cells Using Hierarchical DBSCAN: Clustering and Visualization of CellDrug Interaction at Multiple Doses. In Proceedings of the 2020 42nd Annual International Conference of the IEEE Engineering in Medicine Biology Society (EMBC), Montreal, QC, Canada, 20–24 July 2020; pp. 2425–2428. [CrossRef]
24. Cai, T.T.; Ma, R. Theoretical Foundations of t-SNE for Visualizing High-Dimensional Clustered Data. *arXiv* **2021**, arXiv:2105.07536.
25. Swetha, S.; Kuehne, H.; Rawat, Y.S.; Shah, M. Unsupervised Discriminative Embedding for Sub-Action Learning in Complex Activities. *arXiv* **2021**, arXiv:2105.00067.
26. Lehmann, D.J.; Theisel, H. Orthographic Star Coordinates. *IEEE Trans. Vis. Comput. Graph.* **2013**, *19*, 2615–2624. [CrossRef]
27. Sánchez, A.; Soguero-Ruiz, C.; Mora-Jimenez, I.; Rivas-Flores, F.J.; Lehmann, D.J.; Rubio-Sánchez, M. Scaled radial axes for interactive visual feature selection: A case study for analyzing chronic conditions. *Expert Syst. Appl.* **2018**, *100*, 182–196. Available online: https://www.sciencedirect.com/science/article/pii/S0957417418300617 (accessed on 12 May 2021). [CrossRef]
28. Rubio-Sánchez, M.; Sanchez, A.; Lehmann, D.J. Adaptable Radial Axes Plots for Improved Multivariate Data Visualization. *Comput. Graph. Forum* **2017**, *36*, 389–399. [CrossRef]
29. Shao, L.; Mahajan, A.; Schreck, T.; Lehmann, D.J. Interactive Regression Lens for Exploring Scatter Plots. *Comput. Graph. Forum* **2017**, *36*, 157–166. [CrossRef]
30. Wang, Y.; Li, J.; Nie, F.; Theisel, H.; Gong, M.; Lehmann, D.J. Linear Discriminative Star Coordinates for Exploring Class and Cluster Separation of High Dimensional Data. *Comput. Graph. Forum* **2017**, *36*, 401–410. [CrossRef]
31. Lehmann, D.J.; Theisel, H. The LloydRelaxer: An Approach to Minimize Scaling Effects for Multivariate Projections. *IEEE Trans. Vis. Comput. Graph.* **2017**. [CrossRef]
32. Lehmann, D.J.; Theisel, H. General Projective Maps for Multidimensional Data Projection. *Comput. Graph. Forum* **2016**, *35*, 443–453. [CrossRef]
33. Lehmann, D.J.; Theisel, H. Optimal Sets of Projections of High-Dimensional Data. *IEEE Trans. Vis. Comput. Graph.* **2015**. [CrossRef]
34. Karer, B.; Hagen, H.; Lehmann, D. Insight Beyond Numbers: The Impact of Qualitative Factors on Visual Data Analysis. *IEEE Trans. Vis. Comput. Graph.* **2020**. [CrossRef]
35. Rubio-Sánchez, M.; Lehmann, D.; Sanchez, A.; Rojo Álvarez, J. Optimal Axes for Data Value Estimation in Star Coordinates and Radial Axes Plots. *Comput. Graph. Forum* **2021**, *40* . [CrossRef]
36. Pezzotti, N.; Lelieveldt, B.; van der Maaten, L.; Höllt, T.; Eisemann, E.; Vilanova, A. Approximated and User Steerable tSNE for Progressive Visual Analytics. *IEEE Trans. Vis. Comput. Graph.* **2016**, *23*, 1739–1752. [CrossRef] [PubMed]
37. Hinton, G.; Roweis, S. Stochastic Neighbor Embedding. *Neural Inf. Process. Syst.* **2002**, 857–864.
38. McInnes, L.; Healy, J.; Melville, J. UMAP: Uniform Manifold Approximation and Projection for Dimension Reduction. *arXiv* **2020**, arXiv:1802.03426.
39. MacQueen, J. Some Methods for Classification and Analysis of Multivariate Observations. *Proc. Berkeley Symp. Math. Stat. Probab.* **1967**, *1*, 281–297.
40. Ester, M.; Kriegel, H.P.; Sander, J.; Xu, X. *A Density-Based Algorithm for Discovering Clusters in Large Spatial Databases with Noise*; Simoudis, E.; Han, J.; Fayyad, U.M., Eds.; KDD; AAAI Press: Palo Alto, CA, USA , 1996; pp. 226–231.
41. Ankerst, M.; Breunig, M.M.; peter Kriegel, H.; Sander, J. *OPTICS: Ordering Points To Identify the Clustering Structure*; ACM Press: New York, NY, USA, 1999; pp. 49–60.
42. Hoffman, P.; Grinstein, G.; Marx, K.; Grosse, I.; Stanley, E. DNA visual and analytic data mining. In Proceedings of the Visualization '97 (Cat. No. 97CB36155), Phoenix, AZ, USA, 19–24 October 1997; pp. 437–441. [CrossRef]
43. Shamsuddin, M.R.; Rahman, S.; Mohamed, A. Exploratory Analysis of MNIST Handwritten Digit for Machine Learning Modelling. In Proceedings of the 4th International Conference on Soft Computing in Data Science, SCDS 2018, Bangkok, Thailand, 15–16 August 2018; Springer: Singapore, 2019; pp. 134–145. [CrossRef]
44. Schott, L.; Rauber, J.; Brendel, W.; Bethge, M. Robust Perception through Analysis by Synthesis. *arXiv* **2018**, arXiv:1805.09190.
45. Tralie, C.J.; Perea, J.A. (Quasi)Periodicity Quantification in Video Data, Using Topology. *arXiv* **2017**, arXiv:1704.08382.
46. Ali, M.; Jones, M.W.; Xie, X. TimeCluster: Dimension reduction applied to temporal data for visual analytics. *Vis. Comput.* **2019**, *35*, 1013–1026. [CrossRef]
47. Ali, M.; Alqahtani, A.; Jones, M.W.; Xie, X. Clustering and Classification for Time Series Data in Visual Analytics: A Survey. *IEEE Access* **2019**, *7*, 181314–181338. [CrossRef]
48. Rauber, P.E.; Falcão, A.X.; Telea, A.C. *Visualizing Time-Dependent Data Using Dynamic t-SNE*; Bertini, E., Elmqvist, N., Wischgoll, T., Eds.; EuroVis 2016—Short Papers; The Eurographics Association: Aire-la-Ville, Switzerland, 2016; [CrossRef]
49. Vernier, E.F.; Garcia, R.; da Silva, I.P.; Comba, J.L.D.; Telea, A.C. Quantitative Evaluation of Time-Dependent Multidimensional Projection Techniques. *arXiv* **2020**, arXiv:2002.07481.

engineering proceedings

MDPI

Proceedings

Kernel Two-Sample and Independence Tests for Nonstationary Random Processes †

Felix Laumann [1,*], Julius von Kügelgen [2,3] and Mauricio Barahona [1]

1 Department of Mathematics, Imperial College London, London SW7 2AZ, UK; m.barahona@imperial.ac.uk
2 MPI for Intelligent Systems, Max-Planck-Ring 4, 72076 Tübingen, Germany; julius.von.kuegelgen@tuebingen.mpg.de
3 Department of Engineering, University of Cambridge, Cambridge CB2 1TN, UK
* Correspondence: f.laumann18@imperial.ac.uk

† Presented at the 7th International Conference on Time Series and Forecasting, Gran Canaria, Spain, 19–21 July 2021.

Abstract: Two-sample and independence tests with the kernel-based MMD and HSIC have shown remarkable results on i.i.d. data and stationary random processes. However, these statistics are not directly applicable to nonstationary random processes, a prevalent form of data in many scientific disciplines. In this work, we extend the application of MMD and HSIC to nonstationary settings by assuming access to independent realisations of the underlying random process. These realisations—in the form of nonstationary time-series measured on the same temporal grid—can then be viewed as i.i.d. samples from a multivariate probability distribution, to which MMD and HSIC can be applied. We further show how to choose suitable kernels over these high-dimensional spaces by maximising the estimated test power with respect to the kernel hyperparameters. In experiments on synthetic data, we demonstrate superior performance of our proposed approaches in terms of test power when compared to current state-of-the-art functional or multivariate two-sample and independence tests. Finally, we employ our methods on a real socioeconomic dataset as an example application.

Keywords: two-sample test; independence test; random process; nonstationary; kernel methods

Citation: Laumann, F.; von Kügelgen, J.; Barahona, M. Kernel Two-Sample and Independence Tests for Nonstationary Random Processes. *Eng. Proc.* **2021**, 5, 31. https://doi.org/10.3390/engproc2021005031

Published: 30 June 2021

1. Introduction

Nonstationary processes are the rule rather than the exception in many scientific disciplines such as epidemiology, biology, sociology, economics, or finance. In recent years, there has been a surge of interest in the analysis of problems described by large sets of interrelated variables with few observations over time, often involving complex nonlinear and nonstationary behaviours. Examples of such problems include the longitudinal spread of obesity in social networks [1], disease modelling from time-varying inter- and intracellular relationships [2], behavioural responses to losses of loved ones within social groups [3], and the linkage between climate change and the global financial system [4]. All such analyses rely on the statistical assessment of the similarity between, or the relationship amongst, noisy time series that exhibit temporal memory. Therefore, the ability to test the statistical significance of homogeneity and dependence between random processes that cannot be assumed to be independent and identically distributed (i.i.d.) is of fundamental importance in many fields.

Kernel-based methods provide a popular framework for homogeneity and independence tests by embedding probability distributions in RKHS [5] (Section 2.2). Of particular interest are the kernel-based two-sample statistic MMD (MMD) [6], which is used to assess whether two samples were drawn from the same distribution, hence testing for *homogeneity*; and the related HSIC (HSIC) [7], which is used to assess dependence between two random variables, thus testing for *independence*. These methods are nonparametric, i.e., they do not make any assumptions on the underlying distribution or the type of dependence.

However, in their original form, both MMD and HSIC assume access to a sample of i.i.d. observations—an assumption that is often violated for temporally-dependent data such as random processes.

Extensions of MMD and HSIC to random processes have been proposed [8,9]. Yet, these methods require the random process to be *stationary*, meaning that its distribution does not change over time. While it is sometimes possible to approximately achieve stationarity with preprocessing techniques such as (seasonal) differencing or square root and power transformations, such approaches become cumbersome and notoriously difficult, particularly with large sets of variables. The stationarity assumption can therefore pose severe limitations in many application areas where multiple nonstationary processes must be taken into consideration. When studying the relationships of climate change to the global financial system, for example, factors such as greenhouse gas emissions, stock market indices, government spending, and corporate profits would have to be transformed or assumed to be stationary over time.

In this paper, we show how the kernel-based statistics MMD and HSIC can be applied to *nonstationary* random processes. At the heart of our proposed approach is the simple, yet effective idea that realisations of a random process in the form of temporally-dependent measurements (i.e., the observed time series) can be viewed as independent samples from a multivariate probability distribution, provided that they are observed at the same points in time, i.e., over the same temporal grid. Then, MMD and HSIC can be applied on these distributions to test for homogeneity and independence, respectively.

The remainder of this paper is structured as follows. After discussing related work in Section 2, we introduce our applications of two-sample and independence testing with MMD and HSIC to nonstationary random processes in Section 3. We then carry out experiments on multiple synthetic datasets in Section 4 and demonstrate that the proposed tests have higher power compared with current functional or multivariate two-sample and independence tests under the same conditions. We provide an example application of our proposed methods to a socioeconomic dataset in Section 5 and conclude the paper with a brief discussion in Section 6.

2. Related Work

Two-sample and independence tests on stochastic processes have been widely studied in recent years. Under the stationarity assumption, ref. [8] investigate how the kernel cross-spectral density operator may be used to test for independence, and [9] formulate a wild bootstrap-based approach for both two-sample and independence tests, which outperforms [8] in various experiments. The wild bootstrap in [9] approximates the null hypothesis H_0 by assuming there exists a time lag τ such that a pair of measurements at any point in time t, $(x_i, y_i)_t$, is independent of $(x_i, y_i)_{t\pm s}$ for $s \geq \tau$. This method is applicable to test for instantaneous homogeneity and independence in stationary processes but requires further assumptions to investigate noninstantaneous cases: a maximum lag $M \leq \tau$ must be defined as the largest absolute lag for the test. This results in multiple hypothesis testing requiring adjustment by a Bonferroni correction. Further, ref. [10] have applied *distance correlation* [11], a HSIC-related statistic, to independence testing on stationary random processes.

Beyond the stationarity assumption, two-sample testing in the functional data analysis literature has mostly focused on differences of mean [12] or covariance structures [13,14]. However, ref. [15] have developed a two-sample test for *distributions* based on generalisations of a finite-dimensional test by utilising functional principal component analysis, and [16] have derived kernels over functions to be used with MMD for the two-sample test. Independence testing for functional data using kernels was recently proposed in [17] but assumes the samples lie on a finite-dimensional subspace of the function space—an assumption not required in our work. Moreover, ref. [18] have developed computationally efficient methods to test for independence on high-dimensional distributions and large sample sizes by using eigenvalues of centred kernel matrices to approximate the distribution under the null hypothesis H_0 instead of simulating a large number of permutations.

3. MMD and HSIC for Nonstationary Random Processes

3.1. Notation and Assumptions

Let $\{\mathbf{X}_t\}$ and $\{\mathbf{Y}_t\}$ denote two nonstationary stochastic processes with probability laws $\mathbb{P}_\mathbf{X}$ and $\mathbb{P}_\mathbf{Y}$, respectively. We assume that we observe m independent realisations of $\{\mathbf{X}_t\}$ and n independent realisations of $\{\mathbf{Y}_t\}$ in the form of time series measured at $T_\mathbf{X}$ and $T_\mathbf{Y}$ time points, respectively. Said differently, the data samples $\mathbf{X} = \{\mathbf{x}_i\}_{i=1}^m \overset{\text{i.i.d.}}{\sim} \mathbb{P}_\mathbf{X}$ are a set of nonstationary time series, $\mathbf{x}_i = \{x_{i,1}, \dots, x_{i,T_\mathbf{X}}\}$, arriving over the same temporal grid, and similarly for $\mathbf{Y} = \{\mathbf{y}_i\}_{i=1}^n \overset{\text{i.i.d.}}{\sim} \mathbb{P}_\mathbf{Y}$ with $\mathbf{y}_i = \{y_{i,1}, \dots, y_{i,T_\mathbf{Y}}\}$. Note that the measurements $x_{i,t}$ and $y_{i,t}$ are not independent across time (we use the terms 'sample' and 'realisation' interchangeably to denote $\mathbf{x}_i$ and $\mathbf{y}_i$ and the term 'measurement' to denote the temporally dependent vectors $x_{i,t}$ and $y_{i,t}$).

We may view the realisations $\mathbf{x}_i$ and $\mathbf{y}_i$ as samples of multivariate probability distributions of dimension $T_\mathbf{X}$ and $T_\mathbf{Y}$, respectively, which are independent at any given point in time, i.e., $x_{i,t} \perp\!\!\!\perp x_{j,t}$ and $y_{i,t} \perp\!\!\!\perp y_{j,t}$ $\forall t$ and $\forall i \neq j$. Consequently, we can represent these distributions by their mean embeddings $\mu_\mathbf{X}$ and $\mu_\mathbf{Y}$ in reproducing kernel Hilbert spaces (RKHSs) and use these to conduct kernel-based two-sample and independence tests. Given a characteristic kernel k, i.e., the mean embedding μ captures all information of a distribution $\mathbb{P}$ [19], the dependence between measurements in time is captured by the ordering of the variables, and the fact that any characteristic kernel k is injective, thus guaranteeing a unique mapping of any probability distribution into a RKHS [20].

For homogeneity testing ($\mathbb{P}_\mathbf{X} \overset{?}{=} \mathbb{P}_\mathbf{Y}$), we use the kernel-based MMD statistic and require equal number of measurements $T = T_\mathbf{X} = T_\mathbf{Y}$ but allow different sample sizes, $m \neq n$. For independence testing ($\mathbb{P}_\mathbf{XY} \overset{?}{=} \mathbb{P}_\mathbf{X}\mathbb{P}_\mathbf{Y}$), we employ the related HSIC, and in this case number of measurements can differ, but we require the same number of realisations, $m = n$. We now describe how two-sample and independence tests can be performed under these assumptions.

3.2. MMD for Nonstationary Random Processes

Let $k : \mathbb{R}^T \times \mathbb{R}^T \to \mathbb{R}$ be a characteristic kernel, such as the Gaussian kernel $k(x,y) = \exp\left(-\|x-y\|^2/\sigma^2\right)$, which uniquely maps $\mathbb{P}_\mathbf{X}$ and $\mathbb{P}_\mathbf{Y}$ to their associated RKHS $\mathcal{H}_k$ via the mean embeddings $\mu_\mathbf{X} := \int k(\mathbf{x}, \cdot)\, d\mathbb{P}_\mathbf{X}(\mathbf{x})$ and $\mu_\mathbf{Y} := \int k(\mathbf{y}, \cdot)\, d\mathbb{P}_\mathbf{Y}(\mathbf{y})$ [5] (Section 2.1). The MMD between $\mathbb{P}_\mathbf{X}$ and $\mathbb{P}_\mathbf{Y}$ in $\mathcal{H}_k$ is defined as [6]:

$$\text{MMD}^2(\mathcal{H}_k, \mathbb{P}_\mathbf{X}, \mathbb{P}_\mathbf{Y}) := \|\mu_\mathbf{X} - \mu_\mathbf{Y}\|^2_{\mathcal{H}_k} \geq 0, \quad \text{with equality iff} \quad \mathbb{P}_\mathbf{X} = \mathbb{P}_\mathbf{Y}. \tag{1}$$

Given samples $\mathbf{X}$ and $\mathbf{Y}$, $\text{MMD}^2(\mathcal{H}_k, \mathbb{P}_\mathbf{X}, \mathbb{P}_\mathbf{Y})$ can then be approximated by the following unbiased estimator [6]:

$$\widehat{\text{MMD}}_u^2(\mathcal{H}_k, \mathbf{X}, \mathbf{Y}) = \sum_{i=1}^{m}\sum_{j\neq i}^{m} \frac{k(\mathbf{x}_i, \mathbf{x}_j)}{m(m-1)} + \sum_{i=1}^{n}\sum_{j\neq i}^{n} \frac{k(\mathbf{y}_i, \mathbf{y}_j)}{n(n-1)} - 2\sum_{i=1}^{m}\sum_{j=1}^{n} \frac{k(\mathbf{x}_i, \mathbf{y}_j)}{mn}. \tag{2}$$

Henceforth, we drop the implied $\mathcal{H}_k$ for ease of notation.

Using $\widehat{\text{MMD}}_u^2(\mathbf{X}, \mathbf{Y})$ as a test statistic, one can construct a statistical two-sample test for the null hypothesis $H_0 : \mathbb{P}_\mathbf{X} = \mathbb{P}_\mathbf{Y}$ against the alternative hypothesis $H_1 : \mathbb{P}_\mathbf{X} \neq \mathbb{P}_\mathbf{Y}$ [21].

Let α be the significance level of the test, i.e., the maximum allowable probability of falsely rejecting H_0 and hence an upper bound on the type-I error. Given α, the threshold c_α for the test statistic can be approximated with a permutation test as follows. We first generate P randomly permuted partitions of the set of all realisations $\mathbf{X} \cup \mathbf{Y}$ with sizes commensurate with $(\mathbf{X}, \mathbf{Y})$, denoted $(\mathbf{X}_p, \mathbf{Y}_p)$, $p = 1, \dots, P$. We then compute $\widehat{\text{MMD}}_u^2(\mathbf{X}_p, \mathbf{Y}_p)$, $\forall p$, and sort the results in descending order. Finally, we select the statistic at position $(1-\alpha) \times P$

as our empirical threshold $\hat{c}_\alpha$. The null hypothesis H_0 is then rejected if $\widehat{\text{MMD}}_u^2(\mathbf{X},\mathbf{Y}) > \hat{c}_\alpha$. For a computationally less expensive (but generally less accurate) option, the inverse cumulative density function of the Gamma distribution can be computed to approximate the null distribution [22].

3.3. HSIC for Nonstationary Random Processes

Let $\mathbb{P}_{\mathbf{XY}}$ denote the joint distribution of $\{\mathbf{X}_t\}$ and $\{\mathbf{Y}_t\}$, and let $\mathcal{H}_k$ and $\mathcal{G}_l$ be separable RKHSs with characteristic kernels $k : \mathbb{R}^{T_\mathbf{X}} \times \mathbb{R}^{T_\mathbf{X}} \to \mathbb{R}$ and $l : \mathbb{R}^{T_\mathbf{Y}} \times \mathbb{R}^{T_\mathbf{Y}} \to \mathbb{R}$, respectively. HSIC is then defined as the MMD between $\mathbb{P}_{\mathbf{XY}}$ and $\mathbb{P}_\mathbf{X}\mathbb{P}_\mathbf{Y}$ [7]:

$$\begin{aligned}\text{HSIC}(\mathcal{H}_k,\mathcal{G}_l,\mathbb{P}_{\mathbf{XY}}) &:= \text{MMD}^2(\mathcal{H}_k\otimes\mathcal{G}_l,\mathbb{P}_{\mathbf{XY}},\mathbb{P}_\mathbf{X}\mathbb{P}_\mathbf{Y}) \\ &= \|\mu_{\mathbf{XY}}-\mu_\mathbf{X}\otimes\mu_\mathbf{Y}\|^2_{\mathcal{H}_k\otimes\mathcal{G}_l} \geq 0, \text{with equality iff} \quad \mathbb{P}_{\mathbf{XY}} = \mathbb{P}_\mathbf{Y}\mathbb{P}_\mathbf{Y}.\end{aligned} \tag{3}$$

Here, $\otimes$ denotes the tensor product. Recall that we assume an equal number of realisations m for both processes, and let $\mathbf{K},\mathbf{L}\in\mathbb{R}^{m\times m}$ be the kernel matrices with entries $k_{ij} = k(\mathbf{x}_i,\mathbf{x}_j)$ and $l_{ij} = l(\mathbf{y}_i,\mathbf{y}_j)$, respectively. Given i.i.d. samples $(\mathbf{X},\mathbf{Y})$, an unbiased empirical estimator of $\text{HSIC}(\mathcal{H}_k,\mathcal{G}_l,\mathbb{P}_{\mathbf{XY}})$ is given by [23] (Theorem 2):

$$\widehat{\text{HSIC}}_u(\mathcal{H}_k,\mathcal{G}_l,\mathbf{XY}) = \frac{1}{m(m-3)}\left[\text{trace}(\widetilde{\mathbf{K}}\widetilde{\mathbf{L}}) + \frac{\mathbf{1}^\top\widetilde{\mathbf{K}}\mathbf{1}\mathbf{1}^\top\widetilde{\mathbf{L}}\mathbf{1}}{(m-1)(m-2)} - \frac{2}{m-2}\mathbf{1}^\top\widetilde{\mathbf{K}}\widetilde{\mathbf{L}}\mathbf{1}\right], \tag{4}$$

where $\widetilde{\mathbf{K}} = \mathbf{K} - \text{diag}(\mathbf{K})$ and $\widetilde{\mathbf{L}} = \mathbf{L} - \text{diag}(\mathbf{L})$, and $\mathbf{1}$ is the $m\times 1$ vector of ones. To ease our notation, we henceforth omit the implied $\mathcal{H}_k$ and $\mathcal{G}_l$.

To test $\widehat{\text{HSIC}}_u(\mathbf{XY})$ for statistical significance, we define the null hypothesis $H_0 : \mathbb{P}_{\mathbf{XY}} = \mathbb{P}_\mathbf{X}\mathbb{P}_\mathbf{Y}$ and the alternative $H_1 : \mathbb{P}_{\mathbf{XY}} \neq \mathbb{P}_\mathbf{X}\mathbb{P}_\mathbf{Y}$. We broadly repeat the procedure outlined in Section 3.2 by bootstrapping the distribution under H_0 via permutations, with the distinction that we only permute the samples $\{\mathbf{y}_i\}_{i=1}^m$, resulting in $\mathbf{Y}_p, p\in[1,P]$, whilst the $\{\mathbf{x}_j\}_{j=1}^m$ are kept unchanged [7]. $\widehat{\text{HSIC}}_u(\mathbf{XY})$ is then computed for each permutation $(\mathbf{X},\mathbf{Y}_p)$ and the empirical threshold $\hat{c}_\alpha$ is taken as the statistic at position $(1-\alpha)\times P$. The null hypothesis H_0 is rejected, if $\widehat{\text{HSIC}}_u(\mathbf{XY}) > \hat{c}_\alpha$.

3.4. Maximising the Test Power

The power of both MMD-based two-sample and HSIC-based independence tests is prone to decay in high dimensional spaces [24,25], as in our setting where each measurement point in time is treated as a separate dimension. Hence, we describe here how a kernel k can be chosen to maximise the test power, i.e., the probability of correctly rejecting H_0 given that it is false. First, note that under H_1 both $\widehat{\text{MMD}}_u^2(\mathbf{X},\mathbf{Y})$ [21] (Corollary 16) and $\widehat{\text{HSIC}}_u(\mathbf{XY})$ [7] (Theorem 1) are asymptotically Gaussian:

$$\frac{\widehat{\text{MMD}}_u^2(\mathbf{X},\mathbf{Y}) - \text{MMD}^2(\mathbb{P}_\mathbf{X},\mathbb{P}_\mathbf{Y})}{\sqrt{V_m^{\text{MMD}}(\mathbb{P}_\mathbf{X},\mathbb{P}_\mathbf{Y})}} \xrightarrow{D} \mathcal{N}(0,1) \tag{5}$$

$$\frac{\widehat{\text{HSIC}}_u(\mathbf{XY}) - \text{HSIC}(\mathbb{P}_{\mathbf{XY}})}{\sqrt{V_m^{\text{HSIC}}(\mathbb{P}_{\mathbf{XY}})}} \xrightarrow{D} \mathcal{N}(0,1), \tag{6}$$

where $V_m^{\text{MMD}}(\mathbb{P}_\mathbf{X},\mathbb{P}_\mathbf{Y})$ and $V_m^{\text{HSIC}}(\mathbb{P}_{\mathbf{XY}})$ denote the asymptotic variance of $\widehat{\text{MMD}}_u^2(\mathbf{X},\mathbf{Y})$ and $\widehat{\text{HSIC}}_u(\mathbf{XY})$, respectively [26] (Section 5.5.1 (A)).

Given a significance level α, we define the test thresholds c_α^{MMD} and c_α^{HSIC} and reject H_0 if $\widehat{\text{MMD}}_u^2(\mathbf{X},\mathbf{Y}) > c_\alpha^{\text{MMD}}$ or $\widehat{\text{HSIC}}_u(\mathbf{XY}) > c_\alpha^{\text{HSIC}}$. Following [27], the test power is defined in terms of $\mathbb{P}_1$, the distributions under H_1, with equal sample sizes $m = n$ as:

$$\mathbb{P}_1\left(\widehat{\mathrm{MMD}}_u^2(\mathbf{X},\mathbf{Y}) > \frac{\hat{c}_\alpha^{\mathrm{MMD}}}{m}\right) \xrightarrow{D} \Phi\left(\frac{\mathrm{MMD}^2(\mathbb{P}_\mathbf{X},\mathbb{P}_\mathbf{Y}) - c_\alpha^{\mathrm{MMD}}/m}{\sqrt{V_m^{\mathrm{MMD}}(\mathbb{P}_\mathbf{X},\mathbb{P}_\mathbf{Y})}}\right) \quad (7)$$

$$\mathbb{P}_1\left(\widehat{\mathrm{HSIC}}_u(\mathbf{XY}) > \frac{\hat{c}_\alpha^{\mathrm{HSIC}}}{m}\right) \xrightarrow{D} \Phi\left(\frac{\mathrm{HSIC}(\mathbb{P}_\mathbf{XY}) - c_\alpha^{\mathrm{HSIC}}/m}{\sqrt{V_m^{\mathrm{HSIC}}(\mathbb{P}_\mathbf{XY})}}\right), \quad (8)$$

where Φ is the cumulative density function of the standard Gaussian distribution and where $\hat{c}_\alpha \to c_\alpha$ with increasing sample size. To maximise the test power, we maximise the argument of Φ, which we approximate by maximising $\widehat{\mathrm{MMD}}_u^2(\mathbf{X},\mathbf{Y})/\sqrt{\hat{V}_m^{\mathrm{MMD}}(\mathbf{X},\mathbf{Y})}$ and minimising $\hat{c}_\alpha^{\mathrm{MMD}}/\left(m\sqrt{\hat{V}_m^{\mathrm{MMD}}(\mathbf{X},\mathbf{Y})}\right)$ for (7), and similarly for (8). The empirical unbiased variance $\hat{V}_m^{\mathrm{MMD}}(\mathbf{X},\mathbf{Y})$ in (7) was derived in [27], and we use [23] (Theorem 5) for $\hat{V}_m^{\mathrm{HSIC}}(\mathbf{XY})$ in (8).

We perform this optimisation by splitting our samples $(\mathbf{X},\mathbf{Y})$ into training and testing sets, of which we take the former to learn the kernel hyperparameters and the latter to conduct the final hypothesis test with the learnt kernel.

4. Experimental Results on Synthetic Data

To evaluate our proposed tests empirically, we first apply our homogeneity and independence tests to various nonstationary synthetic datasets. We report test performance using $\hat{\mu}$, the percentage of rejection of the null hypothesis H_0, which becomes the test power once H_0 is false, by repeating the experiments on 200 trials (i.e., 200 independently generated synthetic datasets). We provide 95% confidence intervals computed as $\hat{\mu} \pm 1.96\sqrt{\hat{\mu}(1-\hat{\mu})/200}$.

4.1. *Homogeneity Tests with* MMD

4.1.1. Setup

We evaluate our MMD-based homogeneity test against shifts in mean and variance of two nonstationary stochastic processes $\{\mathbf{X}_t\}$ and $\{\mathbf{Y}_t\}$ by establishing if they are correctly accepted or rejected under the null hypothesis $H_0 : \mathbb{P}_\mathbf{X} = \mathbb{P}_\mathbf{Y}$. For ease of comparison, we adopt the experimental protocol of [15] and consider two stochastic processes based on a linear mixed effects model. We generate independent samples $\mathbf{X} = \{\mathbf{x}_i\}_{i=1}^m$ and $\mathbf{Y} = \{\mathbf{y}_i\}_{i=1}^n$ on an equally spaced temporal grid of length $T_\mathbf{X} = T_\mathbf{Y} = T$ in the interval $\mathcal{I} = [0,1]$,

$$x_{i,t} = \mu_\mathbf{X}(t) + \sum_{k=1}^{K} \xi_{\mathbf{X}i,k}\,\phi_k(t) + \epsilon_{\mathbf{X}i,t} \quad \text{and} \quad y_{i,t} = \mu_\mathbf{Y}(t) + \sum_{k=1}^{K} \xi_{\mathbf{Y}i,k}\,\phi_k(t) + \epsilon_{\mathbf{Y}i,t}, \quad (9)$$

where we set $K = 2$ with Fourier basis functions $\phi_1(t) = \sqrt{2}\sin(2\pi t)$ and $\phi_2(t) = \sqrt{2}\cos(2\pi t)$. The coefficients $\xi_{\mathbf{X}i,k}$ and $\xi_{\mathbf{Y}i,k}$ and the additive noises $\epsilon_{\mathbf{X}i,t}, \epsilon_{\mathbf{Y}i,t}$ are all independent Gaussian-distributed random variables with means and variances specified below.

We evaluate the test power against varying values of shifts in mean and variance as follows:

- *Mean shift:* $\mu_\mathbf{X}(t) = t$ and $\mu_\mathbf{Y}(t) = t + \delta_\mu t^3$. The basis coefficients are sampled as $\xi_{\mathbf{X}i,1}, \xi_{\mathbf{Y}i,1} \sim \mathcal{N}(0,10)$ and $\xi_{\mathbf{X}i,2}, \xi_{\mathbf{Y}i,2} \sim \mathcal{N}(0,5)$, and the additive noises are sampled as $\epsilon_{\mathbf{X}i,t}, \epsilon_{\mathbf{Y}i,t} \sim \mathcal{N}(0,0.25)$.
- *Variance shift:* We take $\mu_\mathbf{X}(t) = \mu_\mathbf{Y}(t) = 0$, and introduce a shift in variance in the first basis function coefficients via $\xi_{\mathbf{X}i,1} \sim \mathcal{N}(0,10)$ and $\xi_{\mathbf{Y}i,1} \sim \mathcal{N}(0,10+\delta_\sigma)$. The second coefficients are sampled as $\xi_{\mathbf{X}i,2}, \xi_{\mathbf{Y}i,2} \sim \mathcal{N}(0,5)$, and the noises as $\epsilon_{\mathbf{X}i,t}, \epsilon_{\mathbf{Y}i,t} \sim \mathcal{N}(0,0.25)$.

The coefficients δ_μ and δ_σ for mean and variance shifts, respectively, determine the departure from the null hypothesis. Setting $\delta_\mu, \delta_\sigma = 0$ means H_0 is true, whereas $\delta_\mu, \delta_\sigma > 0$

means H_0 is false. Although this is not a necessity, we set the number of independent samples of $\{\mathbf{X}_t\}$ and $\{\mathbf{Y}_t\}$ to be equal, $m = n$. To test for statistical significance, we follow the procedure described in Section 3.2 and perform permutation tests of $P = 5000$ partitions for varying values of δ_μ and δ_σ and different sample sizes $m = 100, 200, 300, 500$.

4.1.2. Baseline Results without Test Power Optimisation

Our baseline results are obtained with a Gaussian kernel $k(x, y) = \exp(-\|x - y\|^2/\sigma^2)$ with bandwidth σ equal to the median distance between observations of the aggregated samples. Figure 1 shows how our method (solid lines) compares to [15] (dashed lines) for $T = 100$ discrete time points. For all sample sizes, the type-I error rate lies at or below the allowable probability of false rejection α, and our method significantly outperforms [15] for nearly all levels of mean and variance shifts. Both shifts become easier to detect for larger sample sizes. Particularly strong improvements are achieved for mean shifts: our method makes no type-II errors for $\delta_\mu \geq 3$ on $m = 100$ samples, whereas [15] only reach such performance with $m = 500$ samples and $\delta_\mu \geq 4.5$. We obtain similar test power results (see Appendix A.1) for coarser realisations with $T = 5, 10, 25, 50$ over the same interval $\mathcal{I} = [0, 1]$.

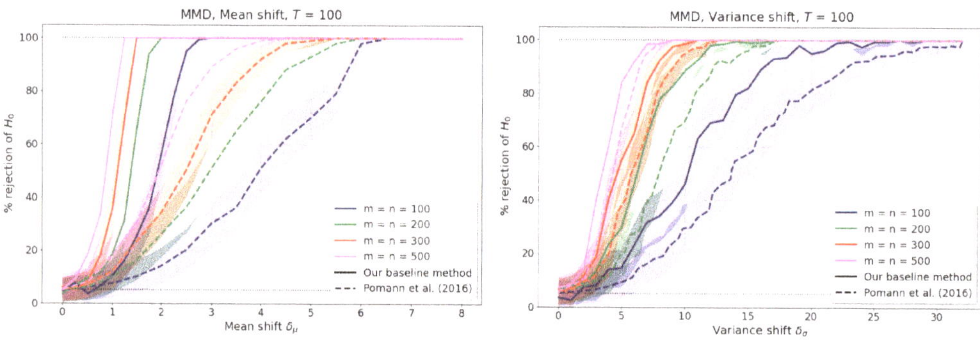

Figure 1. Results of our MMD-based homogeneity test for nonstationary random processes: percentage of rejected H_0 as mean shift (**left**) and variance shift (**right**) are varied. Our baseline method (solid lines) is compared to [15] (dashed lines) for different sample sizes $m = n = 100, 200, 300, 500$ and $T = 100$ discrete time points.

4.1.3. Results of the Optimised Test

Next, we apply the method described in Section 3.4 to maximise the test power. Specifically, we search for the Gaussian kernel bandwidth σ (over spaces defined in Table A1 in Appendix A.2), that maximises the argument of Φ in our approximations of (7) on our training samples. For demonstrative purposes, we choose to split our dataset equally into training and testing sets although other ratios may lead to higher test power. Figure 2 shows the results of the optimised test (dotted lines) against the baseline results (solid lines) and the results of [15] (dashed lines) for $m = 100$ and $m = 200$ samples and $T = 100$ discrete points in time. We find that the test power is significantly improved by our optimisation for the detection of mean shifts. For instance, test power rises fourfold for $\delta_\mu = 1$ and $m = 200$ compared to our baseline method. Furthermore, we have no type-II errors once $\delta_\mu \geq 2$ for $m = 100$, as compared to $\delta_\mu \geq 3$ for our baseline test and $\delta_\mu \geq 6.5$ for [15]. In its current form, however, our optimisation does not yield higher test power for the detection of variance shifts, a fact that we discuss in Section 6.

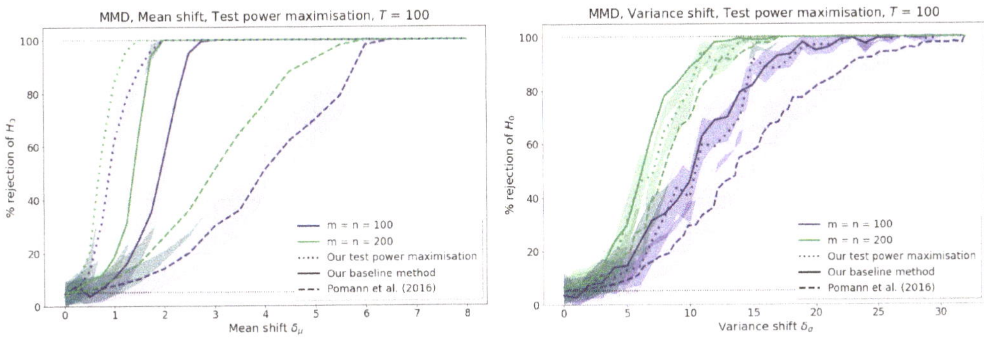

Figure 2. Results of homogeneity test with optimising for test power: percentage of rejected H_0 for mean shift (**left**) and variance shift (**right**) for sample sizes $m = n = 100, 200$ and $T = 100$ discrete time points. Our optimised test power method (dotted lines) is compared to our baseline method (solid lines) and [15] (dashed lines).

4.2. Independence Tests with HSIC

4.2.1. Setup

To test for independence, the null hypothesis is $H_0 : \mathbb{P}_{\mathbf{XY}} = \mathbb{P}_{\mathbf{X}}\mathbb{P}_{\mathbf{Y}}$. We assume we observe measurements $x_{i,t}$ and $y_{i,t}$ over temporal grids of length $T_{\mathbf{X}}$ and $T_{\mathbf{Y}}$ in the interval $\mathcal{I} = [0,1]$, respectively. To measure type-I and type-II error rates, we use the following experimental protocols, partly adopted from [7,18,28]:

- *Linear dependence:* $\mathbf{X}$ is generated as in (9) with $\mu_{\mathbf{X}}(t) = t$, basis coefficients $\xi_{\mathbf{X}i,1} \sim \mathcal{N}(0,10)$, $\xi_{\mathbf{X}i,2} \sim \mathcal{N}(0,5)$, and noise $\epsilon_{\mathbf{X}i,t} \sim \mathcal{N}(0,0.25)$. The samples of the second process are $\mathbf{Y} = \{x_{i,1} + \epsilon_i\}_{i=1}^m$ where $\epsilon_i \sim \mathcal{N}(0,1)$, as in [18].
- *Dependence through a shared coefficient:* $\mathbf{X}$ and $\mathbf{Y}$ are generated as in (9) with $\mu_{\mathbf{X}}(t) = \mu_{\mathbf{Y}}(t) = t$ and independently sampled $\xi_{\mathbf{X}i,1}$, $\xi_{\mathbf{Y}i,1}$, $\epsilon_{\mathbf{X}i,t}$, $\epsilon_{\mathbf{Y}i,t}$ as in the mean shift experiments of Section 4.1, but where the stochastic processes now share the second basis function coefficient: $\xi_{\mathbf{X}i,2} = \xi_{\mathbf{Y}i,2}$.
- *Dependence through rotation:* We start by generating independent $\mathbf{X}^{(0)}$ and $\mathbf{Y}^{(0)}$ as in (9) with $\mu_{\mathbf{X}}(t) = \mu_{\mathbf{Y}}(t) = t$ and $\epsilon_{\mathbf{X}i,t}, \epsilon_{\mathbf{Y}i,t} \sim \mathcal{N}(0,0.25)$, but with $\xi_{\mathbf{X}i,k}$ and $\xi_{\mathbf{Y}i,k}$ drawn from: (i) student-t, (ii) uniform, or (iii) exponential distributions [28] (Table 3). We next multiply $(\mathbf{X}^{(0)}, \mathbf{Y}^{(0)})$ by a 2×2 rotation matrix $R(\theta)$ with $\theta \in [0, \pi/4]$ to generate new rotated samples $(\mathbf{X}, \mathbf{Y})$, which we then test for independence. Clearly, for $\theta = 0$ our samples $(\mathbf{X}, \mathbf{Y})$ are independent and as θ is increased their dependence becomes easier to detect (see [7] (Section 4) and Figure A3 for implementation details).

Statistical significance is computed using $P = 5000$ permutations of $\mathbf{Y}$ whilst $\mathbf{X}$ is kept fixed to approximate the distribution under H_0. Test power is calculated for varying $T = [5, 10, 25, 50, 100]$ and different sample sizes $m = n$.

4.2.2. Baseline Results without Test Power Optimisation

Our baseline results are computed using a Gaussian kernel with σ equal to the median distance between measurements in the corresponding sample. Figure 3 (left) shows the results of our test on the linear dependence experiments, which demonstrate, due to $T_{\mathbf{Y}} = 1$, how dependencies between individual points in time and an entire time series can be detected. We compare our method to: (i) a statistic explicitly aimed at linear dependence, $\text{SubCorr} = \frac{1}{T_{\mathbf{X}}} \sum_{t=1}^{T_{\mathbf{X}}} \text{Corr}(\{x_{i,t}\}_{i=1}^m, \mathbf{Y})$, where $\text{Corr}(\cdot,\cdot)$ is the Pearson correlation coefficient; and (ii) $\text{SubHSIC} = \frac{1}{T_{\mathbf{X}}} \sum_{t=1}^{T_{\mathbf{X}}} \widehat{\text{HSIC}}_u(\{x_{i,t}\}_{i=1}^m, \mathbf{Y})$. For both of these methods, the distribution under H_0 is also approximated via permutations. We find that SubCorr outperforms the other methods in experiments with sample sizes $m < 20$, and SubHSIC achieves comparable results to our method. The results for $T_{\mathbf{X}} = [25, 50, 100]$ (see Appendix A.1) are similar.

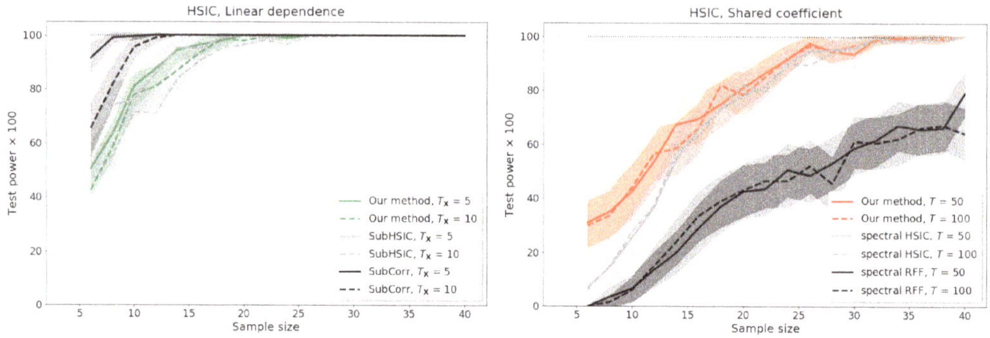

Figure 3. Results of the HSIC-based independence test: Test power for linear dependence (**left**) and dependence through shared coefficients (**right**) as sample size is varied for various numbers of time points. For the linear dependence, we compare our baseline results to SubCorr and SubHSIC; for the shared coefficient, we compare against two spectral approximations [18] (Section 5.1).

Figure 3 (right) displays the power of our independence test for the case of dependent samples through a shared coefficient for varying sample sizes m and measurements T. We compare our results to two spectral methods [18] that approximate the distribution under H_0 using eigenvalues of the centred kernel matrices of **X** and **Y**: spectral HSIC uses the unbiased estimator (4) as the test statistic with the eigenvalue-based null distribution; and spectral RFF uses a test statistic induced by a number of random Fourier features (RFFs) (set here to 10) that approximate the kernel matrices of **X** and **Y**. Our method and spectral HSIC achieve $20 - 50\%$ improvement in test power compared to spectral RFF. For small numbers of samples ($m < 15$), our method outperforms spectral HSIC, which converges to the performance of our method with increasing sample size, as we would expect it [22] (Theorem 1).

Figure 4 shows the rotation dependence experiments, where $\theta = 0$ corresponds to the null hypothesis (independence) and $\theta > 0$ to the alternative. The distribution hyperparameters for $\xi_{\mathbf{X}i,k}$ and $\xi_{\mathbf{Y}i,k}$ are detailed in Appendix A.3, and we set $T_\mathbf{X} = T_\mathbf{Y} = T$, although equality is not required. As expected, dependence is easier to detect with increasing θ. We observe that denser temporal measurements do not result in enhanced test power. Note that the test power is highly dependent on the distribution of the coefficients of the basis functions $\xi_{\mathbf{X}i,k}, \xi_{\mathbf{Y}i,k}$.

4.2.3. Results of the Optimised Test

The test power maximisation was applied to the rotation dependence experiments by searching for optimal Gaussian kernel bandwidths $\sigma_\mathbf{X}$ and $\sigma_\mathbf{Y}$ over predefined intervals (specified in Appendix A.2). Figure 4 shows that the test power is improved when the basis function coefficients are drawn from uniform distributions. In this case, the percentage of rejected H_0 is $20 - 40\%$ higher for θ between 0.2 and $0.75 \times \pi/4$, but it levels off at 95% once $\theta \geq 0.75 \times \pi/4$, which is the same level achieved by our baseline method for $\theta \geq 0.85 \times \pi/4$. With our current test-train split, our optimised test does not improve the test power if the basis function coefficients $\xi_{\mathbf{X}i,k}$ and $\xi_{\mathbf{Y}i,k}$ are drawn from student-t or exponential distributions.

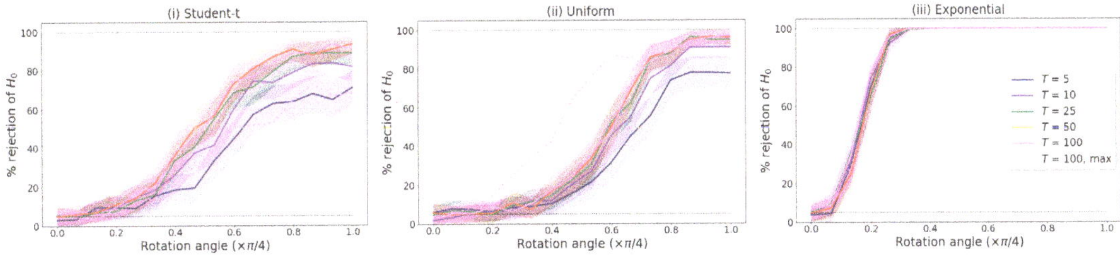

Figure 4. Results of the HSIC-based independence test: Percentage of rejected H_0 in rotation dependence experiments for different number of discrete time points T and coefficients $\xi_{\mathbf{X}i,k}$ and $\xi_{\mathbf{Y}i,k}$ drawn from three distributions: (**i**) student-t, (**ii**) uniform, and (**iii**) exponential (see Appendix A.3). The sample size is $m = 200$. The violet dotted lines are the results of our test power maximisation.

5. Application to a Socioeconomic Dataset

As a further illustration, we apply our method to the United Nations' socioeconomic Sustainable Development Goals (SDGs) (see Appendix A.4 for details). Specifically, we investigate whether some so-called Targets of the 17 SDGs have been homogeneous over the last 20 years across low- and high-income countries and whether certain SDGs in African countries exhibit dependence over the same period. In both settings, we assume countries are independent.

For our homogeneity tests, we classify countries into low- and high-income according to [29]. We use temporal data of 76 Targets for which [30] provides data collected over the last $T = 20$ years for $m = 30$ low-income countries and $n = 55$ high-income countries. Applying our baseline method without test power optimisation, we find that, out of the 76 Targets we have data available for, only 38 have had homogeneous trajectories in low- and high-income countries. For instance, whereas the 'death rate due to road traffic injuries' (Target 3.6) has been homogeneous between these two groups, the 'fight the epidemics of AIDS, tuberculosis, malaria and others' (Target 3.3) has not been homogeneous in low- and high-income countries.

For our independence tests, we consider temporal data from $m = n = 49$ African countries over $T = 20$ years and test any two Targets for pairwise independence. Of the total 2850 possible pairwise combinations, the null hypothesis of independence is rejected for 357. As an illustration, we examine the dependencies of 'implementation of national social protection systems' (Target 1.3) with 'economic growth' (Target 8.1) and the 'proportion of informally employed workers' (Target 8.3). Applying our baseline method, we accept the null hypothesis of independence between Target 1.3 and 8.1, i.e., we find that the 'implementation of national social protection systems' has been independent of economic growth. In contrast, we find that Target 1.3 has been dependent on the 'proportion of informally employed workers' (Target 8.3).

6. Discussion and Conclusions

Building on ideas from functional data analysis, we have presented approaches to testing for homogeneity and independence between two nonstationary random processes with the kernel-based statistics MMD and HSIC. We view independent realisations of the underlying processes as samples from multivariate probability distributions to which MMD and HSIC can be applied. Our tests are shown to outperform current state-of-the-art methods in a range of experiments. Furthermore, we optimise the test power over the choice of kernel and achieve improved results in most settings. However, we also observe that our optimisation procedure does not always yield an increase in test power. We leave the investigation of this behaviour open for future research with the possibility of defining search spaces and step sizes over kernel hyperparameters differently or of choosing a gradient-based approach for optimisation [27]. Our results show that small sample sizes of less than 40 independent realisations can already achieve high test power and that denser measurements over the same time period do not necessarily lead to enhanced test power.

The proposed tests can be of interest in many areas where nonstationary and nonlinear multivariate temporal datasets constitute the norm, as illustrated by our application to test for homogeneity and independence between the United Nations' Sustainable Development Goals measured in different countries over the last 20 years.

Institutional Review Board Statement: Not applicable

Informed Consent Statement: Not applicable

Data Availability Statement: The socioeconomic dataset is freely available at [30].

Appendix A

Appendix A.1. Results for Realisations with Varying Number of Time Points, T

MMD We show here the results for mean and variance shifts for $m = n = 100$, but the results are similar for all tested sample sizes $m = n = 100, 200, 300, 500$,

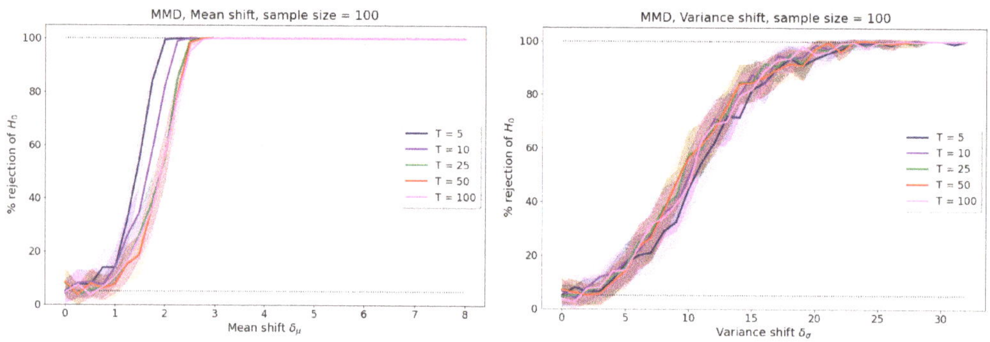

Figure A1. Results of MMD-based homogeneity test with $T = [5, 10, 25, 50, 100]$: Percentage of rejected H_0 for mean shift (**left**) and variance shift (**right**) for sample sizes $m = n = 100$ and T discrete time points in $d = 1$ dimensions.

HSIC Experiments for linear dependence and dependence through shared second basis function coefficient for various T. We find that the granularity of measurements over time does not influence the text power significantly.

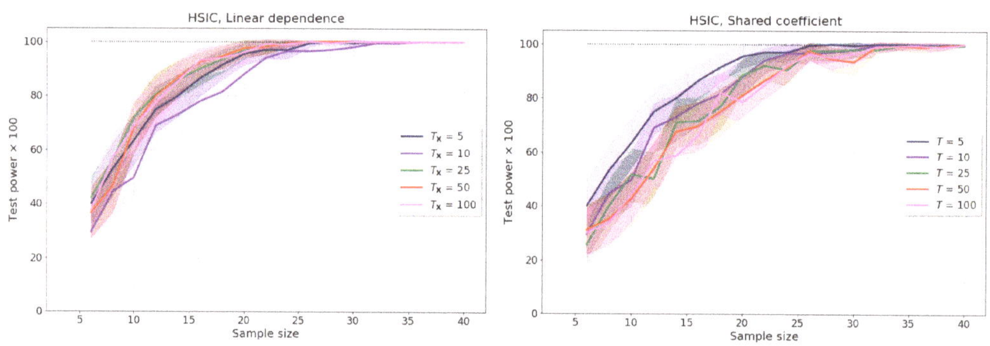

Figure A2. Results of the HSIC-based independence test: Test power for linear dependence (**left**) and dependence through shared coefficient (**right**) as sample size is varied for various numbers of time points $T = [5, 10, 25, 50, 100]$.

A.2. Test Power Maximisation

MMD For mean shift experiments for MMD, we predefine a linear search space with 11 values for the Gaussian kernel bandwidth σ due to the dependence on δ_μ and similarly for variance shift experiments (both stated in Table A1). These search spaces resulted from extensive manual explorations for all shifts and sample sizes. We acknowledge that the test power may be further improved with search spaces of finer granularity.

HSIC We define search intervals of both σ_X and σ_Y across all angles θ but different for the student-t, uniform, and exponential distributions. For student-t and exponential distributions, both σ_X and σ_Y were chosen as 20 evenly spaced numbers on a linear scale between 1 and 20. For uniform distributions, both σ_X and σ_Y were chosen as 40 evenly spaced numbers on a linear scale between 1 and 40. These search spaces resulted from extensive manual explorations for all angles and distributions. We acknowledge that the test power may be further improved with search spaces of finer granularity.

Table A1. Linear search spaces for bandwidth σ in MMD mean (**left**) and variance (**right**) shift experiments.

δ_μ	0–2	2.25–3	3.25–5	5.5–8	δ_σ	0–4	5–14	15–32
		Step Size = 0.25		Step Size = 0.5			Step Size = 1	
search space for σ	1	6	11	16	search space for σ	10	20	30
	3	8	13	18		12	22	32
	5	10	15	20		14	24	34
	7	12	17	22		16	26	36
	9	14	19	24		18	28	38
	11	16	21	26		20	30	40
	13	18	23	28		22	32	42
	15	20	25	30		24	34	44
	17	22	27	32		26	36	46
	19	24	29	34		28	38	48
	21	26	31	36		30	40	50

Appendix A.3. Distribution Specifications for Basis Function Coefficients in Rotation Mixing

Table A2. Specifications of distributions for the rotation mixing. They are a subset of the distributions in [28] (Table 3), and **Z** is a proxy for both **X** and **Y**.

Distribution	Fourier Basis Function Coefficients	
	ξ_{Zi1}	ξ_{Zi2}
Exponential	$\lambda = 1.5$	$\lambda = 3$
Student-t	$\nu = 3$	$\nu = 5$
Uniform	$\mathcal{U}[-10, 10]$	$\mathcal{U}[-5, 5]$

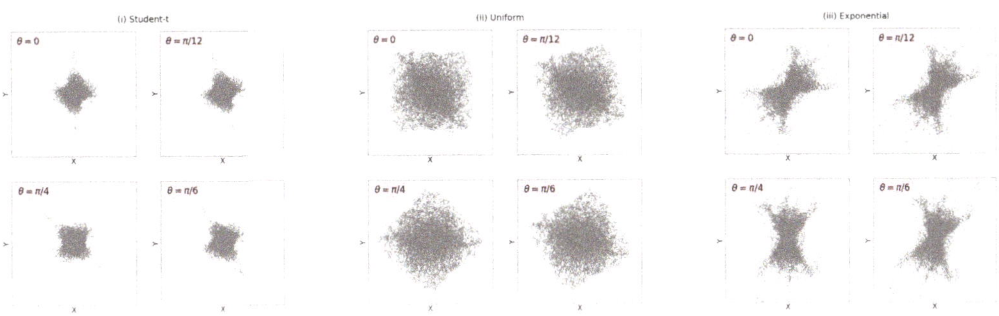

Figure A3. Illustration of **X** and **Y** with (**i**) student-t, (**ii**) uniform, and (**iii**) exponential basis function coefficients being mixed by different rotation angles θ, ordered clockwise by increasing θ.

A.4. SDG Dataset

Data of the Indicators measuring the progress of the Targets of the SDGs can be found at [30]. Each of these Indicators measures the progress towards a specific Target.

For instance, an Indicator for Target 1.1, *'by 2030, eradicate extreme poverty for all people everywhere, currently measured as people living on less than $1.90 a day'*, is the *'proportion of population below the international poverty line, by gender, age, employment status and geographical location (urban/rural)'*. Each of the Targets belongs to one specific Goal (e.g., Target 1.1 belongs to Goal 1, *'end poverty in all its forms everywhere'*). There are 17 such Goals, which are commonly referred to as the Sustainable Development Goals (SDGs). We compute averages over all Indicators belonging to one Target for our analyses in Section 5.

The dataset of [30] has many missing values, especially for the time span 2000–2005. We impute these values using a weighted average across countries (where data is available) with weights inversely proportional to the Euclidean distance between indicators.

References

1. Christakis, N.A.; Fowler, J.H. The spread of obesity in a large social network over 32 years. *N. Engl. J. Med.* **2007**, *357*, 370–379. [CrossRef] [PubMed]
2. Barabási, A.L.; Gulbahce, N.; Loscalzo, J. Network medicine: a network-based approach to human disease. *Nat. Rev. Genet.* **2011**, *12*, 56–68. [CrossRef]
3. Bond, R. Complex networks: network healing after loss. *Nat. Hum. Behav.* **2017**, *1*, 1–2. [CrossRef]
4. Battiston, S.; Mandel, A.; Monasterolo, I.; Schütze, F.; Visentin, G. A climate stress-test of the financial system. *Nat. Clim. Chang.* **2017**, *7*, 283–288. [CrossRef]
5. Muandet, K.; Fukumizu, K.; Sriperumbudur, B.; Schölkopf, B.; others. Kernel mean embedding of distributions: A review and beyond. *Found. Trends Mach. Learn.* **2017**, *10*, 1–141. [CrossRef]
6. Gretton, A.; Borgwardt, K.; Rasch, M.; Schölkopf, B.; Smola, A.J. A kernel method for the two-sample-problem. *arXiv* **2008**, arXiv:0805.2368.
7. Gretton, A.; Fukumizu, K.; Teo, C.H.; Song, L.; Schölkopf, B.; Smola, A.J. A kernel statistical test of independence. *NIPS* **2008**, *20*, 585–592.
8. Besserve, M.; Logothetis, N.K.; Schölkopf, B. Statistical analysis of coupled time series with Kernel Cross-Spectral Density operators. In *Advances in Neural Information Processing Systems 26*; Curran Associates, Inc.: Red Hook, NY, USA, 2013; pp. 2535–2543.
9. Chwialkowski, K.; Sejdinovic, D.; Gretton, A. A wild bootstrap for degenerate kernel tests. In *Advances in Neural Information Processing Systems*; Curran Associates, Inc.: Red Hook, NY, USA, 2014; pp. 3608–3616.
10. Davis, R.A.; Matsui, M.; Mikosch, T.; Wan, P.; others. Applications of distance correlation to time series. *Bernoulli* **2018**, *24*, 3087–3116. [CrossRef]
11. Székely, G.J.; Rizzo, M.L.; Bakirov, N.K.; others. Measuring and testing dependence by correlation of distances. *Ann. Stat.* **2007**, *35*, 2769–2794. [CrossRef]
12. Horváth, L.; Kokoszka, P.; Reeder, R. Estimation of the mean of functional time series and a two-sample problem. *J. R. Stat. Soc. Ser. B* **2012**, *75*, 103–122. [CrossRef]
13. Fremdt, S.; Steinbach, J.G.; Horváth, L.; iotr Kokoszka. Testing the Equality of Covariance Operators in Functional Samples. *Scand. J. Stat.* **2012**, *40*, 138–152. [CrossRef]
14. Panaretos, V.M.; Kraus, D.; Maddocks, J.H. Second-Order Comparison of Gaussian Random Functions and the Geometry of DNA Minicircles. *J. Am. Stat. Assoc.* **2010**, *105*, 670–682. [CrossRef]
15. Pomann, G.M.; Staicu, A.M.; Ghosh, S. A two-sample distribution-free test for functional data with application to a diffusion tensor imaging study of multiple sclerosis. *J. R. Stat. Soc. Ser. C* **2016**, *65*, 395–414. [CrossRef] [PubMed]
16. Wynne, G.; Duncan, A.B. A kernel two-sample test for functional data. *arXiv* **2020**, arXiv:2008.11095.
17. Górecki, T.; Krzyśko, M.; Wołyński, W. Independence test and canonical correlation analysis based on the alignment between kernel matrices for multivariate functional data. *Artif. Intell. Rev.* **2018**, *53*, 475–499. [CrossRef]
18. Zhang, Q.; Filippi, S.; Gretton, A.; Sejdinovic, D. Large-scale kernel methods for independence testing. *Stat. Comput.* **2018**, *28*, 113–130. [CrossRef]
19. Sriperumbudur, B.K.; Gretton, A.; Fukumizu, K.; Schölkopf, B.; Lanckriet, G.R. Hilbert space embeddings and metrics on probability measures. *J. Mach. Learn. Res.* **2010**, *11*, 1517–1561.
20. Sriperumbudur, B.K.; Fukumizu, K.; Lanckriet, G.R. Universality, Characteristic Kernels and RKHS Embedding of Measures. *J. Mach. Learn. Res.* **2011**, *12*, 2389–2410.
21. Gretton, A.; Borgwardt, K.M.; Rasch, M.J.; Schölkopf, B.; Smola, A.J. A kernel two-sample test. *J. Mach. Learn. Res.* **2012**, *13*, 723–773.
22. Gretton, A.; Fukumizu, K.; Harchaoui, Z.; Sriperumbudur, B.K. A fast, consistent kernel two-sample test. *NIPS* **2009**, *23*, 673–681.
23. Song, L.; Smola, A.J.; Gretton, A.; Bedo, J.; Borgwardt, K. Feature selection via dependence maximization. *J. Mach. Learn. Res.* **2012**, *13*, 1393–1434.

24. Ramdas, A.; Reddi, S.J.; Póczos, B.; Singh, A.; Wasserman, L. On the decreasing power of kernel and distance based nonparametric hypothesis tests in high dimensions. In Proceedings of the Twenty-Ninth AAAI Conference on Artificial Intelligence, Austin, TX, USA, 25–30 January 2015.
25. Reddi, S.; Ramdas, A.; Póczos, B.; Singh, A.; Wasserman, L. On the high dimensional power of a linear-time two sample test under mean-shift alternatives. Artif. Intell. Stat. **2015**, *38*, 772–780.
26. Serfling, R.J. *Approximation Theorems of Mathematical Statistics*; Wiley Series in Probability and Mathematical Statistics; Wiley: New York, NY, USA, 2002.
27. Sutherland, D.J.; Tung, H.Y.; Strathmann, H.; De, S.; Ramdas, A.; Smola, A.J.; Gretton, A. Generative models and model criticism via optimized maximum mean discrepancy. *arXiv* **2016**, arXiv:1611.04488.
28. Gretton, A.; Herbrich, R.; Smola, A.J.; Bousquet, O.; Schölkopf, B. Kernel methods for measuring independence. *J. Mach. Learn. Res.* **2005**, *6*, 2075–2129.
29. World Bank. World Bank Country and Lending Groups. 2020. Available online: https://datahelpdesk.worldbank.org/knowledgebase/articles/906519-world-bank-country-and-lending-groups (accessed on 28 January 2020).
30. World Bank. Sustainable Development Goals. 2020. Available online: https://datacatalog.worldbank.org/dataset/sustainable-development-goals (accessed on 28 January 2020).

engineering proceedings

Proceedings

Improved Output Gap Estimates and Forecasts Using a Local Linear Regression †

Marlon Fritz

Department of Economics, Paderborn University, Warburger Str. 100, 33098 Paderborn, Germany; marlon.fritz@uni-paderborn.de; Tel.: +49-5251-602115

† Presented at the 7th International Conference on Time Series and Forecasting, Gran Canaria, Spain, 19–21 July 2021.

Abstract: The output gap, the difference between potential and actual output, has a direct impact on policy decisions, e.g., monetary policy. Estimating this gap and its further analysis remain the subject of controversial debates due to methodological problems. We propose a local polynomial regression combined with a Self-Exciting Threshold AutoRegressive (SETAR) model and its forecasting extension for a systematic output gap estimation. Furthermore, local polynomial regression is proposed for the (multivariate) OECD production function approach and its reliability is demonstrated in forecasting output growth. A comparison of the proposed gap to the Hodrick–Prescott filter as well as to estimations by experts from the FED and OECD shows a higher correlation of our output gap with those from those economic institutions. Furthermore, sometimes gaps with different magnitude and different positions above or below the trend are observed between different methods. This may cause competing policy implications which can be improved with our results.

Keywords: business cycles; nonparametric methods; output gap; trend identification

JEL Classification: C14; C22; E31; E52

Citation: Fritz, M. Improved Output Gap Estimates and Forecasts Using a Local Linear Regression . *Eng. Proc.* **2021**, *5*, 32. https://doi.org/10.3390/engproc2021005032

Published: 30 June 2021

1. Introduction

Since the influential work of [1], the output gap and its reliability have been widely discussed. Also, the importance of gap estimations for "conjunctural and monetary policy analysis" ([2], p. 2) is undisputed. The difficulties in the estimation of potential output are summarized by [3]. They distinguish three methods for its estimation: (i) statistical, (ii) production function and (iii) structural approaches. Ref [2] show that some statistical methods produce unreliable real-time estimates of the gap. These unreliable output gap estimates have induced unfavorable monetary policy activities, as [4] demonstrate for the UK. Thus, monetary policy recommendations need to be treated carefully as they depend heavily on the estimation method used for potential output.

The following four reasons for instable output gap estimations: (i) influence of first estimates on policy decisions, (ii) forecast errors, (iii) data revisions and (iv) varying decompositions of trend and cycle are identified by [5]. We focus on reducing the effects of (iv) by applying a new decomposition method and of (ii) by using more information, e.g., a regime-switching SETAR model. Furthermore, higher correlations of the recent proposal with output gaps from policy institutions and an improvement in the accuracy of output growth forecasts using the new output gap underline its reliability.

Since no true output gap exists one must rely in accordance with [6] on estimates without having an unambiguous definition from theory. Their paper summarizes numerous methods used to estimate the output gap, e.g., the Hodrick–Prescott (HP) filter, the [7] filter, and the [8] (BN) decomposition (see [9]). They distinguish between univariate time series methods and multivariate methods. Although multivariate methods process information from additional explanatory variables, ref [6] conclude that no multivariate model outperforms its univariate competitors. One of the most widely used methods for output gap estimations, employed e.g., by the European Commission (EC) and (indirectly) by the OECD, is the HP filter introduced by [10]. Using this penalized spline smoother results in very different gaps depending on the arbitrarily selected smoothing parameter λ ([1]), yielding somewhat arbitrary, either negatively or positively, output gaps [11]. Recently, the HP filter is criticized by [12] for causing problems at an unknown amount of boundary points. This becomes obvious once new data is available, which results in significant output gap revisions and reduces real-time reliability.

As mentioned by [13], many detrending methods perform poorly at time series endpoints, which results in output gaps that are sensitive to large revisions. This is also proposed by [1], who argue that the vast majority of output gap revisions are attributable to the boundary problems of detrending methods. This view contrasts with the expectation that data revisions are the primary source of uncertainty, whereas in line with [2] model and estimation uncertainty are much smaller. Further discussions on improvements of the FED output gap estimates and purely statistical methods over the last decades can be found in [13]. Nevertheless, they also confirm the poor reliability of solely statistical methods over the whole period and add that the accuracy of output gap estimates depends on the period under investigation, while the last decade eases gap estimation.

In order to estimate a reliable output gap accurately, an identification of trend and cycle is a prerequisite ([5]) and needs to be combined with the systematic analysis of the gap component. Therefore, any analyses need to provide additional information that can be used to estimate a more precise output gap. To use all available information in the sense of applying a two-sided filter, the local polynomial regression of [14] may be a better alternative for estimating the output gap. This method in its local linear (LLR) version also improves the estimation quality at boundary points, since an asymmetric boundary kernel is introduced to enhance the estimation quality at time series endpoints (real-time reliability). Moreover, the LLR allows for short-range dependence between trend and cyclical movements as required by [9] who analyze the revision properties of the BN decomposed output gap. The use of a (semi-)SETAR model provides additional information for the output gap. We then extend the method to forecast output gaps. Moreover, the univariate LLR of [14] is extended to multivariate analysis to examine the contribution of multivariate methods. Besides introducing this methodology, we also compare the gaps produced using the LLR with those using the HP filter for (i) statistically based estimations and (ii) the production function approach used by the OECD. Finally, the output gap estimated by experts from the FED and OECD is used as a benchmark. However, since no original gap exists, the comparison with external criteria on the appropriateness of the gap is difficult.

Section 2 presents the nonparametric LLR. Section 3 shows its application and compares it to the HP gap. Section 4 combines output gaps and semi-SETAR models by comparing different methodologies to those of the FED and the OECD by extending the univariate LLR approach to a multivariate method. Section 5 shows the predictive power of the new gap for output growth. Section 6 concludes.

2. Local Linear Regression

In the introduction, the HP filter is criticized for its suboptimality at boundary points. The LLR has automatic boundary correction [15], ensuring that asymptotic properties of the estimators in the interior still hold at boundary points. We focus on the estimation quality at these points and use an asymmetric boundary kernel to obtain stable boundary estimates, which are the key to obtaining reliable real-time output gap estimates. Ref [14] use an additive component model:

$$Y_t = m(x_t) + \xi_t, \tag{1}$$

where Y_t is a sequence of macroeconomic time series with time $t = 1, \ldots, T$, $x_t = t/T$ denotes the rescaled time, $m(x)$ is some smooth function and ξ_t denotes a zero mean stationary process.

Thus, a data-driven local polynomial estimator for the smooth trend function is used in line with [14] without any parametric assumptions on ξ_t. Under the assumption of short-range dependence the authors use the following Equation (2) for estimating the trend $m(x_t)$ by minimizing the locally weighted least squares:

$$Q = \sum_{t=1}^{T} \left\{ y_t - \sum_{j=0}^{p} \beta_j (x_t - x)^j \right\}^2 W\left(\frac{x_t - x}{h} \right), \quad (2)$$

where $W(u) = C_\mu (1 - u^2)^\mu 1_{[-1,1]}(u)$, $\mu = 0, 1, \ldots$ is the weight function (a second order kernel on $[-1, 1]$) and h is the (relative) bandwidth. In Equation (2) the bandwidth determines the smoothness of the trend and is the counterpart to HP's λ. Minimizing Equation (2) yields any v-th derivative of $m(x)$, defined as $m^{(v)}(x)$ $(v \leq p)$. If $p - \nu$ is odd, the linear smoother $\hat{m}^{(v)}(x)$ has automatic boundary correction and the bias is of order $k - \nu$. We use the Epanechnikov kernel as the weight function, which is optimal in the MSE sense. The resulting trend estimates are $\hat{m}^{(v)}(x) = v!\hat{\beta}_v$, where $v = 0, 1, \ldots, p$. Since the local linear estimator, where $p = 1$, results in the most stable boundary estimates (for two-sided filters), it seems a logical choice for estimating the output gap. In order to estimate the bandwidth in a data-driven manner, we follow [14], where the bandwidth is estimated by minimizing the asymptotic mean integrated squared error (AMISE):

$$AMISE(h) = h^{2(k-v)} \frac{I\left[m^{(k)}\right] \beta^2}{[k!]^2} + \frac{2\pi c_f (d_b - c_b) R(K)}{Th^{2v+1}}. \quad (3)$$

The corresponding optimal bandwidth h for estimating $m(x)$ on $[0, 1]$ is chosen using:

$$h_A = \left(\frac{2v+1}{2(k-v)} \frac{2\pi c_f [k!]^2 (d_b - c_b) R(K)}{I\left[m^{(k)}\right] \beta^2_{(v,k)}} \right)^{\frac{1}{(2k+1)}} T^{-1/(2k+1)}, \quad (4)$$

where $I\left[m^k\right] = \int_{c_b}^{d_b} \left[m^{(k)}(x)\right]^2 dx$, $\beta_{(v,k)} = \int_{-1}^{1} u^k K(u) du$, and $R(K) = \int_{-1}^{1} K^2(u) du$, and K is the asymptotically equivalent kernel in the interior. Furthermore, v is the order of the derivative and $k = p + 1$, so that m is k-times continuously differentiable. $c_f = f(0)$ is the value of the spectral density of ξ_t at the origin, with $f(\lambda) = 1/2\pi \sum_{l=-\infty}^{\infty} \gamma_\xi(l) e^{-il\lambda}$, $-\pi \leq \lambda \leq \pi$. The dependence structure is fully captured by the bandwidth. The values c_b and d_b can be chosen to select the bandwidth using only observations between these bounds. Details of the data-driven IPI are described in [15]. To address the criticism of [12,16], an asymmetric boundary kernel is used to weight the boundary points and the bandwidth at the boundary is kept constant such that the asymptotic properties at the boundary are the same as in the interior [17].

3. Output Gap Estimation Using the LLR

In this section, the LLR is used to estimate the output gap for the US economy without any parametric model assumptions of the output gap component. Therefore, quarterly US GDP vintages from 1947.1 to 2018.3 and annual US GDP from 1790 to 2018 by [18] are used. To contrast our results with those of [1,19], we follow their definitions. Thus, the final estimate of the output gap is defined as the detrended last available vintage (2018.3). Using the LLR for every vintage and collecting each endpoint estimation delivers a new time series that is defined as "the real-time estimate of the output gap" ([1], p. 571). As in [19], the last 2 years are not used to ensure that the comparison is not biased by the last vintages.

Figure 1 shows the real-time output gap estimates of the LLR compared to those of the HP filter ($\lambda = 1600$) for quarterly US GDP data. The HP filter and the LLR could be quite similar if λ is chosen correctly, which can also be detected in the resulting output gaps. Nevertheless, these approaches sometimes yield very different output gaps. In some cases, only the magnitude of the gaps differs, whereas in others the sign is contradicting. The HP gap is slightly smaller for the period from 1966.1 to 2018.3. This is obvious especially since the 2000s. These observations confirm the analysis of [1], where different detrending procedures yield various output gaps.

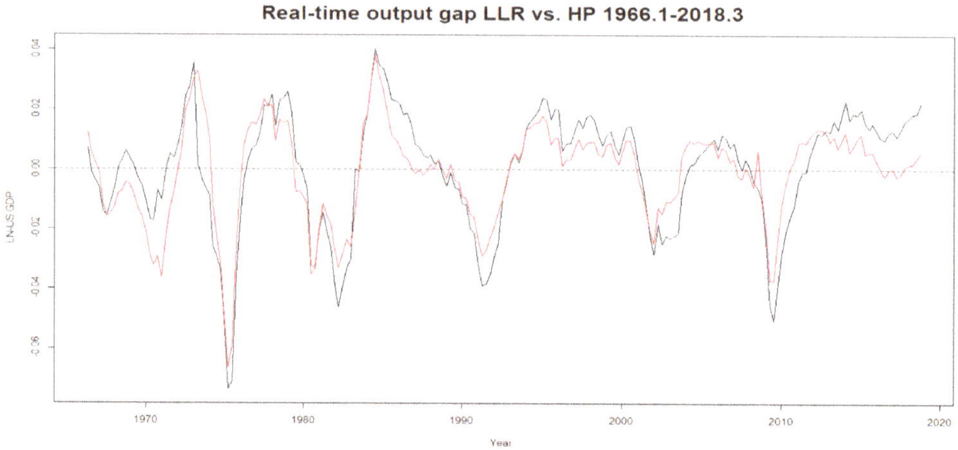

Figure 1. Real-time output gap estimation for quarterly US GDP data from 1966.1 to 2018.3 using the local linear regression (black) and the HP filter (red).

An even more stable real-time estimation of the gap is possible by using the LLR, since the trend is estimated appropriately with regard to the data-driven degree of smoothing and the introduced boundary correction increases the reliability of the output gap. The poor performance of the HP filter during periods of increased cyclical variation is examined by [20,21]. They conclude that using an unreliable detrending method such as the HP filter results in crises that are shown to be less intense than they actually are because most changes are attributed to trend movements. This presumable underestimation of the output gap using the HP filter is evident in Figure 2, where the gaps are shown for the Great Depression using data from 1790 to 2018. In this figure, the LLR (black) and the HP trend with $\lambda = 6.25$ (red) are shown for annual observations (grey line) from 1920 to 1960. The HP filter gap (red area) is significantly smaller than that estimated with the LLR (blue area). This may be a hint for the underestimation properties of the gap proposed by [20,21]. It is important to note that the amplitude of the HP filter can be adjusted using different values of λ. Nevertheless, for the LLR the bandwidth estimation is data-driven, so the arbitrary choice of λ is not necessary. To summarize, the data-driven selection and the stable and automatic boundary correction demonstrate the advantages of the LLR. The effects due to parameter, model and data uncertainty in the sense of [2] are per definition lower using the HP filter, but these smaller effects may not reflect the true output gap.

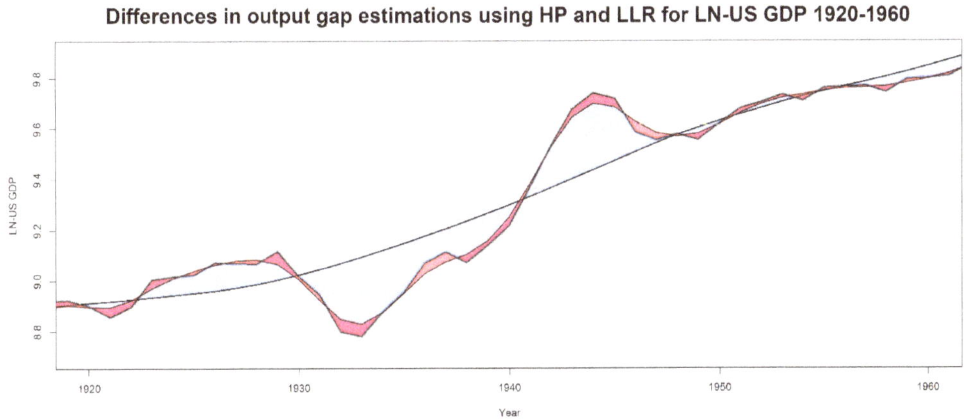

Figure 2. Estimation of the LLR (black) and the HP trend (red) with its corresponding gap estimations (blue area for LLR gap and red area for HP gap) and the original observations (grey) for US GDP from 1920 to 1960.

Since crises are unusually volatile transitory events, it is expected that the HP filter, which assumes a constant signal-to-noise ratio, performs less reliable in those periods. Although the IPI captures heteroscedastic events due to minimization of the AMISE, we improve the LLR by implementing a version that is able to leave those periods out for bandwidth selection.

The possible underestimation heavily influences monetary policy by, e.g., central banks, that in turn under- or overshoot with their interventions. Moreover, the different gap estimations influence the timing of policy actions. The HP filter has a similar disadvantage as one-sided filters. [16] argues in the setting of bandpass filtering that the underestimation of the output gap using these methods is a substantial error. A similar analysis for the period of the financial crisis around 2007/2008 shows that the estimated output gaps get smaller after the 2000s. As expected, the gap is significantly smaller than that estimated during the Great Depression, independent of the detrending approach, with neither method showing a significant gap for the recent period.

Various sources of uncertainty for gap estimation are identified by [2]. To analyze parameter uncertainty and parameter instability, we compare the final estimates and the real-time estimates of the output gap in Figure 3, which compares the real-time LLR gap (black) to the final LLR gap (green). Further, the real-time gap estimated with the HP filter (red) is compared to the final HP gap (blue). The differences between the real-time LLR gap (black) and the final LLR gap (green) partly reflect these different uncertainties. It is argued that a higher correlation between final and real-time estimation shows a lower level of revisions [2]. The calculated correlation for the LLR is 0.2949 and that for the HP filter is 0.5083. This discrepancy can be explained by the data-driven nature of the LLR, where the bandwidth changes slightly with every new observation point because the bandwidth depends on the sample size T (Equation (3)). By contrast, the smoothing parameter for the HP filter is fixed at $\lambda = 1600$, which causes no additional revisions to the gap estimates. Consequently, the correlation for the HP filter gap is higher per definition. However, the revision properties show that the LLR is appropriate for the ex post analysis of the output gap.

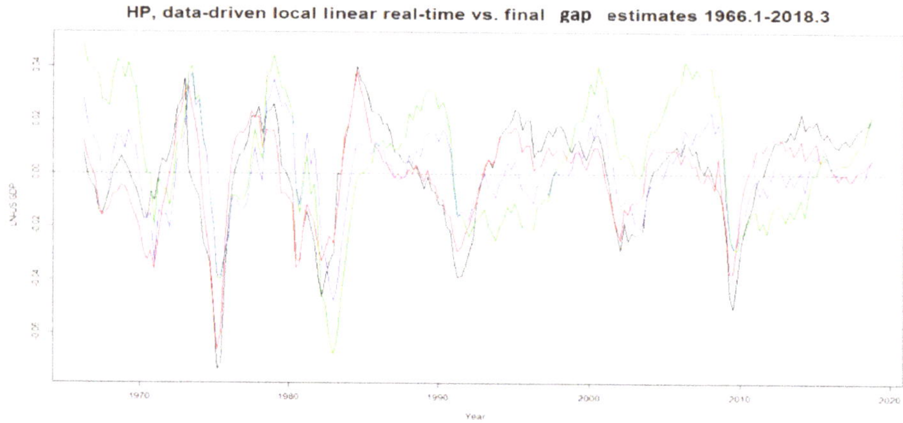

Figure 3. Real-time LLR (black) and final LLR (green) output gaps compared to real-time HP (red) and final HP (blue) output gaps estimation for quarterly US GDP data from 1966.1 to 2018.3.

4. Models for the Output Gap Component

The proposed AMISE-optimal decomposition may identify a more systematic cyclical component that needs to be analyzed during the further estimation of the output gap. Therefore, SETAR is used to further analyze the characteristics of the two different (LLR vs. HP) output gap series.

4.1. Semi-SETAR Model

To verify that the deterministic LLR combined with a further model for the gap component produces a more stable output gap, we fit different SETAR(k,p,d) models, as introduced by [22,23], to the residuals and their growth rates (LLR and HP gaps):

$$\hat{\xi}_t = \phi_0^{(j)} + \phi_1^{(j)}\xi_{t-1} - \ldots - \phi_p^{(j)}\xi_{t-p} + a_t^{(j)}, \text{ if } \gamma_{j-1} \leq \xi_{t-d} < \gamma_j \quad (5)$$

Residuals $\hat{\xi}_t$ are estimated by past realizations ξ_{t-p} and autoregressive coefficients $\phi_p^{(j)}$ such that the threshold variable ξ_{t-d} with d depicting the delay parameter lies in the range of γ_{j-1} up to γ_j dividing the domain of ξ_{t-d} into j regimes. $a_t^{(j)}$ are white noise errors. Trend and cycle are estimated using the LLR and HP filter and gaps are further analyzed with a SETAR model (We focus on the results for annual data as they are mostly used for cyclical analysis, see [2]). This modified and more systematic output gap identification has its merits for accurately timed policy actions, as additional information reduces problems affiliated with unsuitable policy activities [4].

In line with [14], we allow for two different regimes ($j = 2$) which are separated by the threshold zero in a high regime (HR) for expansions and a low regime (LR) for recessions. Moreover, different orders $p = 1, 2, 3$ of the AR part are tested and the delay parameter is set to $d = 0$. The results are displayed in Table S1 in the supplement material. It is evident that the coefficients are larger for the SETAR models fitted to the LLR output gap. Both coefficients are significantly different from zero. Whereas $\phi_{1,LLR}^{LR} = 0.9235$ implies that the next observation will be roughly the same within the same regime, $\phi_{1,HP}^{LR} = 0.3345$, which is much lower in magnitude, implies a much lower probability of similarity to the last observation Y_{t-1}. Thus, the LLR shows more systematic and larger gaps. By contrast, the HP filter gap implies more short-lived differences between actual and potential output. Using the -0.04 gap observed in 2010 leads to a gap of LLR that is three times the magnitude of that calculated using the HP based SETAR model (in absolute terms) and it lasts for a longer period when calculated with the LLR. The growth rates are analyzed in Table S2 in the supplements. The additional information provided by the LLR shows a long-lasting and significant expansion regime resulting in more accurately timed intervention of monetary policy makers (central banks). This drawback of the HP filter is ascribed to the arbitrarily selected smoothing parameter.

4.2. The OECD Approach and the Multivariate LLR

Since the true output gap is not observable, a valuation is difficult but a comparison with a methodological framework used by experts is straightforward. To demonstrate the performance of the LLR output gap estimations, we compare it to the output gap calculated by the OECD Economics Department. Using a Cobb–Douglas production function approach, [24] calculate potential output by using the trend components of labor efficiency (LE), a population between 15 and 74 (POP), and labor force participation rate between 15 and 74 (LFPR) obtained with a cyclical adjustment and the HP filter, with $\lambda = 100$. The unemployment rate is considered and filtered through the Kalman filter, where the productive capital stock (PK) enters the estimation without detrending. Following Equation (4) of [24], potential GDP (PGDP) is estimated by:

$$PGDP = \left[LE \cdot POP \cdot \frac{LFPR}{100} \cdot \left(1 - \frac{UNR}{100}\right)\right]^{\alpha} \cdot (PK)^{(1-\alpha)}. \quad (6)$$

To compare the OECD gap to the LLR gap, we adjust the estimation method of [24] by replacing the HP components in their Cobb–Douglas production function in our Equation (6) by the trend obtained using the LLR. Therefore, we extend the LLR to a multivariate approach. Afterwards, we determine potential output and finally the output gap. Figure 4 displays the OECD output gap approach using the LLR for detrending (green) together with the OECD gap using the HP filter (blue). Again, both estimated output gaps are quite similar and the magnitude is not significantly different, except during the Great Recession, where the OECD gap shows a much larger cycle. From 2008 onwards, the amplitude of the HP-filtered output gap is much larger than that of the LLR-based gap. Surprisingly, these results show that the LLR seems to have a higher variability than the HP trend since the Great Recession, which may be explained by additional cyclical adjustment used in [24]. However, the LLR needs no cyclical adjustment before detrending, is fully data-driven and more stable at boundary points in real-time applications.

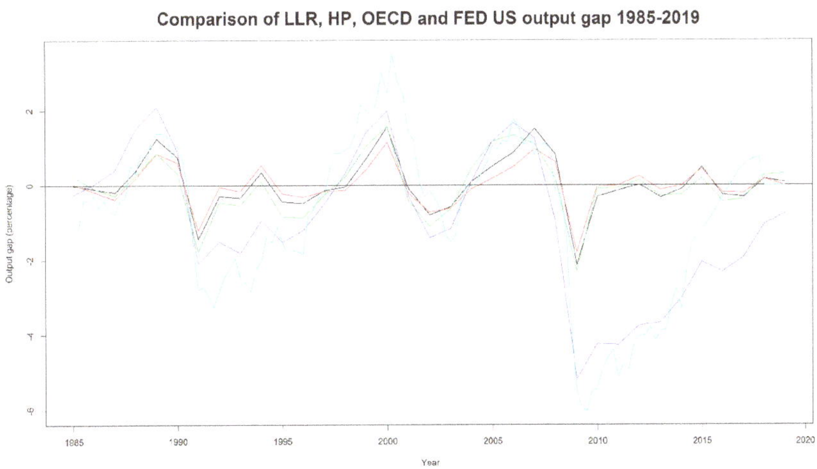

Figure 4. Comparison of univariate LLR (black) and HP filter (red) output gap with the OECD output gap using the LLR (green) and the HP filter (blue) together with the FED output gap (turquoise).

4.3. Comparison of Univariate EC and Multivariate OECD Approach

Multivariate methods, like the OECD production function approach, do not significantly enhance the quality of output gap estimation but impose additional structural assumptions [6]. Univariate time series methods, as used by the EC and the LLR, perform quite reasonably. To compare univariate and multivariate methods and to show the performance of the LLR in both types of applications, Figure 4 displays the gaps using the LLR (black) and the HP filter (red) for the final time series and the output gaps using the production function approach with the LLR (green) and the HP filter (blue). The output gap estimates from the FED (turquoise) are displayed as a benchmark in accordance with [13]. Obviously, both final univariate output gaps yield quite similar results (as for quarterly data). The gaps produced using the OECD approach are larger, although both multivariate gaps show similar dynamics. Those dynamics are also in line with the FED output gap. Surprisingly, the differences between the FED output gap and the OECD HP gap increase after 2014. It is important to mention that the frequency of the FED data is quarterly, thus the variability is larger than in the other series. However, using the HP filter and the additional cyclical adjustments of [24] produces a significantly larger gap than using the LLR. The LLR delivers output gap results in univariate and multivariate approaches that are in between those using the HP filter. Thus, the HP time series method may underestimate the gap while the HP OECD approach may overestimate it. This could be an argument in favor of the data-driven LLR, which is less arbitrary than the HP filter with regard to the degree of smoothness and produces more stable boundary estimates. As mentioned by [19], output gaps estimated by policy institutions provide a good benchmark to compare gaps. Thus, gaps from economic experts may be more reliable compared to purely statistical approaches [13]. They demonstrate that the FED use an evaluated and weighted average of statistical and structural methods. Table 1 shows the correlation coefficients for the LLR and HP based output gaps with those of the FED and the OECD. The correlation coefficients between the proposed LLR approach and the expert gaps are significantly higher than using the real-time HP filter gap. Using the LLR depicts the gaps estimated with economic expertise more reliably than using the HP filter. In other words, the LLR reflects the output gap benchmark provided by policy institutions more precisely than the HP filter.

Table 1. Correlation between real-time LLR and HP gaps with ex post gaps from policy institutions.

Output Gap	FED	OECD
LLR	0.6488	0.7071
HP	0.5071	0.5678

5. Forecasting and Evaluating Output Gaps

Among others, ref [25] use forecasting methods to estimate revisions in potential growth. Since the semi-SETAR model is able to reproduce cyclical features in recessions and expansions, extending it to forecast gaps is straightforward.

5.1. Output Gap Forecasting Using the Semi-SETAR Model

We use the LLR and the SETAR model to forecast the output gap. Firstly, the trend is estimated and removed from the original observations (Equation (1)). Secondly, a SETAR model is fitted to the residuals (Equation (5)). Finally, the SETAR model is used for forecasting using the SETAR(2,3,0) model and quarterly US GDP data. The forecast horizon is set to five quarters ($k = 5$), so the training set ranges from 1947.1 to 1966.1. The series are forecasted (in sample) by recursively updating the sample by one observation starting in 1966.1. To capture different uncertainties, we use a bootstrap method with $n = 10,000$ to forecast the future paths of US output [23]. The forecast results are depicted exemplarily for the sample ending 2017.3 (last in sample forecast) in Figure S1 in the supplements. Compared to the original observations, the semi-SETAR model is able to forecast the output gap quite well for the first year. Original observations and forecasts are nearly identical from 2017.3 to 2018.2 (The detailed results for every forecast value between 1966.1–2018.3 are available upon request). To validate the forecast performance, we calculate the mean absolute scaled error (MASE), which is $MASE = 0.5952$ in this case. Thus, the semi-SETAR model improves forecast performance and delivers a reliable output gap forecast by including information from two regimes.

5.2. Predictive Power of the LLR Output Gap for Output Growth

Problems in evaluating the output gap due to a missing true gap can be overcome by the evaluation of the forecasting performance of the output gap for output growth [26]. use This idea is used by [19] by arguing for a negative correlation between gaps and future growth. They use the following equation to predict output growth using the estimated real-time gap:

$$y_{t+k} - y_t = \alpha + \beta \hat{c}_t + \epsilon_{t+k\,|t}, \tag{7}$$

where $y_{t+k} - y_t$ is output growth, $\hat{c}_t$ is the estimated real-time output gap using either the LLR or HP filter and $\epsilon_{t+k\,|t}$ displays the forecast error. The forecast horizon is $k = 1, \ldots, 8$. Due to the trend-reverting properties, $\beta < 0$ is expected [19,26]. The OLS estimates show the expected signs for LLR- and HP-filtered gaps and are significantly negative for $k = 1, \ldots, 8$.

By comparing the relative RMSEs in Table 2, a small improvement in forecast accuracy is found using the LLR real-time gap. Surprisingly, the gains are higher the larger the forecast horizon is. Compared to the forecasting performance when using the HP gap, the LLR gap improves the forecast accuracy of output growth. A similar exercise can be carried out for inflation. However, in accordance with [13] output gaps usually do not improve inflation forecasts and are hence omitted.

Table 2. Relative RMSEs for output growth forecasts evaluation using LLR and HP real-time gaps.

Horizon	1	2	3	4	5	6	7	8
LLR/HP	0.9999	1.0031	1.0074	1.0058	0.9958	0.9842	0.9717	0.9612

6. Policy Implications and Conclusions

We argue for a more detailed and systematic output gap analysis by combining output gap estimation and SETAR models. Using this additional information, the improved estimation quality at boundary points and the LLR result in an improved estimation of the output gap compared to the standard HP-filtered gap. This is demonstrated by a comparison of both statistically based methods with those estimated with economic expertise by the OECD and the FED. The LLR output gap shows a higher correlation with the OECD and FED gap than the HP filter does. This is partly attributable to the data-driven selection of the bandwidth, which improves the disadvantage of the arbitrary selection in the HP filter. In addition, the HP time series filter attributes more originally cyclical fluctuations to the trend and leaves a too-small gap component, a misspecification that may impede an appropriate real-time gap estimation. Within the OECD approach, we observe the other extreme of a large output gap using the HP filter in combination with the production function approach. While the LLR is successfully extended to the multivariate production function approach, it performs similarly to the OECD method. Extending the semi-SETAR model improves output gap forecasts using additional information from different growth regimes. Using the proposed output gap for forecasting output growth, the LLR real-time gap performs better compared to the HP gap, in the sense that it has a larger predictive power for output growth. These results modify the timing and magnitude of monetary policy decisions as the new model allows a more reliable output gap estimation than the HP filter.

Supplementary Materials: The following are available online at https://www.mdpi.com/article/10.3390/engproc2021005032/s1. Table S1: Estimated model parameters for SETAR(k,p,d) model US GDP 1790-2018 using LLR and HP filter for the residuals. Table S2: Estimated model parameters for SETAR(k,p,d) model US GDP 1790-2018 using LLR and HP filter for the residual growth rates. Figure S1: SETAR(2,3,0) point forecast of the output gap (red dashed) together with the original data (black dotted) for 2017.3-2018.4 using the LLR for quarterly US GDP (black solid) 1947.1-2018.3.

Funding: This research did not receive any specific grant from funding agencies in the public, commercial or not-for-profit sectors.

Institutional Review Board Statement: Not applicable.

Informed Consent Statement: Not applicable.

Data Availability Statement: The data used in this study can be accessed through the following links:

1. Federal Reserve Bank of Philadelphia. Quarterly Real GDP Vintages; Historical Data Files for the Real-Time Data Set by D. Croushore and T. Stark. 2019Available online: https://www.philadelphiafed.org/research-and-data/real-time-center/real-time-data/data-files/routput (accessed on 29 May 2019).
2. Federal Reserve Bank of Philadelphia. Greenbook Output Gap DH Web FED. 2021. Available online: https://www.philadelphiafed.org/surveys-and-data/real-time-data-research/gap-and-financial-data-set (accessed on 20 January 2021).
3. US Bureau of Economic Analysis. Real Gross Domestic Product; 2019. Accessed from FRED, Federal Reserve Bank of St. Louis. Available online: https://research.stlouisfed.org/fred2/series/GDPC1 (accessed on 24 April 2019).
4. Johnston, L.; Williamson, S.H. "What Was the U.S. GDP Then?". Measuring Worth 2019. Available online: http://www.measuringworth.org/usgdp/ (accessed on 15 June 2019).

Acknowledgments: I thank Thomas Chalaux for providing the data for the comparison and Lukas Wiechers for providing excellent research assistance. I also thank Frédérique Bec, Francisco Serranito, Camelia Turcu and the participants of the 7th Applied Macro-Econometrics Workshop in Paris and the 22nd INFER annual conference for their helpful comments and suggestions.

Conflicts of Interest: The author declares no conflict of interest.

References

1. Orphanides, A.; van Norden, S. The Unreliability of Output-Gap Estimates in Real Time. *Rev. Econ. Stat.* **2002**, *84*, 569–583. [CrossRef]
2. Marcellino, M.; Musso, A. The Reliability of Real Time Estimates of the Euro Area Output Gap. *Econ. Model.* **2010**, *28*, 1842–1856. [CrossRef]
3. Coibion, O.; Gorodnichenko, Y.; Ulate, M. *The Cyclical Sensitivity in Estimates of Potential Output*; National Bureau of Economic Research Working Paper No. w23580; National Bureau of Economic: Cambridge, MA, USA, 2017.
4. Nelson, E.; Nikolov, K. UK Inflation in the 1970s and 1980s: The Role of Output Gap Mismeasurement. *J. Econ. Bus.* **2003**, *55*, 353–370. [CrossRef]
5. Grigoli, F.; Herman, A.; Swiston, A.; Di Bella, G. Output Gap Uncertainty and Real-Time Monetary Policy. *Russ. J. Econ.* **2015**, *1*, 329–358. [CrossRef]
6. Álvarez, L.J.; Gómez-Loscos, A. A Menu on Output Gap Estimation Methods. *J. Policy Model.* **2018**, *40*, 827–850. [CrossRef]
7. Baxter, M.; King, R.G. Measuring Business Cycles: Approximate Band-Pass Filters for Economic Time Series. *Rev. Econ. Stat.* **1999**, *81*, 573–593. [CrossRef]
8. Beveridge, S.; Nelson, C.R. A New Approach to Decomposition of Economic Time Series into Permanent and Transitory Components with Particular Attention to Measurement of the Business Cycle. *J. Monet. Econ.* **1981**, *7*, 151–174. [CrossRef]
9. Kamber, G.; Morley, M.; Wong, B. Intuitive and Reliable Estimates of the Output Gap from a Beveridge-Nelson Filter. *Rev. Econ. Stat.* **2018**, *100*, 550–566. [CrossRef]
10. Hodrick, R.J.; Prescott, E.C. Postwar U.S. Business Cycles: An Empirical Investigation. *J. Money Credit Bank.* **1997**, *29*, 1–16. [CrossRef]
11. De Brouwer, G. *Estimating Output Gaps*; Research Discussion Paper 9809; Reserve Bank of Australia: Sydney, Australia, 1998.
12. Hamilton, J.D. Why You Should Never Use the Hodrick-Prescott Filter. *Rev. Econ. Stat.* **2018**, *100*, 831–843. [CrossRef]
13. Edge, R.M.; Rudd, J.B. Real-Time Properties of the Federal Reserve's Output Gap. *Rev. Econ. Stat.* **2016**, *98*, 785–791. [CrossRef]
14. Fritz, M.; Gries, T.; Feng, Y. Growth Trends and Systematic Patterns of Booms and Busts—Testing 200 Years of Business Cycle Dynamics. *Oxf. Bull. Econ. Stat.* **2019**, *81*, 62–78. [CrossRef]

15. Feng, Y.; Gries, T.; Fritz, M. Data-driven local polynomial for the trend and its derivatives in economic time series. *J. Nonparametric Stat.* **2020**, *32*, 1–24. [CrossRef]
16. Watson, M.W. How Accurate Are Real-Time Estimates of Output Trends and Gaps? *Econ. Q.* **2007**, *93*, 143–161.
17. Fritz, M.; Gries, T.; Feng, Y. Secular Stagnation? Is there Statistical Evidence of an Unprecedented, Systematic Decline in Growth? *Econ. Lett.* **2019**, *181*, 47–50. [CrossRef]
18. Johnston, L.; Williamson, S.H. "What Was the U.S. GDP Then?". *Measuring Worth.* 2019. Available online: http://www.measuringworth.org/usgdp/ (accessed on 15 June 2019).
19. Quast, J.; Wolters, M.H. Reliable Real-Time Output Gap Estimates Based on a Modified Hamilton Filter. *J. Bus. Econ. Stat.* **2020**, 1–17. [CrossRef]
20. McCallum, B.T. *Alternative Monetary Policy Rules: A Comparison with Historical Settings for the United States, the United Kingdom, and Japan*; National Bureau of Economic Research Working Paper No. w7725; National Bureau of Economic: Cambridge, MA, USA, 2000.
21. Pollock, D.S.G. Trend Estimation and De-Trending via Rational Square-Wave Filters. *J. Econom.* **2000**, *99*, 317–334. [CrossRef]
22. Tong, H. *Threshold Models in Nonlinear Time Series Analysisl*; Time Series Analysis; Springer: Berlin, Germany, 1983.
23. Grabowski, D.; Staszewska-Bystrova, A.; Winker, P. Generating Prediction Bands for Path Forecasts from SETAR Models. *Stud. Nonlinear Dyn. Econom.* **2017**, *21*, 1–18. [CrossRef]
24. Chalaux, T.; Guillemette, Y. *The OECD Potential Output Estimation Methodology*; OECD Economics Department Working Papers No. 1563; OECD Publishing: Paris, France, 2019. [CrossRef]
25. Blanchard, O.; Lorenzoni, G.; L'Huillier, J.P. Short-Run Effects of Lower Productivity Growth. A Twist on the Secular Stagnation Hypothesis. *J. Policy Model.* **2017**, *39*, 639–649. [CrossRef]
26. Nelson, C.R. The Beveridge-Nelson Decomposition in Retrospect and Prospect. *J. Econom.* **2008**, *146*, 202–206. [CrossRef]

engineering proceedings

MDPI

Proceedings

Rényi Transfer Entropy Estimators for Financial Time Series †

Petr Jizba ‡, Hynek Lavička ‡,§ and Zlata Tabachová *,‡

Faculty of Nuclear Sciences and Physical Engineering, Czech Technical University in Prague, Břehová 7, 115 19 Praha 1, Czech Republic; p.jizba@fjfi.cvut.cz (P.J.); hynek.lavicka@fjfi.cvut.cz (H.L.)
* Correspondence: Zlata.Tabachova@fjfi.cvut.cz (Z.T.); Tel.: +420-775-317-309
† Presented at the 7th International conference on Time Series and Forecasting, Gran Canaria, Spain, 19–21 July 2021.
‡ These authors contributed equally to this work.
§ Current address: Department of Institutional, Environmental and Experimental Economics, University of Economics in Prague, 130 67 Prague, Czech Republic.

Abstract: In this paper, we discuss the statistical coherence between financial time series in terms of Rényi's information measure or entropy. In particular, we tackle the issue of the directional information flow between bivariate time series in terms of Rényi's transfer entropy. The latter represents a measure of information that is transferred only between certain parts of underlying distributions. This fact is particularly relevant in financial time series, where the knowledge of "black swan" events such as spikes or sudden jumps is of key importance. To put some flesh on the bare bones, we illustrate the essential features of Rényi's information flow on two coupled GARCH(1, 1) processes.

Keywords: Rényi's transfer entropy; financial time series; GARCH processes

Citation: Jizba, P.; Lavička, H.; Tabachová, Z. Rényi Transfer Entropy Estimators for Financial Time Series. *Eng. Proc.* **2021**, *5*, 33. https://doi.org/10.3390/engproc2021005033

Published: 30 June 2021

1. Introduction

The linear framework for measuring and testing causality has been widely applied in a number of fields. In finance, one typically uses Granger's linear regression model to study the internal cross-correlations between various market activities. The correlation functions have, however, at least two limitations. First, they measure only linear relations, although it is clear that linear models do not faithfully reflect real market interactions. Second, all they determine is whether two time series (e.g., two stock-index series) have correlated movement. They, however, do not indicate which series affects which, or in other words, they do not provide any directional information about cause and effect. However, there is extensive literature on causality modeling that goes beyond the linear regression model, e.g., applying and combining mathematical logic, graph theory, Markov models, Bayesian probability, etc. (for an extensive review see, e.g., [1]). We will focus here on the information-theoretic approaches, which understand causality as a phenomenon that can be not only detected or measured but also quantified. A particularly important quantifier of the information flow between two time series is the so-called *transfer entropy* (TE).

In his 2000 seminal paper, Schreiber [2] used Shannon's information measure to formulate the concept of TE, which is a version of mutual information operating on conditional probabilities. TE is designed to detect the directed exchange of information between two stochastic variables, conditioned to common history and inputs. An advantage of information-theoretic measures, in comparison with, say, the standard Granger causality, is that they are sensitive to nonlinear signal properties, as they do not rely on linear regression models. A limitation of TEs is that they are, by their very formulation, restricted to bivariate situations. In addition, information-theoretic measures often require substantially more data than regression methods. It can be also shown that for Gaussian variables, Granger causality and transfer entropy are entirely equivalent [3]. For a comparison of TEs with other causal measures, including various implementations of the Granger causality, see, e.g., Ref. [4].

Shannonian TE was generalized to the class of α-Rényi transfer entropies by Jizba et al. in Ref. [5]. The corresponding Rényi TE can be defined in much the same way as its Shannonian counterpart. In particular, one can utilize the concept of mutual information of the order α to quantify the directed exchange of information. Because Rényi's entropy (RE) works, unlike Shannon's entropy, with rescaled distributions, it allows addressing information flow between different parts of underlying distributions in bivariate time series. Consequently, Rényi's TE provides more detailed information concerning the excess (or lack) of information in various parts of the underlying distribution resulting from updating the distribution on the condition that a second time series is known. This is particularly relevant in the context of financial time series, where the knowledge of tale-part (or "black swan") events such as spikes or sudden jumps bears direct implications, e.g., in various risk-reducing formulas in portfolio theory.

In order to quantify the strength of Rényian information flow and its directionality from high-quality time series data, special care has to be taken to select suitable estimators of Rényi's entropy. The aim of this paper is to demonstrate that the estimator introduced by Leonenko et al. [6] is an appropriate instrument for this task. We illustrate this by analyzing Rényi's information flow between two coupled GARCH(1,1) processes.

The paper is organized in the following way. In Section 2, we discuss some essentials from Rényi's entropy and the ensuing Rényi transfer entropy. Section 3 introduces the concept of effective transfer entropy and briefly discusses the pros and cons of Leonenko et al.'s Rényi entropy estimator. In Section 4, we set up our model system, namely a system of two coupled GARCH processes, that will serve as generating processes for bivariate time series to be analyzed. Section 5 is dedicated to the analysis of the effective Rényi's TE for coupled GARCH(1, 1) processes. Finally, in Section 6, we provide some concluding remarks and propose some further generalizations.

2. Rényi Transfer Entropy

In this section, we briefly review some essentials of Rényi's entropy and ensuing directional information flow that will be needed in following sections.

2.1. Rényi Entropy

Rényi's information measure (also known as Rényi's entropy) was introduced by Rényi in his seminal 1961 paper [7] as a one-parameter generalization of *Shannon's entropy*. Let $\alpha \geq 0$, then Rényi's entropy of a probability distribution function $\mathcal{P}$ associated with a discrete random variable X is defined as

$$H_\alpha[\mathcal{P}] = \frac{1}{1-\alpha}\log_2 \sum_{x \in X} p^\alpha(x)\,. \tag{1}$$

In particular, for $\alpha = 0$, we obtain the so-called *Hartley entropy*, while the cases with $\alpha = 2$ and $\alpha = +\infty$ yield the *collision entropy* (that is closely related to *correlation dimension*) and the *Min-entropy*, respectively. Note that for $\alpha = 1$, Rényi's entropy converges to Shannon's entropy by the L'Hospital rule. It can be shown [8] that Rényi's entropy is a non-negative, monotonically decreasing function of α, thus

$$H_0 \geq H_1 \geq H_2 \geq H_{+\infty} \geq 0\,. \tag{2}$$

Particularly important in our following considerations will be the so-called *conditional* Rényi entropy that is defined as [8,9]

$$H_\alpha[\mathcal{P}|\mathcal{Q}] = \frac{1}{1-\alpha}\log_2\frac{\sum_{x\in X,y\in Y}p^\alpha(x,y)}{\sum_{y\in Y}p^\alpha(y)}$$
$$\equiv \frac{1}{1-\alpha}\log_2\sum_{y\in Y}\rho_\alpha(y)\sum_{x\in X}p^\alpha(x|y), \quad (3)$$

where $\mathcal{Q}$ is a probability distribution function of a random variable Y, and ρ_α, defined as

$$\rho_\alpha(x) = \frac{p^\alpha(x)}{\sum_{x\in X}p^\alpha(x)}, \quad (4)$$

is known as *escort distribution* [10]. The latter is also termed as a "zooming" distribution because it scales (deforms and re-emphasizes) different parts of an underlying distribution function $\mathcal{P}$. In particular, for $\alpha < 1$, the central part of the distribution is flattened, i.e., high-probability events are suppressed, and low-probability events are emphasized. This effect is more pronounced for smaller α. In the opposite situation when $\alpha > 1$, low-probability events are suppressed, and high-probability events are emphasized. Thus, for $\alpha \to 0$, the escort distribution tends to a uniform-like shape and $\alpha \to +\infty$ to a very platykurtic (Dirac's-δ function like) distribution. Because this behavior is also true for the conditional RE, it will be seen that REs are instrumental in the understanding of (directional) information flow between bivariate time series.

2.2. Shannon's and Rényi's Transfer Entropies

The concept of *transfer entropy* was introduced by Schreiber in Ref. [2] and independently under the name *conditional mutual information* by Paluš in Ref. [11]. According to these, TE represents a measure of a directional (Shannonian) information flow defined by means of *Kullback–Leibler divergence* on conditional transition probabilities of two Markov processes X and Y as

$$T_{X\to Y}(k,l) = \sum_{x\in X,y\in Y} p(y_{n+1},y_n^{(l)},x_n^{(k)})\log_2\frac{p(y_{n+1}|y_n^{(l)},x_n^{(k)})}{p(y_{n+1}|y_n^{(l)})}. \quad (5)$$

Here l and k denote Markov orders of Y and X processes, respectively, e.g., $x_n^{(k)} \equiv (x_n,\ldots,x_{n-k+1})$. For independent processes, TE is equal to zero. It is also not a symmetric measure as mutual information is; therefore, $T_{Y\to X} \neq T_{X\to Y}$, which becomes clear if we rewrite (5) as

$$T_{X\to Y}(k,l) = H(y_{n+1}|y_n^{(l)}) - H(y_{n+1}|y_n^{(l)},x_n^{(k)}). \quad (6)$$

Now we can use Equation (6) to define *Rényi's transfer entropy* (RTE). Substituting H_α instead of H, we obtain [5]

$$T^R_{\alpha,X\to Y}(k,l) = \frac{1}{1-\alpha}\log_2\frac{\sum\rho_\alpha(y_n^{(l)})p^\alpha(y_{n+1}|y_n^{(l)})}{\sum\rho_\alpha(y_n^{(l)},x_n^{(k)})p^\alpha(y_{n+1}|y_n^{(l)},x_n^{(k)})}. \quad (7)$$

It can be checked that Definition (5) is a special case of (7) for $\alpha = 1$, which we will refer to as *Shannon's transfer entropy* (STE). Most of the aforementioned properties of STE are still valid also for RTE. The most important difference is that the zero values of RTE for $\alpha \neq 1$ do not imply the independence of processes X and Y (i.e., that all order cross-correlations are zero); however, if X and Y are independent, RTE is zero for any α. In addition, in contrast to the Shannonian case, RTE can also have negative values. The reason for this is not difficult to understand. For instance, for, $\alpha < 1$, the negativity of $T^R_{\alpha,X\to Y}(k,l)$ simply means that the knowledge of historical values of both X and Y flattens the tail part of the anticipated distribution function for the price value y_{n+1} more than the historical values of Y alone would do. In other words, extra knowledge of the historical values of X shows that there is a greater risk in the next time step of Y than one would expect by only knowing the historical data of Y. In this sense, $T^R_{\alpha,X\to Y}(k,l)$ represents a *rating factor* that quantifies a gain or loss in the risk concerning the behavior of Y at the future time y_{n+1} after the historical values of X until x_n were accounted for [5]. This information can be further used in financial decisions concerning risk analysis, portfolio selection, or in derivative pricing.

3. Financial Data Processing and Rényi Entropy Estimation

Financial data recorded on stock markets are typically nonstationary, discrete, and with periods of no trading activity. The last two factors can be dealt with technically by means of pertinent data processing methods. The nonstationarity might be problematic for the Markov assumption that we used in the definition of the transfer entropy; however, this is typically solved by introducing a new variable that can be thought of as asymptotic stationary [12].

Following Samuelson's work on geometric Brownian motion [13], it has become clear that the asset *log-return* (rather than raw return) is the relevant financial variable. So, for our future convenience, it is suitable to define the log-return associated with X as

$$R_{X,\tau} = \log\left(\frac{x_{t+\tau}}{x_t}\right), \tag{8}$$

where x_t is the value of the process X at time t.

Another problem that might hinder the correct estimation of the transfer entropy is a limited number of the recorded data. To this end, Marchinski et al. introduced in [12] *effective transfer entropy*, which, for Rényi's TE, can be rewritten in the form

$$T^{R,\text{eff}}_{\alpha,X\to Y} = T^R_{\alpha,X\to Y} - T^R_{\alpha,X_{\text{shuffled}}\to Y}. \tag{9}$$

Effective RTE is thus a difference between two RTEs, where the second one is computed on the *shuffled* X series. Here the shuffling is performed in terms of the *surrogate data technique* [14]. In essence, a surrogate data series has the same mean, the same variance, the same autocorrelation function, and, therefore, the same power spectrum as the original series, but phase relations are destroyed. Consequently, all the potential correlations between X and Y are removed, which implies that $T^R_{\alpha,X_{\text{shuffled}}\to Y}$ should be zero. In practice, this is typically not the case, despite the fact that there is no obvious structure in the data. The nonzero value of $T^R_{\alpha,X_{\text{shuffled}}\to Y}$ must then be a consequence of the finite dataset. Definition (9) then simply ensures that pseudo-effects caused by finite values of k and l are removed.

By its very definition, effective RTE is not symmetric in X and Y. So, in order to visualize and quantify the disparity between the $X \to Y$ and $Y \to X$ flow, it is convenient to define the *balance of flow* or *net flow* of effective RTE as

$$T^{R,\text{bal eff}}_{\alpha,X\to Y} = T^{R,\text{eff}}_{\alpha,X\to Y} - T^{R,\text{eff}}_{\alpha,Y\to X}. \tag{10}$$

The concept of the balance of flow of effective RTE will be employed in Section 5.

In their original work, Marchinski et al. [12] employed effective STE to compute the information transfer between two financial time series. They also used a partitioning method to discretize their financial data. This is a good first approach to data processing that was also employed in our earlier work [5]. However, partitioning can cause the loss of valuable correlations present in the data. That is why, in the following, we test another method of data processing and evaluate RTE using estimators for Rényi's entropy introduced by Leonenko et al. [6].

3.1. Rényi's Entropy Estimation

Estimators of Shannon's entropy based on the k-nearest-neighbor search in one-dimensional spaces were studied in statistics already almost 60 years ago by Dobrushin [15] and a short while later by Vašíček [16]. Unfortunately, they cannot be generalized directly to higher-dimensional spaces, and hence, they are inapplicable to TEs. Presently, there exists a number of suitable entropy estimators—most of them in the Shannonian framework (for a review, see, e.g., [17]). Here we will present the k-nearest-neighbor entropy estimator for higher-dimensional spaces introduced by Leonenko et al. [6]. The latter is not only suitable for RE evaluation but it can also easily be numerically implemented so that RTE can be computed in real time, which is clearly relevant in finance, for instance, in various risk-aversion decisions. An explicit empirical analysis based on this estimator will be presented in Section 5.

Leonenko et al.'s [6] approach is based on an estimator of RE from a finite sequence of N points, and it is defined as

$$\widehat{H}_{N,k,\alpha} = \frac{1}{1-\alpha}\left[\log_2\left(\frac{\Gamma(k)}{\Gamma(k+1-\alpha)}\frac{\pi^{\frac{m(1-\alpha)}{2}}}{\Gamma^{1-\alpha}\left(\frac{m}{2}+1\right)}(N-1)\sum_{i=1}^{N}\left(\rho_k^{(i)}\right)^{m(1-\alpha)}\right) - \log_2 N\right]. \quad (11)$$

Here $\Gamma(x)$ is Euler's gamma function, m is the dimension of the dataset space, and $\rho_k^{(i)}$ is the distance from data i to the k-th nearest data counterpart using the Euclidean metric. The estimator thus depends on the number of data in a dataset N and the rank of the nearest neighbor used k.

Advantages of Estimator (11) in contrast to the standard bin method that estimates probability within a range are:

- Relative accuracy for small datasets;
- Applicability for high-dimensional data;
- Combining the set estimators provides statistics for estimation.

The disadvantage of the method is the computational complexity of the algorithm and the complicated data container. The algorithm can, however, be optimized so that it can run in real time. We can also stress that in contrast to other RE estimators, such as the *fixed-ball* estimator [17], Estimator (11) is not confined to only a certain range of α values.

Estimators of the average and standard deviation for a dataset of size N and the parameter of RE α with Bessel correction are defined, respectively, as

$$\overline{H}_{N,\alpha} = \frac{\sum_{k=1}^{n}\widehat{H}_{N,k,\alpha}}{n},$$

$$\sigma_{H_{N,\alpha}} = \sqrt{\frac{\sum_{k=1}^{n}\left(\widehat{H}_{N,k,\alpha} - \overline{H}_{N,\alpha}\right)^2}{N-1}}, \quad (12)$$

where n is the highest order of the nearest data counterpart. Theoretically, we should use $n = N$, but such a set up would require an enormous amount of computer memory to hold the distances. So, in our calculations, we used $n = 50$, which turned out to be a good compromise between accuracy and computer time.

4. Model Setup: Coupled GARCH Processes

Assuming independent events, as it is typically done in the financial context, is not very realistic. To capture time dependence between different log-returns, it is often convenient to assume that volatility (the square root of the variance of log-returns) for a given financial asset is a time-dependent stochastic process. This assumption is called *heteroskedasticity*, where each term in the time series is described with a generally different variance. Typically, asset returns are not even close to being independent and identically distributed, and their distributions are often heavy-tailed. Another observed fact is a tendency that large changes in prices are followed by large changes and small changes by small changes. This is known as volatility clustering. A stochastic process that is able to capture distributional stylized facts (such as heavy tails or high peakedness), as well as the time series stylized facts (such as volatility clustering), was introduced by Engle in 1982 [18] under the name *autoregressive conditional heteroskedasticity* or simply ARCH.

In Engle's original ARCH(q) model, the conditional variance at time t, i.e., σ_t^2, was postulated to be a linear function of the squares of past q observations modeled by a stochastic process $x_t \sim N(0, \sigma_t^2)$. Unfortunately, in many applications of the ARCH model, a long lag length, and therefore a large number of parameters, is required. This makes the parameter estimation quite impractical. To circumvent this problem, Bollerslev [19] generalized the ARCH process to the so-called *generalized* ARCH (or simply GARCH) process. The latter is (similar to ARCH) a linear function of the squares of past observations plus the linear combination of the past values of variances. For instance, the GARCH(q, p) process specifies the conditional variance as follows:

$$\sigma_t^2 = \alpha_0 + \alpha_1 x_{t-1}^2 + \cdots + \alpha_q x_{t-q}^2 + \beta_1 \sigma_{t-1}^2 + \cdots + \beta_p \sigma_{t-p}^2, \tag{13}$$

where $\alpha_0 > 0, \alpha_i \geq 0$ $(i = 1, \ldots, q)$ and $\beta_k \geq 0$ $(k = 1, \ldots, p)$ are control parameters. It is also common to require that the covariance stationarity condition holds, which implies that $\sum_{i=1}^{q} \alpha_i + \sum_{k=1}^{p} \beta_k < 1$ (see, e.g., [19]). In most empirical applications, it turns out that the simple choice $p = q = 1$ is already able to correctly grasp the volatility dynamics of financial data.

To test RTE with the RE estimator (11), we will examine two GARCH$(1,1)$ processes with unidirectional coupling. The stationary coupling parameter ϵ will allow us, in turn, to probe how the information flows between the two GARCH$(1,1)$ processes change with the coupling strength. To be more specific, let us consider two coupled GARCH$(1,1)$ processes. In particular, let $x_t \sim N(0, \sigma_t^2)$ be a GARCH$(1,1)$ process with $\alpha_0 = 0.2$, $\alpha_1 = \beta_1 = 0.4$ and $y_t \sim N(0, \eta_t^2)$ be a GARCH$(1,1)$ process with $\alpha_0 = 0.3$, $\alpha_1 = \beta_1 = 0.35$ such that

$$\eta_t^2 = \alpha_0 + \alpha_1 y_{t-1}^2 + \beta_1 \eta_{t-1}^2 + \epsilon x_{t-1}^2 . \tag{14}$$

In this way, the coupling between stochastic variables is not direct but mediated through variance. Nonlinear coupling in variances of the GARCH processes is a good approximation of the possible couplings in real-world market data in which one asset, say X, can cause volatility $\eta = \eta(X)$ that will in turn influence another asset, say Y. For future convenience, we will refer to process X as the *master* process and to Y as the *slave* process.

5. Analysis of Effective RTE for Coupled GARCH$(1,1)$ Processes

In this section, we use Estimator (11) for Rényi's entropy to compute effective RTE (9) between coupled GARCH$(1,1)$ processes (13) and (14). The calculations are performed directly on log-returns (8) rather than on amplitudes. A typical dependence of RTE (7) on both α and the coupling strength ϵ is depicted in Figure 1. Ensuing effective RTEs (9) with different Markov parameters are presented in Figure 2. In order to quantify the master–slave relationship in terms of information flow, the balance of flow (10) is calculated (see Figure 3). Respective standard deviations are depicted in Figure 4.

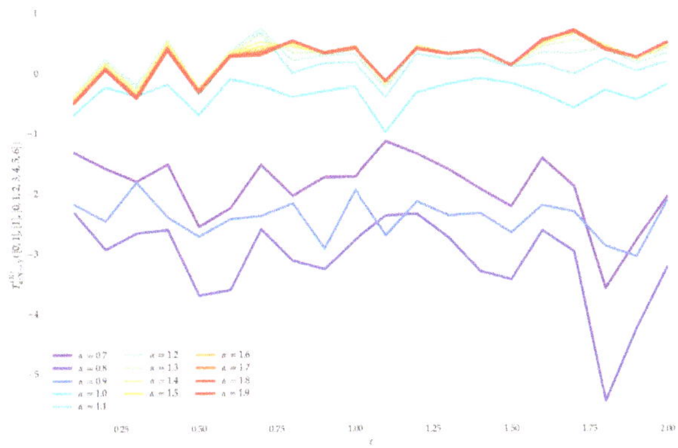

Figure 1. Transfer entropy $T^{R}_{\alpha,X\to Y}(2,7) \equiv T^{(}R)_{\alpha,X\to Y}([0,1],[1],[0,1,2,3,4,5,6])$, where X and Y are GARCH processes described in Section 4. The coupling strength $\epsilon \in \{0.1, 0.2, ..., 2\}$ is on the horizontal axis, and the transfer entropy is on the vertical axis. Each graph represents a different value of $\alpha \in \{0.7, 0.8, ..., 1.9\}$.

Figure 2. Effective transfer entropy $T^{R,\text{eff}}_{\alpha,X\to Y}(k,l) \equiv T^{(}R)_{\alpha,X\to Y}([0,1,...,k-1],[1],[0,1,...,l-1])$, where $(k,l) = (2,2), (2,4), (4,2), (4,4), (2,11), (11,11)$ from left to right and from top to bottom. The coupling parameter ϵ is on the x-axis.

Figure 3. Balance of effective transfer entropy $T^{R,\text{bal eff}}_{\alpha,X\to Y}(k,l)$ for $(k,l) = (2,2), (2,4), (4,2), (4,4), (2,11), (11,11)$ from left to right and from top to bottom. Coupling parameter ϵ is on the x-axis.

Based on experience with STE, one could anticipate that RTE $T^{R,\text{eff}}_{\alpha,X\to Y}$ (see also Figure 1) should increase with the growing value of the coupling parameter ϵ. This expectation is indeed confirmed for $\alpha \geq 1$. In fact, even though the trend is noisy, we can detect a clear upward drift. This can be interpreted as an increase in the information flow between central sectors of the underlying empirical distributions. On the other hand, for $\alpha < 1$, the drift seems to be missing or even decreasing. This fact will be commented on shortly. The aforementioned type of behavior persists even when larger Markov parameters are considered, cf. Figure 2. The smallness of the $X \to Y$ information flow can be attributed to the indirect (nonlinear) coupling between X and Y.

From the results depicted in Figure 2, we can study how the increase in values of the Markov parameters in the respective time series influences the value of effective RTE. It can be observed that when historical values of the slave process Y are included, information flows improve (stabilizes) significantly. On the other hand, conditioning on the additional historical values of the master process X does not seem to change the information flows notably. However, for large histories, it stabilizes the results, as can be seen from the comparison of $(2,11)$ and $(11,11)$ in Figure 2.

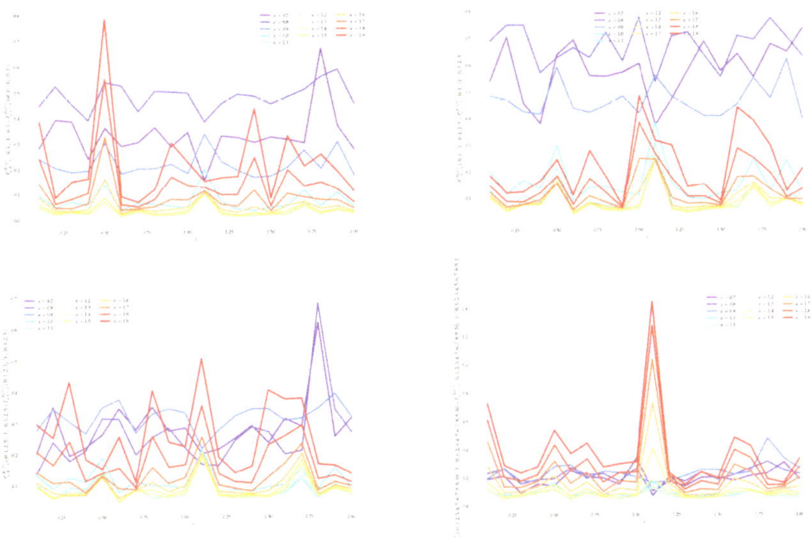

Figure 4. Standard deviation of the balance of effective transfer entropy $T^{R,\text{bal eff}}_{\alpha,X\to Y}(k,l)$ for $(k,l) = (2,2), (2,4), (4,4), (11,11)$ from left to right and from top to bottom. With ϵ on the x-axis.

The balance of effective transfer entropy presented in Figure 3 exhibits an increasing function of the coupling parameter ϵ for $\alpha \geq 1$. Contrary to that, for $\alpha < 1$, it is a stagnating or decreasing function with large fluctuations. The trend is most distinguishable for $(k,l) = (11,11)$. Thus, one can conclude that the deterministic behavior is bared by $\alpha > 1$ and is stochasticity captured by $\alpha < 1$. Positive values of the balance of ERTE suggest higher information flows from the master process to the slave process than in the opposite direction. Therefore, $(11,11)$ confirms the omnipresent master–slave relationship.

Standard deviations of the balance of effective transfer entropy in Figure 4 reveal statistical stability of results for $\alpha > 1$ in contrast to regime $\alpha < 1$. The typical size of fluctuations in the latter case is about $5 - 10$ times larger than in the previous one. However, the results show instability in the results for various strengths of interaction among the time series ϵ. We admit that this is an effect of the insufficient size of datasets or missing statistics of datasets.

6. Conclusions

6.1. Summary

In this paper, we performed extensive computer calculations of the novel method to calculate Rényi's transfer entropy that was applied to a coupled GARCH time series. We performed analogous analysis on the surrogate datasets. Based on that, we calculated the statistics of the balance of effective Rényi transfer entropy.

Analysis of the two-dimensional GARCH process where one dimension influences the latter using Rényi's transfer entropy using the nearest order estimators provides insight into the flow of information within the system. Transfer entropy expectably increases with the increasing strength of the interaction, and the rate increases with the increasing indices of memory. Particularly, it increases with the memory of the time series where the interaction is heading. This observation follows for effective transfer entropy and the balance of effective transfer entropy.

6.2. Perspectives and Generalizations

Transfer entropy is a powerful tool to reveal the strength and direction of information flow in a time series. In combination with the effective use of computer power and modern advancements in the mathematical theory of entropy calculation, it is a better tool to investigate nonlinear causality than the Granger test that is limited to Gaussian time series with linear causality. The advantage of using Rényi's entropy with parameter α is its ability to detect informational flow during extreme events, such as sudden jumps. This is because Rényi's entropy gives an emphasis on the tails or the center of the probability distribution.

Using the complex algorithm on financial datasets can be a potent tool to reveal information flows among, e.g., different stocks or other valuable assets. It can be also used as a precursor of instability or critical behavior in international markets.

Author Contributions: Conceptualization, P.J.; Formal analysis, H.L. and Z.T.; Methodology, P.J., H.L. and Z.T.; Validation, H.L. and Z.T.; Software design, data structures, computer calculation, and visualization, H.L.; Writing—original draft, P.J.; Writing—review editing, P.J., H.L., and Z.T. All authors have read and agreed to the published version of the manuscript.

Funding: P.J., H.L., and Z.T. were supported by the Czech Science Foundation Grant No. 19-16066S.

Institutional Review Board Statement: Not applicable.

Informed Consent Statement: Not applicable.

Data Availability Statement: Not applicable.

Conflicts of Interest: The authors declare no conflicts of interest.

References

1. Pearl, J. *Causality*; Cambridge University Press: Cambridge, UK, 2009.
2. Schreiber, T. Measuring Information Transfer. *Phys. Rev. Lett.* **2000**, *85*, 461–464. [CrossRef] [PubMed]
3. Barnett, L.; Barrett, A.B.; Seth, A.K. Granger Causality and Transfer Entropy Are Equivalent for Gaussian Variables. *Phys. Rev. Lett.* **2009**, *103*, 238701. [CrossRef] [PubMed]
4. Lungerella, M.; Ishigoro, K.; Kuniyoshi, Y.; Otsu, N. Methods for quantifying the causal structure of bivariate time series. *Prog. Neurobiol.* 2007, *77*, 1–37. [CrossRef]
5. Jizba, P.; Kleinert, H.; Shefaat, M. Rényi's information transfer between financial time series. *Physica A* **2012**, *391* , 2971–2989. [CrossRef]
6. Leonenko, N.; Pronzato, L.; Savani, V. A class of Rényi information estimators for multidimensional densities. *Ann. Stat.* **2008**, *36*, 2153–2182. [CrossRef]
7. Rényi, A. On measures of entropy and information. *Proc. Fourth Berkeley Symp. on Math. Statist. Prob.* **1961**, *1*, 547–561.
8. Rényi, A. *Selected Papers of Alfred Rényi*; Akademia Kiado: Budapest, Hungary, 1976; Volume 2.
9. Jizba, P.; Arimitsu, T. World According to Rényi: Thermodynamics of Multifractal Systems. *Ann. Phys.* **2004**, *312*, 17–57. [CrossRef]
10. Beck, C.; Schlögl, F. *Thermodynamics of Chaotic Systems*; Cambridge Nonlinear Science Series (Book 4); Cambridge University Press: Cambridge, UK, 1995.
11. Paluš, M.; Hlaváčkovxax-Schindler, K.; Vejmelka, M.; Bhattacharya, J. Causality detection based on information-theoretic approaches in time series analysis. *Phys. Rep.* **2007**, *441*, 1–46.
12. Marschinski, R.; Kantz, H. Analysing the Information Flow Between Financial Time Series. *Eur. Phys. J. B* **2002**, *30*, 275–281. [CrossRef]
13. Samuelson, P.A. Rational Theory of Warrant Pricing. *Ind. Manag. Rev.* **1965**, *6*, 13–31.
14. Keylock, C.J. Constrained surrogate time series with preservation of the mean and variance structure. *Phys. Rev. E* **2006**, *73*, 036707. [CrossRef] [PubMed]
15. Dobrushin, R.L. A simplified method of experimentally evaluating the entropy of a stationary sequence. *Teor. Veroyatnostei Primen.* **1958**, *3*, 462–464. [CrossRef]
16. Vašíček, O. A test for normality based on sample entropy. *J. Roy. Statist. Soc Ser. B Methodol.* **1976**, *38*, 54–59.
17. Kantz, H.; Schreiber, T. *Nonlinear Time Series Analysis*; Cambridge University Press: Cambridge, UK, 2010.
18. Engle, R.F. Autoregressive Conditional Heteroscedasticity with Estimates of the Variance of United Kingdom Inflation. *Econometrica* **1982**, *50*, 987–1007. [CrossRef]
19. Bollerslev, T. Generalized Autoregressive Conditional Heteroskedasticity. *J. Econom.* **1986**, *31*, 7–327. [CrossRef]

engineering proceedings

MDPI

Proceedings

Revisiting Structural Breaks in the Terms of Trade of Primary Commodities (1900–2020)—Markov Switching Models and Finite Mixture Distributions †

Armand Taranco * and Vincent Geronimi

Cemotev, University of Versailles St Quentin Paris Saclay, 78047 Guyancourt, France; vincent.geronimi@uvsq.fr
* Correspondence: armand.taranco@uvsq.fr

† Presented at the 7th International conference on Time Series and Forecasting, Gran Canaria, Spain, 19–21 July 2021.

Abstract: This paper presents an analysis of the long-term dynamics of the terms of trade of primary commodities (TTPC) using an extended data set for the whole period 1900–2020. Following our original contribution, we implement three approaches of time series—the finite mixture of distributions, the Markov finite mixture of distributions, and the Markov regime-switching model. Our results confirm the hypothesis of the existence of a succession of three different dynamic regimes in the TTPC over the 1900–2020 period. It seems that the uncertainty characterising the long-term dynamic analysis of TTPC is better taken into account with a Markov hypothesis in the transition from one regime to another than without this hypothesis. In addition, this hypothesis improves the quality of the time series segmentation into regimes.

Keywords: commodity prices; terms of trade; long-term fluctuations; structural breaks; finite mixture of distributions; finite Markov mixture of distributions; Markov switching models

Citation: Taranco, A.; Geronimi, V. Revisiting Structural Breaks in the Terms of Trade of Primary Commodities (1900–2020)—Markov Switching Models and Finite Mixture Distributions . *Eng. Proc.* **2021**, *5*, 34. https://doi.org/10.3390/engproc2021005034

Published: 30 June 2021

1. Introduction

One of the main conclusions emerging from the abundant literature dedicated to the study of the long-term evolution of primary commodities' prices is that structural breaks constitute an essential characteristic for the comprehension of the long-term dynamics of terms of trade of primary commodities. Empirical studies of price volatility assess a high level of uncertainty, especially in the post-2008 boom research [1]. However, this literature appears inconclusive on the question of the identification of structural breaks. In this paper, we explore this question by implementing three time series approaches—that have not been, to our knowledge, considered in this literature—for detecting these breaks. We identify structural breaks as the endpoints of the time periods obtained by clustering the data (mixture distributions) or as the endpoints of the regimes (Markov switching regimes). Following our original contribution [2] to the empirical literature on the Prebisch–Singer hypothesis [3,4] of a secular decline in the terms of trade of primary commodities (TTPC), in this paper, we consider an extension of our approaches to the whole period of 1900–2020. The data correspond to the Grilli and Yang Index, here after $\{GY_t\}_{t=t_1,\ldots,t_N}$, see [5,6]. The three approaches of time series we implement—the finite mixture of distributions, the Markov finite mixture of distributions and the Markov regime-switching model—converge in the detection of three different regimes over the 1900–2020 period.

The three following sections of the paper present the methodology and results of, respectively, a finite mixture of distributions approach (Section 2) a finite Markov mixture of distributions approach (Section 3) and a Markov switching model approach (Section 4). The last section is dedicated to the discussion and conclusion (Section 5).

2. A Finite Mixture Distributions Approach

To investigate the hypothesis that the time series $\{GY_t\}_{t=t_1,\ldots,t_N}$ follows different periods over 1900–2020, we first used a finite mixture of distributions with normal components, as a way of putting similar data points (years) together into clusters (which we call regimes). Clusters are represented by the components' distributions of the mixture. The idea is that the years that exhibit the same behaviour belong to the same cluster and come from the same distribution.

A very detailed account of the practical aspects of Markov Chain Monte Carlo (MCMC) for mixture of distributions is given in Frühwirth-Schnatter [7]. The Handbook of mixture analysis [8] provides an overview of the methods of mixture modelling.

2.1. Methodology

A finite mixture of normal distributions can be defined as follows:

$$f(y) = \sum_{i=1}^{K} \eta_i f_i\left(y,\ \mu_i,\ \sigma_i^2\right) \text{ with } \sum_{i=1}^{K} \eta_i = 1,$$

where:

K is the number of components,

η_i is the mixing weight of the ith component, f_i is a normal component distribution of mean μ_i and variance σ_i^2.

In this approach, three kinds of statistical inference problems have to be considered:

- The specification of the number of components K,
- The component parameters $(\mu_i,\ \sigma_i^2)$ and the weight distribution $(\eta_1,\ \ldots,\ \eta_K)$ should be estimated from the data, Finally, we must assign each observation of the time series, $\{GY_t\}_{t=t_1,\ldots,t_N}$, to a certain component of the mixture model by making inference on a hidden vector indicator $\mathbf{S} = (S_{t_1},\ \ldots,\ S_{t_N})$.

To estimate the parameters of the components and the weights, we use Bayesian estimation [9] with MCMC [10] and a two block Gibbs sampling algorithm [7]:

(1) Parameter simulation conditional on the classification $\mathbf{S} = (S_{t_1},\ \ldots,\ S_{t_N})$:

a. Sample the weights $\boldsymbol{\eta} = (\eta_1,\ \ldots,\ \eta_K)$ from a Dirichelet posterior $p(\boldsymbol{\eta}|\mathbf{S})$,

b. Sample the variances σ_i^2 in each group i, from an inverted Gamma distribution $G^{-1}(c_i(\mathbf{S}), C_i(\mathbf{S}))$,

c. Sample the means μ_i in each group i, from an inverted Gamma distribution $G^{-1}(b_i(\mathbf{S}), B_i(\mathbf{S}))$

The precise form of $b_i(\mathbf{S}),\ B_i(\mathbf{S}), c_i(\mathbf{S}),\ C_i(\mathbf{S})$ depends upon the chosen prior distribution family.

(2) Classification of each observation y_i conditional on knowing $\boldsymbol{\mu} = (\mu_1, \ldots,\ \mu_K)$,

$$\boldsymbol{\sigma}^2 = \left(\sigma_1^2, \ldots,\ \sigma_K^2\right) \text{ and } \boldsymbol{\eta} = (\eta_1,\ \ldots,\ \eta_K):$$

$$P\left(S_i = k \middle| \boldsymbol{\mu},\ \boldsymbol{\sigma}^2, \boldsymbol{\eta},\ y_i\right) \propto \frac{1}{\sqrt{2\pi\sigma_k^2}} exp\left\{-\frac{(y_i - \mu_k)^2}{2\sigma_k^2}\right\}\eta_k$$

The number of components may be known or unknown. In our case, the number of components is unknown, and our model selection is based on marginal likelihood [11]. In the academic literature, the unknown number of regimes taken into account is three at most. To determinate the best model, we expand the number of potential regimes to five. Thus, we chose the model with the largest marginal likelihood, approximated by three estimators [7]:

- RI is the estimator obtained by reciprocal importance sampling,
- IS is the estimator obtained by importance sampling,

- BS is the estimator obtained by bridge sampling techniques.

For computing purposes, we use the Matlab library Bayesf 2.0 in this publication.

2.2. Results

The results are presented in the following four sections. First, we confirm the existence of three different components in the mixture. Then, we present the statistical parameters (mean, standard deviation, and weight) of each distribution, associated to the correspondent regime (regime1: 1900–1921; regime 2: 1922–1985 and 2006–2020; regime 3: 1986–2005). The third sub-section presents a point representation of posterior draws. The fourth sub-section clusters the data based on MCMC draws.

In all Monte Carlo simulations using posterior draws, we use 1,000,000 draws after a burn-in of 100,000 draws.

2.2.1. The Choice of the Number of Components

If K is not too large, the different estimators should approximatively agree. As K increases, we observe that, the reciprocal importance sampling and the importance sampling estimators are less precise than the bridge sampling estimator, although all three select the same number of components: among the considered models (number of components ≤ 5), the model with the largest marginal likelihood is a mixture of three normal distributions.

Thus, the results (Table 1) for the mixture of distribution models confirm the accuracy of the hypothesis of the existence of three different components, as already established in our previous analysis for the 1900–2016 period.

Table 1. The choice of the number of components according to three estimators—Source: authors.

Estimators	$K = 1$	$K = 2$	$K = 3$	$K = 4$	$K = 5$
RI	−20.6488	−21.8398	−16.9335	−17.3682	−21.2979
Standard error	8.2511×10^{-5}	1.0581×10^{-3}	3.2659×10^{-3}	6.9481×10^{-2}	5.1914×10^{-1}
IS	−20.6488	−21.8456	−16.9402	−17.1843	−19.2823
Standard error	8.0611×10^{-5}	2.6576×10^{-3}	4.2801×10^{-3}	1.0072×10^{-1}	1.199×10^{-1}
BS	−20.6489	−21.8402	−16.9316	−17.1489	−17.7954
Standard error	5.581×10^{-5}	7.2275×10^{-4}	9.0533×10^{-4}	2.4614×10^{-3}	6.2907×10^{-3}

2.2.2. The Parameters of the Mixture of Three Normal Distributions

The components of the mixture differ mainly in the mean. Components 2 and 3 have nearly the same variance, whereas the first component has a variance that is slightly higher (Table 2).

Table 2. Weight, mean, and standard deviation of each distribution—Source: authors.

Parameters of the kth Component	Distribution 1	Distribution 2	Distribution 3
Weight	0.3238	0.4945	0.1817
Mean	4.9091	4.5876	4.1829
Standard deviation	0.0246	0.0096	0.0153

2.2.3. The Point Process Representation of Posterior Draws

To produce sampling representations of the posterior draws, $\left\{\mu^{(m)}\right\}_{m=1,\ldots,M}$ (M, the number of draws) is plotted against $\left\{\sigma^{2(m)}\right\}_{m=1,\ldots,M}$. This scatter plot is closely related to the point process representation of the underlying mixture distribution. A finite mixture distribution from a fixed parametric family has a representation as a marked point process [12]. Here, we use point process representation (Figure 1) of draws from the posterior density. Three clusters of draws are distinguished, they will scatter around the three points corresponding to the true point process representation, with the spread of the clouds representing the uncertainty of estimating the points (Figure 1).

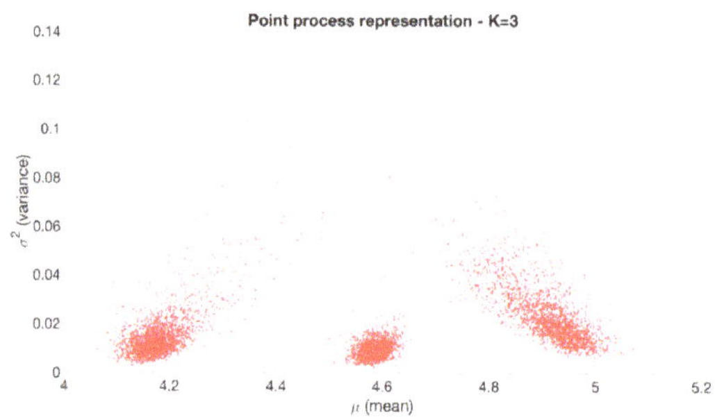

Figure 1. Point process representation for K = 3—Source: authors.

2.2.4. Clustering the Data

We perform clustering of the data into three groups (Figure 2) based on the MCMC draws. Three criteria are used:

- The Bayesian maximum a posteriori (MAP),
- The similarity matrix based on the posterior similarity,
- The misclassification rate.

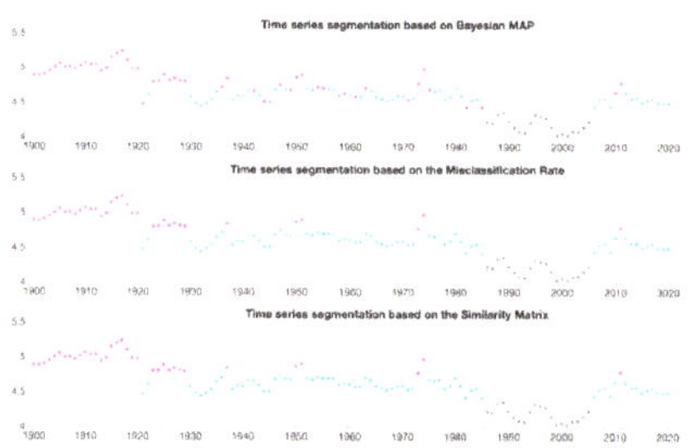

Figure 2. Time-series segmentation according to three methods.

Three regimes are confirmed, (1900–1921; 1922–1985 and 1986–2020; 1986–2005). The second regime is interrupted by the regime 1986–2005, which represents the lowest level in the terms of trade of primary commodities (see Section 5). However, we observe that some years have an ambiguous cluster membership.

3. A Finite Markov Mixture Distributions Approach

3.1. Methodology

In the finite mixture models approach, we assign each observation of the time series $\{GY_t\}_{t=t_1, \ldots, t_N}$ to a certain component of the mixture model by making inference on a hidden vector indicator $\mathbf{S} = (S_{t_1}, \ldots, S_{t_N})$. Now, we suppose that this allocation vector is a hidden Markov chain, $GY_t = \mu_{S_t} + \varepsilon_t$ where ε_t is a zero-mean white noise process with variance σ^2, which is a special case of interest of finite Markov mixture of distributions. Now, the transition probability matrix $\boldsymbol{T}$ of the hidden Markov chain $\mathbf{S} = (S_{t_1}, \ldots, S_{t_N})$ is unknown and need to be estimated from the data. We suppose that the Markov chain is aperiodic and starts from its ergodic distribution $\boldsymbol{\eta} = (\eta_1, \ldots, \eta_K)$:

$$P(S_N = k|\boldsymbol{T}) = \eta_k$$

and the transition probability matrix $\boldsymbol{T}$ is defined by:

$T_{ji} = P(S_{t+1} = j|S_t = i)$ for $i, j = 1, \ldots, K$ and $t = t_1, \ldots, t_N - 1$.

What is the relation between finite mixture distributions and finite Markov mixture distributions? In fact, every finite mixture of distributions may be considered of as a limiting case of a finite Markov mixture of the same family of distributions where $S = (S_{t_1}, \ldots, S_{t_N})$ is an i.i.d. random sequence and where the transition probabilities are all equal to η_k.

3.2. Results

The results are presented in the following three sub-sections. We present the statistical parameters (mean and standard deviation) of each distribution, associated to the correspondent regime (regime1: 1900–1921; regime 2: 1922–1985 and 2006–2020; regime 3: 1986–2005) and transition probabilities from one regime to another one. The second sub-section presents a point representation of posterior draws. The third sub-section clusters the data based on MCMC draws.

3.2.1. The Parameters of the Markov Mixture of Three Normal Distributions

The components of the mixture differ mainly in the mean but have nearly the same variance (Table 3).

Table 3. Mean and standard deviation of each distribution.

	Distribution 1	Distribution 2	Distribution 3
Mean	5.0099	4.6265	4.1822
Standard deviation	0.0110	0.0142	0.0134

The transition probabilities T_{11}, T_{22}, T_{33} are high (Figure 3, Table 4), which indicate that is difficult to change from on regime to the other.

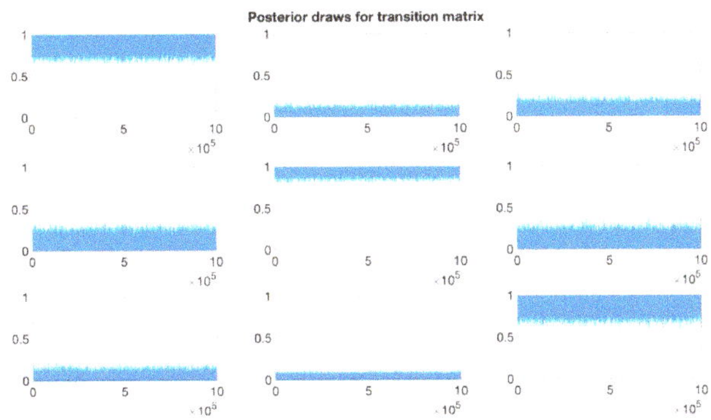

Figure 3. Posterior draws for transition probability matrix T—Source: authors.

Table 4. Transition probability matrix T from posterior draws—Source: authors.

	Regime 1, t	Regime 2, t	Regime 3, t
Regime 1, $t+1$	0.9384	0.0167	0.0284
Regime 2, $t+1$	0.0483	0.9637	0.0571
Regime 3, $t+1$	0.0133	0.0196	0.9146

3.2.2. Point Process Representation of Posterior Draws

We observe that this time, the clusters obtained with the point process representation of posterior draws in the case of a Markov finite mixture (Figure 4) are well-separated and have less dispersion compared with that of the clusters obtained in the case of a finite mixture of distributions. The shapes of the clusters are also different.

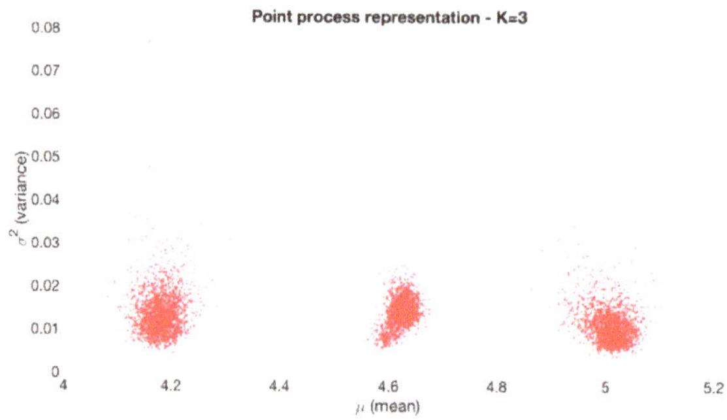

Figure 4. Point process representation for $K = 3$—Source: authors.

3.2.3. Clustering the Data

We confirm the existence of the three regimes previously found (1900–1921; 1922–1985 and 1986–2020; 1986–2005). This time, all the years have a perfect cluster membership. The periods of the regimes are well defined (Figure 5).

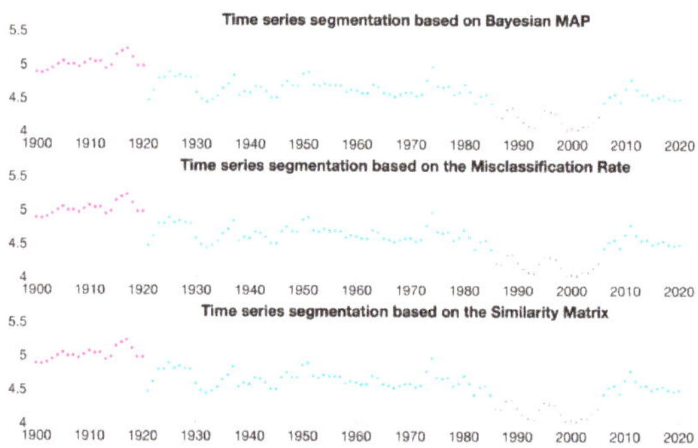

Figure 5. Time-series segmentation according to three methods.

4. A Markov Switching Model Approach

4.1. Methodology

A finite Markov switching (MS) model assumes that the dynamics of a data series, $\{y_t\}_{t=t_1,\ldots,t_N}$, depend on a discrete latent variable S_t, postulated to follow a Markov chain with realizations in $\{1, \ldots, K\}$. This model was popularized by Hamilton [13,14] who applied the Markov-switching approach to model the probability of a recession in the U.S. economy. In this model, the economy alternates between two unobserved states of economic expansion and recession according to a Markov chain process. The model assumes constant transition probabilities for the unobserved states, which, in turn, imply constant expected durations in the various regimes. A general representation is given by:

$$y_t = C_{S_t} + \sum_{i=1}^{p} \alpha_i X_t^{f,i} + \sum_{i=1}^{q} \beta_i(S_t) X_t^{r,i} + \sum_{i=1}^{r} \gamma_i(S_t) y_{t-i} + \varepsilon_t$$

where:

y_t denotes the series observed,

$X_t^{f,i}$ are the independent regressors with fixed effects,

$X_t^{r,i}$ are the independent regressors with random effects,
y_{t-i} these variables represent the autoregressive part of model,
ε_t are independent variables with N (0, $\sigma^2_{\varepsilon,S_t}$) distribution,
S_t is modelled by a homogeneous Markov chain with K states.

The transition probabilities verify:

$P(S_{t+1} = j|S_t = i)= P(S_2 = j|S_1 = i)$, for $t = t_1, \ldots, t_N - 1$ and for $i, j = 1, \ldots, K$ (homogeneity of the chain).

For $i = 1, \ldots, K$:

$$\sum_{j=1}^{K} P(S_{t+1} = j|S_t = i) = 1.$$

We consider only the case where there is no fixed or random effects and no autoregressive part in the model.

Essentially, two computational approaches can be used for the estimation of Markov-switching models. One approach involves maximising the log-likelihood, a function of the transition probabilities, subject to the constraint that the probabilities lie between 0 and 1 and sum to unity. This can be done with the EM algorithm [15], but the non-linear programming approach [16] can also be used. We mobilise this last approach implemented in Oxmetrics. An alternative approach involves using Bayesian estimators with MCMC methods.

4.2. Results

The results of a three-regime model based on the terms of trade of commodities are shown in Tables 5 and 6, and Figure 6. There is a perfect match with the previous results, notably concerning the identification of three regimes over the exact same sub-periods.

Table 5. Statistical characteristics of regimes—Source: authors.

Regimes	Coefficient	Standard Error	*t*-Value	*p*-Value
Regime 1	5.01765	0.02014	249.	0.000
Regime 2	4.63002	0.01365	339.	0.000
Regime 3	4.18067	0.02514	166.	0.000

Table 6. Transition probability matrix.

	Regime 1, *t*	Regime 2, *t*	Regime 3, *t*
Regime 1, *t* + 1	0.95451	0.0000	0.0000
Regime 2, *t* + 1	0.045485	0.98722	0.048562
Regime 3, *t* + 1	0.0000	0.012781	0.95144

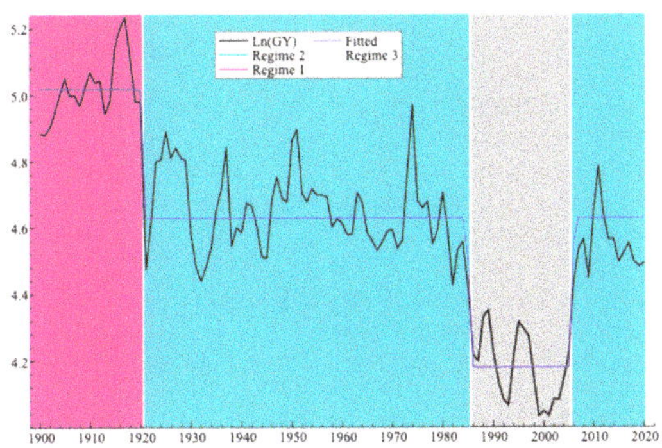

Figure 6. Change of regime in the evolution of the terms of trade for primary commodities (Neperian logarithm, 1900–2020)—Source: authors.

5. Discussion and Conclusions

The existence of different regimes appears robust to various changes in the data span. Indeed, considering data from 1900 to 2010, or 1900 to 2014, or 1900 to 2016 or 1900 to 2020 on the same $\{GY_t\}_{t=t_1, \ldots, t_N}$ index leads to the same representation, with the same break dates (1921, 1986, 2006). The approach using a Markov finite mixture of distributions and the approach using a Markov switching model give very similar results. These two methods differ essentially in the computational aspects. The former uses Bayesian estimation with MCMC and the later involves maximising the log-likelihood. The fact that each observation of the time series $\{GY_t\}_{t=t_1, \ldots, t_N}$ is assigned to a certain component of the Markov mixture model by making inference on a hidden Markov vector indicator, improves the results obtained with the finite mixture model. This time, all years have a perfect cluster membership.

These three approaches applied to the extended 1900–2020 data set confirm the identification of a succession of three different dynamic regimes in the TTPC over the 1900–2020 period. The third regime (1986–2005) is still characterized by the lowest level of terms of trade of the whole period, and the return to the second regime after 2005 is associated with a price significantly higher (56.7% higher). Such an upward shift in primary commodities' prices is unprecedented at the scale of the 20th century and questions more specifically the hypothesis of a secular decline in the terms of trade of primary commodities. Indeed, the entry into a higher level of prices contradicts the hypothesis of a secular decline. However, from 1900 to 2006, this decline manifested itself through the succession of regimes with a lower average level of primary commodity terms of trade, but not in a continuous way. Moreover, data from 2020 for TTPC do not exhibit a specific pattern, leaving open the question of the effect of COVID on the long-term dynamics of primary commodity prices. Therefore, the dynamics behind the evolution of primary commodities in the long-run call for alternative explanations and a change of perspective.

This paper contributes to this change of perspective by considering (and confirming) the existence of three changes in regime in the long term (121 years). Yet, an operational theory of long-term dynamic regime change in primary commodities' terms of trade is still to be constructed.

Following the methodologies used in this present paper, a promising perspective appears to be the introduction of explanatory variables (such as the GDP of main countries, the share of emerging countries in the global GDP, and various indices of real interest rate and exchange rates) in a Markov switching model, in order to identify the incidence of these covariates on the dynamic regimes.

Institutional Review Board Statement: Not applicable.

Informed Consent Statement: Not applicable.

Data Availability Statement: Data Available at https://www.cemotev.uvsq.fr/base-de-donnees.

References

1. Prakash, A. *Safeguarding Food Security in Volatile Global Markets*; FAO: Rome, Italy, 2011.
2. Geronimi, V.; Taranco, A. Revisiting the Prebisch-Singer hypothesis of a secular decline in the terms of trade of primary commodities (1900–2016). A dynamic regime approach. *Resour. Policy* **2018**, *59*, 329–339. [CrossRef]
3. Prebisch, R. The Economic Development of Latin America and its Principal Problems. *Econ. Bull. Lat. Am.* **1962**, *7*, 1–22.
4. Singer, H.W. U.S. Foreign Investment in Underdeveloped Areas: The distribution of gains between Investing and Borrowing countries. *Am. Econ. Rev. Pap. Proc.* **1950**, *40*, 473–485. [CrossRef]
5. Grilli, E.R.; Yang, M.C. Primary commodity prices, manufactured goods prices, and the terms of trade of developing countries: What the long run shows. *World Bank Econ. Rev.* **1988**, *2*, 1–47. [CrossRef]
6. Geronimi, V.; Anani, E.T.G.; Taranco, A. Notes on updating prices indices and terms of trade for primary commodities (No. 3–2017). *Cahier CEMOTEV* **2017**, *2017*, 3.
7. Frühwirth-Schnatter, S. *Finite Mixture and Markov Switching Models; Springer Series in Statistics*; Springer: New York, NY, USA, 2006. [CrossRef]
8. Celeux, G.; Frühwirth-Schnatter, S.; Robert, C.P. (Eds.) *Handbook of Mixture Analysis*; Chapman and Hall/CRC: Boca Raton, FL, USA, 2018. [CrossRef]

9. Diebolt, J.; Robert, C. *Bayesian Estimation of Finite Mixture Distributions, Part I: Theoretical Aspects*; Rapport Technique LSTA; Université Paris VI: Paris, France, 1990; Volume 110.
10. Robert, C.; Casella, G. A short history of Markov chain Monte Carlo: Subjective recollections from incomplete data. *Stat. Sci.* **2011**, *26*, 102–115. [CrossRef]
11. Frühwirth-Schnatter, S. Keeping the balance—Bridge sampling for marginal likelihood estimation in finite mixture, mixture of experts and Markov mixture models. *Braz. J. Probab. Stat.* **2019**, *33*, 706–733. [CrossRef]
12. Stephens, M. Bayesian analysis of mixture models with an unknown number of components-an alternative to reversible jump methods. *Ann. Stat.* **2000**, *28*, 40–74. [CrossRef]
13. Hamilton, J.D. A new approach to the economic analysis of nonstationary time series and the business cycle. *Econometrica* **1989**, *57*, 357. [CrossRef]
14. Hamilton, J.D. Regime-switching models. In *Palgrave Dictionary of Economics*; Durlauf, S., Blume, L., Eds.; Palgrave McMillan Ltd.: London, UK, 2005. [CrossRef]
15. Dempster, A.P.; Laird, N.M.; Rubin, D.B. Maximum likelihood from incomplete data via the EM algorithm (with discussion). *J. R. Statist. Soc. Ser. B* **1977**, *39*, 1–38. [CrossRef]
16. Lawrence, C.T.; Tits, A.L. A computationally efficient feasible sequential quadratic programming algorithm. *Siam J. Optim.* **2001**, *11*, 1092–1118. [CrossRef]

engineering proceedings

MDPI

Proceedings

Forecasting the Spread of the COVID-19 Pandemic Based on the Communication of Coronavirus Sceptics †

Melinda Magyar *, László Kovács and Dávid Burka

Department of Computer Science, Corvinus University of Budapest, Fővám tér 13-15, 1093 Budapest, Hungary; laszlo.kovacs2@uni-corvinus.hu (L.K.); david.burka@uni-corvinus.hu (D.B.)

* Correspondence: melinda.magyar@uni-corvinus.hu

† Presented at the 7th International Conference on Time Series and Forecasting, Gran Canaria, Spain, 19–21 July 2021.

Abstract: The COVID-19 pandemic has left a mark on nearly all events since the start of the year 2020. There are many studies that examine the medical, economic, and social effects of the pandemic; however, only a few are concerned with how the reactions of society affect the spread of the virus. The goal of our study is to explore and analyze the connection between the communication of pandemic sceptics and the spread of the COVID-19 pandemic and its caused damages. We aim to investigate the causal relationship between communication about COVID-19 on social media, anti-mask events, and epidemiological indicators in three countries: the USA, Spain, and Hungary.

Keywords: COVID-19; sceptics; social media; Twitter; sentiment; VAR; Granger causality; government stringency

check for updates

Citation: Magyar, M.; Kovács, L.; Burka, D. Forecasting the Spread of the COVID-19 Pandemic Based on the Communication of Coronavirus Sceptics . *Eng. Proc.* **2021**, *5*, 35. https://doi.org/10.3390/engproc2021005035

Published: 1 July 2021

1. Introduction

Coronavirus is the latest of many infectious diseases affecting humanity throughout history that have reached the state of a pandemic. Pandemics, by definition, affect large regions across continents or even the whole world; thus, even in case of a low mortality rate, the number of casualties can reach millions in a relatively short time period. The COVID-19 outbreak is among the deadliest pandemics of the last hundred years, only outdone by HIV/AIDS (human immunodeficiency virus infection and acquired immune deficiency syndrome) [1].

However, the COVID-19 pandemic is the first to occur since social media became widespread. The swine flu (H1N1) outbreak, being the most recent one, happened between 2009 and 2010 [2], but at that point, Facebook had just started its rise in popularity, and other platforms that are well known today (i.e., Twitter, Reddit, Instagram) had barely started to gain popularity [3]. HIV is an exception, as it still costs around 800 thousand lives per year because of its high mortality rate, but it has infected far fewer people than the other mentioned pandemics [4]. Additionally, HIV was the focus of attention in the 1980s and 1990s, but it has not been covered in the media too often in recent years.

This means that COVID-19 is the first pandemic about which an immense volume of online written communication exists, which can be analyzed with the help of different text mining solutions. Never before has the opportunity been presented to examine the opinion of the masses regarding such events; thus, this is a completely new field of research, and in this relatively short time period, there have not been many investigations exploiting its potential. There are many studies about social communication during the pandemic, including false news and its impact on the pandemic and vice versa [5,6]; however, these usually focus on a single conspiracy theory, a set of news, or a small group of events instead of long-running time series.

Our research aims to examine the connection between social responses and pandemic-related events in the USA, Spain, and Hungary. We examined the most prevalent social platforms of each country and collected a large volume of COVID-19-related comments and their timestamps. Sentiment analysis was used to process this text-based data source; thus, it was possible to create a sentiment time series for each language group.

Reliable corona-related pandemic data are available on the *Our World in Data* (OWID) site in a research-friendly form [7]. Regarding the activity of deniers, we manually collected a list of significant demonstrations and assemblies from different news sources. We only considered "offline" events as these are the ones that could have directly influenced the number of infections.

We compared the sentiment time series with the events and corona-related time series by applying an augmented vector autoregression (VAR) model according to the Toda–Yamamoto procedure [8] on the examined time series in each country separately. Granger causality models have been successfully applied in order to assess the economic and financial effects of the COVID-19 pandemic, for example, by [9] and [10]. We show that the volume of the online comments and the sentiment index had a significant mutual relationship with the official epidemiological indicators. The characteristics of these relationships differed along countries and waves of the pandemic. In Spain, the antimask events had a significant effect on the volume of comments during the first wave and on sentiment in the second wave.

2. Data Sources

For constructing sentiment time series, we need textual data obtained from representative sources. Every target country has some preferred social media sites, such as forums, microblogging sites, or even comment sections of their leading news sites. The most important social media site is Facebook, and Twitter is also in the top 20 in every country except in Hungary, according to Similarweb [11]. The contents of these platforms could be a good starting point to examine social reactions about pandemic events and vice versa. As the most widely used search service in the world, Google cannot be ignored either: not only do the topics searched show an increased interest in the COVID-19 pandemic, but they can give us an idea of the focal points of interest. These platforms together are appropriate sources for text mining research studies, which can transform human sentiments into data, map the topics, and find the most influential ones.

In the examined countries, for data source, the common ground could have been Facebook [11]. However, Facebook is not an easy option for text mining research studies since the Cambridge Analytica scandal [12], so Twitter was chosen as a source for mining sentiments for the English and Spanish languages. Because Twitter is not so popular in Hungary, gyakorikerdesek.hu (hereinafter referred to as FAQ) was used for this country as a text mining source. This is a Q&A-type website, which is the 31st most visited site in Hungary.

Twitter provides an API for researchers under friendly conditions, and there is a project named Twitter Stream Grab by Archive Team that allowed us to download all tweets for the examined period [13]. FAQ does not provide API for grabbing data, so we developed an application for scraping purposes. During scraping, the software collects questions and answers from two relevant categories: health and politics [14].

A series of corona-sceptic events were collected manually based on the collections of national Wikipedia pages related to coronavirus and on the Google Labs search terms related to coronavirus [15].

From the times series published on the website OurWorldInData.org, three are used to describe the pandemic situation. The first time series is the rate of positive coronavirus tests. It is used to describe the spread of the virus. This is in line with WHO recommendations [16]. The severity of the pandemic is described by the daily number of deaths per million people. The daily values of the government stringency index are also considered to examine whether the sentiment of the online public is reacting to government measures or vice versa. The index is calculated by the Oxford Coronavirus Government Response Tracker (OxCGRT) project. This is a composite measure based on nine response indicators, including school closures, workplace closures, and travel bans, rescaled to a value between 0 and 100 (100 = strictest) [17].

The time periods examined were different for each country to ensure an adequate level of variance in each time series as the start of the pandemic differed for each examined country. For example, in the US, the number of deaths was 0 on most days until 13 March 2020, and testing data were only available since 7 March 2020, so 13 March 2020 was used as a starting point. The number of daily deaths per million people was quite scarce for Spain. There were two negative values on 25 May and 12 August that were imputed as 0. There was a weekly seasonality for 0 entries. Therefore, we took a 7-day moving average of daily new deaths per million people for Spain. The end point for all these time series was 31 December 2020, as the focus of our investigation was the past year.

Descriptive statistics for each examined time series are available in Table 1. To check for outlier effect, a mean trimmed off the bottom and upper 10% was used. Outliers had no great effect on the examined time series.

Sentiment in US tweets was the most negative on an average day with low standard deviation, while the mean sentiment in Spain seemed to be the highest, though still a negative value. Hungary had the greatest standard deviation in its sentiment index.

Table 1. Descriptive statistics for all of our examined time series.

Variables	No of Obs.	Mean	St. Dev.	Tr. Mean
Positive rate USA	282	0.08	0.04	0.07
Deaths per million USA	282	3.61	2.30	3.33
Stringency USA	282	68.63	5.41	69.15
Entry count USA	282	3580.59	2166.54	3199.61
Sentiment USA	282	−0.46	0.12	−0.46
Positive rate ESP	282	0.06	0.04	0.06
Deaths per million ESP	246	2.59	2.41	2.31
Stringency ESP	246	66.37	9.48	66.39
Entry count ESP	246	644.04	269.15	627.54
Sentiment ESP	246	−0.09	0.11	−0.09
Positive rate HUN	246	0.08	0.09	0.06
Deaths per million HUN	284	3.47	5.51	2.26
Stringency HUN	284	59.45	12.58	59.62
Entry count HUN	284	88.25	60.45	81.93
Sentiment HUN	284	−0.13	0.19	−0.14

3. Methods

The data on Twitter Stream Grab are available on a monthly basis, and there is one compressed JSON file for every minute, so to examine a whole year, more than half a million files must be processed. A time frame between 01/03/2020 and 31/12/2020 was chosen according to the availability of pandemic data from OWID. Datasets contained time data, text, detailed user data, and language index. There were two important limitations: we did not have data about the specific followers for a given user, and there was no precise location data; we could only rely on user-supplied information. In order to reduce the data size, we filtered out relevant tweets based on a few selected keywords, which were grabbed from Google Labs Corona search terms [15]. English-language tweets were narrowed down to the United States based on user-defined location, and a 10% random sample was taken for Spanish-language tweets. The extracted data were transformed into comma-separated files, which can be easily imported into other systems. The texts scraped from FAQ for Hungarian-language analysis did not needed further preprocessing, as the scraper software was designed specifically for this research and had taken the necessary steps.

After extracting tweets and comments, the texts were cleaned and prepared for sentiment analysis. For stemming and lemmatization, the *hunspell* package was utilized, which is a spell checker and morphological analyzer originally designed for the Hungarian language, but it performs well in English and Spanish also [18]. For examining sentiments regarding the pandemic, collected text entries should be labelled with polarity: negative or positive. To do this, a dictionary-based sentiment analysis was applied.

There could be some structural breaks in each time series due to the different characteristics of the first and second waves of the pandemic. Therefore, we should identify possible structural breaks in each examined time series that best separated the first and second waves of the pandemic.

When investigating Granger causality, it is advisable to fit a model separately on sections defined by structural breaks to ensure stability [8]. To identify structural breaks, the breakpoint function from the *strucchange* R package was utilized [19]. If we assume that the number of breakpoints in a linear trend for a time series is b, then the breakpoint function estimates the location of b breakpoints by minimizing the residual sum of squares (RSS) of a linear model where the slope of the trend can change b times. The optimal b is chosen by the Bayes–Schwarz information criterion (BIC) as this IC prefers the sparsest models. This was preferable for us as we had a relatively small number of observations for each country already, so we should avoid overparameterization.

After determining the breakpoints, we fit VAR models for each country and each wave separately to discover the Granger causality between the time series in both waves of the pandemic. A vector autoregression (VAR) process with k endogenous and m exogenous variables can be considered a system of equation with k equations. Model parameters are estimated by OLS. See [20] for details. Maximum lag of the endogenous variables is denoted by p.

However, the typical Granger causality test based on the classical VAR model cannot be relied on when one or both time series are nonstationary, which could lead to spurious causality [21]. Thus, an augmented Dickey–Fuller (ADF) test was employed. Besides, a Kwiatkowski–Phillips–Schmidt–Shin (KPSS) test, in which the null hypothesis is stationarity, was also conducted as a cross-check. To handle the possible integration in our time series, the VAR models were set up according to the Toda–Yamamoto (TY) procedure [8] using the levels of the data without differencing and adding q extra lags if the maximum order of integration was q. The advantage of the TY procedure is it saves the cointegration test and prevents pretest bias. However, there was a need to ensure that the VAR models of each country were specified in a way that there was no serial correlation in the residual values. This was tested by the portmanteau test.

In the optimal VAR models, Wald tests of Granger causality were applied. The null hypothesis is that the coefficients of the first p lagged values of endogenous variables in each equation are 0 after being tested. The reason for including the coefficient of the lags from $p + 1$ to q is that the additional lagged values are to fix the asymptotic so that the Wald test statistics under the null hypothesis follow asymptotical chi-square distribution. Rejection of the null hypothesis of the Wald test implies a Granger causality.

4. Models

For our investigations, three countries were considered. The English-language tweets were narrowed down to tweets originating from the USA, so epidemiological and government stringency indicators of the US were considered here. For the Spanish-language tweets, the indicators of Spain were considered as during the first wave of the pandemic, Spain was the hardest-hit Spanish-speaking country. By 30 June 2020, the cumulative number of deaths per million was 606 in Spain and 297 and 215 in Chile and Mexico, respectively. During the second wave, the pandemic situation in Latin America became more serious, so the effects of COVID-related tweets from other Spanish-speaking countries could act as confounders. Managing these issues is part of our further research. The indicators of Hungary were considered for the Hungarian language.

Sentiment dictionaries were gathered from different sources: Bing for English, TASS for Spanish, and PrecoSenti for Hungarian [22–24]. Further processing was performed with R using the *tidytext* and *dplyr* packages.

Figure 1 shows that the basic sentiment in Spanish-language tweets was more positive than in English-language ones. The daily count of tweets followed the usual trend of a scandal: at the beginning of the pandemic, we could experience a large volume of comments about corona-related topics, and the numbers started to fall during the year even at the time of the second wave.

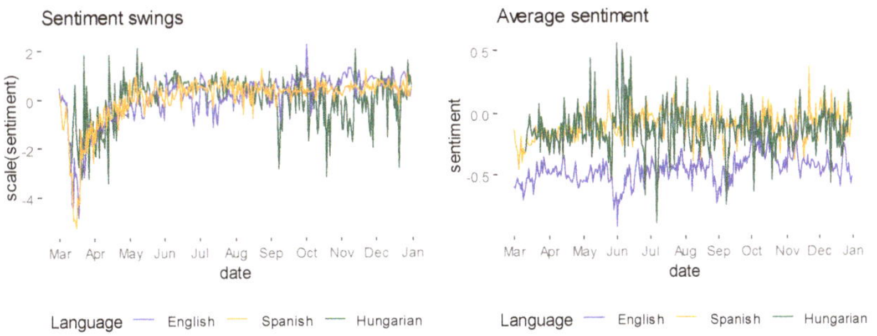

Figure 1. The tendencies in sentiment time series are relatively similar in the three examined datasets; however, the average sentiment is higher in Spanish- and Hungarian-language tweets than English-language contents.

The breakpoint function from the *strucchange* package identified two to four breaks in the time series based on the *BIC*. These breakpoints needed to be narrowed down, as four breakpoints would partition our sample into parts with very small sizes. To select the breakpoints that best separated the two waves, the breakpoints of the positive rate in each county were examined in more detail as this was the measure describing the spread of the pandemic in line with WHO recommendations [16].

The breakpoints of the positive rate in each county are examined in more detail in Figure 2 to define sections on which the Granger causality between the time series is examined by fitting VAR models.

We can see that in Spain and Hungary, we could easily select the structural breakpoint that best separated the start of the second wave of the pandemic. It is also noticeable that Hungary had quite a long period in the summer where the positive rate stagnated on a lower level before the second wave started in September. However, we did not wish to separate this period from the first wave as three breakpoints would result in small subsamples. That is why we also ignored the break that marked the peaking of the second wave. In Spain, the second wave started around the middle of summer, much earlier than in Hungary. We disregarded the other breakpoints marking different periods in the first and second waves as splitting along these would result in small subsamples just like in the case of Hungary.

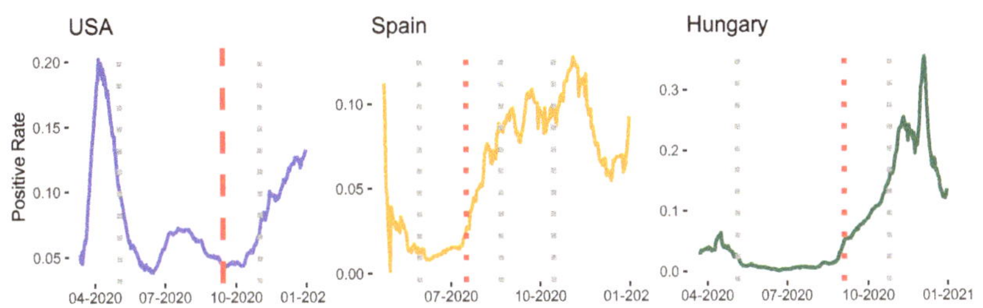

Figure 2. The positive rate for the three examined countries. Structural breakpoints are marked with dotted lines. The breaks marked with red are the ones that best separate the first and second waves of the pandemic. In the US, a custom breakpoint is added to separate the two waves marked with a dashed red line.

The case of the US was more complicated as it had a short flare of the pandemic in the middle of summer and the second wave started in late October. To preserve the sample size, we considered the short flare in positive rate in the summer as an aftershock of the first wave and defined a custom breakpoint on 20/09/2020, marked by the dashed red line in Figure 2. We separated every examined time series into two parts, representing the first and second waves of the pandemic according to the country-specific breakpoints selected as shown in Figure 2.

As we had five time series for each country, we had $k = 5$ endogenous variables. Dummy variables were used as exogenous variables to account for day-of-the-week effect. One more dummy exogenous variable represented whether there was an antimask event with at least 100 participants at time t for each country, making $m = 6 + 1 = 7$. The number of p lags will be chosen later.

Based on the results of the ADF and KPSS tests, taking the first difference of each time series mostly eliminated the unit root. The only exceptions were the stringency time series in the US and the positive rate for Spain and Hungary during the first wave, according to the KPSS test, but only on $\alpha = 10\%$, not on $\alpha = 5\%$. The ADF test rejected the H_0 of the unit root on all common significance levels in these cases. Thus, the maximum order of integration was set to 1.

The VAR models were set up according to the TY procedure to account for the first-order integration. First, we determined the appropriate lag length for the endogenous variables. Based on the Akaike information criterion, Hannan–Quinn information criterion, Bayes–Schwarz criterion, and final prediction error, lags $p = 1$ and $p = 2$ were recommended.

From the results of a portmanteau test controlling for dynamic stability, it was observed that lag 2 removed residual serial autocorrelation at 1% for all VAR models except for Hungary during the first wave. As accepting the H_0 of no serial correlation in the residuals was not convincing on all common significance levels, adding more lags could be considered, but we already had a larger parameter–sample size ratio with the dummies and the two lags for each variable (17 + 1 parameters for each equation, which is slightly less than fifth of the number of observations (circa 160 and 120 for each wave) in all three countries). The VAR models could be considered stable, again except for Hungary during the first wave, as all roots of the characteristic polynomials were inside the unit circle. Detailed diagnostic results for each VAR model are shown in Table 2.

Table 2. Model diagnostic results for the examined VAR(1) and VAR(2) models.

Setup	Lag = 1		Lag = 2	
	Portmanteau Test *p*-Value	**Range of Roots of Characteristic Polynomials**	**Portmanteau Test *p*-Value**	**Range of Roots of Characteristic Polynomials**
USA-1st wave	0.0213	0.508–0.940	0.0596	0.196–0.948
USA-2nd wave	0.8364	0.565–0.902	0.9043	0.038–0.901
Spain-1st wave	0.8667	0.154–0.971	0.9108	0.129–0.962
Spain-2nd wave	0.0369	0.095–0.945	0.0849	0.189–0.936
Hungary-1st wave	0.0005	0.093–1.014	0.0001	0.070–0.992
Hungary-2nd wave	0.1067	0.053–0.980	0.2559	0.094–0.959

Lag $p = 2$ was chosen for the VAR models, and one more lag into each variable was added to every equation, given that the maximum order of integration was 1. Therefore, the augmented VAR models proposed by the TY procedure were constructed, and the Granger causality tests were executed.

5. Results

Results of the Granger causality tests are shown in Table 3. Granger causality in Hungary during the first wave was not investigated as the underlying VAR model was not stable, and it had significant residual serial autocorrelation.

Table 3. Significant Granger causalities found in each examined VAR(2). For each causal relationship, the most significant lag in the appropriate VAR equation and the sign of this lag's coefficient are given in brackets.

Setup	Significant Granger Causalities
USA-1st wave	Stringency -> entry count * (lag = 1; sgn = +) Sentiment -> entry count * (lag = 2; sgn = +) Positive rate -> deaths per million ** (lag = 2; sgn = +) Entry count -> deaths per million ** (lag = 1; sgn = +)
USA-2nd wave	Positive rate -> stringency ** (lag = 2; sgn = +)
Spain-1st wave	Deaths per million -> sentiment * (lag = 1; sgn = −) Deaths per million -> entry Count * (lag = 1; sgn = +) Entry count -> deaths per million ** (lag = 1; sgn = +) Deaths per million -> stringency *** (lag = 2; sgn = +)
Spain-2nd wave	Entry count -> stringency * (lag = 2; sgn = +)
	Entry count -> deaths per million ** (lag = 1; sgn = −)
Hungary-1st wave	-
Hungary-2nd wave	Entry count -> stringency ** (lag = 1; sgn = +)
	Deaths per million -> entry count ** (lag = 1; sgn = −)

* Significant at 10%, ** significant at 5%, *** significant at 1%.

Table 3 shows that more significant Granger causal relationships could be found during the first wave of the pandemic than during the second. This is not surprising as the novelty of the virus posed more challenge during the first wave as decision makers and health professionals had to operate under limited information. Therefore, it is logical that we can find a higher number of relationships between our examined time series during the first wave. Unfortunately, owing to lack of a well-specified model for Hungary, this conclusion can only be made for Spain and the US.

In the US, the two most significant relationships were those between Twitter entry or post count and deaths per million and between positive rate and deaths. It seems that if the test positive rate increased, mortality usually followed 2 days later. This relationship was not significant at any of the common significance levels during the second wave, which suggests that the situation had improved by that time. During the second wave, we could also find that the increase in the rate of positive tests caused a stricter government response. This suggests that by the second wave, the US government started to react faster to changes in the pandemic situation. In the first wave, an increase in government stringency caused the count of Twitter entries to rise a day later. This can confirm that the US population was quite concerned with government response, so the measures were debated on Twitter. This finding is further supported by the fact that 2020 was election year in the US, so it is natural that government actions were under more scrutiny. These debates happened during the hardest days of the pandemic in the US, which is reflected in the significant Granger causality of Twitter entry count on mortality. Lastly, we observed that an increase in Twitter sentiment caused an increase in the number of posts 2 days later. It can be theorized that some positive messages about the pandemic could spread fast in the US, where the population grew frustrated with the lockdowns [25]. The antimask event exogenous variable had no significant effect on any of the endogenous time series in the US.

In Spain, during the first wave, the most significant Granger causality was the one that showed government stringency increasing 2 days after the deaths per million people increased. Therefore, the Spanish government reacted based on mortality, not on the rate of positive tests as the US government did. The less significant relationships showed that the number of tweets increased, and Twitter sentiment declined 1 day after an increase in mortality. Therefore, the increase in government stringency can be also considered an indirect reaction to public sentiment. This seems to suggest that in Spain, the public had some effect on stringency measures, namely, triggering a stricter response. The significant Granger causality of Twitter entry count on mortality suggests that the increased Twitter traffic happened during the hardest days of the pandemic in Spain, similar to the US. These findings seem to confirm the findings of [26,27], who suggest that public opinion had a part in reintroducing strict government measures during the summer of 2020. During the second wave, the effect of Twitter entry count on government stringency remained with a lag of 2 days, although the rest of the Granger causalities in the first wave had become insignificant except for the relationship of Twitter entry count and deaths per million. However, the directions of this relationship changed. It now shows the decrease of deaths per million a day after the number of tweets increases. This might be because Twitter activity concentrated on the peak of the second wave, after which mortality decreased somewhat. In Spain, antimask events had an echo on Twitter, as their exogenous variable had a significant positive effect on Twitter entry count in the first wave and a significant negative effect on Twitter sentiment in the second wave—however, in both cases only at 10%. Therefore, it can be theorized that during the first wave, the increased Twitter entry count that had a significant effect on mortality was partly due to these antimask events.

We only had a stable and well-specified VAR model for Hungary during the second wave, so only the results of this model are discussed. We had two significant Granger causalities—both effects significant at 5%, but not at 1%. The number of posts on Hungary's FAQ page seemed to be followed by an increase in government stringency a day later. This effect is something similar experienced in Spain, as public opinion was critical of the late government response during the second wave in Hungary [28]. We also found that there was a decrease in the number of FAQ posts a day after deaths per million increased. This is something similar to Spain's second wave: posting activity was concentrated on the peak of the second wave where mortality was highest, after which posting activity somewhat decreased. The antimask event exogenous variable had no significant effect on any of the endogenous variables in Hungary.

These VAR models can also be used to make short-period forecasts for any of the endogenous time series based on the other variables in the model. Therefore, for example, government stringency and mortality in Spain can be estimated based of Twitter entry counts of the previous day. However, this direction was not investigated further due to page limits.

6. Summary

Based on our results, the relationships between social media communication and epidemiological indicators were stronger during the first waves of the pandemic than during the later ones.

The US results were heavily influenced by the presidential election throughout the whole year, as the volume of Twitter comments reacted to government stringency in the first wave, but the sentiment did not seem to be affected. By the second wave, government stringency started to react to changes in the positive rate.

During the first wave of Spain, government stringency along with Twitter volume and sentiment all reacted to changes in the mortality rate. Government stringency lagged 2 days behind the changes, while the Twitter events followed only 1 day late. During the second wave, this relationship was reduced to government stringency reacting to Twitter traffic with a delay of 2 days. It is important to note though that around the second wave of Spain, the first wave of Mexico started as well; thus, Spanish Twitter comments might reflect this. Antimask events also had some influence on Twitter traffic, but mainly around the first wave.

In Hungary, our model was not stable for the first wave. However, the discovered relationships were very similar to what we experienced in Spain, as government stringency reacted to the volume of comments on the Hungarian FAQ. The reason for the lack of stable results in the first wave was probably the fact that even though there was a huge media hype in the spring of 2020, the number of confirmed cases was considerably lower than in the other waves.

A number of opportunities for further development have been identified. We would like to achieve greater heterogeneity across source platforms in order to reduce the effects of Twitter's typical "telegram" style. As an effect of abbreviated and compressed tweet texts, inaccuracies resulting from dictionary- and word-based text mining methods are presumably present. Another problem with Twitter is the unbalanced age distribution: only 10% of Twitter users are above 50 years [29]. It follows from all of this that it would be advisable to conduct the research based on the content of the much more widely used Facebook platform, or if it is not possible, then additional country-specific sources need to be utilized.

To identify corona topics and conspiracy theories, the utilized tool should be topic modelling; then social network analysis (SNA) can be performed along with topic modelling results. With SNA, we will examine how these topics spread. Finally, it will be possible to compare the results with the official WHO data collected during the pandemic; thus, we can analyze the impact of society on the pandemic and the impact of the pandemic on society.

Author Contributions: Conceptualization, M.M.; L.K. and D.B.; methodology, M.M.; L.K. and D.B.; software, M.M.; L.K. and D.B.; validation, M.M. and L.K.; formal analysis, M.M. and L.K.; investigation, M.M. and D.B.; resources, M.M.; L.K. and D.B.; data curation, M.M.; writing—original draft preparation, M.M. and L.K.; writing—review and editing, M.M.; L.K. and D.B.; visualization, M.M. and L.K.; supervision, D.B.; project administration, D.B.; funding acquisition, D.B. All authors have read and agreed to the published version of the manuscript.

Funding: This research was funded by Corvinus University of Budapest, grant number EFOP-3.6.3.-VEKOP-16-2017-00007.

Institutional Review Board Statement: Not applicable.

Informed Consent Statement: Not applicable.

Data Availability Statement: Corona-related pandemic data are collected from the site Our World in Data (OWID). https://ourworldindata.org/coronavirus (accessed on 2 February 2021). Tweets had been collected from the Twitter Stream Grab project by Archive Team. The project provides twitter data in compressed downloadable format. https://archive.org/details/twitterstream (accessed on 31 May 2021). Hungarian-language analyzed texts had been collected from the Hungarian FAQ page. This is a Q&A-type website, which is the 31st most visited site in Hungary. Politics and Health categories had been scraped. https://www.gyakorikerdesek.hu/ (accessed on 31 May 2021). Corona-related search terms had been collected from Google Trends Datastore. http://googletrends.github.io/data/ (accessed on 21 May 2021). Corona-sceptic events had been collected from especially national Wikipedia collections. https://en.wikipedia.org/wiki/Protests_over_responses_to_the_COVID-19_pandemic (accessed on 31 May 2021); https://es.wikipedia.org/wiki/Categor%C3%ADa:Manifestaciones_y_protestas_en_2020 (accessed on 31 May 2021).

Conflicts of Interest: The authors declare no conflict of interest.

References

1. Taskinsoy, J. The Great Pandemic of the 21st Century: The Stolen Lives. 2020. Available online: https://papers.ssrn.com/sol3/papers.cfm?abstract_id=3689993 (accessed on 30 June 2021).
2. Centers for Disease Control and Prevention: Estimated Global Mortality Associated with the First 12 Months of 2009 Pandemic Influenza A H1N1 Virus Circulation: A Modelling Study. 2012. Available online: https://www.cdc.gov/flu/spotlights/pandemic-global-estimates.htm (accessed on 31 May 2021).

3. Ortiz-Ospina, E. The Rise of Social Media. Our World in Data. 2019. Available online: https://ourworldindata.org/rise-of-social-media (accessed on 30 June 2021).
4. Centers for Disease Control and Prevention: Statistics Overview. 2020. Available online: https://www.cdc.gov/hiv/statistics/overview/index.html (accessed on 31 May 2021).
5. Douglas, K.M. COVID-19 conspiracy theories. *Group Process. Intergroup Relat.* **2021**, *24*, 270–275. [CrossRef]
6. Vaezi, A.; Javanmard, H.J. Infodemic and Risk Communication in the Era of CoV-19. *Adv. Biomed. Res.* **2020**, *9*, 10. [CrossRef] [PubMed]
7. Roser, M.; Ritchie, H.; Ortiz-Ospina, E.; Hasell, J. Coronavirus Pandemic (COVID-19). Our World in Data. 2020. Available online: https://ourworldindata.org/coronavirus (accessed on 2 February 2021).
8. Toda, H.Y.; Yamamoto, T. Statistical inference in vector autoregressions with possibly integrated processes. *J. Econom.* **1995**, *66*, 225–250. [CrossRef]
9. Ding, D.; Guan, C.; Chan, C.M.; Liu, W. Building stock market resilience through digital transformation: Using Google trends to analyze the impact of COVID-19 pandemic. *Front. Bus. Res. China* **2020**, *14*, 1–21. [CrossRef]
10. Gherghina, Ş.C.; Armeanu, D.Ş.; Joldeş, C.C. Stock Market Reactions to COVID-19 Pandemic Outbreak: Quantitative Evidence from ARDL Bounds Tests and Granger Causality Analysis. *Int. J. Environ. Res. Public Health* **2020**, *17*, 6729. [CrossRef] [PubMed]
11. Similarweb: Top Websites Ranking. 2021. Available online: https://www.similarweb.com/top-websites/ (accessed on 21 May 2021).
12. Bruns, A. After the 'APIcalypse': Social media platforms and their fight against critical scholarly research. *Inf. Commun. Soc.* **2019**, *22*, 1544–1566. [CrossRef]
13. Archive Team: The Twitter Stream Grab. Available online: https://archive.org/details/twitterstream (accessed on 31 May 2021).
14. GyakoriKerdesek Homepage. Available online: https://www.gyakorikerdesek.hu (accessed on 31 May 2021).
15. Google Trends Datastore. Available online: http://googletrends.github.io/data/ (accessed on 21 May 2021).
16. World Health Organization. Overview of Public Health and Social Measures in the Context of COVID-19: Interim Guidance, 18 May 2020. (No. WHO/2019-nCoV/PHSM_Overview/2020.1). World Health Organization. Available online: https://apps.who.int/iris/handle/10665/332115 (accessed on 8 April 2021).
17. Hale, T.; Petherick, A.; Phillips, T.; Webster, S. Variation in Government Responses to COVID-19. Blavatnik School of Government Working Paper 31. 2020. Available online: https://www.bsg.ox.ac.uk/research/publications/variation-government-responses-covid-19 (accessed on 8 April 2021).
18. Hunspell Homepage. Available online: https://hunspell.github.io/ (accessed on 31 May 2021).
19. Zeileis, A.; Kleiber, C.; Kraemer, W.; Hornik, K. Testing and Dating of Structural Changes in Practice. *Comput. Stat. Data Anal.* **2003**, *44*, 109–123. [CrossRef]
20. Stock, J.H.; Watson, M.W. *Introduction to Econometrics, Third Update, Global Edition*; Pearson Education Limited: London, UK, 2015.
21. He, Z.; Maekawa, K. On spurious Granger causality. *Econ. Lett.* **2001**, *73*, 307–313. [CrossRef]
22. Hu, M.; Liu, B. Mining and summarizing customer reviews. In Proceedings of the Tenth ACM SIGKDD International Conference on Knowledge Discovery and Data Mining, Seattle, WA, USA, 22–25 August 2004.
23. TASS: Workshop on Semantic Analysis at SEPLN. Available online: http://tass.sepln.org/ (accessed on 25 May 2021).
24. Szabó, M. Experiences of Creation of a Hungarian Sentiment Lexicon. Conference "Nyelv, kultúra, társadalom". Precognox, Budapest. 2014. Available online: http://publicatio.bibl.u-szeged.hu/8791/12/cikk_mszny_2015.pdf (accessed on 8 April 2021).
25. Deane, C.; Parker, K.; Gramlich, J. A Year of U.S. Public Opinion on the Coronavirus Pandemic. 2021. Available online: https://www.pewresearch.org/2021/03/05/a-year-of-u-s-public-opinion-on-the-coronavirus-pandemic/ (accessed on 8 April 2021).
26. Royo, S. Responding to COVID-19: The Case of Spain. *Eur. Policy Anal.* **2020**, *6*, 180–190. [CrossRef]
27. Oliver, N.; Barber, J.X.; Roomp, K.; Roomp, K. Assessing the Impact of the COVID-19 Pandemic in Spain: Large-Scale, Online, Self-Reported Population Survey. *J. Med. Internet Res.* **2020**, *22*, e21319. [CrossRef] [PubMed]
28. Szakacs, G.; Dunai, M. Orban Given Special Powers as Hungary Locks Down against COVID Surge. 2020. Available online: https://www.reuters.com/article/uk-health-coronavirus-hungary-casualties-idUKKBN27Q2MZ\T1\textquoteright (accessed on 8 April 2021).
29. Tankovska, H. Statista-Distribution of Twitter Users Worldwide as of January 2021, by Age Group. Available online: https://www.statista.com/statistics/283119/age-distribution-of-global-twitter-users/ (accessed on 31 March 2021).

engineering proceedings

MDPI

Proceedings

Does AutoML Outperform Naive Forecasting? †

Gian Marco Paldino [1,*], Jacopo De Stefani [1], Fabrizio De Caro [2] and Gianluca Bontempi [1]

1 Machine Learning Group, Université Libre de Bruxelles, 1050 Bruxelles, Belgium; jdestefa@ulb.ac.be (J.D.S.); gbonte@ulb.ac.be (G.B.)
2 Dipartimento di Ingegneria, Università degli Studi del Sannio, 82100 Benevento, Italy; fdecaro@unisannio.it
* Correspondence: gpaldino@ulb.ac.be
† Presented at the 7th International conference on Time Series and Forecasting, Gran Canaria, Spain, 19–21 July 2021.

Abstract: The availability of massive amounts of temporal data opens new perspectives of knowledge extraction and automated decision making for companies and practitioners. However, learning forecasting models from data requires a knowledgeable data science or machine learning (ML) background and expertise, which is not always available to end-users. This gap fosters a growing demand for frameworks automating the ML pipeline and ensuring broader access to the general public. Automatic machine learning (AutoML) provides solutions to build and validate machine learning pipelines minimizing the user intervention. Most of those pipelines have been validated in static supervised learning settings, while an extensive validation in time series prediction is still missing. This issue is particularly important in the forecasting community, where the relevance of machine learning approaches is still under debate. This paper assesses four existing AutoML frameworks (AutoGluon, H_2O, TPOT, Auto-sklearn) on a number of forecasting challenges (univariate and multivariate, single-step and multi-step ahead) by benchmarking them against simple and conventional forecasting strategies (e.g., naive and exponential smoothing). The obtained results highlight that AutoML approaches are not yet mature enough to address generic forecasting tasks once compared with faster yet more basic statistical forecasters. In particular, the tested AutoML configurations, on average, do not significantly outperform a Naive estimator. Those results, yet preliminary, should not be interpreted as a rejection of AutoML solutions in forecasting but as an encouragement to a more rigorous validation of their limits and perspectives.

Keywords: AutoML; time series forecasting; benchmarking; frameworks

Citation: Paldino, G.M.; De Stefani, J.; De Caro, F.; Bontempi, G. Does AutoML outperform Naive forecasting? *Eng. Proc.* **2021**, *5*, 36. https://doi.org/10.3390/engproc2021005036

Published: 5 July 2021

1. Introduction

The pervasiveness of electronic devices enables the collection of temporal data (about production, development, sales) at a growing rate. Extracting actionable knowledge from temporal data requires specific technical skills, yet the growing availability of data is not accompanied by an equivalent increase in the number of experts able to analyze them, thus reducing their potential impact.

Automated machine learning (AutoML) [1] aims to fill this gap by automatizing the different phases of data analysis and providing suitable solutions for data scientists, practitioners and final users. AutoML approaches can help obtain a glimpse of knowledge about new data, for example, suggesting the optimal model to use. Data may also be too noisy or of poor quality, in which case AutoML would quickly reflect it, by showing failure in multiple pipelines, saving the data scientist a lot of time.

However, finding a procedure that automates the entire ML process for forecasting is a risky endeavor. Time series data have constraints and peculiarities (e.g., trend and seasonalities, outliers, drifts, abrupt changes) to handle in specific ways, often not compatible with more traditional tabular data. Furthermore, most AutoML approaches rely on the assumption that the higher the degree of search in the hyperparameter space, the better the final result. Now, in large dimensional and noisy settings, pushing the degree of grid search too far leads inevitably to a high degree of variance of the returned solution, which, if not adequately assessed with external validation data, could be prone to overfitting [2]. This is particularly the case of large dimensional settings like the ones that can be encountered in multivariate forecasting. In those cases, embedding, i.e., the transformation in a tabular form for supervised learning, produces high-dimensional input datasets. Automatizing the feature selection phase without accounting for an external validation set can be detrimental, returning over-optimistic assessment of the generalization accuracy of the chosen set of features.

Thus far, the main comparative studies on AutoML solutions are [3–8]. They generally compare various frameworks against each other on standard supervised learning tasks. Results show high variance [3], or no significant difference between models [7], although more recent comparisons appear to favor AutoGluon [8], suggesting the importance of feature preprocessing. However, frameworks tend not to significantly outperform traditional models (e.g., random forest within 4 h [4]) nor humans in easy classification tasks [5].

This paper assesses the capabilities of four AutoML frameworks (AutoGluon, H_2O, TPOT, Auto-sklearn) with respect to conventional statistical forecasting strategies (naive, exponential smoothing, Holt-Winter's). This issue is particularly important in the forecasting community, where the relevance of machine learning approaches is still under debate [9]. In order to provide a fair comparison, we took advantages of the possibility provided by AutoML packages to limit the allowed computational time. The goal is to show experimentally the effectiveness of known AutoML frameworks on time series forecasting, challenging the framework by limiting their computational time and comparing the results with fast conventional forecasting strategies. In particular, the main contribution of this manuscript are:

- A description of several state-of-the-art AutoML frameworks;
- The comparison between several state-of-the-art AutoML frameworks on univariate and multivariate time series forecasting on different horizons;
- The assessment of their effectiveness against conventional forecasting strategies such as naive and exponential smoothing on comparable scale times.

Note that the constraint on the execution time is not simply an experimental decision but it reflects a criticism of the authors about the continuous increase of computing resources required by ML methods (notably deep learning). Since this resource consumption is not necessarily followed by a correspondent improvement of the overall performances, we think it is time for the forecasting community to investigate the trade-off between time (and energy) consumption vs. accuracy. The paper is organized as follows: Section 2 introduces the problem formulation, while Section 3 describes the adopted AutoML frameworks. The benchmarking experiments are described in Section 4, with the discussion and conclusions in Section 5.

2. Machine Learning and Forecasting

A multivariate and multitemporal model f aims at learning the mapping between past values and future values of an N-variate time series. Given a time resolution $\Delta t = t_i - t_{i-1}$ at time instant t, a lag L and a forecasting horizon h, the temporal dependency can be represented in the embedded form:

$$\begin{pmatrix} y_{1,t+1}, \dots, y_{1,t+h} \\ \cdots \\ y_{N,t+1}, \dots, y_{N,t+h} \end{pmatrix} = f \begin{pmatrix} y_{1,t-L+1}, \dots, y_{1,t} \\ \cdots \\ y_{N,t-L+1}, \dots, y_{N,t} \end{pmatrix} \tag{1}$$

The multi-input multi-output nature of (1) suggests the adoption of a multi-output approach (e.g., neural networks). However, since most learning algorithms available in

AutoML frameworks are single-output, we will decompose the MIMO problem in a sequence of N multiple-input single-output (MISO) tasks:

$$\begin{gathered}\left(y_{1,t+1},\ldots,y_{1,t+h} \right) = f_{1.1}\begin{pmatrix} y_{1,t-L+1},\ldots,y_{1,t} \\ \cdots \\ y_{N,t-L+1},\ldots,y_{N,t} \end{pmatrix} \\ \cdots \\ \left(y_{N,t+1},\ldots,y_{N,t+h} \right) = f_{1.N}\begin{pmatrix} y_{1,t-L+1},\ldots,y_{1,t} \\ \cdots \\ y_{N,t-L+1},\ldots,y_{N,t} \end{pmatrix}\end{gathered} \tag{2}$$

If we assume that there is no significant cross-series dependency, we may further decompose (2) into a set of N single-input, single-output (SISO) tasks:

$$\begin{gathered}(y_{1,t+1},\ldots,y_{1,t+h}) = f_{2.1}(y_{1,t-L+1},\ldots,y_{1,t}) \\ \cdots \\ (y_{N,t+1},\ldots,y_{N,t+h}) = f_{2.N}(y_{N,t-L+1},\ldots,y_{N,t})\end{gathered} \tag{3}$$

The above formulations make the natural adoption of supervised learning pipelines [10], which are typically composed of the following steps:

1. Preprocessing : the observations are cleaned, normalized and rescaled. Missing data can be removed or replaced. New features may be produced by means of feature engineering [11].
2. Dimensionality reduction: this step aims at reducing the input dimension, to diminish the computational burden and avoid numerical and statistical issues [12].
3. Model estimation: this step estimates from the available data the input-output relationship.
4. Performance assessment: the model performances are validated by means of a validation set, a subsample of the observed data that is kept aside to verify the ability of the model previously trained to correctly predict new unseen samples. This is followed by an analysis of the distribution of performance measures.

It is important to remark that those steps are either skipped or extremely simplified in conventional forecasting strategies (e.g., exponential smoothing) with an evident gain in terms of computational time.

2.1. Conventional Statistical Approaches

Those methods provide a quick insight to the behavior of a time series and are efficient to compute. The simplest approach is the naive: the time series forecast at time $t+1$ is provided by the last available observation at time t. Another simple technique is the mean model, where the forecast at time $t+1$ is the average of all previous observations up to time t. Exponential smoothing is an approach proposed by [13,14], based on exponentially decaying weighted averages of past observations. The approach favors recent observations, and its speed and reliability made it successful. A basic version is the simple exponential smoothing (4), suitable for data with no clear trend or seasonal pattern. $0 \leq \alpha \leq 1$ is the smoothing parameter that controls the rate at which the weights decrease.

$$\hat{y}_{t+1|t} = \alpha y_t + \alpha(1-\alpha)y_{t-1} + \alpha(1-\alpha)^2 y_{t-2} + \cdots \tag{4}$$

Holt and Winters [15] extended the method to capture trends and seasonality. The Holt-Winters seasonal method comprises the forecast equation and three smoothing equations (level, trend, seasonality). A corresponding multiplicative version exists [16]. When the seasonal variations are roughly constant, the additive method is preferred, and when the seasonal variations are changing proportional to the level of the series, the multiplicative method is chosen. By considering variations in the combinations of the trends and seasonal components, nine exponential smoothing methods are possible [17–19].

3. AutoML Frameworks

This section sketches the AutoML frameworks we selected for the experimental comparison. All of them provide the possibility to limit their execution time, enabling a fair comparison with faster statistical methods (Section 2.1).

H_2O is a distributed machine learning platform. Its AutoML module [20] covers a large selection of candidate models, including two stacked ensembles of all the trained models and of the best model of each family, respectively. It provides a simple and highly customizable interface where the user can specify the maximum time of the AutoML process or the maximum number of models to build in an AutoML run. The available ML algorithms are distributed random forest, including both random forest and extremely randomized trees; generalized linear models, XGBoost gradient boosting machine, H_2O gradient boosting machine, and deep neural network. Preprocessing is limited to automated target encoding of high dimension categorical variables.

Auto-sklearn [21] is a Python library for AutoML built on top of the scikit-learn library [22]. It uses 15 learners (notably k nearest neighbors, gradient boosting, stochastic gradient descent, random forest, AdaBoost), 14 feature preprocessing methods, and 4 data preprocessing methods. It leverages meta-learning by evaluating a set of meta-features (e.g., statistics about the number of data points, features) over hundreds of datasets and storing the most accurate related configurations. It also adopts Bayesian optimization to fit a probabilistic model to capture the relationship between hyperparameter settings and their measured performance. Additionally, it uses ensemble selection, a greedy procedure that, starting with an empty ensemble, iteratively adds the model that maximizes the ensemble effectiveness. Its data preprocessing includes one-hot encoding, imputation of missing values and normalization. Its feature preprocessing performs feature selection via principal component analysis, singular value decomposition and other methods.

AutoGluon [8] is a Python library for AutoML dealing with text, image, and tabular data. The set of learners includes neural networks, LightGBM boosted trees, CatBoost boosted trees, random forests, extremely randomized trees, and k nearest neighbors. Its preprocessing is split into model-agnostic preprocessing, including features categorization and treatment (e.g., encoding of categorical variables), and model-specific preprocessing applied on a copy of the data passed to each model. Multi-layer stack ensembling and repeated k-fold bagging are used to combine the base learners.

TPOT [23] is a tree-based pipeline optimization tool that automatically designs and optimizes ML pipelines using genetic programming (GP) [24]. It wraps the scikit-learn library [22], and offers the following models: decision tree, random forest, eXtreme gradient boosting, logistic regression and k nearest neighbor. The preprocessing and feature selection functionalities include standard scaler, randomized PCA, SelectKBest, and recursive feature elimination. Each ML pipeline is treated as a GP primitive, and GP trees are constructed from them. The process starts by generating 100 random tree-based pipelines and evaluating them, while for every generation, the top 20 are selected to maximize accuracy and minimize the number of operators. Each of the top 20 pipelines produces five copies with cross-overs or random mutations over the individual components of the pipeline. The whole procedure is repeated for 100 generations.

4. Experimental Benchmark

This section introduces the time series benchmarks, the methodology and the evaluation metrics and the results. We consider two public datasets made available in [25]. The format of the dataset has been adapted to ease research related to multivariate time series. A link can be found in the footnotes of Section 5.

- **Electricity consumption:** the original dataset (https://archive.ics.uci.edu/ml/datasets/ElectricityLoadDiagrams20112014, accessed on 30 March 2021) contains electricity consumption of 370 clients recorded every 15 min from 2011 to 2014. The preprocessed dataset contains hourly consumption (in kWh) of 321 clients from 2012 to 2014.

- **Exchange rate**: the dataset (possible source: https://fred.stlouisfed.org/series/EXUSEU, accessed on 3 March 2021) is a collection of the daily exchange rates of eight foreign countries, including Australia, Great Britain, Canada, Switzerland, China, Japan, New Zealand and Singapore. The considered time ranges from 1990 to 2016.

We benchmark four AutoML frameworks against simple and conventional forecasting strategies by adopting a SISO (3) approach for univariate forecasting and a MISO (2) approach for multivariate forecasting. The rationale behind the choice of MISO over MIMO (1) is that not all AutoML frameworks provide the possibility of predicting multiple-output, and this would not have produced a fair comparison. An additional reason is the intrinsic univariate nature of conventional methods such as naive or exponential smoothing: a MISO approach provides a more natural comparison by predicting only one variable. For the univariate case, we decide to work with the first available variable, without loss of generality. The choice of additional variables for the multivariate case is made as follows: starting from the first variable, we pick its N most correlated variables, and we forecast on the first variable.

Preprocessing: we do not perform any data preprocessing for three reasons. First, data are already in a format that does not need particular treatment. Second, some AutoML frameworks, as mentioned in Section 3, include some data preprocessing. If this improves the performances, the corresponding framework should be rewarded. Third, this work aims at benchmarking models, rather than maximizing the correctness of forecasting: as long as all models are provided with the same data, the benchmark is fair.

AutoML: the selected frameworks treat tabular data in a supervised learning setting. They hence require an embedding for the time series (see Section 1). The lag parameter for the embedding was fixed at $L = 5$, and future work will explore other values. Since one of the two time series considered contained at most eight variables, we decided to set the possible number of variables to $v \in [1, 3, 5, 8]$. The values for the horizon have been fixed to $h \in [1, 2]$ and the time allowed for each AutoML framework was limited to $t =$ [60 s, 120 s, 300 s]. These values are low with respect to standard AutoML times for an optimal exploration of the space of parameters, but the comparison with particularly fast methods requires those limitations. Longer time frames will be considered in future work. The combinations of all tested parameters are presented in (5), and one experiment has been carried out for each of them. No additional parameter has been set to the frameworks.

$$\begin{bmatrix} \text{Auto-sklearn} \\ \text{AutoGluon} \\ \text{H2O} \\ \text{TPOT} \end{bmatrix} \times \begin{bmatrix} \text{1 variable} \\ \text{3 variables} \\ \text{5 variables} \\ \text{8 variables} \end{bmatrix} \times \begin{bmatrix} \text{Horizon 1} \\ \text{Horizon 2} \end{bmatrix} \times \begin{bmatrix} \text{Time 60 s} \\ \text{Time 120 s} \\ \text{Time 300 s} \end{bmatrix} \quad (5)$$

Conventional forecasting strategies: we consider the naive predictor as our baseline, and from the exponential smoothing family, we choose the simple exponential smoothing and four variations of Holt-Winters. We focus on Holt-Winters because of its historical effectiveness in forecasting [15]. The mentioned variations are summarized in Table 1.

Table 1. Exponential smoothing-approaches considered in this work, specifying the nature of their trends and seasonal components.

Method	Trend Comp.	Seasonal Comp.
Simple exponential smoothing	None	None
Additive Holt-Winters' method	Additive	Additive
Multiplicative Holt-Winters' method	Additive	Multiplicative
Additive Holt-Winters' damped method	Additive damped	Additive
Multiplicative Holt-Winters' damped method	Additive damped	Multiplicative

Train, validation and metrics: We consider the absolute error, i.e., the absolute difference between the real value y_i and the predicted value $\hat{y}_i$. Our validation strategy divides the time series into three equal fragments, considering one additional third at each iteration. The last 16 points of this set are considered as the validation set, while all the previous points are the train set. This results in a total of 48 test points from 3 different areas of the time series.

Experimental Results

Figures 1–4 show critical distance plots, i.e., a graphical representation of the results of the Friedman statistical test (with post hoc Nemenyi test), as suggested in [26]. The methods are ordered according to their performance from left to right (left is better), while the black bar connects methods that are not significantly different (at $p = 0.05$). The nomenclature chosen follows the pattern *framework_variables_time*. A selection of models is shown in Figures 1–3; in particular, we plot the most and least performing variant for each AutoML framework and all the conventional methods. Figures 1 and 2 present the methods ranking over the electricity and exchange time series, respectively, averaging over different horizons. Figure 3 represents the average over both time series. The same results of Figure 3 are presented in Figure 4, but the 20 most performing models are considered. Table 2 presents the win/losses of all studied approaches with respect to the naive predictor. In all cases, the metric considered is the absolute error. This analysis highlights the following results:

- Short-range training times (in the order of few minutes) are not sufficient for the AutoML frameworks considered to significantly outperform conventional methods (Figure 3). For short-horizons quick forecasting, it might therefore be convenient to rely on the latter.
- In terms of training time, 120 s seems to allow slightly better generalization ability than 60 or 300 s (Table 2). This might indicate that with 60 s, the models tend to underfit and with 300 s to overfit the observations.
- All traditional methods dominate every AutoML method in terms of wins count with respect to the naive (Table 2), which reflects the strong forecasting ability of the exponential smoothing family of methods. It could be appropriate to consider those methods as a baseline.
- Moving from a univariate SISO to multivariate MISO approach does not improve the performances of any method despite that the variables are added by maximizing correlation. This seems to suggest a lack of effectiveness in the feature selection approaches of the AutoML frameworks, when implemented.

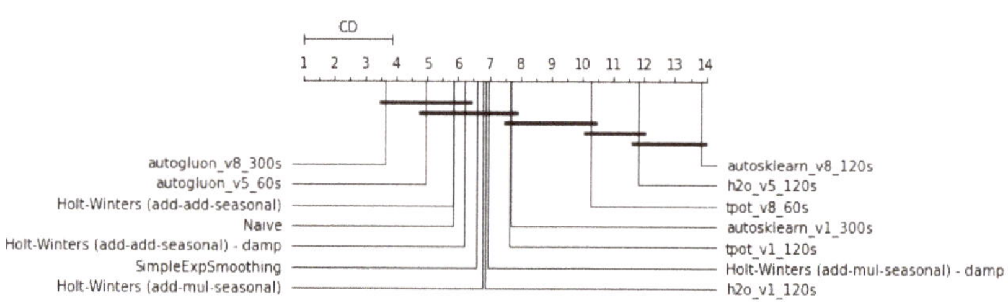

Figure 1. CD plot—selected models comparison of the absolute errors over the validation set for the electricity time series.

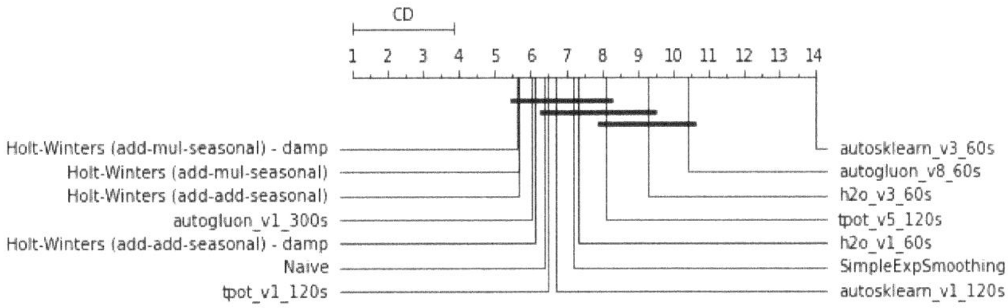

Figure 2. CD plot—selected models comparison of the absolute errors over validation set for the exchange time series.

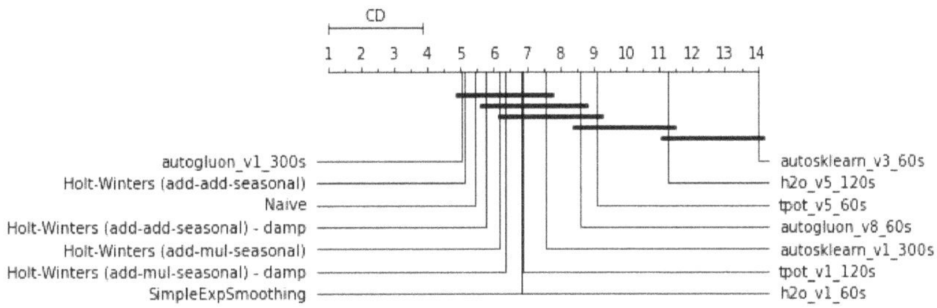

Figure 3. CD plot—selected models comparison of the absolute errors over validation set for both time series.

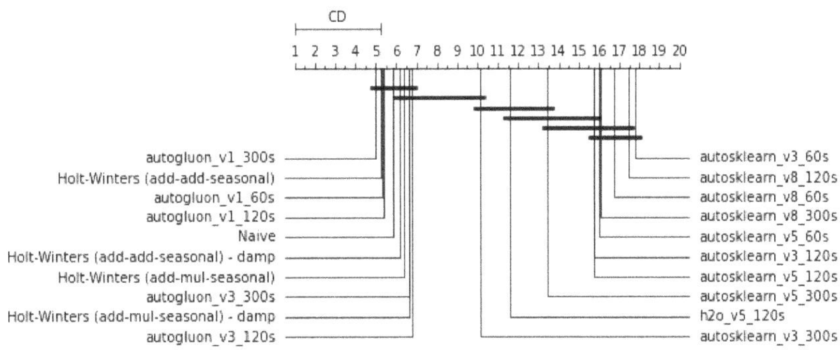

Figure 4. CD plot—top 20 models comparison of the absolute errors over validation set for both time series.

Table 2. Win/loss count with respect to the naive approach presented in Section 2.1. The counts are made over the four case studies considered: electricity and exchange time series for 1-step ahead and 2-steps ahead forecasting. The nomenclature chosen follows the pattern *framework_variables_time*, and the metric considered is the absolute error.

Model	Wins	Losses	Model	Wins	Losses
Holt-Winters (add_d-add)	146	46	h2o_v3_300s	66	126
Holt-Winters (add-add)	146	46	autogluon_v3_300s	65	127
SimpleExpSmoothing	139	53	tpot_v3_60s	65	127
Holt-Winters (add_d-mul)	138	54	h2o_v5_60s	65	127
Holt-Winters (add-mul)	138	54	h2o_v3_60s	63	129
h2o_v1_120s	80	112	h2o_v8_60s	62	130
h2o_v1_300s	79	113	tpot_v5_60s	62	130
autosklearn_v1_300s	75	117	autosklearn_v1_60s	61	131
h2o_v1_60s	75	117	tpot_v8_60s	60	132
h2o_v8_120s	74	118	autogluon_v5_300s	60	132
autogluon_v1_300s	74	118	tpot_v8_120s	60	132
tpot_v1_120s	74	118	autogluon_v8_300s	59	133
tpot_v8_300s	74	118	autogluon_v8_120s	59	133
autogluon_v1_60s	73	119	autogluon_v5_120s	59	133
tpot_v1_60s	72	120	autogluon_v8_60s	57	135
tpot_v1_300s	71	121	autogluon_v5_60s	53	139
h2o_v5_300s	69	123	autosklearn_v3_300s	51	141
autogluon_v1_120s	69	123	h2o_v5_120s	50	142
autogluon_v3_120s	69	123	autosklearn_v5_300s	36	156
h2o_v8_300s	68	124	autosklearn_v3_120s	18	174
tpot_v5_120s	67	125	autosklearn_v5_120s	17	175
autogluon_v3_60s	67	125	autosklearn_v5_60s	14	178
autosklearn_v1_120s	67	125	autosklearn_v8_300s	14	178
tpot_v5_300s	66	126	autosklearn_v8_60s	9	183
tpot_v3_300s	66	126	autosklearn_v8_120s	5	187
tpot_v3_120s	66	126	autosklearn_v3_60s	2	190
h2o_v3_120s	66	126	Naive	Base	Base

5. Conclusions, Recommendations and Future Work

Automated machine learning is a promising research direction aiming to support practitioners in unleashing the potential of ML for data science. Various frameworks currently exist, and they differ by their feature selection, model selection and parameter optimization approaches. With sufficient time and resources, they have been showing excellent results in several learning problems.

This paper supports the idea that it is probably too soon to consider them as a full-fledged solution for time series forecasting. In particular, we deem that most solutions hang more on the complexity and comprehensiveness side than on the one of a rigorous validation of the added value with respect to simpler, yet less prone to overfitting, solutions. This is particularly delicate in forecasting settings where the high noise, the large dimension and the small number of samples would advise for a more cautious attitude with respect to complex automatic solutions. Our conclusion is supported by a benchmark of selected AutoML frameworks against simple statistical methods like naive and Holt-Winter's. The obtained results suggest that, in the short term, AutoML frameworks do not significantly outperform traditional methods, and relying exclusively on them might not be the optimal solution.

On the basis of the results obtained, we would like to make some recommendations to the AutoML community. It is important that any automatic selection strategy is supported by an external validation dataset, including significance tests with respect to simple and naive strategies. In the case of limited data, permutation strategies may be adopted to assess the added value of complex ML pipelines, as well. Last but not least, we deem that AutoML tools should have a pedagogical role with respect to end users by educating them in terms of the trade-off between accuracy and computational resource (and energy) consumption. For instance, a graphical representation of the cost–benefit ratio could help in that sense.

Further work will assess the impact of longer computational time allowed to the AutoML models (in the order of hours or days) and repeat the tests for larger horizons, where traditional methods might suffer. AutoML frameworks also offer deep customization to improve their performance, which has not been considered in this work and will be studied. Additionally, an analysis of other frameworks that offer time-series-specific treatments is foreseen.

Acknowledgments: Gianluca Bontempi was supported by Service Public de Wallonie Recherche under grant n° 2010235–ARIAC by DIGITALWALLONIA4.AI. The authors declare they have no conflicts of interest. Code is available at the following url: https://github.com/gmpal/AutoML-vs-Naive-benchmark (accessed on 30 March 2021). Data are available at the following url: https://github.com/laiguokun/multivariate-time-series-data (accessed on 30 March 2021).

References

1. He, X.; Zhao, K.; Chu, X. AutoML: A Survey of the State-of-the-Art. *Knowl.-Based Syst.* **2021**, *212*, 106622. [CrossRef]
2. Bontempi, G. A blocking strategy to improve gene selection for classification of gene expression data. *IEEE/ACM Trans. Comput. Biol. Bioinform.* **2007**, *4*, 293–300. [CrossRef] [PubMed]
3. Balaji, A.; Allen, A. Benchmarking automatic machine learning frameworks. *arXiv* **2018**, arXiv:1808.06492.
4. Gijsbers, P.; LeDell, E.; Thomas, J.; Poirier, S.; Bischl, B.; Vanschoren, J. An open source automl benchmark. *arXiv* **2019**, arXiv:1907.00909.
5. Hanussek, M.; Blohm, M.; Kintz, M. Can AutoML outperform humans? An evaluation on popular OpenML datasets using AutoML Benchmark. *arXiv* **2020**, arXiv:2009.01564.
6. Guyon, I.; Sun-Hosoya, L.; Boullé, M.; Escalante, H.J.; Escalera, S.; Liu, Z.; Jajetic, D.; Ray, B.; Saeed, M.; Sebag, M.; et al. Analysis of the AutoML Challenge Series 2015–2018. In *Automated Machine Learning*; Hutter, F., Kotthoff, L., Vanschoren, J., Eds.; The Springer Series on Challenges in Machine Learning; Springer: Cham, Switzerland, 2019; doi:10.1007/978-3-030-05318-5_10. [CrossRef]
7. Zöller, M.A.; Huber, M.F. Benchmark and survey of automated machine learning frameworks. *arXiv* **2019**, arXiv:1904.12054.
8. Erickson, N.; Mueller, J.; Shirkov, A.; Zhang, H.; Larroy, P.; Li, M.; Smola, A. Autogluon-tabular: Robust and accurate automl for structured data. *arXiv* **2020**, arXiv:2003.06505.
9. Makridakis, S.; Spiliotis, E.; Assimakopoulos, V. Statistical and Machine Learning forecasting methods: Concerns and ways forward. *PLoS ONE* **2018**, *13*, e0194889. [CrossRef] [PubMed]
10. Bontempi, G.; Taieb, S.B.; Le Borgne, Y.A. Machine learning strategies for time series forecasting. In *European Business Intelligence Summer School*; Springer: Berlin, Germany, 2012; pp. 62–77.
11. Christ, M.; Braun, N.; Neuffer, J.; Kempa-Liehr, A.W. Time series feature extraction on basis of scalable hypothesis tests (tsfresh—A python package). *Neurocomputing* **2018**, *307*, 72–77. [CrossRef]
12. Bermingham, M.L.; Pong-Wong, R.; Spiliopoulou, A.; Hayward, C.; Rudan, I.; Campbell, H.; Wright, A.F.; Wilson, J.F.; Agakov, F.; Navarro, P.; et al. Application of high-dimensional feature selection: Evaluation for genomic prediction in man. *Sci. Rep.* **2015**, *5*, 1–12. [CrossRef] [PubMed]
13. Brown, R.G. *Statistical Forecasting for Inventory Control*; McGraw/Hill: New York, NY, USA, 1959.
14. Holt, C.C. Forecasting seasonals and trends by exponentially weighted moving averages. *Int. J. Forecast.* **2004**, *20*, 5–10. [CrossRef]
15. Goodwin, P. The holt-winters approach to exponential smoothing: 50 years old and going strong. *Foresight* **2010**, *19*, 30–33.
16. Hyndman, R.; Koehler, A.B.; Ord, J.K.; Snyder, R.D. *Forecasting with Exponential Smoothing: The State Space Approach*; Springer Science & Business Media: Berlin/Heidelberg, Germany, 2008.
17. Pegels, C.C. Exponential forecasting: Some new variations. *Manag. Sci.* **1969**, *15*, 311–315.
18. Gardner, E.S., Jr. Exponential smoothing: The state of the art. *J. Forecast.* **1985**, *4*, 1–28. [CrossRef]
19. Taylor, J.W. Exponential smoothing with a damped multiplicative trend. *Int. J. Forecast.* **2003**, *19*, 715–725. [CrossRef]
20. LeDell, E.; Poirier, S. H2O AutoML: Scalable Automatic Machine Learning. In Proceedings of the 7th ICML Workshop on Automated Machine Learning (AutoML), Online, 18 July 2020.
21. Feurer, M.; Klein, A.; Eggensperger, K.; Springenberg, J.; Blum, M.; Hutter, F. Efficient and Robust Automated Machine Learning. 2015. Available online: http://papers.nips.cc/paper/5872-efficient-and-robust-automated-machine-learning (accessed on 30 March 2021).
22. Pedregosa, F.; Varoquaux, G.; Gramfort, A.; Michel, V.; Thirion, B.; Grisel, O.; Blondel, M.; Prettenhofer, P.; Weiss, R.; Dubourg, V.; et al. Scikit-learn: Machine Learning in Python. *CoRR* **2012**, abs/1201.0490. [CrossRef]

23. Olson, R.S.; Bartley, N.; Urbanowicz, R.J.; Moore, J.H. Evaluation of a Tree-based Pipeline Optimization Tool for Automating Data Science. In Proceedings of the Genetic and Evolutionary Computation Conference (GECCO '16), Denver, CO, USA, 20–24 July 2016; ACM: New York, NY, USA, 2016; pp. 485–492. [CrossRef]
24. Banzhaf, W.; Nordin, P.; Keller, R.E.; Francone, F.D. *Genetic Programming: An Introduction*; Morgan Kaufmann Publishers: San Francisco, CA, USA, 1998; Volume 1.
25. Lai, G.; Chang, W.C.; Yang, Y.; Liu, H. Modeling long-and short-term temporal patterns with deep neural networks. In Proceedings of the 41st International ACM SIGIR Conference on Research & Development in Information Retrieval, Ann Arbor, MI, USA, 8–12 July 2018; pp. 95–104.
26. Demšar, J. Statistical comparisons of classifiers over multiple data sets. *J. Mach. Learn. Res.* **2006**, *7*, 1–30.

engineering proceedings

Proceedings

On the Family of Covariance Functions Based on ARMA Models †

Till Schubert *, **Jan Martin Brockmann**, **Johannes Korte** and **Wolf-Dieter Schuh**

Institute of Geodesy and Geoinformation, University of Bonn, 53115 Bonn, Germany; brockmann@geod.uni-bonn.de (J.M.B.); korte@geod.uni-bonn.de (J.K.); schuh@uni-bonn.de (W.-D.S.)
* Correspondence: schubert@geod.uni-bonn.de

† Presented at the 7th International conference on Time Series and Forecasting, Gran Canaria, Spain, 19–21 July 2021.

Abstract: In time series analyses, covariance modeling is an essential part of stochastic methods such as prediction or filtering. For practical use, general families of covariance functions with large flexibilities are necessary to model complex correlations structures such as negative correlations. Thus, families of covariance functions should be as versatile as possible by including a high variety of basis functions. Another drawback of some common covariance models is that they can be parameterized in a way such that they do not allow all parameters to vary. In this work, we elaborate on the affiliation of several established covariance functions such as exponential, Matérn-type, and damped oscillating functions to the general class of covariance functions defined by autoregressive moving average (ARMA) processes. Furthermore, we present advanced limit cases that also belong to this class and enable a higher variability of the shape parameters and, consequently, the representable covariance functions. For prediction tasks in applications with spatial data, the covariance function must be positive semi-definite in the respective domain. We provide conditions for the shape parameters that need to be fulfilled for positive semi-definiteness of the covariance function in higher input dimensions.

Keywords: ARMA processes; covariance function; stochastic modeling; time series analysis; Matérn covariance function; positive definiteness

Citation: Schubert, T.; Brockmann, J.M.; Korte, J.; Schuh, W.-D. On the Family of Covariance Functions Based on ARMA Models. *Eng. Proc.* **2021**, *5*, 37. https://doi.org/10.3390/engproc2021005037

Published: 5 July 2021

1. Introduction and Related Work

Signal covariance modeling is an important part of stochastic methods [1]. In covariance modeling, the choice of the type of covariance function is commonly separated from the actual estimation of its shape parameters. Thus, the estimated covariance model quite strongly depends on the assessed basis functions. From this standpoint, it is desirable to have a very general class of covariance functions that can represent very different shapes with a single functional model and thus includes a large set of possible basis functions. A drop towards negative correlations, i.e., the so-called hole effect [2], is a widespread phenomenon in real-world datasets.

The Matérn family of covariance functions [3] finds application in many fields such as machine learning [4], environmental sciences, and geostatistics [5,6]. Simultaneously, a very similar class is known as Markov models, e.g., [6–8]. For instance, the combination of a degree-two polynomial and an exponential function is known as the third-order Markov model.

In geodetic time series analysis, many standard covariance models have been introduced early. For instance, the authors of [9] provided an application of a simple case of the Matérn covariance function to describe the stochastics of the gravity field. The authors of [10] and [11] introduced second- and third-order Markov models in the geodetic context; see also [12]. The author of [13] used the exponentially damped cosine in a geodetic application. Later, the second-order Markov model was applied to altimetry data [14,15].

On the other hand, Markov models are also referenced by different names. e.g., the respective models can be derived from Radon transforms of the exponential model, cf. [16] (p. 85). In the literature, these models are also denoted as second-order autoregressive (SOAR) models and third-order autoregressive (TOAR) models; see e.g., [8,17–21]. Despite the uncertain terminology in the literature, we distinguish between second-order Gauss–Markov (SOGM) models as in [22,23] and second-order Markov (SOM) models [7,8,10], whereas both models share the property of being second-order autoregressive (SOAR) or second-order ARMA models.

In [23], it was shown that the covariance function of AR and ARMA processes with unique poles corresponds to a sum of SOGM process covariance functions, which are a combination of an exponential function and cosine and sine terms. However, if the autoregressive poles are repeated, the correspondence to SOGM models does not hold anymore. Instead, a higher pole-multiplicity introduces polynomials as basis functions into the family of covariance functions. This more general family is commonly related back to [24] (p. 543) where a family of covariance functions constructed by polynomial functions and exponential damping terms is derived from ARMA models. Whilst the family is mostly introduced in the literature only for real poles, it has a complete set of covariance functions of oscillating type, which is discussed in this work. Examples of this general class appear very sparse in the literature, e.g., in [2] or in a short note on oscillatory Matérn covariance functions in [25] (Section 2.3.3) but never the complete variety of this class. In this work, we merge many known covariance functions to a combined family of covariance functions, namely the ARMA models.

Next to the variety of basis functions involved in the construction of a covariance function, it is essential for the function's flexibility to allow all shape parameters to vary. By this requirement, one can define a family of covariance function, e.g., the class of Markov models. The reference with the most complete variety of functions belonging to this class is [5] (known as Buell's function of index 3; see also [6]) who provides that model with enhanced variability of parameters, which is the general idea in this paper.

In this work, these two extensions to the standard covariance models are introduced as part of the family of non-repeated and repeated poles ARMA models. Hence, starting from the Matérn-type covariance models, it is intended to provide both a variety of basis functions and variability of the shape parameters to achieve the most general family of covariance functions.

Another aspect is the necessity of covariance functions being positive semi-definite. For applications with data in higher dimensions, e.g., spatial data, the reduction to a one-dimensional distance-like norm (e.g., Euclidean) does not guarantee positive definiteness of the covariance function in the higher dimension. Instead, the Bochner theorem extends to the Hankel transform being positive [1,26]. We derived the conditions among the shape parameters that ensure positive semi-definiteness of the covariance function in higher input dimensions.

2. The Family of Non-Repeated Poles ARMA Models

Reference [23] presents elegant parametrizations and fitting procedures for the family of covariance functions defined by autoregressive moving average (ARMA) models. The family is based on covariance functions defined by SOGM processes given in one of the two following parametrizations:

$$\gamma(\tau) = \frac{\sigma^2}{\cos(\eta)}\, \mathrm{e}^{-c\,\tau} \cos(a\,\tau - \eta) \qquad \text{with } a, c \geq 0 \text{ and } |\eta| < \pi/2 \tag{1}$$

$$= \frac{\sigma^2}{\cos(\eta)}\, \mathrm{e}^{-\zeta\omega_0\tau} \cos\Big(\sqrt{1-\zeta^2}\,\omega_0\,\tau - \eta\Big) \quad \text{with } 0 \leq \zeta \leq 1,\ \omega_0 > 0\,. \tag{2}$$

In more detail, the interpolating function to the discrete covariances of an AR(p) process is given by the following finite weighted sum of exponentiated (unique) autoregressive poles $p_1, p_2, \ldots, p_p$:

$$\gamma(\tau) = A_1 p_1^{\tau} + A_2 p_2^{\tau} + \ldots + A_p p_p^{\tau} \qquad \text{with} \quad p_i \in \mathbb{C}, A_i \in \mathbb{C}\,, \tag{3}$$

cf. [27] (Equation (5.2.44)) and [28] (Equation (3.5.44)), which can be mathematically converted to the representation of Equation (2); see [23]. Equation (3) corresponds to either a pure AR(p) process or an ARMA(p,q) process, depending on whether the weights A_i are purely, i.e., uniquely, determined by the autoregressive poles p_i. The two-step approach in [23] starts with an estimation of the autoregressive process parameter and concludes with the fitting of weighting coefficients of the interpolating function.

Positive Definiteness in Higher Dimensions

The application in spatial domains requires positive semi-definiteness of the covariance function in higher dimensions $\mathbb{R}^d$, which is derived here.

Starting from the simple exponentially damped cosine, e.g., [16] (p. 92), the SOGM covariance function is a generalization with three parameters, i.e., additional phase, see [23] for details on the parametrization. Similar to [29] (p. 26), positive semi-definiteness constraints on the parameters can be followed from [30] and amount to

$$\eta \geq -\frac{\pi}{2} + \operatorname{acos}(\zeta) \cdot d \tag{4}$$

as an additional condition to the requirement $\eta \leq \operatorname{asin}(\zeta)$, cf. [23]. The permissible area of parameters is illustrated in Figure 1a and is visibly restricted more and more with increasing dimension.

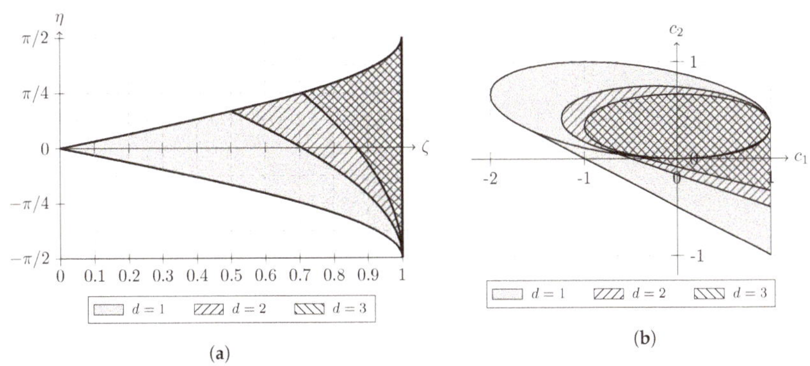

Figure 1. Permissible areas for parameters. For each dimension, the permissible area becomes a subset of that of the lower dimension. (**a**) Permissible areas of the parameters ζ and η in different dimensions 1 to 3. (**b**) Permissible areas of the weights c_1 and c_2 in different dimensions 1 to 3 shown for fixed parameter $c = 1$.

3. Generalization to Repeated Poles ARMA Models

Prior to providing the methodology of repeated poles ARMA processes, we introduce the basics of the Matérn family of covariance functions. The Matérn family of covariance functions can be parameterized in a way such that similarities to the ARMA models become clear.

3.1. The Half-Integer Matérn Covariance Function

The Matérn class of covariance Functions [3,4] defines a covariance functions with the two shape parameters c (scale of correlation length) and order ν. The Matérn covariance function is defined as

$$\gamma_{\text{Mat},\nu}(\tau) = \sigma^2\, 2^{1-\nu}\, \frac{(c\,\tau)^{\nu}}{\Gamma(\nu)}\, \mathrm{K}_{\nu}(c\,\tau) \tag{5}$$

and, in the case of half-integers ν, simplifies to a combination of a polynomial of degree $p = \nu - 1/2$ and an exponential function [4,6]. For the first four half-integers, we have

$$\begin{aligned} \gamma_{\text{Mat},1/2}(\tau) &= \sigma^2\, \mathrm{e}^{-c\,\tau}, \quad \gamma_{\text{Mat},3/2}(\tau) = \sigma^2 (1 + c\,\tau)\, \mathrm{e}^{-c\,\tau}, \\ \gamma_{\text{Mat},5/2}(\tau) &= \sigma^2 \left(1 + c\,\tau + \frac{c^2}{3}\,\tau^2\right) \mathrm{e}^{-c\,\tau} \text{ and} \\ \gamma_{\text{Mat},7/2}(\tau) &= \sigma^2 \left(1 + c\,\tau + \frac{2c^2}{5}\,\tau^2 + \frac{c^3}{15}\,\tau^3\right) \mathrm{e}^{-c\,\tau}. \end{aligned} \tag{6}$$

Note that the attenuation factor c also builds the coefficients of the polynomial.

3.2. Repeated Poles ARMA Models

Equation (3) holds only for the simple case assuming that the autoregressive process has distinct roots. When there are repeated real (positive or negative) poles or repeated complex conjugate poles, special cases have to be considered. Derived from the solution to the difference equation of the autoregressive relation for repeated poles, cf. e.g., [31] (Chap. 3.7), the required basis functions are summarized as one of the following cases of covariance sequences γ_k at discrete lags k, either

$$\gamma_k = \left(c_0 + c_1\, k + \ldots + c_{m-1}\, k^{m-1}\right) \bar{p}^k \tag{7}$$

for $\bar{p} := p_1 = p_2 = \ldots = p_m \in \mathbb{R}^+$, or

$$\gamma_k = \left(c_0 + c_1\, k + \ldots + c_{m-1}\, k^{m-1}\right) |\bar{p}|^k \cos(\pi k), \tag{8}$$

for the case $\bar{p} := p_1 = p_2 = \ldots = p_m \in \mathbb{R}^-$, or finally

$$\gamma_k = \left(c_0 + c_1\, k + \ldots + c_{l-1}\, k^{l-1}\right) |\bar{p}|^k \cos(ak - \eta) \tag{9}$$

for $\bar{p} := p_1 = \ldots = p_l = p^*_{l+1} = \ldots = p^*_{2l} \in \mathbb{C}$. m represents the multiplicity of real roots, l represents the pairwise complex conjugate roots, and c_j is the weights. As a result, these formulae correspond to multiplication and exponentiation of complex-valued weights A_i and poles $\bar{p}$ similar to Equation (3) and with the same correspondences $c = -\ln(|\bar{p}|)$, $a = |\arg(\bar{p})|$, and $|\eta_i| = |\arg(A_i)|$; see [23] (Sections 4.3 and 5.1). However, for repeated poles, e.g., as visible from Equation (7), the summation is performed in the following way

$$\gamma_k = A_1\, \bar{p}^k + A_2\, k\, \bar{p}^k + \ldots + A_m\, k^{m-1}\, \bar{p}^k\,. \tag{10}$$

Although the solution to the difference equation holds for discrete γ_k, we pursue a reinterpretation as a continuous covariance function $\gamma(\tau)$; see [23] (Section 4.3), and use the mathematical elegance of Equation (10) also for the analytical covariance function defined by AR or ARMA models.

From Equation (7), it is evident now that the Matérn covariance functions of order $\nu = p + 1/2$ correspond to ARMA models with $m = p$ repeated real poles $\bar{p} = \mathrm{e}^{-c}$. As known, from the Matérn family, with increasing order ν, the squared-exponential (Gauss-type) covariance function is asymptotically reached. Hence, with increasing pole multiplicity, an increasingly lower slope at the origin is realized.

3.3. Bounds for the Polynomial Coefficients of Markov Models

For the purpose of increasing the flexibility, we adopt the half-integer Matérn covariance function but with arbitrary polynomial coefficients. This approach is followed in [5] with his function of index 3; see also [6]. Similar to [5], we intend to construct a general model with arbitrary weights c_j

$$\gamma(\tau) = \sigma^2 \left(1 + c_1\,\tau + c_2\,\tau^2 + \ldots + c_{m-1}\,\tau^{m-1}\right) \mathrm{e}^{-c\,\tau}\,, \tag{11}$$

e.g., with third-order $\gamma_{\mathrm{TOM}}(\tau) = \sigma^2 (1 + c_1\,\tau + c_2\,\tau^2)\,\mathrm{e}^{-c\,\tau}$, which we denote as a third-order Markov (TOM) model.

As known from [23] (Section 5), allowing arbitrary weights between the basis functions creates a correspondence to ARMA models, i.e., introducing a moving average part. Hence, the covariance functions of index 3 of [5] as well as $\gamma_{\mathrm{TOM}}(\tau)$ also have triple real poles, but they correspond to ARMA(3,q) processes with triple real poles but with unknown order of the moving average part here.

Note that, due to the fixed polynomial coefficients, the Matérn covariance functions determined by c are automatically positive definite for $c > 0$, which makes them simple and easy to handle. However, Markov models with adjustable coefficients exhibit greater flexibility, and they are viable for practical use if the bounds of the coefficients for positive (semi)-definiteness are known.

As in [6] (Equation (14)), we can construct the general model with arbitrary weights c_1 and c_2 from a combination of the half-integer Matérn models Equation (6). The correspondence is

$$\begin{aligned}\left(1 + c_1\,\tau + c_2\,\tau^2\right) \mathrm{e}^{-c\,\tau} = &\left(1 - \frac{c_1}{c}\right) \gamma_{\mathrm{Mat},1/2}(\tau) + \\ &\left(\frac{c_1}{c} - \frac{3c_2}{c^2}\right) \gamma_{\mathrm{Mat},3/2}(\tau) + \frac{3c_2}{c^2}\,\gamma_{\mathrm{Mat},5/2}(\tau)\,.\end{aligned} \tag{12}$$

In the d-dimensional space, the general Matérn covariance function has the Fourier transform

$$F(s) = \frac{\Gamma\!\left(\frac{d}{2} + \nu\right) c^{2\nu}}{\Gamma(\nu)\,\pi^{-\frac{d}{2}}\,(c^2 + s^2)^{-\left(\frac{d}{2}+\nu\right)}}\,, \tag{13}$$

cf. [32] (Equation (4.130)), which, weighted as in Equation (12) and simplified (cf. [6]), yields

$$F(s) = \frac{\Gamma(1/2+d/2)}{\Gamma(1/2)\,\pi^{d/2}\,(c^2+s^2)^{5/2+d/2}} \left(\begin{array}{llll} \left(1 - \frac{c_1}{c}\right) & & c & (c^2+s^2)^2 \; + \\ \left(\frac{c_1}{c} - \frac{3c_2}{c^2}\right) & (1+d) & c^3 & (c^2+s^2) \; + \\ \left(\frac{c_2}{c^2}\right) & (1+d)(3+d) & c^5 & \end{array}\right). \tag{14}$$

From this, bounds for the non-negativity conditions can be derived. In detail, c_2 can lie within the bounds defined by the functions

$$c_2 = \frac{c(2cd + 6c + c_1 d - 3c_1)}{9(d+1)} \pm \frac{2c\sqrt{(c_1 - c)(cd + 3c + 2c_1 d)(d+3)}}{9(d+1)}\,, \tag{15}$$

which form the shape of an ellipsis added to a straight line. If c_1 is larger than $c_1 \geq -c(2d+3)/(d(d+2))$, the domain extends to the straight line lower bound $c_2 \geq -(c(c + c_1 d))/(d(d+1))$ and up to $c_1 < c$ and $c_2 \leq c^2/3$; see Figure 1b.

3.4. Oscillatory Repeated Poles ARMA Models

It is intuitive to combine the Matérn covariance function with an oscillating function in order to create a more versatile function; see [25] (Section 2.3.3). Hence, by multiplying Equation (11) with a cosine of frequency a and phase η, we have the following:

$$\gamma(\tau) = \sigma^2 \left(1 + c_1\,\tau + c_2\,\tau^2 + \ldots + c_{l-1}\,\tau^{l-1}\right) \mathrm{e}^{-c\,\tau} \cos(a\,\tau - \eta) \tag{16}$$

We define the general class of repeated poles ARMA covariance functions with $l = \nu + 1/2$ times repeated complex-conjugate pairs of poles given by

$$p_{i,i+l} = \mathrm{e}^{-c}(\cos(a) \pm i\,\sin(a)) \tag{17}$$

and thus autoregressive order $p = 2l$. Again, the moving average parameters, i.e., also the dependence of weights c_j on poles and zeros of the ARMA process, are not derived here, cf. [23].

When combining the covariance models of Sections 2 and 3.3, the conditions of positive definiteness are the joint requirements of both types, i.e., Figure 1a,b.

4. Application to Altimetry Data: A Demonstration

The following empirical covariance function of a two-dimensional geodetic application shall serve as a small example to demonstrate the necessity of different covariance functions presented in this work. Here, we interpret a time series of sea level anomalies (SLA) along the altimeter track as a stationary stochastic field in planar approximation, i.e., two-dimensional domain. To obtain SLA, sea surface heights observed by the Envisat satellite launched and operated from 2002 until 2012 by the European Space Agency were reduced by a long-term mean sea surface model (in this case, CNES-CLS11, [33]) interpolated along the satellites ground track. For the demonstration example, we extracted a subset of 10,905 observations in a local area of the North Atlantic ocean of cycle 13 (13 January 2003 to 17 February 2003); see Figure 2. We computed empirical estimates of the isotropic covariance function averaged for equidistant lags ($\Delta\tau = 0.2°$) and by using the biased estimator (see the black dots in Figure 3).

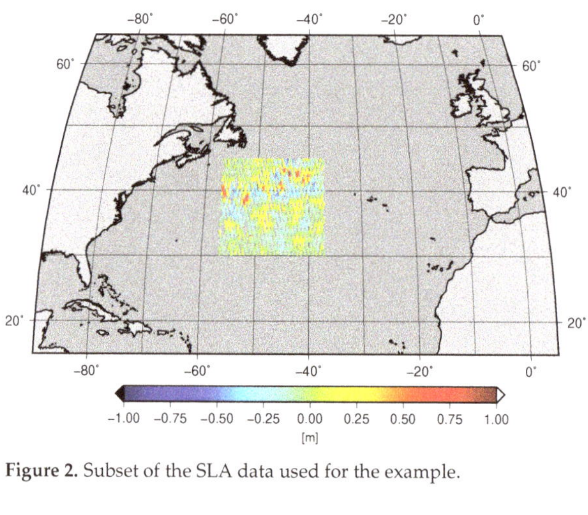

Figure 2. Subset of the SLA data used for the example.

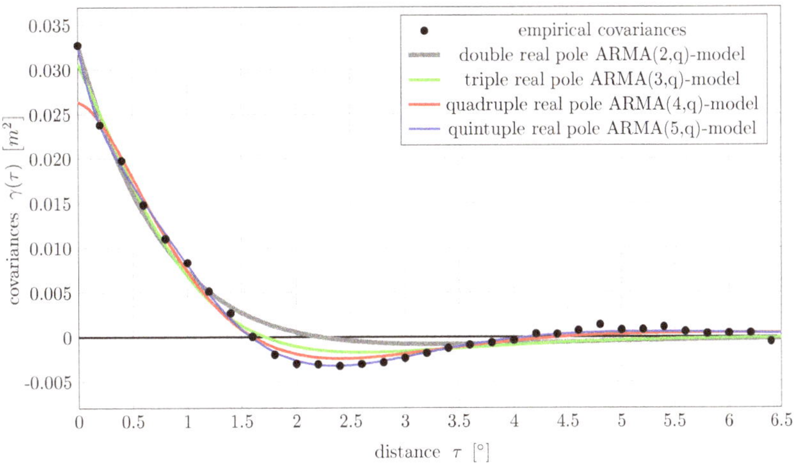

Figure 3. Functions of the type in Equation (11) with different orders fitted to the empirical covariances.

For Figure 4, we fit functions of the type in Equation (16) with different orders to the empirical covariances. These ARMA models do not experience an improvement from the higher pole multiplicity because the oscillatory nature of the complex poles ARMA model already nicely captures the hole effect. The higher-order models slightly improve the very long-range correlations.

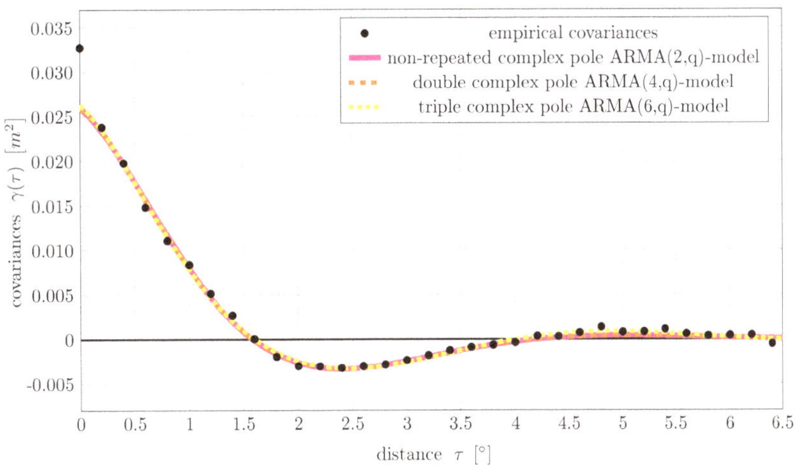

Figure 4. Functions of the type in Equation (16) with different orders fitted to the empirical covariances.

For demonstration purposes, different types of covariance functions belonging to the family of (repeated pole) ARMA models are fitted to the empirical estimates g_k of the covariances γ_k from lag $k = 1$ up to $k = 34$ using non-closed-form solvers. We used the GNU Octave's nonlinear minimization routine `fmincon`, cf. [34] and implemented a constrained least squares fitting.

In a first plot, we fit repeated real pole ARMA models, i.e., Equation (11), of orders $p = 2, 3, 4$ and 5. These correspond to linear combinations of Matérn covariance functions, where some Matérn functions can even be subtracted in the combination. The number of fitted parameters, i.e., σ, c, and c_j, are 3, 4, 5, and 6. The order q of the moving average part is not determined here. All results are estimated to be positive semi-definite in $\mathbb{R}^2$.

Figure 3 shows that the polynomial component of the covariance function can successfully capture the negative correlations. The quality of fit gets better with increasing order and is sufficient for the fifth-order model. We are aware that the nugget $\gamma(0) - g_0$ (white noise variance component) is quite different for the estimated models, but that is because we did not restrict it by a priori knowledge.

5. Summary and Conclusions

The example demonstrates that relatively complex correlations structures can also be captured by simple covariance models such as Markov models. Enhanced flexibility is achieved by adjustable polynomial coefficients, which makes them favorable to the Matérn covariance function, especially for modeling negative correlations as in the example. The underlying methodology of ARMA processes builds the general family for all of these covariance functions and thus also holds out the prospect of suited optimization methods such as the Yule–Walker equations, cf. [23].

In addition, we provide bounds for all parameters of the ARMA covariance models in order to ensure positive semi-definiteness in the respective domain of the data. In general, this work demonstrates the necessity for a large variety of basis functions collected in a family of covariance functions as well as suited fitting procedures. Tailored optimization problems for the repeated poles ARMA models are still an open research field.

Author Contributions: Conceptualization, methodology, formal analysis, investigation, software, validation, visualization, writing—original draft preparation: T.S., writing—review and editing: T.S., J.K., J.M.B., and W.-D.S., data curation: J.M.B., funding acquisition, project administration, resources, supervision: W.-D.S. All authors have read and agreed to the published version of the manuscript.

Funding: This research is funded by the Deutsche Forschungsgemeinschaft (DFG, German Research Foundation) grant No. 435703911 (SCHU 2305/7-1 'Nonstationary stochastic processes in least squares collocation—NonStopLSC'). The second author acknowledges the financial support of the DFG in the context of the PARASURV (grant BR5470/1-1) project.

Data Availability Statement: The used Delayed Time CorSSH products were processed with support from CNES (by CLS Space Oceanography Division) and distributed by AVISO+ (http://www.aviso.altimetry.fr/, accessed on 2 October 2019).

Conflicts of Interest: The authors declare no conflict of interest.

Abbreviations

The following abbreviations are used in this manuscript:

AR	Autoregressive
ARMA	Autoregressive moving average
MA	Moving average
SOGM	Second-order Gauss–Markov
SLA	Sea level anomalies

References

1. Moritz, H. *Covariance Functions in Least-Squares Collocation*; Number 240 in Reports of the Department of Geodetic Science; Ohio State University: Columbus, OH, USA, 1976.
2. Journel, A.G.; Froidevaux, R. Anisotropic Hole-Effect Modeling. *J. Int. Assoc. Math. Geol.* **1982**, *14*, 217–239. [CrossRef]
3. Matérn, B. Spatial Variation: Stochastic Models and Their Application to Some Problems in Forest Surveys and Other Sampling Investigations. Ph.D. Thesis, University of Stockholm, Stockholm, Sweden, 1960. [CrossRef]
4. Rasmussen, C.; Williams, C. *Gaussian Processes for Machine Learning*; Adaptive Computation and Machine Learning; MIT Press: Cambridge, MA, USA, 2006.
5. Buell, C.E. Correlation Functions for Wind and Geopotential on Isobaric Surfaces. *J. Appl. Meteorol.* **1972**, *11*, 51–59. [CrossRef]
6. Gneiting, T. Correlation Functions for Atmospheric Data Analysis. *Q. J. R. Meteorol. Soc.* **1999**, *125*, 2449–2464. [CrossRef]

7. Gelb, A. *Applied Optimal Estimation*; The MIT Press: Cambridge, MA, USA, 1974.
8. Moreaux, G. Compactly Supported Radial Covariance Functions. *J. Geod.* **2008**, *82*, 431–443. [CrossRef]
9. Shaw, L.; Paul, I.; Henrikson, P. Statistical Models for the Vertical Deflection from Gravity-Anomaly Models. *J. Geophys. Res. (1896–1977)* **1969**, *74*, 4259–4265. [CrossRef]
10. Kasper, J.F. A Second-Order Markov Gravity Anomaly Model. *J. Geophys. Res. (1896–1977)* **1971**, *76*, 7844–7849. [CrossRef]
11. Jordan, S.K. Self-Consistent Statistical Models for the Gravity Anomaly, Vertical Deflections, and Undulation of the Geoid. *J. Geophys. Res. (1896–1977)* **1972**, *77*, 3660–3670. [CrossRef]
12. Moritz, H. Least-Squares Collocation. *Rev. Geophys.* **1978**, *16*, 421–430. [CrossRef]
13. Vyskočil, V. On the Covariance and Structure Functions of the Anomalous Gravity Field. *Stud. Geophys. Geod.* **1970**, *14*, 174–177. [CrossRef]
14. Andersen, O.B.; Knudsen, P. Global Marine Gravity Field from the ERS-1 and Geosat Geodetic Mission Altimetry. *J. Geophys. Res. Ocean.* **1998**, *103*, 8129–8137. [CrossRef]
15. Andersen, O.B. Marine Gravity and Geoid from Satellite Altimetry. In *Geoid Determination: Theory and Methods*; Sansò, F., Sideris, M.G., Eds.; Lecture Notes in Earth System Sciences; Springer: Berlin/Heidelberg, Germany, 2013; pp. 401–451. [CrossRef]
16. Chilès, J.P.; Delfiner, P. *Geostatistics: Modeling Spatial Uncertainty*; Wiley Series in Probability and Statistics; John Wiley & Sons: Hoboken, NJ, USA, 1999. [CrossRef]
17. Julian, P.R.; Thiébaux, H.J. On Some Properties of Correlation Functions Used in Optimum Interpolation Schemes. *Mon. Weather. Rev.* **1975**, *103*, 605–616. [CrossRef]
18. Thiébaux, H.J. Anisotropic Correlation Functions for Objective Analysis. *Mon. Weather. Rev.* **1976**, *104*, 994–1002. [CrossRef]
19. Franke, R.H. *Covariance Functions for Statistical Interpolation*; Technical Report NPS-53-86-007; Naval Postgraduate School: Monterey, CA, USA, 1986.
20. Weber, R.O.; Talkner, P. Some Remarks on Spatial Correlation Function Models. *Mon. Weather. Rev.* **1993**, *121*, 2611–2617. [CrossRef]
21. Gaspari, G.; Cohn, S.E. Construction of Correlation Functions in Two and Three Dimensions. *Q. J. R. Meteorol. Soc.* **1999**, *125*, 723–757. [CrossRef]
22. Maybeck, P.S. *Stochastic Models, Estimation, and Control*; *Mathematics in Science and Engineering*; Academic Press: New York, NY, USA, 1979; Volume 141-1. [CrossRef]
23. Schubert, T.; Korte, J.; Brockmann, J.M.; Schuh, W.D. A Generic Approach to Covariance Function Estimation Using ARMA-Models. *Mathematics* **2020**, *8*, 591. [CrossRef]
24. Doob, J.L. *Stochastic Processes*; Wiley Series in Probability and Mathematical Statistics; Wiley: New York, NY, USA, 1953.
25. Li, Z. Methods for Irregularly Sampled Continuous Time Processes. Ph.D. Thesis, University College London, London, UK, 2014.
26. Wackernagel, H. *Multivariate Geostatistics: An Introduction with Applications*; Springer: Berlin/Heidelberg, Germany, 1995. [CrossRef]
27. Jenkins, G.M.; Watts, D.G. *Spectral Analysis and Its Applications*; Holden-Day: San Francisco, CA, USA, 1968.
28. Priestley, M.B. *Spectral Analysis and Time Series*; Academic Press: London, UK; New York, NY, USA, 1981.
29. Gelfand, A.E.; Diggle, P.; Guttorp, P.; Fuentes, M. *Handbook of Spatial Statistics*; Handbooks of Modern Statistical Methods; Chapman & Hall/CRC: Boca Raton, FL, USA, 2010. [CrossRef]
30. Zastavnyi, V.P. Positive Definiteness of a Family of Functions. *Math. Notes* **2017**, *101*, 250–259. [CrossRef]
31. Goldberg, S. *Introduction to Difference Equations*; Dover Publications: New York, NY, USA, 1986.
32. Yaglom, A.M. *Correlation Theory of Stationary and Related Random Functions: Volume I: Basic Results*; Springer Series in Statistics; Springer: New York, NY, USA, 1987.
33. Schaeffer, P.; Faugére, Y.; Legeais, J.F.; Ollivier, A.; Guinle, T.; Picot, N. The CNES_CLS11 Global Mean Sea Surface Computed from 16 Years of Satellite Altimeter Data. *Mar. Geod.* **2012**, *35*, 3–19. [CrossRef]
34. Eaton, J.W.; Bateman, D.; Hauberg, S.; Wehbring, R. *GNU Octave; Version 5.2.0 Manual: A High-Level Interactive Language for Numerical Computations*; Free Software Foundation: Boston, MA, USA, 2020.

engineering proceedings

MDPI

Proceedings

Learning Curves: A Novel Approach for Robustness Improvement of Load Forecasting †

Chiara Giola *, Piero Danti * and Sandro Magnani

Yanmar R&D Europe, viale Galileo 3/A, 50125 Firenze, Italy; sandro_magnani@yanmar.com
* Correspondence: chiara_giola@yanmar.com (C.G.); piero_danti@yanmar.com (P.D.)
† Presented at the 7th International conference on Time Series and Forecasting, Gran Canaria, Spain, 19–21 July 2021.

Abstract: In the age of AI, companies strive to extract benefits from data. In the first steps of data analysis, an arduous dilemma scientists have to cope with is the definition of the 'right' quantity of data needed for a certain task. In particular, when dealing with energy management, one of the most thriving application of AI is the consumption's optimization of energy plant generators. When designing a strategy to improve the generators' schedule, a piece of essential information is the future energy load requested by the plant. This topic, in the literature it is referred to as load forecasting, has lately gained great popularity; in this paper authors underline the problem of estimating the correct size of data to train prediction algorithms and propose a suitable methodology. The main characters of this methodology are the Learning Curves, a powerful tool to track algorithms performance whilst data training-set size varies. At first, a brief review of the state of the art and a shallow analysis of eligible machine learning techniques are offered. Furthermore, the hypothesis and constraints of the work are explained, presenting the dataset and the goal of the analysis. Finally, the methodology is elucidated and the results are discussed.

Keywords: learning curves; energy load forecasting; time series; training-set size

Citation: Giola, C.; Danti, P.; Magnani, S. Learning Curves: A Novel Approach for Robustness Improvement of Load Forecasting. *Eng. Proc.* **2021**, *5*, 38. https://doi.org/10.3390/engproc2021005038

Published: 13 July 2021

1. Introduction

The advent of electricity markets and the progress in Renewable Energy Sources (RES) have changed the nature of electricity production and consumption [1]. In order to increase the RES share and to use energy more effectively, energy system flexibility needs to be improved, for example, by means of enabling the demand-side management [2]. In this framework, electricity load prediction is required as an essential part in the energy industry to manage load fluctuations and aleatory RES [3]. Load forecasting is a useful and practical tool for efficient energy management, safer grid operation, and optimal maintenance planning. An accurate load forecasting is a key element to improve the environmental impact, sustainability, and cost-effectiveness of smart grids.

Electricity load prediction is vitally essential for the industries in deregulated economics [3]; load forecasting is necessarily implemented in Energy Management Systems (EMS) that optimally control appliances. An increasing number of numerical approaches has been proposed for energy prediction. A lot of models have been used and a coarse clustering is usually adopted in review articles: Statistical models, time series analysis, Machine Learning (ML), and Deep Learning (DL) [1,3]. Wei et al. [4] propose a review of data-driven approaches for the prediction and classification of buildings energy consumption; a comparison among white-box, grey-box, and black-box approaches for predicting consumption is described. White-box models lean on a complete knowledge of the physics of the systems while black-box models are completely data-driven and require historical data. Grey-box models are a hybrid solution between the other two. The paper focuses on data-driven models and, above all, describes Artificial Neural Networks

(ANNs), Support Vector Machines (SVMs), and statistical regression. Yildiz et al. [5] cover the same subject, indeed they emphasize the broad application of ANNs and SVMs but, moreover, also Auto-Regressive (AR) models are cited; in particular the Auto-Regressive Integrated Moving Average (ARIMA) is defined as one of the most used techniques in load forecasting. In the latest years, by means of a greater computational power, many researchers started to apply Deep Learning techniques in order to improve load forecasting precision: Zhang et al. [6] details a variety of DL models like Restricted Boltzmann Machines (RBMs), Deep Belief Network (DBN), and RNN (Recurrent Neural Networks). Another novel algorithm taken into consideration is Extreme Gradient Boosting (XGBoost) that couples great performance with low execution time.

A plethora of researchers focused on model selection and hyper-parameters optimization; Khalid et al. [7] classify optimization methods for algorithm hyper-parameters in two groups: Nature-inspired approaches and statistical methods.

Recently, many companies have developed EMS whose services are based on data collection and artificial intelligence algorithms; load forecasting represents one of the most implemented service. Hence when an EMS business model is developed, a crucial point is the available amount of data. Once the model for prediction is selected, finding the trade-off between the volume of data and goodness of forecast is still a challenge. If the appropriate volume of training data can be coupled with the forecasting algorithm, the EMS has a robust load forecasting model. This aspect could be of great interest among companies developing services based on ML routines; indeed, when implementing an intelligent platform in a customer plant, estimating the monitoring period needed to collect data is significant to build an efficient business model. A powerful tool to tackle this estimation is to build Learning Curves (LCs).

A learning curve shows the measure of predictive performance on a given domain as a function of the training sample size. Reviewing learning curves of models can be used to diagnose problems with learning, such as underfitting or overfitting, as well as whether the training and validation datasets are suitably representative. Building an overly complex model leads to high variance error in prediction, but a too simple model has a high bias error. The opportunity of training the model with the proper number of observations leads to finding the architecture with an optimal trade off between variance and bias errors [8]. Although the learning curves are promising, in the literature they have been mainly applied to other types of data, with non correlated observations [9–12]. Hence in this context, the present work tries to bridge this research gap applying the learning curves procedure to time series.

The remainder of the paper is structured as follows. Section 2 presents a state of the art on Learning Curves and a background for this analysis. Proposed methodology is described in Sections 3 and 4 presents results and discussion. Finally, conclusions are provided in Section 5.

2. State of the Art

Learning curves aim to compare the generalization performance of an algorithm as a function of training-set size. A learning curve shows the validation and training score of an estimator for different numbers of training samples. It is a tool to find out how much the estimator benefits from adding more training data and whether it suffers more from a variance error or a bias error [13]. Over two decades ago in machine learning research, the analysis of learning curves was a widespread tool for comparisons of Machine Learning techniques [14]; nowadays, it is rarely presented. Moreover, time-series LCs are not commonplace mainly because procedure presents some issues [15].

A common procedure for building LCs is implemented in the function named *learning_curve* of scikit-learn [13], a mainstream Python library. The just mentioned *learning_curve* function needs an estimator, the number of training observations that will be used to generate the LC, and the number k of fold to split data while using the k-fold validation strategy. This strategy splits the whole dataset k times, each time a different train-set and validation-set are extrapolated (Figure 1).

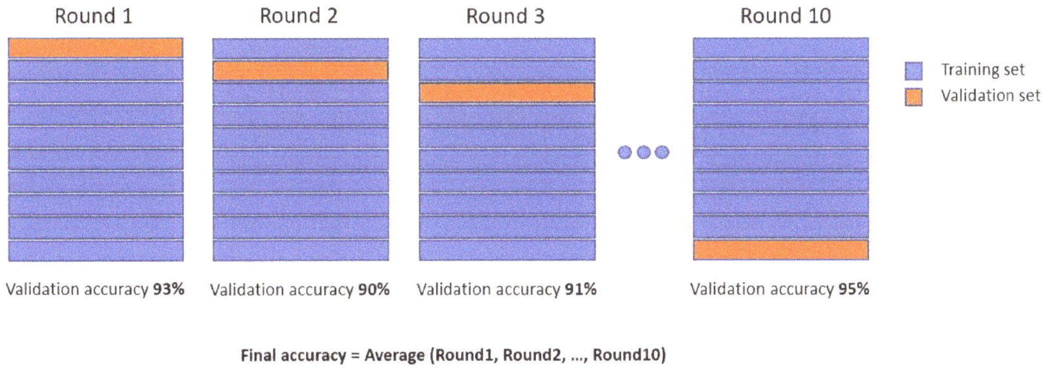

Figure 1. Example of k-fold validation with $k = 10$.

k-fold validation is a particular type of cross-validation (CV), a validation technique for assessing how the results of a statistical analysis will generalize on an independent dataset. Subsets of the training-set, whose size will be incremented after each k-fold validation, will be used to train the estimator and a score for each training subset size and for the test-set will be computed. Afterwards, the scores will be averaged over all k runs; in the end, two over-k-runs averaged score (both for train and test) will be obtained for each training subset size [13].

In the literature, there is not an extensive discussion of the LC subject. Most of the articles dealing with it refer to different fields of application. Ning et al. [12] test how the performance of Deep Convolutional Neural Networks (DCNNs) are affected by the size of the training-set in an image segmentation task: Six training-sets are considered and the performance of the DCNN trained with the larger dataset is used as the baseline. Zhu et al. [16] investigate the question of whether existing object recognition detectors will continue to improve as data grows, or saturate in performance due to limited model complexity. Beleites et al. [9], Figueroa et al. [10], and Hess and al. [11] study the importance of LCs in classification problems applied to the biomedical field where it is very difficult to obtain big datasets for training the estimator. All these analyses take into account independent samples, this means that the training-set can be enlarged, shuffled, and split without considering the samples order. However, this hypothesis is not valid when dealing with time-series.

Several strategies have been proposed in the literature for performance estimation of time-series and currently there is no consensual approach [17]. Out-of-Sample (OOS) approaches hold out a test-set in order to test a model on a never-seen portion of data. Train/test split can be faced with a different procedure: Sliding window or growing window [18]. OOS methods always retain the temporal order to guarantee the preservation of correlation among observations. In order to produce a robust estimation of predictive performance, Tashman [19] recommends applying OOS strategies in multiple test periods. Thus, by using OOS, the benefits of CV, especially for small datasets, cannot be exploited [20]. In general, CV is a common strategy both for model selection and for testing the generalization performance of an algorithm [21]. A fundamental hypothesis of CV is independence and identical distribution (i.i.d.) among observations. However, time-series has serial correlation in the data, possible non-stationarities, and time ordering, which forces not to use future data to predict the past; consequently the application of CV to time-series is not straightforward. There are several revised CV approaches designed for time-series; a wide review is presented from Bergmeir et al. [22]. Most common procedures are blocked

CV and hv-block CV. Blocked CV has no initial random shuffling of data, and divides observation in K blocks as in k-fold CV; time order is kept within each block, although is broken across them [18]. h-block CV is a non-dependent cross-validation, as it leaves out the possibly dependent observations and only considers data points that can be considered to be independent [20].

Cerqueira et al. [18] compare different approaches on both stationary and non-stationary time-series. They conclude that CV procedures are suitable for stationary time-series but are not compliant with real data and with potential non stationarities; thus, OOS applied in multiple testing periods is recommended. Süzen et al. [15] present a procedure for time-series learning curves based on reconstructive CV; it combines OOS estimation and imputation of missing data at random by means of techniques like Kalman filtering [23].

3. Proposed Methodology

In this section, the building of learning curves is explained. As mentioned above, a cross-validation method is applied in a learning curves procedure in order to select the optimal number of observations needed for training a forecast algorithm. The aim of the proposed procedure is to present an adaptable methodology that can cope with all types of algorithms and all types of time-series data. The proposed methodology consists of three main steps: Data collection, algorithm selection, and building of learning curves. All these stages are described in the following sections.

3.1. Data Collection

When approaching a problem of energy load forecasting, the first activity to be performed is represented by data collection. Even if, lately, the words artificial intelligence and big data are mainstream, this does not mean that every facility manager arranges a data storage routine. Often data are monitored by means of a local Human-Machine-Inteface (HMI) by maintenance operators, whose goal is to check real-time behavior of the plant without a compulsory need of heaping data in an accessible structure.

Usually, many kinds of features can affect the energetic behavior of a plant and they can be grouped in the following short list:

- Field measurements like energy consumption or plant temperatures. These signals are collected by a field device (e.g., a PLC or a remote I/O);
- Management details like hotel reservations or hospital occupants. These numbers are collected by ERP softwares or, in the worst case, by hand-written registers;
- Weather measurements and forecasts like external temperature or wind speed. These values are collected by weather stations or directly downloaded from the web.

All the above-mentioned data must be aggregated in a central entity whose task is to forward an average value to a database located in cloud or in a local server. The real importance of each measurement and its correlation with the load to be predicted is strictly dependent on the plant's use case; when the monitored plant satisfies the energetic needs of a hotel then it is very useful to acquire for example the rooms reservation, the meeting room usage, and the external weather. Otherwise, when the building under investigation is a parking lot, it is helpful to know the period of the year and the parking spots occupation. A third example is represented by a manufacturing factory where the most important Key Performance Indicator (KPI) is the produced quantity of goods. In the real world, the machine learning engineer in charge of developing the ad-hoc model to predict energy load forecasting will not have access to all these information; most of the time model inputs will consist of the date and external temperature. Another important feature of data shape is granularity: In Italy, the energy market regulator [24] imposes to work with values averaged every hour or, in some cases, every 15 min. In order to maintain generality, in this paper measured signals are sampled every hour and the considered features are the most likely to be available: Date, external temperature and, of course, energy load consumption.

3.2. Algorithms Selection

When discussing forecasting, it is crucial to define the time interval to be forecasted; as Hammad et al. list in [25] there are four types of forecasting horizon:

- Long-Term Load Forecasting (LTLF), time interval ranges from one year to 20 years ahead;
- Medium-Term Load Forecasting (MTLF), time interval ranges from a week up to a year;
- Short-Term Load Forecasting (STLF), time interval ranges from one hour to a week;
- Ultra/Very Short-Term Load Forecasting (VSTLF), time interval ranges from a few minutes to an hour ahead and is used for real-time control.

In this paper, the goal is to predict the energy load forecasting of the next day, so it is a STLF problem. This assumption is not a required hypothesis for the methodology proposed in Section 3.4. As briefly introduced in Section 2, in the latest 40 years many methods have been developed and used for time-series forecasting and, in particular, for energy STLF. Makridakis et al. in [26] make a coarse division between statistical and ML methods; this kind of grouping method is widely used and, more in-depth forecasting model can be detailed as follows: Statistical Methods, ML Methods, and DL Methods.

- Statistical Methods are historically the most used because of their easy implementation and fast execution, and among these ARIMA and Holt–Winter methods are very popular. These approaches usually work better when dealing with low-frequency signals and when the target variable understays the hypothesis of time-invariance: Both statements are not compliant with the object of this paper.
- Machine Learning Methods had great promise at the beginning of 21st century and represent a good trade-off between performance and computational costs. Among the ML group, in this paper three techniques have been selected: Support Vector Regressor (SVR) because it is the most simple and understandable algorithm, Extreme Gradient Boosting (XGBoost) [27] because it is a novel algorithm able to outperform state-of-the-art techniques in many competitions, and Multi-Layer Perceptron (MLP) because it is often used as a load forecasting benchmark.
- Deep Learning Methods and in particular Recurrent Neural Networks (RNNs) could act as a central character in the short-term energy load forecasting because of their affinity with time-series and their well-known high performance; on the other hand, the hard hyper-parameter tuning phase risks a change in the focus of the work. Indeed, in order to face the LCs subject, it is important to train models with pre-selected hyper-parameters whose value can be considered correct by the authors with a high confidence degree.

3.3. Hyper-Parameters Selection

When building ML models to proceed with the LCs methodology, a strict hypothesis must be met: All hyper-parameter's values must be tuned and then fixed to a defined value. In other words, optimization routines like randomized search [28] or grid search are not compliant.

In Section 3.2, the selected techniques used in this paper have been introduced: SVR, XGBoost, and MLP. Below, an extensive description of the settled hyper-parameter is reported.

The first algorithm selected is SVR; the SVR Scikit-learn library [29] has been used and four parameters have been tuned:

- $C = 1$, the regularization parameter;
- $epsilon = 0.1$, the epsilon-tube within which no penalty is associated;
- $kernel =$ 'rbf', the kernel type to be used in the algorithm;
- $\gamma = 0.08$, the kernel coefficient.

The second algorithm taken into analysis is XGBoost; the Scikit-learn Wrapper interface for XGBoost [30] has been implemented and four parameters have been tuned:

- *max_depth* = 4, maximum tree depth for base learners;
- *learning_rate* = 0.1, boosting learning rate;
- $\lambda = 1$, regularization term on weights;
- *n_estimators* = 100, the number of gradient boosted trees.

The third developed algorithm is a two-layer MLP; the Scikit-learn library for MLP [31] has been exploited and four parameters have been tuned:

- *hidden_layer_sizes* = 8, one hidden layer with 8 neurons;
- *activation* = 'relu', activation function for the hidden layer;
- $\alpha = 10^{-7}$, regularization term;
- *batch_size* = 1, size of minibatches for stochastic optimizers.

3.4. Building of Learning Curves: The Proposed Methodology

As underlined in Section 3.1, a multivariate time-series with n independent variables and only one dependent variable is analyzed with each observation of an independent time-series being $x \in \mathbb{R}^n$ and observations of dependent target variable are $y \in \mathbb{R}$. At time $t_i \in \mathbb{R}_0^+$, $x(t_i)$, and $y(t_i)$ represent the observations of independent and dependent variables; t_0 is considered as the first time sample available in the dataset. In this specific case, y is the time-series for energy load. A list of training-set size to test have to be defined since the generalization performance have to be shown as a function of the training-set size. The training-set size list has q elements; each p_j with $j \in [1, ..., q]$ is a training-set size. The test-set size d is fixed to a constant value and does not vary during the whole learning curves procedure. As cited in Section 3.2, the testing period considered in this work has a one-day length because the aim is a day-ahead load prediction. Moreover, the day of test immediately following the training-set is adopted. This is not a lack of generality, rather a different size for test-set can be applied and can be shifted from the end of training-set, as long as time order is retained.

By means of an OOS approach, a part of available data is used to fit the model, a different part to test it and assess the performance of the prediction algorithm. This procedure is repeated for each training-set size in the aforementioned list.

If p_j is the training length, a set of p_j consecutive observations is used for training the model and the following set of length d is used for testing purposes. The analyzed sets are:

$$x(t_i) \leq x \leq x(t_i + p_j)$$
$$y(t_i) \leq y \leq y(t_i + p_j)$$

and the test-set, if d is length for testing, is:

$$x(t_i + p_j + 1) \leq x \leq x(t_i + p_j + 1 + d)$$
$$y(t_i + p_j + 1) \leq y \leq y(t_i + p_j + 1 + d).$$

In order to produce a robust estimation of forecasting performance, for the same p_j length of training, this strategy is applied in multiple test periods with a sliding window approach (Figure 2). It is worth underlining that, as the methodology is conceived, whenever the test-set is shifted, the training-set slides.

A statistically significant k number of tested days has to be chosen. In the period from $t_0 + p_j$ to the end of the multivariate time-series, k tested days are chosen in a uniformly distributed and random way. The selection of k should be a trade off between the maximum p_j training size and the possibility of testing an heterogeneous number of data portions also according to seasonality and trend in the time-series. Since in the present work one year of data is available, $k = 30$ is enough to evaluate the algorithm generalized performance; this number of tested days allows to mitigate the sensitivity of error to different phases of a business cycle. Every time a sliding window is tested a metric has to be evaluated in order to compute y forecasting error for both the training-set and test-set. Different metrics can be used as a performance indicator, i.e. Root Mean Squared Error (RMSE) or Mean

Absolute Error (MAE); they all have different shortcomings and merits. Mean Absolute Percentage Error (MAPE) has been used since it is scaled to the original y value and gives an intuitive interpretation of error. MAPE, for training of the m-th day is expressed by the formula:

$$MAPE_{m,train} = \sum_{i=1}^{p_j} \frac{|y_i^m - \hat{y}_i^m|}{y_i^m} \quad (1)$$

where y_i^m is the actual value and $\hat{y}_i^m$ is the forecast value. For training, for example, it is computed over all the p_j observations of the training set. The aforementioned procedure is performed k times for each p_j training set size obtaining k MAPE error values for testing and k reconstruction errors of training. The average value e_k of MAPEs for testing and training is reported in a learning curve graph. For training MAPE is computed as:

$$e_{k,train} = \frac{1}{k} \sum_{m=1}^{k} MAPE_{m,train}. \quad (2)$$

For the testing MAPE, the procedure is the same, but it is computed over d time-steps of the testing set as follows:

$$e_{k,test} = \frac{1}{k} \sum_{m=1}^{k} \sum_{i=1}^{d} \frac{|y_i^m - \hat{y}_i^m|}{y_i^m}. \quad (3)$$

To plot the learning curves, the mean value of training errors and the mean value of test errors are taken; accordingly only two error scores for each training set size are plotted. Moreover, in order to show the scatter of data, the variance of error for both the training and testing curve is depicted by means of a colored shade.

Further details are reported in the implementation code available at [32].

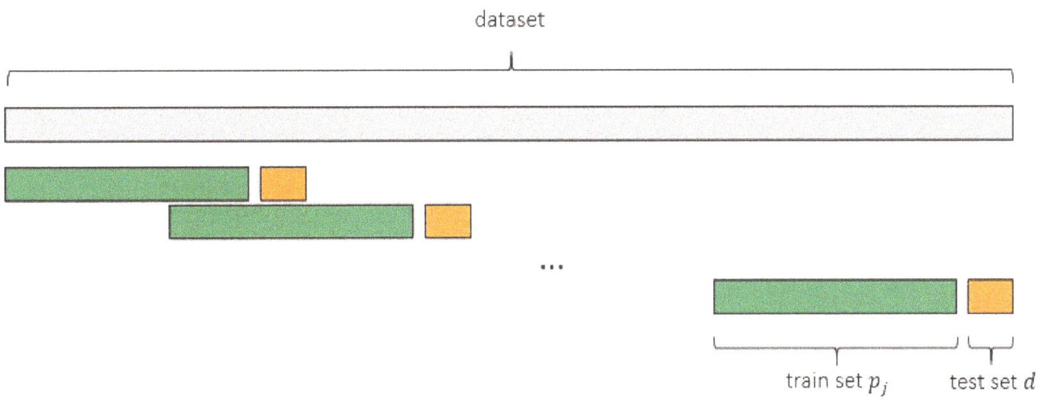

Figure 2. Scheme of proposed methodology.

4. Discussion and Results

In this work the proposed procedure is applied to the "ASHRAE-Great Energy Predictor III" competition data [33]; in particular, one year of hourly sampled data of a parking building (building id: 1215) has been selected. The target dependent variable is the electric load and the independent variables are:

- The time of the day as a cyclical variable (sine and cosine);
- The day of the week one-hot encoded;
- The month of the year as a cyclical variable (sine and cosine);

- The outdoor air temperature.

The training sizes p_j range from 2 weeks to 36 weeks within a span of 2 weeks.

Figure 3 shows LCs obtained with the XGBoost algorithm. When the training-set size is small, the training error is lower since the information variance to be learnt is tiny. As a consequence, the model has no generalization capability and its test error is high. When the training set size increases, training MAPE increases and test MAPE decreases. Adding training data helps to reduce bias error. Training and test curves are very close between 20–28 training weeks. This narrow gap shows a low variance error: Training data are fitted well and the algorithm can generalize on unseen data. The gap increases for a training size higher than 28 weeks, which may indicate an overfitting problem. In this case, a training size of 24 weeks seems to be a good compromise for XGBoost.

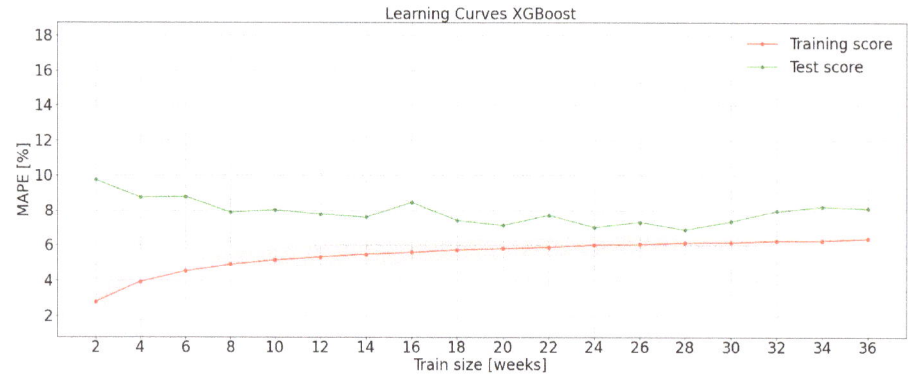

Figure 3. Learning curves obtained with the XGBoost algorithm.

LCs for the SVR algorithm are shown in Figure 4. Bias error slightly but progressively decreases with a training set size; variance error reaches its minimum between 20 and 28 weeks. Hence, an appropriate training set size is 24 weeks.

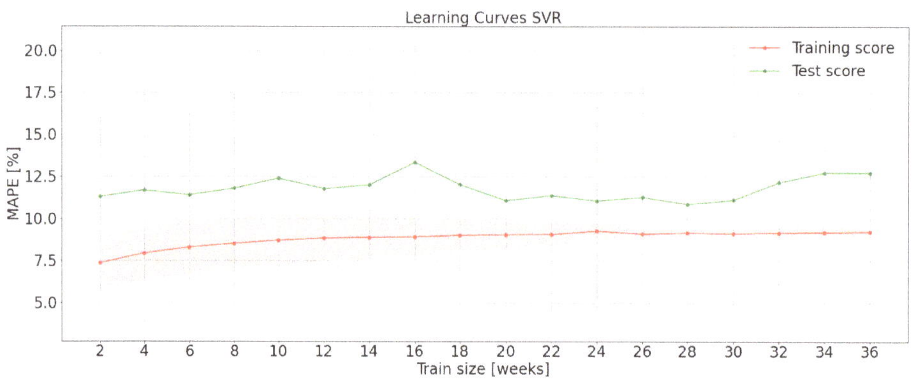

Figure 4. Learning curves obtained with a SVR algorithm.

The same performance is presented from MLP whose LCs are plotted in Figure 5. Adding training data to small train dataset leads to increase the training error and decrease test error. This is mainly due to a reduction of bias error. The bias-variance dilemma is settled between 20 and 24 weeks of training. The appropriate training set size could be 20 weeks for MLP with two layers and hyperparameters as described in Section 3.3.

If learning curves are characterized by high training and test errors according to domain knowledge, the model may seem to suffer of a high bias error. Getting more training data will not help much. In this particular case, the desired MAPE is around 6%; while XGBoost reaches this target, SVR and MLP seem to suffer from underfitting. This problem highlights that SVR and MLP models have been tuned with simple architectures.

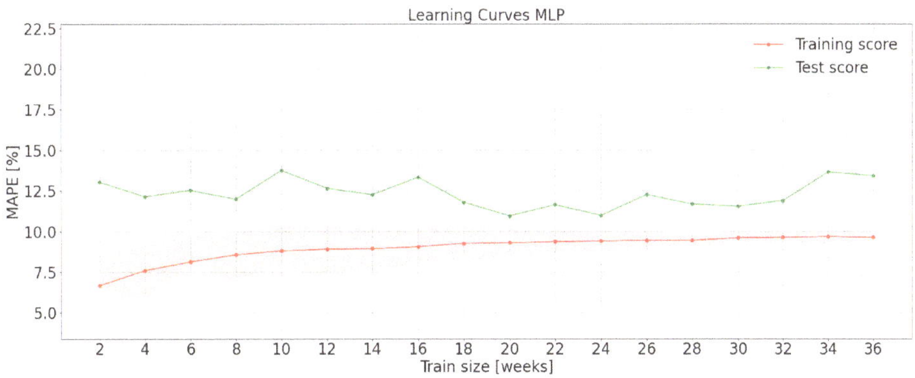

Figure 5. Learning curves obtained with the MLP algorithm.

5. Conclusions

In this paper a procedure to analyze learning curves for time-series is presented; it aims to show a generalized performance of an algorithm with different training set sizes. The procedure retained time order and allowed to test heterogeneous samples for each training-set size. The performance estimation was analyzed in an Out-of-Sample approach with a sliding window. This methodology is suitable for real world data with potential non-stationarities. The developed procedure could be applied to any kind of data or algorithm.

The proposed methodology was applied to electrical load forecasting of a parking building. Learning curves were obtained with three different regression algorithms; namely XGBoost, SVR, and MLP. This analysis underlines how learning curves could give information about training and test as a function of a training set size and how to choose an appropriate size of data to cope with the bias-variance problem.

The full code is available at [32] in order to guarantee the reproducibility of the presented procedure.

As a next step, this research could be used as a tool for evaluating the estimator architecture by using different sets of hyperparameters to build LCs guides to understand their impact on the learning process.

Author Contributions: Conceptualization, C.G. and P.D.; methodology, C.G. and P.D.; software, C.G. and P.D.; validation, C.G. and P.D.; formal analysis, C.G. and P.D.; investigation, C.G. and P.D.; resources, C.G., P.D. and S.M.; data curation, C.G. and P.D.; writing—original draft preparation, C.G. and P.D.; writing—review and editing, C.G., P.D. and S.M.; visualization, C.G. and P.D.; supervision, S.M.; project administration, S.M. All authors have read and agreed to the published version of the manuscript.

References

1. El-Hawary, M.E. *Advances in Electric Power and Energy Systems-Load and Price Forecasting*; IEEE Press Wiley: Piscataway, NJ, USA, 2017.
2. D'Ettorre, F.; De Rosa, M.; Conti, P.; Testi, D.; Finn, D. Mapping the energy flexibility potential of single buildings equipped with optimally-controlled heat pump, gas boilers and thermal storage. *Sustain. Cities Soc.* **2019**, *50*, 101689. [CrossRef]
3. Ahmad, T.; Zhang, H.; Yan, B. A review on renewable energy and electricity requirement forecasting models for smart grid and buildings. *Sustain. Cities Soc.* **2020**, *55*, 102052. [CrossRef]

4. Wei, Y.; Zhang, X.; Shi, Y.; Xia, L.; Pan, S.; Wu, J.; Han, M.; Zhao, X. A review of data-driven approaches for prediction and classification of building energy consumption. *Renew. Sustain. Energy Rev.* **2018**, *82*, 1027–1047. [CrossRef]
5. Yildiz, B.; Bilbao, J.I.; Sproul, A.B. A review and analysis of regression and machine learning models on commercial building electricity load forecasting. *Renew. Sustain. Energy Rev.* **2017**, *73*, 1104–1122. [CrossRef]
6. Zhang, L.; Wen, J.; Li, Y.; Chen, J.; Ye, Y.; Fu, Y.; Livingood, W. A review of machine learning in building load prediction. *Appl. Energy* **2021**, *285*, 116452. [CrossRef]
7. Khalid, R.; Javaid, N. A survey on hyperparameters optimization algorithms of forecasting models in smart grid. *Sustain. Cities Soc.* **2020**, *61*, 102275. [CrossRef]
8. Würsch, C. Bias-Variance-Tradeoff: Crossvalidation & Learning Curves. Available online: https://stdm.github.io/downloads/courses/ML/V06_BiasVariance-LearningCurves.pdf (accessed on 5 October 2020).
9. Beleites, C.; Neugebauer, U.; Bocklitz, T.; Krafft, C.; Popp, J. Sample Size Planning for Classification Models. *Anal. Chim. Acta* **2013**, *760*, 25–33. [CrossRef] [PubMed]
10. Figueroa, R.L.; Zeng-Treitler, Q.; Kandula, S.; Ngo, L.H. Predicting sample size required for classification performance. *BMC Med. Inform. Decis. Mak.* **2012**, *12*, 8. [CrossRef]
11. Hess, K.R.; Wei, C. Learning Curves in Classification With Microarray Data. *Semin. Oncol.* **2010**, *37*, 65–68. [CrossRef]
12. Ning, H.; Li, Z.; Wang, C.; Yang, L. Choosing an appropriate training-set size when using existing data to train neural networks for land cover segmentation. *Ann. Gis* **2020**, *26*, 329–342. [CrossRef]
13. Pedregosa, F.; Varoquaux, G.; Gramfort, A.; Michel, V.; Thirion, B.; Grisel, O.; Blondel, M.; Prettenhofer, P.; Weiss, R.; Dubourg, V.; et al. Scikit-learn: Machine Learning in Python. *J. Mach. Learn. Res.* **2011**, *12*, 2825–2830.
14. Perlich, C.; Provost, F.; Simonoff, J. Tree Induction vs. Logistic Regression: A Learning-Curve Analysis. *J. Mach. Learn. Res.* **2003**, *4*, 211–255.
15. Süzen, M.; Yegenoglu, A. Generalised Learning of Time-Series: Ornstein-Uhlenbeck Processes. *arXiv* **2020**, arXiv:1910.09394.
16. Zhu, X.; Vondrick, C.; Fowlkes, C.C.; Ramanan, D. Do We Need More Training Data? *Int. J. Comput. Vis.* **2016**, *119*, 76–92. [CrossRef]
17. Cerqueira, V.; Torgo, L.; Smailovič, J.; Mozetixcx, I. A comparative study of performance estimation methods for time series forecasting. In Proceedings of the 2017 IEEE International Conference on Data Science and Advanced Analytics (DSAA), Tokyo, Japan, 19–21 October 2017; pp. 529–538.
18. Cerqueira, V.; Torgo, L.; Smailovič, J.; Mozetixcx, I. Evaluating Time Series Forecasting Models: An Empirical Study on Performance Estimation Methods. *arXiv* **2019**, arXiv:1905.11744.
19. Tashman, L.J. Out-of-sample tests of forecasting accuracy: An analysis and review. *Int. J. Forecast.* **2000**, *16*, 437–450. [CrossRef]
20. Bergmeir, C.; Hyndman, R.J.; Koo, B. A note on the validity of crossvalidation for evaluating autoregressive time series prediction. *Comput. Stat. Data Anal.* **2018**, *120*, 70–83. [CrossRef]
21. Arlot, S.; Celisse, A. A survey of cross-validation procedures for model selection. *Stat. Surv.* **2010**, *4*, 40–79. [CrossRef]
22. Bergmeir, C.; Benítez, J.M. On the use of cross-validation for time series predictor evaluation. *Inform. Sci.* **2012**, *191*, 192–213. [CrossRef]
23. Kalman, R.E. A New Approach to Linear Filtering and Prediction Problems. *J. Basic Eng.* **1960**, *82 (Series D)*, 35–45. [CrossRef]
24. Gestore Mercati Energetici. Available online: https://www.mercatoelettrico.org/en/ (accessed on 5 April 2021).
25. Hammad, M.A.; Jereb, B.; Rosi, B.; Dragan, D. Methods and Models for Electric Load Forecasting: A Comprehensive Review. *Logist. Sustain. Transp.* **2020**, *11*, 51–76. [CrossRef]
26. Makridakis, S.; Spiliotis, E.; Assimakopoulos, V. The M4 Competition: 100,000 time series and 61 forecasting methods. *Int. J. Forecast.* **2020**, *36*, 54–74. [CrossRef]
27. Chen, T.; Guestrin, C. XGBoost: A Scalable Tree Boosting System. In Proceedings of the 22nd ACM SIGKDD International Conference on Knowledge Discovery and Data Mining, San Francisco, CA, USA, 13–17 August 2016.
28. Bergstra, J.; Bengio, Y. Random Search for Hyper-Parameter Optimization. *J. Mach. Learn. Res.* **2012**, *13*, 281–305.
29. Support Vector Regression (SVR) Scikit-Learn Library. Available online: https://scikit-learn.org/stable/modules/generated/sklearn.svm.SVR.html (accessed on 5 April 2021).
30. Scikit-Learn Wrapper interface for XGBoost. Available online: https://xgboost.readthedocs.io/en/latest/python/python_api.html#module-xgboost.sklearn (accessed on 5 April 2021).
31. Multi-Layer Perceptron (MLP) Regressor Scikit-Learn Library. Available online: https://scikit-learn.org/stable/modules/generated/sklearn.neural_network.MLPRegressor.html (accessed on 5 April 2021).
32. Giola, C.; Danti, P. Learning-Curves. 2020. Available online: https://github.com/jolachi/learning-curves/ (accessed on 5 April 2021).
33. ASHRAE-Great Energy Predictor III. Available online: https://www.kaggle.com/c/ashrae-energy-prediction (accessed on 5 April 2021).

engineering proceedings

MDPI

Proceedings

Decomposition-Based Hybrid Models for Very Short-Term Wind Power Forecasting †

Juan Manuel González Sopeña [1,*], Vikram Pakrashi [2,3,4] and Bidisha Ghosh [1,5]

1 QUANT Group, Department of Civil, Structural and Environmental Engineering, Trinity College Dublin, D02 PN40 Dublin, Ireland; bghosh@tcd.ie
2 UCD Centre for Mechanics, Dynamical Systems and Risk Laboratory, School of Mechanical & Materials Engineering, University College Dublin, D04 V1W8 Dublin, Ireland; vikram.pakrashi@ucd.ie
3 SFI MaREI Centre, University College Dublin, D04 Dublin, Ireland
4 The Energy Institute, University College Dublin, D04 V1W8 Dublin, Ireland
5 CONNECT: SFI Research Centre for Future Networks & Communications, Trinity College Dublin, D02 Dublin, Ireland
* Correspondence: gonzlezj@tcd.ie

† Presented at the 7th International Conference on Time Series and Forecasting, Gran Canaria, Spain, 19–21 July 2021.

Abstract: Wind power forecasting is a tool used in the energy industry for a wide range of applications, such as energy trading and the operation of the grid. A set of models known as decomposition-based hybrid models have stood out in recent times due to promising results in terms of performance. As many publications on this matter are found in the literature, a comparison of these models is difficult, because they are tested under different conditions in terms of data, prediction horizon, and time resolution. In this paper, we provide a comparison unifying these parameters using the main decomposition algorithms and a set of artificial neural network-based models for very short-term wind power forecasting (up to 30 min ahead). For this purpose, a case study using data from an Irish wind farm is performed to analyze the models in terms of accuracy and robustness for a variety of wind power generation scenarios.

Keywords: very short-term wind power forecasting; decomposition-based hybrid models; artificial neural networks; data-driven forecasting models

Citation: Sopeña, J.M.G.; Pakrashi, V.; Ghosh, B. Decomposition-Based Hybrid Models for Very Short-Term Wind Power Forecasting. *Eng. Proc.* **2021**, 5, 39. https://doi.org/10.3390/engproc2021005039

Published: 7 July 2021

1. Introduction

Wind power forecasting (WPF) is a tool of importance for practitioners in the wind energy industry, and it accomplishes different tasks depending on the time horizon, from reserve requirement decisions [1] to energy trading [2].

Several standards are found in the literature to classify WPF models with respect to the forecast horizon. One of the most well-known conventions is presented in [3], in which four time horizons are defined: very short-term (up to 30 min ahead), short-term (from 30 min to 6 h ahead), medium-term (up to 1 day ahead), and long-term (more than 1 day ahead) horizons. For medium- and long-term forecasts, physical models are preferred, whereas statistical models are used for very short- and short-term horizons, as they are easier to model and less computationally expensive than physical-based approaches. Among the statistical models, a family of models known as decomposition-based hybrid models has gained the attention of wind forecasting practitioners, with more than 100 papers on this topic having been published [4]. These models have a preprocessing step in which the complexity of wind power time series is avoided by decomposing the signal into a set of more stationary components (usually known as modes). However, as the literature on this type of models is already very extensive, it is difficult to determine which of these models are more suitable for very short-term and short-term forecasts, as they are tested under datasets of different nature, length, and resolution. In addition, the resulting components are usually fit using a broad variety of artificial neural networks (ANNs), whose capacity to identify and model the features of wind power time series differ depending on the intrinsic characteristics of the type of ANN. Taking all these aspects into consideration, the aim of this article is to provide a case study where (1) the state-of-the-art decomposition techniques are considered to decompose wind power time series, (2) a set of ANN models are used to train the resulting modes, and (3) a time-scale classification of the models for very short-term wind power forecasts using common criteria is presented.

The paper is organized as follows: Section 2 introduces the main elements of decomposition-based hybrid models; Section 3 describes the data used in this study; Section 4 presents the results; and Section 5 provides the concluding remarks of this paper.

2. Methodology

In this section, the main decomposition algorithms and ANN-based forecasting models are described, as well as the metrics used to analyze the performance of the models.

2.1. Decomposition-Based Hybrid Models

Decomposition-based hybrid models decompose the original time series into a set of more stationary modes that are easier to handle. In terms of forecasting, ANNs allow us to exploit diverse features of the data, such as recurrent neural networks (RNN) or convolutional neural networks (CNN). The main structure for this family of models is shown in Figure 1: (1) the wind power time series is decomposed into a set of modes; (2) a forecasting model is built independently for every mode; and (3) the wind power forecast is estimated by adding the values of all modes.

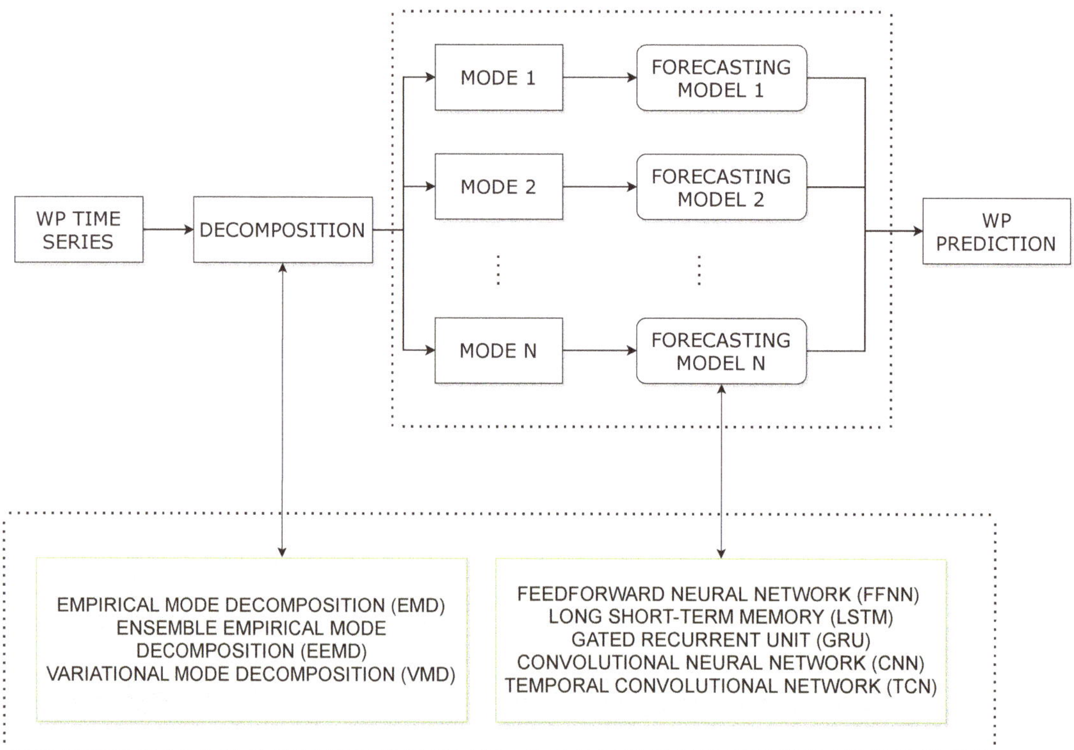

Figure 1. Flowchart for decomposition-based hybrid models.

Two of the most common decomposition techniques are empirical mode decomposition (EMD) [5] and variational mode decomposition (VMD) [6]. The wind power time series are decomposed into modes as

$$y(t) = \sum_{k=1}^{K} y_k(t) \quad (1)$$

where $y_k(t)$ is the k-mode extracted from the data. The modes, also known in the literature as intrinsic mode functions (IMFs), can be expressed as amplitude-modulated–frequency-modulated (AM–FM) signals [6,7]:

$$y_k(t) = A_k(t)\cos\phi_k(t), \quad A_k(t), \phi'_k > 0 \; \forall t \quad (2)$$

where $\phi_k(t)$ is a non-decreasing function. The main assumption is that the variation of A_k and ϕ'_k is slower than the variation of $\phi_k(t)$. Thus, every mode $y_k(t)$ can be considered as a harmonic signal with amplitude A_k and frequency ϕ'_k for a sufficiently long time interval $[t-\delta, t+\delta]$ where $\delta \approx 2\phi/\phi'_k$ [7].

The EMD algorithm extracts these modes as described in the following four steps: (1) local maxima and minima are located in the time series data $y(t)$ and then interpolated to build an upper and a lower envelope, respectively; (2) the mean value $m(t)$ of these envelopes is determined, and the first component H_1 is built by subtracting this value from the original time series $y(t)$; (3) these two steps are repeated until the stopping criterion is satisfied, and in this case, H_1 will be equivalent to the first mode $y_1(t)$ and the residue to $y(t) - H_1$, the difference between the original time series and the first mode; and (4) steps 1–3 are repeated with the residues until all of the modes and the last residue are computed.

Mode mixing and aliasing can occur when the EMD algorithm is applied to decomposed the data [8]. A variation of the original EMD approach known as ensemble empirical mode decomposition (EEMD) [9] was proposed to overcome this: a set of trials following the EMD algorithm are performed, but mixing the original time series $y(t)$ with Gaussian white noise. Thus, the EEMD algorithm is developed in four steps: (1) Gaussian white noise is added to the original data, (2) the EMD algorithm is applied to the data mixed with white noise, (3) steps 1–2 are repeated using different white noise series, and (4) the final decomposition is obtained calculating the mean value of all trials. This way, the white noise series cancel each other, and the risk of mode mixing is significantly reduced.

On the other hand, VMD is a non-recursive signal processing method designed for decomposing non-stationary signals. The decomposition takes place by means of a constrained variational problem to calculate the bandwidth of each mode. This process consists of three steps: (1) the Hilbert transform is used to obtain the unilateral frequency spectrum for each mode, (2) an exponential tuned to the estimated center frequencies is used to shift every mode's frequency spectrum to baseband, and (3) the bandwidth of each mode is identified using the H^1 Gaussian smoothness of the demodulated signal. As suggested in the original paper [6], the constrained variational problem can be transformed into an unconstrained problem by introducing a quadratic penalty term and Lagrangian multipliers λ as follows:

$$L(\{y_k\},\{\omega_k\},\lambda) = \alpha \sum_{k=1}^{K} \left\| \partial_t \left[\left(\delta(t) + \frac{j}{\pi t} \right) * y_k(t) \right] e^{-j\omega_k t} \right\|_2^2 + \\ + \left\| y(t) - \sum_{k=1}^{K} y_k(t) \right\|_2^2 + \left\langle \lambda(t), y(t) - \sum_{k=1}^{K} y_k(t) \right\rangle \quad (3)$$

where $y(t)$ is the original time series, $\{y_k\}$ is the set of all modes, $\{\omega_k\}$ is the set of the respective center frequencies, $\delta(t)$ is the Dirac function, $*$ denotes a convolution, $\|\|_2^2$ denotes a squared L^2-norm, and α denotes the balancing parameter of the data fidelity constraint. Then, this unconstrained problem is solved by using a technique known as the alternate direction method of multipliers (ADMM) [10,11], which allows one to obtain the modes y_k and the center frequencies ω_k with the following expressions:

$$\hat{y}_k^{n+1}(\omega) = \frac{\hat{y}(\omega) - \sum_{i \neq k} \hat{y}_i(\omega) + \frac{\hat{\lambda}(\omega)}{2}}{1 + 2\alpha(\omega - \omega_k)^2} \tag{4}$$

$$\omega_k^{n+1} = \frac{\int_0^\infty \omega |\hat{y}_k(\omega)|^2 dw}{\int_0^\infty |\hat{y}_k(\omega)|^2 dw} \tag{5}$$

where $\hat{y}(\omega)$, $\hat{y}_k(\omega)$, and $\hat{\lambda}(\omega)$ are the Fourier transformations of $y(t)$, $y_k(t)$, and $\lambda(t)$ respectively.

Regarding the forecasting models, the basic ANN structure is known as a feedforward neural network (FFNN), which is composed of a set of three layers (input, hidden, and output layers), and the information is propagated forward to the output layers using the backpropagation algorithm [12]. Given an input $\mathbf{x} = \{x_1, \ldots, x_t\}$ and a hidden layer with h neurons, the output is of the form

$$\sum_{i=1}^{h} \beta_i \phi(\omega_i \mathbf{x} + b_i) \tag{6}$$

where β_i represents the weights resulting from connecting the hidden and output layers (*output weights*), ω_i the weights connecting the input and hidden layers (*input weights*), b_i the biases of the neurons in the hidden layer, and ϕ the activation function.

Other types of ANNs can learn spatial and temporal features of time series data. For instance, RNNs take into consideration temporal patterns by maintaining an internal state in order to process a sequence of inputs. In order to process long-term dependencies, advanced RNN structures, such as long-short term memory (LSTM) [13], and gated recurrent units (GRU) [14] should be implemented, as basic RNNs experience vanishing and exploding gradients in this scenario [15]. On the other hand, spatial features can be extracted using CNNs. Both temporal and spatial features can be considered simultaneously by combining RNN and CNN structures [16], resulting, for instance, in CNN-GRU and CNN-LSTM models. Temporal and spatial features are also taken into consideration in temporal convolutional networks (TCN) [17], in which the convolutions are causal, meaning that the outputs are only related to the current and previous inputs.

All of the decomposition algorithms and ANN-based models can be combined to build any decomposition-based hybrid model. To make this study as comprehensive as possible, 21 models in total are considered for the simulations, resulting from the combination of the 3 decomposition algorithms (EMD, EEMD, and VMD) and the 7 forecasting models (FFNN, GRU, LSTM, CNN, CNN-GRU, CNN-LSTM, and TCN) introduced in this section.

2.2. Performance Evaluation

The performance of the models is measured using one of the most widespread metrics in the WPF literature [18], the mean absolute error (*MAE*):

$$MAE = \frac{1}{N} \sum_{i=1}^{N} |\hat{y}_i - y_i| \tag{7}$$

where N indicates the number of samples over the testing set, y_i the value of wind power measurements, and $\hat{y}_i$ the value of the forecasts. To facilitate the understanding of the error measures, *MAE* values are normalized by the total capacity of the farm and, therefore, the normalized *MAE* (NMAE) is used from here onwards to report the results.

3. Data

A dataset containing historical wind power measurements from an Irish wind farm is used to carry out the simulations. Data were collected from 1 January 2017 to 30 June 2019 at a 10 min resolution. In order to benchmark the models in the most comprehensive manner, the data are divided into one-year long sets, where the first eleven months are used for training and validation and the last month as the testing set.

Figure 2 shows all of the testing sets, in which the fluctuating nature of wind power can be observed clearly from DS-1 to DS-8. This variety of wind power generation scenarios allow us to examine the performance of the models not only in terms of accuracy but in terms of robustness. Furthermore, large periods where the wind farm has been halted can be observed in the testing sets corresponding to DS-9 and DS-10.

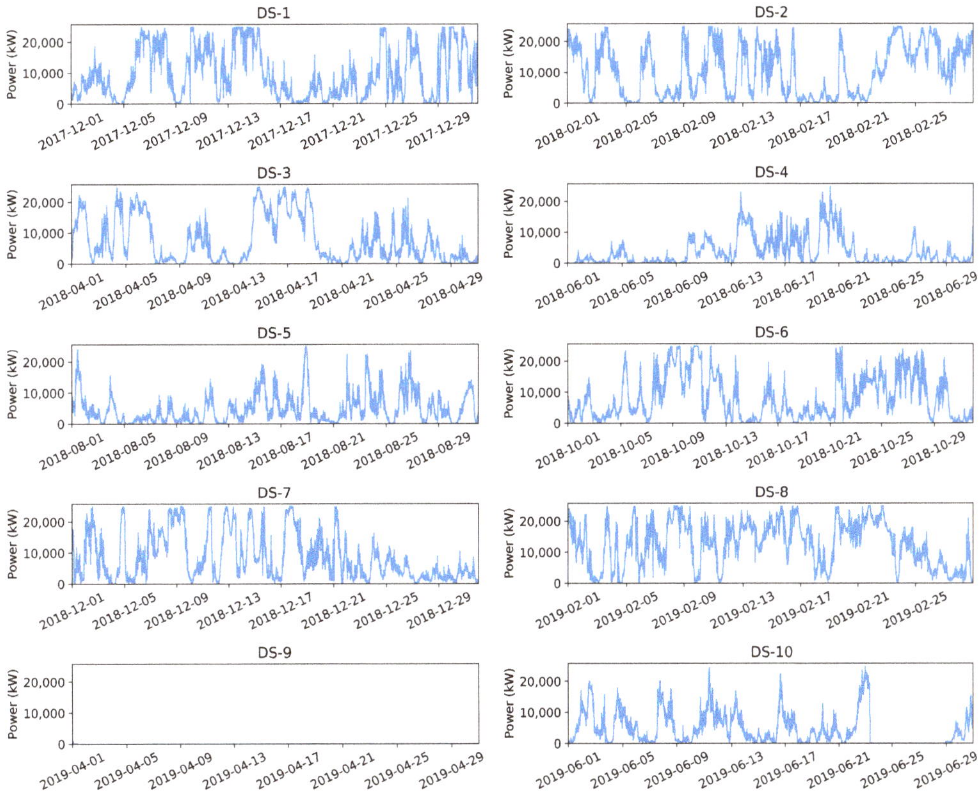

Figure 2. Testing sets considering different periods of the dataset.

4. Results

Every model was run five times for every dataset, although the results coming from training the models using the datasets DS-9 and DS-10 are omitted in this Section, as the corresponding testing sets contain large periods where the wind farm is halted, which may bias the evaluation of model performance. Thus, a total of 40 simulations were performed for all models, meaning that the models were trained 40 times, yielding different numerical results every time due to (1) the use of different subsets of data and (2) random initialization of the weights of the ANN structures, which influences the training process [19]. This way, the parameters learned by the model in the training stage vary even if the same training data are fed to the model.

Using either VMD, EMD or EEMD, data were divided into six modes, which were all trained under the same conditions. Regarding the parameters, the models were trained using a batch of size 64 for 100 epochs, using early stopping [20] to halt the training if necessary to avoid overfitting. The hidden layer of the FFNN and RNN-based models have 50 neurons in total; the CNN layers are set with 50 filters with a kernel size = 6; and the TCN layers are formed by 50 filters with dilation factors d = 1, 2, and 4 and a filter size k = 6. The MIMO (multiple-input multiple-output) strategy [21] was implemented to output a vector with the whole sequence of forecasts, so only one model needed to be trained for all horizons. In this case, the models took the previous 72 steps as the input, representing the previous 12 h using 10 min data resolution, and produced a vector containing 36 values, which are equivalent to the next 6 h in 10 min intervals.

As only very short-term WPFs were considered in this study, the results reported correspond to 10-, 20-, and 30-min-ahead forecasts, which are equivalent to output forecasts 1, 2, and 3 steps ahead with the 10 min resolution data used in this study. Some examples of these simulations are shown in Figure 3, where 30-min-ahead forecasts are shown for DS-1, DS-2, DS-5, DS-6, and DS-8 using two of the models with better performance: the VMD-GRU and the VMD-CNN-LSTM models.

The average value of the NMAE over all the simulations is shown in Table 1. In terms of the decomposition algorithm, VMD proves to be the better than EEMD and EMD at decomposing wind power time series, as the performance using VMD is higher than that of the others in terms of accuracy, regardless of the ANN model used. Among these, the models where the temporal patterns of data are considered exhibit the best performance: an average NMAE value of 0.42 with the VMD-CNN-GRU model for 10-min-ahead forecasts; 0.59 with the VMD-GRU model for 20-min-ahead forecasts; and 0.91 with the VMD-GRU, VMD-CNN-GRU, and VMD-CNN-LSTM models for 30-min-ahead WPFs. Thus, adding the CNN layer prior to either the LSTM or GRU layer does not result in any significant improvement of performance.

Figure 4 provides additional information with respect to model performance, showing the distribution of the NMAE values over the simulations for 10-min-ahead WPFs. The combination of VMD with both GRU and LSTM structures, including the CNN-GRU and CNN-LSTM structures, appears to be the more robust among all models, as the variability is very low in terms of model performance. Furthermore, it proves the adaptability of these four models to different training and testing sets of wind power. On the other hand, EMD- and EEMD-based models not only show lower accuracy but also higher variability, which indicates a lower degree of robustness for these models.

The simulations performed in this study indicate that decomposition-based hybrid models based on the VMD algorithm for the purpose of decomposing wind power time series and RNN-based forecasting models are the most adequate for WPFs up to 30 min ahead, both in terms of forecast accuracy and robustness to different testing sets. The nature of LSTM and GRU structures is reflected in the predictions, which are able to adequately capture the temporal patterns present in the data.

Table 1. Average NMAE for very short-term forecasts.

Model	10 min	20 min	30 min
VMD-FFNN	0.77	0.97	1.13
VMD-GRU	0.43	0.59	0.91
VMD-LSTM	0.46	0.66	0.92
VMD-CNN	0.82	0.91	1.1
VMD-CNN-GRU	0.42	0.61	0.91
VMD-CNN-LSTM	0.43	0.61	0.91
VMD-TCN	0.57	0.8	1.05
EMD-FFNN	1.69	2.13	2.58
EMD-GRU	1.35	1.8	2.18
EMD-LSTM	1.31	1.71	2.1
EMD-CNN	1.62	2.02	2.31
EMD-CNN-GRU	1.3	1.72	2.08
EMD-CNN-LSTM	1.3	1.69	2.07
EMD-TCN	1.38	1.7	2.04
EEMD-FFNN	1.38	1.75	1.93
EEMD-GRU	1.23	1.56	1.71
EEMD-LSTM	1.21	1.54	1.69
EEMD-CNN	1.37	1.69	1.85
EEMD-CNN-GRU	1.22	1.54	1.7
EEMD-CNN-LSTM	1.23	1.57	1.74
EEMD-TCN	1.28	1.59	1.75

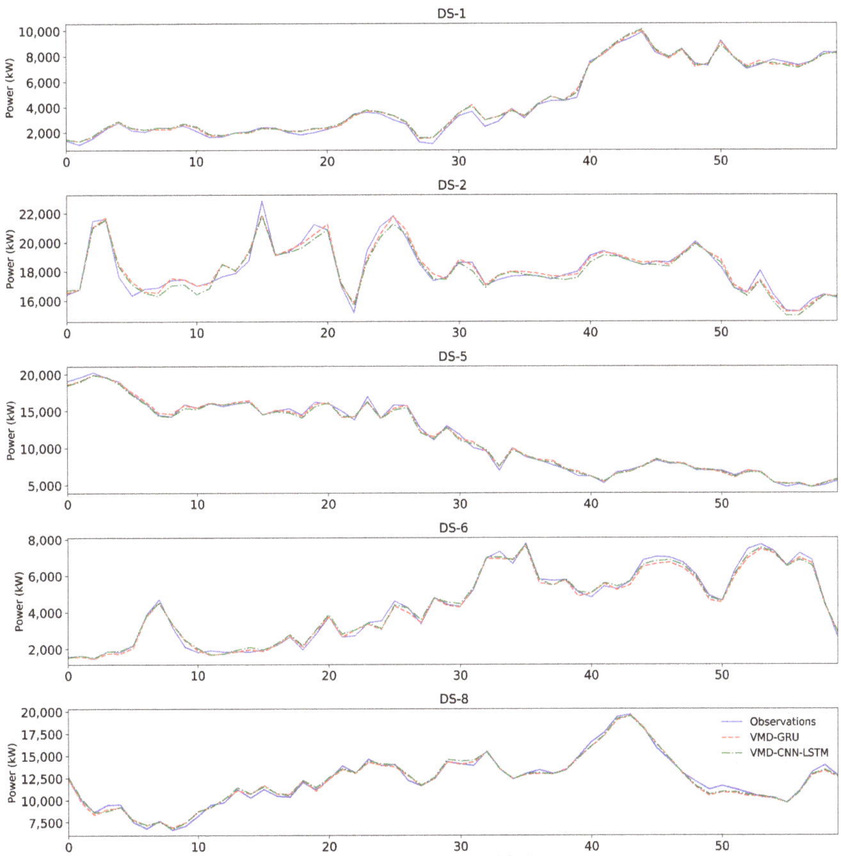

Figure 3. Examples of predictions for 30-min-ahead forecasts using the models VMD-GRU and VMD-CNN-LSTM.

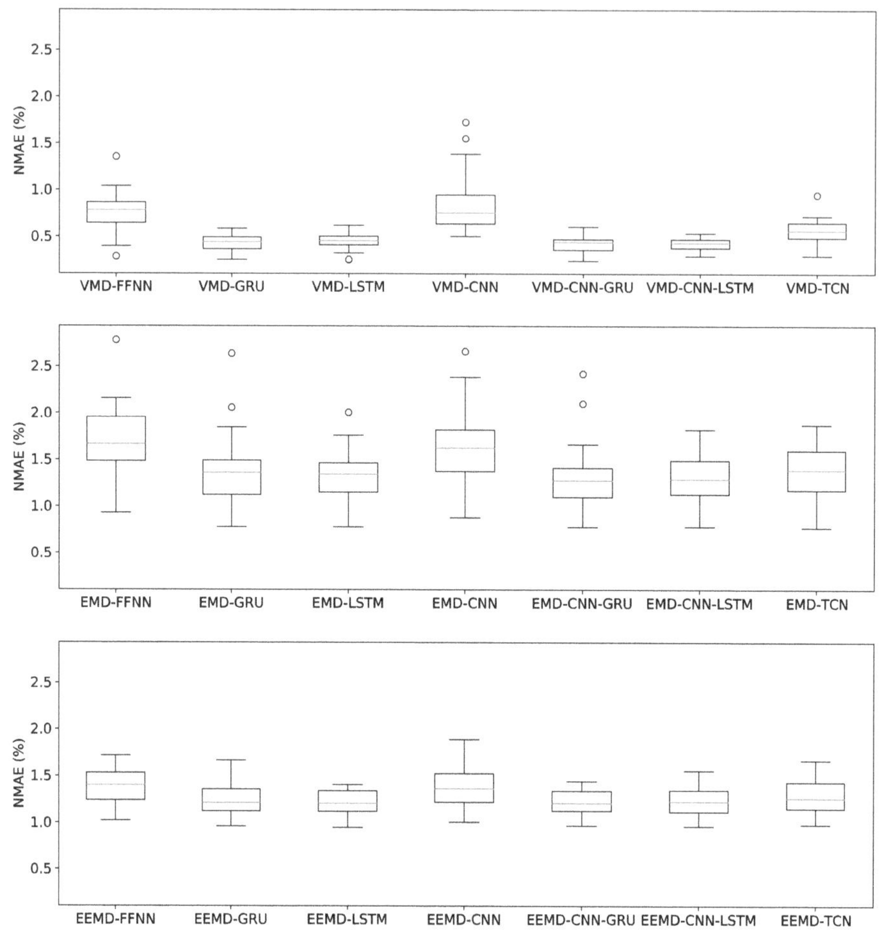

Figure 4. NMAE distribution for 10-min-ahead forecasts.

5. Conclusions

In recent times, decomposition-based hybrid models have shown promising results for very short- and short-term wind power forecasting. As the number of papers published on the topic is considerable, comparing them is a strenuous task, because the models are tested under different conditions, such as the prediction horizon, the time resolution of the data, or the amount of data used to train the model. To bring some light to this issue, this paper provides a classification of decomposition-based hybrid models for very short-term wind power forecasting, where the main state-of-the-art decomposition algorithms and the main ANN-based forecasting models are combined and benchmarked under the same conditions.

A set of simulations was performed using data from an Irish wind farm. The data were divided into several subsets to analyze the data under different training and testing conditions, as wind power time series show very high variability. As such, this study does not only identify the model performance in terms of accuracy but also their robustness to different wind power generation situations. The results indicate that using variational mode decomposition together with advanced RNN structures (LSTM and GRU) provides the most accurate and robust WPFs for very short-term horizons, showing lower average NMAE values over all the simulations and lower variability in the NMAE distribution when compared to those of the other models. To further validate the scalability of these results, additional wind power datasets can be considered following the same experimental design shown in this paper.

Author Contributions: Conceptualization, J.M.G.S., V.P. and B.G.; methodology, J.M.G.S. and B.G.; software, J.M.G.S.; validation, J.M.G.S., V.P. and B.G.; formal analysis, J.M.G.S.; investigation, J.M.G.S., V.P. and B.G.; resources, J.M.G.S., V.P. and B.G.; data curation, J.M.G.S., V.P. and B.G.; writing—original draft preparation, J.M.G.S. and B.G.; writing—review and editing, J.M.G.S., V.P. and B.G.; visualization, J.M.G.S.; supervision, V.P. and B.G.; project administration, V.P. and B.G.; funding acquisition, V.P. and B.G. All authors have read and agreed to the published version of the manuscript.

Funding: The authors acknowledge the funding of SEAI WindPearl Project 18/RDD/263.

Data Availability Statement: Relevant computed data are available from authors upon reasonable request.

Acknowledgments: Bidisha Ghosh would like to acknowledge the support of ENABLE (Grant number 16/SP/3804) and Connect Center (Grant number 13/RC/2077_P2). Vikram Pakrashi would like to acknowledge the support of SFI MaREI centre (Grant number RC2302_2) and UCD Energy Institute.

Conflicts of Interest: The authors declare no conflict of interest.

Abbreviations

The following abbreviations are used in this manuscript:

ADMM	Alternate direction method of multipliers
ANN	Artificial neural network
CNN	Convolutional neural network
EEMD	Ensemble empirical mode decomposition
EMD	Empirical mode decomposition
FFNN	Feedforward neural network
GRU	Gated recurrent unit
IMF	Intrinsic mode function
LSTM	Long short-term memory
MAE	Mean absolute error
NMAE	Normalized mean absolute error
RNN	Recurrent neural network
TCN	Temporal convolutional network
VMD	Variational mode decomposition
WPF	Wind power forecasting

References

1. Bessa, R.J.; Matos, M.A.; Costa, I.C.; Bremermann, L.; Franchin, I.G.; Pestana, R.; Machado, N.; Waldl, H.P.; Wichmann, C. Reserve setting and steady-state security assessment using wind power uncertainty forecast: A case study. *IEEE Trans. Sustain. Energy* **2012**, *3*, 827–836. [CrossRef]
2. Pinson, P.; Chevallier, C.; Kariniotakis, G.N. Trading wind generation from short-term probabilistic forecasts of wind power. *IEEE Trans. Power Syst.* **2007**, *22*, 1148–1156. [CrossRef]
3. Soman, S.S.; Zareipour, H.; Malik, O.; Mandal, P. A review of wind power and wind speed forecasting methods with different time horizons. In Proceedings of the North American Power Symposium 2010, Arlington, TX, USA, 26–28 September 2010; pp. 1–8.
4. Qian, Z.; Pei, Y.; Zareipour, H.; Chen, N. A review and discussion of decomposition-based hybrid models for wind energy forecasting applications. *Appl. Energy* **2019**, *235*, 939–953. [CrossRef]
5. Huang, N.E.; Shen, Z.; Long, S.R.; Wu, M.C.; Shih, H.H.; Zheng, Q.; Yen, N.C.; Tung, C.C.; Liu, H.H. The empirical mode decomposition and the Hilbert spectrum for nonlinear and non-stationary time series analysis. *Proc. R. Soc. Lond. Ser. A* **1998**, *454*, 903–995. [CrossRef]
6. Dragomiretskiy, K.; Zosso, D. Variational mode decomposition. *IEEE Trans. Signal Process.* **2013**, *62*, 531–544. [CrossRef]
7. Daubechies, I.; Lu, J.; Wu, H.T. Synchrosqueezed wavelet transforms: An empirical mode decomposition-like tool. *Appl. Comput. Harmon. Anal.* **2011**, *30*, 243–261. [CrossRef]
8. Mandic, D.P.; ur Rehman, N.; Wu, Z.; Huang, N.E. Empirical mode decomposition-based time-frequency analysis of multivariate signals: The power of adaptive data analysis. *IEEE Signal Process. Mag.* **2013**, *30*, 74–86. [CrossRef]

9. Wu, Z.; Huang, N.E. Ensemble empirical mode decomposition: A noise-assisted data analysis method. *Adv. Adapt. Data Anal.* **2009**, *1*, 1–41. [CrossRef]
10. Boyd, S.; Parikh, N.; Chu, E. *Distributed Optimization and Statistical Learning via the Alternating Direction Method of Multipliers*; Now Publishers Inc.: Delft, The Netherlands, 2011.
11. Ghadimi, E.; Teixeira, A.; Shames, I.; Johansson, M. Optimal parameter selection for the alternating direction method of multipliers (ADMM): Quadratic problems. *IEEE Trans. Autom. Control* **2014**, *60*, 644–658. [CrossRef]
12. Rumelhart, D.E.; Hinton, G.E.; Williams, R.J. Learning representations by back-propagating errors. *Nature* **1986**, *323*, 533–536. [CrossRef]
13. Hochreiter, S.; Schmidhuber, J. Long short-term memory. *Neural Comput.* **1997**, *9*, 1735–1780. [CrossRef] [PubMed]
14. Chung, J.; Gulcehre, C.; Cho, K.; Bengio, Y. Empirical evaluation of gated recurrent neural networks on sequence modeling. *arXiv* **2014**, arXiv:1412.3555.
15. Pascanu, R.; Mikolov, T.; Bengio, Y. On the difficulty of training recurrent neural networks. In Proceedings of the International Conference on Machine Learning, Atlanta, GA, USA, 16–21 June 2013; pp. 1310–1318.
16. Chen, Y.; Zhang, S.; Zhang, W.; Peng, J.; Cai, Y. Multifactor spatio-temporal correlation model based on a combination of convolutional neural network and long short-term memory neural network for wind speed forecasting. *Energy Convers. Manag.* **2019**, *185*, 783–799. [CrossRef]
17. Gan, Z.; Li, C.; Zhou, J.; Tang, G. Temporal convolutional networks interval prediction model for wind speed forecasting. *Electr. Power Syst. Res.* **2021**, *191*, 106865. [CrossRef]
18. González-Sopeña, J.; Pakrashi, V.; Ghosh, B. An overview of performance evaluation metrics for short-term statistical wind power forecasting. *Renew. Sustain. Energy Rev.* **2020**, *138*, 110515. [CrossRef]
19. Pollack, J.B. Backpropagation is sensitive to initial conditions. *Complex Syst.* **1990**, *4*, 269–280.
20. Prechelt, L. Early stopping-but when? In *Neural Networks: Tricks of the Trade*; Springer: Berlin/Heidelberg, Germany, 1998; pp. 55–69.
21. Taieb, S.B.; Sorjamaa, A.; Bontempi, G. Multiple-output modeling for multi-step-ahead time series forecasting. *Neurocomputing* **2010**, *73*, 1950–1957. [CrossRef]

engineering proceedings

MDPI

Proceedings

If You Like It, GAN It—Probabilistic Multivariate Times Series Forecast with GAN †

Alireza Koochali [1,2,3,*], Andreas Dengel [2,3] and Sheraz Ahmed [2]

1 Ingenieurgesellschaft Auto und Verkehr (IAV) GmbH, 10587 Berlin, Germany
2 Deutsches Forschungszentrum für Künstliche Intelligenz (DFKI) GmbH, 67663 Kaiserslautern, Germany; andreas.dengel@dfki.de (A.D.); sheraz.ahmed@dfki.de (S.A.)
3 Department of Computer Science, University of Kaiserslautern, 67663 Kaiserslautern, Germany
* Correspondence: alireza.koochali@iav.de

† Presented at the 7th International Conference on Time Series and Forecasting, Gran Canaria, Spain, 19–21 July 2021.

Abstract: The contribution of this paper is two-fold. First, we present ProbCast—a novel probabilistic model for multivariate time-series forecasting. We employ a conditional GAN framework to train our model with adversarial training. Second, we propose a framework that lets us transform a deterministic model into a probabilistic one with improved performance. The motivation of the framework is to either transform existing highly accurate point forecast models to their probabilistic counterparts or to train GANs stably by selecting the architecture of GAN's component carefully and efficiently. We conduct experiments over two publicly available datasets—an electricity consumption dataset and an exchange-rate dataset. The results of the experiments demonstrate the remarkable performance of our model as well as the successful application of our proposed framework.

Keywords: time-series; generative adversarial networks; forecasting; probabilistic; prediction

Citation: Koochali, A.; Dengel, A.; Ahmed, S. If You Like It, GAN It—Probabilistic Multivariate Times Series Forecast with GAN. *Eng. Proc.* **2021**, *5*, 40. https://doi.org/10.3390/engproc2021005040

Published: 8 July 2021

Publisher's Note: MDPI stays neutral with regard to jurisdictional claims in published maps and institutional affiliations.

1. Introduction

Many sectors, such as health care, the automotive industry, the aerospace industry and weather forecasting, deal with time-series data in their operations. Knowledge about what will happen in the future is essential for making genuine decisions, and accurately forecasting future values is key to their success. A huge body of research is therefore dedicated to addressing the forecasting problem. An overview of various studies on the forecasting problem is provided in [1]. Currently, the field is dominated by point prediction methods, which are easy to understand. However, these deterministic models report the mean of possible outcomes and cannot reflect the inherent uncertainty that exists in the real world. The probabilistic forecast models are devised to rectify these shortcomings. These models try to quantify the uncertainty of the predictions by forming a probability distribution over possible outcomes [2].

In this paper, we propose ProbCast, a new probabilistic forecast model for multivariate time series based on Conditional Generative Adversarial Networks (GANs). Conditional GANs are a class of NN-based generative models that enable us to learn conditional probability distribution given a dataset. ProbCast is trained using a Conditional GAN setup to learn the probability distribution of future values conditioned on the historical information of the signal.

While GANs are powerful methods for learning complex probability distributions, they are notoriously hard to train. The training process is very unstable and quite dependent on careful selection of the model architecture and hyperparameters [3]. In addition to ProbCast, we suggest a framework for transforming an existing deterministic forecaster—which is comparatively easy to train—into a probabilistic one that exceeds the performance of its predecessor. By using the proposed framework, the space for searching the GAN's architecture becomes considerably smaller. Thus, this framework provides an easy way to adapt highly accurate deterministic models to construct useful probabilistic models, without compromising the accuracy, by exploiting the potential of GANs.

In summary, the main contributions of this article are as follows:

- We introduce ProbCast, a novel probabilistic model for multivariate time-series forecasting. Our method employs a conditional GAN setup to train a probabilistic forecaster.
- We suggest a framework for transforming a point forecast model into a probabilistic model. This framework eases the process of replacing the deterministic model with probabilistic ones.
- We conduct experiments on two publicly available datasets and report the results, which show the superiority of ProbCast. Furthermore, we demonstrate that our framework is capable of transforming a point forecast method into a probabilistic model with improved accuracy.

2. Related Work

Due to the lack of a standard evaluation method for GANs, initially they were applied to domains in which their results are intuitively assessable, for example, images. However, recently, they have been applied to time-series data. Currently, GANs are applied to various domains for generating realistic time-series data including health care [4–8], finance [9,10] and the energy industry [11–13]. In [14], the authors combine GAN and auto-regressive models to improve sequential data generation. Ramponi et al. [15] condition a GAN on timestamp information to handle irregular sampling.

Furthermore, researchers have used conditional GANs to build probabilistic forecasting models. Koochali et al. [16] used a Conditional GAN to build a probabilistic model for univariate time-series. They used Long Short Term Memory (LSTM) in the GAN's component and tested their method on a synthetic dataset as well as on two publicly available datasets. In [17], the authors utilized LSTM and Multi-Layer Perceptron (MLP) in a conditional GAN structure to forecast the daily closing price of stocks. The authors combined the Mean Square Error (MSE) with the generator loss of a GAN to improve performance. Zhou et al. [18] employed LSTM and a convolutional neural network (CNN) in an adversarial training setup to forecast the high-frequency stock market. To guarantee satisfying predictions, this method minimizes the forecast error in the form of Mean absolute error (MAE) or MSE during training in conjunction with the GAN value function. Lin et al. [19] proposed a pattern sensitive forecasting model for traffic flow, which can provide accurate predictions in unusual states without compromising its performance in its usual states. This method uses conditional GAN with MLP in its structure and adds two error terms to the standard generator loss. The first term specifies forecast error and the second term expresses reconstruction error. Kabir et al. [20] make use of adversarial training for quantifying the uncertainty of the electricity price with a prediction interval. This line of research is more aligned with the method we presented in this article; however, the methods suggested in [17–19] include a point-wise loss function in the GAN loss function. Minimizing suggested loss functions would decrease the statistical error values such as RMSE, MAPE and MSE. However, they encourage the model to learn the mean of possible outcomes instead of the probability distribution of future value. Hence, their probabilistic forecast can be misleading despite the small statistical error.

3. Background

Here, we work with a multivariate time-series $X = \{X_0, X_1, ..., X_T\}$, where each $X_t = \{x_{t,1}, x_{t,2}, ..., x_{t,f}\}$ is a vector with size f equal to the number of features. In this paper, $x_{t,f}$ refers to the data point at time step t of feature f and X_t points to the feature vector at time step t. The goal is to model $P(X_{t+1}|X_t, .., X_0)$, the probability distribution for X_{t+1} given historical information $\{X_t, .., X_0\}$.

3.1. Mean Regression Forecaster

To address the problem of forecasting, we can take the predictive view of regression [2]. Ultimately, the regression analysis aims to learn the conditional distribution of a response given a set of explanatory variables [21]. The mean regression methods are deterministic methods, which are concerned with accurately predicting the mean of the possible outcome, that is, $\mu(P(X_{t+1}|X_t, .., X_0))$. There is a broad range of mean regression methods available in the literature; however, all of them are unable to reflect uncertainty in their forecasts. Hence, their results can be unreliable and misleading in some cases.

3.2. Generative Adversarial Network

In 2014, Goodfellow et al. [22] introduced a powerful generative model called the Generative Adversarial Network (GAN). GAN can implicitly learn probability distribution, which describes a given dataset, that is, $P(data)$ with high precision. Hence, it is capable of generating artificial samples with high fidelity. The GAN architecture is composed of two neural networks, namely generator and discriminator. These components are trained simultaneously in an adversarial process. In the training process, first, a noise vector z is sampled from a known probability distribution $P_{noise}(z)$ and is fed into a generator. Then, the generator transforms z from $P_{noise}(z)$ to a sample, which follows P_{data}. On the other hand, the discriminator checks how well the generator is performing by comparing the generator's outputs with real samples from the dataset. During training, this two-player minimax game is set in motion by optimizing the following value function:

$$\begin{aligned} \min_G \max_D V(D,G) = \mathbb{E}_{x \sim P_{data}(x)}[log(D(x))] + \\ \mathbb{E}_{z \sim P_{noise}(z)}[log(1 - D(G(z)))]. \end{aligned} \tag{1}$$

However, GAN's remarkable performance is not acquired easily. The training process is quite unstable and the careful selection of GAN's architecture and hyperparameters is vital for stabilizing the training process [3]. Since we should search for the optimal architecture of the generator and discriminator simultaneously, it is normally a cumbersome and time-consuming task to find a perfect combination of structures in a big search space.

3.3. Conditional GAN

Conditional GAN [23] enables us to incorporate auxiliary information, called the condition, into the process of data generation. In this method, we provide an extra piece of information, such as labels, to both the generator and the discriminator. The generator must respect the condition while synthesizing a new sample because the discriminator considers the given condition while it checks the authenticity of its input. The new value function $V(G, D)$ for this setting is:

$$\begin{aligned} \min_G \max_D V(D,G) = \mathbb{E}_{x \sim P_{data}(x)}[log(D(x|y))] + \\ \mathbb{E}_{z \sim P_{noise}(z)}[log(1 - D(G(z|y)))]. \end{aligned} \tag{2}$$

After training a Conditional GAN, the generator learns implicitly the probability distribution of the given condition of the data, that is, $P(data|condition)$.

4. Methodology

4.1. ProbCast: The Proposed Multivariate Forecasting Model

In this article, we consider Conditional GAN as a method for training a probabilistic forecast model using adversarial training. In this perspective, the generator is our probabilistic model (i.e., ProbCast) and the discriminator provides the required gradient for optimizing ProbCast during training. To learn $P(X_{t+1}|X_t, .., X_0)$, the historical information $\{X_t, .., X_0\}$ is used as the condition of our Conditional GAN and the generator is trained to generate X_{t+1}. Hence, the probability distribution, which is learned by the generator, corresponds to $P(X_{t+1}|X_t, .., X_0)$, that is, our target distribution. The value function, which we used for training the ProbCast (indicated as PC), is:

$$\min_{PC} \max_{D} V(D, PC) = \mathbb{E}_{X_{t+1} \sim P_{\text{data}}(X_{t+1})}[\log D(X_{t+1}|X_t, .., X_0)] + \mathbb{E}_{z \sim P_z(z)}[\log(1 - D(PC(z|X_t, .., X_0)))]. \quad (3)$$

4.2. The Proposed Framework for Converting Deterministic Model to Probabilistic Model

By stepping into the realm of multivariate time-series, other challenges also need to be addressed. In the multivariate setting, we require more complicated architecture to figure out dependencies between features and to forecast the future with high accuracy. Furthermore, as previously mentioned, GANs require precise hyperparameter tuning to have a stable training process. Considering required network complexity for handling multivariate time-series data, it is very cumbersome, or in some cases impossible, to find suitable generator and discriminator architecture concurrently which performs accurately. To address this problem, we propose a new framework for building a probabilistic forecaster based on a deterministic forecaster using GAN architecture.

In this framework, we build the generator based on the architecture and hyperparameters of the deterministic forecaster and train it using appropriate discriminator architecture. In this fashion, we can perform the task of finding an appropriate generator and discriminator architecture separately, which results in simplification of the GAN architecture search process. In other words, by using this framework, we can transform an existing accurate deterministic model into a probabilistic model with increased precision and better alignment with the real world.

4.3. Train Pipeline

Figure 1 demonstrates the proposed framework, as well as the conditional GAN setup, for training ProbCast. First, we build an accurate point forecast model by searching for the optimal architecture of the deterministic model. In the case that a precise point forecast model exists, we can skip the first step and use an existing model. Then, we need to integrate the noise vector z into the deterministic model architecture. In our experiments, we obtain the best results when we insert the noise vector into the later layers of the network, letting earlier layers of the network learn the representation of the input window. Finally, we train this model using adversarial training to acquire our probabilistic forecast model, that is, ProbCast.

With the generator architecture at hand, we only need to search for an appropriate time-series classifier to serve as the discriminator during the training of GAN. By reducing the search space of GAN architecture to the discriminator only, we can efficiently find a discriminator structure that is capable of training the ProbCast with a superior performance in comparison to the deterministic model. The following steps summarize the framework:

1. Employ an accurate deterministic model;
    (a) Either use an existing model;
    (b) Or search for an optimal deterministic forecaster;
2. Structure the generator based on deterministic model architecture and incorporate the noise vector into the network, preferably into later layers;
3. Search for an optimal discriminator structure and train the generator using it.

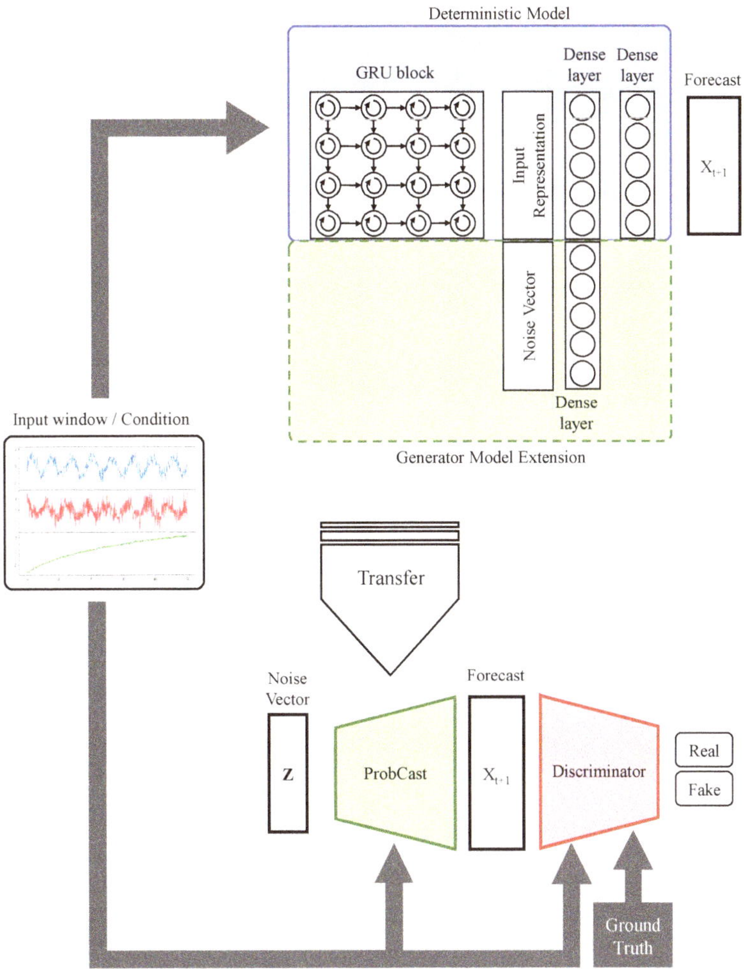

Figure 1. The demonstration of the proposed framework and adversarial training setup. The pipeline is followed from top to bottom. First, we search for the optimal architecture of the deterministic model. The deterministic model consists of a GRU block for the learning input window representation and two dense layers to map the representation to the forecast. Then, the noise vector z is integrated into the deterministic model to build the generator. Finally, the generator is trained using a suitable discriminator in a conditional GAN setup to obtain ProbCast.

5. Experiment

5.1. Datasets

We tested our method on two publicly available datasets—electricity and exchange-rate datasets (we used datasets from https://github.com/laiguokun/multivariate-time-series-data as they were prepared by the authors of [24]). The electricity dataset consists of the electricity consumption of 321 clients in KWh, which was collected every 15 min between 2012 and 2014. The dataset was converted to reflect hourly consumption. The exchange-rate dataset contains the daily exchange-rate of eight countries, namely Australia, Great Britain, Canada, Switzerland, China, Japan, New Zealand, and Singapore, which was collected between 1990 and 2016. Table 1 lists the properties of these two datasets.

For our experiments, we used 75% of the datasets for training, 5% for validation and 20% for testing.

Table 1. The properties of datasets.

	Electricity Dataset	Exchange-Rate Dataset
Dataset length	62,304	7588
Number of feature	321	8
Sample rate	1 h	1 day

5.2. Setup

In each of our experiments, first we ran an architecture search to find an accurate deterministic model. For training the deterministic model, we employed MAE as a loss function. In Figure 1, the architecture of the deterministic model is indicated. We used a Gated recurrent unit (GRU) [25] block to learn the representation of the input window. Then, the representation passed through two dense layers to map from the representation to the forecast. We adopted the architecture of the most precise deterministic model, which we found in order to build the ProbCast by concatenating the noise vector to the GRU block output (i.e., representation) and extending the MLP block as shown in Figure 1. Finally, we searched for the optimal architecture of the discriminator (Figure 2) and trained the ProbCast. The discriminator concatenated X_{t+1} to the end of the input window and constructed $\{X_{t+1}, X_t, .., X_0\}$. Then it utilized a GRU block followed by two layers of MLP to inspect the consistency of this window. We used the genetic algorithm to search for the optimal architecture. We coded our method using Pytorch [26].

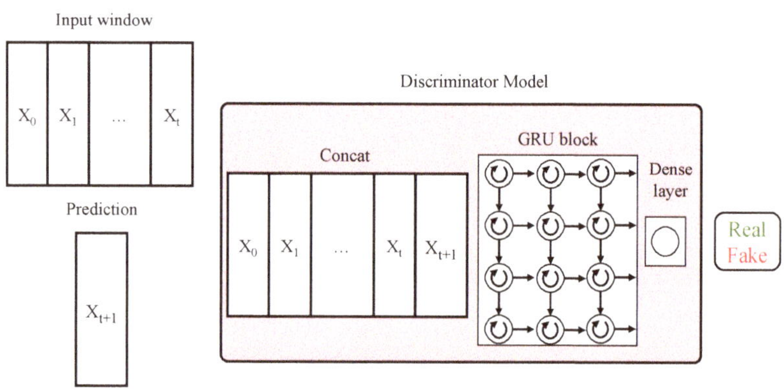

Figure 2. The discriminator architecture of our conditional GAN. The number of layers and cells in the GRU block are hyperparameters.

5.3. Evaluation Metric

To report the performance of the ProbCast, we used the negative form of the Continuous Ranked Probability Score [27] (denoted by $CRPS^*$) as the metric. The $CRPS^*$ reflects the sharpness and calibration of a probabilistic method. It is defined as follows:

$$CRPS^*(F, x) = E_F|X - x| - \frac{1}{2}E_F|X - X'|, \quad (4)$$

where X and X' are independent copies of a random variable from probabilistic forecaster F and x is the ground truth. The $CRPS^*$ provides a direct way to compare deterministic and probabilistic models. In the case of the deterministic forecaster, the $CRPS^*$ reduces to Mean Absolute Error (MAE), which is a commonly used point-wise error metric.

In other words, in a deterministic setting, the $CRPS^*$ is equivalent to MAE:

$$MAE(x, \hat{x}) = E|\hat{x} - x|, \quad (5)$$

where x is the ground truth and $\hat{x}$ is the point forecast. After the GAN training concluded, we calculated the $CRPS^*$ of the ProbCast and the deterministic model. To calculate $CRPS^*$ for ProbCast using Equation (4), we sampled it 200 times (100 times for each random variable).

6. Results and Discussion

Table 2 presents the optimal hyperparameters we found for each dataset using our framework during the experiments, and Table 3 summarizes our experiments' results presenting the $CRPS^*$ of the best deterministic model and the ProbCast for each dataset.

Table 2. List of the hyperparameters alongside their optimal values for each experiment.

Generator Hyperparameters	Electricity	Exchange-Rate
Input windows size	174	170
Noise size	303	183
Number of GRU layers	1	1
Number of GRU cells in each layer	119	119
Discriminator Hyperparameters		
Number of GRU layers	3	1
Number of GRU cells in each layer	146	149

In the experiment with the electricity dataset, the ProbCast was more accurate than the deterministic model despite having an almost identical structure. Furthermore, this experiment showed that our model can provide precise forecasts for multivariate time-series even when the number of features is substantial. In the exchange-rate experiment, the ProbCast outperforms its deterministic predecessor, again despite structural similarities. We can also observe that our method works well even though the dataset is considerably smaller in comparison to that of the previous experiment.

Furthermore, it confirms that our framework is capable of transforming a deterministic model to a probabilistic model that is more accurate than its predecessor. The question now arises: Considering the sensitivity of GAN to the architecture of its components, why does employing the deterministic model architecture to define the ProbCast work well, when it is borrowed from a totally different setup? We think that the deterministic model provides us an architecture that is capable of learning a good representation from the input time window. Since the model is trained to learn the mean of possible outcomes, these representations contain a distinctive indicator of where the target distribution is located. With the help of these indicators, the MLP block learns to accurately transform the noise vector z to the probability distribution of future values.

Table 3. The results of the experiments for the deterministic model and the ProbCast reported in $CRPS^*$.

Dataset	Deterministic Model	ProbCast
Electricity	235.96	232.00
Exchange-rate	1.04×10^{-2}	8.66×10^{-3}

7. Conclusion and Future Works

In this paper, we present ProbCast, a probabilistic model for forecasting one step ahead of a multivariate time-series. We employ the potential of conditional GAN in a learning conditional probability distribution to model the probability distribution of future values given past values, that is, $P(X_{t+1}|X_t, .., X_0)$.

Furthermore, we propose a framework to efficiently find the optimal architecture of GAN's components. This framework builds the probabilistic model upon a deterministic model to improve its performance. Hence, it enables us to search for the optimal architecture of a generator and a discriminator separately. Furthermore, it can transform an existing deterministic model into a probabilistic model with increased precision and better alignment with the real world.

We assess the performance of our method on two publicly available datasets. The exchange-rate dataset is a small dataset with few features, while the electricity dataset is bigger with a considerably larger number of features. We compare the performance of the ProbCast with its deterministic equivalent. In both experiments, our method outperforms its counterpart. The results of the experiments demonstrate that the ProbCast can learn patterns precisely from a small set of data and at the same time, it is capable of figuring out the dependencies between many features, and can forecast future values accurately in the presence of a big dataset. Furthermore, the results of the experiments indicate the successful application of our framework, which paves the way for a systematic and straightforward approach to exchanging currently used deterministic models with a probabilistic model to improve accuracy and obtain realistic forecasts.

The promising results of our experiments signify great potential for probabilistic forecasting using GANs and suggest many new frontiers for further pushing the research in this direction. For instance, we employ vanilla GAN for our research and there have been a lot of modifications suggested for improving GANs in recent years. One possible direction is to apply these modifications and inspect the improvement in the performance of the ProbCast. The other direction is experimenting with more sophisticated architectures for the generator and the discriminator. Finally, we only use the knowledge from the deterministic model to shape the generator. It would be interesting to push this direction and try to incorporate more knowledge from the deterministic model into the GAN training process to improve and optimize the probabilistic model.

Data Availability Statement: The datasets used in this study can be found at https://github.com/laiguokun/multivariate-time-series-data.

References

1. Mahalakshmi, G.; Sridevi, S.; Rajaram, S. A survey on forecasting of time series data. In Proceedings of the 2016 International Conference on Computing Technologies and Intelligent Data Engineering (ICCTIDE'16), Kovilpatti, India, 7–9 January 2016; pp. 1–8.
2. Gneiting, T.; Katzfuss, M. Probabilistic forecasting. *Annu. Rev. Stat. Its Appl.* **2014**, *1*, 125–151. [CrossRef]
3. Goodfellow, I.; Bengio, Y.; Courville, A. *Deep Learning*; MIT Press: Cambridge, MA, USA, 2016. Available online: http://www.deeplearningbook.org (accessed on 1 June 2019).
4. Esteban, C.; Hyland, S.L.; Rätsch, G. Real-valued (medical) time series generation with recurrent conditional GANs. *arXiv* **2017**, arXiv:1706.02633.
5. Golany, T.; Radinsky, K. PGANs: Personalized Generative Adversarial Networks for ECG Synthesis to Improve Patient-Specific Deep ECG Classification. In Proceedings of the AAAI Conference on Artificial Intelligence, Honolulu, HI, USA, 27 January–1 February 2019; Volume 33, pp. 557–564.
6. Haradal, S.; Hayashi, H.; Uchida, S. Biosignal data augmentation based on generative adversarial networks. In Proceedings of the 2018 40th Annual International Conference of the IEEE Engineering in Medicine and Biology Society (EMBC), Honolulu, HI, USA, 18–21 July 2018; pp. 368–371.
7. Nikolaidis, K.; Kristiansen, S.; Goebel, V.; Plagemann, T.; Liestøl, K.; Kankanhalli, M. Augmenting Physiological Time Series Data: A Case Study for Sleep Apnea Detection. *arXiv* **2019**, arXiv:1905.09068.
8. Ye, F.; Zhu, F.; Fu, Y.; Shen, B. ECG Generation With Sequence Generative Adversarial Nets Optimized by Policy Gradient. *IEEE Access* **2019**, *7*, 159369–159378. [CrossRef]
9. Wiese, M.; Bai, L.; Wood, B.; Buehler, H. Deep Hedging: Learning to Simulate Equity Option Markets. 2019. Available online: https://ssrn.com/abstract=3470756 (accessed on 1 December 2019).
10. Wiese, M.; Knobloch, R.; Korn, R.; Kretschmer, P. Quant GANs: deep generation of financial time series. *Quant. Financ.* **2020**, *20*, 1–22. [CrossRef]

11. Chen, Y.; Wang, Y.; Kirschen, D.; Zhang, B. Model-free renewable scenario generation using generative adversarial networks. *IEEE Trans. Power Syst.* **2018**, *33*, 3265–3275. [CrossRef]
12. Fekri, M.N.; Ghosh, A.M.; Grolinger, K. Generating Energy Data for Machine Learning with Recurrent Generative Adversarial Networks. *Energies* **2020**, *13*, 130. [CrossRef]
13. Zhang, C.; Kuppannagari, S.R.; Kannan, R.; Prasanna, V.K. Generative Adversarial Network for Synthetic Time Series Data Generation in Smart Grids. In Proceedings of the 2018 IEEE International Conference on Communications, Control, and Computing Technologies for Smart Grids (SmartGridComm), Aalborg, Denmark, 29–31 October 2018.
14. Yoon, J.; Jarrett, D.; van der Schaar, M. Time-series Generative Adversarial Networks. In Proceedings of the 33rd Conference on Neural Information Processing Systems (NeurIPS 2019), Vancouver, BC, Canada, 8–14 December 2019; pp. 5509–5519.
15. Ramponi, G.; Protopapas, P.; Brambilla, M.; Janssen, R. T-cgan: Conditional generative adversarial network for data augmentation in noisy time series with irregular sampling. *arXiv* **2018**, arXiv:1811.08295.
16. Koochali, A.; Schichtel, P.; Dengel, A.; Ahmed, S. Probabilistic Forecasting of Sensory Data with Generative Adversarial Networks–ForGAN. *IEEE Access* **2019**, *7*, 63868–63880. [CrossRef]
17. Zhang, K.; Zhong, G.; Dong, J.; Wang, S.; Wang, Y. Stock market prediction based on generative adversarial network. *Procedia Comput. Sci.* **2019**, *147*, 400–406. [CrossRef]
18. Zhou, X.; Pan, Z.; Hu, G.; Tang, S.; Zhao, C. Stock market prediction on high-frequency data using generative adversarial nets. *Math. Probl. Eng.* **2018**. [CrossRef]
19. Lin, Y.; Dai, X.; Li, L.; Wang, F.Y. Pattern sensitive prediction of traffic flow based on generative adversarial framework. *IEEE Trans. Intell. Transp. Syst.* **2018**, *20*, 2395–2400. [CrossRef]
20. Kabir, H.M.D.; Khosravi, A.; Nahavandi, S.; Kavousi-Fard, A. Partial Adversarial Training for Neural Network-Based Uncertainty Quantification. *IEEE Trans. Emerg. Top. Comput. Intell.* **2019**, 1–12. [CrossRef]
21. Hothorn, T.; Kneib, T.; Bühlmann, P. Conditional transformation models. *J. R. Stat. Soc. Ser. B (Stat. Methodol.)* **2014**, *76*, 3–27. [CrossRef]
22. Goodfellow, I.; Pouget-Abadie, J.; Mirza, M.; Xu, B.; Warde-Farley, D.; Ozair, S.; Courville, A.; Bengio, Y. Generative adversarial nets. In Proceedings of the 28th Annual Conference on Neural Information Processing Systems, Montreal, QC, Canada, 8–13 December 2014; pp. 2672–2680.
23. Mirza, M.; Osindero, S. Conditional generative adversarial nets. *arXiv* **2014**, arXiv:1411.1784.
24. Lai, G.; Chang, W.C.; Yang, Y.; Liu, H. Modeling long-and short-term temporal patterns with deep neural networks. In Proceedings of the 41st International ACM SIGIR Conference on Research & Development in Information Retrieval, Ann Arbor, MI, USA, 8–12 July 2018; pp. 95–104.
25. Cho, K.; Van Merriënboer, B.; Gulcehre, C.; Bahdanau, D.; Bougares, F.; Schwenk, H.; Bengio, Y. Learning phrase representations using RNN encoder-decoder for statistical machine translation. *arXiv* **2014**, arXiv:1406.1078.
26. Paszke, A.; Gross, S.; Chintala, S.; Chanan, G.; Yang, E.; DeVito, Z.; Lin, Z.; Desmaison, A.; Antiga, L.; Lerer, A. Automatic Differentiation in PyTorch. In Proceedings of the 31st Conference on Neural Information Processing Systems (NIPS 2017), Long Beach, CA, USA, 4–9 December 2017.
27. Gneiting, T.; Raftery, A.E. Strictly proper scoring rules, prediction, and estimation. *J. Am. Stat. Assoc.* **2007**, *102*, 359–378. [CrossRef]

engineering proceedings

MDPI

Proceedings

Unemployment and COVID-19 Impact in Greece: A Vector Autoregression (VAR) Data Analysis †

Christos Katris

Department of Accounting and Finance, Athens University of Economics and Business, 76, Patission Street, GR-10434 Athens, Greece; chriskatris@aueb.gr

† Presented at the 7th International Conference on Time Series and Forecasting, Gran Canaria, Spain, 19–21 July 2021.

Abstract: In this paper, the scope is to study whether and how the COVID-19 situation affected the unemployment rate in Greece. To achieve this, a vector autoregression (VAR) model is employed and data analysis is carried out. Another interesting question is whether the situation affected more heavily female and the youth unemployment (under 25 years old) compared to the overall unemployment. To predict the future impact of COVID-19 on these variables, we used the Impulse Response function. Furthermore, there is taking place a comparison of the impact of the pandemic with the other European countries for overall, female, and youth unemployment rates. Finally, the forecasting ability of such a model is compared with ARIMA and ANN univariate models.

Keywords: unemployment; COVID-19; Greece; VAR; Impulse Response function; forecasting

1. Introduction

The scope of this paper is to examine the impact of COVID-19 on the Greece unemployment rate. To achieve this goal, we designed and implemented an econometric analysis. It is a quite common debate whether and to what scale the pandemic, will cause unemployment problems to society. This is an attempt to answer this question using econometric analysis. Another major question is whether female unemployment will be impacted further than that of males and another interesting question is if this situation affects more the youth (under the age of 25) in terms of unemployment. Such answers uncover whether these specific groups (i.e., women and young people) are more vulnerable to a pandemic situation. The answers to the above questions are particularly important in terms of the designing of economic policies.

The country of interest of this study is Greece. Thus, is important to investigate the impact of COVID-19 on Greece in comparison with other countries and in this study we considered the European Union of 27 countries (EU27). We can extract some conclusions about how much time should we expect the impact of COVID-19 to the unemployment rate of the country to last. The comparison with the EU27 allows the direct comparison of the impact in terms of time. Whether the impact of this situation is similar, then the same policies are expected to be effective both for Greece and the EU27. The impact of the pandemic on female and youth unemployment unveils increased vulnerabilities for these specific groups and the economic measures should be directed more to them in order to gain increased efficiency, i.e., smaller effects of the COVID-19 to the unemployment rate. A final question under examination is the forecasting ability of such an approach (VAR model) compared to some other approaches. The target of this question is to answer if this approach is suitable for both impact measuring (and maybe for deciding forecasting horizon) and for forecasting or some other approach should be used for forecasting purposes. The core of this econometric analysis is the Vector Autoregression (VAR) model. The unemployment rate is expressed in monthly data and the COVID-19 cases in daily terms. To create a time series of equal length for the unemployment series, we use interpolation while for

Citation: Katris, C. Unemployment and COVID-19 Impact in Greece: A Vector Autoregression (VAR) Data Analysis . *Eng. Proc.* **2021**, *5*, 41. https://doi.org/10.3390/engproc2021005041

Published: 8 July 2021

the COVID-19 series, we considered the number of new cases per five days. An essential feature of this model is the Impulse response function which allows for observation of the future impact of the situation per unit (in this analysis per 5 days).

There are already some attempts with the aim to describe and explain the impact of COVID to the dynamics of macroeconomic variables. Examples are these of [1] in which the author studies the social and economic responses to the COVID-19 pandemic in a large sample of countries, of [2] in which the authors study the influences of the COVID-19 pandemic on unemployment in five selected European economies and of [3] in which the author investigate the impact of globalization to the speed of initial transmission and on the scale of initial infections to a country. Moreover, there are mentioned some additional relevant studies whose analyses share the common characteristic of the usage of VAR models. These studies include [4] in which the author study the impact of fear sentiment caused by the coronavirus pandemic on Bitcoin price dynamics using Google search queries, ref. [5] in which the author investigates the impact of COVID-19 in the stock market (specifically in Dow Jones and S&P 500 returns), ref. [6] in which the authors consider several indicators of economic uncertainty for the US and the UK before and during the COVID-19 situation and study the impact of the pandemic to these indicators, ref. [7] in which the authors study the assumptions which are needed for forecasting of the evolution of the U.S. economy following the outbreak of COVID-19, ref. [8] in which the author study the effect of the virus outbreak on the economic output of New York state. There are also several papers about the impact of macroeconomic variables to unemployment using VAR models such as the following: ref. [9] in which the authors study the influence of Foreign Direct Investment on Unemployment, ref. [10] in which the author analyzes the dynamic effects of different macroeconomic shocks on unemployment in Germany, ref. [11] in which the authors use Bayesian SVAR models to analyze the role of oil price movements in the evolution of unemployment in the UK, ref. [12] in which the author uses a Structural VAR (SVAR) approach to study the effects of shocks to the Austrian unemployment, ref. [13] in which the authors review the main causes of Spanish unemployment using the structural VAR methodology [14] in which the author uses a bivariate VAR model with to describe output–unemployment dynamics. Attempts which are related to forecasting are that of [15] who use three time series methods to forecast the Swedish unemployment rate, and a recent attempt for forecast youth unemployment in Italy in the aftermath of the COVID-19, using an artificial neural network (ANN) model, in [16].

The scope of this paper is the exploration of impact of COVID-19 in the unemployment in Greece and the comparison with the rest EU countries overall, for females and for young people. This is performed through the fitting of Vector Autoregression (VAR) models. Finally, we studied the contribution of such model to the forecasting ability for the unemployment rate. Specifically, there is an attempt to answer the following question: is the usage of such model for forecasting purposes a suitable approach or is preferable the usage of some other approach? Some important conclusions could be derived from such an analysis. The rest of the manuscript is organized as follows: in Section 2 we analyzed the VAR model, in Section 3 we discuss the Impulse Response function, in Section 4 we discuss the forecasting ability of VAR model and the detection of a suitable forecasting approach, in Section 5 we perform the data analysis and Section 6 contains the conclusions of the paper.

2. VAR Model

The Vector Autoregression model is a statistical model which describes the evolution of multivariate linear time series with k endogenous variables. The evolution of these endogenous variables in the system is considered not only as function of their own history, but as a function of the lagged values of all endogenous variables. In essence, this model is a generalization of ARIMA models for univariate time series. This is the simplest and most used model for multivariate time series forecasting.

The VAR model introduced in [17] where the author explains the usefulness of VAR models and show their use through applications. All variables in this approach are endogenous and are functions of the lagged values of all the considered variables. A brief review of the illustration of such a model follows.

In terms of characterization the order of the model, i.e., the number of previous periods that the model will use, has crucial role. For example a VAR(3) model is a model where each variable is linear combination of the last three periods (lags) of all the variables of the system.

The general form of a VAR(p) model with k variables and p lags in terms of a matrix follows:

$$\begin{bmatrix} y_{1,t} \\ y_{2,t} \\ \vdots \\ y_{k,t} \end{bmatrix} = \begin{bmatrix} c_1 \\ c_2 \\ \vdots \\ c_k \end{bmatrix} + \begin{bmatrix} a_{1,1} & a_{1,2} & \cdots & a_{1,k} \\ a_{2,1} & a_{2,2} & \cdots & a_{2,k} \\ \vdots & \vdots & \cdots & \vdots \\ a_{k,1} & a_{k,2} & \cdots & a_{k,k} \end{bmatrix} \begin{bmatrix} y_{1,t-1} \\ y_{2,t-1} \\ \vdots \\ y_{k,t-1} \end{bmatrix} + \cdots + \begin{bmatrix} a_{1,1} & a_{1,2} & \cdots & a_{1,k} \\ a_{2,1} & a_{2,2} & \cdots & a_{2,k} \\ \vdots & \vdots & \cdots & \vdots \\ a_{k,1} & a_{k,2} & \cdots & a_{k,k} \end{bmatrix} \begin{bmatrix} y_{1,t-p} \\ y_{2,t-p} \\ \vdots \\ y_{k,t-p} \end{bmatrix} + \begin{bmatrix} e_{1,t} \\ e_{2,t} \\ \vdots \\ e_{k,t} \end{bmatrix} \quad (1)$$

or simpler $Y_t = c + \sum_{i=1}^{p} A_i Y_i + e$

where each Y_i represents a vector of length k and each A_i is a $k \times k$ matrix. The vector of residuals (e) has expected value of zero and the error terms ($e_{i,t}$) are not autocorrelated. The validity of the previous properties offers consistent and efficient estimators through the method of Least Squares (LS).

The interpretation of VAR models is especially important. We should be careful that these models do not allow us to extract any inference about causality between the variables (only Granger causality could be examined, i.e., if a time-series contribute to the prediction of another time-series, this property is obviously much weaker compared to the normal causality). On the other hand, VAR models allow interpretations about the dynamic relationship between variables of the system. Detailed information about VAR models can be found in [18].

3. Impulse Response Function

Impulse response functions are used frequently in macroeconomic modeling to describe how the economy reacts over time to economic shocks, which are considered to be exogenous impulses. These functions are often used in the context of a VAR model. Additionally, these functions describe the reaction of endogenous macroeconomic variables to the economic shock (of one or more standard deviations) both at the time of the shock and in future points in time. In other words, the major purpose is the description of the evolution of a variable in the model when it reacts to a shock in other variables of the system, and this makes them a very useful tool for policy makers for assessing alternative economic policies.

The idea of the impulse response is that we look at the adjustment of the endogenous variables over time, after a hypothetical shock in t, and we compare this adjustment with the time series process without the shock, i.e., the actual process. The impulse response sequences plot this difference. The impulse response function is obtaining through the consideration of the moving average (MA) representation for a linear VAR model. The discrepancy between the expected value of the variable with and without the considered shock is the forecast error impulse response (FEIR) function. The FEIR function for the ith period after the shock is expressed as

$$\Phi_i = \sum_{j=1}^{i} \Phi_{i-j} A_j \quad (2)$$

where $\Phi_0 = I_k$ and $A_j = 0$ for $j > p$, k is the number of exogenous variables and p is the lag order of the VAR model.

In this work, orthogonal impulse responses are used. The reason for the use of such functions is that we assume that the other impulse remains constant, i.e., to isolate a concurrent effect to the variable which is arising solely because of an impulse in the same equation. The basic idea is that the variance-covariance matrix (Σ) is decomposed (usually with a Choleski decomposition) in a way that $\Sigma = PP'$, where P is a lower triangular matrix with positive diagonal elements.

Additional and detailed information about impulse response functions one can find in [18]. In [19] is discussed the identification of shocks for studying specific economic problems. Moreover, have been suggested asymmetric impulse response functions that separate the impact of a positive shock from a negative one in [20].

4. Forecasting Using the VAR Approach

The VAR model is certainly useful for studying the impact of COVID-19 cases on unemployment. One question is whether this approach is useful for forecasting. The answer is not straightforward, in the sense that accurate forecasting is a different task than studying the impact of a factor. Could a VAR model be used for both of these tasks effectively, or could the consideration of an alternative model for forecasting be advantageous in terms of forecasting accuracy? This question is explored in this section of this paper.

As benchmark model for unemployment forecasting is considered the plain ARIMA model (i.e., using only past values of the series). The other considered forecasting approaches are: the ARIMA model using COVID-19 cases as external regressor (the lags are decided from the corresponding VAR model), Feed-Forward Artificial Neural Networks (ANN) and Feed-Forward Artificial Neural Networks (ANN) using the COVID-19 cases as external input (the lags are decided from the corresponding VAR model). Two questions are explored here, with the aim to specify a suitable model: (i) whether the insertion of COVID-19 cases can improve a forecasting approach and (ii) whether a machine learning approach (in our case ANN) can offer additional forecasting accuracy.

The details of our analysis are as follows. The training set are 60 observations and the test set 30 observations. The forecasting task in this study is as follows: models are fitted using the first 60 observations. In every step, the models are refitted with all the available data up to this point. The forecasts are 1-step ahead and finally, there are 30 forecasts with each model which are compared with the actual values of the test set using the Root Mean Squared Percentage Error (RMSPE) and the Mean Absolute Percentage Error (MAPE) multiplied by 1000 for easier direct comparison of the models. We consider that we have N observations with y_i the actual values of the time-series and $\hat{y}_i$ the forecasts of the values of the time series, then the formulas for the RMSPE and MAPE are following:

$$RMSPE = \frac{1}{N} \cdot \sqrt{\sum_{i=1}^{N} \left(\frac{y_i - \hat{y}_i}{y_i} \right)^2} \times 100\%, \; MAPE = \frac{1}{N} \cdot \sum_{i=1}^{N} \frac{|y_i - \hat{y}_i|}{y_i} \times 100\%.$$

A basic question, which we try to answer in this work, is whether such a model is the suitable approach for forecasting or should be accompanied by a model for gaining additional forecasting accuracy. In this study, the considered approaches are the ARIMA model (1,0,0) from the univariate time series domain and the Feed-forward Artificial Neural Networks (ANN) with one (1) hidden layer and with one (1) to ten (10) nodes in this layer (the model which gives the lowest RMSPE is considered). Using both frameworks, we consider the insertion of COVID-19 cases (as external regressor in the ARIMA model and as a second input in ANN model) and study if such an insertion offers additional accuracy. Details about ARIMA models can be found on [21] and details about Artificial Neural Nets can be found in [22].

5. Data Analysis

The source of the data used in this study is Eurostat. The reported data describe monthly unemployment by sex and age: https://ec.europa.eu/eurostat/web/products-datasets/-/une_rt_m (accessed on 6 March 2021). On the other hand, the COVID-19 data are freely available from the European Centre for Disease Prevention and Control and are downloaded from the https://ourworldindata.org/coronavirus site (download the complete database) (accessed on 6 March 2021). The analysis in this work is performed through the statistical software R. Specifically, the following packages are used: *moments* [23] for calculating skewness, kurtosis and for performing Jarque-Bera test, *vars* [24–26]) for VAR model estimation and prediction, *forecast* [27,28] for ARIMA models, *AMORE* ([29]) for the Feed-Forward Fully Completed Artificial Neural Network Models and *MLmetrics* [30] for the computation of RMSPE and MAPE metrics.

5.1. Data Overview

The cases of COVID-19 are considered. The data cover the period from November 2019 until 31 January of 2021. For both Greece and EU, the considered unemployment data are from November 2019 until January of 2021. To achieve equal length of the datasets, we consider five days as a time point (each month has six time points—day 5, day 10, day 15, day 20, day 25, end of month, i.e., day 30 or 31 or 29 for February 2020). For the COVID-19 data, we consider the number of new cases at the end of the five days as a single point, while for the unemployment series we consider constant interpolation to fill the gaps (until the start of November 2020 for Greece and until the start of January 2021 for EU). For the additional time points until the finish of January 2021, the data for the unemployment series are filled using exponential smoothing models. The analysis performed in R package forecast. The mean value, the standard deviation (SD), the Coefficient of variation (CV) and the Jarque-Bera test for normality (statistic and p-value in parenthesis) of the data are shown in Table 1.

Table 1. Statistical Characteristics of the Data.

Variable	Mean	SD	Skewness	Kurtosis	CV	Normality Test (Jarque-Bera)
Total Unemployment Greece	16.47	0.55	0.732	4.258	0.033	13.980 (0.001)
Total Unemployment EU27	7.07	0.47	0.050	1.443	0.066	9.133 (0.010)
Female Unemployment Greece	20.03	0.72	−0.857	3.112	0.036	11.080 (0.004)
Female Unemployment EU27	7.45	0.55	0.091	1.446	0.074	9.179 (0.010)
Youth Unemployment Greece	34.77	2.33	0.859	2.591	0.067	11.701 (0.003)
Youth Unemployment EU27	16.37	1.19	−0.121	1.617	0.072	7.396 (0.025)
COVID-19 New Cases Greece	2.08	1.42	−0.357	1.766	0.683	7.621 (0.022)
COVID-19 New Cases EU27	3.85	2.19	−0.937	2.233	0.570	15.382 (0.01)

Greece displays unemployment over EU27 countries for all categories (Total, Female and Under 25 years old—"Youth"). The most severe case—for both Greece and EU27 countries—seems to be the youth unemployment because it is on a higher level according to mean value and is more dispersed according to SD. Normality can be rejected at any case for Greece's unemployment, while for the EU27 unemployment series cannot be rejected at 0.01 level. With the aim to perform our analysis to find the impact of COVID-19 new cases on unemployment, the data are transformed to natural log values (COVID-19 new cases are transformed to natural log (values+1) because there are zeros in the sample). We observe mainly that Greece unemployment display higher kurtosis than EU27 for all types of unemployment and the rejection of Normality at the 0.01 level only for all types of Greece unemployment.

5.2. Unemployment Analysis

5.2.1. Overall Unemployment

The first scope of this paper is to explore the effects of COVID-19 to the overall unemployment of Greece and to compare this impact to the rest EU27 countries (EU27). To achieve this, a VAR model is applied for two variables, i.e., unemployment and COVID-19 new cases both for Greece and for EU27. The lags are decided through the BIC criterion or is selected the model with the minimum lags which leads to no autocorrelated or heteroskedastic residuals. The residuals of the models are checked for autocorrelation with the Pertmanteau Multivariate test, for heteroskedasticity using the ARCH-LM test, for normality with the Jarque-Bera test and for stationarity with the ADF test. Additionally, we observe in a CUSUM graph if there is evidence of the existence of a structural break. Table 2 displays the results of the fitting for these models.

Table 2. VAR Models for Total Unemployment (Y_t) and COVID-19 Cases [1].

Coefficients	Greece	EU27
Constant	0.2372	0.0733
Y_{t-1}	0.9528 [2]	0.9608 [2]
COVID-19 $cases_{t-1}$	−0.0090	0.0031
Y_{t-2}	−0.0047	
COVID-19 $cases_{t-2}$	0.0321	
Y_{t-3}	−0.0336	
COVID-19 $cases_{t-3}$	−0.0211	
Model Evaluation		
Multiple R^2	0.8499	0.9757
Adjusted R^2	0.8317	0.9751

[1] Selection with Schwarz criterion gave the model with one lag for both Greece and EU27, respectively. For Greece we choose three lags to avoid heteroskedasticity of the residuals. [2] Significant at 0.05 level.

For both Greece and the EU27, autocorrelation and no heteroskedasticity of residuals can be assumed at 5%, while there is no graphical evidence of the existence of a structural break. Additionally, the residuals of the regression with Unemployment Rates as (y-variables) can be assumed stationary at 1%. Finally, the assumption of normality of the residuals is rejected at almost every level of significance which led to the use of bootstrap both for the construction of confidence intervals for Impulse Responses and for the calculation of the p-value for Granger causalities (the p-values of the tests are calculated by considering 10,000 bootstrap replicates). Using Granger causalities, only the EU27 shows that it can be assumed at 10% that COVID-19 cases cause Granger Unemployment.

With the aim to directly compare Greece with the EU27 case, we construct a table which shows the cumulative impulse response in terms of Unemployment Rates for the next seven months. These results are shown in Table 3. The impact of COVID-19 cases is expected to raise unemployment more in EU27 countries than in Greece. This situation can be considered less in Greece as a factor of deterioration in unemployment.

Table 3. Cumulative Impulse Response of COVID-19 cases to Unemployment Rate.

Months Ahead	Greece	EU27
1	0.00772	0.00896
2	0.01962	0.02838
3	0.02986	0.05327
4	0.03820	0.08037
5	0.04478	0.10757
6	0.04986	0.13360
7	0.05372	0.15773

5.2.2. Female and Youth Unemployment

An additional aim is the exploration of the effect of COVID-19 to the female and to the youth unemployment in Greece and EU27. Again, VAR models are fitted for COVID-19 cases and unemployment of these specific groups and the lags are decided through the BIC criterion. For female unemployment, Table 4 displays the results of the fitting and Table 5 the cumulative impulse responses for Greece and EU27 respectively. For youth unemployment the results are shown in Table 6 and the cumulative impulse responses for Greece and EU27 respectively in Table 7. Again, the lags are decided through the BIC criterion or the model with the minimum lags is selected, which leads to no autocorrelated or heteroskedastic residuals and the residuals of the models are checked for autocorrelation with Pertmanteau Multivariate test, for heteroskedasticity using the ARCH-LM test, for normality with Jarque-Bera test and for stationarity with ADF test, and we observe whether there is evidence of the existence of a structural break in a CUSUM graph.

Table 4. VAR Models for Female Unemployment (Y_t) and COVID-19 Cases [1].

Coefficients	Greece	EU27
Constant	0.2731 [2]	0.0690
Y_{t-1}	0.9076 [2]	0.9643 [2]
COVID-19 $cases_{t-1}$	0.0037	0.0028
	Model Evaluation	
Multiple R^2	0.8471	0.9716
Adjusted R^2	0.8436	0.9710

[1] Selection with Schwarz (SC) criterion gave the model with one lag for Greece and EU27. [2] Significant at 0.05 level.

Table 5. Cumulative Impulse Response of COVID-19 cases to Female Unemployment Rate.

Months Ahead	Greece	EU27
1	0.00555	0.00800
2	0.01665	0.02544
3	0.02956	0.04795
4	0.04217	0.07265
5	0.05340	0.09768
6	0.06283	0.12187
7	0.07042	0.14454

Table 6. VAR Models for Youth Unemployment (Y_t) and COVID-19 Cases [1].

Coefficients	Greece	EU27
Constant	0.8555 [2]	0.2785 [2]
Y_{t-1}	0.9455 [2]	0.8837 [2]
COVID-19 $cases_{t-1}$	−0.0495	−0.0038
Y_{t-2}	−0.0867	−0.0004
COVID-19 $cases_{t-2}$	0.1787 [2]	−0.0004
Y_{t-3}	0.0210	−0.0007
COVID-19 $cases_{t-3}$	−0.2024 [2]	0.0007
Y_{t-4}	−0.0123	0.0010
COVID-19 $cases_{t-4}$	0.0759	0.0017
Y_{t-5}	−0.0044	0.0001
COVID-19 $cases_{t-5}$	−0.0293	0.0020
Y_{t-6}	−0.1057	0.4826 [2]
COVID-19 $cases_{t-6}$	0.0320	0.0022
Y_{t-7}		−0.4661 [2]
COVID-19 $cases_{t-7}$		0.0054
	Model Evaluation	
Multiple R^2	0.7650	0.9768
Adjusted R^2	0.7253	0.9721

[1] Selection with Schwarz (SC) criterion gave the model with 6 and 7 lags for Greece and EU27 respectively, to avoid autocorrelation of the residuals. [2] Significant at 0.05 level.

Table 7. Cumulative Impulse Response of COVID-19 cases to Youth Unemployment Rate.

Months Ahead	Greece	EU27
1	0.00177	−0.00556
2	0.00327	0.00731
3	0.02219	0.04319
4	0.03688	0.08766
5	0.04438	0.12623
6	0.05107	0.15274
7	0.05696	0.16842

For female unemployment, both for Greece and the EU27, autocorrelation (for Greece, the residuals cannot be considered autocorrelated at 1% level) and no heteroskedasticity of residuals can be assumed at 5% level, while there is no graphical evidence of the existence of a structural break and the residuals of the regression with Unemployment Rates as (y-variables) can be assumed stationary at 0.05 level. However, normality of residuals is rejected at almost any level of statistical significance. The same applies for youth unemployment. The rejection of normality of the residuals again lead to the use of bootstrap for the construction of confidence intervals for Impulse Responses.

With the aim to directly compare the Greece with EU27 case, we construct a table which show the cumulative impulse response in terms of Unemployment Rates for the next seven months. These results are shown to Table 5. The impact of COVID-19 to both Greece and EU27 is not observable the first month, but in the end of the seventh month the affection is clear and female unemployment is expected to rise more in EU27 countries than in Greece.

What follows are the results for youth unemployment. Table 6 displays the fitted VAR model. With the aim to directly compare the Greece with EU27 case, we construct a table which show the cumulative impulse response in terms of Unemployment Rates for the next seven months. These results are shown in Table 7. To sum up, Table 8 displays the analysis which presents the results for the cumulative impact of COVID-19 cases both for Greece and EU27. All categories of unemployment are expected to be affected positively from the pandemic. According to the type of unemployment, young people are expected to experience a higher increase of their unemployment, while females are expected to be affected less than the overall population in the EU27 countries and to be affected more heavily than the other categories of unemployment in Greece. According to the country, Greece is expected to be affected less than the EU27 countries for all types of unemployment. Probable reasons are maybe structural, and we point out that the values of Unemployment rates in Greece are already in higher level than the EU27 countries. This fact leads to two main remarks: first that unemployment is expected to rise in all cases due to the COVID-19 situation and the average EU27 country is expected to be affected more than Greece in terms of unemployment rates. This is maybe a sign that it is more urgent for Greece to solve structural problems, while for the average EU27 country it seems more urgent to take measures to protect its economy from this situation. Secondly, female unemployment in Greece and the unemployment of young people in EU27 countries are expected to be affected more heavily by COVID-19, which indicates that the policies should have a different focus, to alleviate from the consequences.

Table 8. Cumulative impact (seven months ahead) of COVID-19 cases to different types of Unemployment.

Type of Unemployment	Greece	EU27
Overall	0.0538	0.1577
Female	0.0704	0.1445
Youth	0.0570	0.1684

5.3. Forecasting the Unemployment Rates

This analysis closes the paper and answers the question whether such a VAR model is better for forecasting purposes of unemployment or whether other approaches should be considered because they could achieve more accurate results. The results are displayed in Table 9 which displays the RMSPE values (multiplied by 1000) and the MAPE values inside parenthesis (multiplied by 1000). To decide about the suitability of the model, we use as alternatives for forecasting, the following approaches: the plain ARIMA model (as benchmark), the ARIMA model with COVID-19 cases as external regressor, a Feed-Forward Multivariate Artificial Neural Network (ANN) based solely on previous cases of the unemployment and the same model but additionally with observations of COVID-19 cases. The models are compared in terms of RMSPE and MAPE. Sixty (60) observations are used for training of the models and thirty (30) for testing the forecasting ability of the models.

Table 9. Forecasting Unemployment—Comparison of Approaches.

Model	Greece	EU27
Overall Unemployment		
VAR	0.618 (0.553)	0.609 (0.572)
ARIMA (benchmark model)	0.437 (0.398)	0.425 (0.420)
ARIMA (with COVID-19 cases as external regressor)	1.220 (1.213)	0.299 (0.181)
ANN: (1,1,1), (1,10,1)/(1,10,1), (1,10,1)	2.026 (0.941)	5.951 (4.931)
ANN (with COVID-19 cases as input): (2,3,1), (2,3,1)/(2,9,1), (2,9,1)	3.108 (2.299)	5.954 (4.860)
Female Unemployment		
VAR	0.439 (0.377)	0.684 (0.653)
ARIMA (benchmark model)	0.295 (0.206)	0.547 (0.537)
ARIMA (with COVID-19 cases as external regressor)	0.283 (0.197)	0.414 (0.262)
ANN (1,1,1), (1,4,1)/(1,10,1), (1,10,1)	1.373 (0.981)	6.307 (5.334)
ANN (with COVID-19 cases as input): (1,8,1), (1,8,1)/(1,9,1), (1,9,1)	3.606 (3.133)	6.301 (5.268)
Youth Unemployment		
VAR	2.328 (1.916)	0.529 (0.400)
ARIMA (benchmark model)	1.902 (1.709)	0.350 (0.346)
ARIMA (with COVID-19 cases as external regressor)	2.131 (1.842)	0.348 (0.344)
ANN (1,1,1), (1,1,1)/(1,10,1), (1,10,1)	7.362 (3.165)	4.775 (3.839)
ANN (with COVID-19 cases as input): (1,1,1), (1,1,1)/(1,9,1), (1,9,1)	9.456 (6.432)	5.517 (4.944)

The main conclusions from Table 9 are as follows. First, the VAR model is not the best approach for forecasting for the EU27 nor for Greece, for all the considered subcategories of unemployment. Next, the ANN approach displays lower performance than VAR and ARIMA models. Finally, under the ARIMA framework, the insertion of COVID-19 cases improves the forecasting only for the case of EU27 countries and not in the case of Greece (expected due to the Granger causality).

6. Conclusions

In this work, we constructed and fitted Vector Autoregressive (VAR) models with the aim to explore the impact of COVID-19 cases on Greece's general unemployment and on two more sensitive cases, i.e., Female and the Youth unemployment. Furthermore, the forecasting ability of the VAR model is found to be limited and other univariate approaches appear as preferable. A strategy is to use the VAR model to explore effects of shocks, while it seems advantageous the use of other approaches for forecasting purposes. Additionally, there is evidence that COVID-19 cases Granger cause the overall unemployment rates only for the EU27 countries (the non-causality cannot be rejected at the 0.1 level). Additionally, a shock in COVID-19 cases in Greece will have a lower impact in all considered types of unemployment. For all unemployment types (overall, female and youth) the effect of COVID-19 cases is expected to be lower for Greece compared to the EU27 countries. However, the impact does not appear to stop after seven months for all types of unemployment. In terms of forecasting, a suggestion is that the VAR model can be used to investigate the impact of a shock and should be accompanied by an ARIMA model for forecasting purposes.

Data Availability Statement: The overview of the data is analyzed in Section 5.1.

Conflicts of Interest: The author declares no conflict of interest.

References

1. Milani, F. COVID-19 outbreak, social response, and early economic effects: A global VAR analysis of cross-country interdependencies. *J. Popul. Econ.* **2021**, *34*, 223–252. [CrossRef] [PubMed]
2. Su, C.W.; Dai, K.; Ullah, S.; Andlib, Z. COVID-19 pandemic and unemployment dynamics in European economies. *Econ. Res. Ekon. Istraživanja* **2021**, 1–13. [CrossRef]
3. Zimmermann, K.F.; Karabulut, G.; Bilgin, M.H.; Doker, A.C. Inter-country distancing, globalisation and the coronavirus pandemic. *World Econ.* **2020**, *43*, 1484–1498. [CrossRef] [PubMed]
4. Chen, C.; Liu, L.; Zhao, N. Fear sentiment, uncertainty, and bitcoin price dynamics: The case of COVID-19. *Emerg. Mark. Financ. Trade* **2020**, *56*, 2298–2309. [CrossRef]
5. Onali, E. COVID-19 and Stock Market Volatility. 2020. Available online: https://ssrn.com/abstract=3571453 (accessed on 6 March 2021). [CrossRef]
6. Altig, D.; Baker, S.; Barrero, J.M.; Bloom, N.; Bunn, P.; Chen, S.; Davis, S.J.; Leather, J.; Meyer, B.; Mihaylov, E. Economic uncertainty before and during the COVID-19 pandemic. *J. Public Econ.* **2020**, *191*, 104274. [CrossRef] [PubMed]
7. Primiceri, G.E.; Tambalotti, A. *Macroeconomic Forecasting in the Time of COVID-19*; Manuscript; Northwestern University: Evanston, IL, USA, 2020.
8. Gharehgozli, O.; Nayebvali, P.; Gharehgozli, A.; Zamanian, Z. Impact of COVID-19 on the Economic Output of the US Outbreak's Epicenter. *Econ. Disasters Clim. Chang.* **2020**, *4*, 561–573. [CrossRef] [PubMed]
9. Balcerzak, A.P.; Żurek, M. Foreign Direct Investment and Unemployment: VAR Analysis for Poland in the Years 1995–2009. *Eur. Res. Stud.* **2011**, *14*, 3–14.
10. Linzert, T. Sources of German Unemployment: Evidence from a Structural VAR Model. ZEW Discussion Paper No.01-41. 2001. Available online: https://ssrn.com/abstract=358349 (accessed on 6 March 2021).
11. Cuestas, J.C.; Ordóñez, J. Oil prices and unemployment in the UK before and after the crisis: A Bayesian VAR approach. A note. *Phys. A Stat. Mech. Its Appl.* **2018**, *510*, 200–207. [CrossRef]
12. Maidorn, S. The effects of shocks on the Austrian unemployment rate—A structural VAR approach. *Empir. Econ.* **2003**, *28*, 387–402. [CrossRef]
13. Dolado, J.J.; Jimeno, J.F. The causes of Spanish unemployment: A structural VAR approach. *Eur. Econ. Rev.* **1997**, *41*, 1281–1307. [CrossRef]
14. Evans, G.W. Output and unemployment dynamics in the United States: 1950–1985. *J. Appl. Econ.* **1989**, *4*, 213–237. [CrossRef]
15. Edlund, P.O.; Karlsson, S. Forecasting the Swedish unemployment rate VAR vs. transfer function modelling. *Int. J. Forecast.* **1993**, *9*, 61–76. [CrossRef]
16. Fenga, L.; Son-Turan, S. *Forecasting Youth Unemployment in the Aftermath of the COVID-19 Pandemic: The Italian Case*; Research Square: Durham, NC, USA, 2020.
17. Sims, C.A. Macroeconomics and reality. *Econ. J. Econ. Soc.* **1980**, *48*, 1–48. [CrossRef]
18. Lütkepohl, H. *New Introduction to Multiple Time Series Analysis*; Springer Science & Business Media: Berlin, Germany, 2005.
19. Lütkepohl, H. Impulse Response Function. In *The New Palgrave Dictionary of Economics*; Macmillan Publishers Ltd., Ed.; Palgrave Macmillan: London, UK, 2018.
20. Hatemi-J, A. Asymmetric generalized impulse responses with an application in finance. *Econ. Model.* **2014**, *36*, 18–22. [CrossRef]
21. Hyndman, R.J.; Athanasopoulos, G. *Forecasting: Principles and Practice*, 2nd ed.; OTexts: Melbourne, Australia, 2018; Available online: https://otexts.com/fpp2/ (accessed on 6 March 2021).
22. Lippmann, R. An introduction to computing with neural nets. *IEEE ASSP Mag.* **1987**, *4*, 4–22. [CrossRef]
23. Komsta, L.; Novomestky, F. Moments: Moments, Cumulants, Skewness, Kurtosis and Related Tests. R Package Version 0.14. 2015. Available online: https://CRAN.R-project.org/package=moments (accessed on 6 March 2021).
24. Pfaff, B. VAR, SVAR and SVEC Models: Implementation within R Package Vars. *J. Stat. Softw.* **2008**, *27*. Available online: http://www.jstatsoft.org/v27/i04/ (accessed on 6 March 2021). [CrossRef]

25. Pfaff, B. *Analysis of Integrated and Cointegrated Time Series with R*, 2nd ed.; Springer: New York, NY, USA, 2008; ISBN 0-387-27960-1.
26. Pfaff, M.B.; Stigler, M. Package 'Vars'. 2018. Available online: https://cran.r-project.org/web/packages/vars/vars.pdf (accessed on 6 March 2021).
27. Hyndman, R.J.; Athanasopoulos, G.; Bergmeir, C.; Caceres, G.; Chhay, L.; O'Hara-Wild, M.; Petropoulos, F.; Razbash, S.; Wang, E.; Yasmeen, F. forecast: Forecasting Functions for Time Series and Linear Models. R Package Version 8.13. Available online: https://pkg.robjhyndman.com/forecast/ (accessed on 6 March 2021).
28. Hyndman, R.J.; Khandakar, Y. Automatic time series forecasting: The forecast package for R. *J. Stat. Softw.* **2008**, *27*, 1–22. [CrossRef]
29. Limas, M.C.; Meré, J.B.O.; Marcos, A.G.; de Pisón Ascacibar, F.J.M.; Espinoza, A.V.P.; Elías, F.A.; Ramos, J.M.P. A MORE Flexible Neural Network Package. R Package Version 0.2-16. 2020. Available online: https://cran.r-project.org/web/packages/AMORE/AMORE.pdf (accessed on 6 March 2021).
30. Yan, Y. MLmetrics: Machine Learning Evaluation Metrics. R Package Version 1.1.1. 2016. Available online: https://cran.r-project.org/web/packages/MLmetrics/MLmetrics.pdf (accessed on 6 March 2021).

engineering proceedings

MDPI

Proceedings

STL Decomposition of Time Series Can Benefit Forecasting Done by Statistical Methods but Not by Machine Learning Ones †

Zuokun Ouyang *, Philippe Ravier and Meryem Jabloun

Laboratoire Pluridisciplinaire de Recherche en Ingénierie des Systèmes, Mécanique, Energétique, Université d'Orléans, 12 rue de Blois, 45067 Orléans, France; philippe.ravier@univ-orleans.fr (P.R.); meryem.jabloun@univ-orleans.fr (M.J.)

* Correspondence: zuokun.ouyang@univ-orleans.fr

† Presented at the 7th International Conference on Time Series and Forecasting, Gran Canaria, Spain, 19–21 July 2021.

Abstract: This paper aims at comparing different forecasting strategies combined with the STL decomposition method. STL is a versatile and robust time series decomposition method. The forecasting strategies we consider are as follows: three statistical methods (ARIMA, ETS, and Theta), five machine learning methods (KNN, SVR, CART, RF, and GP), and two versions of RNNs (CNN-LSTM and ConvLSTM). We conduct the forecasting test on six horizons (1, 6, 12, 18, and 24 months). Our results show that, when applied to monthly industrial M3 Competition data as a preprocessing step, STL decomposition can benefit forecasting using statistical methods but harms the machine learning ones. Moreover, the STL-Theta combination method displays the best forecasting results on four over the five forecasting horizons.

Keywords: time series forecasting; ARIMA; ETS; Theta method; STL decomposition; machine learning; RNN

Citation: Ouyang, Z.; Ravier, P.; Jabloun, M. STL Decomposition of Time Series Can Benefit Forecasting Done by Statistical Methods but Not by Machine Learning Ones. *Eng. Proc.* **2021**, *5*, 42. https://doi.org/10.3390/engproc2021005042

Published: 8 July 2021

1. Introduction

Time series forecasting is a subdomain of time series analysis in which the historical measurements are modeled to describe the underlying characteristics of the observable and extrapolated into the future [1].

For a few decades, the domain of time series analysis has been dominated by statistical methodology. One of the most important and generally used models is the AutoRegressive Integrated Moving Average (ARIMA), which can be quickly built thanks to the Box–Jenkins methodology [2]. The ARIMA family has excellent flexibility for presenting varying time series. Nevertheless, it has limits due to its assumption of linearity of the time series [1,3].

Another dominating and widely used statistical method is the ExponenTial Smoothing (ETS) method, which was proposed in the 1950s [4–6]. It is often considered as an alternative to the ARIMA models. While the ARIMA family develops a model where the prediction is a weighted linear sum of recent past observations or lags, forecasts produced by ETS are weighted averages of past observations, with the weights decaying exponentially as the observations get older [7].

The M Competitions were initiated by professor Spyros Makridakis. There are different open competitions (M1–M5) dedicated to the performance comparison of different forecasting methods [8]. Particularly, the Theta method had impressive success in the M3 Competition and thus was used in the M4 Competition as a benchmark. It was first proposed by Assimakopoulos et al. [9] and then extended to forecast multivariate macroeconomic and financial time series [10]. Hyndman demonstrated the Theta method applied in the M3 Competition is equivalent to the simple exponential smoothing with a drift [11].

Time series can also have many underlying patterns, and decomposition can reveal them by splitting a time series into several components. In our study, we focus on STL decomposition. STL stands for Seasonal-Trend decomposition using LOESS, where LOESS is LOcal regrESSion, which was proposed by Robert et al. in 1990 [12], contributing to a decomposition method robust to anomalies.

Artificial Intelligence (AI) has gained significant prominence over the last decade, especially in object recognition [13], natural language processing (NLP) [14], and autonomous driving [15]. Convolutional Neural Networks (CNNs) have revolutionized the field of computer vision [16]. Recurrent Neural Networks (RNNs) benefited from lots of NLP tasks, such as machine translation [17] and speech recognition [18]. In the field of time series, many machine learning methods such as support vector regression (SVR), neural networks, classification and regression tree (CART), and k-nearest neighbor (kNN) regression were proven able to model and forecast time series as well [19,20].

There are some discussions on comparing the performance of different forecasting approaches. Ahmed et al. [21] performed an empirical comparison of eight machine learning models over the 1045 monthly series involved in the M3 Competition, but only one-step-ahead forecasting was considered. Makrirdakis et al. did some similar works [22], comparing statistical and machine learning methods, but without any decomposition method being introduced as a preprocessing step. Using the M1 Competition dataset, Theodosieu [23] compared a new STL-based method with some common benchmarks but without combining STL with them, and only up to an 18-month forecasting was considered.

As the preprocessing step often plays an integral part in prediction tasks and substantially impacts the results, we propose to conduct a new comparison work to identify its benefit: (1) by exploring STL decomposition when using it as a preprocessing step for all methods; and (2) by considering multiple forecasting horizons.

The rest of this paper is organized as follows. In the next section, we present a concise description of all the involved models and the decomposition methods. Section 3 presents how we organized and conducted the experiments. In Section 4, we present the comparison results and discussions based on these results. Section 5 gives the conclusion.

2. Methods

Although there are many different variations of each model, we considered only the primitive versions of each model in our experiments. As this paper is inspired heavily by the M3 and M4 Competitions, we kept the six benchmark methods used in these two competitions by the organizer [24].

2.1. Existing Benchmarks in M4 Competitions

Below is a list of descriptions of the benchmarks utilized in the M4 Competition.

- **Naïve 1.** Naïve 1 assumes future values are identical to the last observation.
- **Naïve S.** Naïve S assumes future values are identical to the values from the last known period, which, in our case, is 12 months.
- **Naïve 2.** Naïve 2 is similar to Naïve 1, except the data are seasonally adjusted by a conventional multiplicative decomposition if tested seasonal. We performed a 90% autocorrelation test at lag 12 for each series.
- **Simple Exponential Smoothing (SES).** SES forecasts future values as exponentially decayed weighted averages of past observations [7].
- **Holt.** Holt's linear trend method extends SES for data with a trend [7].
- **Damped.** The damped model dampens the trend in Holt's method [7].

2.2. Conventional Decomposition and STL Decomposition

Here, we introduce two commonly used decomposition methods.

2.2.1. Conventional Decomposition

The conventional multiplicative classical decomposition algorithm for a series with seasonal period m has four steps [7]:

1. Compute the trend component $\hat{T}_t$ using a simple moving average method.

2. Detrend the time series: $y_t/\hat{T}_t$.
3. Compute the seasonal component $\hat{S}_t$ by averaging the corresponding season's detrended values and then adjusting to ensure that they add to m.
4. Compute the remainder component $\hat{R}_t$: $\hat{R}_t = y_t/(\hat{T}_t\hat{S}_t)$.

2.2.2. STL Decomposition

STL decomposition consists of two recursive procedures: an inner loop and an outer loop. The inner loop fits the trend and calculates the seasonal component. Every inner loop consists of six steps in total:

1. Detrending. Calculate a detrended series $y_v - T_v^{(k)}$. For the first pass, $T_v^{(0)} = 0$.
2. Cycle-Subseries Smoothing. Use LOESS to smooth the subseries of values at each position of the seasonal cycle. The result is marked as $C_v^{(k+1)}$.
3. Low-Pass Filtering of Smoothed Cycle-Subseries. This procedure consists of two MA filters and a LOESS smoother. The result is marked as $L_v^{(k+1)}$.
4. Detrending of Smoothed Cycle-Subseries. $S_v^{(k+1)} = C_v^{(k+1)} - L_v^{(k+1)}$.
5. Deseasonalizing. $y_v - S_v^{(k+1)}$.
6. Trend Smoothing. Use LOESS to smooth the deseasonalized series to get the trend component of this pass $T_v^{(k+1)}$.

If any anomaly is detected, an outer loop will be applied and replace the LOESSs at the second and sixth steps of the inner loop with the robust LOESS.

2.3. ARIMA, ETS, and Theta

- **ARIMA.** An ARIMA model assumes future values to be linear combinations of past values and random errors, contributing to the AR and MA terms, respectively [2]. SARIMA (Seasonal ARIMA) is an extension of ARIMA that explicitly supports time series data with a seasonal component. Once STL decomposition is applied, SARIMA models degenerate into regular ARIMA models as STL handles the seasonal part.
- **ETS.** The ETS models are a family of time series models with an underlying state space model consisting of a level component, a trend component (T), a seasonal component (S), and an error term (E). Forecasts produced using exponential smoothing methods are weighted averages of past observations, with the weights decaying exponentially as the observations get older [7]. After concatenating STL on the ETS model, the full ETS model degenerates into Holt's method [7] as the seasonal equation is handled by STL.
- **Theta Method.** The Theta method, initially proposed in 2000 by Assimakopoulos et al. [9], performed exceptionally well in the M3 Competition and was used as a benchmark in the M4 Competition. The Theta method is based on the concept of modifying the local curvature of the time series through a coefficient θ, which is applied directly to the second difference of the data [9]. Hyndman demonstrated that the h-step-ahead forecast obtained by the Theta method is equivalent to an SES with drift depending on the smoothing parameter value of SES, the horizon h, and the data [11].

2.4. Machine Learning Methods

It is interesting to closely examine how machine learning methods perform in time series forecasting tasks. Using the embedding strategy to transform this task into a supervised learning problem [25], we can apply machine learning techniques to time series forecasting tasks. The following briefly introduces the machine learning methods used in this experimentation.

- ***k*-NN.** k-NN is a non-parametric method used for classification and regression. In both cases, the input consists of the k closest training examples in the feature space.

In k-NN regression, the output is the property value for the object. This value is the average of the values of k nearest neighbors based on the Euclidian distances.

- **SVR.** Support Vector Machine (SVM) is a successful method that tries to find a separation hyperplane to maximize the margin between two classes, while SVR seeks a hyperplane to minimize the margin between the support vectors and the hyperplane.
- **CART.** CART is one of the most generally used machine learning methods and can be used for classification and regression. CART dichotomizes each feature recursively and divides the input space into several cells. CART computes the probability distributions of the corresponding prediction in it.
- **RF.** RF is an ensemble learning algorithm based on the Decision Tree [26]. Similar to CART, Random Forest can be used for both classification and regression. It operates by constructing many decision trees at training time and calculating the average predictions from the individual trees.
- **GP.** A GP is a generalization of the Gaussian probability distribution [27]. It uses a measure of homogeneity between points as a kernel function to predict an unknown point's value from the input training data. The result of its prediction contains the value of the point and the uncertainty information, i.e., its one-dimensional Gaussian distribution [22].

2.5. Deep Learning Methods

For the promising capacity of RNNs to memorize the long-term values, we decided to test the deep learning models. Here, we present two structures of RNNs implemented in our experimentation. The first one is the well-known CNNs stacked with the Long Short-Term Memory (LSTM) cells, and the other one is the ConvLSTM structure proposed by Xingjian Shi et al. in NeurIPS 2015 [28].

- **CNN-LSTM.** We use a 1D CNN to handle univariate time series. It has a hidden convolutional layer iterating over a 1D sequence and follows a pooling layer to extract the most salient features, which is then interpreted by a fully connected layer. Then, we stack it with some LSTM layers, which is a widely used RNNs model that provides a solution to the vanishing gradient problem for RNNs. It was proposed by Sepp Hochreiter et al. in 1997 [29].
- **Convolutional LSTM (ConvLSTM).** ConvLSTM is an RNNs with convolutional structures in both the input-to-state and state-to-state transitions. It determines the future state of a certain cell in the grid by its local neighbors' inputs and past states. This is achieved using a convolution operator in the state-to-state and input-to-state transitions [28]. Rather than reading and extracting the features with a CNN and then interpreting them by an LSTM, ConvLSTM reads and interprets them at a time.

3. Experimentation Setup

This section presents how we organized and performed our experimentation.

3.1. Dataset

We selected 332 monthly series from the industry category which contains the highest number of points per series from the M3 Competition dataset. We set 84 as the length of the historical data and tested five different forecast horizons, i.e., 1, 6, 12, 18, and 24 months. Thus, the total length required for an appropriate series is 108. The two series N2011 (78 points) and N2118 (104 points) were thus removed from the original 334-series dataset.

3.2. Pipeline for Machine Learning and Deep Learning Methods

3.2.1. Data Preprocessing

In our experimentation, three preprocessing techniques were conducted on all the series:

1. **Deseasonalizing**: A 90% autocorrelation test at lag 12 is performed to decide whether the series is seasonal. We perform a conventional multiplicative decomposition or an STL decomposition if the series is seasonal and extract the seasonal part.

2. **Detrending**: A one-order differencing is performed to eliminate the trend.
3. **Scaling**: A standardization step is applied to remove the mean and scale the features to unit variance.

3.2.2. Supervised Learning Setting

A time series prediction problem can be transformed into a supervised learning task that machine learning and deep learning methods can do. A commonly used approach is to formulate a training set by lagging and stacking the historical series several times, which is often referred to as the embedding technique in the R implementation [30].

Typically, for an h-step-ahead prediction problem, we can construct a training set $\{X, Y\}$ as follows:

$$X = \begin{bmatrix} y_1 & y_2 & \cdots & y_n \\ y_2 & y_3 & \cdots & y_{n+1} \\ \vdots & \vdots & \vdots & \vdots \\ y_{N-n-h+1} & y_{N-n-h+2} & \cdots & y_{N-h} \end{bmatrix}, Y = \begin{bmatrix} y_{n+1} & y_{n+2} & \cdots & y_{n+h} \\ y_{n+2} & y_{n+3} & \cdots & y_{n+h+1} \\ \vdots & \vdots & \vdots & \vdots \\ y_{N-h+1} & y_{N-h+2} & \cdots & y_N \end{bmatrix}, \quad (1)$$

where N is the total length of the series, n is the number of times we lag the series, often referred to as the window length. Each row in X represents a training example, while its label corresponds to the vector in the same row in Y.

3.2.3. Results Post-Processing

The post-processing part comprises the inverted operations of the three preprocessing steps:

1. **Rescaling**: A rescaling step is performed by inverting the standardization.
2. **Retrending**: A cumulated summing is conducted to bring back the trend.
3. **Reseasonalizing**: A reseasonalization step is executed to integrate the seasonal component into the prediction.

3.3. Pipeline for Statistical Methods

Statistical methods require no preprocessing or post-processing as the machine learning and deep learning methods demand. However, the same deseasonalization and reseasonalization steps are necessary for the STL-based methods.

In our experimentation, we performed an STL decomposition and constructed the ARIMA, ETS, and Theta models upon the seasonally adjusted series to compute the point forecasts. It comprises the following three procedures:

1. **Deseasonalizing.** Compute the deseasonalized series by extracting the seasonal component calculated by STL decomposition.
2. **Point forecasting.** Construct the ARIMA, ETS, and Theta models on the seasonally adjusted data and calculate the forecasting values.
3. **Reseasonalizing.** Integrate the seasonal component back to calculate the final forecasting results.

One effect of applying the STL decomposition on statistical methods is that it cancels these statistical methods' intrinsic seasonality handlers.

3.4. Implementation and Parameters Tuning

3.4.1. Statistical Methods

All of the statistical methods, as well as their STL-based versions, were conducted using the `forecast-8.13` package [31] in R 4.0.2.

3.4.2. Machine Learning Methods

The machine learning methods and their STL-based versions were tested exploiting the `statsmodels-0.12.1` module [32] and the `scikit-learn-0.23.2` [33] and `sktime-0.4.2` [34] packages in Python 3.8.5.

3.4.3. Deep Learning Methods

The two deep learning models were constructed in Python 3.8.5 with the `Keras-2.4.0` framework [35] under `TensorFlow-2.3.1` [36]. The hyperparameters were empirically tuned.

For CNN-LSTM, we stacked one CNN layer by two LSTM layers and three dense layers. The CNN uses a `ReLU` activation function and has 16 filters, where each filter has a kernel size of 5. Each LSTM layer has 128 units, and the two following dense layers have 32 and 16 units, respectively. The number of units of the last dense layer is identical to the forecast horizons.

For ConvLSTM, we stacked two ConvLSTM layers, followed by three dense layers. Each ConvLSTM layer has 128 filters where each filter has a shape of `[1, 2]`. The three dense layers are identical to those of CNN-LSTM.

3.5. Evaluation Metrics

Three evaluation metrics were used in this experimentation.

We used the symmetric Mean Absolute Percentage Error (sMAPE) [37]. It has the following formula: $\mathrm{sMAPE} = \frac{2}{k}\sum_{t=1}^{k}\frac{|y_t-\hat{y}_t|}{|y_t|+|\hat{y}_t|} \times 100\%$, where k is the forecasting horizon, y_t is the actual values at time t, and $\hat{y}_t$ is the forecast produced by the model.

We also used the Mean Absolute Scaled Error (MASE) introduced by Rob Hyndman [38]: $\mathrm{MASE} = \frac{1}{k}\frac{\sum_{t=1}^{k}|y_t-\hat{y}_t|}{\frac{1}{n-m}\sum_{t=m+1}^{n}|y_t-y_{t-m}|}$, where n is the number of the observations and m is the number of periods per season.

The Overall Weighted Average (OWA) to Naïve 2 was also adopted [24]:

$$\mathrm{OWA} = \frac{1}{2}\left(\frac{\mathrm{sMAPE}_{\text{Model X}}}{\mathrm{sMAPE}_{\text{Naïve 2}}} + \frac{\mathrm{MASE}_{\text{Model X}}}{\mathrm{MASE}_{\text{Naïve 2}}}\right). \quad (2)$$

4. Results and Discussion

4.1. Results

The results of our experimentation are presented in Table 1, Figures 1–3, and the following contents.

Table 1 represents the forecast results of different methods on different forecast horizons. Note that Naïve 2 method was chosen as the reference method for the OWA indicator, meaning that OWA equals 1 whatever the horizon value h. At first glance, in Table 1, most of the statistical methods give better forecasting results with respect to naive methods (OWA < 1) than the machine learning methods (OWA > 1). This result confirms the conclusion from the M3 Competition that sophisticated machine learning methods do not assure a more accurate prediction than simple statistical methods.

This result becomes obvious in Figure 1, showing OWA ≤ 0.910 performance results for the three advanced statistical methods (ARIMA, ETS, and Theta), by comparison with Figure 2, showing OWA ≥ 0.914 performance results for the five machine learning methods. Above all, Figures 1 and 2 show the impact of STL decomposition as a preprocessing step of statistical and ML methods on the forecasting performance results.

Significant improvement from STL decomposition was found for statistical methods. Among all the tested STL-based methods, the STL-Theta method outperforms the other methods on almost all forecast horizons. The STL-Theta method can even give a lower OWA on a 24-month forecast horizon than the other methods on the 18-month one.

In Figure 2, we can find that the SVR model gives the best result. No significant improvement from STL preprocessing was detected.

Figure 3 shows the mean and standard deviation of the gain brought by STL decomposition. On average, STL improves the OWA of ARIMA by 1.5%, ETS by 0.9%, and Theta by 5%, but it conducts a loss of OWA for machine learning methods. It harms SVR by 2.3%, RF by 3.3%, GP by 2.3%, KNN by 2.2%, and CART by 1.1%.

Table 1. Forecast results of different methods on different forecast horizons.

Statistical	h = 1			h = 6			h = 12			h = 18			h = 24		
	sMAPE	MASE	OWA	sMAPE	MASE	OWA	sMAPE	MASE	OWA	sMAPE	MASE	OWA	sMAPE	MASE	OWA
Naive	12.536	1.006	1.071	16.011	1.280	1.177	16.238	1.312	1.152	17.480	1.395	1.153	18.044	1.456	1.143
sNaive	12.464	0.882	1.002	12.001	0.874	0.842	12.726	0.925	0.857	14.088	1.033	0.891	14.689	1.094	0.894
Naive2	11.704	0.939	1.000	13.813	1.071	1.000	14.374	1.118	1.000	15.431	1.189	1.000	16.053	1.254	1.000
SES	9.277	0.723	0.781	11.386	0.844	0.806	12.376	0.931	0.847	13.640	1.017	0.870	14.397	1.092	0.884
Holt	9.734	0.741	0.810	11.669	0.865	0.826	13.522	1.004	0.920	15.710	1.161	0.997	17.197	1.293	1.051
Damped	9.288	0.720	0.780	11.388	0.844	0.806	12.572	0.942	0.859	13.985	1.036	0.889	14.740	1.110	0.902
ARIMA	8.643	0.623	0.701	10.037	0.730	0.704	11.824	0.873	0.802	13.581	1.015	0.867	14.794	1.127	0.910
ETS	7.805	0.591	0.648	9.875	0.716	0.692	11.718	0.849	0.787	13.608	0.987	0.856	14.751	1.085	0.892
Theta	8.645	0.640	0.710	10.668	0.749	0.736	11.862	0.854	0.794	13.403	0.962	0.839	14.399	1.047	0.866
STL-ARIMA	8.245	0.604	0.674	9.915	0.717	0.693	11.755	0.856	0.792	13.457	0.993	0.854	14.481	1.093	0.887
STL-ETS	7.760	0.569	**0.635**	9.882	0.704	0.686	11.728	0.845	0.786	13.433	0.969	0.843	14.552	1.074	0.882
STL-Theta	7.963	0.580	0.649	9.502	0.678	**0.660**	11.177	0.801	**0.747**	12.817	0.921	**0.803**	13.891	1.011	**0.836**
ML & DL	**h = 1**			**h = 6**			**h = 12**			**h = 18**			**h = 24**		
	sMAPE	**MASE**	**OWA**	**sMAPE**	**MASE**	**OWA**	**sMAPE**	**MASE**	**OWA**	**sMAPE**	**MASE**	**OWA**	**sMAPE**	**MASE**	**OWA**
KNN	13.636	0.965	1.096	16.070	1.166	1.126	17.781	1.326	1.212	20.421	1.497	1.291	21.741	1.610	1.319
STL-KNN	15.318	1.077	1.228	18.359	1.390	1.313	17.980	1.306	1.210	22.513	1.698	1.444	22.154	1.610	1.332
SVR	11.253	0.855	**0.936**	12.732	0.971	**0.914**	14.712	1.116	**1.011**	17.485	1.284	**1.107**	19.526	1.429	1.178
STL-SVR	12.978	1.006	1.090	15.285	1.225	1.125	14.919	1.109	1.015	19.338	1.484	1.251	19.589	1.410	1.172
CART	14.080	1.025	1.147	19.081	1.377	1.334	25.490	1.930	1.750	30.934	2.314	1.975	35.956	2.596	2.155
STL-CART	15.820	1.191	1.310	21.715	1.660	1.561	25.157	1.862	1.708	32.285	2.446	2.075	35.715	2.537	2.124
RF	11.756	0.898	0.980	13.668	1.027	0.974	15.432	1.186	1.067	17.831	1.369	1.153	19.692	1.496	1.210
STL-RF	13.667	1.054	1.145	16.401	1.289	1.195	15.880	1.177	1.079	20.237	1.581	1.321	19.947	1.465	1.205
GP	12.540	0.972	1.053	14.268	1.093	1.027	15.528	1.195	1.075	17.395	1.313	1.116	18.720	1.418	**1.148**
STL-GP	14.163	1.120	1.201	16.950	1.351	1.244	15.782	1.187	1.080	19.624	1.526	1.278	18.974	1.408	1.152
CNN-LSTM	13.105	0.985	1.084	15.439	1.176	1.108	16.233	1.213	1.107	17.811	1.332	1.137	18.821	1.423	1.154
ConvLSTM	12.976	0.929	1.049	16.257	1.235	1.165	17.121	1.283	1.169	18.926	1.399	1.202	19.372	1.441	1.178

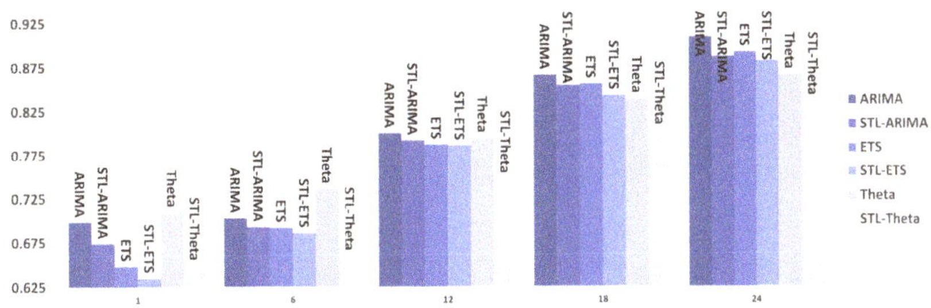

Figure 1. OWAs for STL decomposition on statistical models.

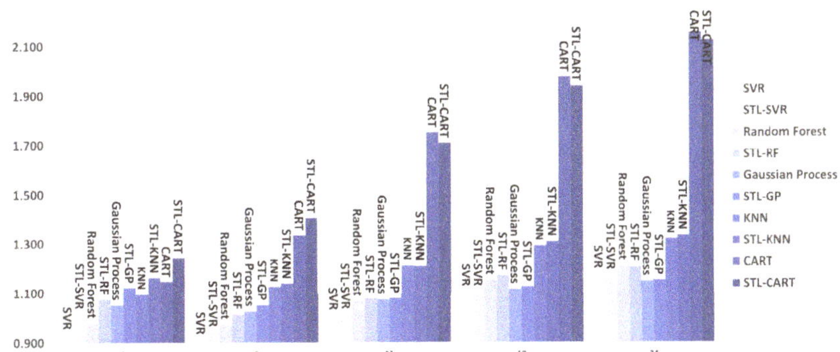

Figure 2. STL decomposition on ML models.

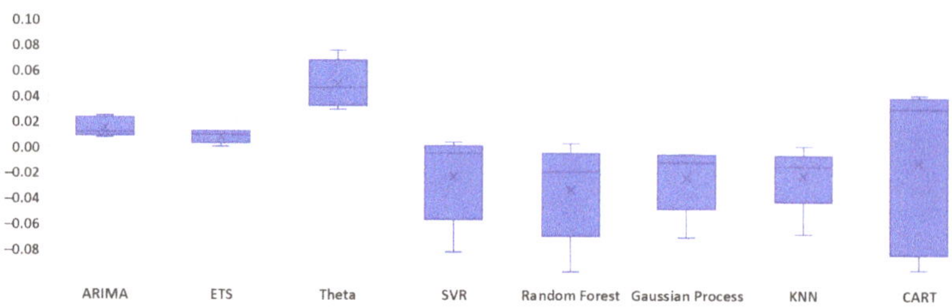

Figure 3. Boxplot of OWA gain from STL for each method.

4.2. Discussion

It is interesting to note from the results in Figure 3 that CART performs the worst among all these methods, which is easy to understand as CART is a single forecaster. Its ensemble method Random Forest performs much better in terms of the precision of forecasting. At the same time, it consumes the most time.

The initial objective of this study was to determine whether STL decomposition can be helpful as a preprocessing step for time series forecasting methods. Our results confirm using STL decomposition as a preprocessing method can effectively improve the statistical methods' performance, which is consistent with Theodosiou [23] using M1 Competition data, but, for machine learning methods, it can be harmful.

A possible explanation for this might be extracting the seasonal information from the series can affect the features to be modeled. For statistical models, their intrinsic ability for handling the seasonality might be worse than the STL decomposition. For the machine learning models, it could be easier to model seasonal data. Further research is required to confirm this hypothesis.

5. Conclusions

The present study was designed to determine the effect of using STL decomposition as a preprocessing step on different forecasting strategies. The results show some vast differences between these methods. Among all tested models, the STL decomposition-based Theta method is the best one. In the meantime, the STL decomposition can benefit the statistical methods by providing a more robust decomposition procedure than their intrinsic mechanism. The machine learning methods tested in this experimentation failed to outperform most statistical methods but still have some potentials for improvement. We can perform other preprocessing methods without harming the natural feature of the time series. More research is required in the future. For deep learning methods, as there are so many architectures and combinations of hyperparameters for neural networks, the two tested architectures in this experimentation may not be the optimal solutions. At the same time, there are many architectures more suitable for short sequence learning and worthy of further research.

Author Contributions: Conceptualization, Z.O., P.R. and M.J.; methodology, Z.O., P.R. and M.J.; software, Z.O.; validation, Z.O., P.R. and M.J.; formal analysis, Z.O., P.R. and M.J.; investigation, Z.O. and P.R.; resources, Z.O.; data curation, Z.O.; writing—original draft preparation, Z.O.; writing—review and editing, Z.O., P.R. and M.J.; visualization, Z.O.; supervision, P.R. and M.J.; and project administration, Z.O. and P.R. All authors have read and agreed to the published version of the manuscript.

Funding: This research was funded by Association Nationale de la Recherche et de la Technologie (CIFRE 2019/0551).

Data Availability Statement: Data is contained within the article.

Conflicts of Interest: The authors declare no conflict of interest.

References

1. Zhang, G. Time series forecasting using a hybrid ARIMA and neural network model. *Neurocomputing* **2003**, *50*. doi:10.1016/S0925-2312(01)00702-0. [CrossRef]
2. Box, G.E.P.; Jenkins, G.M.; Reinsel, G.C.; Ljung, G.M. *Time Series Analysis: Forecasting and Control*, 5th ed.; John Wiley & Sons, Inc.: Hoboken, NJ, USA, 2004; Volume 20.
3. Kihoro, J.; Otieno, R.; Wafula, C. Seasonal time series forecasting: A comparative study of ARIMA and ANN models. *Afr. J. Sci. Technol.* **2004**, *5*. [CrossRef]
4. Brown, R.G. *Statistical Forecasting for Inventory Control*; McGraw/Hill: New York, NY, USA, 1959.
5. Holt, C.C. Forecasting seasonals and trends by exponentially weighted moving averages. *Int. J. Forecast.* **2004**, *20*. doi:10.1016/j.ijforecast.2003.09.015. [CrossRef]
6. Winters, P.R. Forecasting sales by exponentially weighted moving averages. *Manag. Sci.* **1960**, *6*, 324–342. [CrossRef]
7. Hyndman, R.J.; Athanasopoulos, G. *Forecasting: Principles and Practice*; OTexts: Melbourne, Australia, 2018.
8. Hyndman, R.J. A brief history of forecasting competitions. *Int. J. Forecast.* **2020**, *36*, 7–14. [CrossRef]
9. Assimakopoulos, V.; Nikolopoulos, K. The theta model: A decomposition approach to forecasting. *Int. J. Forecast.* **2000**, *16*, 521–530. [CrossRef]
10. Thomakos, D.D.; Nikolopoulos, K. Forecasting multivariate time series with the theta method. *J. Forecast.* **2015**, *34*, 220–229. [CrossRef]
11. Hyndman, R.J.; Billah, B. Unmasking the Theta method. *Int. J. Forecast.* **2003**, *19*, 287–290. [CrossRef]
12. Cleveland, R.B.; Cleveland, W.S.; McRae, J.E.; Terpenning, I. STL: A seasonal-trend decomposition. *J. Off. Stat.* **1990**, *6*, 3–73.
13. Szegedy, C.; Toshev, A.; Erhan, D. Deep neural networks for object detection. In *Proceedings of the 26th International Conference on Neural Information Processing Systems—Volume 2*; Curran Associates Inc.: Red Hook, NY, USA, 2013.
14. Graves, A. Supervised sequence labelling. In *Supervised Sequence Labelling with Recurrent Neural Networks*; Springer: Berlin/Heidelberg, Germany, 2012.
15. Yurtsever, E.; Lambert, J.; Carballo, A.; Takeda, K. A survey of autonomous driving: Common practices and emerging technologies. *IEEE Access* **2020**, *8*. [CrossRef]
16. Krizhevsky, A.; Sutskever, I.; Hinton, G.E. Imagenet classification with deep convolutional neural networks. In *Proceedings of the 25th International Conference on Neural Information Processing Systems—Volume 1*; Curran Associates Inc.: Red Hook, NY, USA, 2012.
17. Sutskever, I.; Vinyals, O.; Le, Q.V. Sequence to sequence learning with neural networks. In *Proceedings of the 27th International Conference on Neural Information Processing Systems—Volume 2*; Curran Associates Inc.: Red Hook, NY, USA, 2014.
18. Hinton, G.; Deng, L.; Yu, D.; Dahl, G.E.; Mohamed, A.R.; Jaitly, N.; Senior, A.; Vanhoucke, V.; Nguyen, P.; Sainath, T.N.; et al. Deep neural networks for acoustic modeling in speech recognition: The shared views of four research groups. *IEEE Signal Process. Mag.* **2012**, *29*, 82–97. [CrossRef]
19. Lapedes, A.; Farber, R. *Nonlinear Signal Processing Using Neural Networks: Prediction and System Modelling*; Technical Report (No. LA-UR-87-2662; CONF-8706130-4); Los Alamos National Laboratory: Los Alamos, NM, USA, 1987.
20. Hastie, T.; Tibshirani, R.; Friedman, J. *The Elements of Statistical Learning: Data Mining, Inference, and Prediction*; Springer Science & Business Media: Berlin/Heidelberg, Germany, 2009.
21. Ahmed, N.K.; Atiya, A.F.; Gayar, N.E.; El-Shishiny, H. An empirical comparison of machine learning models for time series forecasting. *Econom. Rev.* **2010**, *29*. doi:10.1080/07474938.2010.481556. [CrossRef]
22. Makridakis, S.; Spiliotis, E.; Assimakopoulos, V. Statistical and Machine Learning forecasting methods: Concerns and ways forward. *PLoS ONE* **2018**, *13*, e0194889. [CrossRef]
23. Theodosiou, M. Forecasting monthly and quarterly time series using STL decomposition. *Int. J. Forecast.* **2011**, *27*, 1178–1195. doi:10.1016/j.ijforecast.2010.11.002. [CrossRef]
24. Makridakis, S.; Spiliotis, E.; Assimakopoulos, V. The M4 Competition: 100,000 time series and 61 forecasting methods. *Int. J. Forecast.* **2020**, *36*, 54–74. [CrossRef]
25. Bontempi, G.; Taieb, S.B.; Le Borgne, Y.A. Machine learning strategies for time series forecasting. In *European Business Intelligence Summer School*; Springer: Berlin/Heidelberg, Germany, 2012.
26. Breiman, L. Random forests. *Mach. Learn.* **2001**, *45*, 5–32. [CrossRef]
27. Rasmussen, C.E. Gaussian processes in machine learning. In *Summer School on Machine Learning*; Springer: Berlin/Heidelberg, Germany, 2003.
28. Shi, X.; Chen, Z.; Wang, H.; Yeung, D.Y.; Wong, W.K.; Woo, W.C. Convolutional LSTM network: A machine learning approach for precipitation nowcasting. In *Proceedings of the 28th International Conference on Neural Information Processing Systems—Volume 1*; Curran Associates Inc.: Red Hook, NY, USA, 2015.
29. Hochreiter, S.; Schmidhuber, J. Long short-term memory. *Neural Comput.* **1997**, *9*, 1735–1780. [CrossRef] [PubMed]
30. R Core Team. *R: A Language and Environment for Statistical Computing*; R Foundation for Statistical Computing: Vienna, Austria, 2020.
31. Hyndman, R.J.; Khandakar, Y. Automatic time series forecasting: The forecast package for R. *J. Stat. Softw.* **2008**, *27*. doi:10.18637/jss.v027.i03. [CrossRef]
32. Seabold, S.; Perktold, J. Statsmodels: Econometric and statistical modeling with python. In Proceedings of the 9th Python in Science Conference (SciPy 2010), Austin, TX, USA, 28 June–3 July 2010.
33. Pedregosa, F.; Varoquaux, G.; Gramfort, A.; Michel, V.; Thirion, B.; Grisel, O.; Blondel, M.; Prettenhofer, P.; Weiss, R.; Dubourg, V.; et al. Scikit-learn: Machine learning in Python. *J. Mach. Learn. Res.* **2011**, *12*, 2825–2830.

34. Löning, M.; Bagnall, A.; Ganesh, S.; Kazakov, V.; Lines, J.; Király, F.J. sktime: A unified interface for machine learning with time series. *arXiv* **2019**, arXiv:1909.07872.
35. Chollet, F. Keras. 2015. Available online: https://keras.io (accessed on 13 October 2020).
36. Abadi, M.; Agarwal, A.; Barham, P.; Brevdo, E.; Chen, Z.; Citro, C.; Corrado, G.S.; Davis, A.; Dean, J.; Devin, M.; et al. TensorFlow: Large-Scale Machine Learning on Heterogeneous Systems. 2015. Available online: tensorflow.org (accessed on 13 October 2020).
37. Makridakis, S. Accuracy measures: Theoretical and practical concerns. *Int. J. Forecast.* **1993**, *9*, 527–529. [CrossRef]
38. Hyndman, R.J.; Koehler, A.B. Another look at measures of forecast accuracy. *Int. J. Forecast.* **2006**, *22*, 679–688. [CrossRef]

engineering proceedings

MDPI

Proceedings

Flow and Density Estimation in Grenoble Using Real Data †

Martin Rodriguez-Vega *, **Carlos Canudas-de-Wit** and **Hassen Fourati**

GIPSA-Lab, Université Grenoble Alpes, CNRS, INRIA, 38400 Saint-Martin-d'Hères, France; carlos.canudas-de-wit@gipsa-lab.fr (C.C.-d.-W.); hassen.fourati@gipsa-lab.fr (H.F.)

* martin.rodriguez-vega@gipsa-lab.fr

† Presented at the 7th International conference on Time Series and Forecasting, Gran Canaria, Spain, 19–21 July 2021.

Abstract: This work deals with the Traffic State Estimation (TSE) problem for urban networks, using heterogeneous sources of data such as stationary flow sensors, Floating Car Data (FCD), and Automatic Vehicle Identifiers (AVI). A data-based flow and density estimation method is presented and tested using real traffic data. This work presents a study case applied to the downtown of the city of Grenoble in France, using the Grenoble Traffic Lab for urban networks (GTL-Ville), which is an experimental platform for real-time collection and analysis of traffic data.

Keywords: traffic state estimation; large urban networks; floating car data; turning ratios

Citation: Rodriguez-Vega, M.; Canudas-de-Wit, C.; Fourati, H. Flow and Density Estimation in Grenoble Using Real Data. *Eng. Proc.* **2021**, *5*, 43. https://doi.org/10.3390/engproc2021005043

Published: 8 July 2021

1. Introduction

Traffic state estimation (TSE) is an important stage in the development of Intelligent Transportation Systems (ITS), as the knowledge of the evolution of traffic state variables such as flow and density for each road can be used to implement control strategies, or help in the decision-making stages for network design for better smart cities. TSE refers to the use of partially observed and noisy traffic data to infer the value of traffic indicators such as flow, density, velocity, traveling time, and others [1]. This information can be used to calculate the mean traveling times for users, fuel consumption and vehicle emissions (important for air quality assessment), estimate the life of pavement, and many other applications. Because of this, accurate TSE is an active field in the transportation research literature [2].

Classical TSE methods were initially proposed for the case of highways [3]. Generally, these methods are based on the Lighthill–Whitham [4] and Richards [5] (LWR) model, and its discrete counterpart, the Cell Transmission Model (CTM) [6], which use the empirical flow-density relation known as the Fundamental Diagram. In [7], the authors propose the use of an Extended Kalman Filter (EKF) by linearizing the CTM around a current state to estimate the density of road sections. In [8], the CTM is used to identify observable modes, where a graph-constrained density observer is applied. In [9], semi-analytical solutions to the LWR are coupled with a mixed integer problem to estimate highway density.

The case of networks has received less attention [2], but the need to study this scenario is increasing in the last few years for TSE issues. This relative lack of attention is due to the additional modeling tools required to describe vehicle interactions in intersections [10]. The extended version of the CTM developed in [11] brings a solution to this problem via a flow maximization formulation under constraints provided by the fundamental diagram. This approach is widely used as can be seen in [12,13]. However, the use of the fundamental diagram, especially in urban networks, is challenging as it requires the calibration of many parameters. Furthermore, recent studies have found that the fundamental diagram does not effectively describe vehicle deceleration at intersections [14]. To solve this issue, the authors in [15] proposed a data-based method that collects data from connected vehicles to estimate the exiting flow of each road. Nevertheless, such rich data are not always available, and other methods are required.

Our contribution in this paper is the proposal of a data-based TSE method for general urban networks. We make use of three different data sources: stationary flow sensors, AVI using Bluetooth devices, and Floating Car Data (FCD). Data provided by these sources are used to estimate the external inflows to a traffic network, the turning ratios for a selection of intersections, and the space mean speed of the road sections of the network. Additionally, the method is tested using real traffic data collected from a sensor network in the city of Grenoble, France.

This paper is organized as follows. Section 2 presents the traffic dynamics model used to estimate the flow and density for the road sections of an urban traffic network. Section 3 describes the experimental platform Grenoble Traffic Lab (GTL-Ville), and the available data used to deploy and validate the proposed model. Section 4 describes a method used to estimate some of the parameters of the model that are not measured directly. Section 5 presents the results of the estimation approach and compares them to real data. Finally, Section 6 ends the paper with some conclusions.

2. Density Estimation Model

We consider urban traffic networks which are modeled as a directed graph $\mathcal{G} = \{\mathcal{N}, \mathcal{E}\}$ where the nodes $\mathcal{N}$ correspond to intersections, and the edges $\mathcal{E} \subset \mathcal{N} \times \mathcal{N}$ correspond to road sections. Additionally, let $\mathcal{E}^{\text{in}} \subset \mathcal{E}$ denote the boundary incoming roads which have no upstream neighbors, and $\mathcal{E}^{\text{out}} \subset \mathcal{E}$ denote the boundary outgoing roads that have no downstream neighbors.

For all roads, we consider as state variables the density $\boldsymbol{\rho}$ (veh/km), incoming flow $\boldsymbol{\varphi}^{\text{in}}$ (veh/h), outgoing flow $\boldsymbol{\varphi}^{\text{out}}$ (veh/h), and space–mean velocity $\mathbf{v}$ (km/h), which are all time dependent and have dimensions equal to the number of roads $|\mathcal{E}|$.

To model the traffic dynamics, consider the following conservation law for the traffic density [16],

$$\frac{\mathrm{d}}{\mathrm{d}t}\boldsymbol{\rho}(t) = L^{-1}(\boldsymbol{\varphi}^{\text{in}}(t) - \boldsymbol{\varphi}^{\text{out}}(t)) \tag{1}$$

where L is a diagonal matrix containing the road lengths. Furthermore, the inflows and outflows of adjacent roads are dependent on each other through intersections as shown in Figure 1. Intersections are modeled as 0-dimensional points that do not store vehicles. To model the exchange of inflows and outflows of the different roads at the intersections, we use the parameters called turning ratios. Let $\mathcal{I}(n)$ be the set of incoming roads to some intersection $n \in \mathcal{N}$ and $\mathcal{O}(n)$ be the set of outgoing roads from n. A turning ratio $r_{i,j}$ for $i \in \mathcal{I}(n)$ and $j \in \mathcal{O}(n)$ defines the proportion of vehicles exiting i that enters j.

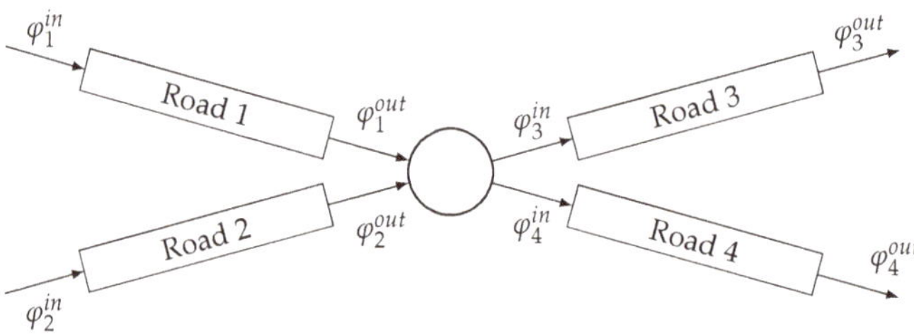

Figure 1. Flow exchange at an intersection.

As intersections do not store vehicles, then the conservation of density implies that

$$\sum_{j \in \mathcal{O}(n)} r_{i,j} = 1\,, \quad \forall n \in \mathcal{N} \quad \forall i \in \mathcal{E} \setminus \mathcal{E}^{\text{out}}. \tag{2}$$

Let $R \in \mathbb{R}^{|\mathcal{E}|\times|\mathcal{E}|}$ be the turning ratio matrix with elements $r_{i,j}$. If there is no connection between roads i, j, then $R_{i,j} = 0$. The input flows of each section can be expressed as a linear combination of the output flows of the preceding sections:

$$\boldsymbol{\varphi}^{\text{in}}(t) = R^\top \boldsymbol{\varphi}^{\text{out}}(t) + B\mathbf{u}(t) \tag{3}$$

where $\mathbf{u}(t)$ is the vector of input demands at the boundaries of the network, and B is a selection matrix which identifies the elements of $\mathcal{E}^{\text{in}}$. Combining Equations (1) and (3), we have

$$\frac{\mathrm{d}}{\mathrm{d}t}\boldsymbol{\rho}(t) = L^{-1}(R^\top - \mathbb{I})\boldsymbol{\varphi}^{\text{out}}(t) + L^{-1}B\mathbf{u}(t) \tag{4}$$

Using the hydrodynamic relation, we can approximate the outflows of each road from the values of density and the space–mean speed as

$$\boldsymbol{\varphi}^{\text{out}}(t) \approx V(t)\boldsymbol{\rho}(t) \tag{5}$$

where $V(t) = \text{diag}(\mathbf{v}(t))$. This relation applies accurately when considering very short distances, or when the spatial variations in vehicle speed and density are negligible. We make the following assumption:

Assumption 1. *The speed and density throughout a road section do not vary significantly in the spatial domain.*

In urban settings, this assumption can be justified as road lengths between intersections are generally on the order of 100 m. Therefore, we rewrite (4) as

$$\frac{\mathrm{d}}{\mathrm{d}t}\boldsymbol{\rho}(t) = L^{-1}(R^\top - \mathbb{I})V(t)\boldsymbol{\rho}(t) + L^{-1}B\mathbf{u}(t) \tag{6}$$

In this work, we consider the use of (6) as an open-loop estimator for the state of the network. To achieve this goal, we require as input data the values of the turning ratios for all intersections, the space–mean road speeds, and the input demands. Denote by $\hat{R}$, $\hat{V}$ and $\hat{\mathbf{u}}$ the estimated or measured values for these variables. Thus, the proposed density estimator is

$$\frac{\mathrm{d}}{\mathrm{d}t}\hat{\boldsymbol{\rho}}(t) = L^{-1}(\hat{R}^\top - \mathbb{I})\hat{V}(t)\hat{\boldsymbol{\rho}}(t) + L^{-1}\hat{\mathbf{u}}(t) \tag{7}$$

In the next section, we describe how the estimated quantities for the input data are obtained.

3. Experimental Platform

In this work, we make use of the Grenoble Traffic Lab for Urban Networks, GTL-Ville (http://gtlville.inrialpes.fr, accessed on 29 June 2021). This is an experimental platform for real-time collection of traffic data coming from a network of sensors installed in the city of Grenoble, France. This platform also provides real-time traffic indicators and analysis oriented towards the users of the city, traffic operators, and researchers. The collected data and computed indicators are available for download at the website.

In this work, we consider a section of Grenoble's downtown of an area of approximately 1.4 km by 1 km (see Figure 2). In this section, we describe the available data for the intersections and roads contained in this section.

3.1. Stationary Counting Sensors

Stationary sensors are placed in a fixed position in a road section, and collect information of the vehicles passing through that point. The collected data vary according to the technology, but generally variables such as length, speed, and time of passage are recorded. Two sensor technologies are available:

- Induction loop sensors, installed under the pavement, detect changes in the inductance due to the passage of vehicles. It provides information about flow and occupancy.
- Microwave radars, located above the ground, emit pulses of radiation and then measure the properties of the reflected beam. It provides information about flow, vehicle speeds, and length.

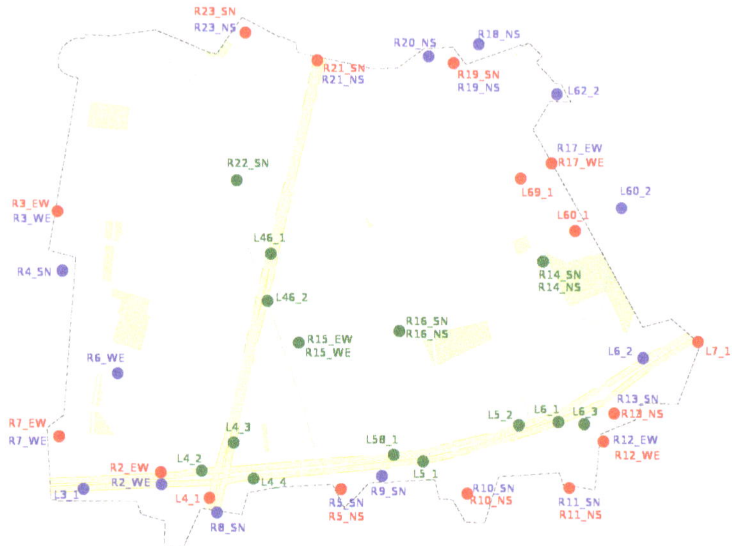

Figure 2. Stationary flow sensors located in downtown Grenoble. The text refers to sensor identifiers. Sensors in blue, correspond to boundary inflows; in red to boundary outflows; and in green to validation flows.

The sensor locations are shown in Figure 2. Induction loop sensors have an identifier starting with "L", whereas microwave radars start with "R". Each dot corresponds to the location of a single sensor. As radars are able to measure in multiple lanes and directions, some locations present two identifiers that correspond to each direction.

Furthermore, according to their locations, sensor data are classified as

- Boundary inflows (blue dots in the figure), providing the values of $\hat{\mathbf{u}}(t)$ in Equation (7).
- Boundary outflows (red dots in the figure). Data from these locations are denoted by $\mathbf{y}(t)$.
- Validation flows (green dots in the figure). Data from these locations will be used to validate estimation results.

3.2. Floating Car Data

Floating Car Data (FCD) are trajectories of individual vehicles collected via devices such as GPS navigators. Due to privacy policies, data from multiple users are aggregated.

Define by $\mathcal{V}_i(t)$ the set of vehicle indexes that provide FCD that are inside road i at time t. Let v_α be the speed of a vehicle indexed by α. We define the aggregated speed for road section i from FCD data by

$$\hat{v}_i(t) = \frac{1}{|\mathcal{V}_i(\tau)|} \sum_{\alpha \in \mathcal{V}_i(t)} v_\alpha(t) \tag{8}$$

which provides the estimates of the space–mean speeds for all roads, $\hat{V}(t)$ in Equation (7). However, this information is not available for roads that have few vehicles during the day, resulting in low precision estimates. For this cases, we use the value of the free-flow velocity, as roads with few vehicles are often under the critical density.

3.3. Turning Ratio Measurements

To measure the values of the turning ratio parameters, Bluetooth vehicle identifiers were used. For a given intersection, these devices are located at the adjacent incoming and outgoing roads. During a time interval, each device is able to detect vehicles that are equipped with another Bluetooth device, and records a unique identifier and its time of passage. By comparing the information across the installed devices, it is possible to assign the origin and destination road of individual vehicles.

As the rate of vehicles equipped with Bluetooth devices is unknown, these measurements cannot provide the total flow. However, this information can be used to compute the relative use of each turn, so the turning ratios can be estimated. Due to economical constraints, only 12 intersections were monitored during a measurement campaign lasting one week. Denote by $\mathcal{B} \subset \mathcal{N}$ the set of intersection monitored by these devices, whose locations are shown in Figure 3. The corresponding turning ratios are computed as

$$r_{i,j}^{\mathrm{BT}} = \frac{\mathrm{Counts}(i,j)}{\sum_k \mathrm{Counts}(i,k)} \tag{9}$$

where $\mathrm{Counts}(i,j)$ is the total number of detected vehicles going from road i to j during the campaign duration. To provide turning ratio estimates for the remaining intersections, a method is described in Section 4 which uses the data presented in Section 3.4.

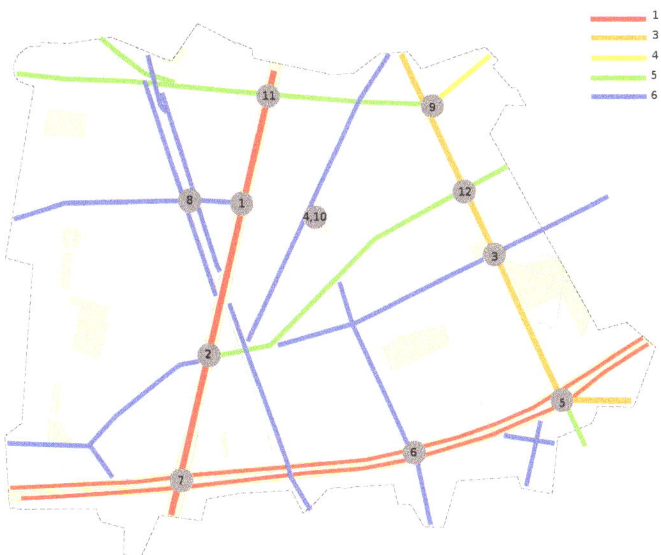

Figure 3. Grey dots: intersections equipped with Bluetooth identifier devices. Colored lines: functional road classification of the network.

3.4. Functional Road Classification

The Functional Road Classification (FRC) is used to classify roads into homogeneous classes depending on their role in a transportation network [17]. This classification determines the type of use of each road, for instance, as it differentiates between major roads that experience heavy traffic from a variety of O/D pairs, and minor roads which are inside of a residential area and experience light traffic only. Figure 3 shows the FRC of each road of the considered zone of Grenoble, and Table 1 shows each class description (Source: https://developer.tomtom.com/traffic-stats/support/faq/what-are-functional-road-classes-frc, accessed on 29 June 2021) . Roads that are not colored in the figure have class 7.

Table 1. Description of the road classes provided by TomTom.

Class	Short Description	Long Description
1	Major roads of high importance	Roads of high importance that are used for international and national traffic.
2	Other major roads	Roads used to travel between neighboring country regions.
3	Secondary roads	Roads used to travel between parts of the same region.
4	Local connecting roads	Roads making settlements accessible or making parts of a settlement accessible.
5	Local roads of high importance	Local roads that are the main connections in a settlement.
6	Local roads	Roads used to travel within a part of a settlement.
7	Local roads of minor importance	Roads that only have a destination function.

4. Parameter Estimation

To estimate the values of the turning ratio parameters for the intersections that have no direct measurement (see Section 3.3), we propose the use of the FRC information. The reasoning for this is that roads with a higher importance are more commonly used than smaller roads, and will therefore present higher turning ratios.

For each FRC class in the set $\{1, 2, \ldots, 7\}$, we define a weight $\theta \in (0, 1]$. Let $\boldsymbol{\theta} \in (0, 1]^7$ be the vector of class weights. Suppose that the turning ratios at each intersection are distributed proportionally to the class weights of each of its outgoing roads. Thus, these parameters are computed as

$$r_{i,j}^{\mathrm{FRC}} = \frac{\theta_{\mathrm{FRC}(j)}}{\sum_{k \in \mathcal{O}(n_i)} \theta_{\mathrm{FRC}(k)}} \tag{10}$$

where $\mathrm{FRC}(i)$ is the FRC class of road i, n_i is the intersection connected to i, and $\mathcal{O}(n_i)$ is the set of outgoing roads from intersection n_i.

To compute the value of $\boldsymbol{\theta}$, we consider the following optimization problem:

$$\begin{aligned} \min_{\boldsymbol{\theta}} \quad & ||\bar{\mathbf{y}} - C(\mathbb{I} - \hat{R}^\top(\boldsymbol{\theta}))^{-1} B\bar{\mathbf{u}}|| \\ \text{subject to} \quad & \boldsymbol{\theta} \in (0,1]^7, \\ & \theta_1 = 1, \\ & \hat{R}_{i,j}(\boldsymbol{\theta}) = \begin{cases} r_{i,j}^{\mathrm{BT}} & \text{if} \quad n_i \in \mathcal{B} \\ r_{i,j}^{\mathrm{FRC}} & \text{if} \quad n_i \notin \mathcal{B} \end{cases} \end{aligned} \tag{11}$$

where C is a selection matrix which identifies the outgoing roads $\mathcal{E}^{\mathrm{out}}$, and

$$\bar{\mathbf{u}} = \frac{1}{T}\int_0^T \mathbf{u}(t) \quad , \quad \bar{\mathbf{y}} = \frac{1}{T}\int_0^T \mathbf{y}(t). \tag{12}$$

are the average flows from the input and output sets, respectively. This optimization problem tries to match the observed outflows of the network with the outflows computed from the measured inflows and the turning ratio estimates,

$$\hat{\mathbf{y}} = C(\mathbb{I} - \hat{R}^{\top})^{-1} B\bar{\mathbf{u}}. \tag{13}$$

The condition $\theta_1 = 1$ is set arbitrarily without loss of generality, as only the relative differences between the weights are important.

Due to the limited number of parameters, this problem can be solved with common optimization solvers. We obtained the values shown in Table 2 (The considered network has no road with FRC 2. Thus, its weight does not affect the calculations). Note that, as the importance of the road decreases, so does the corresponding class weight as is to be expected.

Table 2. Value of FRC weights for the estimation of turning ratio parameters.

Class index	1	2	3	4	5	6	7
Class weight θ	1.00	N/A	1.00	0.50	0.23	0.13	0.03

5. Experimental Results

For evaluation purposes, we considered the data collected for 8 January 2021. Figure 4 shows the time series for the real and estimated flows for a selection of validation sensors. Note that, for most cases, the estimated and real values have a very similar trajectory. The mismatches obtained for some of the sensors can be attributed to several factors. The main source of error is due to deviations between the real and estimated turning ratios. As these parameters were computed using a simplifying hypothesis using the FRC, there are intersections for which the obtained values present error. However, this method provides a simple to use manner to compute these parameters for large networks with easily obtainable information, and provides good initial results for a large number of locations which can be improved with time. Another possible error source is the presence of internal sources and sinks of traffic flow which are not taken into account.

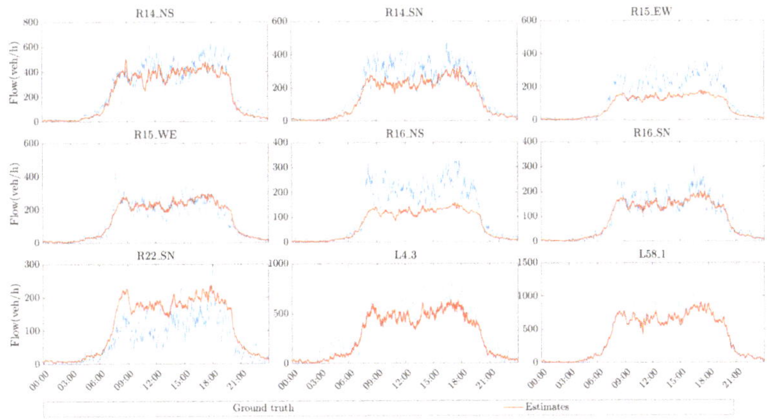

Figure 4. Ground truth flows obtained from cross-validation sensors, and the corresponding estimated flows.

To quantify the error in time for each location, we use as error metrics the Relative Mean Error (RME) and the Relative Absolute Error (RAE), defined as

$$\mathrm{RME}_i = \frac{\left|\int_0^T \varphi_i^{\mathrm{out}}(t) - \hat{\varphi}_i^{\mathrm{out}}(t)\mathrm{d}t\right|}{\int_0^T \varphi_i^{\mathrm{out}}(t)\mathrm{d}t}, \quad \mathrm{RAE}_i = \frac{\int_0^T \left|\varphi_i^{\mathrm{out}}(t) - \hat{\varphi}_i^{\mathrm{out}}(t)\right|\mathrm{d}t}{\int_0^T \varphi_i^{\mathrm{out}}(t)\mathrm{d}t}. \tag{14}$$

Figure 5 shows the obtained error metrics for all the available validation sensors. The RME shows that the proposed estimator provides close estimates to the real values, as half of the validation locations present an error under 20%, and all cases presented an error under 50%. When considering the RAE, the error increases as this metric considers not only the differences between the mean trajectories, but also takes into account the dispersion of the real data. Nevertheless, for half of the locations, the RAE lies between 20% and 30% showing a good agreement of the estimation with the real data. Similarly to the RME, all locations have a RAE under 50%.

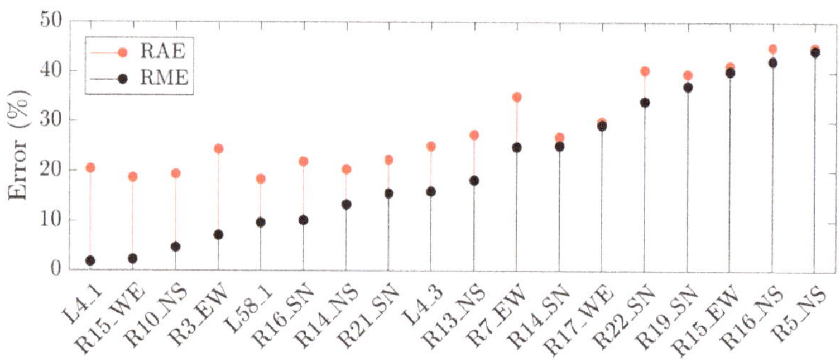

Figure 5. RME and RAE obtained for the available validation and output sensors.

6. Conclusions

In this paper, we proposed a data-based flow and density estimator that uses heterogeneous data sources such as stationary counting sensors, FCD, and Bluetooth devices. The estimator was tested using real data from the city of Grenoble, France, using the sensing infrastructure developed in the project GTL-Ville.

Although the problem of TSE in large networks is challenging, the obtained results are encouraging as the estimated flow for individual roads are very close to the ground truth data provided by sensors. When considering all the available validation locations, more than half of the mean trajectories presented an error below 20%. For some locations, there is a mismatch between the predicted flow and the real one. Nevertheless, even in this case, the obtained errors were below 45%. We identify as the main error source the uncertainty in the values of the turning ratios, as only a few locations are computed using real data. However, this can be improved in the future by performing more measuring campaigns, so the estimation results in the real application are expected to improve significantly.

Author Contributions: M.R.-V.: Conceptualization, Data curation, Formal analysis, Methodology, Software, Visualization, Writing—Original draft; C.C.-d.-W.: Conceptualization, Formal analysis, Funding acquisition, Methodology, Project administration, Supervision, Writing—Review and editing; H.F.: Conceptualization, Formal analysis, Methodology, Supervision, Writing—Review and editing. All authors have read and agreed to the published version of the manuscript.

Funding: This work is supported by the European Research Council (ERC) under the European Unions Horizon 2020 research and innovation program, ERC-AdG No. 694209, Scale-FreeBack (http://scale-freeback.eu/, accessed on 29 June 2021).

Institutional Review Board Statement: Ethics approval was not sought, as the collected information includes no personal data. The used sensors do not allow to identify individuals, and no video, images, or license plates can be obtained. Available information is only an aggregated number of traffic counts at given locations.

Informed Consent Statement: Subject consent was waived as no personal data is collected. This research involves no risk to participants, and only aggregated traffic count and speeds are collected.

Data Availability Statement: Publicly available datasets were analyzed in this study. These data can be found here: http://gtlville.inrialpes.fr/ (accessed on 29 June 2021).

Conflicts of Interest: The authors declare no conflict of interest.

Abbreviations

The following abbreviations are used in this manuscript:

TSE	Traffic State Estimation
FCD	Floating Car Data
AVI	Automatic Vehicle Identifier
ITS	Intelligent Transportation Systems
CTM	Cell Transmission Model
FRC	Functional Road Classification

References

1. Treiber, M.; Kesting, A. *Traffic Flow Dynamics: Data, Models and Simulation*; Springer: Berlin/Heidelberg, Germany, 2013; pp. 1–503.
2. Seo, T.; Bayen, A.M.; Kusakabe, T.; Asakura, Y. Traffic state estimation on highway: A comprehensive survey. *Annu. Rev. Control.* **2017**, *43*, 128–151. [CrossRef]
3. Ferrara, A.; Sacone, S.; Siri, S. *Freeway Traffic Modelling and Control*, 1st ed.; Advances in Industrial Control; Springer International Publishing: Cham, Switzerland, 2018.
4. Lighthill, M.J.; Whitham, G.B. On kinematic waves II. A theory of traffic flow on long crowded roads. *Proc. R. Soc. London. Ser. A Math. Phys. Sci.* **1955**, *229*, 317–345.
5. Richards, P.I. Shock Waves on the Highway. *Oper. Res.* **1956**, *4*, 42–51. [CrossRef]
6. Daganzo, C.F. The cell transmission model: A dynamic representation of highway traffic consistent with the hydrodynamic theory. *Transp. Res. Part B* **1994**, *28*, 269–287. [CrossRef]
7. Tampère, C.M.; Immers, L.H. An extended Kalman filter application for traffic state estimation using CTM with implicit mode switching and dynamic parameters. In Proceedings of the 2007 IEEE Intelligent Transportation Systems Conference, Seattle, WA, USA, 30 September–3 October 2007; pp. 209–216.
8. Canudas-de Wit, C.; Ojeda, L.L.; Kibangou, A.Y. Graph constrained-CTM observer design for the Grenoble south ring. *FAC Proc. Vol.* **2012**, *45*, 197–202. [CrossRef]
9. Canepa, E.S.; Claudel, C.G. Networked traffic state estimation involving mixed fixed-mobile sensor data using Hamilton-Jacobi equations. *Transp. Res. Part B Methodol.* **2017**, *104*, 686–709. [CrossRef]
10. Jabari, S.E. Node modeling for congested urban road networks. *Transp. Res. Part B Methodol.* **2016**, *91*, 229–249. [CrossRef]
11. Daganzo, C.F. The cell transmission model, part II: Network traffic. *Transp. Res. Part B* **1995**, *29*, 79–93. [CrossRef]
12. Lovisari, E.; Canudas-de Wit, C.; Kibangou, A.Y. Density/Flow reconstruction via heterogeneous sources and Optimal Sensor Placement in road networks. *Transp. Res. Part C Emerg. Technol.* **2016**, *69*, 451–476. [CrossRef]
13. Ladino, A.; Canudas-de Wit, C.; Kibangou, A.; Fourati, H.; Rodriguez, M. Density and flow reconstruction in urban traffic networks using heterogeneous data sources. In Proceedings of the 2018 European Control Conference (ECC), Limassol, Cyprus, 12–15 June 2018; pp. 1679–1684.
14. Liou, H.T.; Hu, S.R.; Peeta, S. *Estimation of Time-Dependent Intersection Turning Proportions for Adaptive Traffic Signal Control under Limited Link Traffic Counts from Heterogeneous Sensors*; Technical Report; NEXTRANS Center, Purdue University: West Lafayette, IN, USA, 2017.
15. Rostami Shahrbabaki, M.; Safavi, A.A.; Papageorgiou, M.; Setoodeh, P.; Papamichail, I. State estimation in urban traffic networks: A two-layer approach. *Transp. Res. Part C Emerg. Technol.* **2020**, *115*, 102616. [CrossRef]
16. Bianchin, G.; Pasqualetti, F.; Kundu, S. Resilience of Traffic Networks with Partially Controlled Routing. In Proceedings of the 2019 American Control Conference (ACC), Philadelphia, PA, USA, 10–12 July 2019; pp. 2670–2675.
17. D'Andrea, A.; Cappadona, C.; La Rosa, G.; Pellegrino, O. A functional road classification with data mining techniques. *Transport* **2014**, *29*, 419–430. [CrossRef]

engineering proceedings

MDPI

Proceedings

Predictability of Scrub Typhus Incidences Time Series in Thailand †

Valeria Bondarenko [1,*], Pierre Mazzega [1,2] and Claire Lajaunie [2,3]

1 Solidarity, Societies, Territories, Interdisciplinary Laboratory LISST UMR 5193, CNRS—University of Toulouse Jean-Jaurès, 31000 Toulouse, France; pmazzega@gmail.com
2 Strathclyde Centre for Environmental Law and Governance (SCELG), University of Strathclyde, Glasgow G11, UK
3 INSERM, Laboratory Population Environment Development, IRD UMR 151, Aix Marseille University, IRD, 13331 Marseille, France; claire.lajaunie@gmail.com
* Correspondence: valeria_bondarenko@yahoo.com
† Presented at the 7th International conference on Time Series and Forecasting, Gran Canaria, Spain, 19–21 July 2021.

Abstract: Scrub typhus, an infectious disease caused by a bacterium transmitted by "chigger" mites, constitutes a public health problem in Thailand. Predicting epidemic peaks would allow implementing preventive measures locally. This study analyses the predictability of the time series of incidence of scrub typhus aggregated at the provincial level. After stationarizing the time series, the evaluation of the Hurst exponents indicates the provinces where the epidemiological dynamics present a long memory and are predictable. The predictive performances of ARIMA (autoregressive integrated moving average model), ARFIMA (autoregressive fractionally integrated moving average) and fractional Brownian motion models are evaluated. The results constitute the reference level for the predictability of the incidence data of this zoonosis.

Keywords: infectious disease; predictability; incidence; time series; ARIMA; ARFIMA; fBm; scrub typhus; Thailand

Citation: Bondarenko, V.; Mazzega, P.; Lajaunie, C. Predictability of Scrub Typhus Incidences Time Series in Thailand . *Eng. Proc.* **2021**, *5*, 44. https://doi.org/10.3390/engproc2021005044

Published: 13 July 2021

1. Introduction

Scrub typhus is a relatively underdiagnosed infectious disease, affecting approximately 1 million people worldwide. It is caused by *Orientia tsutsugamushi*, a bacterium transmitted by "chigger" mites that mostly parasitize rodents, humans being only accidentally a host. In the Mekong region of Southeast Asia, the frequency of isolation of pathogens implicated in the aetiology of non-malarial febrile illness has shown that *Orientia tsutsugamushi* was, immediately following dengue virus, the most frequently reported [1,2]. In Thailand, the sporadic incidence in humans reported by the Armed Forces Research Institute of Medical Sciences for 2018 and 2019, respectively, was 7.11 and 5.82 individuals per 100,000 populations [3]. Although scrub typhus is endemic in various countries of Asia and Southeast Asia [4], it remains an underdiagnosed disease and its public health burden is still poorly known [5]. While knowing the distribution and the factors that may contribute to the disease incidence is essential for public health [6], it is also crucial to understand the seasonal trend of scrub typhus in order to be ready to adopt the appropriate public health response. Indeed, the study of time series of incidence of scrub typhus aggregated at the provincial level may help to forecast future incidence [7], to monitor the evolution of the trend of the diseases and thus to prevent and control the outbreaks of the disease by providing a way to facilitate decision-making and determine whether an apparent excess represents an outbreak rather than a random variation [8].

Anticipating the timing and severity of the peak can improve the decision-making in real time [9,10] and help allocating the resources needed during public health emergencies [11]. The analysis of the predictability of the time series of incidence of scrub typhus also presents a double interest from the point of view of the knowledge of this infectious disease and its modelling: (a) it involves evaluating a property intrinsic to its epidemiological dynamics (the long-term dependence of the number of cases being likely to vary significantly from one province to another.); (b) it offers a decisive criterion for testing the performance of the models and their usefulness for prevention. After presenting the data from the Thai Ministry of Public Health and their aggregation in Section 2, the predictive models used are briefly presented in Section 3 as well as the criteria used for their evaluation. The results concerning the predictability of scrub typhus time series are gathered and discussed in Section 4, before concluding in Section 5.

2. Data Time Series of Scrub Typhus Incidences

The analyzed data are provided by the Thai Ministry of Public Health. More than 110,000 individual cases have been diagnosed in Thailand between January 2003 and December 2019. Each data record corresponds to a case and includes the code indicating the geographical position of the hospital which took care of the patient, and other information such as her/his age and occupation.

For our analysis, we retain as the date the one associated with the entry to the hospital. The number of cases considered at too high a spatial scale of resolution leads to unpredictable processes. Aggregating data at the provincial level and for each month is a good compromise between structuring a consistent signal and sufficient length of the time series. Figure 1 shows the series obtained for two provinces in northern Thailand. The patterns of these two time-series are similar to those obtained for the other provinces (as well as for the whole Thailand time series) where scrub typhus is endemic. The number of cases is low until 2006–2007. Then the annual cycle becomes wider, with a strong annual variability. A decrease in the cycle in 2016 and 2017 is not confirmed in the following years. We can also see a non-linear rise in the base level, which we will consider to be a trend. Our analysis focuses on those provinces where the number of cases (which we will sometimes call "incidence" a little improperly but for better readability of the text) is relatively high. In fact, the majority of the 76 provinces only present very sporadic and low incidence episodes. For this same reason, the prediction and the prevention that could make use of it are not suitable for prompting the implementation of dedicated public health measures.

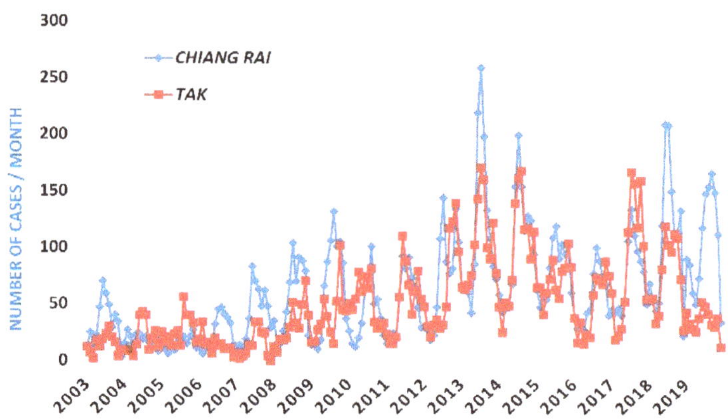

Figure 1. Time series of the number of cases of scrub typhus per month from January 2003 to December 2019 in Chiang Rai and Tak provinces.

3. Incidence Time Series Modelling

The ARIMA (autoregressive integrated moving average model), ARFIMA (autoregressive fractionally integrated moving average) and fractional Brownian motion(fBm) models are briefly described followed by the three criteria used to select the best performing candidate for making predictions. The ARIMA time series prediction method was proposed by

Box and Jenkins in the 1970s [12]. In general, a stationary sequence can establish a metrology model. The unit root test is used to evaluate the stationarity of the sequence. Some non-stationary sequences are convertible in stationary sequences with one or several applications of a difference operation (subtracting the previous datum to the current one). A difference-stationary or unit root I(D) process is a process that makes a sequence stationary by taking D differences, the integer D being the degree of the model. The ARIMA (p, D, q) model is essentially a combination of difference operations and ARMA (p, q) model (linear combination of the previous p terms for the autoregressive polynomial in the lag operator, and updated q terms entering the moving average polynomial). After transforming the original data in a stationary time series (when possible), an ARIMA model is sought by estimating its parameters. The data autocorrelation and partial autocorrelation functions are used to estimate the (p,q) parameters. Then the polynomial coefficients of the model are estimated using a least squares or maximum likelihood method [13,14].

The ARFIMA model [15] is a good candidate for data time series with long-term memory. Recently, Kartikasari et al. [16] successfully applied ARFIMA models to predict the occurrence of new cases of patients dying from coronavirus disease 2019 (COVID-19) in Indonesia over a 3-month period. In contrast with ARIMA model, the differencing parameter d which governs the memory of the process is fractional (not integer; [17]) and reflects a property inherent to the processed time series. The type of time series (coloured noise) and model's degrees of freedom required to ensure an accurate prediction of the time series must be determined (with semi-parametric estimators, [18]). Parameter d is estimated by a nonlinear least squares method using time-domain approach [19]. The (p,q) values and coefficients of the ARIMA (p,D,q) and ARFIMA (p,d,q) model are estimated with the Software EViews 3 [20].

Hurst exponent H is a measure of the long-term memory of the stochastic process underlying a time series. A value of $H = 0.5$ indicates the lack of memory in the data (Wiener process, white noise). The closer the value of H to 1, the higher the persistence of long-term addiction (positive correlation). The data sample exhibits a persistent behavior when $0.5 < H \leq 1$. In such case, if the series increases (decreases) in the previous period, then there is a significant probability that it will maintain this tendency for some time in the immediate future. By contrast, the range $0 \leq H < 0.5$ corresponds to an anti-persistent series (negative correlation), a behaviour all the more marked as H approaches zero. There are several methods for estimating the Hurst exponent of time series with different sizes [21–23]. Figure 2 provides the highest Hurst exponents found at the province level in Thailand.

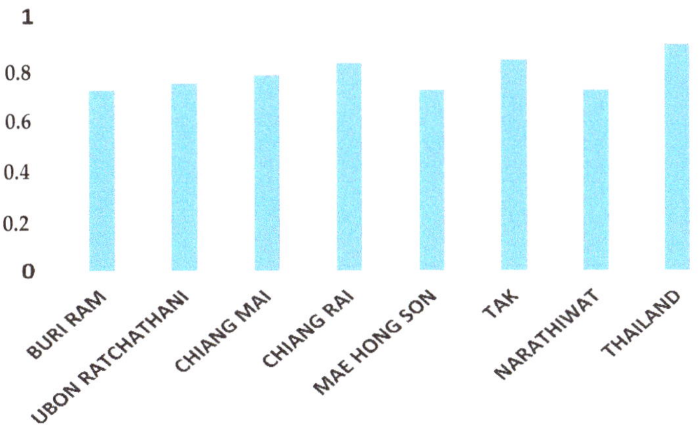

Figure 2. Hurst exponents for Thailand and provinces with $H > 0.7$ and stationary scrub typhus incidence time series.

The forecasting of the values of a Fractional Brownian Motion is possible only for the persistent case. Indeed, the fBm [24,25] generalizes Brownian motion with time increments not necessarily independent. fBm probability distribution is self-similar. It is a function of the Hurst exponent as is fBm autocorrelation. Having a null expectation, its use as a model of empirical time series requires the prior withdrawal of the data trend. We model the data trend with an ARIMA process. The Hurst exponents and fBm models are estimated with a R-package [26–28].

In the test phase of candidate models, predictions of the number of scrub typhus cases are made on dates for which we already have actual observed data. Each model is optimized based on the data of the learning window and the prediction is performed for the following h_p = 1 to h_p = 14 months (h_p denoting a prediction horizon). By doing this over the entire length of a time series of observations, we can estimate the average performance of each model and compare it with that of other models. For this we use the following three criteria. Conventionally, the criterion R2 (or coefficient of determination) measures the proportion of variance of the observed data which is explained by a model (here in predictive mode). The criterion R2 takes the value 1 when the data explained (or predicted) coincide exactly with the observations. The Akaike criterion is a measure of the compromise established between the good fit of the data by a model (using a R2-like criterion) and the parametric frugality of the model. Indeed, too many parameters can lead to an adjustment which integrates noise but too few parameters do not allow us to adjust to all the variability of the signal. When comparing models fitted to the same time series, the one with the lowest Akaike criterion value is to be preferred. The Durbin–Watson test is also applied in order to check that the residues (observed minus predicted data) are not correlated (which would introduce an estimation bias of the R2 performance criterion of the model, while omitting a fraction of the signal). A value of 2 of the test indicates that the residuals are fully uncorrelated. These criteria or tests are applied to all models combined with the learning windows in order to estimate their relative predictive performances. Used in conjunction, they provide a good basis for selecting a model for predicting the number of scrub typhus cases.

4. Incidence Predictability: Results

Three models are in competition: ARIMA, ARFIMA and fBm combined with an ARIMA model to capture the trend of time series (this combined model is denoted fBm*). We rule out the fBm model alone because it does not correctly reproduce the trend of the time series. Each model is optimized to fit the number of cases monthly data in a so-called learning window. The length q of these windows, of 12 (one year), 25 or 50 months (a little more than two and four years respectively), is attached to the label assigned to each model. Adjustments on training sets of q = 6 months have not given satisfactory results and are ruled out. They do not even adequately capture the annual cycle of scrub typhus incidences.

Once adjusted to the monthly data of the training window, a model is used to predict the number of scrub typhus cases over the next 14 months. As the predictions being made in past years, they can be compared to available data of the Thai Ministry of Public Health. The procedure is repeated throughout the data time series. The performances of the different models are thus evaluated and compared. The results are presented here for a prediction horizon h_p ranging from 1 to 3 months, a period of anticipation of incidences useful for the implementation of preventive health measures. Naturally, the performance of the models deteriorates for longer-term predictions. The inspection of the results does not show any exploitable results for longer horizons (h_p $\geq$ 4), which the dominance of the incidences' annual cycle and inter-annual variability could allow one to hope.

National level: each of the three models is applied with the three learning windows to the data aggregated at the Thailand level. To compare the performance of the models and their configuration of use, the values of R2 of the prediction horizons of 1, 2 and 3 months are averaged to give a "R2 indicator". This procedure makes it possible consider the best score—obtained for h_p = 1 month—but also the decrease in this score over the period of interest for health prevention. The same procedure is used to produce an Akaike indicator and a Durbin–Watson indicator. The results are collated in Figure 3 for comparison.

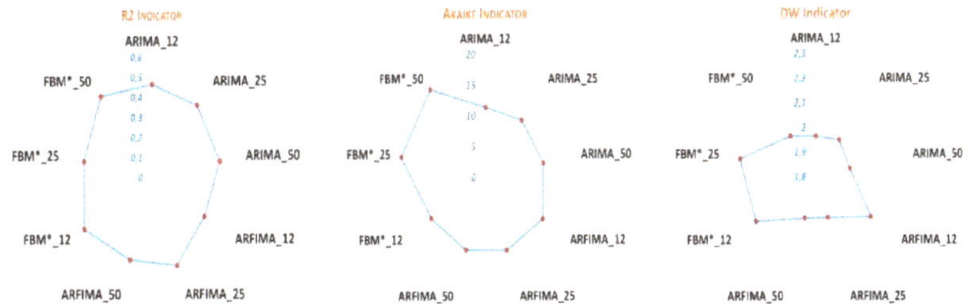

Figure 3. R2, Akaike and Durbin–Watson (DW) indicators of ARIMA (autoregressive integrated moving average model, ARFIMA (autoregressive fractionally integrated moving average) and fbm plus ARIMA-trend" (labelled fBm*) models' performances for 3 learning windows (12, 25 and 50 months) with Thailand level incidence data (see text).

In most cases, the comparison of entities on the basis of three criteria does not constitute an order relation which would unambiguously designate the entity to be retained, but a partial order [29]. Because it directly concerns the prediction of the number of people affected by scrub typhus, we will primarily consider the score of R2 indicator, and use the other two indicators in addition. The sole Akaike's indicator would systematically disqualify the fBm* model which combines the parameters of a fBm model and of an ARIMA model for the trend. However, fBm* gives results of the same order as the other candidates in terms of R2 or Durbin–Watson indicators (whereas ARIMA or ARFIMA models associated with a similar Akaike criterion perform less in terms of R2).

Among the nine configurations tested, fBm*_50 presents the best R2 indicator (and criterion with R2 = 0.57 for h_p = 1) as well as fBm*_25 in second position. Unsurprisingly, these two models have a number of parameters which tend to disqualify them compared to other models, in particular ARFIMA_12. However, with a DW indicator of 2.01, (2.21 for fBm*_25 and 2.13 for ARFIMA_12), it turns out to be our preferred candidate. On the other hand, the configuration using training windows of q = 12 months leads to models that leave correlated prediction residuals (DW indicator furthest from 2). It is excluded from analyses at the province level.

Provincial level: we focus here on four provinces with high Hurst exponents (Table 1). These are also provinces where the number of people infected is the highest each year, while some other provinces have almost no cases. Provincial government bodies enjoy the autonomy necessary to implement preventive health measures in their territory. Therefore, it is at this administrative scale that prediction may be the most useful and effective. By comparing the performance of the models for each province as we did previously (with q = 25 and q = 50 learning windows), we end up with the selection of models presented in Table 1. Except for the province of Tak, the R2 indicators are higher than 0.5. The best score (0.62) is obtained in Chiang Mai, a province of northern Thailand. The R2 associated with the only prediction horizon h_p = 1 month is 0.72.

Table 1. Best models for provinces with Hurst exponents higher than 0.7, and their performance indicators. The R2 score for h_p = 1 is given in parenthesis.

Province	Hurst	Best Model	R2 I Akaike I DW Indicators
U. Ratchathani	0.74	fBm_25	0.52 (0.53) I 16.05 I 2.02
Chiang Mai	0.77	ARFIMA_50	0.62 (0.72) I 9.70 I 1.99
Chiang Rai	0.82	fBm*_50	0.59 (0.63) I 20.10 I 2.02
Tak	0.83	ARFIMA_25	0.47 (0.51) I 10.22 I 1.97
Thailand	0.93	fBm*_50	0.53 (0.57) I 18.55 I 2.01

The other provinces show scores comparable to those of Thailand, although their Hurst exponent, and therefore the memory measure of the number of people infected with scrub typhus, is significantly lower. The Durbin–Watson indicator indicates a lack of correlation of the residuals, and the Akaike indicator primarily reports the number of parameters in each model. Figure 4 shows as an example the observed time series of Tak province, the predicted data with 3 months of antecedence (h_p = 3) as well as the scatter plot of the prediction residues.

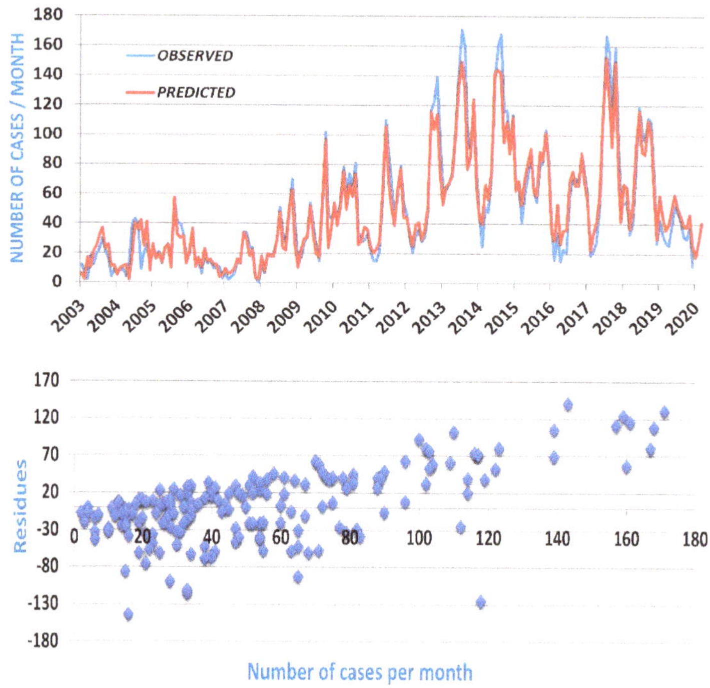

Figure 4. (**Top**) Observed and predicted time series of scrub typhus cases for Tak province with ARFIMA_25 model and hp = 3 months; (**Bottom**) scatter plot of the residues (observed minus predicted data).

Very similar figures are obtained for the other provinces and for Thailand. The number of scrub typhus cases is a positive variable. The models predict fairly well the local minima of the series, associated with the months of December through April of the following year. The scatter diagram (Figure 4 bottom) shows that the maxima are almost systematically underestimated. The models provide a low estimate of the number of scrub typhus cases to come in the following months which, in itself, is useful information for provincial governments. However, this is another observation that catches our attention. It is neither always the same type of model (fBm* or ARFIMA), nor the same training length (q = 25 or q = 50), which offers the best predictive capacity. Added to this is the fact that in some provinces an ARIMA model has a performance only slightly inferior to that of the best model. On the other hand, nothing in Table 1 indicates that a long learning window is preferable when the Hurst exponent is higher, neither that national-level data aggregation is more predictable (compare models' performances in Chiang Mai or Chiang Rai with Thailand). Nevertheless, the results show that models are available to produce a low estimate of the number of scrub typhus cases in the next 2 or 3 months in the provinces where the disease is endemic and has the highest incidence.

Thailand Overview and Discussion

None of our results militate in favour of a "universal" model, in the sense that it would apply to all provinces with equal success and would emerge as the best representation of the epidemic dynamics of scrub typhus, at least in Thailand. It should also be remembered that in most provinces the series are generally not stationary, nor can they even be transformed into stationary series. Our understanding, therefore, tends more in the direction of an out-of-equilibrium dynamic, somewhat latent (weak background signal), which would take off towards a more stable regime in favour of contingent local ecological, social and environmental conditions. It is the establishment of this stable regime and time series regular behaviour that would condition the capacity to predict with a few months of antecedence the number of scrub typhus cases, aggregated at the correct spatial and temporal scales (in this case the province and the month).

Therefore, the most useful strategy is to select a model and a learning window adjusted to each province whose Hurst exponent is greater than a threshold value (e.g., H_threshold = 0.6). The prediction horizon must also be adapted to each provincial context so as to operationalize a compromise between: (1) an acceptable level of performance for the predictions; (2) taking into account the incubation times of the disease and diagnosis in a hospital; (3) taking into account the delays for implementing public health measures aimed at preventing the resurgence of cases and reducing the incidence of scrub typhus.

The preventive public health actions, scaled according to the low estimates of the predictions, could be implemented by the authorities in the districts where the cases are concentrated rather than at the level of an entire province. Data analysis identifies those districts whose relatively persistent ecological and social factors favour the annual recurrence of local scrub typhus epidemics. Let us add that the results presented here give a reference level against which to develop other more efficient modelling approaches (if there are any). Along with other model performance indicators that may obscure their ultimate utility, the production of reliable predictions of scrub typhus incidences remains both an excellent criterion for qualifying models and a challenge for research.

5. Conclusions

In the Thai provinces where scrub typhus is endemic, the prediction of the monthly number of cases is a public health issue of interest for implementing preventive measures. Applying a difference operation makes the observed data time series stationary. The selected series have a Hurst exponent greater than 0.7 and, therefore, a long-term memory making them suitable for a prediction over horizons of several months. The ARIMA, ARFIMA and fBm models of these time series, adjusted on moving training windows of approximately 1, 2 and 4 years, are put in competition in order to identify the best prediction options and evaluate their performance from three criteria (R2, Akaike and Durbin–Watson). Predictions for horizons spanning 1 to 14 months are compared to existing observations (up to 2019). The results obtained allow operational use of the predictions of scrub typhus epidemic events in the provinces concerned with up to three months of antecedent. However, it is neither the same model nor the same size of learning window that give the best predictions from province to province. The performance of the models presented constitutes a benchmark against which several improvement strategies are possible. These analyses also suggest that the ability to predict the number of scrub typhus cases (and even of other infectious diseases) several months in advance is a real challenge that modelling should not avoid.

Author Contributions: Conceptualization, V.B. and P.M.; methodology, V.B. and P.M.; software, validation and formal analysis, V.B.; writing—review and editing, P.M., C.L.; supervision, P.M., C.L. All authors have read and agreed to the published version of the manuscript.

Funding: This research was funded by the Agence Nationale de la Recherche, Research project FutureHealthSEA "Predictive scenarios of health in Southeast Asia: linking land use and climate changes to infectious diseases", ANR-17-CE35-0003-01.

Acknowledgments: We gratefully thank H. Soawapak for giving us the opportunity to access the data and R. Kumlert, for providing the selected data, both at the Department of Disease Control, Ministry of Public Health in Thailand.

Conflicts of Interest: The authors declare no conflict of interest.

References

1. Acestor, N.; Cooksey, R.; Newton, P.N.; Menard, D.; Guerin, P.J.; Nakagawa, J.; Christophel, E.; Gonzalez, I.J.; Bell, D. Mapping the aetiology of non-malarial febrile illness in Southeast Asia through a systematic review—Terra Incognita impairing treatment policies. *PLoS ONE* **2012**, *7*, e44269. [CrossRef]
2. Aung, A.K.; Spelman, D.W.; Murray, R.J.; Graves, S. Rickettsial infections in Southeast Asia: Implications for local populace and febrile returned travelers. *Am. J. Trop. Med. Hyg.* **2014**, *91*, 451–460. [CrossRef] [PubMed]
3. Prompiram, P.; Poltep, K.; Pamonsupornvichit, S.; Wongwadhunyoo, W.; Chamsai, T.; Rodkvamtook, W. Rickettsiae exposure related to habitats of the oriental house rat (Rattus tanezumi, Temminck, 1844) in Salaya suburb, Thailand. *Int. J. Parasitol. Parasites Wildl.* **2020**, *13*, 22–26. [CrossRef] [PubMed]
4. Xu, G.; Walker, D.H.; Jupiter, D.; Melby, P.C.; Arcari, C.M. A review of the global epidemiology of scrub typhus. *PLoS Negl. Trop. Dis.* **2017**, *11*, e0006062. [CrossRef] [PubMed]
5. Bonell, A.; Lubell, Y.; Newton, P.N.; Crump, J.A.; Paris, D.H. Estimating the burden of scrub typhus: A systematic review. *PLoS Negl. Trop. Dis.* **2017**, *11*, e0005838. [CrossRef] [PubMed]
6. Wangrangsimakul, T.; Elliott, I.; Nedsuwan, S.; Kumlert, R.; Hinjoy, S.; Chaisiri, K.; Day, N.; Morand, S. The estimated burden of scrub typhus in Thailand from national surveillance data (2003–2018). *PLoS Negl. Trop. Dis.* **2020**, *14*, e0008233. [CrossRef] [PubMed]
7. Gao, J.; Li, J.; Wang, M. Time series analysis of cumulative incidences of typhoid and paratyphoid fevers in China using both Grey and SARIMA models. *PLoS ONE* **2020**, *15*, e0241217. [CrossRef] [PubMed]
8. Allard, R. Use of time-series analysis in infectious disease surveillance. *Bull World Health Organ.* **1998**, *76*, 327–333. [PubMed]
9. Holloway, R.; Rasmussen, S.A.; Zaza, S.; Cox, N.J.; Jernigan, D.B. Updated preparedness and response framework for influenza pandemics. MMWR. Recommendations and reports. *MMWR Morb. Mortal. Wkly. Rep. Recomm. Rep.* **2014**, *63*, 1–18.
10. Lutz, C.S.; Huynh, M.P.; Schroeder, M.; Anyatonwu, S.; Dahlgren, F.S.; Danyluk, G.; Fernandez, D.; Greene, S.K.; Kipshidze, N.; Liu, L.; et al. Applying infectious disease forecasting to public health: A path forward using influenza forecasting examples. *BMC Public Health* **2019**, *19*, 1659. [CrossRef] [PubMed]
11. Fischer, L.S.; Santibanez, S.; Hatchett, R.J.; Jernigan, D.B.; Meyers, L.A.; Thorpe, P.G.; Meltzer, M.I. CDC Grand Rounds: Modeling and public health decision-making. *MMWR Morb. Mortal. Wkly. Report.* **2016**, *65*, 1374–1377. [CrossRef] [PubMed]
12. Box, G.E.P.; Jenkins, G.M. *Time Series Analysis Forecasting and Control*; Holden-Day: San Francisco, CA, USA, 1970.
13. Beran, J. *Statistics for Long-Memory Processes*; Chapman and Hall: London, UK, 1994.
14. Hyndman, R.; Koehler, A.B.; Ord, J.K.; Snyder, R.D. *Forecasting with Exponential Smoothing: The State Space Approach*; Springer: Berlin, Germany, 2008.
15. Granger, C.W.J.; Joyeux, R. An Introduction to Long-Range Time Series Models and Fractional Differencing. *J. Time Ser. Anal.* **1980**, *1*, 15–29. [CrossRef]
16. Kartikasari, P.; Yasin, H.; Maruddani, D.A. ARFIMA model for short term forecasting of new death cases COVID-19. In *E3S Web of Conferences*; EDP Sciences: Paris, France; London, UK, 2020; Volume 202, p. 13007.
17. Hosking, J.R.M. Fractional differencing. *Biometrika* **1981**, *68*, 165–176. [CrossRef]
18. Robinson, P.M. (Ed.) *Time Series Analysis with Long Memory*; Oxford University Press: Oxford, UK, 2003.
19. Breidt, F.J.; Crato, N.; De Lima, P. The detection and estimation of long memory in stochastic volatility. *J. Econom.* **1998**, *83*, 325–348. [CrossRef]
20. Ma, L.; Hu, C.; Lin, R.; Han, Y. ARIMA model forecast based on EViews software. *IOP Conf. Ser. Earth Environ. Sci.* **2018**, *208*, 012017. [CrossRef]

21. Willinger, W.; Taqqu, M.; Erramilli, A. Bibliographical guide to self-similar traffic and performance modeling for modern high-speed network. In *Stochastic Networks: Theory and Applications*; Kelly, F.P., Zachary, S., Ziedins, I., Eds.; Claredon Press—Oxford University Press: Oxford, UK, 1996; Chapter 20; pp. 339–366.
22. Clegg, R.G. A practical guide to measuring the Hurst parameter. *Int. J. Simul. Syst. Sci. Technol.* **2005**, *7*, 3–14.
23. Coeurjolly, J.-F. Hurst exponent estimation of locally self-similar Gaussian processes using sample quantiles. *Ann. Stat.* **2008**, *36*, 1404–1434. [CrossRef]
24. Mandelbrot, B.; van Ness, J.W. Fractional Brownian motions, fractional noises and applications. *SIAM Rev.* **1968**, *10*, 422–437. [CrossRef]
25. Mishura, Y. *Stochastic Calculus for Fractional Brownian Motion and Related Processes*; Lecture Notes in Mathematics; Springer: Berlin, Germany, 2008.
26. Buhovets, A.G.; Moskalev, P.V.; Bogatova, V.P.; Ya Biryuchinskaya, T. *Statistical Analysis of the Data in the R*; VGAU: Voronezh, Russia, 2010.
27. McLeod, A.I.; Yu, H.; Mahdi, E. Time Series Analysis with R. *Time Ser. Anal. Methods Appl.* **2012**, *30*, 661–712. [CrossRef]
28. Shang, H.L. FTSA: An R package for analyzing functional time series. *R J.* **2013**, *5*, 64–72. [CrossRef]
29. Mazzega, P.; Lajaunie, C.; Leblet, J.; Barros-Platiau, A.F.; Chansardon, C. How to compare bundles of national environmental and development indexes? In *Law, Public Policies and Complex Systems: Networks in Action*; Boulet, R., Lajaunie, C., Mazzega, P., Eds.; Law, Gov. & Tech. Series 16; Springer: Berlin, Germany, 2019; pp. 243–265.

engineering proceedings

MDPI

Proceedings

Financial Time Series: Market Analysis Techniques Based on Matrix Profiles †

Eoin Cartwright *,‡, **Martin Crane** ‡ and **Heather J. Ruskin** ‡

Modelling & Scientific Computing Group (ModSci), School of Computing, Dublin City University, D09Y074 Dublin 9, Ireland; martin.crane@dcu.ie (M.C.); heather.ruskin@dcu.ie (H.J.R.)
* Correspondence: eoin.cartwright3@mail.dcu.ie
† Presented at the 7th International Conference on Time Series and Forecasting, Gran Canaria, Spain, 19–21 July 2021.
‡ These authors contributed equally to this work.

Abstract: The Matrix Profile (*MP*) algorithm has the potential to revolutionise many areas of data analysis. In this article, several applications to financial time series are examined. Several approaches for the identification of similar behaviour patterns (or *motifs*) are proposed, illustrated, and the results discussed. While the *MP* is primarily designed for single series analysis, it can also be applied to multi-variate financial series. It still permits the initial identification of time periods with indicatively similar behaviour across individual market sectors and indexes, together with the assessment of wider applications, such as general market behaviour in times of financial crisis. In short, the *MP* algorithm offers considerable potential for detailed analysis, not only in terms of motif identification in financial time series, but also in terms of exploring the nature of underlying events.

Keywords: financial time series; matrix profile; time series motifs

check for updates

Citation: Cartwright, E.; Crane, M.; Ruskin, H.J. Financial Time Series: Market Analysis Techniques Based on Matrix Profiles. *Eng. Proc.* **2021**, *5*, 45. https://doi.org/10.3390/engproc2021005045

Published: 16 July 2021

1. Introduction

Time series *motifs* (repeated, matched or partially matched sequences) occur both within and between individual time series [1]. Motif discovery is the task of extracting previously unknown recurrent patterns from such data sets [2] with applications in fields ranging from music [3] to seismology [4], and of course, to finance, facilitating attempts to assess the importance of historical events and predict future trends.

In the financial domain, a wide range of motif discovery approaches have been explored to date, including that of piecewise aggregate approximation (*PAA*) [5], used to investigate historical Standards and Poor's *S&P500* index data. In addition, a *motif tracking algorithm* was used to examine motifs in a West Texas intermediate (*WTI*) crude oil daily price time series (a popular indicator of oil prices in general) [6].

A spatio-temporal pattern-mining approach was also applied to the examination of company portfolios, where for each company examined, this was taken to correspond to a moving trajectory over a two-dimensional financial grid (for discretised size and *price-to-book* ratio) [7]. A set of similar financial trajectories taken over the same time period was then considered to be a motif. For a more detailed review of currently available motif discovery and evaluation techniques for financial applications, as can be seen, e.g., [8].

Among the motif discovery algorithms that we have investigated [8,9], a new data construct based upon an efficient *nearest neighbours* discovery method and designated as the *Matrix Profile* (*MP*) [10] has already clearly demonstrated considerable potential for its extension and flexibility of application. Thus, we are less concerned here with the relative superiority of the *MP* on a point-by-point basis as compared to other motif discovery algorithms, but rather a demonstration of how visual tools, namely *MP* plots, can provide insight on single and multiple financial series data and their *macro-economic* interpretation.

While such tools have their limitations (discussed in Sections 3.3 and 3.5), considerable insight can be obtained on series coincidences and responses to events of different types, including the identification of potential hedging opportunities.

To demonstrate relevance, *MP* plots created using *Matlab* were used to identify similar patterns (*motifs*) within a single series. The impact, on the plot evolution of increasing *motif length* was also examined, where this can indicate the persistence of given behaviour over longer timescales. Additionally, histogram plots of *MP* data can illustrate whether the proportion of matches (*motifs*) or mismatches (*discords*) is greater for a given financial time series.

The examination of multi-dimensional *MP* plots for localised minima allows the combination of different measures for a single financial series to be explored. Additionally, periods of similar behaviour both *within* and *across* market sectors can be demonstrated in representative time series, while individual stocks contributing to a given index can also be investigated. *MP* use is illustrated for the financial crisis period, from January 2007 to January 2009, and verified against the relevant raw series.

2. Materials and Methods

The *Matrix Profile* (*MP*) is a novel algorithm (proposed by the Keogh research group) that has proven useful for numerous data mining and time series analysis tasks [11]. As the *MP* is highly scalable for time series *sub-sequence all-pairs-similarity* search [10], it efficiently identifies time series motifs and discords (i.e., mis-matches). Thus, the examination of *MP* plots can aid in the interpretation of distinctive or recurring patterns in financial time series.

The main advantages (amongst others) of the *MP* algorithm are that it:

- Returns an exact solution for motif discovery.
- Requires only one input parameter (sub-sequence length m).
  - For example, a similarity/distance threshold does not need to be specified (unlike for many other similar algorithms).
- Has a time complexity that is constant in sub-sequence length.
  - Thus, it can be constructed in a deterministic timeframe, an important consideration for time-sensitive financial applications.
- Incorporates flexibility.
  - No assumptions are made about the underlying data.
  - Is incrementally maintainable.

For an input time series with a given sub-sequence length m, the *MP* returns four results. These are:

Matrix Profile Index (*MPI*)

- For every index i (or time point) in the examined series, the *MPI* contains a pointer to another index j (in the original series) indicating the start location of the nearest neighbour sub-sequence (or similar behaviour pattern).

Matrix Profile (*MP*)

- For every index i in the examined series, the *MP* contains a record of the Z-normalised Euclidean distance [10] to the nearest neighbour sequence (as indicated by the *MPI*).
  Note: Zero distance implies exact match.

Motif Index (M_i)

- For the given series, M_i records the start location index of the sub-sequence that has the *lowest* sub-sequence distance value of *MP*, i.e., closest match in terms of distance or 'classical' time series motif.

Discord Index (D_i)

- D_i records the start location index of the sub-sequence that has the *highest* sub-sequence distance value of *MP*, i.e., poorest match in terms of distance or 'classical' time series discord.

A sample *MP* plot (*red* line) based on a synthetic input series (*blue* line) is shown in Figure 1. Illustrated is a *MP* with (a) a *matching region*, i.e., low *MP* distance values and (b) a *mis-match region* corresponding to high *MP* distance values. One important feature of the *MP* utilised in the following analysis is that *exact matches* of content are not necessary to obtain meaningful results, as a localised *MP* minimum value can be used to identify a close match even if the *MP* distance value considered is non zero.

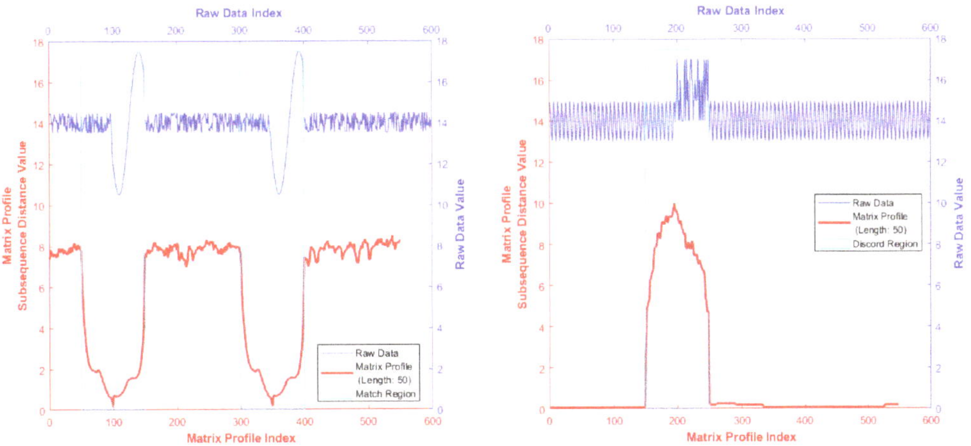

(**a**) (**b**)

Figure 1. Sample synthetic Matrix Profiles [11]: (**a**) *MP* with motif region; and (**b**) *MP* with discord region.

3. Results

When investigating the Matrix Profile of a financial time series, a typical focus is on regions (as highlighted by lower *MP* distance values) indicating similar behaviour at some other point in the data series, as financial markets show evidence of auto-regression [12].

The nature of this behaviour can be characterised by shorter or longer sub-sequences or by common 'shapes', indicative of standard financial features of the original series. Examples include *pennant*, consisting of significant rise or fall in the series followed by a period of consolidation and the *triple bottom*, which occurs when the reduction in series values creates three distinct troughs, at around the same price level, before breaking out and then reversing the trend [13–15].

Constructing a sub-sequence of length m (to create the given *MP*) and starting at the index value indicated by the lowest *MP* distance value (i.e., the closest match), it is possible to explore whether similar regions occur at regular intervals or can be associated with external events such as, for example, an *FED* rate announcement.

3.1. Single Series Motif Identification

Financial data are inherently noisy however, so the *MP* interpretation is inevitably affected to some degree [16]. Figure 2a shows a *MP* for the full *S&P500* time series (available at time of writing [17]) labelled by both date and original series index, while Figure 2b shows a subset of the original *S&P500* series restricted by a given date window.

Figure 2b thus shows *MP* patterns illustrated in greater detail, facilitating the relation of these patterns to market conditions occurring within the given timeframe. The window chosen and used for further analysis reflects the considerable stress experienced in the global marketplace at this time [18] corresponding to initial confidence issues in the American sub-prime property market.

This sparked a global liquidity crisis [19] that caused many financial institutions to collapse and triggered large systemic interventions in the form of bailouts from both governments and global financial institutions such as the *IMF*, in order to re-establish system stability.

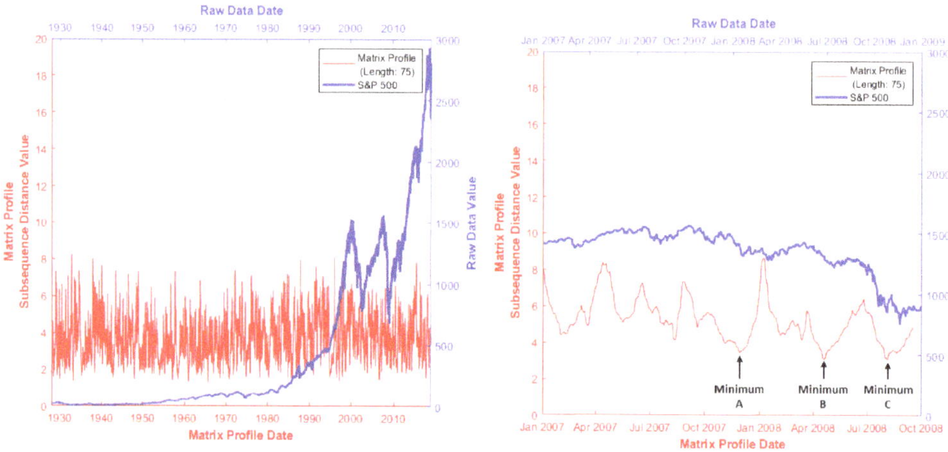

(a) (b)

Figure 2. *S&P500* series and associated *MP* distance values: (**a**) January 1928–March 2019; and (**b**) January 2007–January 2009. Further *MP* minima location detail is contained in Table 1.

Figure 2b (*red* series) illustrates three points of interest highlighted as points **A**, **B** and **C** (with further detail in Table 1). Low *MP* values indicate the similar behaviour of the *S&P500* index (*blue* series) at some other point in the time window examined (obtained from the corresponding *MPI*). Thus, *MP* plots can highlight behavioural similarities which may be less obvious from the raw series data.

To demonstrate in more detail, Figure 3 indicates typical motifs obtained from the raw *S&P500* series (as indicated by the *MP* and *MPI* values of Figure 2b and Table 1). These are constructed by the generation of a sub-sequence of *MP* length (m = 75) to facilitate the display of longer term sub-sequences within the length bounds enforced by the *MP* algorithm (minimum and maximum constraints relative to the series length apply). An initial sub-sequence from the start index of the minimal *MP* distance value (visual inspection) is compared with a second sub-sequence, which starts at the nearest 'matching index', as indicated by the corresponding *MPI* value (Table 1).

Table 1. *MP* minima details of reduced *S&P500* series as highlighted in Figure 2b. *Matrix Profile Index* (*MPI*) values, i.e., locations of the matching index, are also shown.

Local *MP* Minima Location	Identified *MP* Minima Index	Identified *MP* Minima Date	*MPI* Value of Identified Index	*MPI* Date of Identified Index
A	229	28 November 2007	412	20 August 2008
B	326	18 April 2008	401	5 August 2008
C	401	5 August 2008	326	18 April 2008

Note that although several local *MP* minima locations are identified in Figure 2b, only two raw data sequences are displayed in Figure 3 as, in this particular case, the remaining localised *MP* minima form a 'classic' motif (i.e., closest match in terms of distance). This can be seen in Table 1, where for minima locations ***B*** and ***C***, the *MPI* values are reversed, i.e., marking the same sub-sequence.

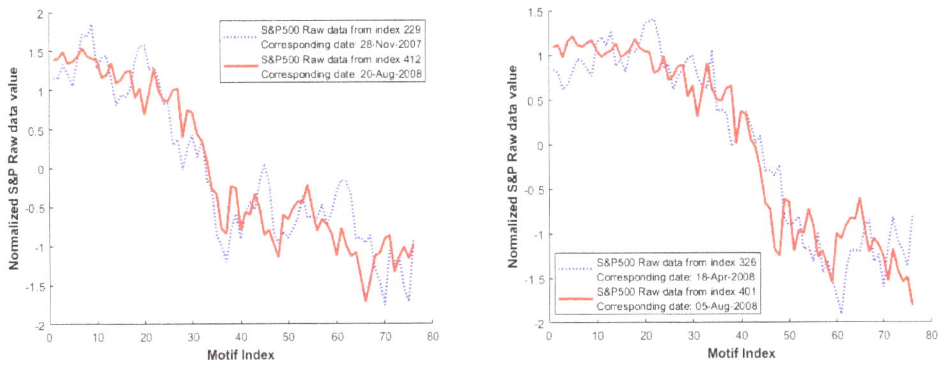

Figure 3. Raw data of *S&P500* series indicated as motif locations by low *MP* values in Figure 2b and Table 1. Here, the blue series indicates the sub-sequence visually identified from low *MP* values, while the red sub–sequence represents the nearest 'match' as indicated by the corresponding *MPI* value: (**a**) location ***A*** (index 229) in Figure 2b; (**b**) location ***B*** (index 326) in Figure 2b.

3.2. Single Series MP Evolution over Length

As the *MP* sub-sequence length increases, the average *MP* distance value for that sub-sequence length also appears to increase, indicating a less-exact match (in terms of average *Z*-normalised Euclidean distance) over the entire length of the *MP* (Figure 4a). This result is intuitive, as the shorter the sub-sequence length is, the more readily it is matched [2].

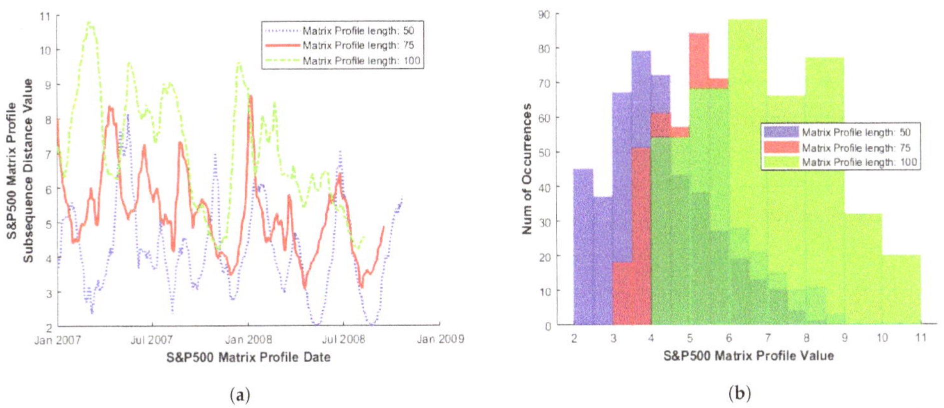

Figure 4. (**a**) *MP* and (**b**) histogram of the *S&P500* series. Illustrated over increasing sub-sequence length, during the period January 2007–January 2009.

As the *MP* sub-sequence length is increased, the frequencies of *MP* motif match and discord values correspondingly decrease, (Figure 4a). However, where found, these large *MP* distance values (occurring at approximately the same index in *MP* plots of shorter and longer sub-sequence lengths) may indicate the existence of longer term trends in the data, despite more volatile behaviour being observed at shorter *MP* sub-sequence lengths.

It should be noted that an increase in *MP* sub-sequence length does not necessarily result in a clearer, 'less noisy' *MP* structure (particularly for the multi-variate cases examined, as shown in Sections 3.3 and 3.4) in individual series. Hence, both a range of sub-sequence lengths and *MP* distance minima are needed for balanced interpretation.

This is further illustrated by a histogram plot of the same *MP* data in Figure 4b. The entire histogram (of overall distance to repeats) was shifted to the right (for given sub-sequence length). The global behaviour of the *MP* can be linked to the distributional morphing. The shorter *MP* length (of 50) here with a higher frequency of occurrence of matches/discords, is closer to the Normal (or Gaussian) form. For higher sub-sequence length (of 100 here) the distribution is flatter, indicating larger variation in the motif and discord distance values. However, an examination of detailed motif shape in these longer sequences may prevent over-reliance on short-term volatility, while capturing longer-term patterns of growth or stability, with a corresponding reduction in transaction costs.

The *MP* distance histogram also highlights the fat-tailed distribution of many financial market series data, (where a right-skew indicates a higher proportion of discords and a left-skew indicates a higher proportion of motifs). Figure 4a shows *MP* line series plots of increasing sub-sequence length while Figure 4b illustrates their corresponding histogram values. These plots are based upon *S&P500* [17] and share value data, again, for the time window of January 2007–January 2009.

The same generalised behaviour as that of Figure 4 is observed in Figure 5 for *Microsoft* [20] series data; however, in this case, with a higher proportion of increased *MP* distance values (or discords) as indicated by a skewed distribution to the right. This occurs for all *MP* sub-sequence lengths examined to date and indicates that series behaviour is consistent over longer timescales.

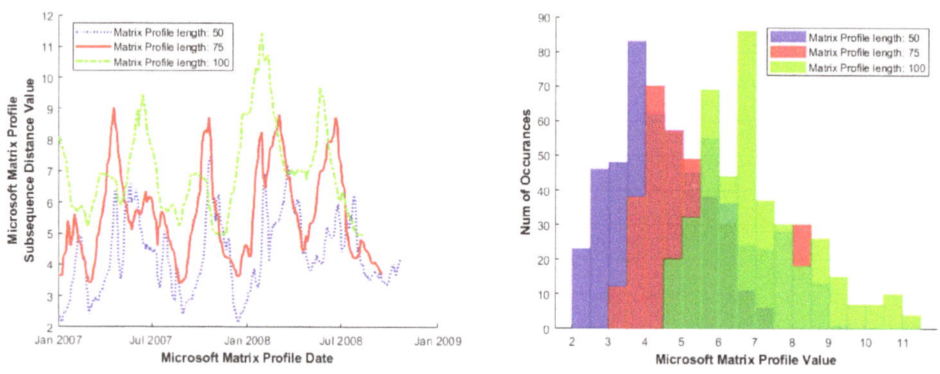

Figure 5. (**a**) *MP* and (**b**) histogram of the *Microsoft* series. Illustrated over increasing sub-sequence length, during the period January 2007–January 2009.

3.3. *Multi-Variate Series*

In an attempt to characterise wider market behaviour, the *MP* single-series approach must be expanded to multi-variate series. Applications for finance include the investigation of multiple companies within the same market sector, as opposed to an individual stock or index considered independently.

3.3.1. Single Sector

Figure 6 illustrates the *MP* plot of stock series for influential companies within (a) technology and (b) pharmaceutical sectors chosen at random from several top 10 lists based on market cap, percentage annual return and market value [21,22]. Although fluctuations in amplitude are large, coherent movements at lower *MP* distances are observed over short time-frames, (i.e., local minima regions correspond across series).

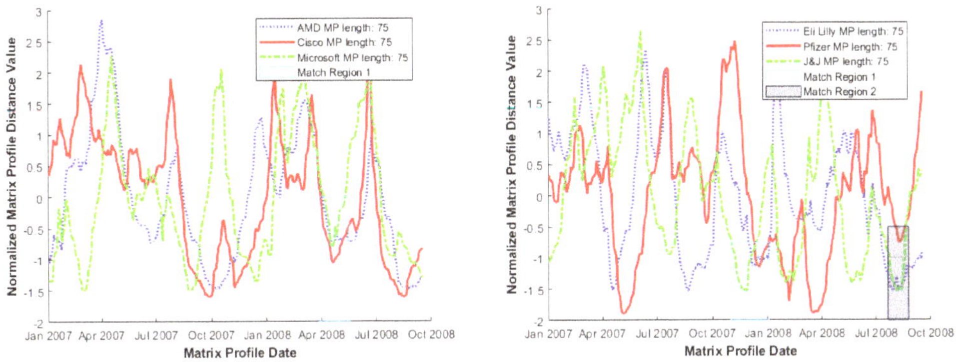

(a) (b)

Figure 6. Sample set of normalised Matrix Profiles across individual market sectors: (**a**) tech sector; and (**b**) pharmaceutical. Localised *MP* minima coherence are indicated by coloured rectangles. During the period January 2007–January 2009.

Clearly, both the occurrence and values of these local *MP* minima over the shortest timeframe are of interest for motif identification and verification. The main considerations are (i) the time duration to when similar behaviour is repeated (i.e., *when* a match occurs); and (ii) distance range (indicating how close a match it is). Thus, a visual choice of the point at which a generalised local minima region occurs in a multi-variate *MP* series plot is made based upon obtaining the best combination of local minima over the shortest timeframe and restricting the *MP* minima spread to be as low as possible. We consider these to be match regions as highlighted by shaded areas in Figure 6a,b for example.

Occurrence of a motif within an identified match region may be slightly shifted from series to series, either with respect to the starting index or by extension, date. In consequence, plots can be constructed to start at a specific index (where a given series feature may slightly overlap with a similar or matching feature in another series) or at a specific date, where shifts between series may be clarified.

It should be noted that, due to total *MP* series variance and the fact that areas of interest are small compared to the overall plot size involved, visual *MP* distance plot analysis is a limited technique. These plots become harder to interpret and sectors of interest more challenging to identify as multiple series are added, so that typically, only a small series set is examined. However, consistent behaviours such as reduced volatility, less precise matching (increased *MP* distance) and better-defined *MP* structure are generally observed for long as well as shorter sub-sequence lengths.

3.3.2. Multi Sector

Expanding the approach to multiple sectors (including indexes) can be useful in illustrating more generalised market behaviour where, for example, large events such as global shocks can generate coherence that is reflected in the behaviour of the corresponding *MPs*. To illustrate this, a range of leading sectoral companies were chosen, again from several top 10 lists based on Market Cap, percentage annual return and market value. These sectors span information technology (*Microsoft*), the pharmaceutical industry (*Merck&Co*) and the finance sector (*Citigroup*) [21–23]. *MP* line plots, constructed for the same time window of January 2007–January 2009, are shown in Figure 7, together with coincident local *MP* minima that occur within narrower time intervals (shaded match regions).

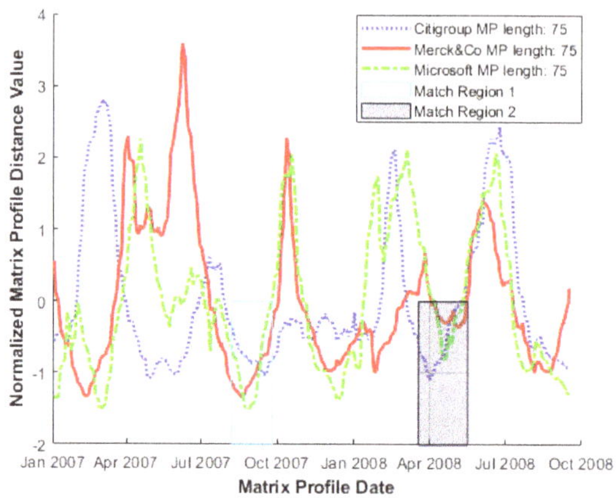

Figure 7. Sample set of normalised Matrix Profiles across multiple market sectors, where local *MP* minima coherence is indicated by coloured rectangles. January 2007–January 2009.

3.4. Stocks within an Index

Matrix Profile plots are also useful in examining the influence of individual stock series on the index to which these contribute. Comparison of *MP* index series against several *MP* plots of individual companies (chosen to cover a wide range of sectors trading within that index) serves to characterise the convergence of lower *MP* distance values (Figure 8).

Within the time window examined, short periods occur where localised *MP* minima coincide with those of the *S&P500*, suggesting coherent behaviour; (for raw data analysis, see Section 3.5). Table 2, moreover, shows the shift in *location* (and by extension *timing*) of *MP* minima occurrence within these series.

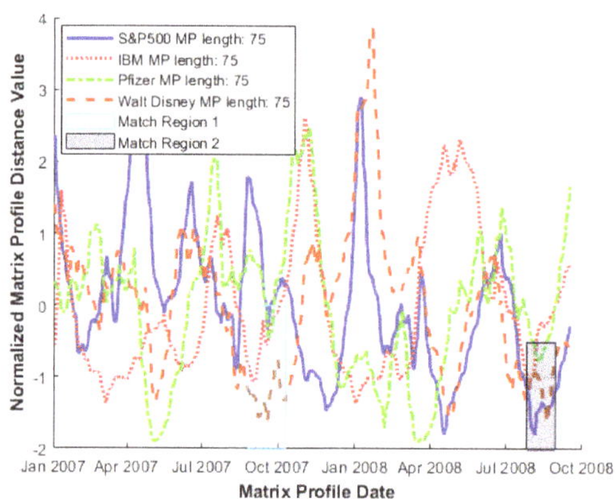

Figure 8. Multi-sector *MP* plots including S&P500 index, local *MP* minima coherence indicated by coloured rectangles: January 2007–January 2009. Dates of occurrence of low minima regions in Figure 8 are summarised in Table 2.

For some series, the *MP* minima occur before the *S&P500* minima, indicating a leading influence upon the index, while others are identified shortly afterwards, indicating that underlying series subsequently reflect index movement. Only *three* sub-series are currently included of course so, given that other stock series may be influential, a comprehensive analysis would need to consider additional index components and combinations thereof.

Table 2. Identified *MP* minima dates and indexes of match regions 1 and 2 (i.e., localised *MP* minima coherence) as highlighted in Figure 8.

Series	Sector	Match Region 1		Match Region 2	
		Identified MP Minima Index	Identified MP Minima Date	Identified MP Minima Index	Identified MP Minima Date
S&P500	Various	179	18 September 2007	401	5 August 2008
IBM	Information Technology	169	4 September 2007	398	31 July 2008
Pfizer	Pharmaceutical	182	21 September 2007	404	8 August 2008
Walt Disney	Entertainment	179	18 September 2007	412	20 August 2008

3.5. Reviewing the Raw Data

In the multi-variate cases examined thus far, a low *MP* value at approximately the same index as for multiple series is taken to be a good indication of similar behaviour. Strictly, however, the *MP* algorithm in its current form examines each series independently so that an extreme *MP* value may indicate either a close match (*motif*) or mismatch (*discord*) within a *single* series. For example, series *X* and *Y* may both have a low *MP* value, coinciding at index *x*, indicating two matches (one within each series) but these are independent, so that event type *motif shapes* may differ. *MP* plots for several series indicate regions of *possible* consistency, so for real behaviour to be characterised, event types in the raw series must be related to *MP* matches.

A motif as a repeated identifiable sub-sequence has a minimum of two parts, namely the initial sequence (as indicated by the index of the localised *MP* minimum) and the corresponding *matching* sequence obtained from the *MPI* (indicating the start point of the nearest neighbour sub-sequence). Figure 3b illustrates the two parts of a sample classic motif of the *S&P500* series found by locating low *MP* distance values in the time window of January 2007–January 2009. However, in the multi-variate case considered here, only one subsection (or motif part) per series is shown for clarity.

The two complementary approaches of the analyses consider: (1) nature of the behaviour of the sub-sequences (indicated by shape), i.e., *event type*, and (2) *timing*. Of interest, with respect to (1) for a set of sub-sequences considered in isolation, is whether such events match in terms of length, magnitude and location. Alternatively, sub-sequences may exhibit amplification or damping over an extended period. In terms of (2), interest centres on whether a motif sub-sequence leads, lags or coincides with other sub-sequences in terms of event *timing*.

Underlying motif sub-sequences in the original series of the *MP* plots (Figure 8) exhibit localised *MP* minima of index-contributing stocks across multiple market sectors. In Figure 9a,c, the motif sequences for each series are plotted according to the *motif* sub-sequence index (i.e., overlapping). Again, illustrative of similar behaviour (in terms of shape), a large drop in value occurs approximately halfway through each of the motif sub-sequences. In Figure 9a, it initially appears that both the *IBM* (red) and *Pfizer* (green) series are reacting at a later point in time to the *S&P500* (blue series). However, when plotted according to date (Figure 9b), it can be seen that the large drop in value actually occurs over the same time window of November 6th–12th 2007 for all series.

To place this in context, this corresponds to a period when a deepening liquidity crisis sparked by issues in the American sub-prime property market [24] began to accelerate globally (as illustrated by the run on the Northern Rock bank in England in September 2007). Despite initial action by the *FED* over 2007 to increase liquidity in short-term money markets through larger open market operation interventions (as described [25]), the peak of market values was reached in October 2007. However, fears of losses at *Citigroup* in combination with poor market sentiment prompted a more generalised sell-off (as reflected in Figure 9a,b).

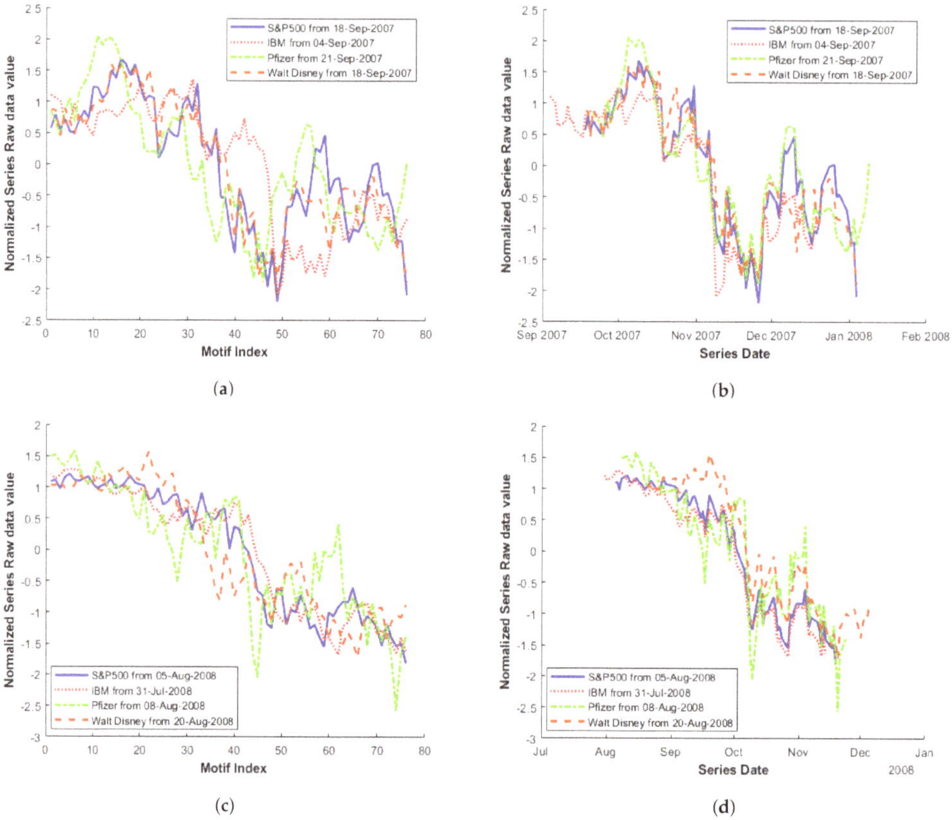

Figure 9. Motif of stocks within an index: i.e., original data sub-sequences with starting indexes obtained from *MP* minima located in *Match Regions 1* and *2* in Figure 8: (**a**) *Match Region 1* overlapping; (**b**) *Match Region 1* by date; (**c**) *Match Region 2* overlapping; and (**d**) *Match Region 2* by date.

Similar behaviour is observed in Figure 9c,d, in this case with the *Pfizer* and *Walt Disney* (brown) series reacting slightly after the *S&P500* (during the period of 1–10 October 2008). This corresponds to the US Congress opening its first hearing on the growing financial crisis when stocks then tumbled further (the *Dow Jones* index dropped below 10,000 for the first time in 4 years [26]), coinciding with the realisation by investors that the credit crisis was spreading around the globe and the recent (29 September) rejection by the US Congress of a proposed USD 700 billion bailout plan would not stabilise the situation. However, as the country's financial system continued to deteriorate, several representatives changed their minds and the legislation was signed off on 3 October 2008 [27]. Overall coherent behaviour was observed for the *S&P500* series and individual stocks, particularly when plotted by date (as initial lag between series is no longer evident).

When examining Figure 8 to identify suitable lowest *MP* minima match regions, an alternative lower index value of the *S&P500 MP* than was initially chosen for *Match Region 1* is also available. This gives a reduced *MP* value (i.e., a closer match in terms of *Euclidean* distance to some other point in the *S&P500* series). Incorporating this alternative *S&P500 MP* minima value (154) occurring on 13 August 2007 into Figure 9a,b gives the plots displayed in Figure 10. Plotting according to date (Figure 10b), the *S&P500* series corresponds quite well with the remaining series in the region where dates overlap. However, Figure 10a illustrates that *event type*, (when considered as motif shape), does significantly differ between the series in question.

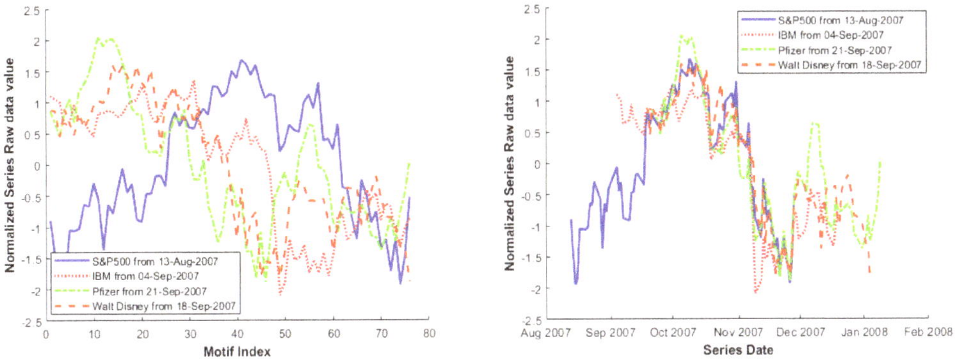

(a) (b)

Figure 10. Motif of stocks within an index, i.e., original data sub-sequences with starting indexes obtained from *MP* minima in Figure 8 *Match Region 1* (using an alternative *S&P500* index). Plotted (**a**) overlapping; and (**b**) by date.

3.6. Multidimensional Analysis of a Single Stock

In addition to utilising the *MP* for the multi-variate analysis of separate series spanning differing market sectors, the approach can also be applied for the combination of series based upon different measures of a single company or index.

In Figure 11a, the *MP* in two measures of *Microsoft* stock (*value* & *volume*) are illustrated (again for the time window of January 2007–January 2009) [20]. A match region (co-incidence of *MP* minima) was identified while raw data sub-sequence values shown in Figure 11b appear to indicate a large increase in both series occurring at approximate dates (26 October 2007 for *volume* and 1 November for share *value*).

Although both series are based upon the same stock, the previous flexibility to display raw data sub-sequences by date of the identified *MP* minima still applies to features identified in both series (in this case, applying to when these occur). Here, it illustrates the timing of the occurrence of features identified in one series relative to another. Figure 11b highlights the reasonable alignment for an increase in both share value and trading volume.

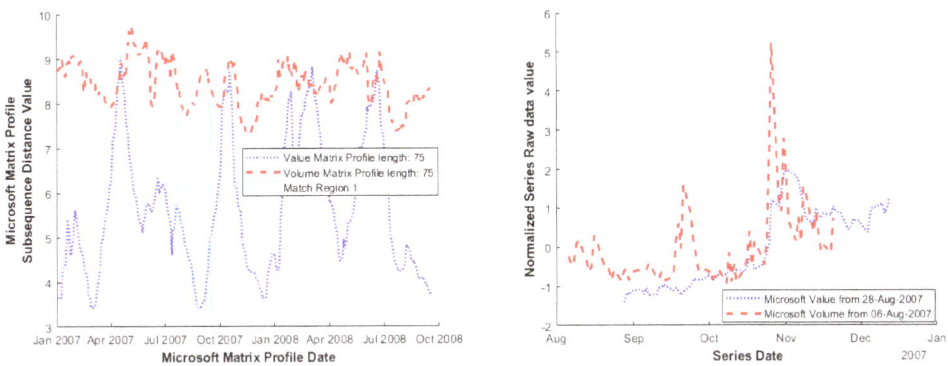

(a) (b)

Figure 11. Motif of differing measures of *Microsoft* stocks during the period of January 2007–January 2009: (**a**) *MP Microsoft* share volume and value; and (**b**) raw data *Microsoft* share volume and value based upon *MP* minima identified in *Match Region 1*.

The examination of other combinations of commonly used company measures such as *price-to-book* and *price-to-earnings* ratios is also possible.

3.7. Motif Length Selection Considerations & Long- vs. Short-Term Behaviour

An important consideration for the selection of the motif or sub-sequence length for analysis is whether interest is focused on short- or long-term behaviour (shorter or longer motif lengths, respectively). The large number of motif locations found for shorter *MP* lengths can obscure particular trends, while the reduced number of motifs returned for longer lengths can facilitate the identification of extended match regions. Recent developments on the length selection process providing an illustration of the motif content (by *MP* length) include the *SKIMP* [28] algorithm.

SKIMP allows the optimised generation of a set of *MPs* for a user-provided length range. The new structure, known as a *Pan Matrix Profile* (*PMP*), can be plotted as a heat-map indicating both the *location* and *length* of motifs in a data set, as illustrated in Figure 12a. Larger motif length locations are indicated by spikes while more frequent motif lengths correspond to areas of increased intensity. *PMP* plots can also provide an indication on common features of financial time-series, i.e., these may contain a large number of smaller length motifs even over a varying time window, as shown in Figure 12a,b. This suggests a shorter *MP* length may be more applicable for financial series analysis.

Thus, a *PMP* can provide an alternative method when obtaining start locations for motif behaviour investigations over reduced timescales. This is important as *MP* plots can become noisy at lower sub-sequence lengths, particularly in the multi-variate case. To illustrate this (within a single series initially), a motif length of 20 was chosen from Figure 12b as a suitable length for probing underlying raw series behaviour.

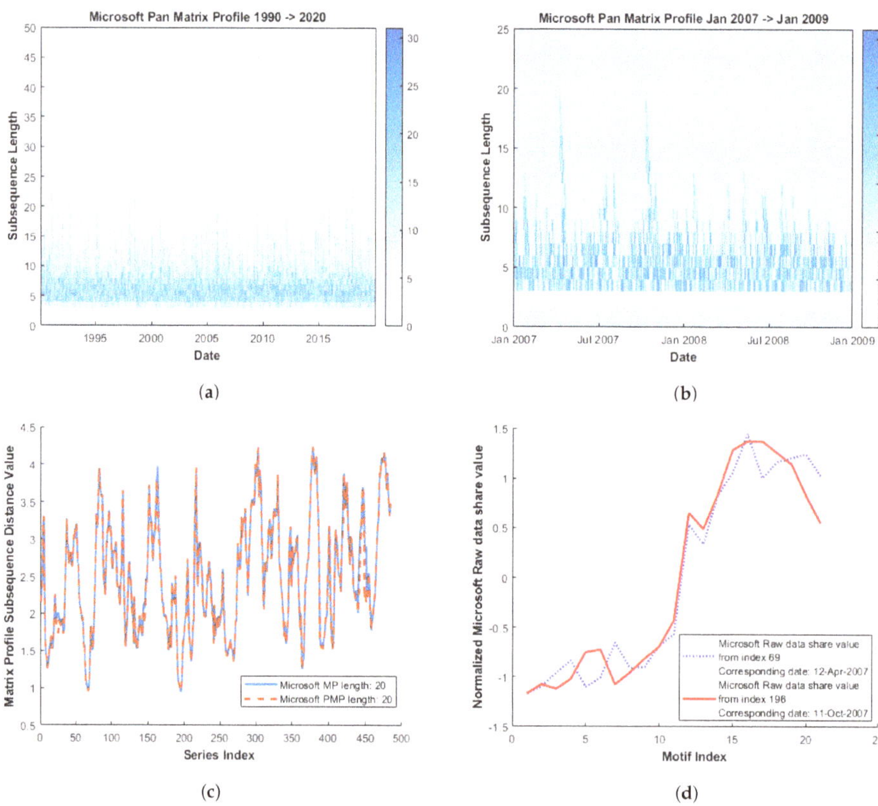

Figure 12. *Microsoft* Pan Matrix Profile (*PMP*) and underlying motif identification. The raw data *Microsoft* motifs in (**d**) were identified by an index of peaks in the reduced *Microsoft* Pan Matrix Profile (**b**) and the corresponding index of low *MP* value locations in (**c**). For context of overall motif length and location, a longer timescale *Microsoft PMP* is also provided in (**a**).

A comparison between the standard *MP* and *PMP* is shown for the given sub-sequence length in Figure 12c illustrating close correlation (as anticipated). The location of peaks within the *PMP* plot (indicating motifs of greater sub-sequence length), are identified by the index which corresponds to localised MP minima values in Figure 12c. The underlying raw data sequences are isolated based upon these indexes and are displayed in Figure 12d. In this case, the two locations correspond to the 'classical' motif as the *MPI* indexes refer to each other.

Expanding this approach for the multi-variate case, the same scenario (and individual series) of stock behaviour within an index was considered. Using an initial *S&P500 Pan Matrix Profile* plot (Figure 13a), a sub-sequence length of 13 was chosen for further analysis. For the *S&P500*, indices 134 and 246 exhibited peaks corresponding to motifs of above average length. These were taken as approximate start locations for finding *MP* minima within the individual *Matrix Profiles* (Figure 13b). The alternative of only examining *Matrix Profiles* for low *MP* minima occurrence was not adopted as they become too noisy at this low sub-sequence resolution. Thus, the indexes chosen from the *PMP* serve as regions previously considered as *local match regions* (Section 3.3) when examining the corresponding *MP* plots generated for this sub-sequence length.

Figure 13b displays the full set of *MP* plots for these series (within the time window examined) with match regions centred on these indexes highlighted. Figure 13b also serves to further illustrate the noisy nature of financial *MP* plots at lower sub-sequence lengths, particularly for the multi-variate case as here. For clarity, identified minima indexes and corresponding dates are shown in Table 3, with sub-sequences of interest for both match regions 1 and 2 illustrated from Figure 13d,f.

Table 3. Identified *MP* minima dates and indexes of proposed match regions 1 and 2 as identified from the *PMP* plot (Figure 13a) and highlighted in Figure 13b.

Series	Sector	Match Region 1		Match Region 2	
		Identified MP Minima Index	Identified MP Minima Date	Identified MP Minima Index	Identified MP Minima Date
S&P500	Various	135	17 July 2007	246	21 December 2007
IBM	Information Technology	131	11 July 2007	245	20 December 2007
Pfizer	Pharmaceutical	136	18 July 2007	253	3 January 2008
Walt Disney	Entertainment	134	16 July 2007	246	21 December 2007

For ***Match Region 1***, when plotted according to the sub-sequence index (Figure 13c), independent raw data sub-sequences are not in particularly good agreement. However, when plotted according to date (Figure 13d), basic behaviour is similar for all series, although the sharp reduction in value from 25 to 27 July 2007 is not as pronounced for *IBM*.

For ***Match Region 2***, raw data sub-sequence shapes appear to correspond quite well when plotted according to the sub-sequence index (Figure 13e). However, when plotted by date in this case (Figure 13f), the *Pfizer* series briefly demonstrates coherent behaviour, however, in general, it lags relative to the other series.

Figure 13f also illustrates that the *MP* minima location has a disproportionately greater effect at these lower resolutions, causing a larger shift (relative to motif length) as seen previously in Section 3.5 for example. Further when plotting by date, there is less likelihood of an overlap region.

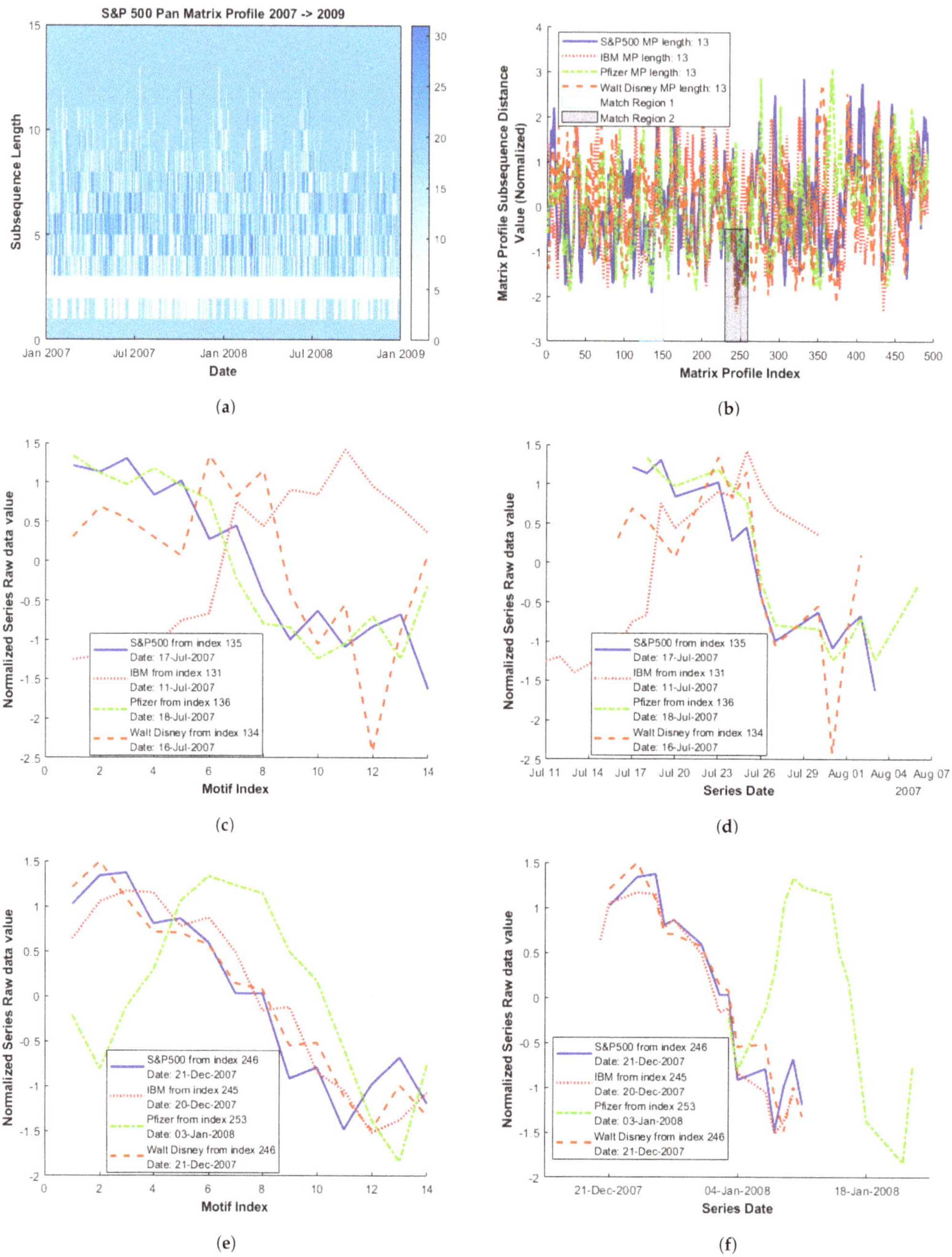

Figure 13. Stock within an index, short-term Pan Matrix Profile (*PMP*) analysis from January 2007 to January 2009: (**a**) *S&P500* Pan Matrix Profile; (**b**) multi-sector *MP* plots with highlighted match regions identified from the *S&P500* Pan Matrix Profile; (**c**) *Match Region 1* overlapping; (**d**) *Match Region 1* by date; (**e**) *Match Region 2* overlapping; and (**f**) *Match Region 2* by date.

4. Discussion

In this work, we explored the potential of the *Matrix Profile* (*MP*) algorithm, to offer additional insight on financial series analysis by the practical demonstration of motif identification and behaviour characterisation. Construction of *MP* series plots *within* a single series can illustrate longer-term trends around a given date (identified from low *MP* values), while *MP* series distributions reflect the percentage of motif matches and discords in the underlying series.

In multiple series analyses, the coincidence of local *MP* minima values can illustrate similar behaviour (i.e., *motif shape*) across single market sectors, as well as more generalised market behaviour (based on a set of companies spanning multiple sectors). The relationship between index data and individual stock data can also be examined using the *MP*. Additionally, the combination of series based upon different measures of a single company or index can be investigated using this approach, providing insight for example on whether a company is under or over valued. The relationship between local *MP* minima and the behaviour of the series they represent is also explored through an examination of raw data sub-sequences (based on the identified *MP* minima location and known *MP* sub-sequence length). This is demonstrated for both the *single* and *multi-variate* case.

The choice of sub-sequence length for analysis is an important consideration. The *Pan Matrix Profile* (*PMP*) algorithm (an extension of the *Matrix Profile*), applied to financial series, demonstrates how this decision can be informed by motif location and length in a given data set. Additionally, it can simplify the interpretation of *MP* plots by using a shortened sub-sequence length range to probe regions of interest. Nevertheless, a more comprehensive automated method for determining localised *MP* minima is clearly desirable, while the robustness of the general methods should be tested on additional time series, such as market rate curves and commodities, for example.

Moreover, while the work presented here has focused on the interpretation of independent *MP* plots for the multi-variate case, recent work on extending the *MP* algorithm, such as *mSTAMP* [29] and *Ostinato* [30], suggests that examining all underlying series simultaneously is within reach. This would facilitate the automation of a process to illustrate occasions where series are conforming with market behaviour, additionally highlighting potential hedging opportunities through the identification of series (within the set examined) that do not exhibit this behaviour.

Author Contributions: Conceptualisation, methodology, software, formal analysis, investigation, visualisation and writing—original draft, E.C.; writing—review and editing, validation, M.C. and H.J.R. All authors have read and agreed to the published version of the manuscript.

Funding: This research received no external funding.

Institutional Review Board Statement: Not applicable.

Informed Consent Statement: Not applicable.

Data Availability Statement: Financial Time Series data available at: https://finance.yahoo.com/lookup (accessed on 29 June 2021), *Matrix Profile* code available at: https://www.cs.ucr.edu/~eamonn/MatrixProfile.html (accessed on 29 June 2021).

Conflicts of Interest: The authors declare no conflict of interest.

Abbreviations

The following abbreviations are used in this manuscript:

MP	Matrix Profile
SKIMP	Scalable Kinetoscopic Matrix Profile
mSTAMP	Multidimensional Scalable Time Series Anytime Matrix Profile
PMP	Pan Matrix Profile
MPI	Matrix Profile Index
PAA	Piecewise Aggregate Approximation

WTI	West Texas Intermediate
FED	Federal Reserve System
IMF	International Monetary Fund
S&P500	Standard and Poor's 500

References

1. Mueen, A.; Keogh, E.; Zhu, Q.; Cash, S.; Westover, B. Exact Discovery of Time Series Motifs. In Proceedings of the SIAM International Conference on Data Mining, Sparks, NV, USA, 30 April–2 May 2009; pp. 35–53, 473–484. [CrossRef]
2. Castro, N.; Azevedo, P. Significant motifs in time series. *Stat. Anal. Data Min.* **2012**, *5*, 35–53. [CrossRef]
3. Silva, D.F.; Yeh, C.M.; Zhu, Y.; Batista, G.; Keogh, E. Fast Similarity Matrix Profile for Music Analysis and Exploration. *IEEE Trans. Multimed.* **2019**, *21*, 29–38. [CrossRef]
4. Senobari, N.S.; Funning, G.; Zimmerman, Z.; Zhu, Y.; Keogh, E. Using the similarity Matrix Profile to investigate foreshock behavior of the 2004 Parkfield earthquake. In Proceedings of the American Geophysical Union Fall Meeting, Washington, DC, USA, 10–14 December 2018; p. S51B-03.
5. Ferreira, P.; Azevedo, P.; Silva, G.; Brito, M. Mining Approximate Motifs in Time Series. In *Discovery Science*; Springer: Berlin/Heidelberg, Germany, 2006; pp. 89–101, ISBN 978-3-540-46493-8.
6. Wilson, W.; Birkin, P.; Aickelin, U. The motif tracking algorithm. *Int. J. Autom. Comput.* **2008**, 32–44. [CrossRef]
7. Xiaoxi, D.; Ruoming, J.; Liang, D.; Lee, V.E.; Thornton, J.H. Migration Motif A Spatial Temporal Pattern Mining Approach for Financial Markets. In Proceedings of the 15th ACM SIGKDD International Conference on Knowledge Discovery and Data Mining, Paris, France, 28 June–1 July 2009; pp. 1135–1144. [CrossRef]
8. Cartwright, E.; Crane, M.; Ruskin, H.J. Abstract: Motif Discovery & Evaluation Focus on Finance. In Proceedings of the Econophysics Colloquium 2018, Palermo, Italy, Online, 2018. Available online: https://sites.google.com/view/econophysics-colloquium-2018 (accessed on 29 June 2021).
9. Cartwright, E.; Crane, M.; Ruskin, H.J. Financial Time Series: Motif Discovery and Analysis Using VALMOD. In Proceedings of the International Conference on Computational Science, Faro, Portugal, 12–14 June 2019; pp. 771–778. [CrossRef]
10. Yeh, C.M.; Zhu, Y.; Ulanova, L.; Begum, N.; Ding, Y.; Dau, H.; Silva, D.F.; Mueen, A.; Keogh, E. Matrix profile I: All pairs similarity joins for time series a unifying view that includes motifs discords and shapelets. *IEEE ICDM* **2016**, *1*, 1317–1322. [CrossRef]
11. Keogh, E. The UCR Matrix Profile Homepage. 2020. Available online: https://www.cs.ucr.edu/~eamonn/MatrixProfile.html (accessed on 29 June 2021).
12. Gao, X.; An, H.; Fang, W.; Huang, X.; Li, H.; Zhong, W. Characteristics of the transmission of autoregressive sub-patterns in financial time series. *Sci. Rep.* **2014**, *4*, 2045–2322. [CrossRef] [PubMed]
13. Investopedia Common Chart Pattern Definitions. Available online: https://www.investopedia.com/articles/technical/112601.asp (accessed on 29 June 2021).
14. Investopedia Pennant Chart Pattern Definition. Available online: https://www.investopedia.com/terms/p/pennant.asp (accessed on 29 June 2021).
15. Investopedia Triple Bottom Chart Pattern Definition. Available online: https://www.investopedia.com/terms/t/triplebottom.asp (accessed on 29 June 2021).
16. Teall, J. *Financial Trading and Investing*, 2nd ed.; Academic Press: Cambridge, MA, USA, 2018; pp. 145–167, ISBN 9780128111161.
17. Yahoo Finance Historical S&P Index. Available online: https://finance.yahoo.com/quote/%5EGSPC?p=%5EGSPC (accessed on 29 June 2021).
18. Meegan, A.; Corbet, S.; Larkin, C. Financial market spillovers during the quantitative easing programmes of the global financial crisis (2007–2009) and the European debt crisis. *J. Int. Financ. Mark. Instit. Money* **2018**, *56*, 128–148. [CrossRef]
19. Bracke, T.; Michael, F. The macro-financial factors behind the crisis: Global liquidity glut or global savings glut?. *N. Am. J. Econ. Financ.* **2012**, *23*, 185–202. [CrossRef]
20. Yahoo Finance Historical *Microsoft* Data. Available online: https://finance.yahoo.com/quote/MSFT (accessed on 29 June 2021).
21. Investopedia Website Technology Companies List. Available online: https://www.investopedia.com/articles/markets/030816/worlds-top-10-technology-companies-aapl-googl.asp (accessed on 29 June 2021).
22. Investopedia Website Pharmaceutical Stocks List. Available online: https://www.investopedia.com/investing/pharmaceutical-stocks/ (accessed on 29 June 2021).
23. Investopedia Website Finance Stocks List. Available online: https://www.investopedia.com/terms/f/financial_sector.asp (accessed on 29 June 2021).
24. OECD Financial Markets Highlights November. *Financ. Mark. Trends* **2007**, *93*, 11–25. Available online: http://www.oecd.org/finance/financial-markets/39654572.pdf (accessed on 29 June 2021).
25. Bernanke, B.S. The Recent Financial Turmoil and its Economic and Policy Consequences. Available online: https://www.federalreserve.gov/newsevents/speech/bernanke20071015a.htm (accessed on 29 June 2021).
26. Yahoo Finance Historical Dow Jones Index 26th October 2004 to 8th October 2008. Available online: https://finance.yahoo.com/quote/%5EDJI/history?period1=1098748800&period2=1223424000&interval=1d&filter=history&frequency=1d&includeAdjustedClose=true (accessed on 29 June 2021).

27. Britannica Financial Crisis of 2007–08 Summary. Available online: https://www.britannica.com/event/financial-crisis-of-2007-2008/Key-events-of-the-crisis (accessed on 29 June 2021).
28. Madrid, F.; Imani, S.; Mercer, R.; Zimmerman, Z.; Shakibay, N.; Mueen, A.; Keogh, E. Matrix Profile XX: Finding and Visualizing Time Series Motifs of All Lengths using the Matrix Profile. *ICBK* **2019**, *1*, 175–182. [CrossRef]
29. Yeh, C.-M.; Kavantzas, N.; Keogh, E. Matrix Profile VI: Meaningful Multidimensional Motif Discovery. *IEEE ICDM* **2017**, *1*, 565–574. [CrossRef]
30. Kamgar, K.; Gharghabi, S.; Keogh, E. Matrix Profile XV: Exploiting Time Series Consensus Motifs to Find Structure in Time Series Sets. *IEEE ICDM* **2019**, *1*, 1156–1161. [CrossRef]

engineering proceedings

MDPI

Proceedings

System for Forecasting COVID-19 Cases Using Time-Series and Neural Networks Models †

Mostafa Abotaleb * and Tatiana Makarovskikh

Department of System Programming, South Ural State University, 454080 Chelyabinsk, Russia; makarovskikh.t.a@susu.ru
* Correspondence: abotalebmostafa@bk.ru

† Presented at the 7th International conference on Time-series and Forecasting, Gran Canaria, Spain, 19–21 July 2021.

Abstract: COVID-19 is one of the biggest challenges that countries face at the present time, as infections and deaths change daily and because this pandemic has a dynamic spread. Our paper considers two tasks. The first one is to develop a system for modeling COVID-19 based on time-series models due to their accuracy in forecasting COVID-19 cases. We developed an "Epidemic. TA" system using R programming for modeling and forecasting COVID-19 cases. This system contains linear (ARIMA and Holt's model) and non-linear (BATS, TBATS, and SIR) time-series models and neural network auto-regressive models (NNAR), which allows us to obtain the most accurate forecasts of infections, deaths, and vaccination cases. The second task is the implementation of our system to forecast the risk of the third wave of infections in the Russian Federation.

Keywords: COVID-19; time-series models; ARIMA; BATS; TBATS; Holt's linear trend; NNAR; forecasting system

Citation: Abotaleb, M.; Makarovskikh, T. System for Forecasting COVID-19 Cases Using Time-Series and Neural Networks Models . *Eng. Proc.* **2021**, 5, 46. https://doi.org/10.3390/engproc2021005046

Published: 9 July 2021

1. Introduction

The COVID-19 (Here and later, all the acronyms are listed at the end of Introduction section, see Table 1) pandemic has become global. While the challenge for the medical sciences where treatments, drugs, vaccines, and test systems are developed are pervasive, the challenges for all fields of knowledge including mathematical and statistical science are also pervasive as they play an important role in modeling and discovering patterns of spread in infection cases and in forecasting COVID-19 infection cases and deaths. Since the first days of the pandemic, various methods for modeling and forecasting the spread of infection cases worldwide and in local regions have developed and appeared. Various articles were devoted to the use of known models, identification of their parameters, and testing on known data.

On 1 March 2020, the virus began to spread in a pattern that resulted in millions of infections in less than a year. Most of the deaths from this virus occur among the elderly and people with chronic heart disease, which is the leading cause of death even in developed countries. Recently, quite a lot of studies have been published on forecasting the number of COVID-19 cases on both a worldwide and regional basis. These studies used mainly the ARIMA model, Holt's linear trend model, and the SIR state transition model. There are also studies devoted to comparing the work of the models, for example, in [1] it is shown that the linear Holt model is better than the ARIMA model for the states considered in it. In this article, we will investigate the performance of these models and provide an analysis of the errors of the forecasts obtained.

Given the similarity of the characteristics of the models in the United States and Italy, it was suggested in [2] that the corresponding forecasting tools can be applied to other countries facing the COVID-19 pandemic, as well as to any pandemics that may arise in the future. However, a general principle for choosing models for forecasting the spread of COVID-19 has not yet been formulated. Moreover, for different states and different conditions of the spread of the epidemic, it is advisable to build a forecast using different models. For example, in [3] it was shown that the LSTM model consistently possessed the lowest rates of forecast errors for tracking the dynamics of infection cases in the four countries considered. There are also studies that show that the ARIMA model and cubic smoothing spline models had lower forecast errors and narrower forecast intervals compared to Holt's and TBATS models. Forecasting time-series data have been around for several decades with techniques such as ARIMA. Recently, recurrent neural networks (LSTM) have been used with much success. The most important advantages of ARIMA include the following: (1) dealing with small data; (2) simple to implement with no parameter tuning; (3) easier to handle multivariate data; (4) quick to run. The advantages of LSTM include the following: (1) no pre-requisites (stationarity, no level shifts); (2) can model non-linear function with neural networks; (3) requires a lot of data (Big data) and so time-series models are considered more appropriate for dealing with COVID-19 data as they have the ability to deal with small data.

Table 1. The list of acronyms.

COVID-19	Corona Virus Disease
ARIMA	Autoregressive integrated moving average
SIR	**S**usceptible-**I**nfected-**R**ecovered
LSTM	**L**ong **S**hort-**T**erm **M**emory
BATS/TBATS	**T**rigonometric, **B**ox-Cox transformation, **A**RMA, **T**rend, **S**easonality
NNAR	**N**eural **n**etwork **a**utoregressive **M**odels
SARIMAX	Seasonal ARIMA
ME	Mean Error
MAE	Mean absolute error
RMSE	Root-mean-square error
MPE	Mean percentage error
MAPE	Mean absolute percentage error
MASE	Mean absolute scaled error
ACF	Autocorrelation function
WHO	World Health Organization
FD	Federal District

The results obtained cannot be generalized to all countries affected by the COVID-19 pandemic due to the different patterns of the virus spreading. At the very beginning of the pandemic, lots of researchers from all over the world tried to forecast the outbreak of COVID-19 by using the models of susceptible-infected-recovered (SIR) family known as the classical epidemiological models [4]. One of the first papers [5] was devoted to the simulation of the COVID-19 in the Isfahan province of Iran for the period from 14 February 2020 to 11 April 2020. The authors of this paper forecasted the remaining infectious cases with three scenarios that differed in terms of the stringency level of social distancing. Despite the prediction of infectious cases in short-term intervals, the constructed SIR model was unable to forecast the actual spread and pattern of the epidemic in the long term. Remarkably, most of the published SIR models developed to predict COVID-19 for other communities suffered from the same conformity. The SIR models are based on assumptions that seem not to be true in the case of the COVID-19 epidemic. Hence, more sophisticated modeling strategies and detailed knowledge of the biomedical and epidemiological aspects of the disease are required to forecast the pandemic.

One more example of using this model is the paper [6] in which the authors predicted that the peak of the second wave of infection cases in Pakistan should have occurred on 25 August 2020; however, the peak of infection in this country was, in fact, in December 2020. The "covid19.analytics" package [7] developed in the R language possesses the same drawbacks. This is evidenced by the results of the SIR model and the prediction of the time of occurrence of the second (and subsequent) wave cycles. Despite these shortcomings, they have been widely accepted. There is also a drawback in that it does not deal with time-series models and neural networks. Due to this deficiency in SIR models, it was important to work on developing time-series models that have been proven effective in modeling and predicting COVID-19 cases. In our paper, we observe that classical SIR model produces greater error than statistical methods.

The purpose of our work is to create an algorithm that allows for the available initial data on the spread of coronavirus infection in a certain region for a given period of time to determine the best model for making a forecast for a given period. The algorithm analyzes forecasts from time-series models (ARIMA, Holt's linear model, BATS, and TBATS), and neural networks model (NNAR) and selects a model that produces a forecast with a minimum mean absolute percentage error (MAPE). The article describes a program in the R language that produces a forecast using the models described above.

2. The Review of Epidemic.TA System

One of the biggest challenges is modeling COVID-19 by using time-series models to obtain very accurate forecasts of infection and death cases. We developed an "Epidemic. TA" system that includes the most important time-series models used for forecasting COVID-19, namely the BATS, TBATS, Holt's Linear trend, ARIMA, and NNAR models. In [1] we concluded that Holt's linear trend model was better than the ARIMA model for forecasting COVID-19 in September 2020. In [8] we show that it is impossible produce a highly accurate forecast without updating the model's parameters during some periods. This pointed to the urgent necessity of developing a system that automatically chooses the best model for forecasting and its best parameters. Figure 1 shows the scheme of the developed software module, which allows choosing the best model with the available initial data, and Figure 2 contains the used global variables. This software module works according to the following algorithm.

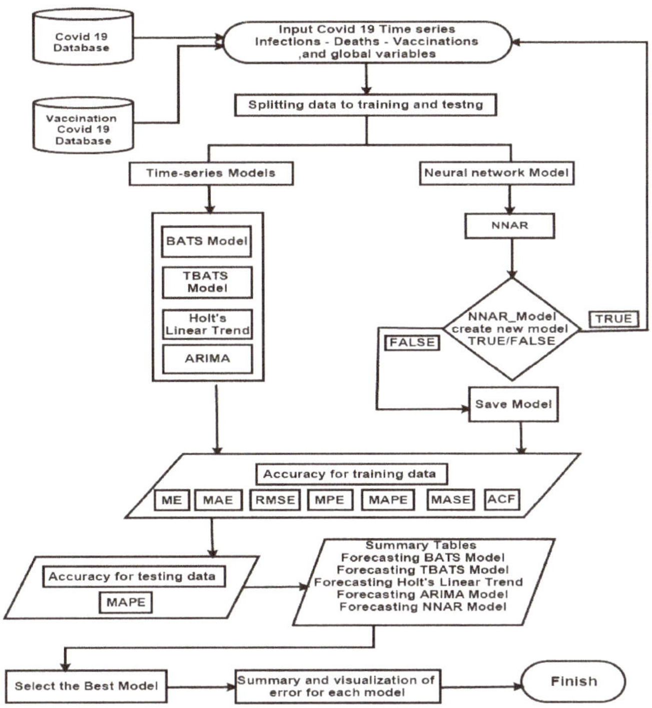

Figure 1. The structure scheme of Epidemic.TA system for forecasting COVID-19 cases.

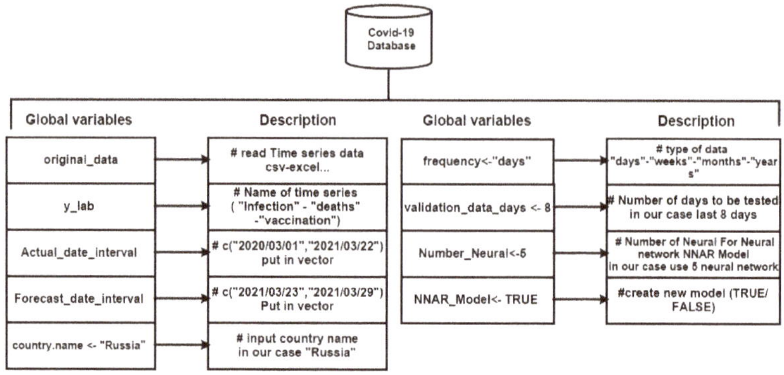

Figure 2. The global variables in Epidemic.TA system.

The source code for "Epidemic.TA" system using this algorithm is published in github [9].

The inferences of time-series forecasting models ARIMA and SARIMAX (taking seasonality into account) were efficient in producing exact and approximate results [10] and so that system selects the best model from five time-series models forecasting COVID-19 with the least error of MAPE in terms of testing data. Note that the considered system can be used to forecast not only the time-series associated with the spread of the epidemic but also for other time-series (for example, to forecast the production volume and the prices of goods, etc.); this could be a topic for further research.

3. Computational Experiments

Let us consider the results of using "Epidemic.TA" system for forecasting the daily infection cases, cumulative infection cases, cumulative deaths cases, and cumulative vaccination cases [9].

3.1. COVID-19 Datasets

The system uses COVID-19 data from the World Health Organization (WHO) [11] related to COVID-19 infection and deaths cases in the Russian Federation for the period from 1 March 2020 to 22 March 2021 and data for vaccinations [12] from 16 December 2020 to 22 March 2021. For our computational experiments, the following data were used [9]:

- COVID-19 data about infection cases in each region of Russia from 12 March 2020 to 22 March 2021;
- Used the last 8 days for testing daily cases (15 March 2021 until 22 March 2021);
- Used the last 8 days for testing cumulative cases and last 4 days for daily cases;
- COVID-19 data in the Russian Federation from 1 March 2020 to 22 February 2021 and used the last 50 days for testing the forecasting of the third wave;
- COVID-19 data in Spain from 1 March 2020 to 31 December 2020 and testing for the last 30 days.
- COVID-19 data in Italy from 1 March 2020 to 28 February 2021 and testing for the last 4 days.
- COVID-19 data in Russian Federation from 1 March 2020 to 22 February 2021 and testing the last 50 days for forecasting the third wave.

3.2. Analysing the Obtained Results

In Table 2, we represent the error of the time-series and neural network models (NNAR) for daily infection cases in the Russian Federation. Our system selects the best model for the simulation of COVID-19 daily infection cases and, for the considered period, the model ARIMA(2,2,3) was chosen. This model has the minimal MAPE for the considered period [13].

Table 2. MAPE (%) for daily COVID-19 infection cases in Russian Federation for testing last 4 days.

Cases	NNAR	BATS	TBATS	Holt's	ARIMA	ARIMA Model	Best Mode
Infections	2.5	5.589	4.29	3.319	1.638	ARIMA(2,2,3)	ARIMA

In Table 3, the MAPE for the last 8 days (testing data) for cumulative data for COVID-19 is presented. We can observe that the ARIMA model is the best one for forecasting infection and vaccinations and the BATS model is the best for death cases for the data we have [9]. This fact once again proves our assumption about choosing the best model for the available time-series.

Table 3. MAPE for cumulative cases of Covid-19 (infection, deaths, and vaccinations) in the Russian Federation for testing last 8 days.

Cases	NNAR	BATS	TBATS	Holt's	ARIMA	ARIMA Model	Best Mode
Infections	0.399	0.063	0.071	0.059	0.009	ARIMA(1,2,4)	ARIMA
Deaths	0.31	0.037	0.038	0.084	0.084	ARIMA(3,2,2)	BATS
Vaccinations	4.747	1.485	2.375	1.752	1.081	ARIMA(1,2,4)	ARIMA

By analyzing the quality of forecasts for different regions, we can observe that different models are chosen to obtain the best result for each region. The choices of models are a consequence of different factors affecting the spreading of the virus and it cannot be obtained without the experiment held.

In order to show the differences in the best obtained model, let us consider eight federal districts of the Russian Federation with different population densities, climates, traditions, and other characteristics. For example, Tables 4 and 5 represent the best chosen models for different federal districts of the Russian Federation either for cumulative data or for daily data, correspondingly [9].

Table 4. Model selection for forecasting cumulative data of COVID-19 infection cases in the Russian Federation Federal Districts (FD) on testing data based on MAPE (%) for last 8 days.

Fed.Distr.	NNAR	BATS	TBATS	Holt's	ARIMA	ARIMA Model	Best Mode
Far Eastern FD	0.192	0.017	0.005	0.042	0.012	ARIMA(2,2,2)	TBATS
Volga FD	0.282	0.003	0.042	0.056	0.004	ARIMA(2,2,3)	BATS
Northwestern FD	0.373	0.036	0.044	0.039	0.002	ARIMA(1,2,1)	ARIMA
North Caucasian FD	0.346	0.044	0.036	0.038	0.039	ARIMA(3,2,2)	TBATS
Siberian FD	0.193	0.004	0.038	0.049	0.006	ARIMA(1,2,2)	BATS
Ural FD	0.458	0.035	0.013	0.026	0.033	ARIMA(1,2,4)	TBATS
Central FD	0.387	0.088	0.084	0.093	0.067	ARIMA(2,2,2)	ARIMA
Southern FD	0.327	0.048	0.045	0.071	0.039	ARIMA(3,2,2)	ARIMA

Table 5. Model selection for the forecasting of daily COVID-19 infection cases in the Russian Federation federal districts on testing data based on MAPE for last 4 days.

Fed.Distr.	NNAR	BATS	TBATS	Holt's	ARIMA	ARIMA Model	Best Mode
Far Eastern FD	9.064	1.614	1.646	3.007	2.038	ARIMA(0,2,3)	BATS
Volga FD	2.503	0.727	1.478	1.376	1.177	ARIMA(4,2,1)	BATS
Northwestern FD	4.641	0.998	3.976	1.711	0.656	ARIMA(4,2,1)	ARIMA
North Caucasian FD	10.626	1.905	2.257	2.452	1.78	ARIMA(4,2,1)	ARIMA
Siberian FD	3.2	1.568	1.182	1.106	1.037	ARIMA(4,2,1)	ARIMA
Ural FD	1.819	2.49	1.71	1.888	1.668	ARIMA(4,2,1)	ARIMA
Central FD	10.352	14.184	3.9	3.708	8.074	ARIMA(2,1,2)	Holt
Southern FD	0.703	4.238	4.174	4.172	4.328	ARIMA(0,2,3)	NNAR(2,5)

The system that allows the definition and utilization of the best forecasting model is expedient, since all the considered forecasting methods work in polynomial time and the automatic use of each of them for time-series with a length of 100–200 elements does not require significant computational resources.

Similar results may be obtained for the whole world and separate countries, continents, and regions, which allows us to classify all the examined regions (or countries) into several clusters with the best model used for forecasting the COVID-19 cases. This approach may become advantageous for the superposition of forecasting results for different regions and different countries. This is an open task and it is not only the statistical but also medical research that is still an open problem: The information on the virus is updated every day and the results of new research are constantly appearing.

3.3. The Risk of the Next Wave Analysis

In March 2021, the third wave of COVID-19 spreading in some countries is one of the main problems in the European Union and in the whole world. As of the end of March 2021, there is a decline in the second wave in the Russian Federation. And now the question arises of lifting the previously introduced restrictions for citizens. It should be understood that weakening of some of the restrictions could result in a new wave of the disease, which is what happened in October 2020. In addition, the study of the likelihood of a new wave of the disease is an urgent and unresearched task not only for the regions of the Russian Federation but also for the whole world.

Undoubtedly, the dynamics of the spread of COVID-19 in each individual country are significantly different, as well as the different models that allow the best forecasts to be obtained. In some countries, the second wave is now occurring (Indonesia and Switzerland) while in other countries the first wave has not yet been completed (India). There are countries that are living in the third wave (Netherlands and Germany), those that have passed the third wave (Israel, Spain, and USA), and there are countries for which data cannot allow, in general, the frequency of the process to be judged (Czech Republic).

Moreover, one more delusion in COVID-19 forecasting is the great number of sophisticated factors, such as the different restrictions of different countries, that affect the spreading of the virus. It seems obvious that these factors must be taken into account. For example, in [14] the authors apply their model to compare several intervention strategies, including restrictions on international air travel, case isolation, home quarantine, social distancing with varying levels of compliance, and school closures. A lot of these factors such as "school closures" are not found to bring decisive benefits unless they are coupled with high levels of social distancing compliance. In our computational experiment, we did not take into account any factors influencing the spreading of virus. The examples are made for the Russian Federation, where the last and the only lockdown ended on 12 May 2020 (Truthfully, it is very hard to call it a lockdown taking into account the Russian attitude of "I don't care") and the strongest restrictions concern the flights between some countries.

Let us consider the application of the forecasting system developed for the prediction of the probability of the next wave in the Russian Federation. The use of the system for medium-term forecasting (NNAR model) predicts the beginning of the next wave (rise in incidence) in mid-July (see Table 6 and Figure 3).

Table 6. Model selection for forecasting the third wave peak in Spain, Italy, and Russia and the obtained data of the third wave peak.

Country	Model	Forecast Date	Actual Date
Italy	NNAR(10, 5)	13 March 2021	13 March 2021
Spain	NNAR(16, 5)	17 January 2021	17 January 2021
Russian Federation	NNAR(8, 50)	19 July 2021	—

As we can observe, Russia, Italy, and Spain have different restrictions and they change these restrictions according to the current situation with virus spreading. Nevertheless, NNAR model allows accurate forecasts to be obtained even without taking into account the existence or absence of these restrictions. Hence, the restrictions do not influence the quality of forecasting using NNAR model.

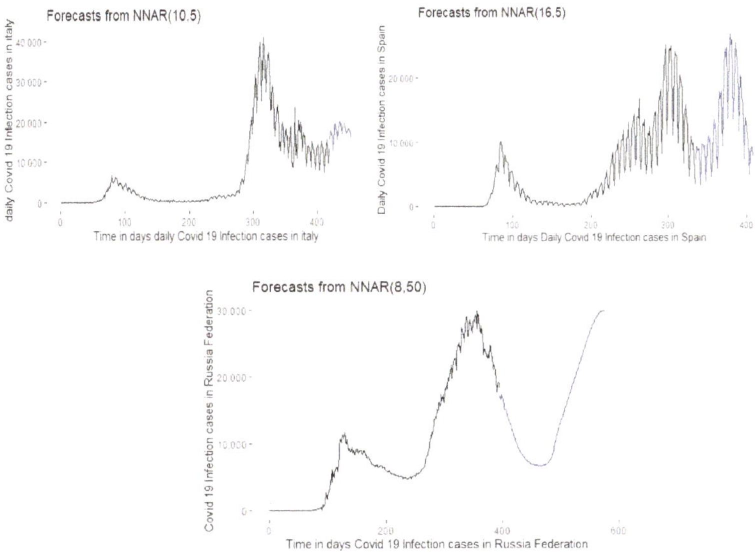

Figure 3. Forecasting the third wave peak of COVID-19 infection cases in Italy, Spain, and the Russian Federation by using NNAR Model from 12 March 2020. The black lines are actual data, and blue ones are forecasted data.

Obviously, this forecast was obtained due to the existing system of restrictions introduced in the considered state. In order to obtain these results, we used NNAR model with five neurons on the hidden level for Italy and Spain for the test periods mentioned before. As for Russia, we needed 50 neurons because the value of testing data had to be increased.

From the WHO data, the inception of the virus in the world is on 1 March 2020, which is represented by time zero on the x-axis in Figure 3.

We used the data for Italy and Spain, since the nature of the spread of coronavirus infection in these countries had clearly defined periods of the rise and fall in infection and there are sufficiently detailed data. We considered the time-series from 1 March 2020 to 28 February 2021 for Italy and the time-series from 1 March 2020 to 31 December 2020 for Spain. The forecast results are also shown in Table 6. For experiments with the peak on the next wave, we take a horizon equal to 45 days for the third wave in Spain, 31 days for Italy, and 129 days for Russian Federation.

Analyzing the results, we note that for the time-series for Italy and in Spain, accurate results were obtained on the date of the onset of the rise in incidence, which coincides with the actual values [9].

Thus, the developed system can be used for medium-term forecasting for up and downtrends in the number of reported cases of COVID-19, which is very important when making management decisions and canceling or introducing various restrictions for citizens.

4. Conclusions

In conclusion, we considered the developed forecasting system and "Epidemic.TA" can automatically select the appropriate model to obtain forecasts with very low MAPE because of the choice of the best model for the time-series used as input data. Surely, the used time-series forecasting can have significant limitations due to time-changing conditions, such as the decisions of the health authorities (e.g., confinements) and vaccine availability, etc. That is, under real circumstances, time-series forecasting can generally be accurate only in the short term. Nevertheless, if we fix the current circumstances (lockdown constraints, vaccine availability and the velocity of vaccination, the capacity of hospitals, etc.) we can observe the scenario of the development situation according to the given circumstances in mid-term or long-term forecasting. This and obtaining the long-term forecasting models are the topics for our future research.

Note that our algorithm for this system is extensible and various modules can be connected to it, providing the construction of forecasts by various methods. Thus, using the considered algorithm scheme it is possible to create a flexible calling function that permits the choice, from the set of implemented methods, of the model with the best result in accordance to a given criterion. This system uses the numbers only, without analyzing any factors influencing the process itself. Hence, the methods used for choosing the best model for forecasting COVID-19 cases may be used for obtaining the forecasts for the other time-series. This topic is an opportunity for further research. To obtain accurate results, it is recommended that the data are updated at least on a weekly basis because there are some factors affecting the process of the virus spreading that can significantly affect the model choice and accuracy of the obtained forecasts.

The open task is testing Epidemic.TA for epidemic data for different countries and the different manners of COVID-19 infections spreading to obtain low MAPE forecasting of peaks for further waves and to define the optimal criteria for choosing the best model while taking into account different exogenous factors (such as lockdown period, vaccination process, etc.).

One of the directions of future research is defining the methods of extending the Epidemic.TA package with deep learning models (LSTM and others), exploring the non-linear models, and the development of our own methods of forecasting that is appropriate for COVID-19 time-series.

Acknowledgments: The work was supported by the Ministry of Science and Higher Education of the Russian Federation (government order FENU-2020-0022).

Conflicts of Interest: The authors declare no conflict of interest.

References

1. Abotaleb, M.S.A. Predicting COVID-19 Cases using Some Statistical Models: An Application to the Cases Reported in China Italy and USA. *Acad. J. Appl. Math. Sci.* **2020**, *6*, 32–40. [CrossRef]
2. Tian, Y.; Luthra, I.; Zhang, X. Forecasting COVID-19 cases using Machine Learning. In Proceedings of the 33rd Conference on Neural Information Processing Systems (NeurIPS 2019), Vancouver, BC, Canada, 8–14 December 2020. [CrossRef]
3. Gecili, E.; Ziady, A.; Szczesniak, R.D. Forecasting COVID-19 confirmed cases, deaths and recoveries: Revisiting established time-series modeling through novel applications for the USA and Italy. *PLoS ONE* **2021**, *16*, e0244173. [CrossRef] [PubMed]
4. Banerjee, M.; Tokarev, A.; Volpert, V. Immuno-epidemiological model of two-stage epidemic growth. *Math. Model. Nat. Phenom.* **2020**, *15*, 27. [CrossRef]
5. Moein, S.; Nickaeen, N.; Roointan, A.; Borhani, N.; Heidary, Z.; Javanmard, S.H.; Ghaisari, J.; Gheisari, Y. Inefficiency of SIR models in forecasting COVID-19 epidemic: A case study of Isfahan. *Sci. Rep.* **2021**, *11*, 1–9. [CrossRef] [PubMed]
6. Hussain, N.; Li, B. Using R-studio to examine the COVID-19 Patients in Pakistan Implementation of SIR Model on Cases. *Int. J. Sci. Res. Multidiscip. Stud.* **2020**, *6*, 54–59.
7. Ponce, M. covid19.analytics: An R Package to Obtain, Analyze and Visualize Data from the Corona Virus Disease Pandemic. *arXiv* **2020**, arXiv:2009.01091v1.
8. Makarovskikh, T.A.; Abotaleb, M.S.A. Automatic Selection of ARIMA Model Parameters to Forecast COVID-19 Infection and Death Cases. *Bull. South Ural State Univ. Ser. Comput. Math. Softw. Eng.* **2021**, *X*, Z1–Z2.
9. Abotaleb, M.; Makarovskikh, T. Epidemic.TA-System. Available online: https://github.com/abotalebmostafa11/Epidemic.TA-System (accessed on 29 June 2021).
10. Bhangu, K.S.; Sandhu, J.K.; Sapra, L. Time-series analysis of COVID-19 cases. *World J. Eng.* **2021**. [CrossRef]
11. World Health Organization. Available online: https://covid19.who.int/info/ (accessed on 29 June 2021).
12. Our World in Data. Available online: https://ourworldindata.org/grapher/daily-COVID-19-vaccination-doses (accessed on 29 June 2021).
13. Abotaleb, M.; Makarovskikh, T. Epidemic.TA System for Forecasting COVID-19 Cases Using Time-series and Neural Networks Models. Available online: https://rpubs.com/abotalebmostafa/744347 (accessed on 29 June 2021).
14. Chang, S.L.; Harding, N.; Zachreson, C.; Cliff, O.M.; Prokopenko, M. Modelling transmission and control of the COVID-19 pandemic in Australia. *Nat. Commun.* **2020**, *11*, 1–13. [CrossRef] [PubMed]

engineering proceedings

MDPI

Proceedings

Cyclic Behavior in the Fumaroles Output Detected by Direct Measurement of Temperature of the Ground †

Iole Serena Diliberto

Istituto Nazionale di Geofisica e Vulcanologia, Sezione di Palermo, 90146 Palermo, Italy; Iole.Diliberto@ingv.it

† Presented at the 7th International conference on Time Series and Forecasting, Gran Canaria, Spain, 19–21 July 2021.

Abstract: On the Island of Vulcano (Aeolian Archipelago, Italy) the temperatures of fumarole emissions, have ranged from about 700 °C to the boiling point. Since the end of the last eruption (1890 A.D.), many periods of increased heating of hydrothermal systems, underlying the La Fossa area have been identified, but an eruptive condition has not yet been reached. The time variation of the high temperature fumaroles has been tracked by the network of sensors located at a few discrete sites on the summit area of La Fossa cone. The same continuous monitoring network has been working for more than 30 years. The time series shows that a natural cyclic modulation has repeated after almost 20 years, and its periodicity yet has to be discussed and interpreted. The statistical approach and the spectral analysis could provide an objective evaluation to reveal the timing, intensity, and general significance of the thermodynamic perturbations that occurred in the hydrothermal circuits of La Fossa caldera, during the study period. The continuous monitoring data series avoid unrealistic interpolations and allow promptly recognizing changes, which perturb the hydrothermal circuits, highlighting—possibly in near real time—the transient phases of energy release from the different sources (hydrologic/magmatic).

Keywords: fumaroles; temperature of the ground; long-term monitoring; close conduit volcano

Citation: Diliberto, I.S. Cyclic Behavior in the Fumaroles Output Detected by Direct Measurement of Temperature of the Ground . *Eng. Proc.* **2021**, *5*, 47. https://doi.org/10.3390/engproc2021005047

Published: 11 July 2021

1. Introduction

In the present technological age, people usually underestimate the danger related to the natural volcanic activity and they tend to quickly forget the negative effects caused by a possible paroxysm. However, volcanic activity still influences the economy, from many points of view, since volcanoes are able to modify air and ground composition, and some eruptions may also pose a risk to people mobility and settlements, for example, in the case of unexpected changes or long-lasting sequences of paroxysms. In general, the temperature monitoring is of interest for the evaluation of geological risks associated with the force acting on active volcanic systems and the main questions addressing the scope of surveillance following the scientific approach, consists of confirming the deterministic models by quantitative interpretation of monitoring results, to forecast the evolution of observed dynamics. The geochemical approach to volcano monitoring has interested, since the eighties, the Island of Vulcano (Aeolian Archipelago, Italy) during its long quiescence. By observing and collecting gases and minerals at the surface, the researchers gathered information on the systems located at depth, which are producing the energy flows transferred by the most mobile components, in the form of fluid releases. By analyzing the gaseous mixtures emitted from fumaroles, steaming grounds, hot springs, and diffuse degassing, the scientific community has thus elaborated information about the composition of hydrothermal systems, recognizing variable influences caused by the neighboring magmatic system. The convective circulation of hot gases, inside hydrothermal systems, causes anomalous surface temperatures, as well. Therefore, at Vulcano—by developing continuous monitoring procedures for the acquisition of some selected parameters (such as surface temperatures or fluxes of gas)—we could show the time variation of the fumarole release, in order to define, the solphataric phenomenon, starting from a lot of systematic observations. I show the longest records of a single parameter, dating back to 1992: The highest temperature of fumaroles. In the Island of Vulcano, the flux of hot fluids expanding from a hot buried source is one of the effects of the quiescent volcanic activity that causes sensible thermal anomalies on the ground surface. On the Island of Vulcano (Aeolian Archipelago, Italy) the temperatures of fumaroles have been recorded continuously, by monitoring stations located at a few discrete sites on the summit area of La Fossa cone. The remote control of these monitoring stations minimizes the hazards associated with data collection and allows a high sampling frequency during long periods. The measured temperatures of fumarole emissions, have ranged from about 700 °C to the boiling point. Since the end of the last eruption (1890 A.D.), many periods of increased heating of hydrothermal systems, underlying the La Fossa area, have been identified, e.g., [1–3]. However, an eruptive condition has not yet been reached. During the time period covered by this work, the main temporal variations of fumarole temperature have revealed many differences in the hydrothermal heat flow, that were originated at depth, as confirmed by other referenced geochemical and geophysical studies. The most likely cause of the main temporal variations in the fumarole temperature was some change in the gaseous input from the magmatic source, e.g., [4,5], but also the local seismic activity that caused episodic increases of hot steam advection from the hydrothermal reservoirs [6].

2. Study Area

Vulcano Island is an active volcanic system located along the southern margin of the Tyrrhenian Sea (Italy, Figure 1a–c) in the southernmost sector of the Aeolian archipelago. The Aeolian archipelago is an arc-shaped structure, formed by seven islands. The Island of Vulcano is located along the Tindari–Letojanni strike-slip tectonic system [7,8] (ATL in Figure 1b), and it is the southernmost island of the NNW–SSE elongated volcanic belt (including also Salina and Lipari, Figure 1b). The release of seismic energy in this area is higher than the regional background [4,7], and the active magmatism is driving the permanent volcanic activity in the island of Stromboli, and is causing the hydrothermal activity at Panarea, Lipari, and Vulcano. The long eruptive history of Vulcano, actually quiescent, has been summarized in eight episodes [8], between 127 ka before the present and historical times (AD 1888–1890). The eruptive episodes are interspersed with variable periods of quiescence, and by volcano-tectonic collapse that gave origin to the calderas of *Il Piano* and *La Fossa* (Figure 1c [9]).

The volcanic activity from *La Fossa* cone consisted of explosive phreatic and phreatomagmatic eruptions, alternated with highly viscous lava flows [8]. The strato-volcano of La Fossa reaches the elevation of 391 m a.s.l. (Figure 1c) and intense fumarolic degassing persists at its top. The main component escaping from the deep system to the surface is water vapor, resulting in the volcanic plume (or hydrothermal cloud) standing over the active cone (Figure 2a,b). The geochemical composition of fumaroles fluids is interpreted as a result of a variable mixing process between magmatic and hydrothermal fluids e.g., [10]. The temperature of fumaroles is one of the variables selected from the general geochemical approach to volcano monitoring. The increasing thermal output through the studied period has been positively correlated to the magmatic component [1,3,11]. More recently, this positive correlation has been less evident. Other effects of the mass and energy flux released by the geothermal system are found in the thermo-mineral aquifers and the diffuse gas emissions, monitored inside the Vulcano Porto village, and in the steam heated ground, spreading on the slopes of the active cone [12]. The thermodynamic evaluations of physic-chemical conditions of the thermal aquifer present at the base of La Fossa cone suggested episodic increases in the equilibrium temperatures and pressures revealing, in some cases, approaches to instability conditions, with increasing risk of phreatic explosions [10]. During periods of enhanced volcanic activity (such as from 1988 to 1992), many water wells receive a larger amount of vapor and the mixture of dissolved gases result similar to the composition of the crater fumaroles [12], and the complex steam heating process affecting the geothermal system has been modelled by Federico et al. [13].

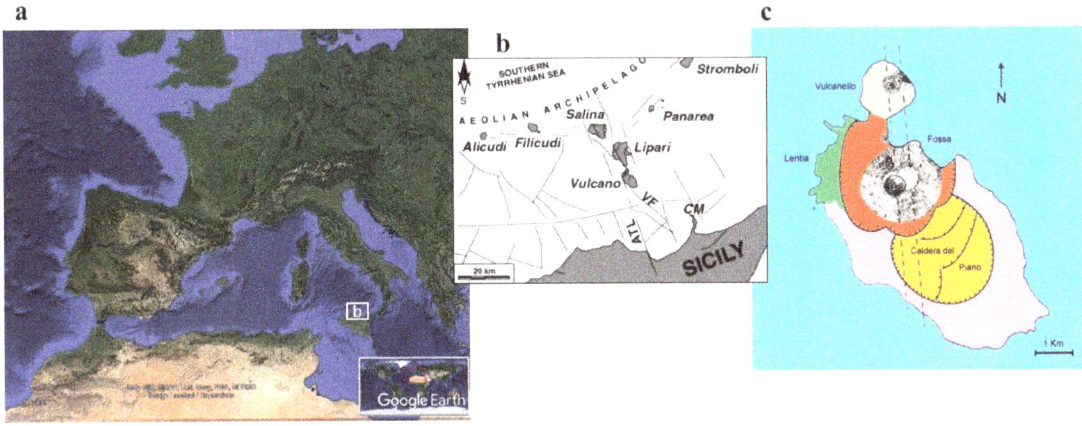

Figure 1. (**a**) Satellite view of the Mediterranean region. (**b**) Location of the Aeolian Archipelago in Southern Tyrrhenian Sea, ATL: Tindari–Letojanni strike-slip tectonic system; VF: Vulcano Faults; CM: Capo Milazzo, from [7]. (**c**) Schematic map of the Island of Vulcano with indication of the main tectonic lines and caldera rims from [9].

Figure 2. (**a**) Panoramic view of the summit zone of La Fossa cone interested by high temperature fumaroles, with indication of the continuous monitoring sites (FA, F5, F5AT). (**b**) Panoramic view of the Island of Vulcano from North-west, with the volcanic plume (or hydrothermal cloud) standing over the active cone.

The northern and southern flanks of La Fossa cone edifice, are covered by hydrothermally altered rocks; here the hydrothermal fluids, which are continuously flowing upwards along the highest permeability zones, influence the hydrological and mechanical properties of the rocks [14], increasing susceptibility to failure of slopes, as indicated by other authors [15–17]. In particular, hydrothermal alteration can interfere with the permeability pattern changing the porosity, and mechanical rock properties. Several temperature anomalies and geochemical crises occurred at La Fossa during the actual quiescent period: 1916–1924, 1977–1993, 1996 2004, 2005, 2006, and 2007 [2,3,18–20]. Many geochemical crises have been accompanied by increases in the number and amplitude of volcano-seismic events at shallow depth (<1–1.5 km) under La Fossa cone [21]. Repeated cycles of thermal volumetric changes increase the probability of landslides and rock falls from slopes interested by the circulation of hot fluid. At La Fossa cone, the relationships of the major temperature increase with the thermal expansion of the active cone and with transient increases of the pore pressure have been already reported, e.g., [6,11,22]. Bonaccorso et al. [22] highlighted the delay between the variations of fumaroles temperature and the deformation rate of the northern slope of the active cone and interpreted the measured displacement as the effect of volumetric changes related to the long-term trending variation of fumaroles temperature that occurred from 1989 to 1999. Since then, other periodical increases of gas emission and seismic activity have occurred, without significant ground deformation [14]. As phreatic explosion is one of the main volcanic risks at La Fossa caldera and it is possibly enhanced by self-sealing processes that occur on the permeable pathways due to the hydrothermal circulation, the temperature trends of fumaroles and thermal ground-waters are considered useful indicators to follow in real time the evolution of the energy and mass fluxes, affecting the equilibrium of this volcanic system.

3. Materials and Methods

Vulcano is constantly monitored by a network of monitoring stations for the temperature monitoring of the ground, tracking changes in the high temperature fumaroles. During a period of enhanced seismic release in the region, that begun in the eighties, the first multidisciplinary monitoring program was set up by the Italian *Gruppo Nazionale di Vulcanologia*. The monitoring program included the automated measurements system, for continuous monitoring of volcanic activity at the Aeolian islands by geochemical parameters, including temperature measurements. Figure 3 shows the location of temperature sensors in the high temperature fumarole network and on some other minor thermal anomalies.

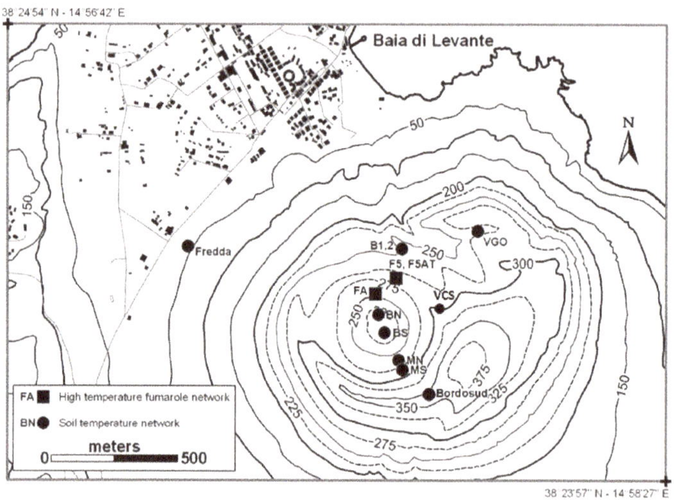

Figure 3. Location of the sensors for the thermal monitoring included in the INGV network.

The time relationships between the thermal signal recorded in the fumaroles vents and other geochemical and geophysical variations have been observed since the beginning of the geochemical monitoring [23], dating back to June 1984 [24]. The extreme environment conditions of La Fossa cone represented a hard challenge: Instrument and people are exposed to uninterrupted fluxes of acidic gases, highest moisture, and the highest temperature of the ground (the thermocouple inserted in the steaming vents measured temperatures higher than 670 °C); heavy rainy periods alternate to very dry seasons; no energy supply is available at the top of the active cone. The data are collected according to conventions aimed to the surveillance of volcanic activity, that have been periodically renewed between the *Dipartimento di Protezione Civile* (DPC) and the *Istituto Nazionale di Geofisica e Vulcanologia* (I.N.G.V.). The outlet temperatures are measured by chromel-alumel thermocouples (sensitivity of 41 $\mu V/°C$), inserted into the vent at a depth of 0.5 m. The measurement accuracy is maximized by applying the cold-junction compensation (CJC) technique. The CJC considers the voltage produced by temperature variations at the cold joint. "Cold" refers to the ambient temperature, in contrast to the "hot" temperature of the fumarole output. Disturbances in the measurement occur when the normally good thermal contact is lost, due to extreme weather conditions around the fumarole. Such uncertain values are removed from the time series data, until the necessary maintenance fieldwork is concluded, because these disturbances may mask the real temperature variation of the fumaroles. Specifically, the most frequent data gaps and technical failures have occurred, due to the frequent condensation of acidic fluids from the vapor release when the temperature output has decreased, reaching temperatures lower than 300 °C. During the technical maintenance of the system, episodic temperature measurements are carried out around the monitored locations with chromel-alumel thermocouples equipped with portable devices (percentage error <±1%). It has been hard to collect temperature data, with the sampling measurement window of 1 h for so many years in the extreme hostile environment created by the hot acidic fluids released in the fumarole field of La Fossa Vulcano. During these years, the monitoring sites were never moved from the former locations, to continue the longest record of data, the positive correlation found with ground deformation, seismic activity, and magmatic gas input has supported this long-term effort. In the last years (2020 and 2021), to observe the restriction imposed by the actual pandemia, we reduced the field work (calibration of the acquisition process and maintenance of the instruments), consequently we could not update all the time series of the monitoring network. From winter 2020 to spring 2021, only two monitoring sites in the high temperature fumaroles and one in the thermal soil have been working.

4. Results and Discussion

The summit area presents a surface thermal anomaly actually ranging from the boiling point to less than 260 °C, due to the presence of many fumarole vents that release high temperature fluids. Moreover, other minor thermal anomalies are present at various distance from the fumarole vents, and they occur where the subterranean steam condensate before reaching the ground surface. These thermal anomalies of minor intensity usually cause ground temperature below the boiling point and reveal high diffuse heat fluxes often associated to high CO_2 fluxes of magmatic origin [12,25]. The Figure 4 show the main thermal zones inside the caldera of La Fossa, as they are remotely sensed by the satellite Landsat 8 [26]. The thermal area in Figure 4b, indicated by the white circle in the summit of the active cone "la Fossa", includes the high temperature fumaroles and the continuous monitoring sites (see Figures 2a and 3 for site locations), and other anomalous surfaces interested by diffuse gas emissions [27]. The other zone indicated in Figure 4b by the white ellipse has the local name of "Levante Bay". The Levante Bay area includes some submarine fumaroles, some subaerial fumaroles, a part of the beach, the "Faraglione" and a "Mud pool" exploited as thermal bathing. The time variation of ground temperature in the Levante Bay is out of the scope of this paper, because the maximum temperature of fumaroles is about 100 °C and is buffered at the boiling point of water.

The Figure 5 is the thermographic photo-mosaic (composed by F. Pisciotta) after a survey made in October 2014 by IR camera [28]. The thermal map has been fitted to the google earth image of the inner flank of la fossa cone. This visualization gives an example of the extension of the thermal anomaly resulting by the combined advection of hot fluids of magmatic and hydrothermal origin in the summit area of the active cone La Fossa [27]. In 1926 the temperature of fumaroles reached more than 600 °C [29], and other intense pulsations have repeated afterwards [3]. The first period of increasing temperature at La Fossa fumarole fields, processed in near real time, lasted from 1988 until 1993, and has been coupled to progressive enlargement of the emissive surface [29] and sensible ground deformation [23]. The dynamics of magmatic and hydrothermal systems has been modelled by interpreting the changes in composition and temperature of the gases released from fumaroles, and suggested some evident interaction between these two systems [1,12,14,30,31]. The highest fumaroles temperature was measured in January 1993 on the inner flank of the La Fossa cone (FA fumarole, T = 670 °C [3,32]. Afterwards, the maximum temperature started to decrease, while the surface of active vents increased and many new vents opened on the active cone [29,32]. The longest time series of continuous temperature data are plotted in Figure 6 (Figures 2a and 3 for site locations). The summary statistics of the continuous monitoring data recorded during the same interval of time is reported in Table 1.

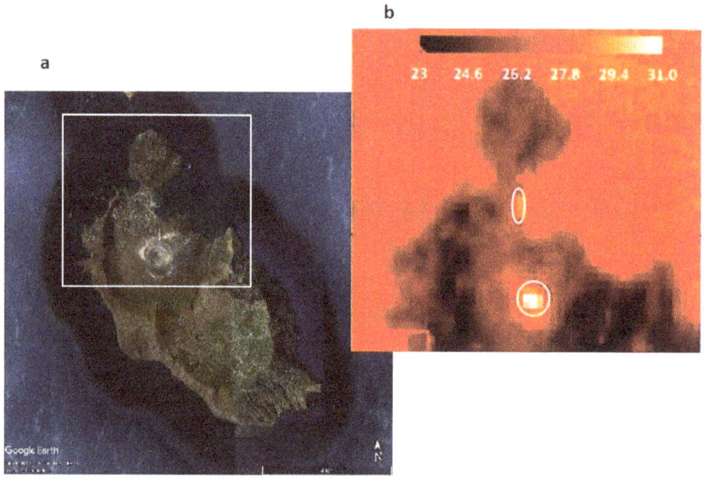

Figure 4. (**a**,**b**) Ground temperature retrieval, sensed on 8 July 2017, by the Landsat 8 TIRS (thermal infrared sensor, spatial resolution over a 190 km swath = 100 m), from [27]. The white circles mark the zones recognized by the thermo-scanning as the highest thermal surface, the scale of values refers the apparent temperature (°C) evaluated for each pixel.

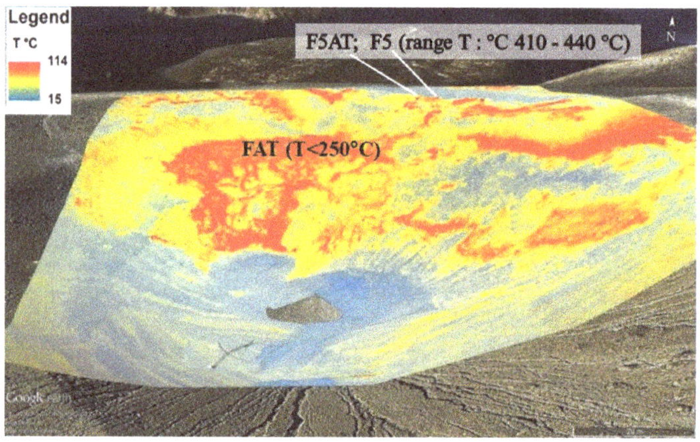

Figure 5. Thermographic photo-mosaic fitted to the google earth image of the inner flank of La Fossa cone modified from [28]. Date of the relief October 2014, author of the image F. Pisciotta. The legend show the scale of apparent temperature sensed by the IR thermo-camera, the labels show the range recorded by the contact sensors in the same month (see Figure 6).

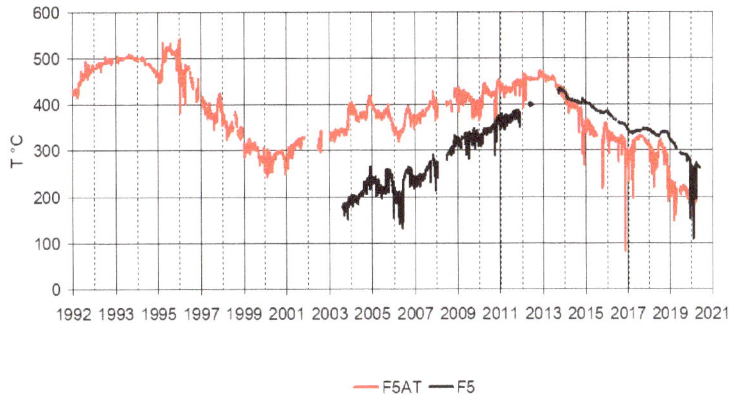

Figure 6. Rough temperature data extracted (daily values at h 12 a.m.) from the fumaroles time series recorded on the upper rim of La Fossa cone.

Table 1. Summary statistics of the continuous monitoring data daily recorded from January 1992 to May 2021.

	F5AT	F5AT2	F5	FAT-0	Tambient
Mean	382.41	354.488	312.1234	228.98	27.47
Standard Error of the Mean	0.83	1.040837	0.986084	1.29608	0.17
Standard Deviation	76.42	68.10	68.84	89.49	8.6
Variance	58403	4637.786	4738.32	8007.723	73.17
Coefficient of Variation	0.1998	0.1921	0.2205	0.3911	0.31
Minimum	82.09	131.24	108.631	12.85	6.4
Maximum	542	458.81	434.37	395	49.52
Sum	3,206,858	1,517,563	1,520,977	1,090,821	70,284
N of records	8386	4281	4873	4764	2559

The F5AT and F5 fumaroles are both located on the rim of the northern slope of La Fossa crater and are periodically sampled to determine the geochemical and the isotopic composition of the mixture of hot gases directly released in air. The set of temperature data, supplied from January 1992 by the continuous monitoring network, shows that in F5AT the temperature has been higher than 500 °C, from November 1993, but it started decreasing in September 1994. From December 2000 to May 2001, the temperature of F5AT (Figure 6) has showed the stable mean value of 275 °C, ranging between 244 and 306 °C. This first recorded period of minimum temperatures has been reached after a negative trending variation lasted 6 years and 3 months. Thereafter, on the upper rim the fumarole temperature progressively increased for more than 12 years, when it reached the second peak value of 473 °C (November 2013). Again in 2021 the maximum temperature probe has recorded the same behavior observed in 2001: A period of minimum temperatures with the stable mean value of 260 °C, ranging between 108 and 284 °C, from December 2020 to May 2021. Now, the F5AT vent ranges between 160 and 215 °C and the highest temperature is measured in the F5 vent (Figure 5), located at a short distance (about 8 m) from the F5AT vent (Figures 2a and 3). During the observation period the maximum temperature fumarole moved from the inner slope (fumarole FA, Figures 2a and 3) to the upper rim (in 1996, fumarole F5AT) and more recently along the upper rim (in 2015 from F5AT to F5). The F5AT time series showed a general negative linear trend, a complete asymmetrical cycle (lasting about 19 years) and several medium-term (lasting from weeks to months) temporal variations (Figure 6). The previous paper [33] decomposed the time series of temperatures recorded between 1998 and June 2012, by the Fast Fourier transform method. The Fast Fourier transform method revealed that: (a) The F5AT temperature variations have been modulated by a mid-term cyclic variation, that repeated every 11 synodic periods (months); (b) the seasonal component (12 synodic periods, nearly corresponding to 356 days) is negligible in the time series of the high temperature fumaroles. The updated time series show that from 1996 to 2018 the medium-term peaks have repeated with comparable amplitudes.

Figure 7 shows the linear and polynomial trends fitting the maximum and minimum values of temperature recorded in F5AT, while Figure 8 shows the results of the linear de-trending applied to the original series of data.

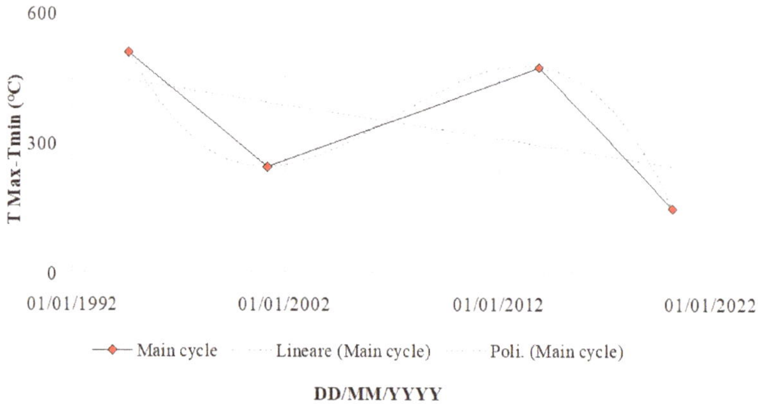

Figure 7. Maximum and minimum temperatures recorded in the F5AT after about 19 years, fitted by the negative linear trend and the polynomial curve of the third order.

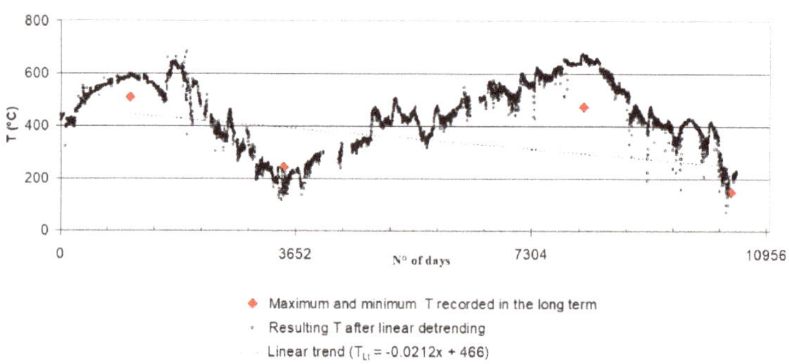

Figure 8. Detrended series of F5AT temperature data ($T_{Lt} = -0.0212x + 466$).

The de-trended time series would show a greater intensity of the asymmetrical cycle, and suggests that the resulting peaks would overcome the temperature values of 600 °C in August 1996 and again in November 2013, reaching more than 670 °C (Figure 7). This is the same temperature that was measured in the FA fumarole vent during the sampling survey performed in January 1993 [3,32], while on the upper rim the output temperature was still showing the real temperature of about 480 °C (Figure 6). In 1996, during the medium term cyclic variation of temperatures, visible in Figures 6 and 8, the geothermal system produced many other anomalous effects measured in the aquifer and in the diffuse gas emissions located at the base of the active cone, e.g., [2,3,19]. Again in 2019, during the medium term cyclic variation of temperatures, visible in Figures 6 and 8, the geochemical monitoring network registered the occurrence of multiple geochemical anomalies, at the base of La Fossa cone. The temperature network consists of a few fumaroles vents, that have been selected more than 30 years ago, when the thermal output was more intense than today, and the vent showing the highest temperature was located differently from today. The extension of the ground surface heated by the advection of high enthalpy fluid has greatly changed from the beginning of this monitoring activity, due to different combinations of many different processes, such as the unstable chemical composition of hydrothermal fluids, weathering processes, and fracturing events. Every monitoring site has shown its own range of temperatures and different local effects, but the overall trend of F5AT temperatures, roughly filtered by the local effect, suggest a major asymmetrical temperature cycle of the hydrothermal fluid expanding from the depth. In the de-trended series, the F5AT temperature pulsated from about 200 to 670 °C, and this maximum value has been already measured only once in the FA fumaroles, during the survey carried out in 1993. Recently, Silvestri et al. [26] and Mannini et al. [33] have tested the remote sensing techniques to follow the evolution of thermal anomalies appearing at the surface, over a larger extension of the ground. The direct measurements collected by the long-term time series presented here, could integrate the results obtained by the remote sensing methodology, to track-back the calibrated ground temperatures surrounding the monitored fumaroles vents and define the time variations of the thermal anomaly over the entire exposed surface. For example, the thermal area surrounding the FA fumaroles is actually extending about 2600 m^2, while the output temperature of this fumarole has been showing the buffered temperature of 120 °C. The intensity of the thermal anomalies, measured by direct monitoring, possibly integrated with the extension of the thermal effects, remotely sensed at the ground surface of active volcanoes could allow tracking of the thermal balance related to the steam advection.

5. Conclusions

In the last 30 years, the temperature of the ground measured in the steaming vents of La Fossa cone (Vulcano Island, Aeolian Archipelago, Italy) has pulsated between more than 650 and 250 °C. The location of the hottest emission has migrated twice, and the extension of the thermal anomalies surrounding the fumarole vents has changed, as well. The monitoring network of temperature tracked the surface heating effects related to the mixture of hot gases, continuously expanding from the shallow hydrothermal source and fed by a variable flux of magmatic gas from a deeper source. The dataset duration (more than 30 years) and temporal discretization (from 12 to 24 measures a day) of the actual records has shown the behaviour of the hidden source of thermal energy, which is useful to understand (hopefully forecast) the possible interaction between the magmatic/hydrothermal system of La Fossa Vulcano. Many different scientific papers have confirmed that this thermal monitoring has tracked the main advection process from the hydrothermal system to the surface during this last quiescent period. The time series methods for analysis (TSA) applied to the temperature data of fumaroles could help unravel the complexity of the hydrothermal system, which makes a deterministic description of the temperature variations difficult or ineffective. At the beginning of the modern volcanic surveillance programs, the deterministic approach has guided the selection of the variables to be monitored on the Island of Vulcano. For example, the thermodynamic approach to evaluate the equilibrium conditions in the geothermal system and the highest mobility of the fluid was based on some theorized behaviour but the quantitative models were lacking the amount of data necessary to confirm the hypothesis aiming to reveal the causative mechanisms of the volcanic processes in act. Therefore, strong approximations of field measurements and frequent data interpolations, were accepted to apply the theories on the fluid thermodynamic, originally based on laboratory experiments.

The frequency content of a volcanic system in a stationary state could be the combination of random variations and the modulation correlated to the external variables, but the surface temperature also reflects transient mass fluxes of hot fluids produced when the volcanic system is excited by the altered condition. This seems a recurrent event at La Fossa caldera where the monitoring evidences have been sometimes interpreted as the progressive accumulation of volatile at the top of an accumulation zone, followed by increased flux of volatiles affecting the hydrothermal fluid budget and the pressurization in the surrounding media, e.g., [13,14,19,20]. At the same time, the size of the thermal surface at the summit of La Fossa cone extended sensibly and also the magmatic gas output in the fumaroles mixture increased [26,31,34].

For many decades, the temperature of the ground has been a variable easy to be interpreted as a proxy of the thermal release from the buried source, and the remote control of the temperature monitoring has resulted technically sustainable. The selection of monitoring sites was fundamental since the monitoring network is highly site sensitive, depending on the permeability distribution that influences—being in turn influenced by—the advection of high enthalpy fluids. Anyway, it is convenient to avoid biases, possibly resulting from a subjective view, to interpret the time series of the measured temperature data. The statistical approach to time series can be used to find the discriminant among the background variations (such as the thermal effects of external origin), the anomalous transients—related to variations of the geothermal flux- and the effects deriving from the natural evolution of hydrothermal alterations. Applied to the geochemical monitoring the TSA data could generally supply many practical benefits to the geological danger management. For example: Assessment of the relative contributions of periodic and aperiodic signal components; cross correlation among different series of data; simulation modelling, and evaluation of a model performance that is based on a sustainable set of reference variables to be monitored in the long-term. The time series analysis could suggest the way to link probability, uncertainty, and randomness with causal dynamics and to incorporate the deterministic controls with random components, which affect the advection processes [35].

The specific result of TSA on this longest set of temperature data seem actually the unique opportunity to explore the thermal behaviour of ground surface on a closed conduit volcano, influenced by an active magmatic system. Moreover, the observed modulation could be interpreted in extensive terms to model the advection processes and to quantify the pulsating energy flowing from the deep system to the surface. Finally, the results and interpretations of this thermal behaviour could be exported in other, more remote, active volcanoes to interpret their thermal trends.

Funding: This work benefited from the agreement between Istituto Nazionale di Geofisica e Vulcanologia and the Italian Presidenza del Consiglio dei Ministri, Dipartimento della Protezione Civile (DPC). This paper does not necessarily represent DPC official opinion and policies.

Data Availability Statement: https://doi.org/10.26022/IEDA/112021, accessed on 11 June 2021.

Conflicts of Interest: The author declares no conflict of interest.

References

1. Chiodini, G.; Cioni, R.; Falsaperla, S.; Montalto, A.; Marini, L.; Guidi, M. Geochemical and seismological investigations at Vulcano (Aeolian Islands) during 1978–1989. *J. Geophys. Res.* **1992**, *97*, 11025–11032. [CrossRef]
2. Diliberto, I.S.; Gurrieri, S.; Valenza, M. Relationships between diffuse CO_2 emissions and volcanic activity on the island of Vulcano (Aeolian Islands, Italy) during the period 1984–1994. *Bull. Volcanol.* **2002**, *64*, 219–228. [CrossRef]
3. Diliberto, I.S. Long-term monitoring on a closed-conduit volcano: A 25-year long time-series of temperatures recorded at La Fossa cone (Vulcano Island, Italy), ranging from 250 °C to 520 °C. *J. Volcanol. Geotherm. Res.* **2017**, *346*, 151–160. [CrossRef]
4. Montalto, A. Seismic signals in geothermal areas of active volcanism: A case study from "La Fossa", Vulcano (Italy). *Bull. Volcanol.* **1994**, *56*, 220–227. [CrossRef]
5. Cannata, A.; Diliberto, I.S.; Alparone, S.; Gambino, S.; Gresta, S.; Liotta, M.; Montalto, P. Multiparametric approach in investigating volcano-hydrothermal systems: The case study of Vulcano (Aeolian Islands, Italy). *Pure Appl. Geophys.* **2012**, *169*, 167–182. [CrossRef]
6. Madonia, P.; Cusano, P.; Diliberto, I.S.; Cangemi, M. Thermal anomalies in fumaroles at Vulcano island (Italy) and their relationship with seismic activity. *J. Phys. Chem. Earth* **2013**, *63*, 160–169. [CrossRef]
7. Mattia, M.; Palano, M.; Bruno, V.; Cannavò, F.; Bonaccorso, A.; Gresta, S. Tectonic features of the Lipari-Vulcano complex (Aeolian archipelago, Italy) from ten years (1996–2006) of GPS data. *Terra Nova* **2008**, *20*, 370–377. [CrossRef]

8. De Astis, G.; Lucchi, F.; Dellino, P.; La Volpe, L.; Tranne, C.A.; Frezzotti, M.L.; Peccerillo, A. Geology, volcanic history and petrology of Vulcano (central Aeolian archipelago). *Geol. Soc. Lond. Mem.* **2013**, *37*, 281–349. [CrossRef]
9. Cortini, M.; Scandone, R. *Un'introduzione alla vulcanologia*; Napoli: Liguori, MO, USA, 1987; pp. 153–161. ISBN 88-207-1596-1.
10. Selva, J.; Bonadonna, C.; Branca, S.; De Astis, G.; Gambino, S.; Paonita, A.; Pistolesi, M.; Ricci, T.; Sulpizio, R.; Tibaldi, A.; et al. Multiple hazards and paths to eruptions: A review of the volcanic system of Vulcano (Aeolian Islands, Italy). *Earth-Sci. Rev.* **2020**, *207*, 103186. [CrossRef]
11. Chiodini, G.; Frondini, F.; Raco, B. Diffuse emission of CO_2 from the Fossa crater, Vulcano Island (Italy). *Bull. Volcanol.* **1996**, *58*, 41–50. [CrossRef]
12. Inguaggiato, S.; Diliberto, I.S.; Federico, C.; Paonita, A.; Vita, F. Review of the evolution of geochemical monitoring, networks and methodologies applied to the volcanoes of the Aeolian Arc (Italy). *Earth-Sci. Rev.* **2018**, *176*, 241–276. [CrossRef]
13. Federico, C.; Capasso, G.; Paonita, A.; Favara, R. Effects of steam-heating processes on a stratified volcanic aquifer: Stable isotopes and dissolved gases in thermal waters of Vulcano Island (Aeolian archipelago). *J. Volcanol. Geotherm. Res.* **2010**, *192*, 178–190. [CrossRef]
14. Currenti, G.; Napoli, R.; Coco, A.; Privitera, E. Effects of hydrothermal unrest on stress and deformation: Insights from numerical modeling and application to Vulcano Island (Italy). *Bull Volcanol.* **2017**, *79*, 28. [CrossRef]
15. Kendrick, J.E.; Smith, R.; Sammonds PMeredith, P.G.; Dainty, M.; Pallister, J.S. The influence of thermal and cyclic stressing on the strength of rocks from Mount St. Helens, Washington. *Bull. Volcanol.* **2013**, *75*, 728. [CrossRef]
16. Hutnak, M.; Hurwitz, S.; Ingebritsen, S.E.; Hsieh, P.A. Numerical models of caldera deformation: Effects of multiphase and multicomponent hydrothermal fluid flow. *J. Geophys. Res.* **2009**, *114*, B04411. [CrossRef]
17. De Natale, G.; Troise, C.; Pingue, F. A mechanical fluid-dynamical model for ground movements at Campi Flegrei caldera. *J. Geodyn.* **2001**, *32*, 487–517. [CrossRef]
18. Italiano, F.; Pecoraino, G. Steam output from fumaroles of an active volcano: Tectonic and magmatic-hydrothermal controls on the degassing system at Vulcano (Aeolian arc). *J. Geophys. Res.* **1998**, *103*, 29829–29842. [CrossRef]
19. Capasso, G.; Favara, R.; Francofonte, S.; Inguaggiato, S. Chemical and isotopic variations in fumarolic discharge and thermal waters at Vulcano Island (Aeolian Islands, Italy) during 1996: Evidence of resumed activity. *J. Volcanol. Geotherm. Res.* **1999**, *88*, 167–175. [CrossRef]
20. Granieri, D.; Carapezza, M.L.; Chiodini, G.; Avino, R.; Caliro, S.; Ranaldi, M.; Tarchini, L. Correlated increase in CO_2 fumarolic content and diffuse emission from La Fossa crater (Vulcano, Italy): Evidence of volcanic unrest or increasing gas release from a stationary deep magma body? *Geophys. Res. Lett.* **2006**, *33*. [CrossRef]
21. Alparone, S.; Cannata, A.; Gambino, S.; Gresta, S.; Milluzzo, V.; Montalto, P. Time-space variation of the volcano seismic events at La Fossa (Vulcano, Aeolian Islands, Italy): New insights into seismic sources in a hydrothermal system. *Bull. Volcanol.* **2010**, *72*, 803–816. [CrossRef]
22. Bonaccorso, A.; Bonforte, A.; Gambino, S. Thermal expansion-contraction and slope instability of a fumaroles field inferred from geodetic measurements at Vulcano. *Bull. Volcanol.* **2010**, *72*, 791–801. [CrossRef]
23. Carapezza, M.; Badalamenti, B.; Valenza, M. *Geochemical Surveillance of the Aeolian Islands by a Radio-Linked Computerized Continuous Monitoring*; Codata Bulletin: Paris, France, 1984.
24. Badalamenti, B.; Falsaperla, S.; Neri, G.; Nuccio, P.M.; Valenza, M. *Confronto Preliminare Tra Dati Sismici e Geochimica Nell'area Lipari–Vulcano*; Consiglio Nazionale Ricerche Gruppo Nazionale Vulcanologia: Roma, Italy, 1986; Volume 1, pp. 37–47.
25. Aubert, M.; Diliberto, S.; Finizola, A.; Chébli, Y. Double origin of hydrothermal convective flux variations in the Fossa of Vulcano (Italy). *Bull. Volcanol.* **2008**, *70*, 743–751. [CrossRef]
26. Silvestri, M.; Rabuffi, F.; Pisciotta, A.; Musacchio, M.; Diliberto, I.S.; Spinetti, C.; Lombardo, V.; Colini, L.; Buongiorno, M.F. Analysis of Thermal Anomalies in Volcanic Areas Using Multiscale and Multitemporal Monitoring: Vulcano Island Test Case. *Remote Sens.* **1998**, *11*, 134. [CrossRef]
27. Revil, A.; Finizola, A.; Piscitelli, S.; Rizzo, E.; Ricci, T.; Crespy, A.; Bolève, A. Inner structure of La Fossa di Vulcano (Vulcano Island, southern Tyrrhenian Sea, Italy) revealed by high-resolution electric resistivity tomography coupled with self-potential, temperature, and CO_2 diffuse degassing measurements. *J. Geophys. Res. Solid Earth* **2008**, *113*. [CrossRef]
28. Diliberto, I.S.; Pisciotta, A.; Vita, F.; Longo, M.; Gagliano, A.L.; Silvestri, M.; Spinetti, C. Vulcano Island from Ground to Space: A focus on changes of thermal release. In Proceedings of the New Space Economy—European Expoforum, Rome, Italy, 10–12 December 2019.
29. Bukumirovic, T.; Italiano, F.; Nuccio, P.M. The evolution of a geological system: The support of a GIS for geochemical measurements at the fumaroles field of Vulcano, Italy. *J. Volcanol. Geotherm. Res.* **1997**, *79*, 253–263. [CrossRef]
30. Capasso, G.; Favara, R.; Inguaggiato, S. Chemical features and isotopic composition of gaseous manifestations on Vulcano Island, Aeolian Islands, Italy: An interpretative model of fluid circulation. *Geochim. Cosmochim. Acta* **1997**, *61*, 3425–3440. [CrossRef]
31. Nuccio, P.M.; Paonita, A.; Sortino, F. Geochemical mixing between magmatic and hydrothermal gases: The case of Vulcano Island, Italy. *Earth Planet. Sci. Lett.* **1999**, *167*, 321–333. [CrossRef]
32. Diliberto, I.S. Time series analysis of high temperature fumaroles monitored on the island of Vulcano (Aeolian Archipelago, Italy). *J. Volcanol. Geotherm. Res.* **2013**, *264*, 150–163. [CrossRef]

33. Mannini, S.; Harris, A.J.L.; Jessop, D.E.; Chevrel, M.O.; Ramsey, M.S. Combining Ground- and ASTER-Based Thermal Measurements to Constrain Fumarole Field Heat Budgets: The Case of Vulcano Fossa 2000–2019. *Geophys. Res. Lett.* **2019**, *46*, 11868–11877. [CrossRef]
34. Paonita, A.; Federico, C.; Bonfanti, P.; Capasso, G.; Inguaggiato, S.; Italiano, F.; Madonia, P.; Pecoraino, G.; Sortino, F. The episodic and abrupt geochemical changes at La Fossa fumaroles (Vulcano Island, Italy) and related constraints on the dynamics, structure, and compositions of the magmatic system. *Geochim. Cosmochim. Acta* **2013**, *120*, 158–178. [CrossRef]
35. Sean, W.; Fleming, A.; Marsh, L.; Alaa, H.A. Practical applications of spectral analysis to hydrologic time series. *Hydrol. Process. Sci. Brief.* **2002**, *16*, 565–574. [CrossRef]

engineering proceedings

Proceedings

Ensemble Precipitation Estimation Using a Fuzzy Rule-Based Model †

O. Burak Akgun and Elcin Kentel *

Department of Civil Engineering, Middle East Technical University, Ankara 06800, Turkey; burak.akgun@metu.edu.tr
* Correspondence: ekentel@metu.edu.tr; Tel.: +90-312-210-5412

† Presented at the 7th International Conference on Time Series and Forecasting, Gran Canaria, Spain, 19–21 July 2021.

Abstract: In this study, a Takagi-Sugeno (TS) fuzzy rule-based (FRB) model is used for ensembling precipitation time series. The TS FRB model takes precipitation predictions of grid-based regional climate models (RCMs) from the EUR11 domain, available from the CORDEX database, as inputs to generate ensembled precipitation time series for two meteorological stations (MSs) in the Mediterranean region of Turkey. For each MS, RCM data that are available at the closest grid to the corresponding MSs are used. To generate the fuzzy rules of the TS FRB model, the subtractive clustering algorithm (SC) is utilized. Together with the TS FRB, the simple ensemble mean approach is also applied, and the performances of these two model results and individual RCM predictions are compared. The results show that ensembled models outperform individual RCMs, for monthly precipitation, for both MSs. On the other hand, although ensemble models capture the general trend in the observations, they underestimate the peak precipitation events.

Keywords: precipitation; fuzzy rule-based models; ensembling; climate

Citation: Akgun, O.B.; Kentel, E. Ensemble Precipitation Estimation Using a Fuzzy Rule-Based Model. *Eng. Proc.* **2021**, *5*, 48. https://doi.org/10.3390/engproc2021005048

Published: 9 July 2021

1. Introduction

Precipitation is one of the main meteorological parameters that affects water resources and it is directly influenced by climate change [1]. General circulation models (GCMs) are used to estimate precipitation under climate change [2]. However, the systematic biases between the simulated and observed values, and coarse resolution (generally a few hundred kilometers), of GCMs prevent their applications for regional-scale climate impact studies [3]. Although regional climate models (RCMs), obtained by the dynamical downscaling of GCMs, provide spatially and physically consistent outputs, with a finer resolution (typical resolution of tens of kilometers), they still have significant biases [3]. Moreover, it has long been recognized that a single model prediction does not provide the range of outcomes that are required to assess the risks of future climate change [4].

One of the alternatives to overcome the above-mentioned issues is to use an ensemble approach (EA). The EA is applied in various fields, such as biology [5], water resources [6], medicine [7], and decision-making [8]. However, the application of the EA for climate predictions is relatively new [9]. The EA is applied by [10] to assess climate change impact on hydrology and water resources. Ref. [11] assessed climate change impact on climate extremes, while [12] carried out future predictions of atmospheric rivers, by using GCMs with the EA. The EA is used by [13] to assess climate change impact on surface winds, and by [14] on extreme rainfall events as well.

Multiple linear regression (MLR) is a simple EA used to generate the superensemble, through minimization of the sum of the squares of the differences between predictors and predictands conclude that MLR has better performance skills compared to single model predictions [15–17]. Machine learning methods are also used for ensembling climate models. Artificial neural networks, random forest, K-nearest neighbor, and support vector machines are examples of machine learning methods for the EA that appeared in the literature [18–20]. In this study, another data-driven method, namely, TS FRB, identified by SC, is applied for ensembling RCM predictions.

The TS FRB model is a powerful practical engineering tool for the modeling and control of complex systems [21]. The main advantage of the TS model is that simple fuzzily defined local models will result in a nonlinear (of high order) global model [22]. The TS FRB model has wide applications in a number of fields, including adaptive nonlinear control, fault detection, performance analysis, forecasting, knowledge extraction, and behavior modeling [21]. In this study, the relation between individual RCM's predictions and observed precipitation is represented using fuzzy rules, which are formulated through SC. Clustering is used in grouping, pattern recognition, data mining, and machine learning [23].

Fuzzy models are used for statistical downscaling of precipitation [24–26]; nevertheless, to the best of our knowledge, this is the first application of the TS FRB model as an EA for climate models in Turkey. The TS FRB model with one rule is equivalent to the MLR; this provides a benchmark for the evaluation of model performance in this study area. In addition to MLR, the performance of the TS FRB model is compared to that of simple ensemble mean (SEM), in terms of the prediction of monthly precipitation in the study area.

2. Methodology

2.1. Data

Daily precipitation observations from two MSs are obtained from the Turkish State Meteorological Service. One of the MSs (Afyon: MS17190) is located inland while the other (Anamur: MS17320) is located close to the shoreline (Figure 1). Observational data are subjected to a two-step quality check (QC). First, the months with more than 10 days missing data are eliminated. Then observed data is examined for the whole historical period and commonly dry months are identified as July, August, and September. Months other than these with an average precipitation less than 0.1 mm are considered unreliable and removed from the time series. For the remaining months, the monthly average time series is obtained and named as the final dataset for observations (FDO).

To build an ensemble model, monthly average precipitation simulations from eight different CORDEX RCMs are obtained for the corresponding months of FDO. RCM predictions are extracted by using the code developed by [27]. Information about the CORDEX RCMs used in this study and their long-term monthly mean and standard deviations (μ/σ) for the grid closest to the MS17190 (~17190) and MS17320 (~17230) are shown in Table 1. The long-term monthly μ/σ for the observed precipitation at MS17190 and MS17320 are 1.16/0.85 and 2.88/3.45, respectively.

Table 1. RCMs used in the study.

General Circulation Models (GCM)	Regional Climate Models (RCMs)	Model Number	μ/σ (~17190)	μ/σ (~17320)
CNRM-CM5 (CNRM-CERFACS)	CCLM4-8-17	RCM1	1.85/1.20	2.89/3.38
	ALADIN53	RCM2	2.50/1.63	1.65/1.89
EC-EARTH (ICHEC)	CCLM4-8-17	RCM3	1.34/1.12	2.26/2.82
	RACMO22E	RCM4	0.95/0.71	1.52/1.75
	HIRHAM5	RCM5	1.14/0.95	2.01/2.59
CM5A-MR (IPSL)	WRF331F	RCM6	2.21/1.63	1.98/2.52
HadGEM2-ES (MOHC)	CCLM4-8-17	RCM7	1.51/1.28	3.33/3.90
	RACMO22E	RCM8	1.19/0.92	2.50/2.64

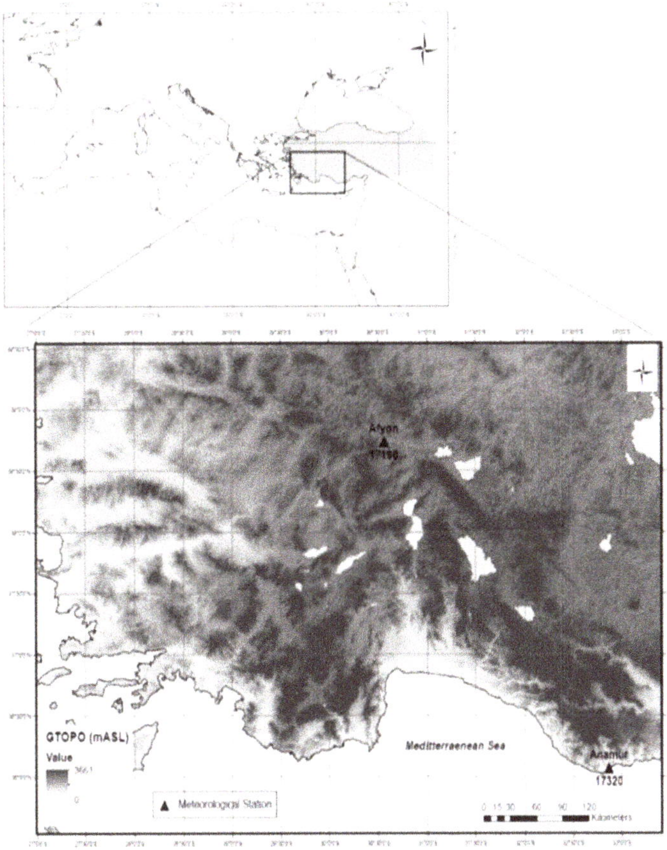

Figure 1. Study area and the location of the meteorological stations (MSs).

2.2. Takagi-Sugeno Fuzzy Rule-Based Model

In this study, a TS FRB model is developed to obtain an ensemble precipitation time series for each MS. Mathematical representation of the TS FRB model is as follows:

$$FRB_i^t = f\left(RCM_{i,1}^t, RCM_{i,2}^t, \ldots, RCM_{i,8}^t\right) \quad (1)$$

where i is the index for the MS (here $i = 1, 2$), t is the index for time, FRB_i^t is the ensembled precipitation value for MS i in month t, $RCM_{i,1}^t, RCM_{i,2}^t, \ldots, RCM_{i,8}^t$ are the precipitation predictions of 8 different RCMs at the grid closest to the MS i in month t.

The rule-based structure of the TS FRB model is identified by using SC. TS FRB takes precipitation predictions of eight RCMs as inputs to predict ensembled precipitation time series at the corresponding MS as the output. In SC, together with the input data, output data is included in the clustering process. Before clustering, log-transformation is applied to all data sets and the feature space is normalized to bind all data in a unit hypercube. In SC, each data point is treated as a candidate to be a cluster center (cc), and the potential of each data point to be a cc is calculated using [28], as follows:

$$P_m = \sum_{k=1}^{n} e^{-(4/{r_a}^2)||X_m - X_k||^2} \quad (2)$$

where X_m is the normalized data point m, P_m is the potential of the X_m, r_a is a user-defined positive constant identifying cluster radius and n is the number of data points.

The potential of a data point exponentially decays with the square of the distance between that data point and all other data points. In this way, the data points with many neighboring data have higher chances to be cluster centers. The data point having the highest potential value is assigned as the first cc. Then, the potentials are updated using [28], as follows:

$$P_m \leftarrow P_m - P_z^* e^{-(4/r_b{}^2)||X_m - X_z^*||^2} \quad (3)$$

where X_z^* is the cc z, P_z^* is the potential of the X_z^* and r_b is a user-defined positive constant. In this study, r_b is taken $1.5r_a$ to avoid cluster centers being spaced closely.

In the updating process, the potential decays exponentially with the square of the distance from each data point to the previously assigned cc. Hence, the updating process ensures the potential of data points that are close to the previously assigned cc drop significantly compared to the data points distant to it; specifically, the potential of the previous cc becomes zero. Note that the predecessor cc is used for the updating process. The data point having the highest potential after the updating process is assigned as the next cc. Updating procedure is repeated until a user-defined number of cluster centers are identified.

Each cc is, in essence, a prototypical data point that exemplifies a characteristic behavior of the dataset. Therefore, each cc can be used as the basis of a fuzzy rule that describes the system behavior [28]. To convert 9-dimensional (8 input RCMs and one output) cc to the fuzzy rules, each cc is decomposed into two vectors (first with eight and second with 1 element). In this study, each fuzzy rule has the following form:

$$\begin{array}{ll}\text{Rule } a: & \text{IF } x_1 \text{ is } M_1^a \& \ x_2 \text{ is } M_2^a \ \& \ldots \& \ x_8 \text{ is } M_8^a \\ & \text{THEN } y = N_1^a x_1 + N_2^a x_2 + \cdots + N_8^a x_8 + N_9^a\end{array} \quad (4)$$

where $x_1,\ x_2, \ldots, x_8$ are the input variables and y is the output variable, $M_1^a, M_2^a, \ldots,\ M_8^a$ are antecedent fuzzy sets for rule a, which are defined by Gaussian membership functions and $N_1^a,\ N_2^a, \ldots, N_8^a,\ N_9^a$ are the parameters to be optimized for rule a. After obtaining fuzzy rules, well-known TS FRB [29] model is constructed and parameters are optimized by recursive least square estimation. For further details, the reader may refer to [28,30].

To select the best combination of the clustering parameters (e.g., the number of cc and r_a), a trial–error procedure is applied. Limiting the study space with 20 cc and maximum r_a of 1, 400 FRB models are built using the FDO as the output and corresponding RCMs as inputs for each MSs. Dataset is randomly sampled and 75% of the dataset is used for training while the rest is used for validation. Combination resulting in the best performance in the validation set is selected and the selected number of cc and r_a are used in ensembling.

The framework of the TS FRB is given in Figure 2. In the EA, 5-fold validation is used. The folds are combined together to form the validation time series (VTS) of precipitation. Note that each data point is used in the validation dataset once.

2.3. Simple Average of the Models for Ensembling

The SEM is formed to predict monthly precipitation values for the closest grid to the selected MS. Formulation of SEM is as follows [31]:

$$SEM_i^t = \overline{OBS}_i + \overline{RCM}_i^t - \overline{RCM}_i \quad (5)$$

where SEM_i^t is the precipitation prediction at grid i in time t, $\overline{OBS}_i$ is the climatology of the precipitation observation at grid i, $\overline{RCM}_i^t$ is the average of the eight RCMs for grid i at time t and $\overline{RCM}_i$ is the climatology of the eight RCMs.

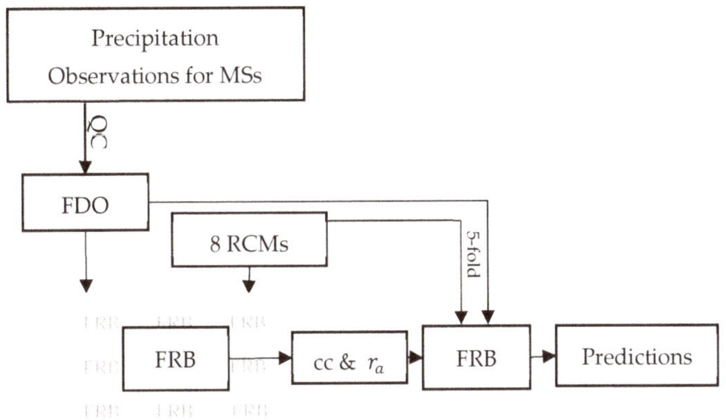

Figure 2. Framework of the ensemble Takagi-Sugeno (TS) fuzzy rule-based (FRB).

3. Results

In this section, the prediction performances of the SEM and TS FRB models are analyzed and compared for the VTS. In the clustering parameter selection process for the TS FRB, it is observed that as the number of cluster centers increases, the models tend to overfit. On the other hand, the performance of the models with relatively low numbers of cluster centers is very similar to those of the models with one cc (e.g., equivalent to the MLR). As a result of the trial-and-error procedure, for MS17190, the TS FRB model with 2 cc and r_a of 0.45 is selected, while 2 cc and r_a of 0.65 is selected for MS17320. Using the selected parameters, ensembling is carried out, and the VTS of each MS is formed. The performances of these models, together with those of the TS FRB model with one cluster center (e.g., MLR) and SEM, are given in Table 2, in terms of correlation (corr), root mean square error (RMSE) and percent bias (PBias).

Table 2. Performance of the EA and RCM.

Meteological Stations (MS)	Criteria	Best RCM		Worst RCM		Takagi-Sugeno (TS) Fuzzy Rule-Based (FRB)	MLR	SEM
		Model	Value	Model	Value			
17190	**Correlation (corr)**	RCM3	**0.29** [1]	RCM2	0.12	**0.40** [1]	**0.40** [1]	**0.40** [1]
	Root mean square error (RMSE)	RCM4	0.96	RCM2	2.21	**0.78** [1]	0.80	0.85
	Percent bias (PBias)	RCM8	**−2.88** [1]	RCM2	−116	10.16	7.16	10.34
17320	corr	RCM5	0.53	RCM2	0.26	**0.65** [1]	0.62	0.62
	RMSE	RCM8	3.12	RCM7	3.86	**2.67** [1]	2.81	2.81
	PBias	RCM8	**3.12** [1]	RCM7	3.86	17.7	23.91	16.88

[1] The blods referred results of the trial-and-error procedure, for MS17190, the TS FRB model with 2 cc and r_a of 0.45 is selected, while 2 cc and r_a of 0.65 is selected for MS17320.

As shown in Table 2, where the best performance is given in bold, the prediction skills of TS FRB, MLR and SEM are very similar for both stations. On the other hand, the ensembled results have higher prediction skills in terms of corr and RMSE, and lower prediction skills in terms of PBias, compared to the best-performing RCM. The time series of the observations and predictions of TS FRB and SEM for MS17190 and MS17320 are given in Figures 3 and 4, respectively. In Figures 3 and 4, to increase visibility, transparent colors are preferred for the TS FRB and SEM predictions.

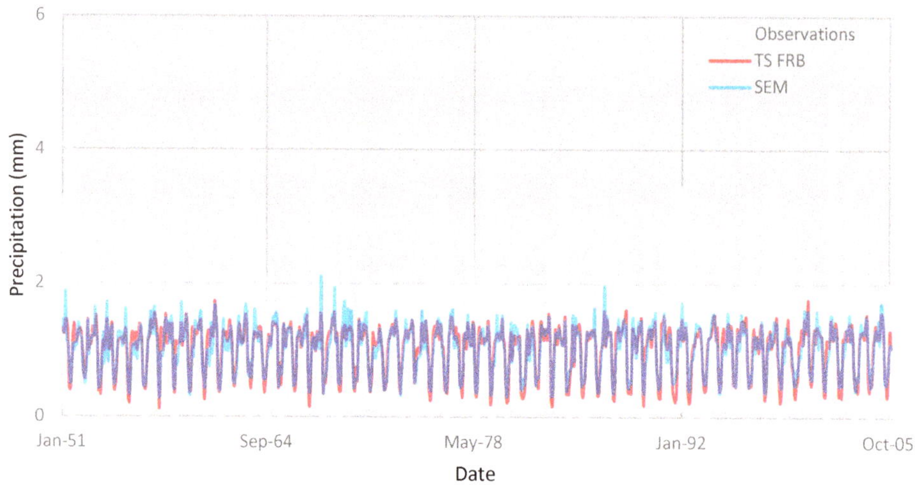

Figure 3. Time series of VTS obtained from TS FRB and simple ensemble mean (SEM) for MS17190.

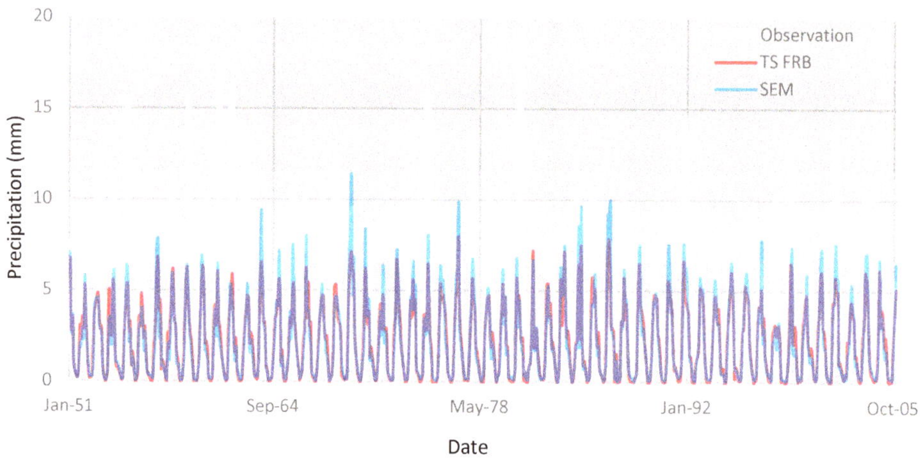

Figure 4. Time series of VTS obtained from TS FRB and SEM for MS17320.

Although TS FRB and SEM follow the general trend in the observations for both MSs, both models underestimate the peak precipitations, as can be seen in Figures 3 and 4. However, despite its simplicity, SEM has higher skill in the prediction of peak precipitations compared to the nonlinear TS FRB. On the other hand, although the peak precipitations for MS17320 are much larger than those of MS17190, the EAs have higher prediction skills for MS17320, especially in terms of corr.

4. Conclusions

In this study, the ensembling performance of a TS FRB model for monthly precipitations is compared to those SEM and individual RCMs.

- The analysis shows that the performance, in terms of corr and RMSE, of the EA is better, compared to the individual RCMs for both MSs. However, the PBias values of the best-performing RCM are much better than those of SEM and TS RFB;

- The nonlinear TS FRB model has very similar prediction skills to the simple SEM model. So, when the effort to select the best cc and r_a combination is taken into account, SEM is more efficient, compared to the TS FRB model for ensembling;
- Both models failed to predict peak precipitation events.

Acknowledgments: This work is supported by the Scientific and Technological Research Council of Turkey (TÜBİTAK) (grant number 118Y365).

Conflicts of Interest: The authors declare no conflict of interest.

References

1. Trenberth, K.E. Changes in precipitation with climate change. *Clim. Res.* **2011**, *47*, 123–138. [CrossRef]
2. Eden, J.M.; Widmann, M.; Maraun, D.; Vrac, M. Comparison of GCM-and RCM-simulated precipitation following stochastic postprocessing. *J. Geophys. Res. Atmos.* **2014**, *119*, 11-040. [CrossRef]
3. Turco, M.; Llasat, M.C.; Herrera, S.; Gutiérrez, J.M. Bias correction and downscaling of future RCM precipitation projections using a MOS-Analog technique. *J. Geophys. Res. Atmos.* **2017**, *122*, 2631–2648. [CrossRef]
4. Buontempo, C.; Mathison, C.; Jones, R.; Williams, K.; Wang, C.; McSweeney, C. An ensemble climate projection for Africa. *Clim. Dyn.* **2015**, *44*, 2097–2118. [CrossRef]
5. Gårdmark, A.; Lindegren, M.; Neuenfeldt, S.; Blenckner, T.; Heikinheimo, O.; Müller-Karulis, B.; Niiranen, S.; Tomczak, M.T.; Aro, E.; Wikström, A.; et al. Biological ensemble modeling to evaluate potential futures of living marine resources. *Ecol. Appl.* **2013**, *23*, 742–754. [CrossRef]
6. Choubin, B.; Moradi, E.; Golshan, M.; Adamowski, J.; Sajedi-Hosseini, F.; Mosavi, A. An ensemble prediction of flood susceptibility using multivariate discriminant analysis, classification and regression trees, and support vector machines. *Sci. Total Environ.* **2019**, *651*, 2087–2096. [CrossRef]
7. Smith, T.; Ross, A.; Maire, N.; Chitnis, N.; Studer, A.; Hardy, D.; Brooks, A.; Penny, M.; Tanner, M. Ensemble Modeling of the Likely Public Health Impact of a Pre-Erythrocytic Malaria Vaccine. *PLoS Med.* **2012**, *9*, e1001157. [CrossRef]
8. Marzocchi, W.; Taroni, M.; Selva, J. Accounting for epistemic uncertainty in PSHA: Logic tree and ensemble modeling. *Bull. Seismol. Soc. Am.* **2015**, *105*, 2151–2159. [CrossRef]
9. Kotlarski, S.; Keuler, K.; Christensen, O.B.; Colette, A.; Déqué, M.; Gobiet, A.; Goergen, K.; Jacob, D.; Lüthi, D.; van Meijgaard, E.; et al. Regional climate modeling on European scales: A joint standard evaluation of the EURO-CORDEX RCM ensemble. *Geosci. Model Dev.* **2014**, *7*, 1297–1333. [CrossRef]
10. Christensen, N.S.; Lettenmaier, D.P. A multimodel ensemble approach to assessment of climate change impacts on the hydrology and water resources of the Colorado River Basin. *Hydrol. Earth Syst. Sci.* **2007**, *11*, 1417–1434. [CrossRef]
11. Tegegne, G.; Melesse, A.M.; Worqlul, A.W. Development of multi-model ensemble approach for enhanced assessment of impacts of climate change on climate extremes. *Sci. Total Environ.* **2020**, *704*, 135357. [CrossRef]
12. Massoud, E.C.; Espinoza, V.; Guan, B.; Waliser, D.E. Global climate model ensemble approaches for future projections of atmospheric rivers. *Earth's Future* **2019**, *7*, 1136–1151. [CrossRef]
13. Najac, J.; Boé, J.; Terray, L. A multi-model ensemble approach for assessment of climate change impact on surface winds in France. *Clim. Dyn.* **2009**, *32*, 615–634. [CrossRef]
14. Padulano, R.; Reder, A.; Rianna, G. An ensemble approach for the analysis of extreme rainfall under climate change in Naples (Italy). *Hydrol. Process.* **2019**, *33*, 2020–2036. [CrossRef]
15. Krishnamurti, T.N.; Kishtawal, C.M.; LaRow, T.E.; Bachiochi, D.R.; Zhang, Z.; Williford, C.E.; Gadgil, S.; Surendran, S. Improved weather and seasonal climate forecasts from multimodel superensemble. *Science* **1999**, *285*, 1548–1550. [CrossRef]
16. Krishnamurti, T.N.; Kishtawal, C.M.; Zhang, Z.; LaRow, T.; Bachiochi, D.; Williford, E.; Gadgil, S.; Surendran, S. Multimodel Ensemble Forecasts for Weather and Seasonal Climate. *J. Clim.* **2000**, *13*, 4196–4216. [CrossRef]
17. Yun, W.T.; Stefanova, L.; Mitra, A.K.; Vijaya Kumar, T.S.V.; Dewar, W.; Krishnamurti, T.N. A multi-model superensemble algorithm for seasonal climate prediction using DEMETER forecasts. *Tellus A Dyn. Meteorol. Oceanogr.* **2005**, *57*, 280–289. [CrossRef]
18. Okkan, U.; Inan, G. Statistical downscaling of monthly reservoir inflows for Kemer watershed in Turkey: Use of machine learning methods, multiple GCMs and emission scenarios. *Int. J. Climatol.* **2015**, *35*, 3274–3295. [CrossRef]
19. Wang, B.; Zheng, L.; Liu, D.L.; Ji, F.; Clark, A.; Yu, Q. Using multi-model ensembles of CMIP5 global climate models to reproduce observed monthly rainfall and temperature with machine learning methods in Australia. *Int. J. Climatol.* **2018**, *38*, 4891–4902. [CrossRef]
20. Ahmed, K.; Sachindra, D.A.; Shahid, S.; Iqbal, Z.; Nawaz, N.; Khan, N. Multi-model ensemble predictions of precipitation and temperature using machine learning algorithms. *Atmos. Res.* **2020**, *236*, 104806. [CrossRef]
21. Angelov, P.P.; Filev, D.P. An approach to online identification of Takagi-Sugeno fuzzy models. *IEEE Trans. Syst. Manand Cybern. Part B* **2004**, *34*, 484–498. [CrossRef] [PubMed]
22. Angelov, P. An approach for fuzzy rule-base adaptation using on-line clustering. *Int. J. Approx. Reason.* **2004**, *35*, 275–289. [CrossRef]
23. Jain, A.K.; Murty, M.N.; Flynn, P.J. Data clustering: A review. *ACM Comput. Surv.* **1999**, *31*, 264–323. [CrossRef]

24. Najafi, M.R.; Moradkhani, H.; Wherry, S.A. Statistical downscaling of precipitation using machine learning with optimal predictor selection. *J. Hydrol. Eng.* **2011**, *16*, 650–664. [CrossRef]
25. Wetterhall, F.; Bárdossy, A.; Chen, D.; Halldin, S.; Xu, C.Y. Statistical downscaling of daily precipitation over Sweden using GCM output. *Theor. Appl. Climatol.* **2009**, *96*, 95–103. [CrossRef]
26. Bárdossy, A.; Pegram, G. Downscaling precipitation using regional climate models and circulation patterns toward hydrology. *Water Resour. Res.* **2011**, *4*. [CrossRef]
27. Kentel, E.; Akgun, O.B.; Mesta, B. User-Friendly R-Code for Data Extraction from CMIP6 outputs. In *AGU Fall Meeting Abstracts*; PA33C-1098; American Geophysical Union: Washington, DC, USA, 2019.
28. Chiu, S.L. Fuzzy model identification based on cluster estimation. *J. Intell. Fuzzy Syst.* **1994**, *2*, 267–278. [CrossRef]
29. Takagi, T.; Sugeno, M. Fuzzy identification of systems and its applications to modeling and control. *IEEE Trans. Syst. Man Cybern.* **1985**, *1*, 116–132. [CrossRef]
30. Mesta, B.; Akgun, O.B.; Kentel, E. Alternative solutions for long missing streamflow data for sustainable water resources management. *Int. J. Water Resour. Dev.* **2020**, 1–24. [CrossRef]
31. Cane, D.; Milelli, M. Multimodel SuperEnsemble technique for quantitative precipitation forecasts in Piemonte region. *Nat. Hazards Earth Syst. Sci.* **2010**, *10*, 265–273. [CrossRef]

engineering proceedings

MDPI

Proceedings

Automatic Hierarchical Time-Series Forecasting Using Gaussian Processes †

Luis Roque [1,*], Luis Torgo [2,3] and Carlos Soares [1,4]

1 Faculdade de Engenharia, Universidade do Porto, 4099-002 Porto, Portugal; csoares@fe.up.pt
2 Faculty of Computer Science, Dalhousie University, Halifax, NS B3H 1W5, Canada; ltorgo@dal.ca
3 Faculdade de Ciências, Universidade do Porto, 4099-002 Porto, Portugal
4 Fraunhofer Portugal, AICOS and LIACC, 4099-002 Porto, Portugal
* Correspondence: luis_roque@live.com

† Presented at the Seventh International Conference on Time-Series and Forecasting, Gran Canaria, Spain, 19–21 July 2021.

Abstract: Forecasting often involves multiple time-series that are hierarchically organized (e.g., sales by geography). In that case, there is a constraint that the bottom level forecasts add-up to the aggregated ones. Common approaches use traditional forecasting methods to predict all levels in the hierarchy and then reconcile the forecasts to satisfy that constraint. We propose a new algorithm that automatically forecasts multiple hierarchically organized time-series. We introduce a combination of additive Gaussian processes (GPs) with a hierarchical piece-wise linear function to estimate, respectively, the stationary and non-stationary components of the time-series. We define a flexible structure of additive GPs generated by each aggregated group in the hierarchy of the data. This formulation aims to capture the nested information in the hierarchy while avoiding overfitting. We extended the piece-wise linear function to be hierarchical by defining hyperparameters shared across related time-series. From our experiments, our algorithm can estimate hundreds of time-series at once. To work at this scale, the estimation of the posterior distributions of the parameters is performed using mean-field approximation. We validate the proposed method in two different real-world datasets showing its competitiveness when compared to the state-of-the-art approaches. In summary, our method simplifies the process of hierarchical forecasting as no reconciliation is required. It is easily adapted to non-Gaussian likelihoods and multiple or non-integer seasonalities. The fact that it is a Bayesian approach makes modeling uncertainty of the forecasts trivial.

Keywords: Gaussian processes; forecasting; hierarchical time-series; Bayesian statistics

Citation: Roque, L.; Torgo, L.; Soares, C. Automatic Hierarchical Time-Series Forecasting Using GPs. *Eng. Proc.* **2021**, *5*, 49. https://doi.org/10.3390/engproc2021005049

Published: 9 July 2021

1. Introduction

The problem of automatically forecasting large numbers of univariate time-series is commonly found in different businesses [1]. In this setting, the selection of the appropriate time-series model, estimation of its parameters, and computation of the forecasts have to be done without human intervention. These large collections of time-series often involve multiple time-series aggregated by groups, such as geography. In this case, the forecasts for the bottom-level-series are required to add up to the forecasts of the aggregated ones. This constraint is referred to as *coherence*, and the process of adjusting forecasts to make them coherent is called forecast *reconciliation* [2]. Our work focuses on automatically forecasting a set of these hierarchically organized time-series. The goal is to generate accurate predictions for both the individual series and for each of the aggregation levels. This is to be achieved while ensuring that the forecasts are *coherent*. Finally, our algorithm is intended to be applied to any time-series domain.

A common approach for this type of problem is to produce forecasts for all aggregation levels and then *reconcile* the forecasts using linear or non-linear models (e.g., [2–4]). This strategy is highly dependent on the forecasting method used, and it is a two-step process.

We follow a different approach where our forecasting algorithm takes into account the hierarchical information when predicting the individual series. This means that we do not need to *reconcile* the forecasts afterwards.

We introduce a combination of additive Gaussian processes (GPs) with a hierarchical piece-wise linear function to estimate, respectively, the stationary and non-stationary components of the time-series. We define a flexible structure of additive GPs that are summed coherently for each element of a group. GPs allow us to model relevant time-series patterns, such as seasonality or noise as their additive components. Its additive nature contributes to capturing the potential nested information in the hierarchy while avoiding overfitting (one common problem with GPs). In the non-stationary component, we are interested in modeling the trend of the data and trend changes over time while also capturing potential hierarchical relationships (e.g., similar trend patterns in the same group). With these goals in mind, we define a hierarchical piece-wise linear function. We adapt the idea of multilevel models from Bayesian statistics to define hyperparameters shared across related time-series. This way, we are estimating parameters for each group and using them to inform the estimation of the individual ones, that is, the individual parameters are partially pooled towards the group mean. This results in forecasts for the trend of the bottom-level series that already take into account the behavior of the aggregated ones.

From our experiments, our algorithm is able to scale to, at least, hundreds of time-series. The estimation of the posterior distribution of the parameters can be challenging for datasets with this size. To be able to perform it at that scale, we used mean-field approximation [5], which is an automatic algorithm to perform Variational Inference (VI). We validated the proposed algorithm in two different real-world datasets, showing its ability to work with small and large datasets with typical trend and seasonal patterns varying between groups.

In summary, our contributions are:

- A new algorithm for hierarchical time-series forecasting that does not require any type of *reconciliation;*
- The definition of a flexible structure of hierarchical additive GPs. Additive GPs are used in statistical analysis [6], whereas we propose a formulation to adapt it to automatic hierarchical time-series forecasting;
- The combination of additive GPs with a hierarchical piece-wise linear function to model, respectively, the stationary and non-stationary components of hierarchical time-series;
- An automatic method that does not require expert intervention to be fitted to new data.

We start by outlining the related work in Section 2. Section 3 introduces the algorithm, covering the main contributions. In Section 4, we present our findings and results followed by conclusions and future directions, in Section 5.

2. Related Work

We work with a collection of s-related univariate time-series $\{z^i_{1:T}\}^s_{i=1}$, where $z^i_{1:T} = [z^i_1, z^i_2, \cdots, z^i_T]$ and $z^i_t \in \mathbb{R}$ denote the value of time-series i at time t. The time-series are aggregated by groups; in fact, each time-series is associated with an element l of every group g present in the dataset. We use vector $\mathbf{q}^{g,l}$ of size s to encode this information. It has value 1, when the series s belongs to the element l of the group g, and 0, otherwise. To give a simple example, consider a dataset with two groups (g1 and g2), each one with two different elements (a and b, and x and y, respectively). We start with the most aggregated level of the data z. We can aggregate the individual series z^i_t by group g1, forming the series z^{g1}_t, and by its elements, forming z^{a}_t and z^{b}_t. We can do the same for the second group g2, forming series z^{g2}_t, or by its elements, z^{x}_t and z^{y}_t. At the bottom level, this would generate four different series (i.e., z^{ax}_t, z^{ay}_t). Figure 1 illustrates a particular example of this dataset.

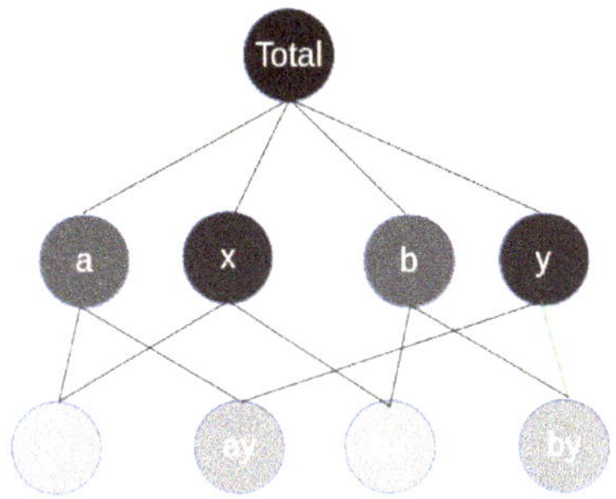

Figure 1. Example of time-series aggregated by group.

The goal of forecasting is to predict the next τ time-steps for all time-series, that is, $\{z^i_{T+1:\tau}\}^s_{i=1}$.

$$p(z^i_{T+1:T+\tau}|z^i_{1:T};\theta), \tag{1}$$

where θ are the parameters of the model. For any given time-series i, we refer to time-series $z^i_{1:T}$ as target time-series, $\{1, 2, ..., T\}$ as the training range, and to time $T+1, T+2, \ldots, T+\tau$ as the prediction range. The time-point $T+1$ is referred to as the forecast start time and $\tau \in \mathbb{N}_{>0}$ is the forecast horizon. Point forecasts for a given time-series, i at time $T+t$ are denoted by $\hat{z}^i_{T+t}$, and the point forecast errors are denoted by $e^i_{T+t} = z^i_{T+t} - \hat{z}^i_{T+t}$.

2.1. Time-Series Forecasting

When we consider the automatic forecasting process of a single time-series and include constraints such as integer or single seasonality, the state-space exponential smoothing (ETS) [7] and automated ARIMA [1] procedures are still considered state-of-the-art approaches. When we extend to non-integer or multiple seasonalities, there are other methods which become relevant, including TBATS [8] or Prophet [9]. In the case of Prophet, there is also additional flexibility on how the trend is modeled. Traditional GPs do not excel in automatic forecasting. An adaptation with benchmark forecasting resulting in single univariate time-series has been proposed [10], but prior knowledge was introduced to achieve that competitiveness. There are several challenges when using GPs to do automatic forecasting, namely, the lack of a criterion for kernel selection and the long time required for training different competing kernels. On the other hand, Recurrent Neural Networks (RNN) are gaining popularity as an alternative to statistical methods. Nevertheless, the settings where they can achieve competitiveness are still very narrow and require user adaptation (see [11] for an extensive study on the topic).

Forecast accuracy is usually measured by summarizing the forecast errors using a scaled metric, such as the Mean Absolute Scaled Error (MASE) (see [7] for an extended overview on forecast error metrics). For seasonal time-series, a scaled error can be computed by:

$$q_j = \frac{e_j}{\frac{1}{T-m}\sum_{t=m+1}^{T}|z_t - z_{t-m}|}, \tag{2}$$

and MASE is defined by $mean(|q_j|)$.

2.2. Hierarchical and Grouped Time-Series

When working with related time-series, the focus is to model cross-series information to improve univariate models. There are cases where the time-series are only related by belonging to the same domain ([12]), while in other cases they can be aggregated in groups or in a hierarchy. There are different methods designed to work with hierarchical time-series. The method initially proposed by [13] and improved in [2,14] consists in optimally combining and reconciling all forecasts at all levels of the hierarchy. A linear regression is used to combine the independent forecasts, guaranteeing that the revised forecasts are as close as possible to the independent forecasts but maintaining coherence. These works were further extended to allow a non-linear combination of the base forecasts [3] and adapted to a Bayesian setting [4]. A Bayesian approach takes into account the uncertainty across all levels of the hierarchy to obtain the revised forecasts.

In a different direction, [15] introduced a Bayesian hierarchical state-space model, which used shared hyperpriors over regression coefficients and latent process characteristics. Thus, the series-level parameters were inherited from global parameters which are shared across all time-series.

2.3. Gaussian Process

In a GP, we directly infer a distribution over functions. Each function can be seen as a random variable assigned to a finite number of discrete training points X and any finite number of these variables have a joint Gaussian distribution. More formally, the GP is completely specified by its mean and covariance functions:

$$f(x) \sim \text{GPs}\big(m(x), k(x_i, x_j)\big). \tag{3}$$

The mean function is usually kept at zero, as the covariance is often flexible enough to model most of the data patterns.

2.3.1. Kernels

The kernels define the types of functions that we are likely to sample from the distribution of functions [16]. We can then draw samples from the distribution of functions evaluated at any number of points, that is, $Cov(f(x_i), f(x_j)) = k_\theta(x_i, x_j)$. They can be separated into stationary kernels, such as the squared exponential kernel (RBF), the periodic kernel (PER) and the white noise kernel (WN) and non-stationary ones, such as the linear kernel (LIN). The stationary kernels can be written as

$$\text{RBF}: \qquad k(x_i, x_j) = \eta_r^2 \exp\big(-\frac{1}{2l_r^2}(x_i - x_j)^T(x_i - x_j)\big) \tag{4}$$

$$\text{PER}: \qquad k(x_i, x_j) = \eta_p^2 \exp\big(-\frac{(2sin^2(\pi|xi - xj|/p)}{l_p^2})\big) \tag{5}$$

$$\text{WN}: \qquad k(x_i, x_j) = \eta_{wn}^2 \delta_{xi,xj}, \tag{6}$$

where $\eta_r, \eta_p, \eta_l, \eta_{wn}$ represent the variances, l_r, l_p are the length-scale parameters which control the smoothness, c defines the offset, and p is the period. The δ_{x_i,x_j} is the Kronecker delta, which has the value of one for $x_i = x_j$, and zero otherwise.

2.3.2. Predictions

When we are predicting using GPs, we are interested in the joint distribution of the training outputs $\mathbf{f}$ and the test outputs $\mathbf{f}_*$. For most of the applications, we are faced with approximate function values, since there is noise to be considered in the form of $z = f(x) + \epsilon$. Assuming additive-independent and identically distributed Gaussian noise with variance σ_n^2, the joint distribution of the observations and the function values at the test positions can be written as

$$\begin{bmatrix} \mathbf{z} \\ \mathbf{f}_* \end{bmatrix} \sim \mathcal{N}\left(0, \begin{bmatrix} K(X,X) + \sigma_n^2 I & K(X, X_*) \\ K(X_*, X) & K(X_*, X_*) \end{bmatrix}\right), \tag{7}$$

where $K(X, X_*)$ denotes the $n \times n_*$ matrix evaluated at all pairs of training (n) and test points (n_*). Finally, we can derive the conditional distribution [16],

$$
\begin{aligned}
\mathbf{f}_* | X_*, X, \mathbf{z} &\sim \mathcal{N}(\bar{\mathbf{f}}_*, \text{cov}(\mathbf{f}_*)) &(8)\\
\bar{\mathbf{f}}_* &\triangleq \mathbb{E}[\mathbf{f}_* | X_*, X] = K(X_*, X)[K(X,X) + \sigma_n^2 I]^{-1}\mathbf{z} &(9)\\
\text{cov}(\mathbf{f}_*) &= K(X_*, X_*) - K(X_*, X)[K(X,X) + \sigma_n^2 I]^{-1}K(X, X_*). &(10)
\end{aligned}
$$

2.4. Variational Inference

Variational inference is an alternative to Markov Chain Monte Carlo for fitting Bayesian models. It provides a deterministic solution computed in a shorter time, potentially less accurate, as it is an approximation. It approximates the true distribution with a simpler distribution (usually Gaussian) $q(\theta;\phi)$, easier to sample from and evaluate. It is called variational density and is parameterized by ϕ. To calculate the distance between these two distributions, the Kullback–Leibler (KL) divergence can be used. Directly minimizing the KL divergence is difficult, but there is an easier and equivalent way to solve this problem, which is by maximizing the evidence lower bound (ELBO). It can be written as

$$
\begin{aligned}
\text{ELBO}(\phi) &= \mathbb{E}_{q(\theta;\phi)}[\log p(z|\theta) - \log q(\theta;\phi)] &(11)\\
\phi^* &= \text{argmax}_\phi\, \text{ELBO}(\phi). &(12)
\end{aligned}
$$

There are several methods to maximize the ELBO that usually require model-specific calculations. The algorithm Automatic Differentiation Variational Inference (ADVI) was proposed by [5] to automate this task. An important thing to notice is that the consequence of choosing to use a mean-field Gaussian for the variational approximation is that it does not capture the correlations between parameters. On the other hand, the full-rank Gaussian variational approximation is able to capture them but the computational cost can be prohibitive.

3. Hierarchical Model

3.1. Hierarchical Structure

As we saw in Section 2.2, our work is focused on datasets that have a hierarchical or grouping structure, and thus, there is potential information nested in those substructures of the data. We defined the two main components of our model: the GP is used to capture seasonality, irregularities and noise, while the trend is modeled using a piece-wise linear model. We denote our algorithm by Hierarchical Piece-Wise Linear GPs (HPLGPs). An illustrative representation is introduced in Figure 2.

To leverage the nested information in each group or hierarchy of our data, we modeled an individual GP per element of each group, where g represents a group from the set of groups G and L_g represents the elements of the specific group g. This way, we were able to model the most important features present in each element of a group, something that we could not do if we have directly modeled each time-series with an individual GP.

Before fitting our hierarchical model, we standardized each time-series to have mean 0 and variance 1. We defined a Normal likelihood for our target variable z_t^s.

$$
z_t^s \sim \mathcal{N}(\mu_t^s, \sigma^s). \tag{13}
$$

For its mean value, we summed the result of the piece-wise linear function p_t^s with the sum of the GPs, defined by $\gamma_t^{g,l}$. Recall that we are using $\mathbf{q}^{g,l}$ to encode the information of what series belong to the element l of group g. We used $\mathbf{q}^{g,l}$ to ensure that we are summing our GPs coherently for each element of a group.

$$
\mu_t^s = p_t^s + \sum_{g \in G} \sum_{l \in L_g} \gamma_t^{g,l} \mathbf{q}^{g,l}. \tag{14}
$$

Thus, we have a number of GPs equal to the number of elements in every group. However, in order to narrow the learning process inside each group, the hyperparameters are only defined per group and not per group element. Notice that we reparameterized $\gamma_t^{g,l}$ following the algorithm described in [16], as it is more efficient. We can denote the GPs as:

$$\gamma_t^{g,l} \sim \text{MvNormal}(0, K^g_{l_r,\eta_r} + K^g_{p,l_p,\eta_p} + K^g_{sigma}), \tag{15}$$

where $K^g_{l_r,\eta_r}, K^g_{p,l_p,\eta_p}, K^g_{sigma}$ are the different kernels defined by group (covered in-depth in Section 3.2). We add a piece-wise linear function to the result of the GPs in the likelihood of our model, defined by:

$$p_t^s = (k^s + A\delta_c^s)\mathbf{x} + (m^s + A(-\mathbf{c}\delta_c^s)), \tag{16}$$

where k is the growth rate, δ is the rate adjustments, m is the offset parameter, and $(-c\delta)$ ensures that the function is continuous (this formulation is covered in-depth in Section 3.3). Notice that we have one δ parameter for each series s and change point c. We will not be focusing on the effect change of using a different number of change points, nor on the automatic selection of the number of change points. Nevertheless, it is possible to address this problem with the current formulation. If one chooses a large number of potential points and uses a sparse prior on the δ parameter, it is the equivalent to performing a L1 regularization.

The approximation of our parameter distributions is performed using ADVI, introduced in Section 2.4.

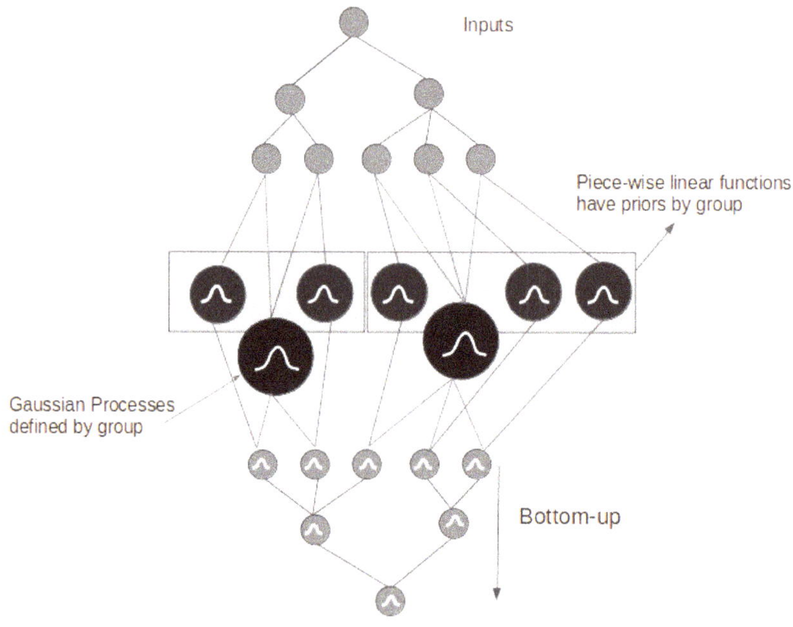

Figure 2. Simplified representation of our proposed algorithm. Notice that the GPs are defined by group, while the piece-wise linear functions are fitted to individual series (with priors defined by group). The model outputs coherent bottom-series forecasts which we can directly sum up (using a bottom-up strategy) to get the forecasts for the higher-level series.

3.2. Gaussian Processes

In designing our GPs, the goal was to define a set of kernels that is flexible enough to be used in different settings. The combination of kernels that we used included the squared exponential kernel (RBF), the period kernel (PER) and the white noise kernel (WN). The equations for each kernel were presented in Section 2.3.1. The RBF kernel was selected to model medium term non-linear irregularities in the data. We could have used a more complex kernel, such as the rational quadratic kernel, but from our experiments, we did not see any relevant improvement and so we avoid adding more parameters to the model. The choice of priors for the length-scale parameters, for both the RBF and PER kernels, needs some attention. First, one important thing to be aware when optimizing the parameters of the kernels are the correlations between them. For instance, the length-scale has a strong interaction with other parameters. The same happens between kernels. The usage of more informative priors on the hyperparameters helps to mitigate (but not erase) that problem. Secondly, the data do not inform length-scales larger or shorter than the maximum or minimum covariate distance. In our case, the distance between points is always one, while the maximum distance is equal to the number of time-points in each series. Thus, we used an inverse gamma with mass inside this interval because it suppresses both zero and infinity. Both l_r and l_p are defined as $l \sim \text{InvGamma}(\alpha = 4, \beta = n)$, where n is the number of data points in the training set.

To model the main seasonal pattern of the data we selected a PER kernel. Since we define a prior over the period p, we just need to be careful with the range of probable values for p. For instance, with weekly data we just need to ensure that our distribution has a significant amount of mass around 52 and we let the algorithm infer the value that best fits our data. In other models, we often need to be precise and define $365.25/7 = 52.18$ weeks in a year. The period is a non-negative parameter, for which we could have chosen a distribution that ensures non-negativity, such as a Gamma distribution. As we defined a very informative prior, we decided to use a Laplace distribution $p \sim \text{Laplace}(\mu = D, \text{b} = 0.1)$, where D is the main seasonality pattern found in the data.

We also specified a noise model with a simple WN kernel. This kernel gives us the capacity to absorb short term irregular behaviors without compromising the fit of the other kernels. It also helps stabilize our covariance function. Once again, we need to use a very informative prior for $\sigma_w \sim \text{HalfNormal}(\sigma = 0.01)$, to avoid losing valid information which could be modeled as noise.

Finally, we can write our covariance function as:

$$K = K_{RBF} + K_{PER} + K_{WN} = K_{l_r,\eta_r} + K_{p,l_p,\eta_p} + K_{\sigma_w}, \tag{17}$$

where l_r, l_p, η_t, η_p, p, and σ_w are all the hyperparameters to learn. As a final note to our kernel design, our algorithm can be trivially extended to have multiple seasonalities. This is useful when there is a weakly seasonality pattern in the data aside from the main pattern. It can be done by adding a new component to our covariance function. A second periodic kernel PER can be added on and the prior for its period p can be defined in the expected range of values for the specific seasonality.

3.3. Trend Model

At this point, it is important to notice that we only used stationary kernels. Our algorithm would not be capable to forecast most of the known time-series datasets, since we would not be able to model the trend component. In the case of ARIMA models, the process consists of performing a first differencing on the data and then fitting an ARMA model. We tested the inclusion of a non-stationary kernel, the linear kernel LIN, to model the trend of the data. Nonetheless, the results were not convincing enough. On one hand, the RBF and LIN kernels were in some cases catching some of the same effects. The necessary regularization to overcome this problem was somewhat customized to the dataset in usage, not easy to generalize. We believe that this was also a consequence of the hierarchical structure that we defined for the GPs. On the other hand, some datasets had non-linear trends or regime changes, which we were unable to catch using a linear kernel. We could also model the trend as the mean of the GPs, but, once again, the results were not as convincing as the option that follows.

We decided to model the trend of the data using a piece-wise linear model, following the definition in [9]. The trend changes are modeled by the definition of a set of change points $\mathbf{c}$ at times $c_j, j = 1, ..., C$, and a vector of trend adjustments $\delta \in \mathbb{R}^c$, where δ_j is the change in rate that occurs at time c_j. The rate at any time-point t is then the base rate k, plus all the adjustments up to that point. This can be represented by a matrix A of size $n \times c$, such that

$$a_j(t) = \begin{cases} 1, & \text{if } t \geq c_j, \\ 0, & \text{otherwise.} \end{cases} \tag{18}$$

$$A = a_j(t)^T \tag{19}$$

$$p = (k + A\delta)\mathbf{x} + (m + A(-\mathbf{c}\delta)). \tag{20}$$

Finally, we cover the prior distributions of the parameters of the piece-wise linear model. To also leverage the group information when estimating the trend of the different series, we defined a hierarchical structure for these parameters. This is also called partial pooling because, while we define individual parameters for each series, they are sharing information through a hyperprior. This has a shrinking effect on the estimations, that is, we assume that the parameters k, m and δ come from a normal distribution centered around their respective group mean μ_k, μ_m and μ_δ, with a certain standard deviation σ_k, σ_m or σ_δ. We only present here the hierarchical structure for δ as it is similar to k and m,

$$\mu_\delta^g \sim \mathcal{N}(0, 0.1) \tag{21}$$

$$\sigma_\delta^g \sim HalfNormal(0.01) \tag{22}$$

$$\delta_c^s \sim \mathcal{N}(\mu_\delta^g, \sigma_\delta^g) \tag{23}$$

4. Results

We chose datasets with different characteristics to ensure that our algorithm is able to capture strong trend and seasonal patterns that vary across groups, while working with either small or large numbers of series.

The first dataset used is fairly small and it represents the quarterly Australia prison population evolution over the period 2005Q1-2016Q4. It has 32 time-series, each one having 48 time-points. We tested with different values for τ but only present here the scenario where $\tau = 2D$ as it was consistent with the other experiments. It comprises three groups: the six states and two territories of Australia (we will refer these two also as states for simplification), the gender of the prisoner (male or female) and the legal status (whether prisoners have already been sentenced or not). Despite its size, there is an interesting change of rate of growth in the data of some groups. We can see that the algorithm behaves rather well even with a small amount of data to be trained on. We used two different models as a benchmark (see Table 1). First, we simply fitted individual GPs to each single series and then aggregate these upwards to produce revised forecasts for the whole hierarchy (usually referred as bottom-up method). The second is the optimal reconciliation algorithm MinT proposed by [2].

Table 1. Results (MASE) for the Australia prison dataset using $\tau = 8$.

Algorithm	Bottom	Total	State	Gender	Legal	All
HPLGPs	2.09	0.244	1.628	0.518	2.682	1.885
BU-GPs	2.319	1.626	1.638	1.396	2.813	2.242
MinT	2.06	0.895	1.698	0.907	1.84	1.96

The second dataset is larger (304 time-series) and comprises the monthly number of visitors in Australia from 1998–2016. We used the first 204 time-points to train our algorithm and the last 24 to evaluate its performance. This dataset can be disaggregated in four different groups. The first three are purely hierarchical and concern the geographical nature of the data: 8 states, 27 zones and 76 regions. The purpose of travelling is a different group that contains four different elements. The total number of elements in groups is 114 which means that we fitted 114 GPs. These data have a strong yearly seasonal pattern and, in specific groups, there are particular trend patterns. Once again, we can see in Table 2 that our algorithm is able to not only capture the relevant patterns in individual series but also to capture the nested ones within groups, specially in the most aggregated level of the hierarchy.

Table 2. Results (MASE) for the Australia tourism dataset using $\tau = 24$.

Algorithm	Bottom	Total	State	Zone	Region	Purpose	All
HPLGPs	1.082	0.779	1.128	1.014	0.981	0.981	1.018
BU-GPs	1.211	1.271	1.312	1.211	1.122	1.121	1.196
MinT	0.906	1.27	1.07	0.893	0.895	1.04	0.897

5. Conclusions and Future Work

We proposed an algorithm for automatic hierarchical time-series forecasting. Our results show that it can compete with its statistical counterparts, being able to effectively capture the behaviors of the aggregated level series while not losing accuracy on the individual ones. Our algorithm is easily extensible, for instance, to work with non-Gaussian likelihoods or multiple seasonality patterns.

As future work, the increase of the scalability of the model can be developed further if we consider approximation methods, such as sparse approximations (e.g., [17]). Another interesting point to address is the posterior distributions correlations of our parameters. There are several works (e.g., [18,19]) on low-rank approximations to the covariance matrix to capture some of these correlations.

In the interest of reproducible science, our proposed algorithm is publicly available (https://github.com/luisroque/automatic_hierarchical_forecaster, last accessed 29 June 2021).

Funding: This work was partially funded by the Canada Research Chairs program, a Discovery Grant from NSERC, and by the project Safe Cities - Inovação para Construir Cidades Seguras, with the reference POCI-01-0247-FEDER-041435, co-funded by the European Regional Development Fund (ERDF), through the Operational Programme for Competitiveness and Internationalization (COMPETE 2020), under the PORTUGAL 2020 Partnership Agreement.

Data Availability Statement: Publicly available datasets were analyzed in this study. The data can be found here: https://github.com/luisroque/automatic_hierarchical_forecaster, last accessed 29 June 2021.

References

1. Hyndman, R.J.; Khandakar, Y. Automatic time-series forecasting: the forecast package for R. *J. Stat. Softw.* **2008**, *26*, 1–22.
2. Wickramasuriya, S.L.; Athanasopoulos, G.; Hyndman, R.J. Optimal Forecast Reconciliation for Hierarchical and Grouped Time Series Through Trace Minimization. *J. Am. Stat. Assoc.* **2019**, *114*, 804–819. [CrossRef]
3. Spiliotis, E.; Abolghasemi, M.; Hyndman, R.J.; Petropoulos, F.; Assimakopoulos, V. Hierarchical forecast reconciliation with machine learning. *arXiv* **2020**, arXiv:2006.02043.
4. Novak, J.; McGarvie, S.; Garcia, B.E. A Bayesian model for forecasting hierarchically structured time series. *arXiv* **2017**, arXiv:1711.04738.
5. Kucukelbir, A.; Tran, D.; Ranganath, R.; Gelman, A.; Blei, D.M. Automatic Differentiation Variational Inference. *J. Mach. Learn. Res.* **2017**, *18*, 1–45.
6. Cheng, L.; Ramchandran, S.; Vatanen, T.; Lietzén, N.; Lahesmaa, R.; Vehtari, A.; Lähdesmäki, H. An additive Gaussian process regression model for interpretable non-parametric analysis of longitudinal data. *Nat. Commun.* **2019**, *10*, 1798. [CrossRef] [PubMed]
7. Hyndman, R.; Athanasopoulos, G. *Forecasting: Principles and Practice*, 3rd ed.; OTexts: Melbourne, Australia, 2021.
8. Livera, A.M.D.; Hyndman, R.J.; Snyder, R.D. Forecasting time-series With Complex Seasonal Patterns Using Exponential Smoothing. *J. Am. Stat. Assoc.* **2011**, *106*, 1513–1527. [CrossRef]
9. Taylor, S.J.; Letham, B. Forecasting at Scale. *Am. Stat.* **2018**, *72*, 37–45. [CrossRef]
10. Corani, G.; Benavoli, A.; Augusto, J.; Zaffalon, M. Automatic Forecasting using Gaussian Processes. *arXiv* **2020**, arXiv:2009.08102.
11. Hewamalage, H.; Bergmeir, C.; Bandara, K. Recurrent Neural Networks for time-series Forecasting: Current status and future directions. *Int. J. Forecast.* **2021**, *37*, 388–427. [CrossRef]

12. Smyl, S. A hybrid method of exponential smoothing and recurrent neural networks for time-series forecasting. *Int. J. Forecast.* **2019**, *36*, 75–85. [CrossRef]
13. Hyndman, R.J.; Ahmed, R.A.; Athanasopoulos, G.; Shang, H.L. Optimal combination forecasts for hierarchical time-series. *Comput. Stat. Data Anal.* **2011**, *55*, 2579–2589. [CrossRef]
14. Athanasopoulos, G.; Ahmed, R.A.; Hyndman, R.J. Hierarchical forecasts for Australian domestic tourism. *Int. J. Forecast.* **2009**, *25*, 146–166. [CrossRef]
15. Chapados, N. Effective Bayesian Modeling of Groups of Related Count time-series. In Proceedings of the 31st International Conference on Machine Learning, Beijing, China, 21–26 June 2014.
16. Rasmussen, C.E.; Williams, C.K.I. *Gaussian Processes for Machine Learning (Adaptive Computation and Machine Learning)*; The MIT Press: Cambridge, MA, USA, 2005.
17. Quiñonero-Candela, J.; Rasmussen, C.E. A Unifying View of Sparse Approximate Gaussian Process Regression. *J. Mach. Learn. Res.* **2005**, *6*, 1939–1959.
18. Ong, V.M.H.; Nott, D.J.; Smith, M.S. Gaussian variational approximation with a factor covariance structure. *arXiv* **2017**, arXiv:1701.03208.
19. Guo, F.; Wang, X.; Broderick, T.; Dunson, D.B. Boosting variational inference. *arXiv* **2016**, arXiv:1611.05559.

engineering proceedings

MDPI

Proceedings

Quantifying Uncertainty for Predicting Renewable Energy Time Series Data Using Machine Learning †

Phil Aupke [1,*], Andreas Kassler [1], Andreas Theocharis [1], Magnus Nilsson [2] and Michael Uelschen [3]

1 Department of Mathematics and Computer Science, Karlstad Universitet, Universitetsgatan 2, 651 88 Karlstad, Sweden; andreas.kassler@kau.se (A.K.); andreas.theocharis@kau.se (A.T.)
2 Glava Energiezentrum, Arvika Näringslivscentrum, 671 29 Arvika, Sweden; magnus.nilsson@arvika.se
3 Faculty of Engineering and Computer Science, University of Applied Science Osnabrück, Albrechtstraße 30, 49076 Osnabrueck, Germany; m.uelschen@hs-osnabrueck.de
* Correspondence: phil.aupke@kau.se

† Presented at the 7th International conference on Time Series and Forecasting, Gran Canaria, Spain, 19–21 July 2021.

Abstract: Recently, there has been growing interest in using machine learning based methods for forecasting renewable energy generation using time-series prediction. Such forecasting is important in order to optimize energy management systems in future micro-grids that will integrate a large amount of solar power generation. However, predicting solar power generation is difficult due to the uncertainty of the solar irradiance and weather phenomena. In this paper, we quantify the impact of uncertainty of machine learning based time-series predictors on the forecast accuracy of renewable energy generation using long-term time series data available from a real micro-grid in Sweden. We use clustering to build different ML forecasting models using LSTM and Facebook Prophet. We evaluate the accuracy impact of using interpolated weather and radiance information on both clustered and non-clustered models. Our evaluations show that clustering decreases the uncertainty by more than 50%. When using actual on-side weather information for the model training and interpolated data for the inference, the improvements in accuracy due to clustering are the highest, which makes our approach an interesting candidate for usage in real micro-grids.

Keywords: machine learning; weather data analysis; clustering; photovoltaic systems forecast; smart grid

Citation: Aupke, P.; Kassler, A.; Theocharis, A; Nilsson, M.; Uelschen, M. Quantifying Uncertainty for Predicting Renewable Energy Time Series Data Using Machine Learning *Eng. Proc.* **2021**, *5*, 50. https://doi.org/10.3390/engproc2021005050

Published: 9 July 2021

1. Introduction

Increasing renewable energy usage is imperative for achieving a climate neutral Europe by 2050, which requires the reduction of greenhouse gas emissions by at least 55% by 2030 [1] according to the Climate Target Plan. Furthermore, refs. [2] and [3] show that the energy capacity of photovoltaic increased from 22 GW in 2009 to 707 GW globally in 2020. To achieve the goal of zero emissions, and to effectively integrate photovoltaic energy into the global energy system, an important cornerstone is the deployment of micro-grids that integrate a large amount of solar power generation photovoltaic panels (PV-Systems). To optimize energy exchanges, recent developments in the area of digitalization, such as the Internet of Things (IoT), Machine Learning and Cloud Computing, aim to develop smart-grids, which provide uninterrupted energy to prosumers while aiming to reduce the stress from the main grid [4,5].

Consequently, there has been growing interest in using machine learning based methods for accurately predicting renewable energy generation for PV-Systems, as such predictions are important for optimal energy management strategies. However, predicting solar power generation is difficult due to the inherent uncertainty of the solar irradiance and weather phenomena that are the most influential factors for the PV output power. Reference [6] measured an uncertainty of 9.5% for the radiation during a period of one year and 8.9% for an increasing span of ten years. As weather stations may not be co-located with each PV-System, interpolated weather and radiance information available from close-by stations may be required for making such predictions, further contributing to the uncertainty of the forecast.

In this paper, we aim to evaluate different factors that impact the uncertainty of PV power forecasting. We use a large dataset available from a Swedish PV power grid and cluster the data according to weather information. Using those clusters, we build different forecasting models using LSTM and Facebook Prophet. We compare the forecasting accuracy of clustered versus non clustered models. Because, in reality, exact weather information from co-located weather stations may not be available, we evaluate the impact of using interpolated data for training and inference on prediction accuracy. Our evaluation shows that clustering decreases the uncertainty by more than 50%. When using actual on-side weather information for the model training and interpolated data for the inference, the improvements in accuracy due to clustering are the highest, which makes our approach an interesting candidate for usage in real PV micro-grids.

The rest of this paper is structured as follows: in Section 2, we review related work. Section 3 presents our methodology and approach. In Section 4, we show our evaluation setup and analyse the results. Finally, Section 5 concludes the paper and lists ideas for potential future work.

2. Related Work

There are several studies regarding the prediction of PV-Station energy outcome based on physical, statistical and machine learning methodologies [7]. Statistical prediction models use historic stationary time-series environmental data, which are not that well applicable to non-stationary weather dependent time-series. References [8,9] proposed the usage of LSTM networks for short-term prediction of PV-Station energy production, while [10,11] improved their machine learning approaches by clustering the dataset into four clusters, each of which represent different weather types and train a specific model for each cluster. The findings indicate that the clustered machine learning models provide more accurate and more stable results for the prediction than a model that was trained with a non-clustered dataset. In our approach, we evaluate how clustered models impact prediction accuracy when using interpolated data for training and/or inference.

3. Methodology

Weather phenomena, such as radiation, temperature and clouds, have a significant effect on PV power production. In order to predict such production, machine learning based methods can be used that either use only historical power production information or use additional weather features. However, precise weather information from a co-located weather station may not be readily available. Consequently, we aim to evaluate whether interpolated weather information can be feasible for the prediction of PV-Station energy outcome within a smart-grid and, if so, what the additional uncertainty is when using interpolated weather features. Interpolated data can be used in machine learning based time series prediction at two different stages: during the training time and during inference (when making the prediction).

We compare the following combinations for assessing the prediction accuracy: training on actual co-located weather data and using actual weather data for prediction (denoted A to A, e.g., a micro-grid operator predicts its PV power using its own weather station), training on interpolated data and using interpolated data for the prediction (denoted I to I, e.g., a micro-grid does not have a weather station and uses interpolated weather data for both model training and inference), training on actual co-located weather data and using interpolated data for the prediction (denoted A to I, e.g., a micro-grid operator trains the model using information available from its weather station and provides the model to close-by users that use the interpolated data for inference). For predicting the PV output power, we cluster both datasets and use the associated model for prediction (Figure 1) while comparing it with a non-clustered model.

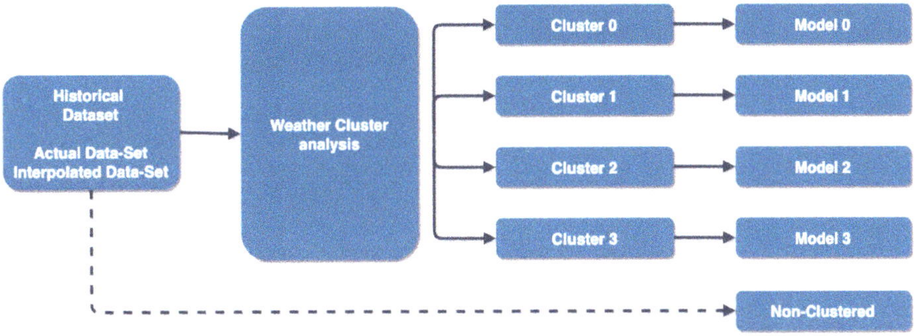

Figure 1. Flowchart for the clustering and the subsequent ML Model.

3.1. dataset

Our dataset contains five consecutive years of weather and PV-Station information from 1 January 2015 to 31 December 2019 for each six seconds, which we average to create one minute time slots. The data are available from a solar park at the Glava Energy Center in Arvika, Sweden (Altitude 220 m, latitude 59.31°N, longitude 12.37°E). The PV panels are oriented to the south with a 40 degr. inclination. The PV modules are comprised of 20xITS 200WP + 20xITS 210WP + 20xITS 210WP + 19xITS 220WP at a total PV Power of 16.580WP (Watt-Peak) using a 4xEltek Valere 4300W inverter. In addition to the Produced Energy (kW) of the solar panels, the Glava Energy Center has a co-located weather station providing the features presented in Table 1.

Table 1. Features of Glava dataset.

Feature	Unit
Wind Direction	Gradient
Precipitation	L/m^2
Temperature	C
Humidity	Percentage
Barometric Pressure	mBar
Wind Speed	m/s
Global Radiation (GR)	W/m^2
Radiation 30 Degrees	W/m^2
Radiation 40 Degrees	W/m^2
Indirect Radiation	W/m^2
Produced Energy	kW

An open source subset of the used dataset is available (https://github.com/AI-4-Energy/Dataset, last accessed: 29 June 2021). For interpolated weather features, we use information available from Meteostat [12] and SMHI [13], using the weather station at the Karlstad Airport (Altitude 107 m, latitude 59.44° N, longitude 13.34° E) at one minute resolution. This dataset contains the same features as the on-site dataset with the exception of radiation, since only the global radiation is available.

3.2. Data Preparation

Both datasets contain several outliers and missing values due to failures in the sensors or systems. For data cleaning, we used the interquartile range (IQR) technique [14], which divides the dataset into quartiles and computes the mean of each segment. Data points that were not within a percentile of the mean were detected as outliers. We also removed data points during the night as they are not as relevant for the prediction [15], because the PV-Systems do not produce energy during this period of the day. Since, especially in Sweden, the night times change throughout the year, the datasets were adapted to consider sunset and sunrise in Sweden, available from [16].

Furthermore, we used the Pearson correlation and the Wrapper selection [17] for feature selection [18]. Our results are inline with [10], which shows that the highest correlations with the energy production of PV-Systems are Wind Speed, Temperature, Humidity, Global Radiation, 30 Degrees Radiation, 40 Degrees Radiation and Indirect Radiation.

3.3. *Clustering*

We applied the density based clustering method DBSCAN [19,20] to define different clusters, which represent different weather characteristics on the co-located and interpolated historical data. We used global radiation (GR) as an input for the DBSCAN clustering and as an indicator of the weather types. DBSCAN clustering requires as input the maximum distance ϵ between two points.

To automatically determine the optimal ϵ, ref. [21] proposes the usage of the nearest-neighbour algorithm to find the distance from each point to its closest neighbour. When applying this technique, we found the optimal value for ϵ_o of 1.6 for both datasets, which resulted in four different clusters. Table 2 presents the results of the clustering, representing the different weather conditions (sunny, partially overcast, overcast and rainy). As can be seen, most of the data points map to the overcast cluster, while the least of the points map to the rainy cluster.

Table 2. DBSCAN–Clustering.

Cluster	Definition	GR Range	Percentage
Cluster 0	Rainy	0–15%	13%
Cluster 1	Partially Overcast	15–35%	28.9%
Cluster 2	Overcast	35–65%	40.5%
Cluster 3	Sunny	65–100%	17.6%

3.4. *Machine Learning Models for PV Power Prediction*

We used both LSTM [22] and Facebook Prophet [23] to build models to predict the Produced Energy of the PV panels. Figure 2 presents the architecture used for the Bi-Directional LSTM model. The eight input features that comprise the weather and PV Power were sent through four *Bi-Directional LSTM Layers*, with the *Return Sequence* set to True. The last layer in the model contains a *Dense Layer*, which is used to compress the result to a singular output value. To evaluate the right amount of neurons within each layer and other parameters, we used hyper-parameter tuning [18].

In addition to the bi-directional LSTM model, we developed a model based on the Facebook Prophet API. In comparison to the LSTM model, the Facebook Prophet model consists of three different types of model—Trend, Seasonality and Holiday model [18,23]—which are combined to form an additive model. As Prophet is designed to only compute univariate inputs, we used the Python library *Multi Prophet* [24] to overcome this restriction and use multivariate input data. This library is a wrapper around the Facebook Prophet interface to handle multiple models for each input feature which can then be used for the prediction of the output value. Prophet creates a model for each input feature as a regressor for the prediction of the PV-Station energy outcome.

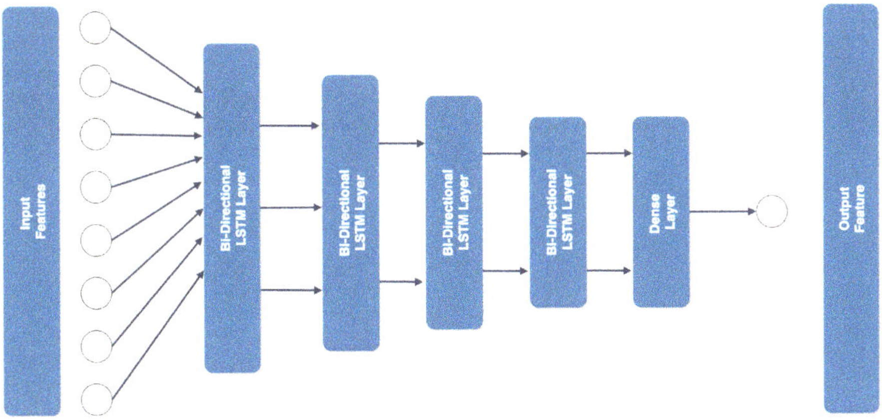

Figure 2. Architecture of the Bi-Directional LSTM model.

4. Experimental Evaluation and Results

In our experimental evaluation, we want to answer the following questions:

- How does the usage of interpolated features impact forecast accuracy?
- How are clustered models impacted by interpolated features during both training and inference?
- How can Facebook Prophet compare with LSTM for predicting PV Power?

To answer these questions, we clustered the first dataset (1 January 2015 to 31 December 2018) into the four categories and created a training (75%) and testing (25%) dataset to train both approaches. The second dataset (1 January 2019 to 31 December 2019) was used for the evaluation. We trained a model for each weather category on the training dataset. To evaluate the results with a baseline model, we also built a model on the combined dataset for both LSTM and Facebook Prophet. Each of our models were trained on actual weather information (A) and on interpolated information (I). The first step of the prediction included a clustering of the input variables into a matching weather type. Once this had been detected, the corresponding model (e.g., as it was rainy, we used the model trained on rainy data for prediction) was chosen for the prediction.

4.1. Prediction Accuracy for On-Site and Interpolated Features

We compared the predicted PV Power with the measured data point and used the normalized Mean Squared Error (MSE) to calculate forecast accuracy. We used both actual on-site (A) and interpolated (I) weather information for the training and prediction. For example, A to A means that the model was trained on actual (A) data and we used actual (A) data for the prediction. The left side of each sub figure in Figure 3 shows violin plots that illustrate the distribution of normalized MSE for the LSTM models when trained on the clustered dataset. The right side (in the red box) shows the normalized error distribution for the baseline LSTM model, trained on the whole dataset, where we evaluated the forecast error separately for each cluster (e.g., we calculated the error distribution over all rainy data points in cluster 0, etc.).

As can be seen from Figure 3a, clustering decreased the normalized MSE for the A to A case. For example, the average normalized MSE decreased from 0.03 to 0.01 in cluster 3 (sunny), which is an improvement of 66%. Furthermore, clustering reduced the standard deviation of the forecast error in general. This can best be observed with cluster 0 (rainy). Without using clustering, the violin graph shows that for some rainy data points the normalized MSE prediction error was as large as 0.26. The forecast accuracy improved when using clustering for rainy days where the maximum observed normalized MSE was around 0.15. When using interpolated data for both training and prediction (I to

I, Figure 3b), a similar improvement in accuracy was observed (both mean normalized MSE reduced as well as standard deviation). However, the normalized MSE is higher when using interpolated data for both training and prediction than when using actual data. While this can be observed for each cluster, clusters 1 and 2, especially, have high inaccuracy due to their more uncertain weather. Figure 3c shows the normalized MSE when training on actual data while using interpolated data for predicting PV Power (A to I). The example use-case is that the micro grid operator has a co-located weather station, and trains the model and provides it to another member of the micro grid that does not have a weather station but uses interpolated weather features. Clearly, using actual data for training reduces the error compared to using interpolated data for training. For example, when using clustering for cluster 1, using I to I resulted in the maximum normalized MSE of 0.31, which reduced to 0.18 when using A to I.

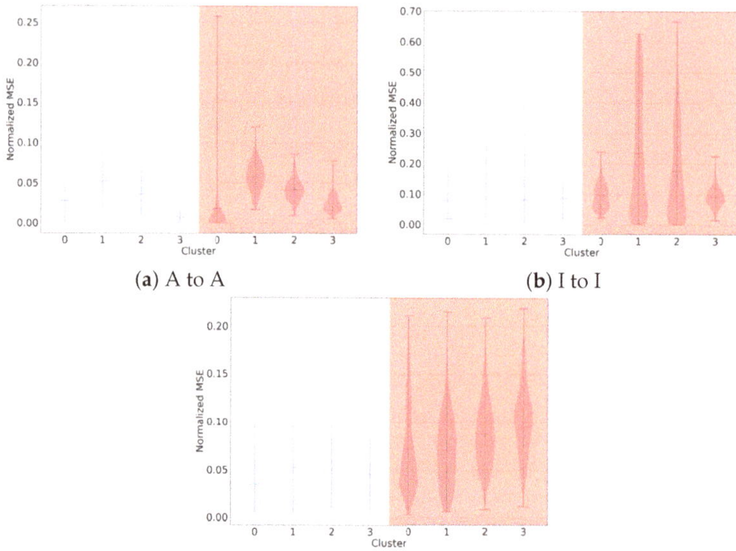

Figure 3. LSTM Accuracy evaluation—normalized MSE for Actual (A) versus Interpolated (I) weather features. Clustered LSTM model (**left**), non-clustered model (**right**).

4.2. *Prediction Uncertainty for Facebook Prophet*

In this section, we evaluate the uncertainty of PV Power prediction when using Facebook Prophet trained on on-site versus interpolated weather features. Compared with the last section, when using the non-clustered model, we used the whole dataset for the prediction instead of the clustered data for the base-model. This was due to the structure of Prophet, which performs best with consecutive data. Figure 4 shows the normalized MSE distribution characterising the prediction error when using both actual on-site weather information and interpolated information. The error characteristics are similar to when using the bi-directional LSTM model, where we can observe an increasing uncertainty while using interpolated weather information. Furthermore, we can see that Prophet handles interpolated weather information better than LSTM models. For example, the mean of the normalized MSE for the clustering based models is below 0.05 in the Prophet model while it fluctuates just below 0.10 when using the LSTM model for the I to I case. Contrary to the mean MSE, Facebook Prophet shows a bigger deviation in the error than the LSTM model, which results in a bigger uncertainty within the prediction.

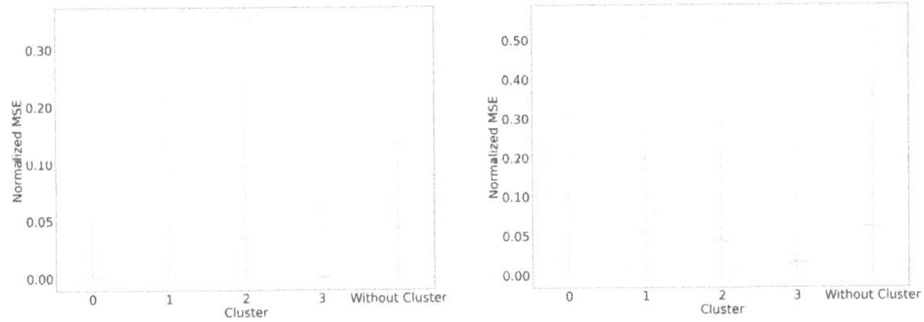

(**a**) Actual Weather Information (**b**) Interpolated Weather Information

Figure 4. Facebook Prophet Accuracy evaluation—normalized MSE for Actual (A) versus Interpolated (I) weather features.

5. Conclusions

In this paper, we evaluate the impact of using interpolated weather information for the prediction of PV Power within smart micro grids, as a co-located weather station may not be readily available. We clustered five years of detailed weather and PV Power information available from a Swedish PV power plant using DBSCAN and evaluate how clustering improves forecast accuracy when using both Facebook Prophet and bi-directional LSTM as predictors. The clustered LSTM model (A to A) performed best of the three scenarios, with an overall normalized MSE of below 0.05 and a small error deviation. The usage of interpolated data increased the error span for both LSTM and Prophet, where mean normalized MSE increased to around 0.10. However, when training on actual data and using interpolated data for inference, the mean normalized MSE dropped, for most cases, to under 0.05, but a significant larger error deviation was determined compared to using on-site data for both training and inference. Consequently, interpolated data are feasible for the prediction of PV-Station energy outcome. Finally, Facebook Prophet outperformed the LSTM model in terms of the average normalized MSE. On the other hand, the LSTM model delivered better results in terms of error standard deviation, which correlates to the uncertainty of the prediction.

Author Contributions: All authors contributed to this work. All authors have read and agreed to the published version of the manuscript.

Funding: Parts of this work have been funded by the Swedish Energy Agency (Energimyndigheten) through the project AI4ENERGY.

Data Availability Statement: A public available subset of the dataset is accessible: https://github.com/AI-4-99Energy/Dataset, last accessed: 29 June 2021.

Conflicts of Interest: The authors declare no conflict of interest.

References

1. European Commission. Energy, Climate Change. Available online: https://www.ec.europa.eu/clima/policies/eu-climate-action/2030_ctp_en (accessed on 4 March 2021).
2. *Grid Integration of Large-Capacity Renewable Energy Sources and Use of Large-Capacity Electrical Energy Storage*; IEC White Paper; Project Team Under Market Strategy Board (MSB) IEC, International Electrotechnical Commission: Geneva, Switzerland, 2012. Available online: https://www.iec.ch/whitepaper/pdf/iecWP-gridintegrationlargecapacityLR-en.pdf(accessed on 17 May 2021).
3. IRENA. *Renewable Capacity Statistics 2021*; International Renewable Energy Agency: Abu Dhabi, United Arab Emirates, 2021.
4. Camarinha-Matos, L.M. Collaborative smart grids—A survey on trends. *Renew. Sustain. Energy Rev.* **2016**, *65*, 283–294. [CrossRef]
5. Bayindir, R.; Hossain, E.; Kabalci, E.; Perez, R. A Comprehensive Study on Microgrid Technology. *Int. J. Renew. Energy Res.* **2014**, *4*, 1094–1107.
6. Jamil, I.; Zhao, J.; Zhang, L.; Rafique, S.F.; Jamil, R. Uncertainty Analysis of Energy Production for a 3 × 50 MW AC Photovoltaic Project Based on Solar Resources. *Int. J. Photoenergy* **2019**, *2019*, 12. [CrossRef]
7. Kostylev, V.; Pavlovski, A. Solar Power Forecasting Performance—Towards Industry Standards. In Proceedings of the 1st International Workshop on the Integration of Solar Power into Power Systems, Aarhus, Denmark, 24 October 2011.

8. Konstantinou, M.; Peratikou, S. Solar Photovoltaic Forecasting of Power Output Using LSTM Networks. *Atmosphere* **2021**, *12*, 124. [CrossRef]
9. Mellit, A.; Pavan, A. Short-term forecasting of power production in a large-scale photovoltaic plant. *Sol. Energy* **2014**, *105*, 401–413. [CrossRef]
10. Lui, L.; Zhao, Y.; Chang, D.; Xie, J.; Ma, Z.; Sun, Q.; Wennersten, R. Prediction of short-term PV power output and uncertainty analysis. *Appl. Energies* **2018**, *228*, 1–711.
11. Acero, J.A.; Koh, E.J.; Pignatta, G.; Norford, L.K. *Clustering Weather Types for Urban Outdoor Thermal Comfort Evaluation in a Tropical Area*; Springer: Friesach, Austria, 2020.
12. Meteostat—Historical Weather and Climate Data. Available online: https://meteostat.net/en (accessed on 18 May 2021).
13. SMHI. Available online: https://www.smhi.se/q/Alster/Karlstad/2726416 (accessed on 18 May 2021).
14. Bajpai, N. *Business Statistics*; Pearson Education India: Gwalior, India, 2009.
15. Isaksson, E.; Conde, M. *Solar Power Forecasting with Machine Learning Techniques*; School of Engineering Sciences: Stockholm, Sweden, 2018.
16. World Data. Sunrise and Sunset in Sweden. Available online: https://www.worlddata.info/europe/sweden/sunset.php (accessed on 18 May 2021).
17. Benesty, J.; Chen, J.; Huang, Y.; Cohen, I. *Pearson Correlation Coefficient*; Springer: Berlin/Heidelberg, Germany, 2009.
18. Aupke, P. Uncertainty in Renewable Energy Time Series Prediction Using Neural Networks. Master's Thesis, Karlstad Universitet, Karlstad, Sweden, 2021.
19. Li, X.; Ramachandran, R.; Movva, S.; Graves, S.; Plale, B.; Vijayakumar, N. *Storm Clustering for Data-Driven Weather Forecasting*; University of Alabama in Huntsville: Huntsville, AL, USA, 2008.
20. Sharma, A.; Chaturvedi, S.; Gour, B. A Semi- Supervised Technique for Weather Condition Prediction using DBSCAN and KNN. *Int. J. Comput. Appl.* **2014**, *95*, 79–183. [CrossRef]
21. Rahmah, N.; Sitanggang, I. Determination of Optimal Epsilon (Eps) Value on DBSCAN Algorithm to Clustering Data on Peatland Hotspots in Sumatra. *Earth Environ. Sci.* **2012**, *31*, 012012. [CrossRef]
22. Hochreiter, S.; Schmithuber, J. Long Short-Term Memory. *Neural Comput.* **1997**, *9*, 1735–1780. [CrossRef] [PubMed]
23. Facebook Prophet. Available online: https://facebook.github.io/prophet/ (accessed on 18 May 2021).
24. Multi Prophet. Available online: https://github.com/vonum/multi-prophet (accessed on 25 May 2021).

engineering proceedings

MDPI

Proceedings

Fuzzy Prediction Intervals Using Credibility Distributions †

Enriqueta Vercher ‡, Abel Rubio ‡ and José D. Bermúdez *,‡

Dept Statistics and Operations Research, University of Valencia, C/Dr. Moliner 50, 46100 Burjassot, Spain; enriqueta.vercher@uv.es (E.V.); abel.rubio@uv.es (A.R.)

* Correspondence: bermudez@uv.es

† Presented at the 7th International conference on Time Series and Forecasting, Gran Canaria, Spain, 19–21 July 2021.

‡ These authors contributed equally to this work.

Abstract: We present a new forecasting scheme based on the credibility distribution of fuzzy events. This approach allows us to build prediction intervals using the first differences of the time series data. Additionally, the credibility expected value enables us to estimate the k-step-ahead pointwise forecasts. We analyze the coverage of the prediction intervals and the accuracy of pointwise forecasts using different credibility approaches based on the upper differences. The comparative results were obtained working with yearly time series from the M4 Competition. The performance and computational cost of our proposal, compared with automatic forecasting procedures, are presented.

Keywords: time series methods; prediction intervals; credibility theory; M4 competition

Citation: Vercher, E.; Rubio, A.; Bermúdez, J.D. Fuzzy Prediction Intervals Using Credibility Distributions. *Eng. Proc.* **2021**, *5*, 51. https://doi.org/10.3390/engproc2021005051

Published: 9 July 2021

1. Introduction

This paper presents a new prediction scheme for time series using fuzzy logic and credibility distributions. Uncertainty about the future behavior of each time series is analyzed through the first and upper differences of the observed data, while the last raw observation is maintained as the level of the time series. That is, we assume that the uncertainty is not in the observed data, but in the underlying processes of the relationship between the data of the time series.

The uncertainty about the future of the time series is approximated using fuzzy variables [1]. Thus, working with fuzzy variables enables us to build prediction intervals with a given credibilistic coverage and provide pointwise forecasts through the credibility expected value, using various fuzzy prediction models. This credibility approach has not been previously proposed in the context of time series, although approximation of the uncertainty of fuzzy events using fuzzy variables is well established in other knowledge areas.

In previous works, we dealt with fuzzy variables to approximate the uncertainty of the future returns on assets or the return of a given portfolio [2,3]. In this paper, we propose to predict the future performance of the times series using LR-type fuzzy variables, whose parameters are estimated by using the sets of differences of several orders of the observed data. Finally, the optimistic values of the fuzzy variable provide us with the prediction intervals for each nominal coverage.

It must be noted that this new approach does not follow the classical fuzzy time series methodology, suggested by Song and Chissom [4], which has subsequently been improved by other authors (for more details about fuzzy time series, see, e.g., [5] and references therein). Recently, we introduced a new weighted fuzzy-trend method to forecast a stock index, which outperforms Chen's methodology [6] for pointwise one-step forecasting [7]. However, the aforementioned approach does not provide suitable tools for building accurate fuzzy prediction intervals and k-step-ahead forecasts.

The main goal of this research is to design new fuzzy prediction models for nonseasonal time series that provide fuzzy variables as forecasts of the future performance of the time series, using the historical datasets. The properties of the fuzzy variables and their credibility distributions are useful to obtain prediction intervals and accurate ex-post forecasts.

To further investigate the predictive behavior of our proposals and the accuracy of the outputs compared to other forecasting approaches, we applied them to a set of yearly time series from the M4 Competition [8]. For the comparison with automatic forecasting procedures, we selected the *exponential smoothing* (*ES*) and *ARIMA* procedures included in the *forecast* package for R [9], which is available in CRAN (https://cran.r-project.org/, accessed on 15 June 2021), by using the R commands *'es'* and *'auto.arima'*, respectively.

The rest of the paper is structured as follows. Section 2 summarizes concepts and definitions of fuzzy theory. The proposed methodology for building prediction intervals and pointwise forecasts is formulated in Section 3. In Section 4, we analyze the numerical results obtained by the fuzzy prediction models for a dataset from the M4 Competition. Finally, we set out our conclusions in Section 5.

2. Uncertainty Estimation with Fuzzy Logic

Let us recall some useful definitions based on fuzzy logic. The possibility and necessity of every fuzzy event (for instance, $\{T \geq x\}$, for any $x \in \mathbb{R}$), where T is a fuzzy number, can be evaluated according to the possibility distribution associated with its membership function ($\mu_T(t)$) as follows [10]:

$$Pos\{T \geq x\} = sup_{t \geq x}\mu_T(t), \text{ and } Nec\{T \geq x\} = 1 - sup_{t<x}\mu_T(t) \tag{1}$$

Thus, an alternative to quantify the uncertainty is the credibility measure introduced in [11]:

$$Cr\{T \geq x\} = \frac{1}{2}Pos\{T \geq x\} + \frac{1}{2}Nec\{T \geq x\}, \tag{2}$$

A fuzzy variable ξ is described by means of the following membership function,

$$\mu_\xi(x) = min(1, 2Cr\{\xi = x\}), x \in \mathbb{R}, \tag{3}$$

whose *credibility distribution* $\Phi_\xi : \mathbb{R} \to [0,1]$ is defined as

$$\Phi_\xi(x) = Cr\{\xi \leq x\} = \frac{1}{2}(sup_{t \leq x}\mu_\xi(t) + 1 - sup_{t>x}\mu_\xi(t)). \tag{4}$$

In particular, ξ is said to be an *L-R power fuzzy variable* if its credibility distribution has the following form [2]:

$$\Phi_\xi(x) = \begin{cases} 0 & \text{if } x \leq l \\ \frac{1}{2}[1 - (\frac{A-x}{A-l})^\alpha] & \text{if } l \leq x \leq A \\ \frac{1}{2}[1 + (\frac{x-A}{u-A})^\beta] & \text{if } A \leq x \leq u \\ 1 & \text{if } x \geq u \end{cases} \tag{5}$$

Throughout the paper, we denote LR-power fuzzy variables of this type by $\xi = (l, A, u)_{\alpha,\beta}$. The estimation of the parameters of ξ is made by means of sample percentiles of the dataset of differences between chronologically consecutive historical data.

Credibility Expected Values and Prediction Intervals

The pointwise estimation of a fuzzy variable ξ is approximated by its credibility expected value, denoted by $E(\xi)$ [11]. For an LR-power fuzzy variable $\xi = (l, A, u)_{\alpha,\beta}$, its expected mean is calculated as follows [2]:

$$E(\xi) = A + [\frac{1}{2}(u - A)\frac{\beta}{\beta+1} - (A - l)\frac{\alpha}{\alpha+1})] \tag{6}$$

Finally, it must be said that the prediction intervals can be built using the credibility distribution (5), calculating suitable γ-optimistic values of the fuzzy variable ξ, defined as follows [1]:

$$\xi_{sup}(\gamma) = \{r : Cr\{\xi \geq r\} \geq \gamma\}. \tag{7}$$

Alternatively, the endpoints of one specific prediction interval, for instance, with a $\gamma 100\%$ of nominal coverage for $\gamma \in [0,1]$, can be directly calculated through the inverse function of the credibility distribution, $\Phi_{\xi}^{-1}(.)$, as follows:

$$IP_{\gamma 100\%}(\xi) = [\Phi_{\xi}^{-1}(\frac{1-\gamma}{2}), \Phi_{\xi}^{-1}(\frac{\gamma}{2})]. \tag{8}$$

3. New Fuzzy Prediction Methods

Let us consider a time series $Y(t)$, whose observed data are $\{y_t\}_{t=1}^{N}$, y_N being the last observation. Then, we can build the k-order differences between consecutive observations, $k \geq 1$.

At time i, $i = 1, \ldots, N-k$, let us define the k-difference, d_i^k, as the difference between the forward observation of k order from time i and the observation at time i, that is

$$d_i^k = y_{i+k} - y_i \quad \forall i = 1, \ldots, N-k, k = 1, \ldots, N-i \tag{9}$$

Thus, for every k-order, $k \geq 1$, we calculate the time series of the k-differences:

$$D^k = \{d_i^k\}_{i=1}^{N-k}. \tag{10}$$

We assume that the uncertainty about the future behavior of the time series $Y(t)$ can be estimated through the credibility approximation of the time series of the k-differences. To do so, we consider the LR-power fuzzy variable $\Delta^k = (l, A, u)_{\alpha,\beta}$, whose credibility distribution is given in Equation (5), which is built using the sample percentiles p_j of the set D^k. The bounded support of Δ^k is approximated by $l = p_0$ and $u = p_{100}$, where $A = p_{50}$. The shape parameters are obtained as $\alpha = ln(0.5)/ln(\frac{A-p_{25}}{A-l})$ and $\beta = ln(0.5)/ln(\frac{p_{75}-A}{l-A})$, assuming that the sample percentiles p_{25} and p_{75} have a 50% possibility of being realistic (see, e.g., [12] for more information about LR fuzzy numbers). The forecaster can choose other sample percentiles, if that decision could improve the approximation of Δ^k, avoiding outliers.

3.1. One-Step Ahead Fuzzy Forecast Model: FFM

For the time series $Y(t) = \{y_1, \ldots, y_N\}$, let us consider the fuzzy variable Δ^1, which is built considering the set of the first differences, $D^1 = \{d_i^1\}_{i=1}^{N-1}$. Since y_N is the last observation, the one-step ahead fuzzy forecast is defined as follows:

$$\hat{F}(N+1) = y_N + \Delta^1 \tag{11}$$

That is, a translation of magnitude y_N is applied to the fuzzy variable Δ^1 to build the one-step-ahead fuzzy forecast. Consequently, by applying the well-known property of the credibility expected mean ($E(a+\xi) = a + E(\xi)$, for $a \in \mathbb{R}$), the pointwise forecast is the credibility expected value $\hat{y}_{N+1} = E(\hat{F}(N+1)) = y_N + E(\Delta^1)$.

In order to provide further fuzzy predictions, we define the fuzzy variables for h steps ahead, $h \geq 2$, using those previously forecasted pointwise as

$$\hat{F}(N+h) = E(\hat{F}(N+h-1)) + \Delta^1, \tag{12}$$

and obtain that $\hat{F}(N+h) = y_N + (h-1)E(\Delta^1) + \Delta^1$. Analogously, we obtain the pointwise forecasts as their credibility expected values: $\hat{y}_{N+h} = E(\hat{F}(N+h)) = y_N + hE(\Delta^1)$, for $h \geq 2$.

The prediction intervals for the h-step-ahead pointwise prediction, $h \geq 1$, are built following Equation (8) for the corresponding fuzzy variables $\hat{F}(N+h)$.

3.2. K-Step-Ahead Fuzzy Forecast Model: FFKM

Let us consider the time series $Y(t) = \{y_1, \dots, y_N\}$, for which k-step-ahead forecasts are requested. For $1 \leq h \leq k$, we build the fuzzy variables Δ^h, using the corresponding set of the h-order differences, $D^h = \{d_i^h\}_{i=1}^{N-h}$, $h \geq 1$. Note that in this modeling approach we need to build k fuzzy variables, $\{\Delta^h\}$ for $1 \leq h \leq k$, depending on the ex-post predictions requested.

Let y_N be the last observation; then, for every $1 \leq h \leq k$, the h-step-ahead fuzzy forecast is defined as follows:

$$\hat{F}(N+h) = y_N + \Delta^h \quad (13)$$

The pointwise h-step-ahead forecast is then $\hat{y}_{N+h} = E(\hat{F}(N+h)) = y_N + E(\Delta^h)$, and the corresponding prediction intervals are built using the credibility distribution of every fuzzy variable $\hat{F}(N+h), 1 \leq h \leq k$.

In the context of fuzzy time series, other authors have also considered the differences of consecutive observations to introduce new forecasting methods [13,14], following the basic scheme introduced in [6]. However, their approaches do not have any element in common with the ones we propose in this paper.

3.3. A Numerical Example

Let us present the performance of our methods FFM and FFkM, using the first yearly time series from the M4 Competition [8]. This time series contains 31 observations, $N = 31$. We make a partition of this observed data, in such a way that 28 first observations belong to the training set (y_{28} = 7651.4), while the last 3 are reserved for the comparison with the pointwise h-step-ahead forecasts provided, h = 1, 2, 3.

3.3.1. Performance of the FFM Method

Using the aforementioned methodology for applying the FFM method, we build the fuzzy variable Δ^1, which is defined as $\Delta^1 = (-102.3, 97.4, 232.3)_{0.41,0.84}$, the credibility expected value being $E(\Delta^1)$=98.93. Thus, $\hat{y}_{29}$ = 7750.33.

Figure 1 shows the plot of the membership function of the fuzzy variable Δ^1. The observed first differences are plotted as small circles. Using the suitable sample percentiles of the first differences in D^1, we calculate the 90% prediction interval of $IP_{90\%}(\Delta^1) = [-79.0, 226.6]$. Note that the credibility expected value of the fuzzy variable $E(\Delta^1)$ is not the middle point of this prediction interval.

The calculations of the corresponding h-step-ahead forecasts and the prediction intervals, for h = 2, 3, are easily obtained by taking into account both the properties of the expected value and the sample percentiles, $p_j(a + D^1) = a + p_j(D^1)$, for $a \in \mathbb{R}$. In particular, $\hat{y}_{30}$ = 7849.26 and $\hat{y}_{31}$ = 7948.20.

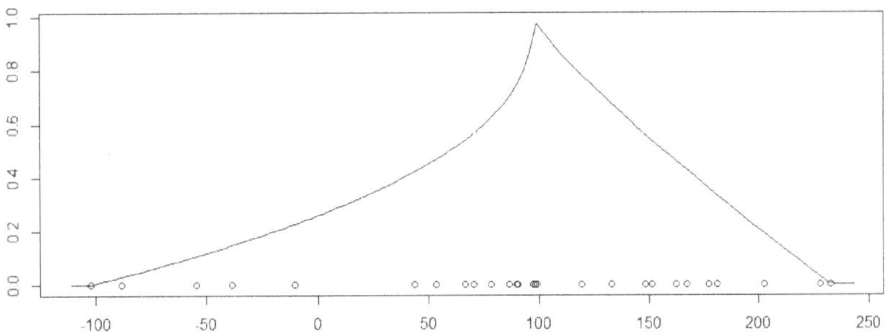

Figure 1. function of the LR-power fuzzy variable $\Delta^1 = (-102.3, 97.4, 232.3)_{0.41,0.84}$. The small circles represent the elements of the set D^1.

3.3.2. Performance of the FFKM Method

In order to determine the out-of-sample forecasts and their prediction intervals, we apply the FFkM method, for $k = 3$. It is easily seen that the first iteration is identical for both methods, FFM and FFkM. However, we need to build the sets of the second- and third-order differences to obtain the fuzzy variables Δ^2 and Δ^3, respectively. Figure 2 shows the membership function of the aforementioned fuzzy variables.

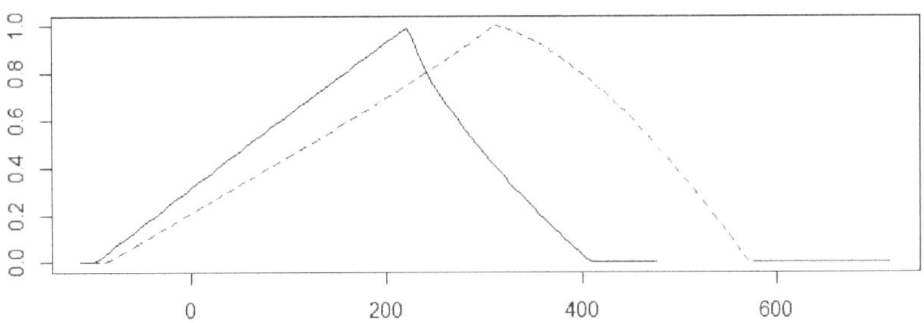

Figure 2. Membership functions of the LR-power fuzzy variables $\Delta^2 = (-98.8, 308.6, 408.0)_{1.02,0.69}$ and $\Delta^3 = (-87.5, 308.6, 570.9)_{0.92,1.51}$. The solid line shows Δ^2, while the dashed line is for Δ^3.

Table 1 shows the raw observation, the pointwise forecasts obtained by applying the FF3M method, the 90% prediction interval and the prediction error, $e_h = \hat{y}(N+h) - y(N+h)$, for h = 1, 2, 3.

Table 1. Pointwise h-step-ahead forecasts obtained by applying the FF3M method, 90% prediction intervals and prediction errors, for h = 1, 2, 3.

	y(28 + h)	$\hat{y}$(28 + h)	IP(90%)	Prediction Error
h = 1	7587.3	7750.3	[7572.4, 7878.0]	163.0
h = 2	7530.5	7831.3	[7568.4, 8053.0]	300.8
h = 3	7261.1	7944.0	[7585.4, 8202.7]	682.9

Note that the trend changes in the observed data have not been suitably estimated by the proposed methods, so that the prediction error has increased when h increases. In fact, only the 90% prediction interval of the first step contains the actual observation. In the next section, we report the results obtained using our proposals for a large set of yearly time series.

4. Results

To investigate the predictive behavior of our fuzzy methods, also in comparison to standard automatic approaches, we applied them to a set of data from the M4 Competition [8]. This competition contains 100,000 seasonal and nonseasonal time series. The results of the M4 Competition have been globally published for each type of subset of time series, depending on their forecasting horizon.

We worked with the nonseasonal yearly time series. Since our proposals need a number of observations to build the fuzzy variables, a filter was applied to this set of 23,000 yearly series, that is, $N \geq 31$. We thus obtained a selection of 9060 yearly time series, with a range of observed sizes: $N \in [31, 700]$. To carry out our study, we used the R language [15], applying various R functions in conjunction with some functions written by the authors.

For the comparison, we followed the specifications established for the M4 Competition and computed forecasts up to six steps ahead. Initially, until 31 May 2018, only the training set was available, but, once it was finished, the values to be forecasted were published in the R package *M4comp2018*. We also selected two automatic forecasting procedures available in the R package forecast, exponential smoothing (*es*) and *auto.arima* [9], for the comparison of both ex-post predictions and coverage of the prediction intervals. We also compared the average computational cost of obtaining these forecasts.

For every model $M_j = \{SE, ARIMA, FFM, FF6M\}$ and each time series, $Y_i(t)$ for $1 \leq i \leq 9060$, the prediction errors were measured as follows:

$$e_{i,M_j}(h) = \hat{y}_{M_j(h,i)} - y_{(h,i)}, h = 1, \ldots, 6, j = 1, \ldots, 4 \tag{14}$$

$\hat{y}_{M_j(h,i)}$ being the forecast at h steps-ahead applying the M_j method and $y_{(h,i)}$ being the observation at time $N + h$. The mean absolute percentage error (MAPE) and the symmetric mean absolute percentage error (sMAPE) were employed to measure forecasting accuracy, respectively:

$$MAPE_{M_j}(h) = 100 * \frac{1}{9060} \sum_{i=1}^{9060} [\frac{e_{i,M_j}(h)}{y_{(h,i)}}], \; h = 1, \ldots, 6 \tag{15}$$

$$sMAPE_{M_j}(h) = 100 * \frac{2}{9060} \sum_{i=1}^{9060} \frac{|e_{i,M_j}(h)|}{|\hat{y}_{M_j(h,i)} + y_{(h,i)}|}, \; h = 1, \ldots, 6 \tag{16}$$

For the comparison of the prediction intervals, wed use their empirical coverage, that is, the percentage of times that the observation in included in those intervals.

4.1. Prediction Intervals

Table 2 shows the results of the empirical coverage obtained by applying the different methods at each h-step ahead, denoted as $E.h$, for $h = 1, \ldots, 6$, on the scale [0,1], for the 9060 yearly time series. Note that the nominal coverage is established in 80% and 95%, respectively. The last column shows the mean of the empirical coverage, as a percentage.

The results in Table 2 show that the prediction intervals provided by the FFM method are wider than those provided by FFkM method, while the coverage of the prediction intervals provided by FFkM is closer to that provided by exponential smoothing. The *ARIMA* procedure provided prediction intervals with lower empirical coverage.

The computational time was calculated in milliseconds for every run, that is, the elapsed time from when time series data were read until the six ex-post forecasts were given. Table 3 shows the descriptive statistics of the computing time for each method.

Table 2. Comparison of the empirical coverage at each step ahead attained for every method. The last column shows the mean of the coverage as a percentage.

IP(80%)	E.1	E.2	E.3	E.4	E.5	E.6	Mean
SE	0.65	0.72	0.76	0.76	0.77	0.78	74%
ARIMA	0.51	0.53	0.56	0.56	0.56	0.57	55%
FFM	0.66	0.79	0.89	0.92	0.93	0.94	86%
FF6M	0.66	0.80	0.86	0.83	0.78	0.73	78%
IP(95%)							
SE	0.81	0.87	0.89	0.89	0.88	0.89	87%
ARIMA	0.49	0.56	0.65	0.68	0.69	0.72	63%
FFM	0.74	0.86	0.93	0.95	0.96	0.96	90%
FF6M	0.74	0.87	0.93	0.91	0.86	0.83	86%

Table 3. Statistics of computing times for forecasting the 9060 yearly time series.

Milliseconds	SE	ARIMA	FFM	FF6M
Mean	10.80	80.91	0.75	2.61
Standard deviation	27.81	53.60	17.51	2.07
TOTAL TIME	97,876.75	73,3054.9	6,788.27	23,605.61

Concerning the computational cost, the new fuzzy proposals are much more competitive than the automatic forecasting procedures. The average time consumed for each run favors the FFM method, although the FFkM method reported better empirical coverage for the prediction intervals.

4.2. Pointwise Forecasts

For the predictive analysis of the performance of a forecasting method, it is also important to verify the accuracy of the out-of-sample forecasts. Table 4 shows the averaged post-sample accuracy of the forecasts obtained for the aforementioned methods, both using MAPE and sMAPE prediction errors. These results are consistent with the best performance of our proposed fuzzy forecast methods, at least with respect to forecasting accuracy. The forecasting errors were calculated for each forecasting horizon (1–6 steps ahead) and for the aggregated periods.

Table 4. MAPE and sMAPE for the 9060 time series of yearly data.

MAPE	E.1	E.2	E.3	E.4	E.5	E.6	Average 1 to 6
SE	7.70	10.80	13.44	16.48	18.80	22.05	14.88
ARIMA	8.14	11.38	13.77	16.70	18,60	22.02	15.10
FFM	7.55	10.26	12.64	15.67	17.63	20.93	14.11
FF6M	7.55	10.08	12.14	14.86	16.53	19.67	13.47
sMAPE							
SE	7.36	9.89	12.10	14.3	16.38	18.08	13.02
ARIMA	7.70	10.25	12.10	14.12	16.02	17.52	12.95
FFM	7.31	9.63	11.69	13.95	15.82	17.62	12.67
FF6M	7.31	9.37	11.17	13.16	14.89	16.63	12.09

The results concerning the accuracy of the ex-post forecasts are consistent with the best performance of our proposed fuzzy forecast methods. It must also be noted that the FFkM method obtained the best results at every step for the averaged MAPE and sMAPE, and it provided these accurate ex-post forecasts in a very competitive computation time.

In addition, a statistical pairwise comparison of the MAPE errors was performed, using the Wilcoxon nonparametric test and adjusting the p-values by Holm's method. All comparisons were statistically significant (adjusted p-values not greater than 0.012) except the *SE* and *ARIMA* comparison for the second-step-ahead forecast (p value = 0.32).

The statistical differences and homogeneities using sMAPE follow similar patterns to those of averaged MAPE. We also found analogous results when averaged ex-post forecast errors were compared using parametric methods.

5. Conclusions

In this paper, we introduce a new forecasting scheme based on fuzzy variables and credibility distributions of the first and upper differences of the data series. Our fuzzy forecasting methods enable us to obtain accurate pointwise forecasts and reliable prediction intervals for nonseasonal time series. Our methods also provide LR-power fuzzy variables as ex-post fuzzy forecasts.

Our approach was tested using yearly time series from the M4 Competition. Concerning the out-of-sample performance of the point forecast in comparison with automatic forecasting packages, we obtained a competitive accuracy with the lowest computation time, for a set of 9060 time series.

Statistical pairwise comparison was used to compare the averaged forecasting errors, showing significant differences between the averaged MAPE and sMAPE forecast errors of our fuzzy methods and the other automatic forecasting methods included in the experiment.

In further studies, we will analyze the performance of our proposed techniques on other type of time series. Those studies could also focus on the analysis of other fuzzy strategies to improve the detection of trend changes, using the upper differences between observed data.

Author Contributions: Conceptualization, J.D.B. and E.V.; methodology, J.D.B., A.R. and E.V.; software, A.R.; validation, J.D.B., A.R. and E.V.; formal analysis, E.V.; investigation, A.R.; resources, A.R.; data curation, A.R.; writing—original draft preparation, E.V.; writing—review and editing, E.V.; supervision, J.D.B.; project administration, J.D.B.; and funding acquisition, J.D.B. and E.V. All authors have read and agreed to the published version of the manuscript.

Funding: This work was supported by the Spanish Ministry of Economy and Competitiveness (Project MTM2017-83850-P), co-financed by FEDER funds.

Data Availability Statement: Datasets available in: Montero-Manso, P., Netto, C. & Talagala, C. (2018). M4comp2018: Data from the M4-Competition. R package version 0.1.0.

Conflicts of Interest: The authors declare no conflict of interest.

References

1. Liu, B. A survey of credibility theory. *Fuzzy Optim. Decis. Mak.* **2006**, *5*, 387–408. [CrossRef]
2. Vercher, E.; Bermúdez, J.D. Portfolio optimization using a credibility mean-absolute semi-deviation model. *Expert Syst. Appl.* **2015**, *42*, 7121–7131. [CrossRef]
3. Ruiz, A.B.; Saborido, R.; Bermúdez, J.D.; Luque, M.; Vercher, E. Preference-based evolutionary multi-objective optimization for portfolio selection: A new credibilistic model under investor preferences. *J. Glob. Optim.* **2020**, *76*, 295–315. [CrossRef]
4. Song, Q.; Chissom, B.S. Fuzzy time series and its models. *Fuzzy Sets Syst.* **1993**, *54*, 269–277. [CrossRef]
5. Wang, L.; Liu, X.; Pedrycz, W.; Shao, Y. Determination of temporal information granules to improve forecasting in fuzzy time series. *Expert Syst. Appl.* **2014**, *41*, 3134–3142. [CrossRef]
6. Chen, S.-M. Forecasting enrollments based on fuzzy time series. *Fuzzy Sets Syst.* **1996**, *81*, 311–319. [CrossRef]
7. Rubio, A.; Bermúdez, J.D.; Vercher, E. Improving stock index forecasts by using a new weighted fuzzy-trend time series method. *Expert Syst. Appl.* **2017**, *76*, 12–20. [CrossRef]
8. Makridakis, S.; Spiliotis, E.; Assimakopoulos, V. The M4 competition: Results, findings, conclusion and way forward. *Int. J. Forecast.* **2018**, *34*, 802–808. [CrossRef]
9. Hyndman, R.; Khandakar, Y. Automatic time series forecasting: the forecast package for R. *J. Stat. Softw.* **2008**, *26*, 1–22.
10. Zadeh, L. Probability measures of fuzzy events. *J. Math. Anal. Appl.* **1968**, *23*, 421–427. [CrossRef]
11. Liu, B.; Liu, Y.-K. Expected value of fuzzy variable and fuzzy expected value models. *IEEE Trans. Fuzzy Syst.* **2002**, *10*, 445–450.

12. León, T.; Vercher, E. Solving a class of fuzzy linear programs by using semi-infinite programming techniques. *Fuzzy Sets Syst.* **2004**, *146*, 235–252. [CrossRef]
13. Singh, S. A computational method of forecasting based on fuzzy time series. *Math. Comput. Simul.* **2008**, *79*, 539–554. [CrossRef]
14. Stevenson, M.; Porter, J. Fuzzy time series forecasting using percentage change as the universe of discourse. *World Acad. Sci. Eng. Technol.* **2009**, *55*, 154–157.
15. R Core Team. *R: A Language and Environment for Statistical Computing*; R Foundation for Statistical Computing: Vienna, Austria, 2014.

engineering proceedings

MDPI

Proceedings

Epidemiology SIR with Regression, Arima, and Prophet in Forecasting Covid-19 †

Pedro Furtado

Departamento de Engenharia Informatica/CISUC, Universidade de Coimbra/Polo II, 3030-790 Coimbra, Portugal; pnf@dei.uc.pt

† Presented at the 7th International Conference on Time Series and Forecasting, Gran Canaria, Spain, 19–21 July 2021.

Abstract: Epidemiology maths resorts to Susceptible-Infected-Recovered (SIR)-like models to describe contagion evolution curves for diseases such as Covid-19. Other time series estimation approaches can be used to fit and forecast curves. We use data from the Covid-19 pandemic infection curves of 20 countries to compare forecasting using SEIR (a variant of SIR), polynomial regression, ARIMA and Prophet. Polynomial regression deg2 (POLY d(2)) on differentiated curves had lowest 15 day forecast errors (6% average error over 20 countries), SEIR (errors 25–68%) and ARIMA (errors 15–85%) were better for spans larger than 30 days. We highlight the importance of SEIR for longer terms, and POLY d(2) in 15-days forecasting.

Keywords: time series forecasting; epidemiology; SIR

Citation: Furtado, P. Epidemiology SIR with Regression, Arima, and Prophet in Forecasting Covid-19 . *Eng. Proc.* **2021**, *5*, 52. https://doi.org/10.3390/engproc2021005052

Published: 13 July 2021

1. Introduction

The typical initial evolution of an epidemic when the population has no immunity and the pathogen has high virulence and high death rate is a frightening exponential curve. Scientists still lack a lot of knowledge about the SARS-CoV-2 virus and variants, and recently new virus variants have increased its contagiousness. Nevertheless, we can say that its reproduction number (rate of growth) should be around 3 and its death rate could be around 0.3%. Since a reproduction number higher than 1 already means exponential growth, a value of 3 indicates a significant virulence and containment is necessary if no vaccination is available or significant acquired immunity in the population. In order to study and predict the evolution of the curve and to decide how to act at each moment, epidemiologists and mathematicians frequently use variants of the Susceptibility Infected Removed (SIR). SIR or the alternative SIR with an additional state called Exposed (SEIR) we use here, is actually a simple model. Given four possible states (S, E, I and R), at any given moment each individual from a population can be in any of those states. At the start, with zero immunity, all population is in state S (Susceptible). In the model individuals transition from S to I (infected) and from there to recovered or dead (R). The model also uses some other parameters to describe rates of transition in three differential equations.

There are many other ways to fit and forecast curves in general. Approaches such as linear regression, polynomial regression, ARIMA [1] or other time series forecasting approaches could in principle be used in this context as in any other context, but the question is whether they would stand any chance when compared with SEIR that integrates epidemic-specific parameters. Lacking all the specific model parameters that SEIR can include, we expect those more generic time series analysis models to miss important information that leads to future changes in the curve, but on the other hand it is a possibility that they could be useful in short-term (few days) forecasting, with more constant conditions.

We review the approaches, setup variants of polynomial regression, ARIMA, Prophet and SEIR, collect the Covid-19 evolution curves of the 20 hardest hit countries at the end of March and compare their performance forecasting the last 15 days of the curves and different lengths as well. This setup allowed us to reach conclusions regarding their relative merits. In this work we were slightly constrained by the fact that, except for China where the outbreak started much earlier, we had only around 45 days of data for most countries (the time from the start of the outbreak in most countries to this study), but on the other hand this study is especially interesting because it deals with an ongoing outbreak of hard consequences. Future work would be interesting in generalizing these results with more diseases and more forecasting evaluation alternatives.

2. Related Work

Epidemiological modelling has been discussed extensively in works such as [2–6]. The Susceptible-Infected-Removed model (SIR) is reviewed in [7] and an analytical solution to it is discussed in [8]. According to the definition, infection transitions between the three states given in the name itself, and a set of three equations describes the transitions between those states. Parameters include the average contacts of individuals (Beta) times transmission probability per contact (I), rate of deaths and recoveries, time a person is infected D and its inverse γ. There are many other models that evolved from SIR, including those in [9].

Polynomial regression is a fitting procedure that tries to approximate a given curve using a polynomial of degree n. Reference [10] describes numerical methods for curve fitting. Regression analysis [11,12] focuses especially on statistical inference related to curve fitting and associated uncertainty. These kinds of approaches can be helpful for abstracting trends and forecasting into the near future in different contexts.

The model Autoregressive Integrated Moving Average is reviewed in [1,13]. It uses differentiation iterations to solve non-stationarity, plus regression, moving averages and integration to successively improve data fitting. A simpler but also effective model [14] was proposed by Facebook^{TM}. In [15] we used both Prophet and a modified ARIMA to predict evolution of business performance indicators in Telecom. In that specific application, we concluded that ARIMA outperformed Prophet.

3. Curve Fitting and Forecasting Models

3.1. The SEIR Model Plus a Social Distance Factor

The SEIR model [5,16] has susceptible (S), exposed (E), infected (I) and recovered (R) states and describes the dynamics of the population successively moving from one of the states to the next. As soon as individuals reach state R they are no longer able to become infected. Initially, the whole population is in state S. The following differential equations model how the individuals of a population evolve between these states in SEIR. For instance, $S'(t)$ is the change in the number of people in state S from moment t to $t + 1$. A social distancing factor is also added to model the degree of distance between people, which has the potential to decrease the rate of contagions:

$$S_{t+1} = \rho \times \beta \times S_t \times I_t \tag{1}$$

$$E_{t+1} = \rho \times \beta \times S_t \times I_t - \propto \times E_t \tag{2}$$

$$I_{t+1} = \propto \times E_t - \gamma \times I_t \tag{3}$$

$$R_{t+1} = \gamma \times I_t \tag{4}$$

In these four equations, α is the inverse of the incubation time ($1/d\alpha$), estimated to be 5 days in average for Covid-19 (varying between 1 and 14 days); $\beta = \tau \times c$ is transmissibility (τ = infection probability with contact with infected) and the average rate of contact between susceptible and infected individuals c, obtained by curve fitting; γ is the inverse of the mean infectious time ($1/d\gamma$), estimated to be 10 days; ρ is the social distancing factor, varying between 0 and 1, observable by curve fitting. We have coded this model together with least squares fitting to find the term "social distancing × Beta ($\rho \times \beta$) that minimizes the average root mean squared error (RMSE) between the SEIR curve and the official country curve. Forecasting was done by assuming that ($\rho \times \beta$) remains the same for the next days.

3.2. Polynomial Regression

Polynomial regression (POLY) uses least-squares fitting to find the coefficients of a polynomial of degree n that best fit a curve. Equation (1) shows the polynomial, where some curve Y is to be approximated by the polynomial function with coefficients C_0 to C_n,

$$Y_a = C_0 \times x^n + C_1 \times x^{n-1} + \ldots + C_{n-1} \times x + C_n \quad (5)$$

In our case the x is the time unit (the forecast is Y_a value for time unit ti. Different polynomial degrees were tested in our experiments. We will show in our experiments that forecasting with POLY over differentiated curves of "number of active patients" instead of the original followed by an inverse transformation to reconstruct the curve yielded best results (ARIMA also uses differentiation to work on stationary curves).

3.3. Time Series Forecasting with ARIMA

For our own review of time series forecasting using ARIMA, please refer to [15], another reference is [9]. In [15] we explain how ARIMA uses Auto-Regressive (AR) and Moving Average (MA) models and how a set of parameters are applied ineach of those component models. Please, refer to [15] for more details on ARIMA.

3.4. Automated Parameterization of ARIMA

In automated parameterization of ARIMA, a set of three parameters are tuned (p, d, q) and seasonality as well are obtained by automatically testing the fit of the curve for each combination of those values. The Akaike criteria (AIC), estimating the prediction error, provides a relative metric for the model quality. Thus, AIC provides a means for model selection. In the following example, the combination with lowest AIC is chosen.

pdq = 0, 0, 0 resulting in AIC = 705.7393610322358

...

pdq = 0, 1, 1 resulting in AIC = 456.58099826482464

...

pdq = 1, 1, 1 resulting in AIC = 304.90034871700635

3.5. Time Series Forecasting with Prophet

Forecasting using Prophet is described in [17] and in detail in [18], or in our own prior work [15]. It uses three sub-models, one analyzing the trend, another one analyzing the seasonality and the third one taking into account festivity periods. Each of those sub-models is modelled by a function (logistic for trend, Fourier for seasonality and an adjustment for festivity periods.

4. Experimental Work

Our experimental setup was created using the pandemics curves (number of active cases) up to a specific day (27 March 2020). This data can be obtained for instance in [19]. Figure 1 illustrates the curves, showing the per-million aligned active cases in 5 European countries (we show only 5 countries to avoid cluttering). The number of active cases of the 20 most hit countries except China (27 March) were used for curve fitting, and Chinas's curve was used for testing longer forecasting spans.

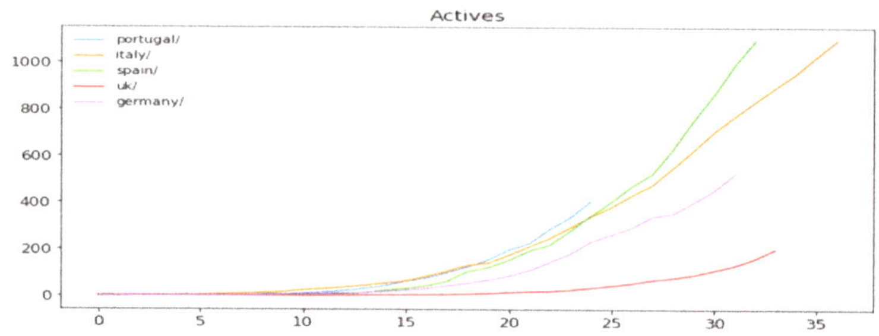

Figure 1. Active cases up to 27 March, 5 Europe countries align start of outbreak (number of cases > 0).

For our experimental setup we implemented each of the forecasting approaches (polynomial regression, ARIMA, Prophet and SEIR) and all necessary data loading and data transformations in python. The full list of countries in our setup included 21 countries. China was used in forecasting larger lengths. We tested polynomial degrees 1 to 4 and differentiation as well. The experimental procedure is simple: extract x days from the curve, fit the model to the truncated curve and finally forecasting the last x days using the model. All results report the average error using *MAE* relative to the values (*MAEr*) for the new forecasted segment, shown in Equation (5):

$$MAEr = \frac{\sum_1^n \left| \frac{Y_i - YY_{ei}}{Y_i} \right|}{n} \tag{6}$$

4.1. Fifteen-Day Forecast over 20 Countries

The 15-day forecast is done on each country by extracting 15 days from the curve, then fitting the model to the truncated curve and finally forecasting the last 15 days using the model. Figure 2 and Table 1 show the results for 20 countries (POLY d(n) = polynomial regression with degree n on differentiated curve, Prophet d = Prophet on differentiated curve).

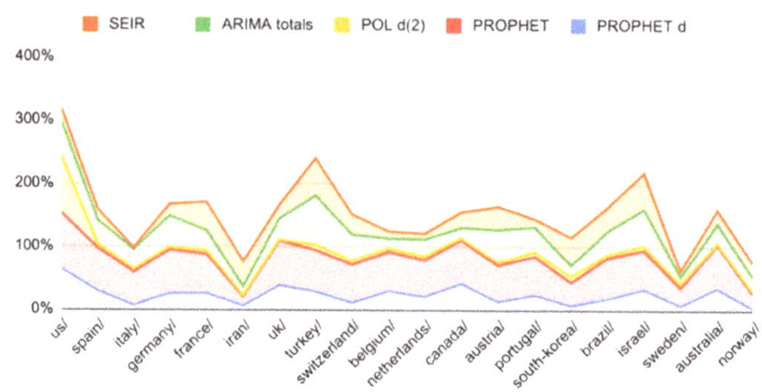

Figure 2. Comparison of methods over 20 countries (MAER, stacked chart).

Note that in the previous results we have chosen POLY d(2) because it had the best forecasting results. Table 2 shows the comparison of forecasting results of different polynomial regression options for the same experimental setup. In that table the number in parenthesis is the polynomial degree and d stands for differentiation.

Table 1. Comparison of methods over 20 countries (MAER).

	PROPHET d	PROPHET	POLY d(2)	ARIMA	SEIR
Countries avg	23%	57%	9%	35%	28%
stdev	15%	18%	19%	16%	16%
US	64%	89%	89%	54%	21%
Spain	29%	68%	6%	37%	18%
Italy	7%	52%	4%	32%	3%
Germany	25%	69%	4%	50%	19%
France	26%	62%	6%	32%	45%
Iran	6%	14%	2%	15%	40%
UK	39%	71%	1%	34%	22%
Turkey	29%	66%	8%	78%	59%
Switzerland	11%	60%	6%	42%	32%
Belgium	29%	62%	5%	16%	11%
Netherlands	21%	58%	5%	28%	9%
Canada	41%	69%	3%	17%	25%
Austria	13%	58%	4%	52%	37%
Portugal	24%	61%	7%	40%	13%
South-Korea	7%	37%	9%	18%	45%
Brazil	18%	65%	4%	38%	38%
Israel	33%	62%	7%	58%	58%
Sweden	7%	28%	5%	12%	10%
Australia	35%	70%	2%	31%	23%
Norway	3%	23%	3%	23%	23%

Table 2. Summary of MAER for polynomial regression, 20 countries 15-day forecast.

POLY 4	POLY 3	POLY(2)	POLY(1)	POLY d(4)	POLY d(3)	POLY d(2)
44%	28%	21%	49%	14%	11%	9%

4.2. Discussion of Results

The results in Figure 2 and Table 1 show that POLY d(2) had the best 15-day forecasts (MAEr 9 $\pm$ 19%), less than half the next competitor, then Prophet d (22 $\pm$ 15%) and SEIR (28 $\pm$ 16%). ARIMA had (35 $\pm$ 16%) and Prophet without differentiation was the worst (55 $\pm$ 18%). All techniques had a similar degree of variation between countries (stdevs 15% to 19%). The experiment with multiple alternatives of polynomial regression show that differentiating was useful and the error was smaller with smaller polynomial degree in the tested interval 2 to 4. In essence, the polynomial regression of degree 2 is fitting the curve by a parabola that mimics the initial steep increase of the number of daily cases, then as confinement and social distancing kicks in, the change first to a stabilization and then a decrease of the number of daily cases. But degree 3 or 4 on the differentiated curve also had relatively small errors.

The fact that SEIR did not achieve the best 15-days forecast may seem surprising, since SEIR is the preferred epidemic modeling approach. However, although its results were still reasonable and we still expect SEIR to be the best for forecasting the whole epidemic curve, some of its parameters are abstractions that mean it may not fit official epidemics curves perfectly. One such parameter is the initial population of Susceptibles (S), and the variability is due to the degree of susceptibility of individuals to the contagion and to transmission varying in reality. There is also a problem with the official account of quantity of infected actually infected, since there are many asymptomatic patients and knowing the actual quantity of infected would require extensive continuous testing. The dynamics of transmission and social distancing at a country level is also a coarse approximation, since there exist high-density, highly populated cities and lower-density zones in every country.

Prophet on the differentiated curve scored 22% error, and both Prophet and POLY were much better if used on the differentiated curve. The best ARIMA results were obtained with the input curve not differentiated (35%), but note that ARIMA itself differentiates through the parameter d for stationarity.

4.3. Testing the Approaches on Larger Spans (China)

China was the only Covid-19 worst-hit countries curve for which there were considerably more days (around 90). The next experiment consisted in removing a variable number of days from that series, building the model and forecasting those removed days using the model. The results are shown in Figure 3 (stacked chart) and Table 3.

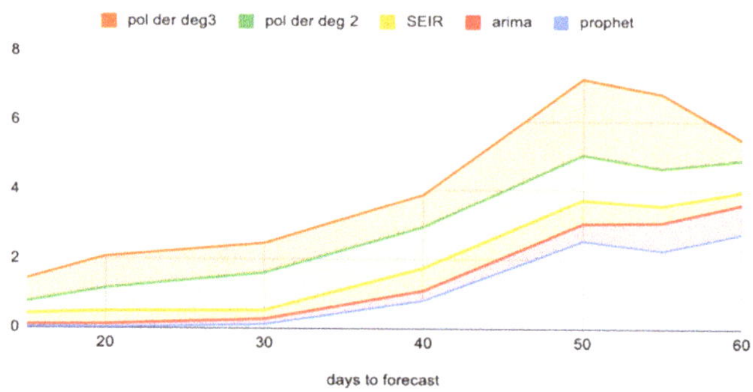

Figure 3. China with different spans (MAER, stacked chart) (der = differentiated, deg = degree).

Table 3. MAER data for China spans.

Days to Forecast	Prophet	ARIMA	SEIR	Poly der Deg 2	Poly der Deg 3
15	4%	7%	32%	34%	67%
20	3%	10%	37%	67%	92%
30	13%	15%	25%	109%	85%
40	82%	28%	65%	119%	94%
50	254%	50%	68%	131%	220%
55	226%	81%	49%	108%	215%
60	275%	85%	37%	91%	59%
AVG	123%	39%	45%	94%	119%
STD	125%	33%	17%	33%	69%

4.4. Discussion on Larger Forecasts

SEIR was clearly superior for this case of longer forecast periods, it had the best scores overall and least variance (error between 32% and 45%). The specific modeling that considers important epidemiology concepts overcame the results of generic models when modeling on longer spans. ARIMA was next (7% to 85% errors), then POLY der deg 2 (34% to 131%), while the polynomial of degree 3 was much worse (59% to 220%). Prophet (4% to 275%) had the smallest errors up to forecast length 30 and then the largest errors for the remaining lengths. The advantage of ARIMA over Prophet on longer spans could be related to ARIMA adjusting its p, d, q parameters.

4.5. Visualization of Some Results

We do not have space to show most visualizations, however, we show a few illustrative examples next. Figure 4 is the daily cases in Spain together with the POLY d(2) forecast, and Figure 5 is the daily cases in China together with the SEIR forecast, both showing the actual values (blue curve) together with the estimations. Figure 6a is Spain's 15 day forecast using ARIMA and Figure 6b is the forecast for the same case using Prophet. We can see that POLY d(2) was near the actual curve in Figure 4, although at the end of the interval the divergence increased, SEIR forecast was also quite good in Figure 5, while in

ARIMA the forecast for the last days was diverging significantly upwards and in Prophet it was diverging downwards.

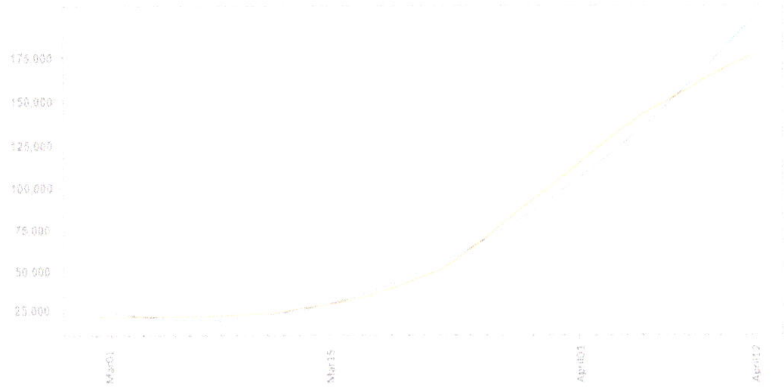

Figure 4. Example of POLY d(2) Spain 15-days forecasting (blue) over original line (orange), daily cases.

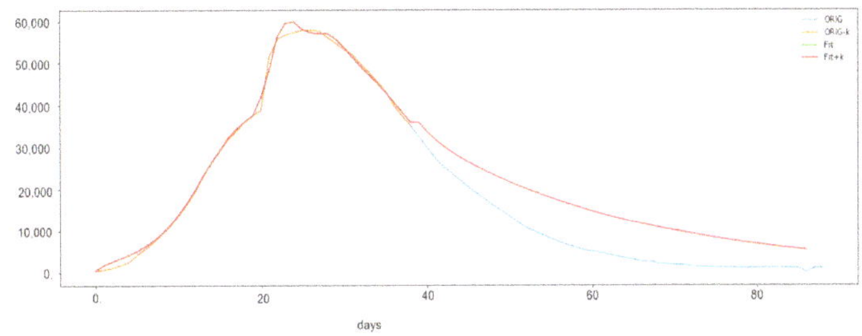

Figure 5. SEIR forecasting (red) and original curve (orange) plus cut (blue) for China, daily cases.

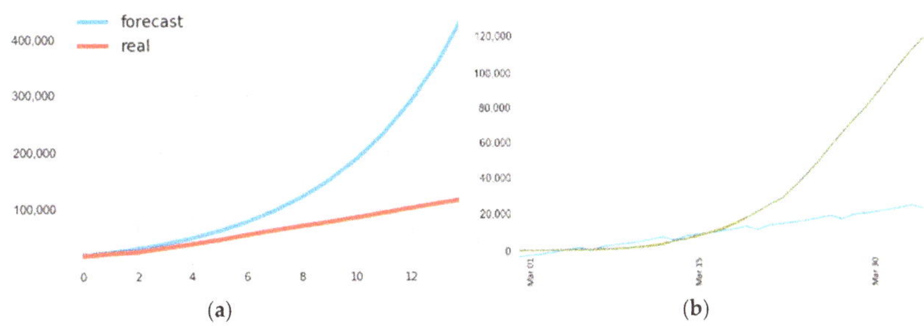

Figure 6. ARIMA and Prophet 15-days forecasting of Spain (days versus number of infected). (**a**) ARIMA (forecast in blue); (**b**) Prophet (forecast in blue).

5. Conclusions

In this article we investigated the issue of short and longer-term forecasting over epidemiological curves of Covid-19 using both generic forecasting approaches and the more specific epidemiological SEIR model, with the objective of confronting the alternatives. After describing the approaches used we created an experimental setup with the alternatives and tested over 20 countries, plus longer-term forecasts on the longest curve (China). We concluded that, in average, polynomial regression of degree 2 was the best for short term (15 days or less), but on longer term SEIR was clearly superior to the competition, which is explained by its use of more specific epidemiological parameters. The use of Covid-19 curves makes the work very up-to-date, but on the other hand we would like to experiment with other epidemics curves and to test different segments on multiple spans. Our own current and future work deals also with automatic fitting, parameter optimization and what-if analysis with the SEIR model.

Funding: The funders had no role in the design of the study; in the collection, analyses, or interpretation of data; in the writing of the manuscript, or in the decision to publish the results.

Institutional Review Board Statement: Not applicable.

Informed Consent Statement: Not applicable.

Data Availability Statement: All the data used in this study is publicly available in https://www.worldometers.info/coronavirus/, accessed on 1 June 2020.

Conflicts of Interest: The authors declare no conflict of interest.

References

1. Brockwell, P.J.; Davis, R.A. *Time Series: Theory and Methods*, 2nd ed.; Springer: New York, NY, USA, 2009; p. 273. ISBN 9781441903198.
2. Hethcote, H. The Mathematics of Infectious Diseases. *SIAM Rev.* **2000**, *42*, 599–653. [CrossRef]
3. Bailey, N.T. *The Mathematical Theory of Infectious Diseases and its Applications*, 2nd ed.; Griffin: London, UK, 1975; ISBN 0-85264-231-8.
4. Altizer, S.; Nunn, C. Infectious diseases in primates: Behavior, ecology and evolution. In *Oxford Series in Ecology and Evolution;* Oxford Univers Press: Oxford, UK, 2006; ISBN 0-19-856585-2.
5. Brauer, F.; Castillo-Chávez, C. *Mathematical Models in Population Biology and Epidemiology;* Springer: New York, NY, USA, 2001; ISBN 0-387-98902-1.
6. Anderson, R.M. *Population Dynamics of Infectious Diseases: Theory and Applications;* Chapman and Hall: London, UK, 1982; ISBN 0-412-21610-8.
7. Kermack, W.O.; McKendrick, A.G. A Contribution to the Mathematical Theory of Epidemics. *Proc. R. Soc.* **1927**, *115*, 772.
8. Is Prophet Really Better than ARIMA for Forecasting Time Series Data. Available online: https://blog.exploratory.io/is-prophet-better-than-arima-for-forecasting-time-series-fa9ae08a5851 (accessed on 16 August 2018).
9. A Comprehensive Beginner's Guide to Create a Time Series Forecast. Available online: https://www.analyticsvidhya.com/blog/2016/02/time-series-forecasting-codes-python (accessed on 15 March 2018).
10. Guest, P.G.; Guest, P.G. *Numerical Methods of Curve Fitting;* Cambridge University Press: Cambridge, UK, 2012.
11. Motulsky, H.; Christopoulos, A. *Fitting Models to Biological Data Using Linear and Nonlinear Regression: A Practical Guide to Curve Fitting;* Oxford University Press: Oxford, UK, 2004.
12. Freund, R.J.; Wilson, W.J.; Sa, P. *Regression Analysis;* Elsevier: Amsterdam, The Netherlands, 2006.
13. Box, G.; Jenkins, G.M.; Reinsel, G.C. *Time Series Analysis: Forecasting and Control*, 3rd ed.; Prentice-Hall: Hoboken, NJ, USA, 1994; ISBN 0130607746.
14. Prophet Forecasting at Scale. Available online: https://facebook.github.io/prophet/ (accessed on 16 April 2019).
15. Pinho, A.; Costa, R.; Silva, H.; Furtado, P. Comparing Time Series Prediction Approaches for Telecom Analysis. In *International Conference on Time Series and Forecasting;* Springer: Cham, Switzerland, 2018; pp. 331–345.
16. Harko, T.; Lobo, F.S.; Mak, M.K. Exact analytical solutions of the Susceptible-Infected-Recovered (SIR) epidemic model and of the SIR model with equal death and birth rates. *Appl. Math. Comput.* **2014**, *236*, 184–194. [CrossRef]
17. Wang, M.; Wang, Y.; Wang, X.; Wei, Z. Forecast and Analyze the Telecom Income based on ARIMA Model. *Open Cybern. Syst. J.* **2015**, *9*, 2559–2564. [CrossRef]
18. Taylor, S.J.; Letham, B. Forecasting at scale. *Am. Stat.* **2018**, *72*, 37–45. [CrossRef]
19. World-o-Meter Site with Worldwide and Per-Country Corona Virus Information. Available online: https://www.worldometers.info/coronavirus/ (accessed on 7 April 2020).

engineering proceedings

Proceedings

Estimation of COVID-19 Dynamics in the Different States of the United States during the First Months of the Pandemic †

Ignacio Rojas-Valenzuela [1,*], Olga Valenzuela [2], Elvira Delgado-Marquez [3] and Fernando Rojas [4]

1 School of Technology and Telecommunications Engineering, University of Granada, 18071 Granada, Spain
2 Department of Applied Mathematics, University of Granada, 18071 Granada, Spain; olgavc@ugr.es
3 Department of Economics and Statistics, University of Leon, 24071 León, Spain; elvira.delgado@unileon.es
4 Department of Architecture and Computer Technology, University of Granada, 18071 Granada, Spain; frojas@ugr.es
* Correspondence: e.ignaciorojas@go.ugr.es
† Presented at the 7th International conference on Time Series and Forecasting, Gran Canaria, Spain, 19–21 July 2021.

Abstract: Estimation of COVID-19 dynamics and its evolution is a multidisciplinary effort, which requires the unification of heterogeneous disciplines (scientific, mathematics, epidemiological, biological/bio-chemical, virologists and health disciplines to mention the most relevant) to work together towards a better understanding of this pandemic. Time series analysis is of great importance to determine both the similarity in the behavior of COVID-19 in certain countries/states and the establishment of models that can analyze and predict the transmission process of this infectious disease. In this contribution, an analysis of the different states of the United States will be carried out to measure the similarity of COVID-19 time series, using dynamic time warping distance (DTW) as a distance metric. A parametric methodology is proposed to jointly analyze infected and deceased persons. This metric allows comparison of time series that have a different time length, making it very appropriate for studying the United States, since the virus did not spread simultaneously in all the states/provinces. After a measure of the similarity between the time series of the states of United States was determined, a hierarchical cluster was created, which makes it possible to analyze the behavioral relationships of the pandemic between different states and to discover interesting patterns and correlations in the underlying data of COVID-19 in the United States. With the proposed methodology, nine different clusters were obtained, showing a different behavior in the eastern zone and western zone of the United States. Finally, to make a prediction of the evolution of COVID-19 in the states, Logistic, Gompertz and SIR models were computed. With these mathematical models, it is possible to have a more precise knowledge of the evolution and forecast of the pandemic.

Keywords: COVID-19; pandemic in the united states; time series; DTW distance; hierarchical clustering; SIR model

Citation: Rojas-Valenzuela, I.; Valenzuela, O.; Delgado-Marquez, E.; Rojas, F. Estimation of COVID-19 Dynamics in the Different States of the United States during the First Months of the Pandemic . *Eng. Proc.* **2021**, *5*, 53. https://doi.org/10.3390/engproc2021005053

Published: 13 July 2021

1. Introduction

The COVID-19 epidemic started in Hubei Province, China, around December 2019. Since then, the disease has spread to all continents and countries of the world, being categorized as a pandemic by World Health Organization on 11 March 2020.

In recent months, contributions have been made that analyze the evolution of the pandemic in different countries, implementing mathematical models to predict their evolution. Traditional predictive models for infectious diseases mainly include models for predicting differential equations and models for predicting time series based on statistics and random processes.

For example, in [1] a methodology with the aim of estimating the actual number of people infected with COVID-19 in France is presented, since according to the authors, the number of screening tests carried out and the methodology do not directly calculate the actual number of cases and infection mortality rate (IFR). A mechanistic–statistical approach was developed that combines an epidemiological SIR model that describes these unobserved epidemiological dynamics, a probabilistic model that describes the data collection process and a method of statistical inference.

The logistic growth model, the generalized logistic growth model, the generalized growth model and the generalized Richards model were used to model the number of infected cases in the 29 provinces of China (and several countries), performing a detailed analysis on the heterogeneous situations by four phases of the outbreak in China [2].

In [3] the Kermack–McKendrick SEIR model (Susceptible, Exposed, Infectious and Recovered) is presented to analyze the effects of behavioral changes on the reduction in community transmission in Mexico. A variable contact rate over time is proposed and the consequences of disease spread in an affected population of non-essential activities is analyzed.

The behavior of the virus in Japan has also been analyzed [4]. By 29 February 2020, in addition to the 619 confirmed cases (passengers and crew members) infected with COVID-19 in a cruise ship (near Tokyo), 215 locally transmitted cases had been also confirmed in Japan. To evaluate the effectiveness of reaction strategies based on avoiding large accumulations or crowded areas and to predict the spread of COVID-19 infections in Japan, in [4] a stochastic transmission model produced by expanding the epidemiological model based on SIR (Susceptible-Infected-Removed) had been presented. The simulation results showed that the number of Infected and Removed patients will increase rapidly if there is no reduction in the time spent in crowded zone.

In [5] using the Maximum-Hasting (MH) parameter estimation method and the SEIR model, the spread of COVID-19 and its prediction in South Africa, Egypt, Nigeria, Senegal, Kenya, and Algeria under three intervention scenarios (suppression, mitigation, mildness) is presented.

In addition to the most relevant epidemiological models used in the literature, models typically based on time series have also been used to analyze the behavior of the pandemic in different countries. The autoregressive integrated moving average (ARIMA) model is a mathematical model widely studied in the context of time series that has been successfully applied in the field of health (estimate the incidence and prevalence of influenza mortality, malaria incidence, hepatitis, and other infectious diseases) as well as in different fields in the past due to its simple structure, fast applicability and ability to explain the data set.

In [6], ten Brazilian states are analyzed using the autoregressive integrated moving average (ARIMA), the cubistic regression (CUBIST), the random forest (RF), ridge regression (RIDGE), the support vector regression (SVR) and the stacking-ensemble learning in the task of time series forecasting of the number of patients infected with COVID-19 with one, three, and six-days ahead. A forecasting model based on ARIMA has also been presented in [7] for Pakistan, presenting the high exponential growth in the number of confirmed cases, deaths and recoveries. In [8], ARIMA time series models were applied to forecast the total confirmed cases of COVID-19 for the next ten days using the model ARIMA (0,2,1), ARIMA (1,2,0) and ARIMA (0,2,1) for Italy, Spain, and France, respectively.

Currently, the analysis of the evolution of COVID-19 in America is of great importance due to the impact of this epidemic on this continent. In this contribution we will focus on the United States. The first patient detected in the United States was a travel-associated case from Washington state on 19 January 2020. The preponderance of initial cases of infected patients with COVID-19 in the United States were correlated with travel to a "high-risk" country or close contacts of previously identified cases corresponding to the testing criteria adopted by the Centers for Disease Control and Prevention (CDC) (https://www.cdc.gov/, accessed March 2020). From 1 to 31 March 2020, the number of reported COVID-19 cases in the United States rapidly increased from 30 to 188,172, the number of deaths rising from 1 to 5531, and the virus being detected in all the states. At the end of April, the number of infected reached 1,069,424 and the number of deceased stood at 62,996. At the time of writing this contribution (14 June 2020) the number of infected is more than 2×10^6 and more than 100.000 deaths, the United States being one of the countries that is suffering the most from COVID-19.

In a recent paper [9], an attempt was made to estimate the actual number of infected people, even if they have not been counted. It was estimated that the true number of COVID-19 cases in the United States is likely in the tens of thousands, suggesting substantial undetected infections and spread within the country [10].

This contribution presents a methodology to analyze the evolution patterns of COVID-19 in the states of United States (including Puerto Rico and the District of Columbia). A parametric similarity measure is presented, based on a robust distance measure between time series, the dynamic time warping distance (DTW), with which the number of infected and dead in each of the states can be compared simultaneously, even though the start of the epidemic originated on different dates in each zone (therefore, the time series that need to be compared have different lengths).

To the best of our knowledge, this contribution is the first study that tries to develop a hierarchical clustering time series algorithm in order to globally compare and classify the behavior of all the states of United State simultaneously in their evolution of infected and deceased patients suffering COVID-19. Carrying out this classification is very useful, since it will allow the establishment of similarities and patterns in the evolution of the pandemic among the states of the United States.

2. Material and Methods

A time series is a sequence of numerical (temporal) data points in successive order, which is naturally high dimensional and large in data size. There are two main operations that could be performed when working with time-series with its sequential data: (a) the analysis of a single time series; (b) the analysis of multiple time series simultaneously. This contribution is concentrated on the analysis of multiple time series for all the states of US suffering COVID-19, with the purpose of finding similarities between multiple time series by performing a clustering time-series methodology.

Clustering such complex objects is particularly advantageous because it may lead to the discovery of interesting patterns in time-series datasets, which contributes to a better understanding of the COVID-19 spread in different regions of the United States.

Clustering of time-series sequences has received noteworthy attention [11,12], not only as a formidable exploratory method and powerful tool for discovering patters, but also as a pre-processing step or subroutine for other tasks [13].

In this section, the database used is presented first (Section 2.1). Subsequently, a review of the most popular distance measures for time series is described (Section 2.2) and a new parametric distance is proposed.

2.1. Data Set

The COVID-19 epidemic data set used in this contribution was collected from the Johns Hopkins University [14]. In this platform, the number of confirmed, deaths and recovered cases until 14 June 2020 for different countries are presented. For the United States, two additional .csv files are provided, in with detail of administration and province/state is reported (including Puerto Rico and District of Columbia). In order to compare countries behavior, the time-series data are divided by state population.

2.2. Similarity/Distance Measure in Time Series

In a simplified way, the similarity of two simple time series with the same number of points (denoted by m), and defined by $X = \{x_1, x_2, \ldots .x_m\}$ and $Y = \{y_1, y_2, \ldots .y_m\}$, can be achieved by simply calculating the Minkowski (or Euclidean) distance (shortest path between two points) between points on both time series that happen at the same time. This distance is the measure of similarity, denoted as $d(X,Y)$, and it is a function that takes both times series (X,Y) as input and calculates their distance "d", defined as:

$$d(X,Y) = \left(\sum_{i=1}^{m} |x_i - y_i|^k\right)^{\frac{1}{k}} \quad (1)$$

when $k = 2$, the distance between two series is called Euclidean Distance. Using the Minkowski distance is a good metric to analyze the similarity of two time series, if these time series are synchronized (that is, all similar events in both time series occur at exactly the same time) and have the same length.

The evolution of time series in the different states of the United States present a different start date, both for the number of confirmed and death cases, and therefore its length is also different. Suppose as an analogy the time series of the sound of a mother's voice when she speaks slowly to her child. If the mother says the same phrase quickly, the child will most likely recognize that she is still his mother. However, if the Euclidean distance between both series were used as a metric, these two time series would have a very low similarity and would not be considered fundamentally equal. This would lead to the conclusion that the two voices did not come from the same person. To solve this problem, the dynamic time warping distance (DTW) method is frequently used in the bibliography [15].

DTW is a technique that can be considered as an extension of the Euclidean Distance between series [16], that calculates an optimal match between two given time series with certain restriction, performing non-linearly in the series (by stretching or shrinking along its time-axis). This distortion (denoted as warping) between two time series is used to find corresponding regions and determine the similarity between them.

The DTW of two series X and Y, defined as $X = \{x_1, x_2, \ldots .x_n\}$ and $Y = \{y_1, y_2, \ldots .y_m\}$ is computed in the following way. An n-by-m matrix D is computed with the $(i.\ j)$th element, defining the local distance of two elements by:

$$d(x_i, y_j) = (x_i - y_j)^2 \quad (2)$$

The point-to-point alignment between series X and Y can be represented by a time warping path W, defined as:

$$W = \begin{pmatrix} w_x(k) \\ w_y(k) \end{pmatrix}, \ k = 1, 2, \ldots p \quad (3)$$

where p is the length of the warping path W, and w_x (k) and w_y (k) represent the indexes in time series X and Y, respectively. The warping path $\begin{pmatrix} w_x(k) \\ w_y(k) \end{pmatrix}$ indicates that the w_x (k)th element in time series X maps to the w_y (k)th element in time series Y. There are some constraints and rules for the construction of the warping path:

- Every index from the first time series must be matched with one or more indices from the other time series (and vice versa)
- The first (the same for the last index) index from the first time series must be matched (not only this match) with the first (last) index from the other time series. That is, the warping path should start at $W(1) = (1,1)$ and end up at $W(p) = (n,m)$.
- The mapping of the indices from the first time series to indices from the other time serie must be monotonically increasing, and vice versa. The adjacent elements of path W, $W(k)$ and $W(k+1)$ must be subject to $w_x(k+1) - w_x(k) \geq 0$ and $w_y(k+1) - w_y(k) \geq 0$.
- The warping path should also have the property of continuity, mathematically expressed as adjacent elements of path W, $W(k)$ and $W(k+1)$ must be subject to $w_x(k+1) - w_x(k) \leq 1$ and $w_y(k+1) - w_y(k) \leq 1$.

The optimal match is denoted by the match that satisfies all the restrictions and the rules and that has the minimal cost, where the cost is computed as the sum of absolute differences, for each matched pair of indices, between their values. The DTW (minimal distance and optimal warping path) could be found using a dynamic programming algorithm:

$$RD(x_i, y_j) = d(x_i, y_j) + \min \begin{cases} RD(x_{i-1}, y_{j-1}) \\ RD(x_{i-1}, y_j) \\ RD(x_i, y_{j-1}) \end{cases} \quad (4)$$
$$DTW(X, Y) = \min\{RD(x_n, y_m)\}$$

where $RD(x_i, y_j)$ is the minimal cumulative distance from (0, 0) to (i, j) in matrix D. In the methodology proposed in this paper, for each of the states analyzed, both the time series of the number of infected and the time series of deaths will be simultaneously taken into account.

If each of these time series needs to be weighted differently, the following parametric metric, $DTW_\propto(S_A, S_B)$ is defined:

$$DTW_\propto(S_A, S_B) = \propto DTW(TSD_A, TSD_B) + (1 - \propto) DTW(TSC_A, TSC_B) \quad (5)$$

that measures the similarity in the evolution of the COVID time series for two states of the United States (S_A y S_B), TSC_A and TSC_B represent the time series of the number of infected, TSD_A and TSD_B represent the time series of the number of deaths for the states S_A and S_B, respectively. The parameter α (with $0 \leq \alpha \leq 1$) indicates the relative relevance given to the similarity measure, taking into account the time series of infected or deaths.

2.3. Clustering Method for Time Series

Clustering is a data mining technique in which similar data are divided into related or homogeneous groups, in an unsupervised way, that is, without a priori advanced knowledge of the data. For the problem presented in this contribution, working with time series of states of the United States suffering COVID-19, given a set of individual time series data, the objective is to group similar time series into the same cluster.

The problem of grouping time series data is formally defined by, given a dataset of N time series data $Q = \{X_1, X_2 \ldots X_N\}$, finding in an unsupervised way a partition of Q into K cluster, denoted as $C=\{C_1, C_2, \ldots C_k\}$, taking into account that homogeneous or similar series are grouped together based on a certain similarity/distance measure. In this paper, the parametric metric $DTW_\propto(S_A, S_B)$ is used and there is no intersection between clusters, therefore:

$$Q = \cup_{i=1}^{K} C_i; \text{with } C_i \cap C_j = \varnothing \ (i \neq j) \quad (6)$$

The methods used in the area of time series clustering [11,17] are usually based in a conventional clustering algorithm by substituting standard distance measurements with a more suitable distance to compare time series (raw methods) or converting series into normal data and using directly classical algorithms (feature-based methods and models).

Among the most popular clustering algorithms, the hierarchical clustering and the k-means algorithm are widely used in time series clustering. In this contribution the hierarchical clustering is used, mainly due to its great visualization power and its simple and intuitive interpretation.

Hierarchical clustering creates a nested hierarchy of similar time series, according to a pair-wise distance matrix of the time series analyzed. The similarity measure $DTW_\propto(S_A, S_B)$ used is therefore essential in this time-series clustering process.

The most widely used linkage criteria, such as single, average and complete linkage variants [18], were analyzed. Hierarchical clustering can be converted into a partitional clustering, with k cluster, by cutting the first k links.

3. Results and Discussion

To evaluate the performance of the proposed method, several experiments are conducted in this section for three values of the parameter α in the distance metric $DTW_{\alpha}(S_A, S_B)$. The time series of the states of the United States was taken from John Hopkins database. Data range from 22 January 2020 to 14 June 2020, covering all the states (including Puerto Rico and District of Columbia). For the computation of the distance metric, a threshold I_{min} was defined, defining the minimum number of infected people to start the time series, being for this study $I_{min} = 5$ (the number of confirmed was greater than 5). Therefore, the length of the time series is different for each state, being on average 101 days (Figure 1). The index of each of the states is presented in Table 1.

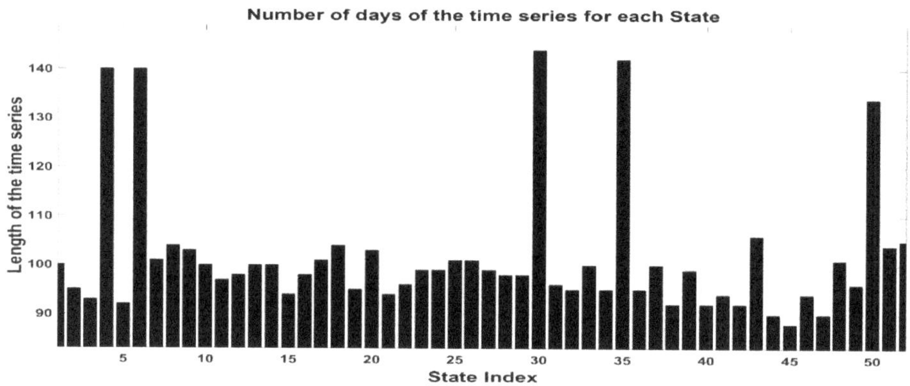

Figure 1. Length of the different time series, corresponding to the states analyzed.

Table 1. Distribution of the states obtained by means of hierarchical clustering with 9 clusters ($\alpha = 0.5$ and, in bold, the state for which the SIR model is calculated). *DCluster* is the distance between the elements that make up a cluster (its value is zero in the case that there is only one element in a cluster).

Cluster Number (C_N)	$D_{Cluster}$	States ($\alpha = 0.5$)
1	0.015	**(1) Nebraska,** (2) South Dakota
2	0.042	(3) Alabama, (4) Arizona, (5) Arkansas, **(6) California,** (7) Colorado, (8) Florida, (9)Georgia, (10) Indiana, (11) Iowa, (12) Kansas, (13) Kentucky, (14) Minnesota, (15) Mississippi, (16) Missouri, (17) Nevada, (18) New Hampshire, (19) New Mexico, (20) North Carolina, (21) North Dakota, (22) Ohio, (23) Oklahoma, (24) South Carolina, (25) Tennessee, (26) Texas, (27) Utah, (28) Vermont, (29) Virginia, (30) Washington, (31) Wisconsin
3	0.019	(32) Michigan, **(33) Pennsylvania**
4	0.031	(34) Delaware, **(35) Illinois,** (36) Louisiana, (37) Maryland
5	0.024	(38) Alaska, (39) Hawaii, (40) Idaho, (41) Maine, **(42) Montana,** (43) Oregon, (44) Puerto Rico, (45) West Virginia, (46) Wyoming
6	0	**(47) District of Columbia**
7	0	**(48) New Jersey**
8	0.033	(49) Connecticut, (50) Massachusetts, **(51) New York**
9	0	**(52) Rhode Island**

The values of the parameter α analyzed will be $\{0, 0.5, 1\}$. This section of results begins with the value of $\alpha = 0.5$, that is, the information of the confirmed patients time series with the same relevance as the time series of deaths for the final computation of the distance

$DTW_{\propto}(S_A, S_B)$. The distance matrix between the different states is presented in Figure 2 (for a better visual representation, the distance matrix was multiplied by a constant and the states are ordered according to the cluster to which each one belongs).

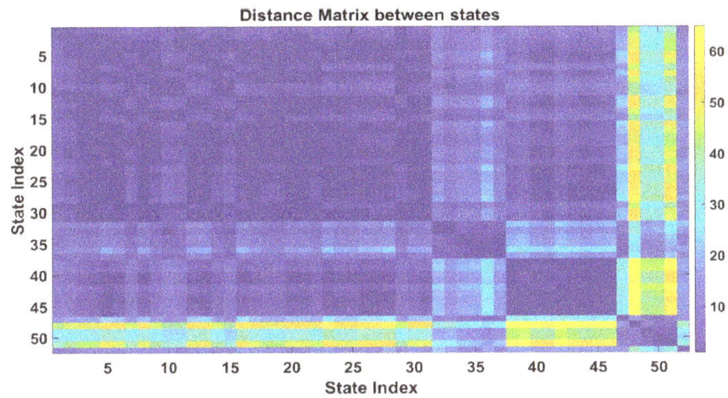

Figure 2. Distance or similarity symmetric matrix to characterize the behavior of the time series for the states of the United States (parameter α = 0.5). The greater the similarity, the smaller the distance between the series (being the diagonal of this matrix of zero value).

It is important to highlight the existence of various clusters with only one state (corresponding with District of Columbia, New Jersey and Rhode Island). Cluster 7 (New Jersey, listed as 48 in Table 1) links directly to cluster 8 ((49) Connecticut, (50) Massachusetts, (51) New York), which denote similar behavior between these states. For cluster 6 (District of Columbia, number 47), the linkage is performed for both cluster 3 ((32) Michigan, (33) Pennsylvania) and cluster 4 ((34) Delaware, (35) Illinois, (36) Louisiana, (37) Maryland). There are two large clusters (cluster 2 and cluster 5) that contain 29 and 9 states, respectively. Its linkage is performed through cluster 1, which contains only two states (Nebraska and South Dakota).

The similarities and distances between the different states and clusters obtained can be analyzed using the results presented in the hierarchical clustering (Figure 3) and distance matrix (Figure 2). Figure 4 presents a geographical representation of all the states according to the cluster grouping obtained in Table 1, using the similarity between the clusters obtained using the dendogram in Figure 3.

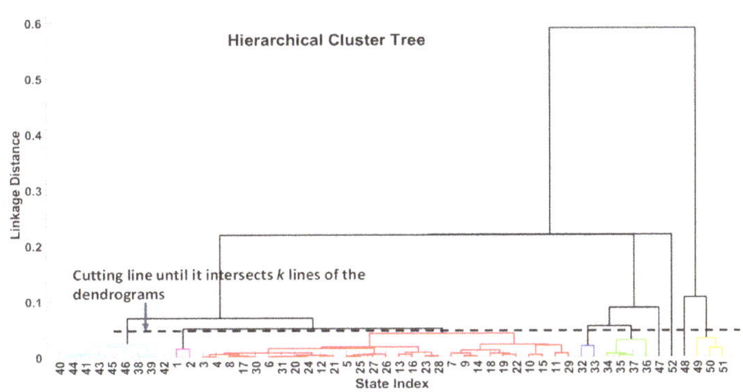

Figure 3. Hierarchical cluster tree obtained using as the distance metric the $DTW_{\propto}(S_A, S_B)$ and α = 0.5.

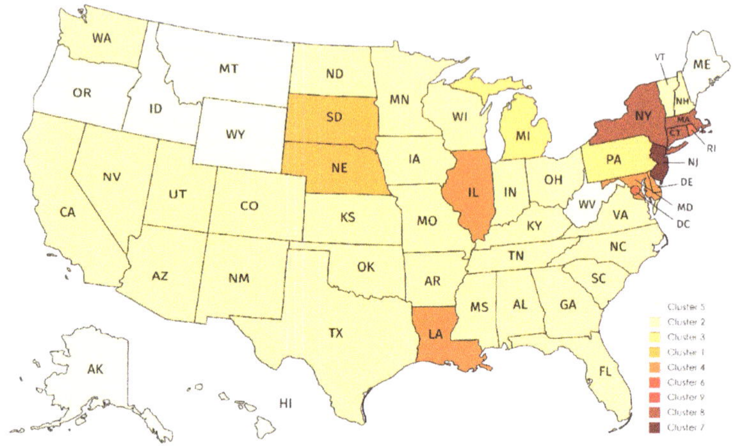

Figure 4. Geographical representation of all the states of the United States according to the cluster grouping obtained in Table 1 using the $DTW_{\propto}(S_A, S_B)$ and $\alpha = 0.5$.

The hierarchical clustering previously presented was obtained taking into account the time series of the number of confirmed and death cases simultaneously ($\alpha = 0.5$).

4. Conclusions

A powerful tool for the analysis of time series is the grouping through clustering. Clustering time series is usually an unsupervised process, with the aim of finding behavioral similarities between the different time series that are analyzed. This article proposed a parametric metric, based on the dynamic time warping distance, in order to measure the distance or similarity between time series corresponding to different states in the United States, taking into account the behavior of the number of COVID-19 confirmed cases and persons deceased due to COVID-19 simultaneously. The proposed parametric metric, named $DTW_{\propto}(S_A, S_B)$, is robust to the different lengths of data sequences (different beginning of the epidemic in the different states of the United States).

- Using the Calinski–Harabasz criterion, the optimal number of clusters in which the different states of United States can be grouped was obtained, taken as a value of $\alpha = 0.5$ (same relevance for the time series of confirmed and death patients). A total of nine heterogeneous clusters were found, in the sense that there are clusters within a large number of states (there are two large clusters, which encompass 29 and 9 countries) and other clusters with only one state (indicating that their behavior was unique, as they do not have excessive similarities with the rest of states).
- With the proposed hierarchical clustering procedure, it is possible to identify and summarize interesting patterns and correlations in the underlying data of the time series of the states of the United States suffering COVID-19 and therefore determine similar behaviors that different states may have.

Author Contributions: Conceptualization, I.R.-V.; O.V.; methodology, I.R.-V.; E.D.-M.; F.R.; software, I.R.-V.; E.; F.R.; validation, O.V.; E.D.-M.; formal analysis, I.R.-V.; O.V.; F.R.; investigation, I.R.-V.; F.R.; resources, I.R.-V.; O.V.; F.R.; data curation, I.R.-V.; O.V.; F.R.; writing—original draft preparation, I.R.-V.; O.V.; F.R.; writing—review and editing, I.R.-V.; O.V.; F.R.; visualization, I.R.-V.; supervision, I.R.-V.; project administration, I.R.-V.; O.V.; funding acquisition, I.R.-V.; O.V.; E.D.-M.; F.R. All authors have read and agreed to the published version of the manuscript.

Funding: This research was funded by projects with reference RTI2018-101674-B-I00 and CV20-64934.

Institutional Review Board Statement: Not applicable.

Informed Consent Statement: Not applicable.

Acknowledgments: This contribution has been partially supported by the projects with reference RTI2018-101674-B-I00 and CV20-64934.

Conflicts of Interest: The authors declare no conflict of interest.

References

1. Roques, L.; Klein, E.; Papaix, J.; Sar, A.; Soubeyrand, S. Using Early Data to Estimate the Actual Infection Fatality Ratio from COVID-19 in France. *Biology* **2020**, *9*, 97. [CrossRef] [PubMed]
2. Wu, K.; Darcet, D.; Wang, Q.; Sornette, D. Generalized logistic growth modeling of the COVID-19 outbreak in 29 provinces in China and in the rest of the world. *Nonlinear Dyn.* **2020**. [CrossRef] [PubMed]
3. Acuña-Zegarra, M.; Santana-Cibrian, M.; Velasco-Hernandez, J. Modeling behavioral change and COVID-19 containment in Mexico: A trade-off between lockdown and compliance. *Math. Biosci.* **2020**, *325*, 108370. [CrossRef] [PubMed]
4. Karako, K.; Song, P.; Chen, Y.; Tang, W. Analysis of COVID-19 infection spread in Japan based on stochastic transition model. *BioSci. Trends* **2020**, *14*, 134–138. [CrossRef] [PubMed]
5. Zhao, Z.; Li, X.; Liu, F.; Zhu, G.; Ma, C.; Wang, L. Prediction of the COVID-19 spread in African countries and implications for prevention and control: A case study in South Africa, Egypt, Algeria, Nigeria, Senegal and Kenya. *Sci. Total Environ.* **2020**, *729*, 138959. [CrossRef] [PubMed]
6. Ribeiro, M.H.D.M.; da Silva, R.G.; Mariani, V.C.; Coelho, L.d.S. Short-term forecasting COVID-19 cumulative confirmed cases: Perspectives for Brazil. *Chaos Solitons Fractals* **2020**, *135*, 109853. [CrossRef] [PubMed]
7. Yousaf, M.; Zahir, S.; Riaz, M.; Hussain, S.M.; Shah, K. Statistical analysis of forecasting COVID-19 for upcoming month in Pakistan. *Chaos Solitons Fractals* **2020**, *138*, 109926. [CrossRef] [PubMed]
8. Ceylan, Z. Estimation of COVID-19 prevalence in Italy, Spain, and France. *Sci. Total Environ.* **2020**, *729*, 138817. [CrossRef] [PubMed]
9. Perkins, A.; Cavany, S.M.; Moore, S.M.; Oidtman, R.J.; Lerch, A.; Poterek, M. Estimating unobserved SARS-CoV-2 infections in the United States. *Proc. Natl. Acad. Sci. USA* **2020**. [CrossRef] [PubMed]
10. Fauver, J.R.; Petrone, M.E.; Hodcroft, E.B.; Shioda, K.; Ehrlich, H.Y.; Watts, A.G.; Vogels, C.B.F.; Brito, A.F.; Alpert, T.; Grubaugh, N.D.; et al. Coast-to-Coast Spread of SARS-CoV-2 during the Early Epidemic in the United States. *Cell* **2020**, *181*, 990–996. [CrossRef] [PubMed]
11. Aghabozorgi, S.; Shirkhorshidi, A.; Teh Ying, W. Time-series clustering—A decade review. *Inf. Syst.* **2015**, *53*, 16–38. [CrossRef]
12. Johnpaul, C.I.; Prasad, M.V.N.K.; Nickolas, S.; Gangadharan, G.R. Trendlets: A novel probabilistic representational structures for clustering the time series data. *Expert Syst. Appl.* **2020**, *145*, 113119.
13. Taoying, L.; Xu, W.; Zhang, J. Time Series Clustering Model based on DTW for Classifying Car Parks. *Algorithms* **2020**, *13*, 57.
14. Dong, E.; Du, H.; Gardner, L. An interactive web-based dashboard to track COVID-19 in real time. *Lancet Infect. Dis.* **2020**, *3099*, 19–20. [CrossRef]
15. Keogh, E.; Ratanamahatana, C.A. Exact indexing of dynamic time warping. *Knowl. Inf. Syst.* **2005**, *7*, 358–386. [CrossRef]
16. Sakoe, H.; Chiba, S. Dynamic programming algorithm optimization for spoken word recognition. *IEEE Trans. Acoust. Speech Sign. Process.* **1978**, *26*, 43–49. [CrossRef]
17. Bandara, K.; Bergmeir, C.; Smyl, S. Forecasting across time series databases using recurrent neural networks on groups of similar series: A clustering approach. *Expert Syst. Appl.* **2020**, *140*, 112896. [CrossRef]
18. Kaufman, L.; Rousseeuw, P. *Finding Groups in Data: An Introduction to Cluster Analysis*; John Wiley & Sons: Hoboken, NJ, USA, 2009; Volume 344.

engineering proceedings

MDPI

Proceedings

The Conflicting Developments of RMB Internationalization: Contagion Effect and Dynamic Conditional Correlation †

Xiangqing Lu and **Roengchai Tansuchat** *

Center of Excellence in Econometric, Faculty of Economics, Chiang Mai University, Chiang Mai 50200, Thailand; xiangqing_lu@cmu.ac.th
* Correspondence: roengchaitan@gmail.com

† Presented at the 7th International Conference on Time Series and Forecasting, Gran Canaria, Spain, 19–21 July 2021.

Abstract: As the world's largest exporter and second-largest importer, China has made exchange rate stability a top priority for its economic growth. With development over decades, however, China now holds excess dollar reserves that have suffered a huge paper loss because of quantitative easing in the United States. In this reality, China has been provoked into speeding RMB internationalization as a strategy to reduce the cost and get rid of the excessive dependence on the US dollar. Thus, this study attempts to investigate the volatility contagion effect and dynamic conditional correlation among four assets, namely China's onshore exchange rate (CNY), China's offshore exchange rate (CNH), China's foreign exchange reserves (FER), and RMB internationalization level (RGI). Considering the huge changes before and after China's "8.11" exchange rate reform in 2015, we separate the period of study into two sub-periods. The Diagonal BEKK-GARCH model is employed for this analysis. The results exhibit large GARCH effects and relatively low ARCH effects among all periods and evidence that, before August 2015, there was a weak contagion effect among them. However, after September 2015, the model validates a strengthened volatility contagion within CNY and CNH, CNY and RGI, and CNH and RGI. However, the contagion effect is weakened between FER and CNY, FER and CNH, and FER and RGI.

Keywords: China; diagonal BEKK; exchange rate; foreign exchange reserve; RMB internationalization; volatility contagion effect

check for updates

Citation: Lu, X.; Tansuchat, R. The Conflicting Developments of RMB Internationalization: Contagion Effect and Dynamic Conditional Correlation. *Eng. Proc.* **2021**, *5*, 54. https://doi.org/10.3390/engproc2021005054

Published: 15 July 2021

1. Introduction

In the last two decades, China's gross domestic product (GDP) has been increasing year by year, from 1.34 trillion US dollars in 2001 to 15.42 trillion US dollars in 2020, and currently ranks as the second largest in the world by nominal GDP and the largest in the world by purchasing power parity. The main source of China's economic growth is the net trade of import and export. The ratio of China's trade to GDP reached a record high of 64.48% in 2006. Since then, although the share fluctuates, the ratio has remained above 35%. As the world's largest exporter and second-largest importer, China still needs to use the US dollar and other international currencies when it participates in international trade, overseas investment, or debt. The large fluctuation of the benchmark exchange rate based on the RMB against the US dollar will bring great risks to the domestic economy. Accordingly, China has made exchange rate policy a top priority for its economic development. From 2005, China has shifted gradually from a fixed exchange rate system to a managed floating exchange rate regime to regulate and control its exchange rate by the change of China's foreign exchange reserves until the exchange rate between the renminbi and the dollar reaches its target level. Although China's intervention in exchange rate fluctuations has long been a controversial topic, the Chinese government have been longing for a balance between the fixed exchange rate system and the floating exchange rate system.

However, stung by the financial crisis of 2008, the US Fed's open up to long years of quantitative easing, with the direct creation of base currency, has led to the huge expansion of the balance sheet and China has the largest dollar-denominated foreign exchange reserve assets in the world, which suffered huge paper losses and damage or loss of opportunity cost. To reduce the transaction cost and exchange rate risk, China proposed to promote the internationalization of the RMB in 2009 as a strategic measure to get rid of the excessive dependence on the US dollar by fundamentally reducing the dollar reserves relatively but holding more special drawing rights (SDRs) currencies and gold. Since then, RMB settlement in the international market has been increasing. For example, the RMB joined SDRs on 1 October 2016 to become one of the top five international currencies officially. On 26 March 2018, renminbi-denominated crude oil futures, namely SC1906, were listed, which means that renminbi is trying to peg to crude oil. Until 2020, China had signed currency swap agreements with 40 countries, including Russia, the European Union, the United Kingdom, Japan, Canada, Brazil, South Korea, Thailand, Australia, and so on.

There are many articles examining exchange rate volatility and its impact on trade flows (Jiang, 2014) [1] only analyzing causality within exchange rate and foreign exchange reserves (Mayuresh et al. 2013) [2], or studying RMB internationalization level and exchange rate fluctuation from the perspective of local currency as a trade settlement currency (Wenbing et al. 2014) [3] and reserve currency (Yanjing, 2012) [4]. McKinnon et al. (2014) [5] pointed out that only once the domestic financial system has been well-capitalized and competently regulated is it safe to open the economy to allow the exchange rate to float, and on that basis to internationalize the currency. However, they did not analyze the contradiction between foreign exchange control and internationalization from the direction of measurement in China. This study aims to fill this gap in the literature, to examine the dynamic correlation and volatility contagion effect of exchange rate volatility concerning currency internationalization under foreign exchange reserve interference in China. To accomplish our goal, four different variables of China's onshore market exchange rate CNY, China's offshore market rate CNH, renminbi globalization index RGI, and China's foreign exchange reserves are considered in this study. Studying the volatility of CNY and CNH markets separately will more directly reflect the contradictions and fragmentation of China's exchange rate system. Selecting the RMB globalization index (RGI) to represent the level of renminbi internationalization (Kelvin et al. 2012) [6] should be more comprehensive than using a single variable, e.g., overseas RMB deposit (Xueceng, 2015) [7] or major currencies accounting for the share of international bond issuance (Zhiwen et al., 2013) [8]. Moreover, the change of China's foreign exchange reserves (FER) is taken as a substitute variable of China's foreign exchange intervention (Qiumin, 2015) [9], which is very typical and representative. The contagion effect is a transmission of volatility from shocks arising in one country to other countries, but from the finance perspective, the contagion effect can explain how the shock of one asset can be transferred to other assets. Indeed, some studies use it to explain the market transmission effects, e.g., in major coal markets (M. Thenmozhi et al. 2020) [10]. In this study, we analyze the contagion effect of the above four assets.

From the methodological point of view, the contagion effect is tested by correlation. Dynamic correlation models include, e.g., the Constant Conditional Correlation GARCH, Dynamic Conditional Correlation GARCH, and Full BEKK-GARCH model. The diagonal BEKK-GARCH model is a restricted form of Full BEKK, but DBEKK is mathematically and statistically preferable to the fatally flawed full BEKK and the DCC model, and thus provides a suitable benchmark (McAleer, 2019) [11,12]. In short, the objective of this study is to investigate the volatility contagion effect of four variables in two periods and analyse the dynamic correlations by the Diagonal BEKK GARCH model.

The rest of the paper is structured as follow. Section 2 reviews the previous literature, and Section 3 describes the data and methodology. Section 4 presents the empirical results. Finally, the conclusions with a discussion of major findings are presented in Section 5.

2. Literature Review

There has been no consensus regarding whether there exists a strong correlation between China's exchange rate, China's foreign exchange reserves, and renminbi globalization. Marggie et al. (2014) [13] have investigated the volatility of CNY and CNH exchange rates by GARCH and EGARCH models. Their results showed that the exchange rate volatility is asymmetric. Zhilai (2019) [14] divided CNY and CNH data from 2012 to 2018 into three different periods to make the DCC-GARCH model and proved that the intensity of the volatility spillover effect was different at different time nodes. Qiumin (2015) [9] estimated the exchange rate volatility under the condition of foreign exchange intervention by the GARCH-VaR model. Their results show that the use of China's foreign exchange reserve intervention effectively reduces the exchange rate fluctuations in the period from 2008 to 2015. Since 2012, the International Monetary Institute of Renmin University of China has regularly released the RMB Internationalization Report every year. Yanjing (2012) [4] undertook an econometric analysis of the historical evolution of the composition of the current international reserve currency (1980–2008), and concludes that RMB has preliminary met the conditions for internationalization. Zhiwen et al. (2013) [8] documented the exchange rate volatility and local currency internationalization on the Australian Dollar, finding that the greater exchange rate volatility of the AUD against the USD has a significantly negative effect on AUD internationalization. This has an important reference value for RMB internationalization. Luyao (2018) [15] pointed out that the development level of China's financial market has a certain degree of influence on RMB internationalization, but it is not significant. Foreign exchange reserves are not only conducive to regulating China's balance of international payments, but also play an important role in stabilizing the RMB exchange rate. The volatility of the exchange rate has a significant negative effect on RMB internationalization. There is no agreement on the effect of the degree of currency internationalization on the size of a country's foreign exchange reserves. Some studies believe that the degree of currency internationalization has a relationship with the scale of foreign exchange reserves of a country, increasing at first and then decreasing (Zhu Guoping et al. 2014) [16], while others believe that there is a negative relationship between foreign exchange reserves and the degree of currency internationalization (Zhang et al., 2011) [17]. Lian et al. (2017) [18] believed that foreign exchange reserves have a phased impact on the internationalization of the local currency. In the initial stage, it promotes the internationalization of local currency, while excessive accumulation hinders it. Mengnan (2017) [19,20] founded that the internationalization degree of the Japanese yen is negatively correlated with the exchange rate level of the Japanese yen against the US dollar. The influence of the scale of Japan's foreign exchange reserves on the internationalization of the yen depends on the relative size of foreign exchange reserves. The restraining effect of the size of foreign exchange reserves on the internationalization of the yen is due to the "insufficient" rather than the "excessive" size of Japan's foreign exchange reserves.

3. Materials and Methods

3.1. Methods

To examine the magnitude of volatility contagion across the provided series, we employ a diagonally restricted BEKK-GARCH model. The estimation of the model involves the joint estimation of both its mean as well as variance equations. We specify the model as follows.

Given the mean equation:

$$R_t = \alpha + DR_{t-1} + \varepsilon_{t-1}. \quad (1)$$

Given the diagonal (or scalar) BEKK model, DBEKK-GARCH (1,1), namely:

$$Q_t = QQ' + A\varepsilon_{t-1}\varepsilon'_{t-1}A' + BQ_{t-1}B', \quad (2)$$

where A and B are diagonal (or scalar) matrices, $\Omega = QQ'$ is positive semidefinite, the H_t matrix can be represented as:

$$H_t = \begin{bmatrix} h_{11,t} & h_{12,t} & h_{13,t} & h_{14,t} \\ h_{12,t} & h_{22,t} & h_{23,t} & h_{24,t} \\ h_{13,t} & h_{23,t} & h_{33,t} & h_{34,t} \\ h_{14,t} & h_{23,t} & h_{33,t} & h_{44,t} \end{bmatrix} = \begin{bmatrix} \Omega_{11,t} & \Omega_{12,t} & \Omega_{13,t} & \Omega_{14,t} \\ \Omega_{12,t} & \Omega_{22,t} & \Omega_{23,t} & \Omega_{24,t} \\ \Omega_{13,t} & \Omega_{23,t} & \Omega_{33,t} & \Omega_{34,t} \\ \Omega_{14,t} & \Omega_{23,t} & \Omega_{33,t} & \Omega_{44,t} \end{bmatrix} + \begin{bmatrix} a_{11} & 0 & 0 & 0 \\ 0 & a_{22} & 0 & 0 \\ 0 & 0 & a_{33} & 0 \\ 0 & 0 & 0 & a_{44} \end{bmatrix}' \begin{bmatrix} \varepsilon_{1t-1} \\ \varepsilon_{2t-1} \\ \varepsilon_{3t-1} \\ \varepsilon_{4t-1} \end{bmatrix} \begin{bmatrix} \varepsilon_{1t-1} \\ \varepsilon_{2t-1} \\ \varepsilon_{3t-1} \\ \varepsilon_{4t-1} \end{bmatrix}' \begin{bmatrix} a_{11} & 0 & 0 & 0 \\ 0 & a_{22} & 0 & 0 \\ 0 & 0 & a_{33} & 0 \\ 0 & 0 & 0 & a_{44} \end{bmatrix} + \begin{bmatrix} b_{11} & 0 & 0 & 0 \\ 0 & b_{22} & 0 & 0 \\ 0 & 0 & b_{33} & 0 \\ 0 & 0 & 0 & b_{44} \end{bmatrix}' \begin{bmatrix} h_{11,t-1} & h_{12,t-1} & h_{13,t-1} & h_{14,t-1} \\ h_{12,t-1} & h_{22,t-1} & h_{23,t-1} & h_{24,t-1} \\ h_{13,t-1} & h_{23,t-1} & h_{33,t-1} & h_{34,t-1} \\ h_{14,t-1} & h_{23,t-1} & h_{33,t-1} & h_{44,t-1} \end{bmatrix} \begin{bmatrix} b_{11} & 0 & 0 & 0 \\ 0 & b_{22} & 0 & 0 \\ 0 & 0 & b_{33} & 0 \\ 0 & 0 & 0 & b_{44} \end{bmatrix} \tag{3}$$

Then, each conditional variance and covariance equation are given as:

$$h_{11,t}^{cny\_cny} = \Omega_{11} + a_{11}^2\varepsilon_{1t-1}^2 + b_{11}^2 h_{11,t-1} \tag{4}$$

$$h_{12,t}^{cny\_cnh} = \Omega_{12} + a_{11}a_{22}\varepsilon_{1t-1}\varepsilon_{2t-1} + b_{11}b_{22}h_{12,t-1} \tag{5}$$

$$h_{13,t}^{cny\_fer} = \Omega_{13} + a_{11}a_{33}\varepsilon_{1,t-1}\varepsilon_{3,t-1} + b_{11}b_{33}h_{13,t-1} \tag{6}$$

$$h_{14,t}^{cny\_rgi} = \Omega_{14} + a_{11}a_{44}\varepsilon_{1,t-1}\varepsilon_{4,t-1} + b_{11}b_{44}h_{14,t-1} \tag{7}$$

$$h_{22,t}^{cnh\_cnh} = \Omega_{22} + a_{22}^2\varepsilon_{2,t-1}^2 + b_{22}^2 h_{22,t-1} \tag{8}$$

$$h_{23,t}^{cnh\_fer} = \Omega_{23} + a_{22}a_{33}\varepsilon_{2t-1}\varepsilon_{3,t-1} + b_{22}b_{33}h_{23,t-1} \tag{9}$$

$$h_{24,t}^{cnh\_rgi} = \Omega_{24} + a_{22}a_{44}\varepsilon_{2t-1}\varepsilon_{4,t-1} + b_{22}b_{44}h_{24,t-1} \tag{10}$$

$$h_{33,t}^{fer\_fer} = \Omega_{33} + a_{33}^2\varepsilon_{3,t-1}^2 + b_{33}^2 h_{33,t-1} \tag{11}$$

$$h_{34,t}^{fer\_rgi} = \Omega_{34} + a_{33}a_{44}\varepsilon_{3t-1}\varepsilon_{4,t-1} + b_{33}b_{44}h_{34,t-1} \tag{12}$$

$$h_{44,t}^{rgi\_rgi} = \Omega_{44} + a_{44}^2\varepsilon_{4t-1}^2 + b_{44}^2 h_{44,t-1} \tag{13}$$

where $h_{12,t}^{cny\_cnh}$, $h_{13,t}^{cny\_fer}$, $h_{14,t}^{cny\_rgi}$, $h_{23,t}^{cnh\_fer}$, $h_{24,t}^{cnh\_rgi}$, and $h_{34,t}^{fer\_rgi}$ represent the conditional covariance between CNY return and CNH return, CNY return and FER return, CNY return and RGI return, CNH return and FER return, CNH return and RGI return, and RGI return and FER return, respectively.

3.2. Materials

The data as Table 1 employed in this study are the monthly data of exchange rate central parity rate on CNY and CNH markets, foreign reserve assets in China (FER), and RMB internationalization index (RGI). Considering the different development periods of China's onshore market (CNY) and offshore market (CNH), the data were selected from August 2010 to January 2021. The monthly returns are computed as the difference of the natural logarithm of all consecutive assets.

Table 1. Variables description.

Data	Description	Source
CNY	In the onshore mainland China market, the Chinese Yuan is called CNY, and the CNY central parity exchange rate is calculated from the average of the buying price plus selling price.	investing.com
CNH	The offshore market includes traditional Yuan centers, such as Hong Kong (a special administrative region of China), Singapore, London, and newly developed centers such as Luxembourg is called CNH. while the CNH central parity exchange rate is calculated from the average of the buying price plus selling price.	investing.com
RGI	Represent the RMB globalization index, which calculate from offshore RMB deposits, trade settle and other international payments, dim sum bonds and foreign exchange turnover.	Standard Chartered Research
FER	Represent China's foreign exchange reserves, the change of China's foreign exchange reserve is the substitution variable of foreign exchange intervention.	People's republic Bank of China

4. Empirical Results

4.1. Descriptive Statistics

In Table 2, the first R for returns, it can be seen that the standard deviation of the four variables indicates fluctuation degree of the variables is small, among which RGI has the largest fluctuation degree, followed by FER, CNH, and CNY. All skewness is not zero, which means all the rates of return are not symmetric. The skewness of CNY, CNH, and RGI is greater than 0, indicating that the sequence is right-skewed and has the characteristic of sharp peak and thick tail. The skewness of FER is less than 0, indicating that the sequence is left-skewed. All kurtosis is larger than 3, indicating that these sequences have excessive kurtosis, which means all assets return steeper than the normal distribution. Besides, all Jarque–Bera test values are large (12.0259, 10.8568, 7.6614, 120.4339) and significant at the 5% level, which means the returns of all series are not a normal distribution. Rather, they are peaked distributions with fat tails. The more obvious the thick tail is, the longer the state lasts, and the historical information is important for the future forecast. Figure 1 exhibits the monthly returns series corresponding to CNY, CNH, FER, and RGI and exposes the differences in their fluctuations.

Table 2. Descriptive statistics for returns in CNY, CNH, FER, and RGI.

	RCNY	RCNH	RFER	RRGI
Mean	−0.0003	−0.0003	0.0016	0.0248
Median	−0.0013	−0.0014	0.0022	0.0128
Maximum	0.0314	0.0300	0.0416	0.2231
Minimum	−0.0253	−0.0242	−0.0383	−0.1024
Std. Dev.	0.0088	0.0092	0.0130	0.0546
Skewness	0.5656	0.6234	−0.0169	1.6294
Kurtosis	4.0238	3.7396	4.2222	6.6424
J-B	12.0259	10.8568	7.6614	120.4339
Prob.	0.0024	0.0044	0.0217	0.0000

Notes: The Jarque-Bera test corresponds to the test statistic for the null hypothesis of normality in the distribution of sample returns.

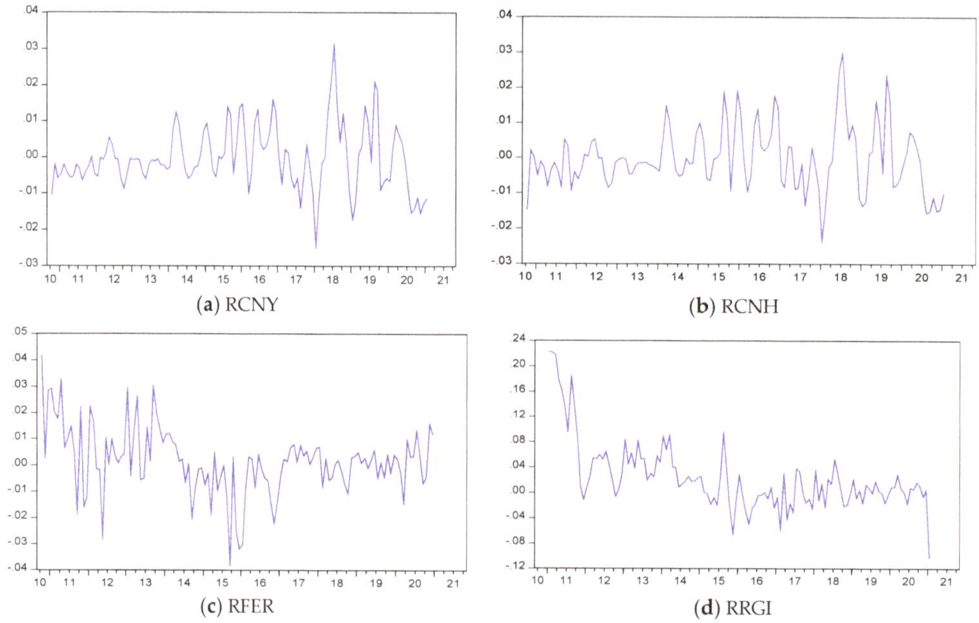

Figure 1. Monthly data of returns in CNY, CNH, FER, and RGI.

4.2. Unit Root Test

As the result of the ADF test has shown in Table 3, the large negative values in all cases. A *p*-value of all sequences less than 0.01 indicate rejection of the null hypothesis at the 1% level. Thus, the analysis shows that the rate of return sequences has no unit root and satisfies the stationary condition, suitable for further analysis.

Table 3. Unit Root Test for sample returns-ADF approach.

	T-Stat	1% Level	5% Level	10% Level	Prob.	Results
RCNH	−5.1187	−3.4856	−2.8857	−2.5797	0.0000	Stationary
RCNY	−5.0032	−3.4856	−2.8857	−2.5797	0.0000	Stationary
RRGI	−3.9669	−3.4870	−2.8863	−2.5800	0.0022	Stationary
RFER	−8.0091	−3.4847	−2.8852	−2.5795	0.0000	Stationary

4.3. Volatility Contagion Effect

As mentioned in the introduction part, the contagion effect is used to measure the transmission of volatility from shocks arising in one country to other countries, and from the finance perspective, can explain how the shock of one asset can be transferred to other assets. The contagion effect is tested by correlation. The diagonal BEKK model is mathematically and statistically preferable to the fatally flawed full BEKK and the DCC model, that's why we use it. Besides, we suspect that the volatility contagion effect of those variables may not be stable over the different periods and obtain underlying long memory and time-varying characteristics. Accordingly, we take China's exchange rate reform, namely "8.11" in 2015, as the boundary, dividing the analysis into two stages, the first period from September 2010 to August 2015, and the second period from September 2015 to January 2021.

The estimated diagonally restricted BEKK-GARCH results as shown in Table 4. We can find that, in the first period, λ values are found to be significant at the 1% level of significance for both CNY and RGI, while CNH is significant at the 1% level, which indicates that the mean value of returns for each of those three variables is influenced by their own returns effect. However, the value for FER is not significantly affected by its own early returns.

Table 4. Estimated Coefficients for Diagonal BEKK GARCH model.

	Period 1: September 2010 to August 2015			**Period 2: September 2015 to January 2021**		
Mean equation	Coeff.	Std. error	z-Stat	Coeff.	Std. error	z-Stat
α_{cny}	−0.0009 *	0.0005	−1.8123	−0.0013	0.0016	−0.8023
λ_{cny}	0.3088 ***	0.0977	3.1605	0.3675 ***	0.0979	3.7534
α_{cnh}	−0.0009 *	0.0005	−1.8581	−0.0014	0.0017	−0.7870
λ_{cnh}	0.3976 ***	0.1018	3.9070	0.3122 ***	0.1057	2.9537
α_{fer}	0.0039 *	0.0020	1.8959	0.0006	0.0015	0.4234
λ_{fer}	0.1624	0.1265	1.2839	0.3881 **	0.1543	2.5148
α_{rgi}	0.0116 **	0.0050	2.3058	0.0043 *	0.0024	1.8130
λ_{rgi}	0.6161 ***	0.0970	6.3521	0.1198	0.1473	0.8131
Variance equation	Coeff.	Std. error	z-Stat.	Coeff.	Std. error	z-Stat.
Ω_{11}	1.24×10^{-6}	8.50×10^{-7}	1.4560	2.94×10^{-5} *	1.60×10^{-5}	1.8317
Ω_{12}	1.62×10^{-6} **	6.30×10^{-7}	2.5646	3.89×10^{-5} **	1.76×10^{-5}	2.2110
Ω_{13}	-7.06×10^{-7}	1.23×10^{-6}	−0.5761	-6.71×10^{-6}	8.51×10^{-6}	−0.7888
Ω_{14}	1.94×10^{-6}	3.72×10^{-6}	0.5220	6.21×10^{-6}	8.95×10^{-6}	0.6938
Ω_{22}	1.99×10^{-6} ***	7.12×10^{-7}	2.8006	4.86×10^{-5} **	2.00×10^{-5}	2.4259
Ω_{23}	-2.78×10^{-6}	2.09×10^{-6}	−1.3290	-7.25×10^{-6}	1.09×10^{-5}	−0.6650
Ω_{24}	2.54×10^{-6}	5.12×10^{-6}	0.4961	8.45×10^{-6}	1.43×10^{-5}	0.5906
Ω_{33}	-1.89×10^{-5}	1.86×10^{-5}	−1.0127	1.40×10^{-5}	1.16×10^{-5}	1.2080
Ω_{34}	7.59×10^{-6}	1.74×10^{-5}	0.4349	-7.02×10^{-6}	5.04×10^{-6}	−1.3937
Ω_{44}	7.66×10^{-5}	0.0001	0.6865	-1.09×10^{-5} ***	2.34×10^{-6}	−4.6486
a_{11}	0.2643 ***	0.1020	2.5912	0.287159 **	0.1258	2.2819
a_{22}	0.4094 ***	0.1010	4.0517	0.4044 ***	0.1321	3.0606
a_{33}	−0.3772 **	0.1532	−2.4612	0.2649 *	0.1601	1.6548
a_{44}	0.0682	0.1770	0.3852	−0.1243	0.1493	−0.8326
b_{11}	0.9153 ***	0.0551	16.6129	0.7684 ***	0.1064	7.2244
b_{22}	0.8412 ***	0.0358	23.4845	0.6249 ***	0.1218	5.1322
b_{33}	0.9855 ***	0.0400	24.6565	0.7945 ***	0.1356	5.8599
b_{44}	0.9366 ***	0.0579	16.1700	0.9801 ***	0.0182	53.9888
LogL	790.1209			879.7569		
AIC	−27.74985			−27.5406		

Note: *** indicates reject the null hypothesis at the 1% significance level; ** indicates reject the null hypothesis at the 5% significance level; * indicates reject the null hypothesis at the 10% significance level.

In the second period, λ values of CNY and CNH were significant at 1% level, and the λ value of FER was significant at 5%. However, at this period, RGI is no longer significantly affected by its own early returns. The coefficients of the ARCH term and GARCH term in the conditional variance and covariance equation indicate a cross-volatility effect and own-volatility effect, respectively. The own-volatility effect is under the dominant influence of its own past shocks and volatility (one lag), and the cross-volatility effect under the influence of the news that came from other exogenous variables. The sum of the coefficients ARCH term and GARCH term in the below conditional variance-covariance equation is less than 1, which indicates the stability condition of the variance. At the same time, as the sum is close to 1, the process slightly oscillates around the mean value, and shows the effects of long memory in the four series. Besides, the conditional covariance model effectively captures the own volatility and cross volatility contagion effects among the four markets. The parameters b_{11}, b_{22}, b_{33}, b_{44} are statistically significant in Table 4, indicating that, at this stage, the offshore exchange rate market, onshore exchange rate market, China's foreign exchange reserve, and RMB internationalization index are significantly affected by their own volatility in the previous period. While the results stand as evidence for strong GARCH effects, we find the presence of ARCH effects to be relatively weak. All GARCH term coefficients in the above equations are greater than the ARCH term coefficients, which means that the volatility of four variables is mainly affected by their own volatility effects.

In the first period, the coefficient $a_{33} < a_{44} < a_{11} < a_{22}$ means that dealers in China's onshore exchange rate market, foreign exchange reserves market, as well as renminbi globalization market should pay attention to the news coming from the Chinese offshore exchange rate market. The coefficients $b_{22} < b_{11} < b_{44} < b_{33}$ mean that the shocks on the China's foreign exchange reserves market last longer than in the renminbi internationalization market, the onshore exchange rate market, and offshore exchange rate market in terms of future market volatility. In the second period, the parameters of b_{11}, b_{22}, b_{33}, b_{44} are statistically significant, indicating that, during this period, the offshore exchange rate market, onshore exchange rate, foreign exchange reserve, and RMB internationalization index are significantly affected by their own volatility effects in the previous period. The second period evidenced the presence of both ARCH and GARCH effects, but the GARCH term still higher than the ARCH term. The volatility sequences are mainly affected by its own previous shocks. $a_{44} < a_{33} < a_{11} < a_{22}$ noted that dealers in China's offshore exchange rate market must pay greater attention to the news coming from China's onshore exchange rate market and then the news from foreign exchange reserves markets. Meanwhile, the impact of news from the offshore market exchange rate to the RGI market is small in the current stage. $b_{22} < b_{11} < b_{33} < b_{44}$ means that the shocks on the renminbi internationalization market last longer than in China's foreign exchange reserves market, onshore exchange rate market, and offshore exchange rate market in terms of future market volatility.

Comparing the two periods, we can see that after 2015, the contagion effect of volatility among CNY and CNH markets has been significantly strengthened. As contrast, the pass-through effect of volatility among FER and RGI are reduced. Besides, the covariance parameters of CNY and FER as well as FER and CNH changed from negative to positive. Meanwhile, the covariance coefficients of CNY with RGI and CNH with RGI shifted from positive to negative. The absolute value shows that, before August 2015, there is a weak contagion effect among them. However, after September 2015, the model validates the presence of a strengthened volatility contagion within CNY and CNH, CNY and RGI, and CNH and RGI. However, the contagion effect is weakened between FER and CNY, FER and CNH, and FER and RGI. The difference in volatility contagion effect within two periods is shown in Figure 2. Moreover, we can observe the volatility dynamic correlation from Figure 3.

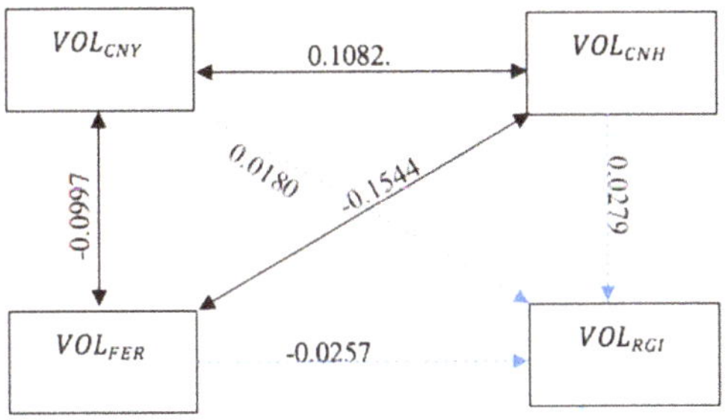

(**a**) the first period (2010m9–2015m8)

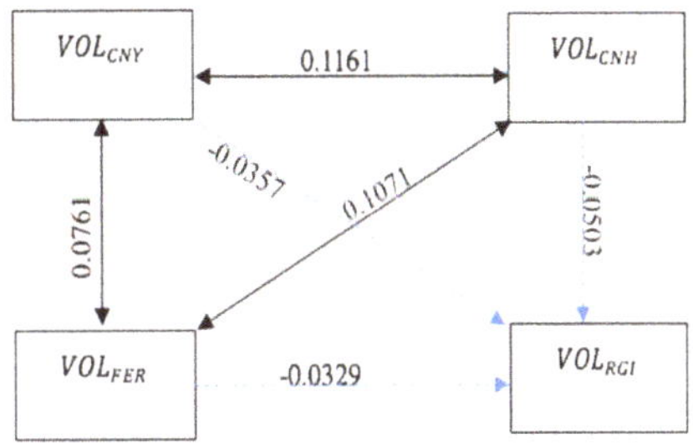

(**b**) the second period(2015m9–2021m1)

Figure 2. The volatility contagion effect in two stages.

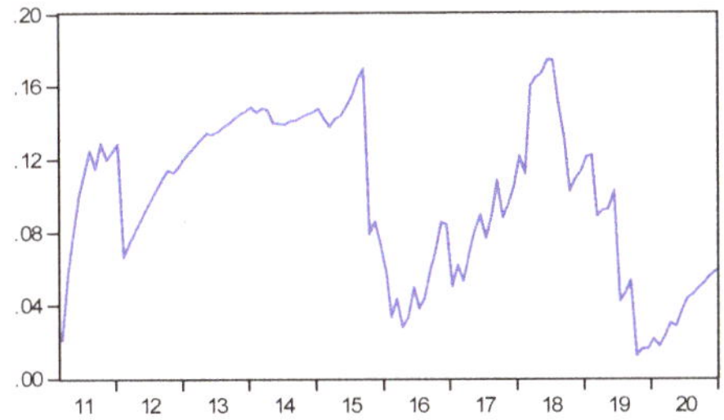

(**a**) Correlation for VOL_{CNY} and VOL_{CNH}

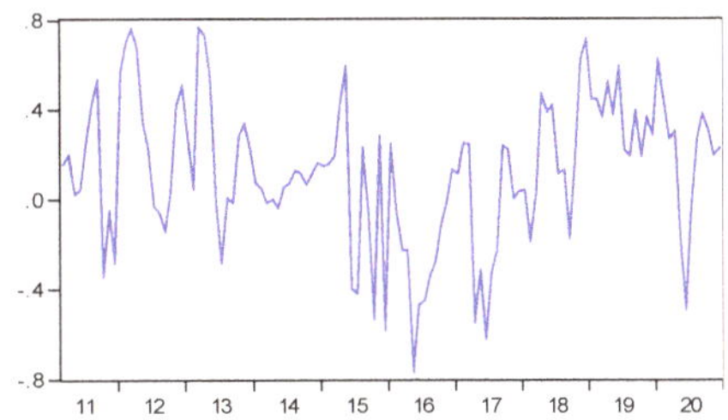

(**b**) Correlation for VOL_{FER} and VOL_{RGI}

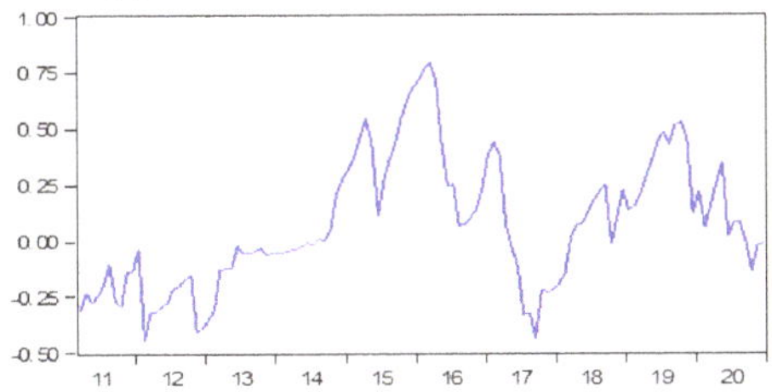

(**c**) Correlation for VOL_{CNY} and VOL_{FER}

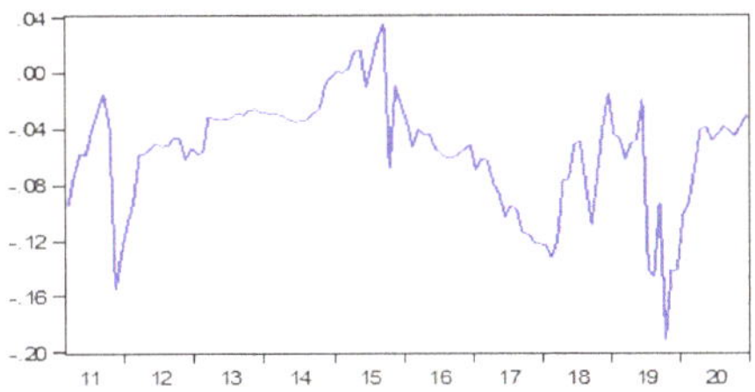

(**d**) Correlation for VOL_{CNH} and VOL_{FER}

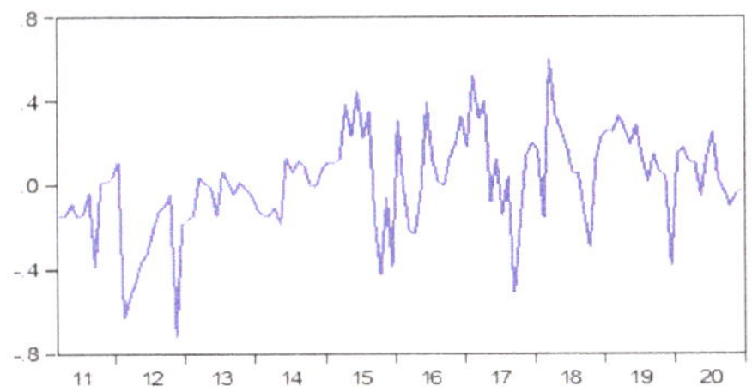

(**e**) Correlation for VOL_{CNY} and VOL_{RGI}

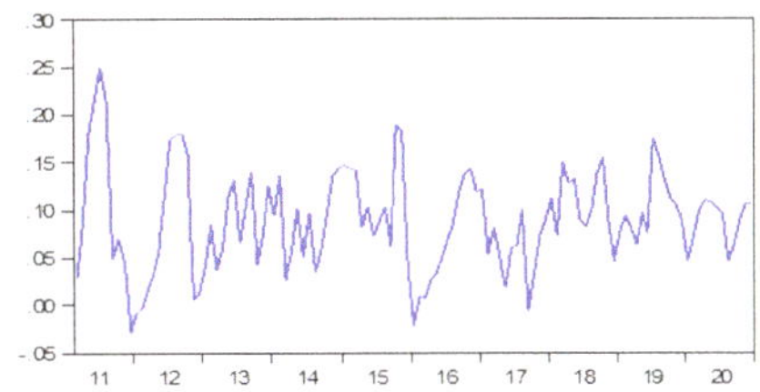

(**f**) Correlation for VOL_{CNH} and VOL_{RGI}

Figure 3. Dynamic conditional correlation for volatilities.

5. Conclusions

From the model to the actual situation, we can argue that, from 2010 to 2015, the renminbi was in a primitive stage of internationalization. Even though China adopted a controlled floating exchange rate system, during this period, China's foreign exchange reserve accumulation grew at a much faster rate than normal. Hence, during this period, RMB gradually appreciated, China's foreign exchange reserves increased rapidly, the government had to start the financial firewall, and the level of RMB globalization was stimulated to initiate development. After 2015, with the reform of China's exchange rate system, the Chinese government gradually opened the foreign exchange market and adopted a relatively flexible managed floating exchange rate system. The RMB was accelerating the internationalization process to an immature stage. However, China's excessive accumulation of foreign exchange reserves at this stage made the huge amount of foreign exchange outstanding lead to currency overissue, which brought about inflation risk, local currency devaluation expectation, and capital flight, which hindered the process of RMB internationalization to a certain extent. This is where the contradiction lies.

Firstly, from this study, there is a strengthened trend of positively bidirectional volatility contagion effect between the offshore and onshore markets of RMB. This is consistent with previous research. The currency volatility is quietly stable, which means that, although China is trying to find a balance between a fixed exchange rate system and floating exchange rate system, it is still a controlled exchange rate market, also leading to less liquidity in China's capital market. From this perspective, we suggest that the Chinese government further open its foreign exchange market and promote foreign exchange circulation. For international trade, enterprises should be able to take advantage of the currently favorable foreign exchange market environment.

Secondly, a large amount of foreign exchange reserves in China was indeed conducive to stabilizing the volatility of exchange rate of China's onshore and offshore markets before 2015. However, nowadays, the increasing of FER will rise the volatility and risk on the RMB exchange rate market and the internationalization level of RMB. We already see a weakened contagion effect trend in the model, but not enough. Thus, we recommend that China still needs to reduce the frequency of intervention in the foreign exchange market and further reduce the proportion of the US dollar to promote the diversification of reserve asset structures.

Thirdly, this study finds that China's foreign exchange reserves have a two-way contagion effect with both CNY and CNH markets, indicating that the management of foreign exchange reserves should be at the top of the list of issues that China needs to face in coming years.

From the research, we came to recognize that the internationalization level of RMB is still relatively low and cannot have a significant effect on China's exchange rate system and China's foreign exchange reserves. As a result, it is necessary to continue to encourage and promote the internationalization process of RMB for China. The contagion effect shows a way to strengthen CNY to RGI as well as CNH to RGI so we can stimulate the internationalization of RMB from these aspects, e.g., by promoting the development and growth of the CNH market, increasing the yuan settlement in trade, accelerating the rollout of digital payments and cryptocurrencies, and so on.

Institutional Review Board Statement: The study did not involve humans or animals.

Informed Consent Statement: The study did not involve humans.

Data Availability Statement: The data sources for this paper are publicly available, the data of CNY and CNH are from investing.com, and the data of RMB globalization index from Standard Chartered Research. Finally, the data of FER came from People's republic Bank of China.

References

1. Jiang, W. The Effect of RMB Exchange Rate Volatility on Import and Export Trade in China. *Int. J. Acad. Res. Bus. Soc. Sci.* **2014**, *4*, 615.
2. Gokhale, M.S.; Raju, J.R. Causality between Exchange Rate and Foreign Exchange Reserves in the Indian Context. *Glob. J. Manag. Bus. Res.* [S.l.]. July 2013. Available online: https://journalofbusiness.org/index.php/GJMBR/article/view/1021 (accessed on 14 July 2021).
3. Wenbing, S.; Hongzhong, L. RMB Internationalization, Exchange Rate Fluctuation and Exchange Rate Expectations. *Stud. Int. Financ.* **2014**, *8*, F832.6.
4. Yanjing, L. The Quantitative Analysis on the Evolution of the International Reserve Currency—Also on the Feasibility of the Internationalization of RMB. *Stud. Int. Financ.* **2012**, *4*.

5. McKinnon, R.; Schnabl, G. China's exchange rate and financial repression: The conflicted emergence of the RMB as an international currency. *China World Econ.* **2014**, *22*, 1–35. [CrossRef]
6. Lau, K.; Mann, D.; Maratheftis, M.; Weng, S. CNH—Introducing the Renminbi Globalisation Index. *Glob. Res. Res.* 2012. Available online: https://research.sc.com/reports/RGI_20121114.pdf (accessed on 14 July 2021).
7. Zha, X. Study on the Relationship between RMB Internationalization and Exchange Rate. *Commer. Econ. Res.* 2015. No. 17. Available online: http://www.cnki.com.cn/Article/CJFDTotal-SYJJ201517032.htm (accessed on 14 July 2021).
8. Zhang, Z.; Bai, Q. Exchange Rate Volatility and Local Currency Internationalization: An Empirical Study on the Australian Dollar (20 August 2012). *Stud. Int. Financ.* April 2013. Available online: https://ssrn.com/abstract=2147416 (accessed on 14 July 2021). (In Chinese)
9. Li, Q. Study on the Trend and Fluctuation of RMB Exchange Rate under the Management Floating Exchange Rate System. Ph.D. Thesis, University of Electronic Science and Technology of China, Beijing, China, 24 April 2015. Available online: http://cdmd.cnki.com.cn/Article/CDMD-10614-1016049477.htm (accessed on 14 July 2021).
10. Thenmozhi, M.; Maurya, S. Crude Oil Volatility Transmission across Food Commodity Markets: A Multivariate BEKK-GARCH Approach. *J. Emerg. Market Financ.* **2020**. [CrossRef]
11. McAleer, M. What They Did Not Tell You about Algebraic (Non-) Existence, Mathematical (IR-)Regularity and (Non-) Asymptotic Properties of the Full BEKK Dynamic Conditional Covariance Model. *J. Risk Financ. Manag.* **2019**, *12*, 61. [CrossRef]
12. McAleer, M. What They Did Not Tell You about Algebraic (Non-) Existence, Mathematical (IR-)Regularity, and (Non-) Asymptotic Properties of the Dynamic Conditional Correlation (DCC) Model. *J. Risk Financ. Manag.* **2019**, *12*, 66. [CrossRef]
13. Marggie, M.; Jiangze, D.; Kin, L. Modeling Volatility Exchange Rate of Chinese Yuan against US Dolla Based on GARCH Models. In Proceedings of the Sixth International Conference on Business Intelligence and Financial Engineering, Hangzhou, China, 14–16 November 2013; Available online: https://ieeexplore.ieee.org/abstract/document/696114 (accessed on 14 July 2021).
14. Liu, Z. Study on the Formation Mechanism of RMB Exchange Rate. Master's Thesis, Shanghai Academy of Social Science, Shanghai, China, 1 June 2019. Available online: https://cdmd.cnki.com.cn/Article/CDMD-87903-1019045437.htm (accessed on 14 July 2021).
15. Shao, L. The Influence of Exchange Rate Fluctuation on RMB Internationalization. Master's Thesis, Soochow University, Suzhou, China, May 2018. Available online: http://cdmd.cnki.com.cn/Article/CDMD-10285-1018102808.htm (accessed on 14 July 2021).
16. Zhu, G.; Liu, L.; Zhang, W. Optimal International Reserve Size in the Process of Currency Internationalization. *Int. Finance Res.* **2014**, *3*. Available online: http://www.cnki.com.cn/Article/CJFDTOTAL-GJJR201403003.htm (accessed on 14 July 2021).
17. Bai, Q.; Zhang, Z. The Size of Foreign Exchange Reserve and Local Currency Internationalization: An Empirical Study on the Japanese Yen (4 May 2013). *Econ. Res. J.* October 2011. Available online: https://ssrn.com/abstract=2260861 (accessed on 14 July 2021). (In Chinese)
18. Lian, P.; Ding, J.; E, Y. Internationalization of RMB and Management of Foreign Exchange Reserves—Based on Theoretical and Empirical Analysis. *Int. Financ.* 2017. Available online: http://www.cnki.com.cn/Article/CJFDTOTAL-JRGJ201706008.htm (accessed on 14 July 2021).
19. Zhu, M.; Yan, S. Foreign Exchange Reserve Size, Exchange Rate and Currency Internationalization: An Empirical Study Based on Yen. *J. Southwest Univ. Natl. (Humanit. Soc. Sci. Ed.)*. 2017. Available online: http://www.cnki.com.cn/Article/CJFDTOTAL-XNZS201703022.htm (accessed on 14 July 2021).
20. Zhu, M.; Cao, C. Currency Globalization, Financial Stability and Reserve Requirement. *Stat. Res.* **2019**, *36*. [CrossRef]

engineering proceedings

Proceedings

Prediction of Consumption and Income in National Accounts: Simulation-Based Forecast Model Selection †

Adusei Jumah [1] and Robert M. Kunst [2,3,*]

1 Department of Economics, Central University, Tema, Accra P.O. Box 2305, Ghana; aduseijumah@gmail.com
2 Institute for Advanced Studies, Josefstadter Strasse 39, 1080 Vienna, Austria
3 Department of Economics, University of Vienna, Oskar Morgenstern Platz, 1090 Vienna, Austria
* Correspondence: robert.kunst@univie.ac.at

† Presented at the 7th International Conference on Time Series and Forecasting, Gran Canaria, Spain, 19–21 July 2021.

Abstract: Simulation-based forecast model selection considers two candidate forecast model classes, simulates from both models fitted to data, applies both forecast models to simulated structures, and evaluates the relative benefit of each candidate prediction tool. This approach, for example, determines a sample size beyond which a candidate predicts best. In an application, aggregate household consumption and disposable income provide an example for error correction. With panel data for European countries, we explore whether and to what degree the cointegration properties benefit forecasting. It evolves that statistical evidence on cointegration is not equivalent to better forecasting properties by the implied cointegrating structure.

Keywords: forecasting; household saving rate; parametric bootstrap

Citation: Jumah, A.; Kunst, R.M. Prediction of Consumption and Income in National Accounts: Simulation-Based Forecast Model Selection. *Eng. Proc.* **2021**, *5*, 55. https://doi.org/10.3390/engproc2021005055

Published: 16 July 2021

1. Introduction

The joint movement of aggregate household consumption and income caught the attention of researchers even in the early days of the research on cointegration in macroeconomic data (see [1]). There are two variants of the long-run relationship of consumption and income. The first variant focuses on the share of consumer spending in the domestic product, a ratio that well exceeds 50% in most developed economies and often yields the impression of being fairly stable in a longer perspective. It is this variant that was studied, among others, by [2,3]. The other variant focuses on household disposable income rather than overall output and on the concept of a stable household saving rate. Representatives of this latter strand are [4] or [5]. It is this latter concept that corresponds to the historical 'great ratios of economics' ([6]), and it is also the focus here.

In the project presented here, we investigate the issue of whether assuming and estimating such a cointegrating relationship benefits forecasting of the two variables, in cases where it really exists and in cases where it remains fictitious. We study forecasting in a finite sample rather than in an asymptotic framework. This implies that 'anything can happen' in the sense that incorrect models may 'defeat' correct ones, whereas in asymptotic comparisons, the generating model always outperforms simplified rival models. Our tool in pursuing the issue is a novel simulation-based technique. For a detailed discussion of the method, we refer to Section 2. Cross-checking prediction models by simulation has been documented in [7] and was used in [8].

We interpret the question as to whether the error-correction mechanism of consumption and household income helps in forecasting consumption and income in an *empirical* or finite-sample Granger causality sense. The original Granger causality definition [9] is an asymptotic concept: a time-series variable causes another variable if and only if it improves the prediction of the effect variable, assuming known coefficient parameters. This is not the empirically relevant question for a forecaster. With empirical Granger causality, a variable causes another one only if it improves forecasting at a specific sample size, assuming estimated coefficients. Whereas empirical Granger causality implies Granger causality, the reverse does not necessarily hold, as our experiments confirm. The concept is related to the conditional predictive ability of [10].

The structure of this article is as follows. Following this introductory section, Section 2 discusses the utilized econometric methodology utilized. Section 3 describes the data. The main experiments are reported in Section 4. Section 5 concludes.

2. Methodology

Our forecasting experiments are based on the concept of simulation-based forecast model selection. This is a computer-intensive method that has been documented in [7] and applied by [8]. In fact, usage of the method may be more widespread and, for example, Ref. [11] uses an identical concept without explicitly naming it. The first subsection motivates and describes the method, the second subsection reports a small Monte Carlo experiment.

2.1. Simulation-Based Forecast Model Selection

The traditional view on model selection is inspired by statistical hypothesis testing. Researchers may consider nested sequences of models and evaluate restriction tests, such as the simple F- and *t*-tests of the Wald type. The more complex model is chosen if the tests reject, otherwise the simpler model is maintained. In many situations, this approach is justified by the fact that there is no clear monetary loss that is suffered if the decision turns out to be non-optimal. A decision for a model is regarded as correct when the generating model class is selected, and it is a requirement that incorrect decisions disappear as the sample size increases.

If the aim of the model selection exercise is specified as prediction, it is difficult to maintain this statistical paradigm. A simple model may be preferred to a complex model when it forecasts better, and this decision may depend on the sample size. Larger samples admit precise estimation of more parameters such that even a small advantage for a complex model may be worth the additional sophistication. A decision for a model is regarded as optimal when it minimizes prediction error whether the selected model is correct or not, and the decision may take the sample size into account such that a decision may be good for small samples and bad for larger samples.

An important difference between the tasks of forecasting and of approximating a true structure is that the former decision problem is symmetric, whereas the hypothesis testing framework is asymmetric. Statistical textbooks often explain hypothesis tests using the metaphor of an accused person in a trial. The accused is regarded as innocent until the evidence is so strong that he or she can be regarded as guilty 'beyond all reasonable doubt'. In practice, this means that a risk of 5% can be accepted for the probability that a convicted person is really innocent. One may take the metaphor further and demand for a risk of 1% if the amount of evidence used before court increases. Some recommend that the null should become harder to reject as the sample size increases.

Forecasting does not need this asymmetry. A small set of potential prediction models is at hand, and the forecaster chooses among less and more sophisticated variants. If the decision is based on an out-of sample forecast evaluation for realized data, concern for simplicity is no longer required. Models are cheap, and the model that comes closest to the realization can be selected even if it is very complex. The winner model, however, is typically not too complex as, otherwise, its performance would be hampered by the sampling variation in parameter estimation.

Thus, the following strategy appears to be informative when the purpose of a model is prediction. All rival models are estimated, i.e., the closest fit to the data at hand within the specified model class is determined, and then all estimated structures are simulated. These simulated pseudo-data are again predicted by all rival models, freshly estimated from the simulated data. For example, the qualitative outcome of this experiment may be as follows:

1. Model A predicts data generated by model A well;
2. Model A predicts data generated by model B satisfactorily and only slightly less precisely than model B;
3. Model B predicts data generated by model B well;
4. Model B predicts data generated by model A poorly.

Given this general impression, a forecaster may prefer model A as a prediction tool rather than model B, unless support for model B by the data is truly convincing. We note that models do not always perform well in forecasting their own data. For example, models containing parameters that are small and estimable only with large standard errors are often dominated by forecast models that set the critical parameters at zero. Ref. [12] also suggested quantitative measures for evaluating the four experiments summarily, but within the limits of this article we will stay with qualitative evaluations, particularly processing the reaction to changing sample sizes. For example, model A may forecast B data well up to 300 observations, when model B would clearly dominate. In this case, the preference may depend on the time range for future applications. If the researcher intends to base such forecasts on 500 observations, model B becomes competitive.

It is worthwhile contrasting the method with alternative concepts that have been suggested in the literature. For example, ref. [13] investigate a related problem, a decision between multivariate and univariate prediction models. They introduce the comparative population-based measure $P_{M|U}(h)$, which is approximated by sample counterparts $\hat{P}_{M|U}(h)$ and $\hat{F}_{M|U}(h)$. The large-sample measure

$$P_{M|U}(h) = 1 - \frac{\sigma_M^2(h)}{\sigma_U^2(h)}$$

depends on the ratio of prediction error variances corresponding to the true best multivariate and the true best univariate prediction model if these are forecasting at a horizon of h steps. By definition, the multivariate model with known coefficients must always outperform the univariate rival, and $P_{M|U}$ is restricted to the interval $[0,1]$. If the prediction error variances are estimated from data in finite samples, the estimate $\hat{P}_{M|U}(h)$ will inherit these properties. Ref. [13] show that, under plausible conditions, the estimate converges to the true value as the sample size grows. Ref. [13] concede, however, that in empirical applications, the multivariate forecast can be genuinely worse than the univariate rival, so they suggest adjusting the ratio for degrees of freedom, following the role model of information criteria. In particular, they consider the final prediction error (FPE) criterion due to [14], which multiplies the empirical prediction error variances by the correction factor $1 + k/N$, with k standing for the number of estimated parameters and N for the sample size. The resulting complexity-adjusted measure can be negative when the multivariate model uses many parameters without delivering better predictive performance.

We see the main differences between this approach and ours in the fact that [13] do not explicitly forecast the data by the two rival models. They basically use the one sample at hand and fit univariate and multivariate models to it. We proceed one step further and simulate the data under the tentative assumption that the forecast models are data generators. This permits explicitly evaluating the reaction of forecast precision to changing sample sizes. On the other hand, we will focus exclusively on one-step forecasting in the following. There is no impediment in principle, however, to generalizing our simulations to multi-step predictions, and we intend to pursue this track in future work.

2.2. A General Simulation Experiment

In order to find out a bit more about the strengths and weaknesses of our suggested procedure, we ran some prediction experiments based on simulated data. Because of the hierarchy of steps, such simulation experiments are time-consuming, so the number of replications remains limited. We simulate time-series data from basic time-series models, such as ARMA(1,1), and consider prediction based on ARMA(1,1) with coefficients estimated from the observations and also the simpler AR(1) model. The AR(1) model omits the MA(1) term of the generating model, but it may be competitive for small samples and for small MA coefficients.

We use a grid for the ARMA(1,1) model

$$X_t = \phi_0 + \phi X_{t-1} + \varepsilon_t - \theta\varepsilon_{t-1}$$

over $\phi = -0.75, -0.5, ..., 0.5, 0.75$ and $\theta = -0.75, -0.5, ..., 0.5, 0.75$. The intercept ϕ_0 is always kept at zero, but all estimated models include an intercept term. We consider two distributions for the *iid* errors ε, a standard $N(0,1)$ and a Cauchy distribution. The simulation-based strategy uses two variants. In the first variant, both the ARMA(1,1) and an AR(1) model are estimated from the data, and both estimated structures are simulated and again predicted based on both models. This delivers four sub-experiments and, finally, the model—either AR(1) or ARMA(1,1)—is selected as a prediction model that *more often* defeats its 'rival' model. In the other variant, the mean squared forecast errors (MSFE) that evolve from both models are added up, and the model with the smaller average MSFE is selected. Forecasts from the thus selected models are then compared with the choice based on a classical AIC, an extremely competitive benchmark that is hard to beat. The experiments are summarized in Figure 1, which shows the optimum strategy for each combination of AR and MA coefficients, with an optimum defined as that strategy that ultimately yields the minimum (absolute) forecast error for the out-of-sample observation at position $t = N + 1$. By construction, the diagonal connecting the southwest and the northeast corners represents white noise, as AR and MA terms cancel in $X_t = \phi_0 + \phi X_{t-1} + \varepsilon_t - \phi\varepsilon_{t-1}$. The heterogeneity visible in the graphs reflects sampling variation.

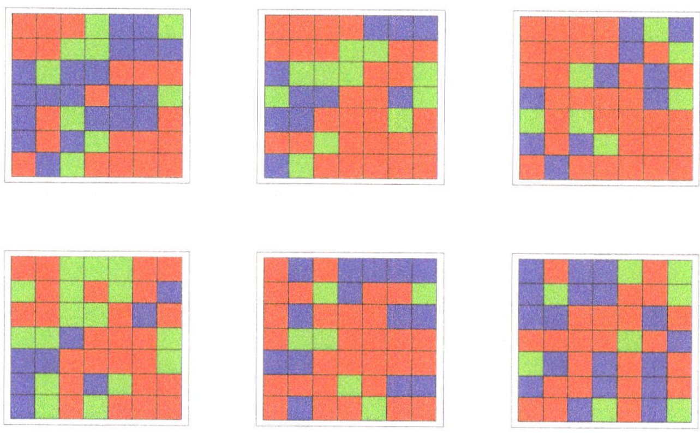

Figure 1. Best selection strategy for forecasting simulated ARMA(1,1) data. ARMA(1,1) models are generated with $T = 25, 50, 100$. Upper row $N(0,1)$ errors, lower row Cauchy errors. $\phi = -0.75, -0.5, \ldots, 0.75$ on the x-axis and $\theta = -0.75, -0.5, \ldots, 0.75$ on the y-axis. Red denotes AIC, green MSE simulator, blue counting simulator.

For Gaussian errors, Figure 1 shows a preponderance of AIC-supporting cases for $T = 50$ and $T = 100$, whereas the contest is more open for $T = 25$. Dominance is less explicit for Cauchy errors. From the two different evaluation methods for the simulation method, counting cases is preferable for most cases, so it may be interesting to reduce the rival strategies to two, the AIC and the simulation method with evaluating case counts. This design results in a quite similar figure, with almost all green dots turning blue. In summary, AIC appears invincible for large samples and Gaussian errors, whereas the simulator deserves consideration for small samples. This simulation experiment may be relevant for the empirical example to be studied in the next section, as the macroeconomic time-series sample remains in the lower region of the Monte Carlo evaluated in this section.

3. Data and Their Time-Series Properties

3.1. The Data

A full intrinsically homogeneous data set for household consumption and corresponding disposable income is not available from the Eurostat database, at least not for the majority of member countries. For this reason, we took the available information and constructed consumption and income series based on these support series. Even this reconstruction, however, was only feasible for a subset of countries, at least for the targeted full time range of 1995 to 2019 (annual data): Austria, Belgium, Cyprus, Czechia, Denmark, Estonia, Finland, France, Germany, Hungary, Ireland, Italy, Latvia, Lithuania, Netherlands, Poland, Slovakia, Slovenia, Spain, Sweden, and the United Kingdom. These are 21 countries that, for a considerable part of the time range, have been members of the European Union: some joined a bit later, whereas the United Kingdom left the EU recently. A visual summary of the consumption and income series is provided in Figure 2, where countries have been sorted in the EU tradition according to the beginning letters in the local languages. The full numerical data are available on request. Figure 2 shows that saving rates were almost always positive, which implies that households have been under some pressure during episodes of fiscal austerity but that aggregates were not often forced to dissave on a large scale. Of course, single households have been and are confronted with quite different situations. An exception to the rule are economies that were confronted with a fierce transition from a socialist to a market economy, such as the Baltic countries.

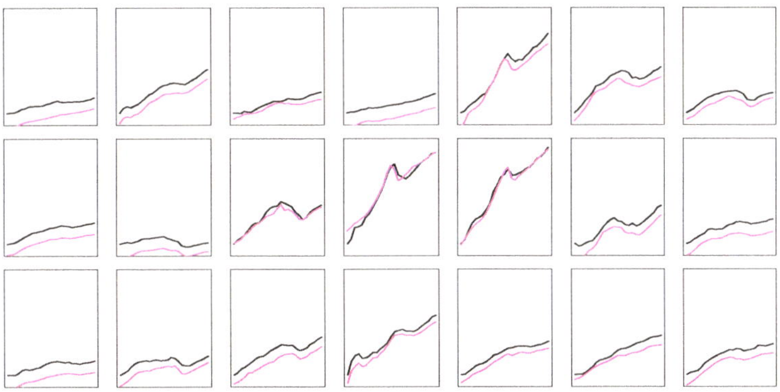

Figure 2. Eurostat data for household consumption (pink) and aggregate disposable income (black) for the following countries: Belgium, Czechia, Denmark, Germany, Estonia, Ireland, Spain, France, Italy, Cyprus, Latvia, Lithuania, Hungary, Netherlands, Austria, Poland, Slovenia, Slovakia, Finland, Sweden, United Kingdom.

A similar heterogeneity can be seen regarding real income growth. The Eastern European economics have started from a much lower level and had to grow faster in order to catch up with Western Europe. It is known that this catching-up process was successful, and the Eastern spearheads such as Czechia have already overtaken the Western laggards such as Portugal. The distance between the two variables appears to be reasonably stable, and eyeballing would support cointegration.

Other disciplines may be surprised at the comparatively short time series that are routinely used by macroeconomists for their forecasts. Because of the deep transformation processes toward the end of the 20th century, longer series are not available for large parts of Europe. Forecasts based on such data sets represent an interesting challenge.

3.2. Time-Series Properties

The series at hand, 21 consumption and 21 income series, were subjected to unit-root tests with the alternative of stationarity. The series are rather short, and the unit-root tests have low power, so results are only summarily reported here. If one augmentation term is added to the basic Dickey–Fuller regression, which is the standard option in R, and a linear time trend is used in the regression, the tests fail to reject for any of the income series and reject only three times for the consumption series. Rejections are recorded for Belgium, France, and the Netherlands, i.e., three countries with some similarities and strong interaction. The time-series graphs, however, do not look very different from other countries, and the rejections may be caused by the smoothness of the curves that permits linear trends to approximate them particularly well, such that the remainder looks stationary. On the whole, it appears reasonable to proceed under the assumption of 42 first-order integrated (I(1)) variables.

If variables are I(1), they may be cointegrated. In modeling consumption and income, the research concentrates on the potential stationarity of the difference in logs between the variables, not on a freely estimated cointegrating vector. The economic motivation is that, for C and Y representing log consumption and income, respectively, any stationary variable of the form $C - Y^{\lambda}$ with $\lambda \neq 1$ would imply an unsustainable long-run relationship, with either consumption systematically exceeding income or a saving rate converging to zero. Figure 3 shows the calculated differences in a single plot. This graph intends to convey a general tendency, with country indicators—for example, the very high value for 1995 belongs to Latvia—deliberately suppressed. The summary visual evidence supports stationarity, but statistical Dickey–Fuller tests are less clear, rejecting the unit-root null only for five cases: Italy, Spain, and three more rather unrelated countries.

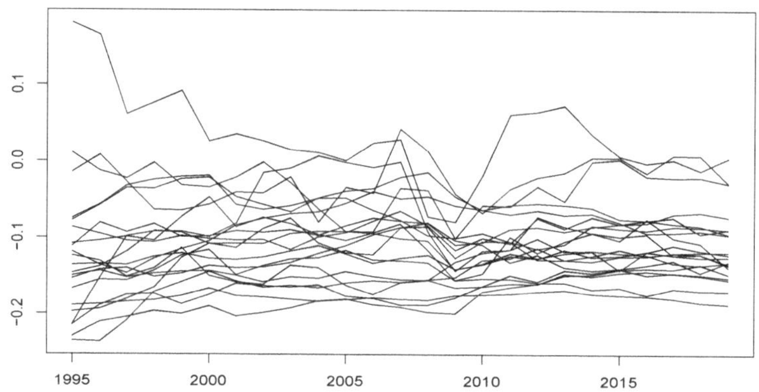

Figure 3. Log consumption minus log income for 21 countries.

Figure 3 also gives an impression of convergence, as the cross-sample volatility appears to be stronger in the 1990s than toward the end of the sample. Such convergence would suggest assigning stronger weights to the late part of the sample or including a nonlinear, converging trend line. Within the limits of this paper, such extensions are ruled out and will be reserved for future work.

4. The Prediction Experiments

Assume a toolbox consisting of two models, a bivariate autoregression with cointegration and a bivariate autoregression in differences without cointegration. Furthermore, we are interested in whether the lag order of two that is recommended for most countries by most criteria is better than a lag order of three that is recommended only for relatively few countries. The two issues can be combined in the sense that cointegration can be studied with more or less lags in differences. Instinctively, we may conjecture that low-order models in differences yield poorer forecasts, as they process less information, and that they only forecast on par with their rivals when the restrictions are all valid, i.e., the parameters of concern are all zero.

It turns out, however, that this assumption is not generally confirmed in simulation experiments. Even if a structure with invalid zero restrictions has generated the data, it is quite often the case that simpler models outperform the 'true' models due to the fact that the entertained true models are not true literally, but they contain coefficient parameters that must be estimated from the data and necessarily involve some sampling variation. This effect has been well documented repeatedly in the forecasting literature (see, e.g., KOLASSA, 2016, for the case of seasonality), although it still puzzles many researchers.

4.1. Models with and without Error Correction

This subsection reports our central experiment. Bivariate models with and without cointegration are fitted to the data. From the estimated parametric structures, pseudo-samples are generated. To these pseudo-data, again, both types of models, with and without error correction, are fitted. Finally, the performance of all four designs is comparatively evaluated.

We note that the seminal contribution by [15] did not consider a comparison between a cointegrating and a pure-difference model but, rather, between a cointegrating and a level autoregression, such that cointegration restricts the model. In practice, this comparison is less natural, as univariate time-series properties are easier to establish. For example, most researchers would agree that consumption and income series across Europe are better represented by first-order integrated than by stationary models. By contrast, whether and to what degree error correction affects collective behavior is less easy to establish statistically or to agree upon. We surmise that the choice taken by [15] was based on the correct observation that a model in differences is mis-specified if cointegration is present.

Figure 4 provides a graphical representation of the results from this experiment. The abscissa axis represents the available sample size, starting at the left with $T = 20$, which roughly corresponds to the actual data from Eurostat. By contrast, the right end $T = 90$ corresponds to the hypothetical situation with 90 years of available data whose dynamics are 'similar' to the observed Eurostat data. Indeed, the sample size of the pseudo-data proves to be a crucial determinant of relative performance. With $T = 20$ and even $T = 30$, the model without error correction dominates even when the generator model is cointegrating. This implies that inserting an error-correction term does not help in forecasting from small samples that are common in applications. For example, the European Monetary Union has a history of barely 20 years. For longer available data, cointegration helps if it is really present. Only for sample sizes beyond $T = 60$ would it make sense to estimate an error-correction term even if there is a chance that it is spurious. It is a bit surprising that this second boundary appears to be at a slightly larger sample size for consumption, although traded wisdom has it that adjustment to the error correction is primarily performed by the consumption variable (the slave) rather than by the income variable (the master).

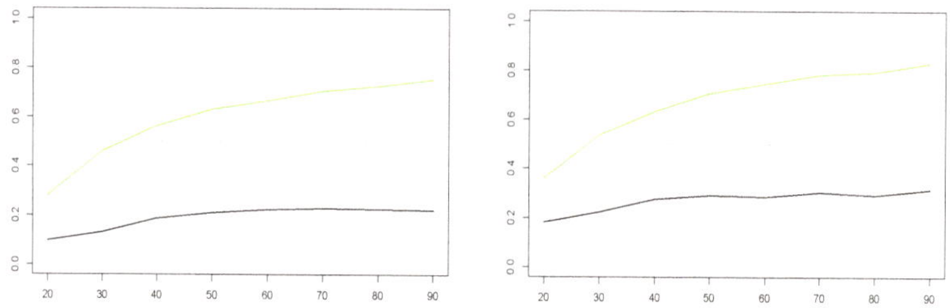

Figure 4. Percentage of error-correction models yielding the better forecast for consumption (**left**) and income (**right**). Green curve for cointegrated generators, black curve for non-cointegrated generators.

4.2. Longer Lags

The implications of lag specification are a traditional centerpiece of time-series model selection. It is well known that optimizing (i.e., minimizing) the AIC criterion due to [14] is tantamount to optimizing forecasting properties in large samples, but also that this strategy does not lead to consistency in the sense of capturing the true lag order as $T \to \infty$. In our panel, a lag order of two, i.e., a VAR(2), has been suggested by most criteria for most countries. In some countries, a lag order of one suffices, and for other countries, three is the recommended order.

In our simulation experiment, we consider first- and second-order vector autoregressions in differences, i.e.,

$$\Delta X_t = \Phi_0 + \Phi_1 \Delta X_{t-1} + \varepsilon_t$$

and

$$\Delta X_t = \Phi_0 + \Phi_1 \Delta X_{t-1} + \Phi_2 \Delta X_{t-2} + \varepsilon_t.$$

Both models are fitted to the data and then used to generate artificial pseudo-data. These pseudo-data are then forecast using both models. The outcome is summarized in Figure 5.

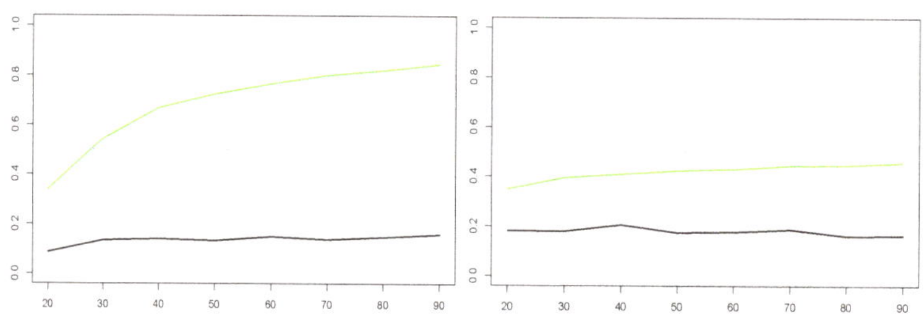

Figure 5. Percentage of VAR(2) models in differences yielding the better forecast for consumption (**left**) and income (**right**). Green curve for VAR(2) generators, black curve for VAR(1) generators.

Figure 5 shows that larger lag orders are not promising for the income series at all. Even at $T = 90$, D-VAR(1) remains the better forecaster than the 'true' D-VAR(2). Ninety years of macroeconomic data are unlikely to accumulate in the foreseeable future. In forecasting consumption, only 50–60 observations would be required to show an advantage for larger lag orders, even if the researcher is 50–50 undecided regarding whether such a lag order is needed.

5. Summary and Conclusions

We explore the technique of simulation-based forecast model selection in a panel data set of European income and consumption data. Variants of the time-series models are formulated, estimated, and simulated, and the relative merits of using each variant as a prediction tool are evaluated. The results confirm that statistical significance is an incomplete guideline for selecting forecasting models.

In detail, whereas the variables display noteworthy error-correction behavior, pure difference VAR models predict best in small samples, and they do so systematically. Similarly, the low lag order chosen for most individuals defines the best forecast model for quite large samples, at least for income, even if we allow for the possibility that a VAR with more lags has generated the sample.

The results not only contribute to the study of dynamic linkages among macroeconomic variables, they also demonstrate the power of the simulation-based selection procedure—it is simple and informative. More such examples will be considered in the future. In particular, it is to be noted that the symmetric approach does not face the usual difficulties in decisions between unit-root and stationary processes, a statistically non-standard problem that violates the regularity conditions of central limit theorems. Information criteria have been shown to be inadequate for this decision (see [16]). By contrast, the simulation-based approach faces no problem with this type of decision. Just like information criteria and in contrast to restriction tests, the approach is also readily applied to all non-nested decision situations.

We are planning to explore the properties of the method further, both in simulation studies based on artificial data and in empirically relevant applications.

Conflicts of Interest: The authors declare no conflict of interest.

References

1. Engle, R.F.; Granger, C.W.J. Co-Integration and Error Correction: Representation, Estimation, and Testing. *Econometrica* **1987**, *55*, 251–276. [CrossRef]
2. Han, H.L.; Ogaki, M. Consumption, income and cointegration. *Int. Rev. Econ. Financ.* **1997**, *6*, 107–117. [CrossRef]
3. Kunst, R.M.; Neusser, K. Cointegration in a Macroeconomic System. *J. Appl. Econ.* **1990**, *5*, 351–365. [CrossRef]
4. Jin, F. Cointegration of Consumption and Disposable Income: Evidence from Twelve OECD Countries. *South. Econ. J.* **1995**, *62*, 77–88. [CrossRef]
5. Ismail, A.; Rashid, K. Determinants of Household Saving: Cointegrated Evidence from Pakistan (1975–2011). *Econ. Model.* **2013**, *32*, 524–531. [CrossRef]
6. Klein, L.R.; Kosobud, R.F. Some Econometrics of Growth: Great Ratios of Economics. *Q. J. Econ.* **1961**, *75*, 173–198. [CrossRef]
7. Kunst, R.M. Cross validation of prediction models for seasonal time series by parametric bootstrapping. *Austrian J. Stat.* **2008**, *37*, 271–284.
8. Jumah, A.; Kunst, R.M. Seasonal prediction of European cereal prices: Good forecasts using bad models? *J. Forecast.* **2008**, *27*, 391–406. [CrossRef]
9. Granger, C.W.J. Investigating Causal Relations by Econometric Models and Cross-spectral Methods. *Econometrica* **1969**, *37*, 424–438. [CrossRef]
10. Giacomini, R.; White, H. Tests of Conditional Predictive Ability. *Econometrica* **2006**, *74*, 1545–1578. [CrossRef]
11. Kolassa, S. Sometimes It's Better to Be Simple than Correct. *Foresight* **2016**, *40*, 20–26.
12. Jumah, A.; Kunst, R.M. *Forecasting Agricultural Product and Energy Prices: A Simulation-Based Model Selection Approach*; Institute for Advanced Studies: Vienna, Austria, 2018.
13. Peña, D.; Sanchez, I. Measuring the advantages of multivariate vs. univariate forecasts. *J. Time Ser. Anal.* **2007**, *28*, 886–909. [CrossRef]
14. Akaike, H. Statistical predictor identification. *Ann. Inst. Stat. Math.* **1970**, *21*, 243–247. [CrossRef]
15. Engle, R.F.; Yoo, B.S. Forecasting and testing in co-integrated systems. *J. Econom.* **1987**, *35*, 143–159. [CrossRef]
16. Phillips, P.C.B.; Ploberger, W. Empirical Limits for Time Series Econometric Models. *Econometrica* **2003**, *71*, 627–673

engineering proceedings

MDPI

Proceedings

Anomaly and Fraud Detection in Credit Card Transactions Using the ARIMA Model †

Giulia Moschini [1], Régis Houssou [1], Jérôme Bovay [2] and Stephan Robert-Nicoud [1,*]

1 HEIG-Vd, Haute Ecole Spécialisée de la Suisse Occidentale (HES-SO), Rue de Galilée 15, CH-1400 Yverdon-les-Bains, Switzerland; giulia.moschini@heig-vd.ch (G.M.); regis.houssou@heig-vd.ch (R.H.)
2 NetGuardians SA, Avenue des Sciences 13, CH-1400 Yverdon-les-Bains, Switzerland; jerome.bovay@netguardians.ch
* Correspondence: stephan.robert@heig-vd.ch
† Presented at the 7th International conference on Time Series and Forecasting, Gran Canaria, Spain, 19–21 July 2021.

Abstract: This paper addresses the problem of the unsupervised approach of credit card fraud detection in unbalanced datasets using the ARIMA model. The ARIMA model is fitted to the regular spending behaviour of the customer and is used to detect fraud if some deviations or discrepancies appear. Our model is applied to credit card datasets and is compared to four anomaly detection approaches, namely, the K-means, box plot, local outlier factor and isolation forest approaches. The results show that the ARIMA model presents better detecting power than that of the benchmark models.

Keywords: anomaly; fraud; ARIMA; isolation forest; K-means

Citation: Moschini, G.; Houssou, R.; Bovay, J.; Robert-Nicoud, S. Anomaly and Fraud Detection in Credit Card Transactions Using the ARIMA Model. *Eng. Proc.* **2021**, *5*, 56. https://doi.org/10.3390/engproc2021005056

Published: 16 July 2021

1. Introduction

In recent years, there has been a dramatic increase in the use of credit cards as a means of payment due to their ease of use and convenience. As a response to this phenomenon, fraudsters are also adapting their malicious activities to take advantage of the situation. The extent of this issue is significant; according to the Fifth report on card fraud published by the European Central Bank [1], the total value of fraudulent transactions in the SEPA area in 2016 amounted to EUR 1.8 billion. According to the Nilson Report, a publication covering global payment systems, the total loss due to frauds in 2018 amounted to USD 27.85 billion, and it is projected to reach USD 35.67 billion in 2023 [2]. More specifically, a transaction is said to be fraudulent when it is committed by an unauthorised party and without the rightful owner and/or relevant institution knowing [3]. In these cases, fraudsters could use the card for their personal interests, depleting its resources or until they are caught or the card is blocked. This issue has sparked the interest of both academia and industry, where individuals are working to identify solutions to this problem and to keep up with the ever-changing approaches adopted by malicious players [4]. Credit card fraud detection is now an active field of research, and it particularly hinges on the concept of automation; it is in fact not always feasible or possible to manually review each transaction in order to establish its nature [5]. In addition to this, it is also important to consider that there is another significant human component that could make or break the attempt of a fraudster to successfully exploit a card: the promptness of the cardholders in reporting a stolen, lost or suspiciously used card [5]. This requires the implementation of automated tools for smarter and faster detection of frauds, which has resulted in machine learning techniques being increasingly tested and implemented [6]. Various popular algorithms have been tested in this context, such as random forest, logistic regression, decision trees, support vector machines (SVM), and neural networks [7–9]. Khare and Sait in [7] compare logistic regression, SVM, decision tree and random forest using the Kaggle dataset for credit cards containing 284,807 transactions, 492 of which are fraudulent. The features of the dataset are obtained using principal component analysis (PCA) on the original data for confidentiality issues. The authors also state that they use the behavioural characteristics of the owner of the card, which are shown by a variable representing the spending habits of the customer as well as the month, hour of the day, geographical location and type of merchant. Experimental results show that random forest is the algorithm with the best performance, with an accuracy score of 98.6% compared to 97.7% of logistic regression, 97.5% of SVM and 95.5% of decision tree. Varmedja et al. in [8] compare the performances of logistic regression, naive Bayes, random forest and multi-layer perceptron on the Kaggle dataset. The number of features is reduced through the application of feature selection and the class imbalance addressed by oversampling with SMOTE. Their results show that random forest is again the best algorithm , with accuracy, precision and recall equalling 99.06%, 96.38% and 81.63%, respectively. Roy et al. in [9] use a deep learning approach to detect frauds in credit card transactions. The dataset used in the study was provided by a financial institution and contains almost 80 million anonymised transactions performed over a period of 8 months. The authors perform feature engineering to apply field knowledge to the problem and add extra features to the original ones. Due to the unbalanced nature of the dataset, the authors also perform under-sampling at the account level for each unique account ID. artificial neural networks (ANN), recurrent neural networks (RNN), long short-term memory (LSTM) and gated recurrent unit (GRU) are compared in this study; the results highlight that GRU presents the best performance with an accuracy score of 91.6%, followed by 91.2% (LSTM), 90.4% (RNN) and 88.9% (ANN). As can be noted, there is a common fundamental issue in these approaches: the unbalanced nature of the datasets. In the context of credit card fraud detection, it is in fact expected that the dataset will be very unbalanced, which greatly hinders the performance of supervised learning techniques [6]. Another issue involves the lack of properly labelled data, which again represents a substantial obstacle. Finally, many models lack the adaptability required to take into account the fact that the spending behaviour of customers is likely to change over time [6]. In order to tackle these problems, we propose a model that does not require the knowledge of ground truths and that is designed to make the spending behaviour of the customer as the main source of information when categorising transactions as either legitimate or fraudulent. More specifically, we frame the problem as an anomaly detection task in time series, where the variable represented by the time series is the daily count of transactions for a given customer. We propose a method making use of the ARIMA model and of a rolling windows approach to flag suspicious number of transactions as anomalies, which are discussed in depth in the following sections.

2. Fraud Detection with Time Series Approach

2.1. ARIMA Model with Time Series Analysis

Two widely used models for time series are the *autoregressive* (AR) and the *moving average* (MA) models, which can be used together as an *autoregressive moving average* (ARMA) model. ARMA(p, q) is the combination of the AR(p) and MA(q) models, and can be used with univariate time series.

- *Autoregressive Model*
  The AR(p) model is defined by the equation below; it assumes that there is a dependent linear relation between the observation and the values of a specified number of lagged (previous) observations plus an error term.

$$X_t = c + \sum_{i=1}^{p} \phi_i X_{t-i} + \omega_t \quad (1)$$

  where $\phi = (\phi_1, \phi_2, ..., \phi_n)$ are the coefficients of the model, p is a non-negative integer, c is a constant and $\omega_t \sim N(0, \sigma^2)$.
- *Moving Average Model*

The MA(p) model is defined by the equation below; it makes use of the dependency between an observation and the residual errors resulting from the application of a moving average model to lagged observations.

$$X_t = \mu + \sum_{j=1}^{q} \theta_j \omega_{t-j} + \omega_t \tag{2}$$

where μ is the mean of the series, $\theta = (\theta_1, \theta_2, ..., \theta_n)$ are the coefficients of the model, and q is the order and $\omega_t \sim N(0, \sigma^2)$.

The ARMA model, resulting from the combination of these two models, is defined as follows:

$$X_t = c + \omega_t + \sum_{i=1}^{p} \phi_i X_{t-i} + \sum_{j=1}^{q} \theta_j \omega_{t-j} \tag{3}$$

where p refers to the order of the AR model and q refers to the order of the MA model. The main assumption in time series analysis is that the time series is stationary, meaning that its mean and variance are constant over time; however, this is not the case in many practical situations [10]. The solution to this can be found in the generalisation of the ARMA model: the autoregressive integrated moving average(ARIMA) model. ARIMA introduces the possibility to apply differencing to the data points of time series in order to make it stationary [10]. ARIMA is now one of the most popular, flexible and simple models to fit a time series [10]; it is defined as ARIMA(p, d, q), where p and q represent the orders of the AR and MA models and d indicates the degree of differencing. In the context of fraud detection, time series can be used as a tool when working with aggregated features. Aggregation is often used to derive new features from the original ones in order to feed to the model some information that is thought of and expected to be more relevant than the features per se. The number of daily transactions or the total amount spent in a week are examples of aggregated features [5].

2.2. Estimation Process of ARIMA

When using ARIMA, care should be taken to identify the combination of parameters that best represents the data; Box–Jenkins is a method proposed by George Box and Gwilym Jenkins in [11] that is frequently used when tuning an ARIMA model. The method is composed of three steps:

1. *Identification*, which refers to the use of all available data and related information to select the model that best represents the time series. This phase should, however, be split into two sub-steps:
   (a) Differencing
   The first step requires the establishment of whether the time series is stationary or not in order to determine whether it requires differencing. The augmented Dickey–Fuller (ADF) test is a technique that can be used to verify if the time series on hand is stationary. The null hypothesis of the ADF test states that the time series can be represented by a unit root, meaning it presents a time-dependent structure and that it is, thus, not stationary; consequently, rejecting the null hypothesis implies that the time series is stationary.
   (b) Configuration of p and q
   During this phase, it is helpful to use the correlogram to visualise the autocorrelation function (ACF) and the partial autocorrelation function (PACF) that can help to determine a suitable choice for the orders p and q. The fundamental difference between the two functions is that the PACF removes the linear dependence between the intermediate variables in order to return only the correlation between the present and lagged value. Briefly, whereas the autocorrelation function of AR(p) tails off, its partial autocorrelation function

cuts off after the lag p. Conversely, the autocorrelation function of MA(q) has a cut-off after the lag q, while its partial autocorrelation function tails off.

2. *Estimation*, which refers to the training phase. Once the values of p, d, q have been established, the ϕ and θ coefficients can be estimated. This method uses the maximum likelihood estimation process, which is solved by non-linear function maximisation; for more details about this phase, the reader is referred to [11,12].
3. *Diagnostics*, which refers to the evaluation of the model and identification of improvements. This step involves the determination of issues in the model to verify whether it is able to effectively summarise the underlying data. The forecast residuals provide an important source of information for diagnostics. In an ideal model, the error will resemble white noise and will be normally distributed with a mean of 0 and a constant variance. In addition to this, an ideal model would also leave no temporal structure in the residuals, as they should have been learned.

2.3. Fraud Detection with ARIMA Model on Daily Counts of Transactions

Our idea is to use ARIMA on time series representing the daily count of transactions for a given customer to detect frauds. This is based on an important point: we assume that the number of daily transactions for a given customer follows a certain pattern [13]. On a high level, the task of fraud detection in this context is based on the assumption that it is possible to recognise, and hence model, the regular spending behaviour of the customer; once this has been learned, any discrepancies and deviations from it would be likely frauds. We can also refer to such deviations as *anomalies*. An anomaly is a point in a dataset whose characteristics are significantly different compared to the other points; building from this, anomaly detection is the process to isolate such points by determining when they are deviating from the expected behaviour [14]. ARIMA is used to try to model the legitimate spending behaviour of the customer and to produce a forecast. The intuition behind this setting can be easily explained graphically. Figure 1 shows the daily transactions of a credit card for a customer chosen in our dataset; more details about this dataset are given in the next section. The number of legitimate transactions occurring each day for such customer is represented by the blue dot, whereas the number of frauds is represented by the red dot. A significant peak is observed at the same day of fraudulent transactions.

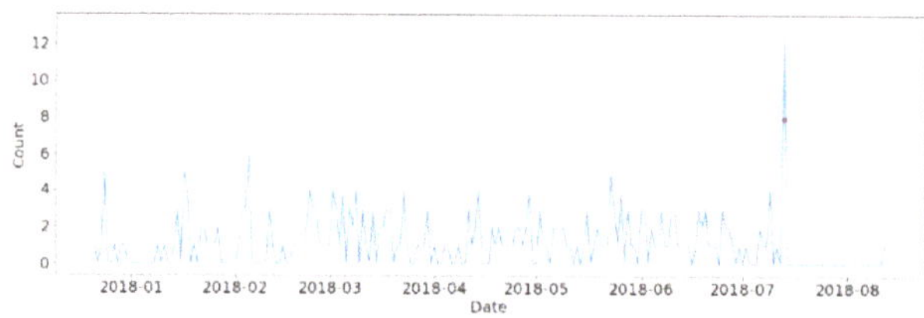

Figure 1. Plot of daily number of transactions for a customer in the dataset. Legitimate transactions are represented by the blue dot, whereas fraudulent transactions are represented by the red dot.

Based on this information, it could be argued that an anomaly detection approach based on the identification of anomalous counts of daily transactions may lead to the detection of frauds. In order to detect frauds, the following steps are proposed:

1. The time series is split into training and testing set; it is important that the training set only contains legitimate transactions so that the model can learn the legitimate behaviour of the customers. This should then allow for the identification of anomalies.
2. In the training set, based on the legitimate transactions, the order of the ARIMA model is identified using the Box–Jenkins method, and, then, the parameters of ARIMA are estimated. During this phase, care is taken to ensure that the estimated coefficients are significant and that there is no temporal structure left in the residuals. Finally, in the testing set, one-step ahead prediction is performed using rolling windows.

3. In order to detect fraud in the testing set, the errors are calculated in terms of difference between the predicted and actual daily count of transactions. Then, the Z-Scores are computed and used to flag the anomalies (i.e., the frauds). The Z-Score is calculated as z-score $= \frac{x-\mu}{\sigma}$, where x is the prediction error on the daily count of transaction in the testing set. μ and σ are the mean and the variance based on the errors of in-sample prediction on the basis of the training set using our model. If the Z-Score is greater than a threshold, the day is flagged as anomalous (i.e., as fraud).

3. Application to Dataset

3.1. Dataset Description

The dataset used for this study was provided by NetGuardians SA and contains information about credit card transactions for 24 customers of a financial institution; it covers the period from June 2017 to February 2019. For reasons of confidentiality, the name of the financial institution is not mentioned. Each row is related to a customer ID and represents a transaction with its various features (i.e., timestamp, amount etc.) including the class label (1 for fraud and 0 for legitimate transaction). An important aspect is that each of the 24 customers presents at least 1 fraud in the whole period. Figure 2 and Table 1 show the number of daily transactions for all customers and the frequency of fraud and legitimate transactions in the whole dataset. We remark that the dataset is highly imbalanced with a proportion of fraud of 0.76%.

Table 1. Frequency of fraud and legitimate transactions in the whole dataset.

	Legitimate	Fraud	Total
Number	11,384	87	11,471
Percentage	99.24%	0.76%	100%

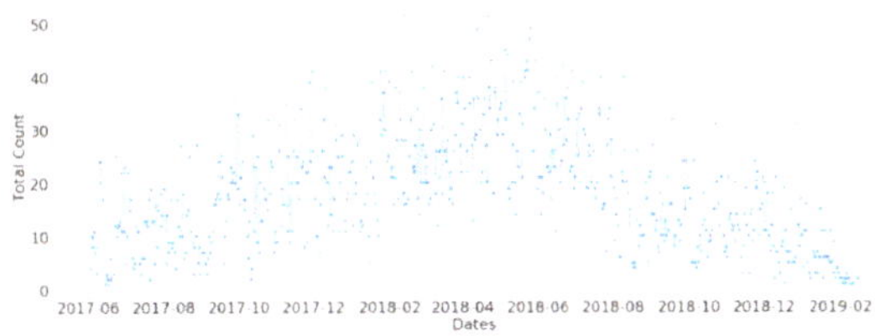

Figure 2. Number of daily transactions summing up all customers.

However, it is important to note that the customers are not necessarily active during the whole period. In fact, as illustrated in Figure 3, some of them perform transactions only in the first part of the considered time frame, others only at the end, and others in the middle. Our approach based on the ARIMA model requires sufficient legitimate transactions in the training set in order to learn the legitimate behaviour of the customers. In addition, our approach requires at least one fraud in the testing set to evaluate the performance of the model. In this context, initially, we propose to split the dataset into the training and testing set with a 70–30 ratio. With this setting, there is at least one fraud in the testing set and no fraudulent transactions in the training set, but, unfortunately, this reduces the number of customers' time series from 24 to 9. Table 2 summarises the composition of the final 9 time series that are used in the next section. The last column indicates the number of frauds over the total number of transactions occurring on the same day; as can be seen, only in one of the time series (number 10) do frauds occur on two different days.

Figure 3. Number of daily transactions: the blue dot represents a specific customer and the red dot represents all customers.

Table 2. Structure of the 9 time series.

Time Series ID	# Days in Train	# Days in Test	Fraud Proportion
0	192	83	1/14
4	193	84	1/3
5	192	83	1/16
7	186	80	1/11
8	131	57	3/15
9	164	71	8/21
10	193	84	4/17
15	191	82	1/11 and 1/2
17	119	51	2/12

3.2. Application of ARIMA Model for Daily Counts of Transactions

The previously outlined steps are performed for each of the 9 time series separately. These are now described in detail for just one of the time series for the sake of clarity and brevity as an illustration. As previously discussed, the first step involves establishing whether the time series is stationary. To do this, we perform the ADF test, whose results are shown in the Table 3. It can be observed that the time series is stationary with significant results.

Next, Figure 4a,b show the PACF and ACF that are used to determine the best values for the order p and q of the ARIMA model. For this time series, there may be a drop-off in the PACF at lag 1 and in the ACF at either lag 1 or 2, suggesting an ARIMA(1,0,1) or ARIMA(1,0,2). The steps for parameter estimation and residual analysis in the training set are carried out to select one of the two models; the model ARIMA(1,0,2) is determined to be a good model for this time series and is used to make forecasts. Figure 4c shows the correlogram of the residuals for the selected model, and this confirms that they have a white noise pattern. These above steps are performed for the remaining eight time series; in some cases, the configuration may require multiple attempts to identify the best parameters. All parameters passed on to the next stage of the study are found to be significant. It is important to mention that for the forecasting in the testing set, we set the threshold to three. So, when the Z-Score is greater than three, there is fraud.

Table 3. Statistics of ADF on the stationarity for one time series.

t-Statistic	−8.73162539099
p-value	$3.180176629 \times 10^{-14}$

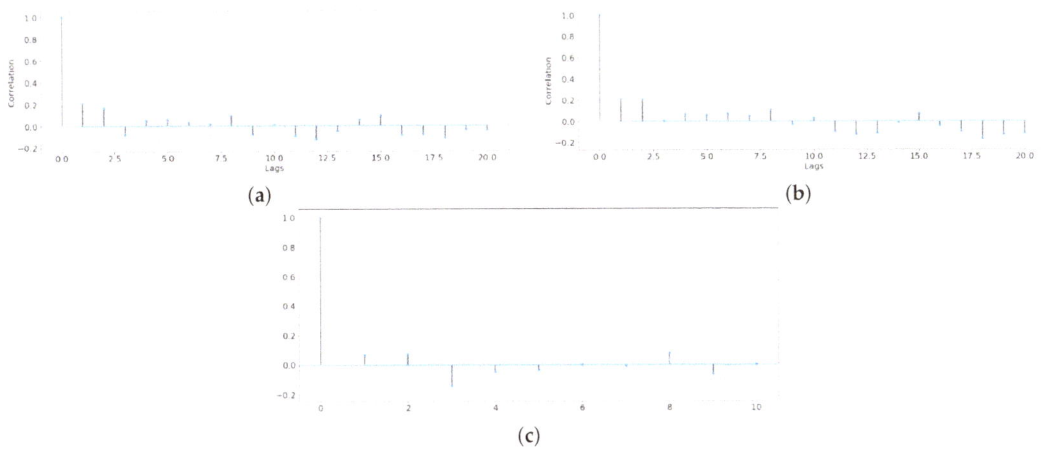

Figure 4. (**a**) Partial autocorrelation plot for sample time series; (**b**) autocorrelation plot for sample time series; (**c**) correlogram of residuals.

3.3. *Benchmark Models*

Our model is compared to four different models of anomaly detection, namely, the box plot, local outlier factor (LOF), isolation forest and the K-means models. Each benchmark model is briefly explained in the following section.

3.3.1. Box Plot

Box plots are used in the context of exploratory data analysis; they can be used to graphically represent data using their descriptive statistics. Box plots do not make any assumptions about the statistical distribution followed by the sample, meaning that potential outliers are identifies solely based on the degree of dispersion of the data points in the sample. Box plots are very useful, because they can be used to effectively identify patterns in groups of numbers that might be invisible to the human eye [15]. Being a visual tool, box plots are often used to increase our understanding of data, thereby allowing for a better interpretation of quantitative data [15]. We apply a box plot to the entire dataset (for each time series); however, only the testing portion of the dataset is considered to calculate the results. This is carried out for consistency reasons in order to ensure a fair comparison of the performances.

3.3.2. Local Outlier Factor (LOF)

The local outlier factor (LOF) is an algorithm that was introduced by Breunig, Kriegel, T. Ng and Sander in 2000 with the aim of identifying anomalous data points based on their local deviation from their neighbours. LOF is a density-based algorithm, and it is centred on the concept of degree of being an outlier [16], as opposed to a binary classification of outliers. The model is local because the anomaly score assigned to each point derives from the degree of isolation of that point compared to the its k neighbours, where k can be specified. More precisely, the locality of a point is given by its k-nearest neighbours. A point is considered to be an outlier when its local density is significantly lower than the densities of its neighbours [17]. For more details about LOF, see [16]. As in the case of the box plot, LOF is applied to the entire dataset only considering the testing set to calculate the results. As previously explained, this is carried out in order to retain consistency across the tests.

3.3.3. Isolation Forest

Isolation forest is an anomaly detection algorithm that implements a new approach compared to other models used for this purpose: rather than focussing on identifying normal points and their deviations (i.e., anomalies), isolation forest directly focuses on the detection of these anomalies without profiling. This can be achieved on the basis of two fundamental properties of outliers; that is, they are few in number and different, which means that they are isolated from the other—regular—points [18]. In order to isolate anomalies, this algorithm makes use of a tree structure, which results in outliers being placed closer to the root of the tree compared to the other points [19]. In isolation forest, each isolation tree isolates the anomalies by randomly selecting features and a split value between the minimum and the maximum values of that feature; the random partitioning should result in anomalies having a shorter path due to both the low number of such instances and to their inherently different characteristics leading to early partitioning. More details about the algorithm of isolation forest can be seen in [19]. Isolation forest does not require labels to work; however, it is trained on the training set comprising only of legitimate transactions and used to classify the data points in the testing set.

3.3.4. K-Means

K-means is an unsupervised learning model used for *clustering*. Clustering is the process by which from a given input, clusters or groupings are identified [20]. The process by which K-means operates can be divided into two parts: given an input comprising of a set of instances $x_1, x_2, x_3, ..., x_n$, and a number of clusters K, the algorithm places the centroids $c_1, c_2, c_3, ..., c_n$ for each cluster J at random locations, and then the steps presented below are followed:

1. For each point x_n:
    (a) Find the nearest centroid c_j. K-means computes the Euclidean distance between each point x_n and centroid c_j. This approach is often called minimising the inertia of the clusters [21] and can be defined as follows:

    $$SS_{w_i} = \sum_n ||x_n - c_j||^2 \forall i \in (1, K)$$

    where n is the number of points x and i is the number of centroids c.
    (b) Assign instance x_n to cluster J.
2. For each cluster $J : 1, 2, ...K$
    (a) Compute the new centroid c_j. This is achieved by calculating the mean from each point x to the centroid x of the cluster J to which is was firstly assigned.
3. Stop when convergence is reached; that is, there are no more changes after the iterations.

For more details on K-means, see also [21,22]. We fit K-means to the entire dataset specifying two clusters (for legitimate and fraudulent daily counts). The cluster containing the smallest number of instances is considered to be the cluster indicating the positive class. As with the box plot and LOF, only the part of the outliers in the testing set is taken into account.

4. Results

The results are presented based on three metrics: precision, recall and F-measure. Precision refers to the ability of the model to be trustworthy in regard to its classified positive points; that is, precision tells us how many of the predicted frauds are actually frauds. High precision means that when the model classifies a point as positive, it is highly likely that it is a correct classification. This metric is defined by the following equation:

Precision = True Positive/(True Positive + False Positive). Recall indicates the ability of the model to detect the positive class. When a model presents a high recall, it means that the majority of positive data points are correctly identified. The equation for recall is as follows: *Recall = True Positive/(True Positive + False Negative)*. Precision and recall indicate two opposite properties of a model, meaning that optimising one implies worsening the other. In order to gain a more comprehensive overview of the performance of the model, we can use the F-measure metric, defined as shown in the following equation: *F-Measure = 2(Precision ∗ Recall)/(Precision + Recall)*. These metrics are calculated for each of the nine time series analysed and are used to obtain the average as described in the previous section. The results are presented in the Table 4. As can be noted, ARIMA presents the best result in terms of precision and F-measure, whereas K-means provides the best performance in terms of recall. The worst-performing model in this setting is the local outlier factor, which presents precision and F-measure scores of 8.4% and 14.04%, respectively. It should be pointed out that LOF was designed to be effective with multidimensional datasets [16], which might explain its bad performance in this particular setting. The box plot model performs the best amongst the benchmarks with a F-measure of 72.22% and is, thus, the only one that is comparable to our model. The advantage of our model that it is based on the concept of modelling the normal behaviour of the customer. In addition, the forecasting by the rolling windows takes into account the dynamic changes in the spending behaviour of the customer. While it can be argued that our model is overall the best one, it underperforms when compared to the box plot, isolation forest and K-means in terms of recall. As previously discussed, only 9 out of the 24 possible time series are retained for analysis due to the lack of frauds in the testing set. Consequently, the results that were presented are highly dependent of that particular set of data. In order to assess the robustness of the model, the time series that were originally discarded are reintegrated through the injection of one fake fraudulent transaction in the testing set. The occurrence of frauds is simulated by the addition of a varying number of counts ranging from 1 to 8 to a random date in the testing set for each time series. The range is set from 1 to 8 as it reflects that observed in the 9 time series previously discussed. It should be noted that the performance of the models highly varies depending on how many counts are added and on which day. In order to account for this randomness, this process is repeated 100 times, and the average of the metrics is computed. In order to gain an overview of the performances over the 24 time series, a global average is computed, which is shown in Table 5.

Table 4. Comparison of the performances of the 5 models using the 9 time series.

METRICS	ARIMA	BOX-PLOT	LOF	IF	K-MEANS
Precision	50%	43.98%	8.4%	25.01%	21.82%
Recall	66.67%	72.22%	66.67%,	72.22%	83.33%
F-Measure	55.56%	52.22%,	14.04%	32.56%	28.95%

Table 5. Global performance of the 5 models using the 24 times series.

METRICS	ARIMA	BOX-PLOT	LOF	IF	K-MEANS
Precision	34.29%	28.96%	6.41%	19.94%	22.51%
Recall	42.03%	60.54%	69.57%,	64.09%	68.16%
F-Measure	36.19%	34.91%,	11.17%	24.82%	26.81%

Despite the fact that all of models under-perform after the injection of fake frauds, the ARIMA presents the best performance in terms of precision and F-measure, whereas the best recall score is achieved by the local outlier factor. The precision of the latter is, however, the worst, which means that in this case too, only the box plot is comparable to ARIMA.

5. Conclusions

This paper addresses the problem of the unsupervised approach of credit card fraud detection using the ARIMA model. The main reason for focussing on time series model is the lack of fraud data due to confidential issues, which could represent a substantial obstacle in the development of machine learning algorithms. In this context, the goal of our approach is to model the regular spending behaviour of the customer, allowing any discrepancies and deviations from it to be deemed potential anomalies. The intuition behind this approach is centred on the assumption that the occurrence of frauds on a given day would cause the daily number of transactions to be altered in such a way that could be detected as suspicious. In the training set, the ARIMA model is first calibrated on the daily number of legitimate transactions in order to learn the regular spending behaviour of the customer. In the second step, the fitted model is used to predict fraud in the testing set by using the rolling windows. The criterion of flagging fraud is based on the Z-score calculated on the prediction errors in the testing set. Our methodology is applied to the dataset that is provided by NetGuardians and is compared with four anomaly detection algorithms, namely, the K-means, box plot, local outlier factor and isolation forest algorithms. It is observed in terms of prediction power that the ARIMA model outperforms the other models following by the box plot method. Among the four benchmark models, the local outlier factor performs the worst. Our model is successful when compared to the benchmark models for two reasons:

1. It works better when there is a significant number of frauds occurring on the same day. This is often the case, as fraudsters are known to take advantage of the time they have before the card is blocked to make several fraudulent transactions in a short time span [13].
2. It presents the best precision; i.e., it reduces the number of false positives compared to the benchmark models.
3. It takes into account the dynamic spending behaviour of the customer by using the rolling windows.

One main problem in our approach is that the ARIMA model assumes that the data come from observations that are equally spaced in time. However, this assumption does not hold in our study since the transaction times are unequally spaced. This issue will be addressed in future research by using advanced approaches, such as the continuous-time autoregressive moving average (CARMA) processes.

Funding: This research was funded by InnoSuisse grant number 34137.1.

References

1. Bank, E.C. *Fifth Report on Card Fraud;* European Central Bank: Frankfurt am Main, Germany, 2019
2. Nilson. The Nilson Report | News and Statistics for Card and Mobile Payment Executives. Available online: Nilsonreport.com (accessed on 1 June 2019)
3. Maniraj, S.P. Credit Card Fraud Detection using Machine Learning and Data Science. *Int. J. Eng. Res. Technol.* **2019**, *8*, 110–115 [CrossRef]
4. Tripathi, D.; Sharma, Y.; Tushar, L.; Shubhra, D. Credit Dard Fraud Detection Using Local Outlier Factor. *Int. J. Pure Appl. Math.* **2018**, *118*, 229–234.
5. Pozzolo, A.D. Learned lessons in credit card fraud detection from a practitioner perspective. *Expert Syst. Appl.* **2014**, *41*, 4915–4928. [CrossRef]
6. Singh, D.; Vardhan, S.; Agrawal, N. Credit Card Fraud Detection Analysis. *Int. Res. J. Eng. Technol. (IRJET)* **2018**, *5*, 1600–1603.
7. Khare, N.; Sait, S.Y. Credit Card Fraud Detection Using Machine Learning Models and Collating Machine Learning Models. *Int. J. Pure Appl. Math.* **2018**, *118–120*, 825–838.
8. Varmedja, D. Credit Card Fraud Detection—Machine Learning methods. In Proceedings of the 18th International Symposium INFOTEH-JAHORINA, Jahorina, Bosnia and Herzegovina, 20–22 March 2019
9. Roy, A. Deep learning detecting fraud in credit card transactions. In Proceedings of the 2018 Systems and Information Engineering Design Symposium, Charlotteville, VA, USA, 27 April 2018; pp. 129–134.
10. Adhikari, R.; Agrawal, R.K. An Introductory Study on Time Series Modeling and Forecasting. *arXiv* **2013**, arXiv:1302.6613.

11. Box, G.E.P.; Jenkins, G.M.; Reinsel, G.C. *Time Series Analysis: Forecasting and Control*, 4th ed.; John Wiley & Sons: Hoboken, NJ, USA, 2008
12. Azrak, R.; Melard, G. *Exact Maximum Likelihood Estimation for Extended ARIMA Models*; Université Libre de Bruxelles Institutional Repository: Brussels, Belgium, 2013.
13. Seyedhossein, L.; Hashemi, M.R. Mining information from credit card time series for timelier fraud detection. *Int. J. Inf. Commun. Technol.* **2010**, *2*, 619–624.
14. Ounacer, S. Using Isolation Forest in anomaly detection: The case of credit card transactions. *Period. Eng. Nat. Sci. (PEN)* **2018**, *6*, 394. [CrossRef]
15. Williamson, D.F. The Box Plot: A Simple Visual Method to Interpret Data. *Ann. Intern. Med.* **1989**, *110*, 916–921. [CrossRef] [PubMed]
16. Breunig, M.M.; Kriegel, H.-P.; Ng, R.T.; Sander, J. LOF: Identifying density-based local outliers. *ACM SIGMOD Rec.* **2000**, *29*, 93–104. [CrossRef]
17. Pedregosa, F.; Varoquaux, G.; Gramfort, A.; Michel, V.; Thirion, B.; Grisel, O.; Blondel, M.; Prettenhofer, P.; Weiss, R.; Dubourg, V.; et al. Scikit-learn: Machine Learning in Python. *J. Mach. Learn.* **2011**, *12*, 2825–2830
18. John, H.; Naaz, S. Credit Card Fraud Detection using Local Outlier Factor and Isolation Forest. *Int. J. Comput. Sci. Eng.* **2019**, *7*, 1060–1064. [CrossRef]
19. Liu, F.T.; Ting, K.M.; Zhou, Z.H. Isolation Forest. In Proceedings of the 8th IEEE International Conference on Data Mining, Pisa, Italy, 15–19 December 2008; Volume 7, pp. 413–422.
20. Kubat, M. *An Introduction to Machine Learning*; Springer: Berlin, Germany, 2015.
21. Dalatu, P.I. Time Complexity of K-Means and K-Medians Clustering Algorithms in Outliers Detection. *Glob. J. Pure Appl. Math.* **2018**, *12*, 4405–4418. .
22. Bonaccorso, G. *Machine Learning Algorithms: Popular Algorithms for Data Science and Machine Learning*; Packt: Birmingham, UK, 2018

engineering proceedings

MDPI

Proceedings

Assessing Statistical Performance of Time Series Interpolators †

Sophie Castel [1,‡] and Wesley S. Burr [2,*,‡,§]

1 Applied Modelling & Quantitative Methods MSc Program, Faculty of Science, Trent University, 1600 West Bank Drive, Peterborough, ON K7L 0G2, Canada; sophiecastel@trentu.ca
2 Department of Mathematics, Faculty of Science, Trent University, 1600 West Bank Drive, Peterborough, ON K7L 0G2, Canada
* Correspondence: wesleyburr@trentu.ca
† Presented at the 7th International conference on Time Series and Forecasting, Gran Canaria, Spain, 19–21 July 2021.
‡ These authors contributed equally to this work.
§ Current address: Trent University, Department of Mathematics, ENW 335, 1600 West Bank Drive, Peterborough, ON K7L 0G2, Canada.

Abstract: Real-world time series data often contain missing values due to human error, irregular sampling, or unforeseen equipment failure. The ability of a computational interpolation method to repair such data greatly depends on the characteristics of the time series itself, such as the number of periodic and polynomial trends and noise structure, as well as the particular configuration of the missing values themselves. The `interpTools` package presents a systematic framework for analyzing the statistical performance of a time series interpolator in light of such data features. Its utility and features are demonstrated through evaluation of a novel algorithm, the Hybrid Wiener Interpolator.

Keywords: time series; interpolation; applied statistics; `interpTools`

Citation: Castel, S.; Burr, W.S. Assessing Statistical Performance of Time Series Interpolators. *Eng. Proc.* **2021**, *5*, 57. https://doi.org/10.3390/engproc2021005057

Published: 16 July 2021

1. Introduction: The Need for Interpolators

Practically-gathered time series data often contain missingness: observations are not known due to a variety of real-world causes, including instrument failure, contamination, data storage losses, or even climate and weather (e.g., for Earth-bound stellar observatories). Many of the best estimation algorithms for time series characteristics assume contiguous samples with no missingness. This contrast is the impetus behind the creation of a number of interpolation (or imputation) methods for time series data [1–5], and the previous development of a test-bench for evaluation of such methods [6].

Recently completed work on a new R package `interpTools` [7,8] provides an additional means of simulating particularly-structured artificial time series, imposing missing observations according to a user-specified gap structure, and repairing the incomplete series via chosen interpolation algorithms, with generous support for evaluating interpolators' statistical performance, and for generating data visualizations. In this paper we discuss the framework developed, and present some results comparing the Hybrid Wiener Interpolator (HWI) [1] to a number of other standard algorithms.

A significant practical challenge when determining the effectiveness of a given interpolator on a particular time series is that the true value of a missing data point at a given index of a stochastic time-ordered process is generally unknown. Performance metrics, such as those described by [2], typically assume the form:

$$C(\hat{x}_i, x_i) = C(\hat{x}_i - x_i), \quad i = 1, ..., I, \tag{1}$$

where C is some function of the deviation between the interpolated data point, $\hat{x}_i$, and the true data point, x_i. Without knowing x_i or at least its probabilistic structure, exactly, it is impossible to determine such a measure of accuracy.

The statistical performance of interpolators depends greatly on the structural nuances of the dataset chosen. Some algorithms are better-suited for time series with high numbers of embedded periodicities, whereas others are more suitable for low-frequency data. Interpolators' performance may also depend on any particular pattern of missing values present in the data, e.g., cubic splines fail exponentially as the gap width increases [8]. Other methods, such as the HWI [1], are more resilient to longer sequences of consecutive missing observations. As such, there are trade-offs to any chosen interpolator, and careful consideration of the research objectives and parameters of the study should be the first step in the selection of an interpolator in any practical setting.

2. Framework for Interpolation Using interpTools

The `interpTools` package allows a user to simulate a 'mock' time series containing similar features to a real-world dataset of interest, such that the original data points are known, and performance metrics of the form $C(\hat{x}_i, x_i)$ can be calculated, following the application of a specified pattern of gaps and a particular interpolation algorithm (Figure 1). Using simulated data enables the user to benchmark performance and make an informed choice regarding which interpolator would be most suitable for use on similar time series outside of the laboratory setting.

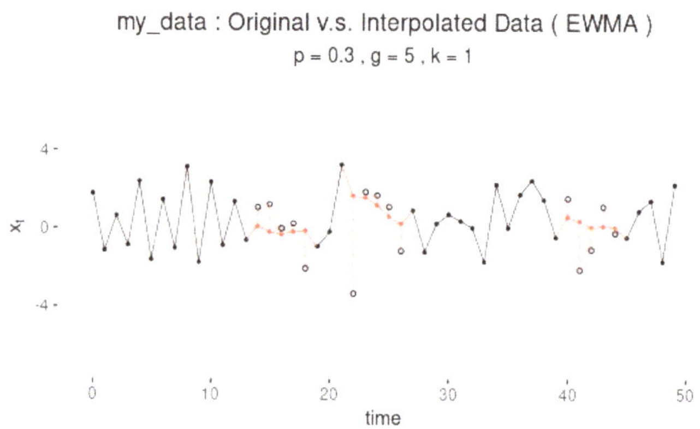

Figure 1. Example interpolation using the Exponential Weighted Moving Average method on the first dataset ($k = 1$) of a series of simulated data with 20% missing values at a minimum gap length of 2, with absolute deviations $|\hat{x}_i - x_i|$ highlighted in red.

The default package model simulates time series x_t based on the classic additive model for time series:

$$x_t = m_t + t_t + \xi_t, \quad t = 0, \cdots, n-1, \tag{2}$$

where m_t is the mean function, t_t is the trend function, and ξ_t is the noise function. The following section provides a brief description of each component, along with its set of defining parameters. The package also supports arbitrary user-generated series for extension beyond this particular model: in particular, it is simple to generate multiplicative models if that is more relevant to the interests of the user (the reason the additive model was used in the development of this algorithm and package is that many astrophysics time series datasets are more accurately modeled using additive structure, with limited or no seasonality to speak of, and the second author has an interest in data of this type).

The mean component, m_t is comprised of a constant, non-varying mean element (a 'grand mean'), μ, and a varying polynomial trend element, μ_t:

$$m_t = \mu + \mu_t, \quad t = 0, ..., n-1 \tag{3}$$

$$= \mu + \sum_{i=1}^{\phi} a_i \left(\frac{t-c}{n} \right)^i, \tag{4}$$

where $\mu \sim u\left(\frac{-n}{100}, \frac{n}{100}\right)$ and c is a randomly sampled integer in the range $[1, n]$. The polynomial coefficients $a_i \sim n(0, \frac{n}{20i})$ and are sampled in this way to facilitate the desirable property that the coefficients 'scale down' (i.e., $a_i \to 0$) as $i \to \phi$. The parameter ϕ is chosen by the user and represents the degree of the polynomial. This could be re-expressed without the μ by allowing the summation to run from $i = 0$, but structurally the code implementation assumes a static non-time-varying mean and time-varying polynomial mean.

The trend component, t_t is considered to be a finite linear combination of sinusoids. The `interptools` package simulates the trend component according to the construction:

$$t_t = \sum_{i=1}^{\psi} b_i \sin(\omega_i t), \tag{5}$$

where $b_i \sim n(1, \frac{n}{200})$ (to allow for variation in relative Signal-to-Noise for individual periodic components, with the normal distribution ensuring extremely high-coefficient sinusoids will be rare) and $\omega = [\omega_1, \omega_2, ..., \omega_\psi]$ with ω_i defining the period of the ith sinusoid. The default is to sample $\psi = 20$ unique values for each $\omega_i \in \left[\frac{2\pi}{N}, \pi\right]$ (using the fundamental Fourier frequency, and bounded by Nyquist). This is user-controllable, and is intended to allow the user a degree of influence over the relative signal-to-noise in the simulations, where "signal" is considered as overall periodic components, and "noise" is the background. Many scientific datasets have dozens to thousands of such periodic signals present (e.g., astrophysics, helioseismology, seismology, oceanography), and they are the object of interest in such fields and analyses, so control of this parameter is of importance for generalizability of the algorithm and package.

The noise component, ξ_t is assumed to be an ARMA$(P^\star, Q^\star)$ stochastic process:

$$\xi_t = \alpha_1 X_{t-1} + ... + \alpha_{P^\star} X_{t-P^\star} + Z_t + \beta_1 Z_{t-1} + ... + \beta_{Q^\star} Z_{t-Q^\star}, \tag{6}$$

with variance $\sigma^2_{\xi_t}$, where $P^\star$ is the autoregressive (AR) order, $Q^\star$ is the moving-average (MA) order, and Z_t is a white noise process.

Each of these components can be generated independently, or simultaneously. Metadata regarding information about the features of each component, such as the polynomial equation of m_t, or the exact frequencies contained in t_t, are saved to memory in list objects of class `simList`. We reiterate here that this is simply the default structure for simulation, and the user is able to specify their own model of interest, and generate their own synthetic time series for testing with ease.

2.1. Imposing a Gap Structure

Once the artificial data (K time series) has been generated, the user can remove observations according to a gap structure defined by parameters p, the proportion of data missing, and g, the gap width, where each observation (except the endpoints) has the same Bernoulli probability of missingness, $p_{\text{omit}} = 1/(N-2)$. These inputs are vectorized such that the user can test any number of different (p, g) combinations, where P and G represent the total number of options for the removal percentage and gap width, respectively. The number of unique gap structures to randomly generate under a particular (p, g) parameterization is specified with K, which gives the user control over the number of iterations: higher K means more replicates, which tends to give more stable estimates, as with most statistics (for the analyses shown here, $K = 100$, which empirically gave stable results on repeated runs and random seeds). An example is shown in Figure 2.

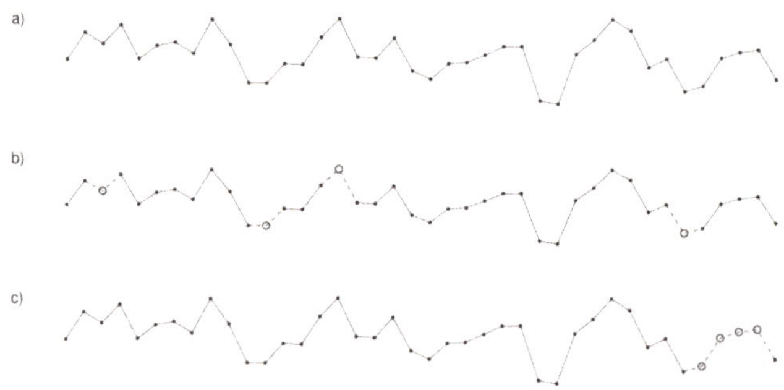

Figure 2. An example of the relationship between the number of holes and gap width for $N = 40$: (**a**) a time-plot of the original series; (**b**) a time-plot of a gappy series with $p = 0.10$ and $g = 1$; (**c**) a time-plot of a gappy series with $p = 0.10$ and $g = 4$.

The result is K 'gappy' series, each with a total of $I = p \cdot N$ missing observations, appearing structurally as $\leq \frac{p \cdot N}{g}$ randomly-spaced non-overlapping holes of width g, where a 'hole' is defined as a sequence of adjacent missing observations. Note that since holes may be placed adjacent to one another, this quantity describes the number of holes visible, at most. Classic Missing Completely At Random (MCAR) can be simulated by setting the gap length to 1.

2.2. Performing Interpolation

Once a set of gap-imposed data has been generated, the user may test interpolation algorithms on those data using `parInterpolate()`, which executes in parallel for efficiency. The package provides a number of built-in interpolation algorithms, though a user may also choose to provide any developed algorithm for flexibility and extension (e.g., [3,5]). The output is a highly nested list with every combination of (m, p, g) having a list of K interpolated time series, each of length N, where m represents the particular interpolation method used, and each $\hat{\mathbf{x}}_k$ for $k = 1, \ldots K$ approximates the original time series.

2.3. Evaluating Statistical Performance

A definition of statistical performance at any time index will be given by some function of the deviation, C (Equation (1)). Generally speaking, C quantifies how well the interpolated series, $\hat{\mathbf{x}} = \{\hat{x}_t\}_{t=0}^{N-1}$, captures the essence of the original series. The package contains 18 such performance metrics, and it is also possible for the user to define their own custom performance statistics. For every $(\mathbf{x}, \hat{\mathbf{x}})$ pair, statistical performance can be calculated via the function `performance()`, with resulting output a list of class `pf` and dimension $M \times P \times G \times K$, where M represents the total number of algorithms tested, and P and G are the total number of different proportion missing and gap width parameterizations applied, respectively. The terminal node of any (m, p, g) branch in this nested list is a vector of all the performance criteria for a particular combination of experimental conditions. Consider that for any (m, p, g) branch, there are a set of K values for each criterion. Thus, each performance metric has a sampling distribution containing K elements. The performance matrices can be condensed by aggregating sample statistics over K to reduce dimensionality, via the function `aggregate_pf()`.

The Hybrid Wiener Interpolator (HWI) [1] is a novel iterative interpolation algorithm based on estimation of sub-components of a time series using robust frequency-domain spectral methods. The essence of the algorithm is the estimation of periodic components and time-varying mean elements using multitaper methods [9], enveloping an embedded

Wiener covariance interpolation step [10] for the approximately stationary noise background. This pair-wise estimation proceeds iteratively until converged, in approximately an Estimation-Maximization model.

As a visual example, imagine a path through such an above list, where $p = 15\%$, $g = 10$, and $m = \text{HWI}$, such that we end with a collection $\{\hat{\mathbf{x}}_k : k \in 1, \ldots, K\}$ (Figure 3a). Imagine applying some performance metric formula, say, $C = \text{MSE}$, to each $(\hat{\mathbf{x}}_k, \mathbf{x}_k)$ pair, such that we have a collection of K values. The sampling distribution of $\{C_k\}$ is shown in Figure 3b. Then, a sample statistic f (e.g., the sample median) can be calculated for this distribution, such that $f = \text{median}(\{C_k, k \in 1, .., K\})$ (Figure 3b). This value represents the aggregated (median) performance of the HWI for a gap structure of $(p = 15\%, g = 10)$, denoted by the value $f(15\%, 10)$ (indicated in Figure 3c).

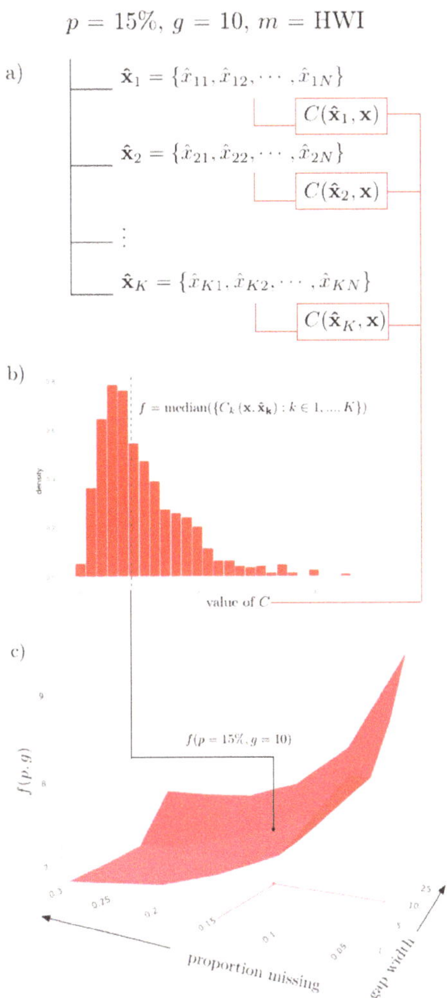

Figure 3. (**a**) Structure diagram of K interpolations performed using the Hybrid Wiener Interpolator for $(p, g) = (15\%, 10)$. (**b**) Example sampling distribution of the median of a selected performance metric C_k. (**c**) An example surface plot f over $\{(p, g) : p \in 1, \ldots, P, g \in 1, \ldots, G\}$, indicating the aggregated performance at $(p, g) = (15\%, 10)$.

3. Data Visualization

3.1. 3D Surfaces

Considering the full set of possible gap combinations $\{(p_i, g_j)\}$, for $i \in 1, ..., P$, $j \in 1, ..., G$ as a discrete mesh in $\mathcal{R}^2$, mapping these aggregated statistics traces out a surface $f(p, g)$ in $\mathcal{R}^3$, where the height of the surface at a point (p, g) represents an interpolator's aggregated performance when subjected to a specific proportion of data missing and gap width. Visualizing performance as a surface helps the user to understand the behavior of an interpolator in light of changes to gap structure. Extreme points on the surface represent gap structures at which performance is exemplified: either optimal, or worst-case. For cross-comparison across interpolation methods, multiple performance surfaces can be graphically layered on top of one another, where the 'best' interpolator for a particular gap structure will be at an extremum of the surfaces at the corresponding (p, g) coordinate point.

This visualization is generated by the `plotSurface()` function, where the user can specify any number of algorithms, the sample statistic to be represented by $f(p, g)$, and a performance metric of choice. It is also possible to select an algorithm to highlight via the argument `highlight`, as well as the colour palette. As the implementation is dynamic, the user can also interact with the surface plot widget by manipulating the camera perspective, adjusting the zoom, and hovering over data points for more precise numerical information. A static export of such a surface is shown in Figure 3c.

3.2. Heatmaps

For more static (and printable) representations of the 3D surface plot, heatmaps are a nice equivalent. Using `heatmapGrid()`, a three-dimensional surface can be collapsed into a heatmap through conversion of the third dimension to colour, to which the value of the metric is proportional. The function `multiHeatmap()` function enables the user to arrange multiple heatmaps into a grid to facilitate cross-comparison between multiple criteria or methods. Demonstration examples are provided in Figure 4.

3.3. Collapsed Cross-Section Plots

Changing the perspective angle on a given surface plot can offer further insights on the relationship between interpolator performance and the gap pattern parameters. Imagine rotating such a surface such that it is viewed perpendicular to either the $p - f$ or $g - f$ plane: this allows a user to examine performance with respect to changes in one variable across all values of the other. The sampling distribution can be visualized as a ribbon, where the upper and lower bounds are the largest and smallest values observed across the set of sample statistics contained in the collapsed variable (the highest and lowest points on the corresponding surface plot), and the central line is the median value of this set. The user can generate these collapsed cross-section plots using the `plotCS()` function.

When `cross_section = 'p'`, the g-axis is collapsed, and the resulting plot is a cross-section of 'proportion missing' (Figure 5, left). Here, we can see how the performance of an algorithm on a particular dataset changes as the total number of missing observations increases ($p \rightarrow P$). When `cross_section = 'g'`, the p-axis is collapsed, and the resulting plot is a cross-section of gap width (Figure 5, right): we can observe how the overall performance of an algorithm on a particular dataset changes as the width of the gaps increases ($g \rightarrow G$). The widths of the ribbons may indicate the sensitivity of an algorithm to a particular gap parameter, where 'thicker' ribbons indicate a greater disparity between the best and worst interpolations, and 'thinner' ribbons correspond to algorithms that seem to perform similarly regardless of the value of the defining axis.

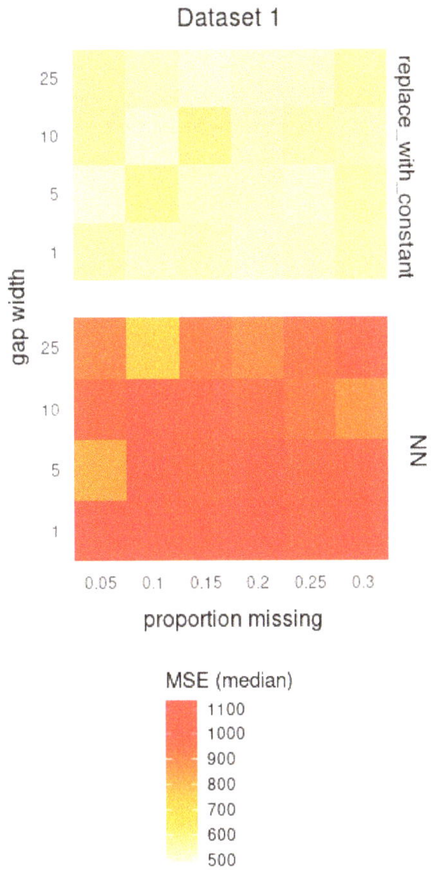

Figure 4. Example of heatmaps generated comparing two interpolation algorithms (Replace with Constant and Nearest Neighbour) on a common set of data. The bottom axis is the proportion of missing points, and the side axis is the designated width of each individual gap, with the colours representing the median value of the Mean Squared Error (MSE) metric.

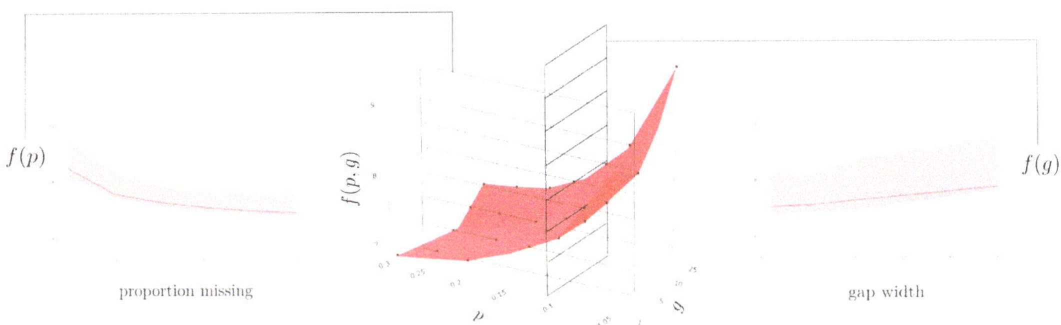

Figure 5. Representation of collapsed cross-section plots derived from a parent performance surface (**center**). The shown bands extract information not shown in the 3D plot, indicating designated percentile intervals across the simulation replicates.

3.4. Uncollapsed Cross-Section Plots

The collapsed cross section plots can be further deconstructed using `plotCS_un()` such that the performance can be assessed with respect to changes in one variable (p or g) across each individual value of the other.

Interpretable as 'slices' of the surface plot generated by `plotSurface()`, these 'uncollapsed' cross-section plots will give insight as to whether there are specific combinations of (p, g) for which the performance of a particular method is particularly sensitive. Each ribbon represents the distribution of a metric across the K simulations, where the upper and lower bounds represent the (e.g., 2.5% and 97.5%) quantiles, respectively, and the interior line is formed by the set of sample statistics from corresponding points on the parent surface plot. An advantage of viewing the data in this way is that it allows us to view the error bars at each (p, g) coordinate point without creating an over-crowded and hard-to-read surface plot visualization. An example is shown in Figure 6.

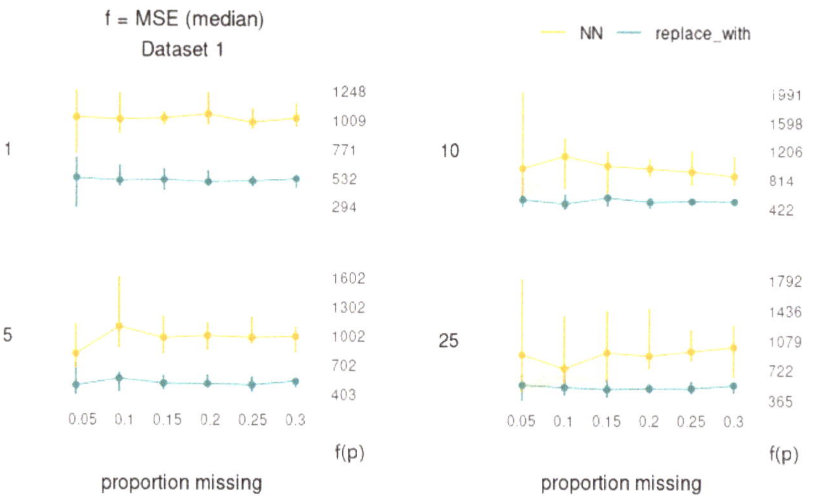

Figure 6. Example of cross-section plots, for the MSE metric, with p and g as shown, showing the algorithms Nearest Neighbours and Replace-With.

4. Analysis of the HWI for $d = 5$ Test Cases

The motivation behind the development of `interpTools` was to test the HWI [1], originally developed to correct missingness in structured astrophysical data. The objective of the research was to audit its statistical performance on various classes of time series following the application of different gap structure parameterizations, and to complete a full comparative analysis against more classically-used interpolation algorithms. The following will present some results from that study demonstrating the robustness of the HWI as well as showcasing the utility of the `interpTools` package.

The analysis was conducted on an (arbitrary) five artificial time series ($d = 1, 2, ..., 5$) of length $N = 1000$, where the mean component m_t was a cubic polynomial function ($\phi = 3$) and was fixed $\forall d$. The periodic trend component t_t was set to vary, where the number of embedded sinusoids, ψ, increased by a factor of ten with each new dataset ($\psi_d = d \times 10$) so as to scale the effective "signal" presence against the "noise" presence. Recall from above that many real-world astrophysical time series have thousands of periodic components present. All five time series were generated against a background of ARMA(0,0) white noise.

The inclusion of a time-varying mean component allowed for the simulation of a particular form of nonstationarity; a property that, in most other conventional interpolation algorithms, would first need to be corrected through estimation and removal of underlying monotonic trends. Often in practice, this correction is either overlooked, or done using crude techniques (such as differencing) that do not preserve the integrity of the data, and are prone to increasing statistical error [8]. One of the major advantages of the HWI is that it can be applied to certain classes of nonstationary data, such that no prior manipulations are necessary [1]. Gap structures were applied with combinations of missingness proportions (p) up to 30% and gap widths (g) up to 25 observations wide, of which each (p, g) parametrization contained $K = 1000$ replicates. In addition to the HWI, other interpolation algorithms were explored: the Kalman filter (KAF), Exponential Weighted Moving Average (EWMA), and cubic splines, with further discussion of other approaches detailed in the appendices of [8].

In assessing the performance metrics, it was clear that in this case the HWI led to significantly more accurate estimation of the missing values, was the most consistent in its estimation, and the most stable when subjected to increasing missingness and data complexity [8]. The HWI maintained its rank, even when compared to the more robust KAF and EWMA algorithms. The cubic splines performed comparably at modest gap structures, but quite poorly when p and g were large, showing particular sensitivity to gap width. Figure 7 provides a summary of the statistical performance of each algorithm (excluding the cubic splines) on the fifth dataset, according to the Normalized Root Mean Squared Deviation (NRMSD) metric (optimal when minimized), which was median-aggregated across the 1000 replicates. The corresponding surface values for the HWI are shown in Table 1.

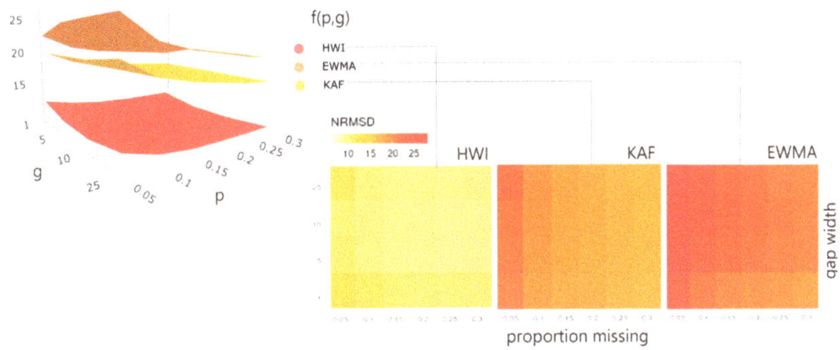

Figure 7. Median Normalized Root Mean Squared Deviation (NRMSD) values across the K simulated interpolations for each (p, g) gap structure imposed on the fifth dataset, with colour proportional to value, and scaled across the set. Each surface and its corresponding heatmap define the performance of a particular interpolation method (HWI, KAF, EWMA). The HWI outperformed the other methods over all parameters.

Note that this should not be considered to be a complete or robust examination of the performance of the HWI against other algorithms. Further analysis was done in [8], but as with many computational algorithms, demonstration of improvements can only be done in limited test cases due to resources. The HWI provides a number of theoretical advantages over other classic algorithms, especially in highly structured time series with large numbers of readily-detected periodic components, which was why it was developed, and these results seem to reinforce that the design was effective for series of this type. The algorithm is being used "in the wild" by several national science agencies for imputation of scientific data sets, with good results.

Table 1. Median NRMSD statistics of the HWI interpolations on the fifth dataset, aggregated across the $K = 1000$ simulations in each (p, g) gap specification. Compare with Figure 6.

		Gap Width (g)			
		1	**5**	**10**	**25**
Proportion missing (p)	5%	11.94	10.62	10.34	11.2
	10%	11.02	9.57	9.51	9.91
	15%	10.42	9.17	9.17	9.42
	20%	10.27	9.09	9.08	9.36
	25%	10.29	9.00	9.01	9.39
	30%	10.31	9.14	9.09	9.52

5. Conclusions

The R package `interpTools` provides a robust set of computational tools for scientists and researchers alike to evaluate interpolator performance on artificially-generated time series data in the presence of various gap structure patterns. Investigating these relationships in the safety of a lab setting with synthetic data allows researchers to benchmark performance and make informed decisions about which interpolation algorithm will be most suitable for a real-world dataset with comparable features. The package also provides a number of data visualization tools that allow a user to distill the resulting copious amounts of performance data into sophisticated, customizable, and interactive graphics.

Through use of this package, we have demonstrated that the Hybrid Wiener Interpolator demonstrates robust performance in the presence of large numbers and lengths of gaps for a selected set of periodic signals against background noise (a relatively broad class of time series often encountered in physical science applications). It is our hope is that by using the framework presented in this paper, interested users will be able to better understand the relationships between interpolators and time series, and minimize the harmful implications of making erroneous inferences from poorly-repaired gappy time series data. The framework also allows for comparison of novel algorithms to accepted standard approaches, novel metrics, and novel time series structural inputs, allowing for a very general support in the development of targeted methods.

Author Contributions: Conceptualization, W.S.B.; methodology, W.S.B. and S.C.; software, S.C.; validation, S.C. and W.S.B.; resources, W.S.B.; writing—original draft preparation, S.C.; writing—review and editing, W.S.B.; visualization, S.C.; supervision, W.S.B.; funding acquisition, W.S.B. All authors have read and agreed to the published version of the manuscript.

Funding: Wesley S. Burr is funded by the National Sciences and Engineering Research Council (NSERC) of Canada, under the Discovery Grant Program. Sophie Castel received an Ontario Graduate Scholarship for the pursuit of this work.

Data Availability Statement: The R package used to develop the simulation results in this paper is available on GitHub [7].

Conflicts of Interest: The authors declare no conflict of interest.

Abbreviations

The following abbreviations are used in this manuscript:

HWI	Hybrid-Wiener Interpolator
NRMSD	Normalized Root Mean Squared Deviation
KAF	Kalman Filter algorithm
EWMA	Exponential Weighted Moving Average algorithm

References

1. Burr, W.S. Air Pollution and Health: Time Series Tools and Analysis. Ph.D. Thesis, Queen's University, Kingston, ON, Canada, 2012.
2. Lepot, M.; Aubin, J.B.; Clemens, F. Interpolation in time series: An introductive overview of existing methods, their performance criteria and uncertainty assessment. *Water* **2017**, *9*, 796. [CrossRef]
3. Andiojaya, A.; Demirhan, H. A bagging algorithm for the imputation of missing values in time series. *Expert Syst. Appl.* **2019**, *129*, 10–26. [CrossRef]
4. Hippert-Ferrer, A.; Yan, Y.; Bolon, P. EM-EOF: Gap-filling in incomplete SAR displacement time series. *IEEE Trans. Geosci. Remote. Sens.* **2020**, 1–18. [CrossRef]
5. Savarimuthu, N.; Karesiddaiah, S. An unsupervised neural network approach for imputation of missing values in univariate time series data. *Concurr. Comput. Pract. Exp.* **2021**, *33*, e6156. [CrossRef]
6. Beck, M.W.; Neeraj, B.; Asencio-Cortés, G.; Kulat, K. R Package `imputeTestbench` to Compare Imputation Methods for Univariate Time Series. *R J.* **2018**, *10*, 218–233. [CrossRef] [PubMed]
7. Castel, S.; Burr, W.S. `interpTools`: Tools for Systematic Testing and Evaluation of Interpolation Algorithms. R Package Version 0.1.0. Available online: https://github.com/castels/interpTools (accessed on 10 July 2021).
8. Castel, S. A Framework for Testing Time Series Interpolators. Master's Thesis, Trent University, Peterborough, ON, Canada, 2020.
9. Thomson, D.J. Spectrum estimation and harmonic analysis. *Proc. IEEE* **1982**, *70*, 1055–1096. [CrossRef]
10. Wiener, N. *Extrapolation, Interpolation, and Smoothing of Stationary Time Series: With Engineering Applications*; MIT Press: Cambridge, UK, 1950.

engineering proceedings

MDPI

Proceedings

Implications of the SARS-Cov-2 Pandemic for Mortality Forecasting: Case Study for the Czech Republic and Spain †

Ondřej Šimpach [1,*] and Marie Šimpachová Pechrová [2]

[1] Department of Statistics and Probability, Faculty of Informatics and Statistics, Prague University of Economics and Business, Winston Churchill sq. 4, 130 67 Prague 3, Czech Republic
[2] Department of Modelling of Impact on Agricultural Policy, Institute of Agricultural Economics and Information, Mánesova 1453/75, 120 00 Prague 2, Czech Republic; simpachova.marie@uzei.cz
* Correspondence: ondrej.simpach@vse.cz; Tel.: +420-737-665-461
† Presented at the 7th International Conference on Time Series and Forecasting, Gran Canaria, Spain, 19–21 July 2021.

Abstract: The current pandemic situation of SARS-Cov-2 is negatively influencing people worldwide, and leading to high mortality and excess mortality, due to more reasons than only the disease itself. Thus, forecasting of the mortality rates and consequent population projections would have been complicated since 2020. Paper models mortality in the Czech Republic and Spain and assesses the possible impact of the COVID-19 on the forecasts. We use a Lee–Carter model and apply it to data from 1981 to 2019 (forecast A) and 1981 to 2020 (forecast B). Our results show differences in forecasts up to 2030 by mean square difference. The highest is in ages above 50 for Spain, where it was observed that the COVID-19 pandemic affected the mortality rates in a way that they were higher, and decreased at a slower pace than they would without taking 2020 into account. In the Czech Republic (CR), the forecast does not seem to be affected yet, but it could be in the future when the number of deaths (not only due to COVID-19, but altogether) increases significantly. Nevertheless, we have to verify our preliminary results on real data as soon as they are available.

Keywords: Lee–Carter model; mortality forecasting; mortality modeling; SARS-Cov-2

Citation: Šimpach, O.; Pechrová, M.Š. Implications of the SARS-Cov-2 Pandemic for Mortality Forecasting: Case Study for the Czech Republic and Spain. *Eng. Proc.* **2021**, *5*, 58. https://doi.org/10.3390/engproc2021005058

Published: 18 July 2021

1. Introduction

"The evolution of the pandemic caused by COVID-19, its high reproductive number and the associated clinical needs, is overwhelming national health systems." (Hierro et al. [1]) Forecasting the outcome of outbreaks as early and as accurately as possible is crucial for decision-making and policy implementations. Much research has tried to predict the development of the pandemic, its spread, and factors that can influence the death rates of COVID-19 or its spatial variations.

For example, Sun et al. [2] investigated the mortality rate across England and assessed the contributions of socioeconomic and environmental factors to spatial variations of COVID-19. They used spatial regression models to estimate the COVID-19 mortality rate and examined the factors that are related to it. They found out that hospital accessibility and relative humidity are negatively related to the COVID-19 mortality rate, but the percentage of Asians and of Blacks, and unemployment rate are related positively to the COVID-19 mortality rate.

Gerli et al. [3] estimated mortality trends in the 27 countries of the European Union (EU), plus Switzerland and the UK, where lockdown dates and confinement interventions have been heterogeneous, and explored its determinants. Hierro et al. [1] employed an OLS and delayed elasticity method to predict mortality for COVID-19 in the US.

Rechtman et al. [4] used gradient-boosting machine learning to predict COVID-19 related mortality based on factors that were identified by logistic regression, i.e., older age, male sex, higher BMI, higher heart rate, higher respiratory rate, lower oxygen saturation, and chronic kidney disease were associated with COVID-19 mortality.

We already know that due to SARS-Cov-2, the life expectancy of people is lower. However, the age-and-sex specific death rates have not been calculated yet. We have to consider and excess mortality (the number of deaths from all causes during a crisis—pandemic situation—above what we would have expected to see under normal conditions). Especially in higher age categories, the cause of death is often the SARS-Cov-2 virus. For example, in Spain, COVID-19 caused deaths even in 44.24% of males and 35.67% females in the age group from 70 to 79 years from 9 March until 9 July 2020 (for detailed data, see Olabi et al. [5]).

Dead rates are changing and have different patterns than in previous years. There are many restrictions, with even complete lockdown, that shall prevent the deaths. Besides, in the Czech Republic, the announcement of the restriction is quite chaotic, and their enforcement relatively weak (in many cases) that forecasting based on econometric models can be misleading. Moreover, it is difficult to predict the situation. As with every modeling apparatus, also mortality forecasting is sensitive to the extraordinary events that are unpredictable and seriously bias the estimate accuracy.

We can agree with Soubeyrand et al. [6] that "discrepancies in population structures, decision making, health systems, and numerous other factors result in various COVID-19-mortality dynamics at the country scale, and make the forecast of deaths in a country under focus challenging."

Ioannidis et al. [7] even proclaimed that forecasting for COVID-19 has failed, due to "poor data input, wrong modeling assumptions, high sensitivity of estimates, lack of incorporation of epidemiological features, poor past evidence on effects of available interventions, lack of transparency, errors, lack of determinacy, consideration of only one or a few dimensions of the problem at hand, lack of expertise in crucial disciplines, groupthink, and bandwagon effects, and selective reporting". Moreover, predicting the future rate of infection is difficult, due to "not knowing the true mechanisms of transmission, infection, and recovery and not having accurate data on who has actually been exposed to the virus and have tested positive to its antibodies." (Allenby [8]).

However, we are concerned with the long forecasting horizon and use real mortality data (regardless of whether the death was caused by the COVID-19 or not), so this gives hope that forecasting is accurate and without above-stated shortcomings. Modern forecasting methods are used.

A similar method—the Lee–Carter model—for mortality forecasting was also used by Diaz-Rojo, Debon, and Mosquera [9]. They found out that changes in mortality patterns might be due to COVID-19. "The mortality changes identified in the control charts pertain to changes in the population's health conditions or new causes of death such as COVID-19 in the coming years." (Diaz-Rojo, Debon, and Mosquera [9]).

The challenges of correct forecasting of COVID-19 spread and mortality of forecasting were also examined by Dayaratna and Michel [10], who showed how different model assumptions change the results of the forecast.

We would like to go further in our analysis and try to project the mortality rates in the future based on available data knowing that COVID-19 might have changed the current situation and can distort the forecasting. We would like to discuss the results of two forecasts by a Lee–Carter model and their differences and to find in what ages they are the highest.

2. Data and Methodology

We chose two heavily hit countries by COVID-19 for a case study: Spain and the Czech Republic (CR). (Besides also Italy and Belgium were also seriously influenced at some time.) Spain had until the end of February over 3.18 mil. cases of disease with total cumulative deaths over 68 thousand. The worst situation was there in March and April 2020. Contrary to that, Czech Republic currently has over 1.23 mil. cases with cumulative deaths over 20 thousand. There have been three peaks so far—October 2020, and January and February 2021. Given that the population of Spain is almost 46.8 mil., and in the Czech Republic, 10.7 mil., the situation is worse in Czechia. (For actual data see [11] https://worldhealthorg.shinyapps.io/covid/ (accessed on 31 May 2021)).

2.1. Data

Data about the Czech Republic and Spain mortality for the period of 1981 to 2018 comes from the *Eurostat database*. We choose 1981 because since that there are full age-and-sex specific data about population and deaths available. (Spain gathered data only up to 85 years until 1980).

Mortality rates were calculated from two Eurostat tables: Deaths by age and sex [demo_magec]—last update 8 February 2021 (extracted 10 February 2021) and Population on 1 January by age and sex [demo_pjan]—last update 8 February 2021 (extracted 10 February 2021). First, the population on 1 January had to be recalculated to mid-year population state in time t ($P_{x,t}$). We took the population state on 1 January in year t and $t+1$, summed it, and divided it by 2 (i.e., simple arithmetic mean). The central mortality rate in time t ($m_{x,t}$) was then calculated as (1):

$$m_{x,t} = \frac{D_{x,t}}{P_{x,t}}, \tag{1}$$

where $D_{x,t}$ is the number of deaths in time t. Data about mortality rates for 2019 in the Czech Republic comes from the *Czech Statistical Office* (Prague, Czech Republic) database, and for Spain for 2019 are taken from *Instituto Nacional de Estadística* (Madrid, Spain). Data for 2020 are available in The *Human Mortality Database*, but they are not age-specific (there are only age categories 0–14, 15–64, 65–74, 75–84, 85+), so the projection cannot include this year directly.

We had to recalculate the number of deaths according to the proportion of dead people in each category. The proportion was calculated as five-year simple arithmetic means of 2015–2019. Particularly the number of deaths in category 0–14 was summed, and the number of deaths in particular age was divided by this sum. This was done for each age category and each year. Then the number of deaths in 2020 was taken from the *Human Mortality Database* (University of California, Berkeley & Max Planck Institute for Demographic Research, Rostock) and recalculated according to the average proportion for each age. The number of inhabitants on 1 January 2020 was taken from Eurostat as the previous data about the number of inhabitants. We could not calculate the mid-term number of inhabitants, because the data for 1 January 2021 are unknown. Finally, death rates were calculated according to (1).

Empirical (and calculated) values of death rates (in logarithms) are displayed in Figure 1 for the Czech Republic and Figure 2 for Spain.

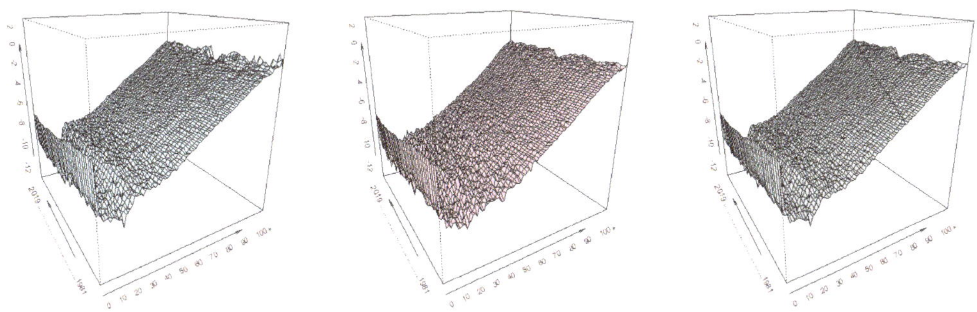

Figure 1. Empirical values of age-specific mortality rates of males (**left**), females (**middle**), and total (**right**)—Czech Republic; x—years, y—ages, z—rates; Source: Human mortality database, own elaboration.

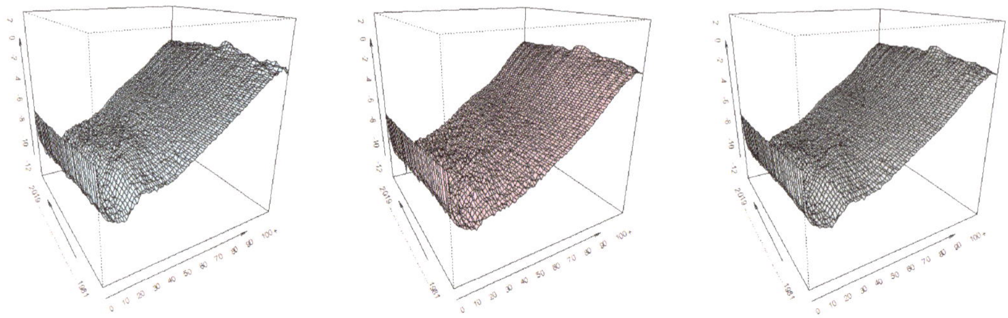

Figure 2. Empirical values of age-specific mortality rates of males (**left**), females (**middle**), and total (**right**)—Spain; x—years, y—ages, z—rates; Source: Human mortality database, own elaboration.

We can identify moments in time and age intervals at which the observed probability of death is substantially different from the pattern of mortality in other periods. Sometimes the patterns are random, e.g., in higher ages, the differences are high, so the mortality rates are modeled in logarithms; sometimes they reflect historical events, for example World War II.

In data from 1981 to 2019, there is nothing significantly different from other periods. Interesting differences are only between 2019 and 2020. Therefore, we take real data from the last 10 years for age categories and display the mortality trend by calculating 5 years moving averages. From Figure 3 (CR) and Figure 4 (Spain), we can clearly see that the COVID-19 pandemic impacted mortality rates. Especially in higher age groups, the increase in the number of deaths is visible. It goes against the trend of decreasing or stagnating mortality. In certain cases, the trend even rapidly changed, due to COVID-19. This was a striking change in Spain in the age category 15 to 39 years. For example, while there were decreasing on average 1,103 of people every week in 2019, it was on average 1187 per week in 2020. Only in age between 0 to 14 years, the number of deaths decreased in line with the long-term trend. So far, the youngest people have stayed reasonably safe from COVID-19, but the situation is changing, and mutation of coronavirus also younger categories can be endangered.

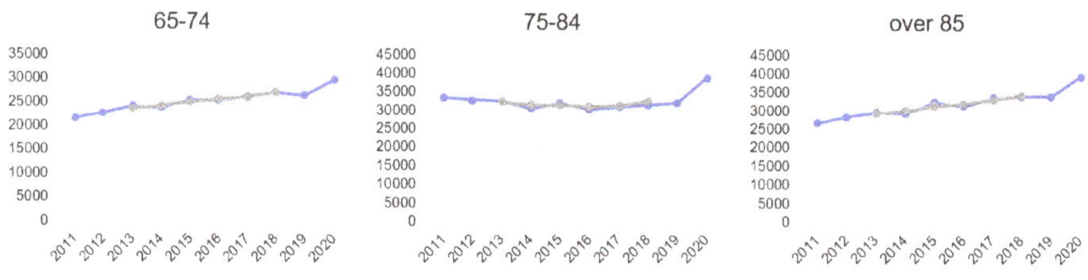

Figure 3. Empirical values of the number of deaths and five-year moving average—Czech Republic; Source: Human mortality database, own elaboration.

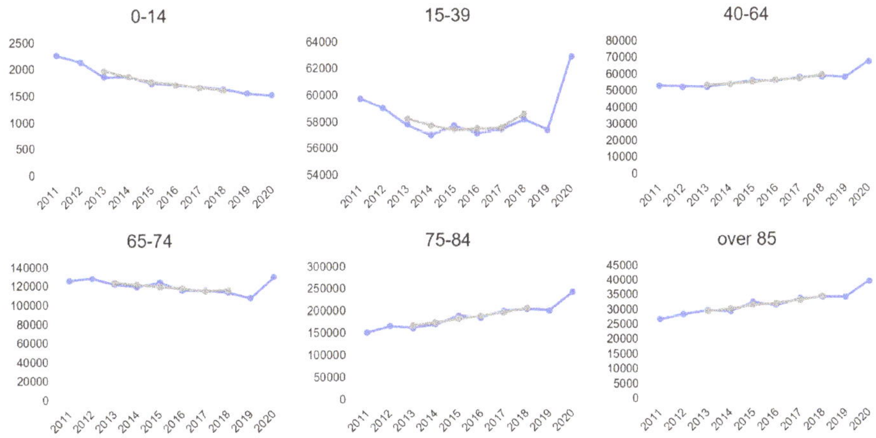

Figure 4. Empirical values of the number of deaths and five-year moving average—Spain; Source: Human mortality database, own elaboration.

In the Czech Republic, there had been an increasing trend of mortality in age categories over 85 and 65–74 already, but in 2020, the increase is more pronounced. While in the age category 75–84 years, there was rather a stagnation in mortality during the observed period, 2020 meant a significant change in the direction to higher mortality. In 2020 a decreasing trend of death in the group 40–64 was broken. In age categories up to 39 years, the number of deaths decreased recently, and even 2020 was not an exception.

Because of the clear difference in mortality patter between 2019 and 2020 we use two basic periods of different length and estimate two predictions of different length. For every population there is: (1) Forecast A used data from 1981 to 2019 ($t = 1, \ldots, 39$) and is elaborated for 2020 to 2030 ($h = 1, \ldots, 11$), and (2) forecast B used data from 1981 to 2020 ($t = 1, \ldots, 40$) and is elaborated for 2021 to 2030 ($h = 1, \ldots, 10$). Both projections are compared by mean squared differences (MSD)—differences between values of two projections that are squared and divided by the length of the projection period (H).

$$MSD_x = \frac{\sum_{h=1}^{H}\left(m_{x,h}^{A} - m_{x,h}^{B}\right)^2}{H}, \tag{2}$$

where $m_{x,h}^{A}$ is the value of mortality rate in time t projected by forecast A and $m_{x,h}^{B}$ by forecast B. The higher is the difference between forecasts, the more COVID-19 pandemic is pronounced and distorts the forecasts. We expect that this will happen, especially for higher ages.

2.2. Methods

This paper aims to model the mortality in the Czech Republic and Spain and assess the possible impact of the COVID-19 on the forecasts. The research brings the discussion about the impact of SARS-Cov-2 on demographic forecasting. Mortality, as one of the demographic processes, is an important part of demographic projections that have a wide impact on policy formulation in every developed country.

Mortality modeling can be based on deterministic or stochastic models. However, currently, more precious stochastic models are rather used than deterministic. We apply the Lee–Carter method that uses the trends and main components of previous development of demographic events.

The level of mortality of younger persons is different in comparison with the oldest, and therefore, it is necessary to correct estimates of mortality at the highest ages, e.g., by Coale–Kisker model, Thatcher model, Kanistö model, or Gompertz–Makeham function (see Boleslawski and Tabeau [12]). Dotlačilová and Šimpach [13] used polynomial functions of the 2nd and 3rd order on the data of Czech and Slovak mortality and concluded that it provided good smoothing in the age from 60 to 80, but have ambivalent results at the ages over 80 years. We expect that due to the SARS-Cov-2 pandemic, the mortality in the highest ages will be affected, and the possible forecast could be distorted, especially in those ages. Thus, there might be a need for correction.

Lee and Carter [14] proposed a "method for modeling and forecasting mortality: A model of age-specific death rates with a time component and a fixed relative age component, and a time series model (an autoregressive integrated moving average—ARIMA) of the time component." (Booth, Maindonald, and Smith [15]) (Diaz-Rojo, Debon, and Mosquera [9]). They defined their model to fit the matrix of mortality rates as (3):

$$\log(\mathbf{m}_{x,t}) = \mathbf{a}_x + \mathbf{b}_x\mathbf{k}_t + \mathbf{e}_{x,t} \text{ or } \mathbf{m}_{x,t} = e^{\mathbf{a}_x+\mathbf{b}_x\mathbf{k}_t+\mathbf{e}_{x,t}}, \quad (3)$$

where $\mathbf{m}_{x,t}$ is the central mortality rate in age x and year t; $\mathbf{a}_x$ and $\mathbf{b}_x$ are the age-specific constants; $\mathbf{a}_x$ is the age specific term which represents the general mortality shape across age; $\mathbf{k}_t$ is time-varying index of the level of mortality for all ages. "The $\mathbf{b}_x$ profile tells us which rates decline rapidly and which rates decline slowly in response to changes in $\mathbf{k}$ $\left(\frac{d\log \mathbf{m}_{x,t}}{dt} = \frac{\mathbf{b}_x dk}{dt}\right)$" (Lee and Carter [14]). The coefficient can be negative for some ages, which indicates that mortality at those ages tends to rise when mortality in other ages is falling. Error term $\mathbf{e}_{x,t} \approx N(0,\sigma^2)$ captures age-specific historical influences not captured by the model and is supposed to be homoscedastic. "As the model written in this way is over parametrized, the two additional constraints are introduced in order to identify the model." (Danesi, Haberman, Millossovich [16]): $\sum_{x=1}^{N} \mathbf{b}_x = 1$ and $\sum_{t=1}^{T} \mathbf{k}_t = 0$. Using these constrains, the least squares estimator for $\mathbf{a}_x$ can be obtained by (4):

$$\hat{\mathbf{a}}_x = \frac{\sum_{x=1}^{N} \log(\mathbf{m}_{x,t})}{N}. \quad (4)$$

Under this normalization, $\mathbf{b}_x$ is the proportion of the change in overall log mortality attributable to age x.

Several advantages of the Lee–Carter model are claimed: "[A] parsimonious demographic model is combined with standard statistical time-series methods; no subjective judgements are involved; forecasting is based on persistent long-term trends; and probabilistic confidence intervals are provided for the forecasts" (Lee and Carter [14]). ARIMA models are used to forecast $\mathbf{k}_t$ and the order of lags is determined based on recommendation by Akaike information criterion. Because the base model estimates the logarithms of rates, thus giving equal weight to ratios of rates, each $\mathbf{k}_t$ was adjusted to reproduce annual total deaths (D_t), while $\mathbf{a}_x$ and $\mathbf{b}_x$ stayed unchanged. This adjustment gives greater weight to the ages at which the numbers of deaths are large. Booth, Maindonald, and Smith [15] modified the method to adjust the time component to reproduce the age distribution of deaths, rather than total deaths, and to determine the optimal fitting period to address non-linearity in the time component.

We used the original Lee–Carter model to elaborate the forecast for the period of 2020 to 2030 (forecast A) and 2021 to 2030 (forecast B). The results were compared according to (2). The model was programmed and fitted in software **R**, and other calculations were done in MS Excel.

3. Results

We present results only for the total population as the data for males and females did not bring meaningful results when they were used for forecasting in model B. It was probably because recalculated data was used. We suppose that 1-year group-specific data are needed to get robust results.

For the Czech Republic, in model A, parameter *k* was estimated by ARIMA(1,1,0) and model B by ARIMA(0,1,0) to predict the level of mortality. In Spain, model A estimated *k* by ARIMA(1,1,0) and model B by ARIMA(1,1,1). The models were selected based on common model selection criteria.

We looked at the average squared differences between forecasts A and B at each age. The differences were relatively low in ages from 1 to 50 in total population in both observed countries. This means that in those ages, adding 2020 to the projection did not bring significant distortion in projections.

No significant distortion is also present in the data for 0-year children in the CR, but in the case of Spain, there is a higher difference. Between years 50 to 90, the mean square difference is still low in the CR, while in raises rapidly in Spain. The results for Spain support our assumption that in higher ages, the forecasts A and B would report higher differences, due to the COVID-19 pandemic.

As an example of the results, we reported the forecasted mortality data for 50, 60, 70, and 80-years-old people because higher ages are more affected, and the differences in forecasts A and B increase after the age of 50. The results are displayed in Figure 5.

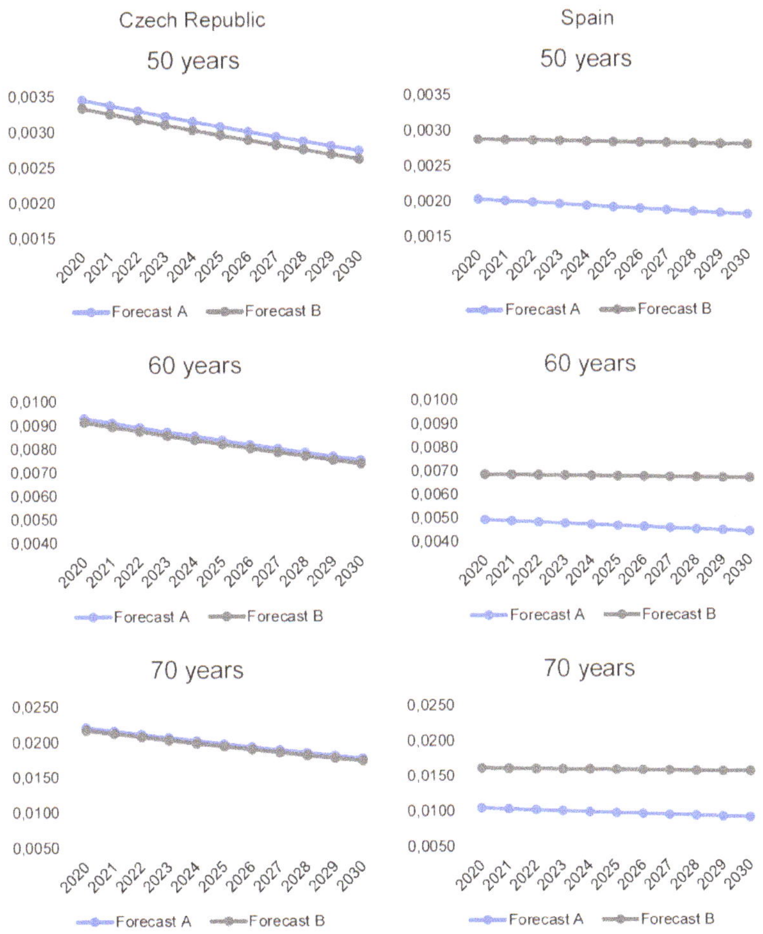

Figure 5. *Cont.*

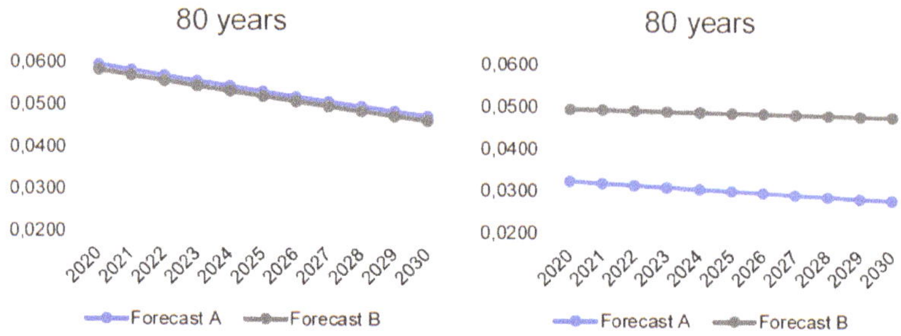

Figure 5. Results of forecasts A and B for the Czech Republic and Spain—example of mortality in higher ages. Source: own elaboration.

Results for the CR are surprising as mortality rates in forecast B are lower than in forecast A. However, the differences are not that significant, and the lines look like parallels. We suppose that it is due to the nature of the data. The population of the CR is relatively small, and the time series is relatively short for forecasting. Besides, the history and living conditions also can play a role. In the CR, since the 1990s, after the end of the communist era, the increase of living standards caused that the mortality rates started to decrease.

This trend is still very strong and rooted in the data. Therefore, despite the COVID-19 pandemic, the mortality rates are still decreasing, and adding 2020 does not change much. Hence, forecast B—i.e., after adding 2020 predicts still lower death rates, than forecast A.

Results for Spain support our assumption. It can be clearly seen that forecast B predicts higher death rates than forecast A. Despite that both projections suggest a decrease, it is milder when 2020 is considered. Hence, the COVID-19 pandemic could influence the decreasing trend of mortality rates negatively. Not only that the death rates are higher, but they are also decreasing at a slower pace.

We also must consider that Spain was hit by COVID-19 already in April 2020, while the CR is suffering heavily only since the end of 2020. Thus, the way the epidemic increasingly influenced how the predictions were affected. Mortality increased rapidly in 2020 in Spain, while in the CR, the differences were not that significant yet. We suppose that in 2021 mortality rates in the CR would be significantly higher, and the mortality forecasts would be affected more.

4. Discussion and Conclusions

Mortality forecasting in the current situation can be disturbed, due to the worldwide SARS-Cov-2pandemic. Because the COVID-19 virus is relatively new, and there are still not available data about age-and-sex specific mortality rates, it is difficult to forecast the consequences and impact of the pandemic on mortality forecasting.

Besides, mortality related to COVID-19 is not caused only by the illness itself, but also due to other psychological issues. Research by Yusuf and Tisler [17] forecasted 388 additional deaths a week in the Netherlands in five years, due to the direct and indirect effects of the lockdown measures. Therefore, they urge that "the additional mortality and increased mental health problem should be considered in evaluating the necessity of lock down and quarantine policy" Yusuf and Tisler [17].

The main problem is that the age-and-sex specific data are not available for 1-year age groups. Only data that are currently obtainable are a weekly number of deaths for large (and not equally length) age groups. The data are divided between males and females, but using them for forecast B created a large bias from forecast A and from expectations. Only forecasts for the total population (especially in CR) were meaningful.

Another problem is that we do not have any information about the mid-year population of 2020, and it can be calculated only from the projection of population in 2021 as the arithmetic mean of those two years. However, any population projection for 2021 without taking the pandemic into account is biased. No institution expected that the mortality rates would increase and developed "higher mortality" scenario, e.g., lower fertility, lower mortality, lower/higher migration. Therefore, the recalculation to one-year mortality rates in 2020 can be inaccurate, so we used data as of 1 January 2020.

Besides, the mortality rates are also influenced by state policy and politicians who try to slow down the pandemic and reduce the number of deaths. Gerli et al. [3] identified a homogeneous distribution of deaths and found a median of 24 days after lockdown adoption to reach the maximum daily deaths. No correlation between lockdown rigidity and population density was observed.

We are aware of the limitations of our results, but consider this topic that important to discuss and draw the path of our future research. We think that it was useful to present at least preliminary results to further elaborate on the topic. We plan to recalculate the model with real data as soon as possible. We believe that using real age-and-sex specific data in 1-year age groups (number of deaths, number of inhabitants, or direct mortality rates) can bring interesting results that will confirm our preliminary results presented in this paper.

Thus far, we can proclaim that the COVID-19 pandemic can affect the mortality rates so that they would be higher and would decrease at a slower pace in higher ages. This can already be the situation in some countries, but in some other countries, this can happen later when the number of deaths (not only due to COVID-19, but altogether) increases significantly. Nevertheless, we must verify our preliminary results on real data when they are available.

Author Contributions: All authors have read and agreed to the published version of the manuscript.

Funding: This research was funded by VŠE IP 400040 project.

Institutional Review Board Statement: Not applicable.

Informed Consent Statement: Not applicable.

Acknowledgments: The paper was supported by long-term institutional support of research activities by the Faculty of Informatics and Statistics of Prague University of Economics and Business.

Conflicts of Interest: The funders had no role in the design of the study; in the collection, analyses, or interpretation of data; in the writing of the manuscript, or in the decision to publish the results.

References

1. Hierro, L.A.; Garzon, A.J.; Atienza-Montero, P.; Marquez, J.L. Predicting mortality for COVID-19 in the US using the delayed elasticity method. *Sci. Rep.* **2020**, *10*, 20811. [CrossRef] [PubMed]
2. Sun, Y.R.; Hu, X.K.; Xie, J. Spatial inequalities of COVID-19 mortality rate in relation to socioeconomic and environmental factors across England. *Sci. Total Environ.* **2021**, *758*, 143595. [CrossRef] [PubMed]
3. Gerli, A.G. COVID-19 mortality rates in the European Union, Switzerland, and the UK: Effect of timeliness, lockdown rigidity, and population density. *Minerva Med.* **2020**, *111*, 308–314. [CrossRef] [PubMed]
4. Rechtman, E.; Curtin, P.; Navarro, E.; Nirenberg, S.; Horton, M.K. Vital signs assessed in initial clinical encounters predict COVID-19 mortality in an NYC hospital system. *Sci. Rep.* **2020**, *10*, 21545. [CrossRef] [PubMed]
5. Olabi, B.; Bagaria, J.; Bhopal, S.S.; Curry, G.D.; Villarroel, N.; Bhopal, R. Population perspective comparing COVID-19 to all and common causes of death during the first wave of the pandemic in seven European countries. *Public Health Pract.* **2021**, *2*, 100077. [CrossRef] [PubMed]
6. Soubeyrand, S.; Ribaud, M.; Baudrot, V.; Allard, D.; Pommeret, D.; Roques, L. COVID-19 mortality dynamics: The future modelled as a (mixture of) past(s). *PLoS ONE* **2020**, *15*, e0238410. [CrossRef] [PubMed]
7. Ioannidis, J.P.A.; Cripps, S.; Tanner, M.A. Forecasting for COVID-19 has failed. *Int. J. Forecast.* **2020**, in press. [CrossRef] [PubMed]
8. Allenby, G. Commentaries: The Bass model: A parsimonious and accurate approach to forecasting mortality caused by COVID-19. *Int. J. Pharm. Healthc. Mark.* **2020**, *14*, 361–365.
9. Diaz-Rojo, G.; Debon, A.; Mosquera, J. Multivariate Control Chart and Lee-Carter Models to Study Mortality Changes. *Mathematics* **2020**, *8*, 2093. [CrossRef]
10. Dayaratna, K.D.; Michel, N.J. The Challenges of Forecasting the Spread and Mortality of COVID-19. *Backgrounder* **2020**, *3486*, 1–27.
11. World Health Organization. COVID-19 Explorer. Available online: https://worldhealthorg.shinyapps.io/covid/ (accessed on 31 May 2021).
12. Boleslawski, L.; Tabeau, E. Comparing Theoretical Age Patterns of Mortality beyond the Age of 80. In *Forecasting Mortality in Developed Countries: Insights from a Statistical, Demographic and Epidemiological Perspective*; Tabeau, E., van den Berg, J.A., Heathcote, C., Eds.; Springer: Berlin/Heidelberg, Germany, 2001; pp. 127–155.

13. Dotlačilová, P.; Šimpach, O. Polynomial function and smoothing of mortality rates: The Czech Republic and Slovakia during their independent development after their separation. In Proceedings of the 14th Conference on Applied Mathematics APLIMAT 2015, Bratislava, Slovakia, 3–5 February 2015; pp. 1–11.
14. Lee, R.D.; Carter, L. Modelling and forecasting US mortality. *J. Am. Stat. Assoc.* **1992**, *87*, 659–675.
15. Booth, H.; Maindonald, J.; Smith, L. Applying Lee-Carter under conditions of variable mortality decline. *Popul. Stud.* **2002**, *56*, 325–336. [CrossRef] [PubMed]
16. Danesi, I.L.; Haberman, S.; Millossovich, P. Forecasting mortality in subpopulations using Lee–Carter type models: A comparison. *Insur. Math. Econ.* **2015**, *62*, 151–161. [CrossRef]
17. Yusuf, E.; Tisler, A. The mortality and psychological burden caused by response to COVID-19 outbreak. *Med. Hypotheses* **2020**, *143*, 110069. [CrossRef] [PubMed]

engineering proceedings

MDPI

Proceedings

Using Least-Squares Residuals to Assess the Stochasticity of Measurements—Example: Terrestrial Laser Scanner and Surface Modeling †

Gaël Kermarrec *, Niklas Schild and Jan Hartmann

Geodetic Institute, Leibniz Universität Hannover, Nienburger Str. 1, 30167 Hannover, Germany; ni.schild@gmx.net (N.S.); jan.hartmann@gih.uni-hannover.de (J.H.)

* Correspondence: kermarrec@gih.uni-hannover.de

† Presented at the 7th International conference on Time Series and Forecasting, Gran Canaria, Spain, 19–21 July 2021.

Abstract: Terrestrial laser scanners (TLS) capture a large number of 3D points rapidly, with high precision and spatial resolution. These scanners are used for applications as diverse as modeling architectural or engineering structures, but also high-resolution mapping of terrain. The noise of the observations cannot be assumed to be strictly corresponding to white noise: besides being heteroscedastic, correlations between observations are likely to appear due to the high scanning rate. Unfortunately, if the variance can sometimes be modeled based on physical or empirical considerations, the latter are more often neglected. Trustworthy knowledge is, however, mandatory to avoid the overestimation of the precision of the point cloud and, potentially, the non-detection of deformation between scans recorded at different epochs using statistical testing strategies. The TLS point clouds can be approximated with parametric surfaces, such as planes, using the Gauss–Helmert model, or the newly introduced T-splines surfaces. In both cases, the goal is to minimize the squared distance between the observations and the approximated surfaces in order to estimate parameters, such as normal vector or control points. In this contribution, we will show how the residuals of the surface approximation can be used to derive the correlation structure of the noise of the observations. We will estimate the correlation parameters using the Whittle maximum likelihood and use comparable simulations and real data to validate our methodology. Using the least-squares adjustment as a "filter of the geometry" paves the way for the determination of a correlation model for many sensors recording 3D point clouds.

Keywords: correlation; Hurst exponent; Whittle maximum likelihood; least-squares; T-splines; surface approximation

Citation: Kermarrec, G.; Schild, N.; Hartmann, J. Using Least-Squares Residuals to Assess the Stochasticity of Measurements—Example: Terrestrial Laser Scanner and Surface Modeling. *Eng. Proc.* **2021**, *5*, 59. https://doi.org/10.3390/engproc2021005059

Published: 19 July 2021

1. Introduction

The widespread use of three-dimensional (3D) laser-scanning technology offers various possibilities to digitize real-world objects (see, e.g., [1]). For instance, the latest generation of terrestrial laser scanners (TLS) is able to record millions of points during a short time period, allowing various applications from standard deformation monitoring [2] to agricultural uses [3]. The noise of the observations is often characterized as being normally distributed; temporal correlations are neglected. This disregard can affect the computation of distances between point clouds recorded at different epochs, as well as test statistics for deformation [4]. Furthermore, the deeper study of correlations can provide precious information about the sensor noise with the aim to decrease its level, and innovative applications such as an analysis of the turbulent atmosphere traveled by the electromagnetic signals (see [5] for an application with GPS observations).

The study of correlations is linked with the filtering of noise from the point cloud, which has been the topic of various publications (see, e.g., [6] for a review). Most of the strategies cannot ensure that the filtered counterpart—also called residuals—will reflect the statistical property of the observation noise. For example, the wavelet chosen will affect the point cloud mode decomposition and, potentially, the correlation structure of the residuals. This is not the case for the least-squares (LS) approximation, provided that the model linking the observations with some parameters to estimate is optimal. In this contribution, we will make use of this property and filter the geometry of a point cloud with the aim of studying the stochasticity of the underlying sensor noise. The surface fitting will be performed with T-splines [7]. This method uses iterative local refinement, as in [8]; it is a powerful and computationally efficient strategy to approximate a point cloud [9], as it is not restricted to predefined forms such as circles or ellipses [10]. Both simulated and real data will highlight the feasibility and potential of our methodology to derive the correlation structure of the underlying noise from the LS residuals. We will focus on correlations modeled as a fractional Gaussian noise (fGn, [11]).

The remainder of the paper is as follows: In the second section, we will present the mathematical background of LS and surface fitting. The third section presents the results of simulations and real data. The fourth section concludes this contribution.

2. Mathematical Background

2.1. Least-Squares

In the following section, we assume that a linear or linearized functional model describes our observations:

$$\mathbf{l} = \mathbf{A}\mathbf{x} + \mathbf{v} \tag{1}$$

where $\mathbf{l}$ is the $n \times 1$ observation vector, $\mathbf{A}$ is the non-stochastic $n \times u$ design matrix with full column rank, $\mathbf{x}$ the $u \times 1$ parameter vector to be estimated, and $\mathbf{v}$ the $n \times 1$ observation noise vector. The error term has zero mean and is normally distributed with $E(\mathbf{v}\mathbf{v}^{\mathbf{T}}) = \boldsymbol{\Sigma}_{\mathbf{vv}}$, where $\boldsymbol{\Sigma}_{\mathbf{vv}}$ is the $n \times n$ positive definite fully populated variance covariance matrix of the error term. $E(.)$ denotes the mathematical expectation.

The system is overdetermined. A solution can be given by minimizing the sum of the squares of the residuals. In that case, the generalized LS estimator $\hat{\mathbf{x}}$ reads:

$$\hat{\mathbf{x}} = \left(\mathbf{A}^{\mathbf{T}}\boldsymbol{\Sigma}_{\mathbf{vv}}^{-1}\mathbf{A}\right)^{-1}\mathbf{A}^{\mathbf{T}}\boldsymbol{\Sigma}_{\mathbf{vv}}^{-1}\mathbf{y} \tag{2}$$

We call $\hat{\mathbf{v}} = \mathbf{y} - \mathbf{A}\hat{\mathbf{x}}$ the estimated $n \times 1$ residual vector of the LS adjustment and $\boldsymbol{\Sigma}_{\hat{\mathbf{v}}\hat{\mathbf{v}}}$ its variance–covariance matrix. We have [12]:

$$\boldsymbol{\Sigma}_{\hat{\mathbf{v}}\hat{\mathbf{v}}} = \boldsymbol{\Sigma}_{\mathbf{vv}} - \mathbf{A}\left(\mathbf{A}^{\mathbf{T}}\boldsymbol{\Sigma}_{\mathbf{vv}}^{-1}\mathbf{A}\right)^{-1}\mathbf{A}^{\mathbf{T}} = \boldsymbol{\Sigma}_{\mathbf{vv}} - \boldsymbol{\Sigma}_{\mathbf{vv}\_est} \tag{3}$$

In many applications, as, for example, in surface approximation, $\boldsymbol{\Sigma}_{\mathbf{vv}}$ is often replaced with the identity matrix, assuming equal variances for the observations. From (3) we have access to the variance and covariance of the noise observations from the residuals of the approximation, if we assume that the terms contained in $\boldsymbol{\Sigma}_{\mathbf{vv}\_est}$ are much smaller than those of $\boldsymbol{\Sigma}_{\mathbf{vv}}$. The knowledge of $\boldsymbol{\Sigma}_{\hat{\mathbf{v}}\hat{\mathbf{v}}}$ allows us to avoid an overestimation of the precision of the LS estimator and potentially biased statistical tests. It can be used to describe and quantify the sources of correlations: the cables, the atmosphere, or the processing at the receiver level. Such information provides precious data for stochastic modeling.

2.2. Surface Approximation Using T-Splines

2.2.1. General Principle of Surface Approximation

In this paper, we focus on the residuals of surface approximation from scattered data using a mesh optimization approach. In the following, we only provide a short introduction on that topic; interested readers should refer to other publications [7,13].

From now on, we consider the observations to be the Cartesian coordinates of a 3D point cloud, recorded by, for example, a TLS. We assume the point cloud to have been parametrized in advance and the observations to be temporally sorted. We use the T-splines surface approximation to estimate a net of control points-sometimes called vertices. The denser the net, the closest the approximation will be to the noisy measurements. We minimize the squared distance between the noisy points and the mathematical surface and focus on the stochasticity of the residuals.

2.2.2. T-Splines Surface

A T-splines surface is defined as $S_T(s,t) = \sum_{i=1}^{m} d_i N_i(s_i,t_i)$, with $N_i(s,t) = N^3_{i,\mathbf{U}b}(s)N^3_{i,\mathbf{V}b}(t)$ and $N^3_{i,\mathbf{U}b}(s)$ as the cubic blending basis functions defined by a recurrence relationship and associated with a knot quintuple $\mathbf{U}b = [u_{i,0}, u_{i,1}, u_{i,2}, u_{i,3}, u_{i,4}]$; $N^3_{j,\mathbf{V}b}(s)$ is defined similarly [7]. We call $\mathbf{x} = \{d_i\}$ the vector matrix, containing the nt control points to be estimated iteratively. The main element of the T-splines surface is the T-mesh, which consists of control points connected by several straight lines. Figure 1 (right) shows an example for the surface under consideration in this paper, depicted in Figure 1 (left).

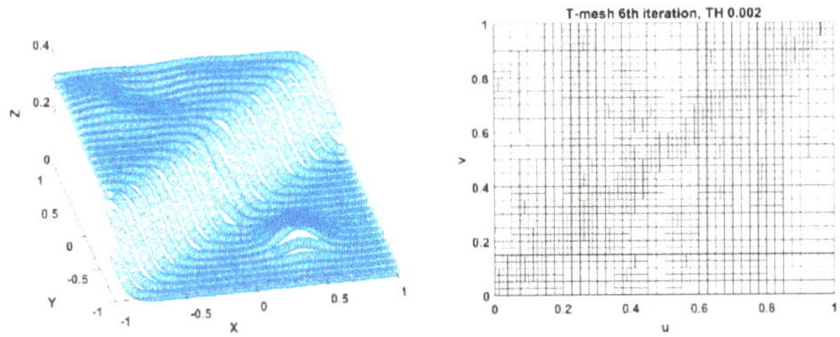

Figure 1. Reference surface under consideration and the corresponding T-mesh after the 6th iteration; threshold (TH) 0.002.

The surface approximation is performed iteratively via LS optimization, and usually starts with a basic rectangular and regular mesh structure. The parameters to be estimated are the nt control points from the T-mesh. The optimization problem can be written in matrix form as $\mathbf{Ax=B}$, where $\mathbf{A}$ is an (nt, nt) matrix that contains the estimation of blending functions at the parameter location: $a_{gi} = \sum_{j=1}^{n} N^3_{g,\mathbf{U}b}(s_j)N^3_{g,\mathbf{V}b}(t_j)$, $g = 1\ldots nt$. Matrix $\mathbf{A}$ depends on the underlying T-mesh via the two local knot vectors $\mathbf{U}b, \mathbf{V}b$. We further define $\mathbf{B} = \{b_g\}$, $b_g = \sum_{i=1}^{nt} \mathbf{l}(s_i,t_i)N_i(s_i,t_i)$. With this notation, we have a unique solution $\hat{\mathbf{x}} = \mathbf{A}^{-1}\mathbf{B}$ which corresponds to (2) by taking $\boldsymbol{\Sigma}_{\mathbf{vv}}$ as the identity matrix. The residuals of the approximation can be computed for each parametrized observation as $e_i = \|\hat{S}_T(s_i,t_i) - \mathbf{l}(s_i,t_i)\|_2, i = 1\ldots n$. We defined TH as being a user-defined threshold. If $e_i > TH$, the cells containing the corresponding data points are refined following the algorithm proposed in [14]. The local refinement allows an economic surface approximation in terms of the parameters to estimate. A wise choice of TH further reduced the risk of overfitting, i.e., fitting the noise of the observations rather than the underlying surface. We propose to set $TH \approx 2\sigma_Z$, with σ_Z the variance in the Z-direction, to prevent such unnecessary refinement. The refinement is ended if the errors do not exceed TH or after a given number of iterations.

2.2.3. Residual Analysis

Once the surface approximation is performed, the residuals can be further analyzed. In this paper, we concentrate on the Z-direction only, as the vertical component is known to be noisier than the horizontal ones [10]. The observations are sorted temporally so that the residuals can be seen as a time series (1D) rather than as a surface (2D).

From physical considerations, it is justifiable to assume the noise to be an fGn, or a combination of them: the high rate of measurements of most sensors induces a long dependency between the observations. More specifically, we are predisposed to think that flicker and white noise (WN, coming from the electronic component and the processing of the raw observations), or atmospheric noise (from the propagation of the signal) will be present. Although they will be found in different bandwidths, a global correlation parameter can be estimated using the WhiE as proposed by [15]. The fGn is fully described by its bounded variance and Hurst exponent H, which is related to the slope of the psd. Following [16], an unbiased likelihood is given by:

$$l_W(H) = -\sum_{\omega \in \Omega} \left[\log\left(\tilde{f}(\omega, H)\right) + \frac{I(\omega)}{\tilde{f}(\omega, H)} \right] \tag{4}$$

with Ω the set of discrete Fourier frequencies, $\tilde{f}(\omega, H)$ the continuous-time process spectral density and $I(\omega)$ the periodogram $I(\omega) \infty \sum_{j=1}^{N} |X_{H,j} e^{-ij\omega}|^2$. The Hurst exponent can be estimated as $\hat{H} = \text{argmax}(l_W(H))$. The slope of the psd β is given by $H = \beta + 1/2$. In this paper, we use the WhiE as implemented in MATLAB by [17]; this function has the main advantage of allowing the estimation of the Hurst parameter when the psd saturates at low frequencies (the so-called Matérn covariance model [18]).

The methodology used in this contribution to derive the analysis of the observation noise from the LS residuals is summarized in a flowchart form in Figure 2.

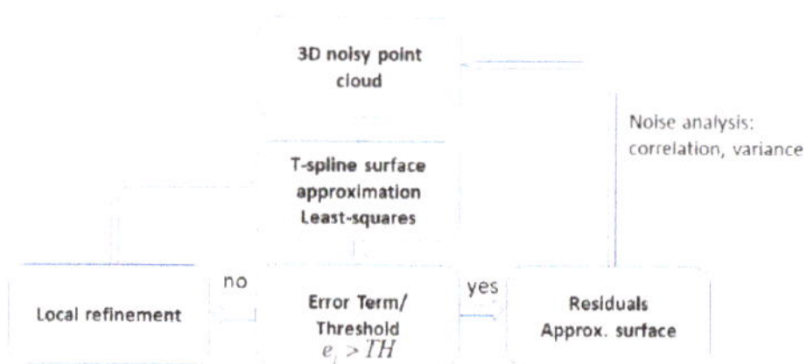

Figure 2. Flowchart explaining the methodology to extract the geometry of a point cloud to analyze the correlation structure of the residuals.

3. Data Analysis

In this section, we will validate the methodology presented in Section 2 to derive the correlation structure of the measurement noise from the residuals of the LS surface approximation. Comparable simulated and real data will be used to highlight the potential of the method.

3.1. Simulated Point Clouds

3.1.1. Reference Point Cloud

We simulated 4000 Terrestrial Laser Scanner observations in Cartesian coordinates from a surface corresponding to Figure 1. This reference surface contains different shapes: it mimics a dam in the middle, a mountain in its side, and a wave structure on the top part of it.

3.1.2. Noise Generation

The reference surface is noised by adding:

- to the X- and Y-components: a Gaussian noise with a standard deviation of 1×10^{-4} m–generated with the Matlab function randn;

- to the Z-component: an fGn or a combination of fGn and WN. We use the MATLAB function ffgn [19].

Two noise vectors are generated: a pure flicker noise (case 1) with $H \approx 0.9$, and a combination of 30% white and 70% flicker noise (case 2). The standard deviation of both is set to $\sigma_Z = 1 \times 10^{-3}$ m, with the aim of approximating the reality of the noise of a TLS sensor [10].

3.1.3. Surface Approximation

The approximation is performed with T-splines using the concept presented in Section 2. We use $TH = 2\sigma_Z$. 6 iterations were performed until the error term did not exceed the threshold. The final T-mesh is shown in Figure 1 (right).

3.1.4. Residual Analysis

The residuals of the approximation are shown for case 1 as an example in Figure 3 (top, red line), together with their power spectral density (psd, Figure 3, bottom). A few outliers caused by the mesh approximation are visible when compared with the reference noise vector (Figure 3, blue line). Fortunately, the shape of the original noise vector is still preserved, which highlights the goodness of the surface approximation. To validate that the correlation structure of the residuals is the same as the one of the measurement noise, we propose to justify empirically that $\mathbf{\Sigma}_{\mathbf{vv}\_est}$ can be discounted (see [3]). To that end, we firstly assume an equal variance of 1 for the error term in the surface approximation. In Figure 3, we plot the diagonal values and the first 100 lines of $\mathbf{\Sigma}_{\mathbf{vv}\_est}$: the diagonal values are more than 5 times smaller than 1; the sparsity of the matrix is evident. The mean value of $\mathbf{\Sigma}_{\mathbf{vv}\_est}$ is found to be of the order of $1 \times 10^{-3} << 1$: this justifies discounting $\mathbf{\Sigma}_{\mathbf{vv}\_est}$ as its impact will be small. The psd of the residuals (Figure 3, right bottom) confirms the validity of this assumption; the slopes of the reference noise psd and the residuals are similar (see Table 1 for the corresponding estimates using the WhiE). This result is valid for cases 1 and 2. We mention that the additional WN component acts to decrease the estimated global slope (0.88 instead of 0.9); This is expected, since the WN was not filtered out. We further point out that low frequencies are deleted in the residuals, which can be interpreted intuitively by considering the T-mesh as a high-pass filter (see Figure 1). Fortunately, the WhiE is not affected by the loss of power at low frequencies, as shown in Table 1. The variance estimated from the residuals is slightly underestimated, which may be due to the aforementioned effect.

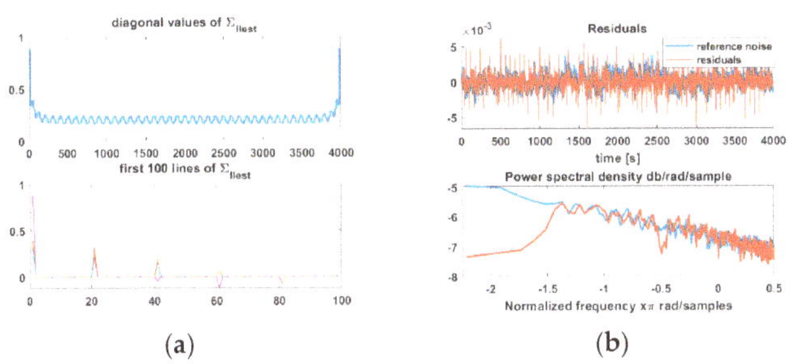

Figure 3. (**a**) Analysis of $\mathbf{\Sigma}_{\mathbf{vv}\_est}$ (diagonal values and first 100 lines). (**b**) (top) residuals (red line) and reference vector (blue line) and their corresponding psd (bottom), loglog plot.

Table 1. Results of the residual analysis for case 1 (no WN) and case 2 (30% WN): slope of the psd and standard deviation (abbreviated as std).

Slope/std [m]	Case 1	Case 2
Original noise	$0.9/1 \times 10^{-3}$	$0.88/1 \times 10^{-3}$
Residuals	$0.89/0.95 \times 10^{-4}$	$0.86/0.91 \times 10^{-4}$

3.2. Real Data Set

3.2.1. Using a 3D Printer

The simulated surface generated in Section 3.1 was created with a high-quality 3D printer (see Figure 4a). To that end, the reference surface was firstly approximated with T-splines to serve as a basis for the printing. To generate a 3D-printable plane from a point cloud, this latter was converted into a solid. A surface model was first created from the points using Delaunay triangulation, and a solid was finally generated by uniformly thickening this model. The final export to a 3D printer-compatible file, such as the .stl format, makes the surface generated from a point cloud 3D-printable. This latter was printed using plastic and stabilized. Grey paint was used in order to avoid strong reflections during the scanning process with a TLS. The additional WN generated by the scanner to generate the surface is considered to be below 1 mm from manufacturer specifications. Consecutively, the printed surface is not exactly the reference one but is superimposed with WN. As the residuals are computed with respect to the generated surface and not the true one, this will have no impact on our conclusions about the correlation structure.

3.2.2. Scanning

The 3D-printed surface was scanned using a phase TLS Zoller+Fröhlich Imager 5016. The scanning configuration was optimal (no angle of incidence, and a surface aligned centrally at the height of the tilting axis of the TLS). The measurements took place indoors at the measuring laboratory of the Geodetic Institute in Hannover (see Figure 4a).

The panel was observed from a total of four standpoints, with distances between 2 and 6 m. No over-radiation occurred. In this contribution, we selected as an example the distance of 2 m, scanned with the angle resolution "high" and the scanning duration "high quality", resulting in a total of 143,490 points. The point cloud was parametrized and approximated with a T-splines surface with $TH = 0.001$, i.e., we assumed a standard deviation of the noise of about 0.5 mm, following the manufacturer specification. The obtained residuals and their psd are depicted in Figure 4 (r, top) and Figure 5, respectively.

3.2.3. Results

The psd of the whole residuals of the approximation is physically barely interpretable (Figure 5, blue line), i.e., it depicts effects coming from the functional model or T-mesh which acts by adding high frequencies—or identically down-weighting low frequencies, see Section 3.1—yielding a serrated psd. For a better understanding of the correlation structure, we thus selected 5000 epochs of the residuals, which are shown in Figure 4 (right, bottom). Other parts were selected: the same results were comparable and are not presented here for the sake of brevity. We eliminated outliers with the mean absolution deviation method [20] and performed a low-pass Butterworth filter of the first order with a normalized cutoff frequency of 0.05, to further eliminate the WN component following the methodology proposed in [21] (see Figures 4 and 5, red line). As expected from [22], we can identify different noises present in different bandwidths: a strong WN component at high frequencies, an fGn with a slope close to $-8/3$, followed by an fGn with a slope close to $-2/3$ and a saturation at low frequencies. The lack of power at low frequencies is to be interpreted following the results of the simulation, and comes from the T-splines surface approximation. The global slope estimated with the WhiE for the non-filtered residuals was found to be -0.8 due to the WN component and close to -2.8 for the filtered residuals. Atmospheric turbulence affects optical signals traveling through a random medium close to the earth's surface and acts to correlate the measurements. This slope is close to the one that is expected from the turbulence theory [21]. The variance of the non-filtered residuals after outlier elimination was 5.8×10^{-4}, and is close to the expected value from the manufacturer specification.

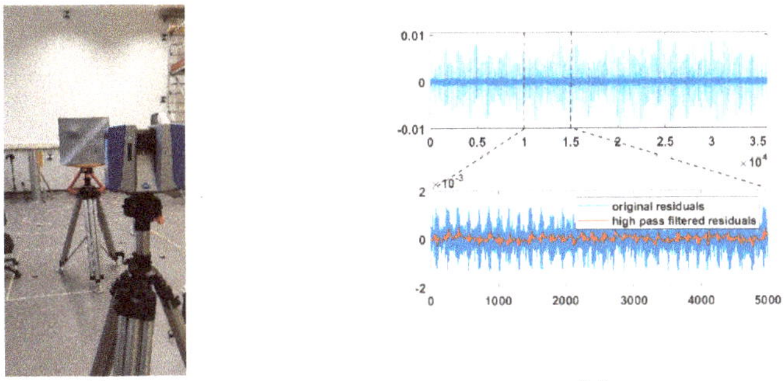

(a) (b)

Figure 4. (**a**) 3D print of the reference surface used for the simulations. (**b**) The residuals of the approximation.

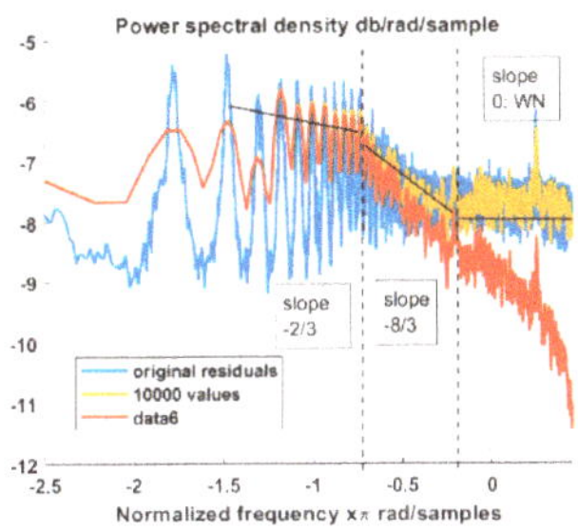

Figure 5. The psd of the whole residuals (blue line), of the selection section (yellow line), and of the filtered section (red line).

4. Conclusions

With the continuous increase of the data rate, the measurement noise of sensors is more likely to become strongly temporally correlated. In this contribution, we have demonstrated how the residuals of the LS approximation can be used to analyze the temporal correlation structure of measurement noise from a TLS. As an example, we performed the extraction of the geometry of a TLS point cloud using an LS approximation with a T-splines surface. This latter has the main advantage of being computationally efficient; it provides an optimal functional model for the further analysis of the correlation structure of the residuals of the approximation. Simulations were set up to validate the methodology with a noised reference surface. We used the unbiased WhiE to estimate the slope parameter of the psd. The same surface was printed in 3D and scanned with a TLS. The residuals could be successfully interpreted based on the results of the simulations. The psd of the real data analysis showed noises present in different bandwidths. Besides a WN component, some of them were identified as coming from the propagation of optical signals through a turbulent medium. This analysis is a validation of the potential of the LS surface approximation to quantify and analyze the correlation structure of sensor noise. Further investigations will focus on its modelization, as well as the impact of functional mismodeling on the spectral decomposition of the residuals.

Author Contributions: Conceptualization, methodology, validation, formal analysis, investigation, writing, funding acquisition: G.K.; writing—review and editing: N.S. and J.H. All authors have read and agreed to the published version of the manuscript.

Funding: This study is supported by the Deutsche Forschungsgemeinschaft under the project KE2453/2-1.

Institutional Review Board Statement: Not applicable.

Informed Consent Statement: Not applicable.

Data Availability Statement: Data can be made available on demand.

Conflicts of Interest: The authors declare no conflict of interest. The funders had no role in the design of the study; in the collection, analyses, or interpretation of data; in the writing of the manuscript, or in the decision to publish the results.

References

1. Borrmann, A. *Building Information Modeling: Technology Foundations and Industry Practice*, 1st ed.; Springer: New York, NY, USA; Berlin/Heidelberg, Germany, 2018; p. 303.
2. Dermanis, A.; Livieratos, E. Applications of deformation analysis in geodesy and geodynamics. *Rev. Geophys.* **1983**, *21*, 41–50. [CrossRef]
3. Friedli, M.; Kirchgessner, N.; Grieder, C.; Liebisch, F.; Mannale, M.; Walter, A. Terrestrial 3D laser scanning to track the increase in canopy height of both monocot and dicot crop species under field conditions. *Plant Methods* **2016**, *12*, 9. [CrossRef] [PubMed]
4. Kermarrec, G.; Kargoll, B.; Alkhatib, H. Deformation Analysis Using B-Spline Surface with Correlated Terrestrial Laser Scanner Observations—A Bridge Under Load. *Remote Sens.* **2020**, *12*, 829. [CrossRef]
5. Kermarrec, G.; Schön, S. On the Mátern covariance family: A proposal for modeling temporal correlations based on turbulence theory. *J. Geod.* **2014**, *88*, 1061–1079. [CrossRef]
6. Han, X.; Jin, J.; Wang, M.; Jiang, W.; Gao, L.; Xiao, L. A review of algorithms for filtering the 3D point cloud. *Signal Process. Image Commun.* **2017**, *57*, 103–112. [CrossRef]
7. Sederberg, T.W.; Zheng, J.; Bakenov, A.; Nasri, A. T-splines and T-NURCCs. *ACM Trans. Graph.* **2003**, *22*, 477–484. [CrossRef]
8. Bracco, C.; Giannelli, C.; Großmann, D.; Sestini, A. Adaptive fitting with THB-splines: Error analysis and industrial applications. *Comput. Aided Geom. Des.* **2018**, *62*, 239–252. [CrossRef]
9. Wang, Y. Free-Form Surface Representation and Approximation Using T-Splines. Ph.D. Thesis, Nanyang Technological University, Singapore, 2009.
10. Wujanz, D.; Burger, M.; Mettenleiter, M.; Neitzel, F. An intensity-based stochastic model for terrestrial laser scanners. *ISPRS J. Photogramm. Remote Sens.* **2017**, *125*, 146–155. [CrossRef]
11. Mandelbrot, B.B.; Van Ness, J.W. Fractional Brownian motion, fractional noises and applications. *SIAM Rev.* **1968**, *10*, 422–437. [CrossRef]
12. Koch, K.R. *Parameter Estimation and Hypothesis Testing in Linear Models*; Springer: Berlin, Germany, 1999.
13. Piegl, L.; Tiller, W. *The NURBS Book*; Springer Science & Business Media: Berlin, Germany, 1997.
14. Morgenstern, P.; Peterseim, D. Analysis-suitable adaptive T-mesh refinement with linear complexity. *Comput. Aided Geom. Des.* **2015**, *34*, 50–66. [CrossRef]
15. Kermarrec, G.; Lösler, M.; Hartmann, J. Analysis of the temporal correlations of TLS range observations from plane fitting residuals. *ISPRS J. Photogramm. Remote Sens.* **2021**, *171*, 119–132. [CrossRef]
16. Sykulski, A.M.; Olhede, S.C.; Guillaumin, A.P.; Lilly, J.M.; Early, J.J. The debiased Whittle likelihood. *Biometrika* **2019**, *106*, 251–266. [CrossRef]
17. Lilly, J. jLab: A Data Analysis Package for Matlab, v. 1.6.6. 2019. Available online: http://www.jmlilly.net/jmlsoft.html (accessed on 28 June 2021).
18. Lilly, J.M.; Sykulski, A.M.; Early, J.J.; Olhede, S.C. Fractional Brownian motion, the Matérn process, and stochastic modeling of turbulent dispersion. *Nonlinear Process. Geophys.* **2017**, *24*, 481–514. [CrossRef]
19. Stoev, S. Simulation of Fractional Gaussian Noise *EXACT* MATLAB Central File Exchange. Available online: https://www.mathworks.com/matlabcentral/fileexchange/19797-simulation-of-fractional-gaussian-noise-exact (accessed on 28 June 2021).
20. Hoaglin, D.C.; Mosteller, F.; Tukey, J.W. *Understanding Robust and Exploratory Data Analysis*; John Wiley & Sons: Hoboken, NJ, USA, 1983; pp. 404–414.
21. Kermarrec, G. On Estimating the Hurst Parameter from Least-Squares Residuals. Case Study: Correlated Terrestrial Laser Scanner Range Noise. *Mathematics* **2020**, *8*, 674. [CrossRef]
22. Wheelon, A.D. *Electromagnetic Scintillation Part I Geometrical Optics*; Cambridge University Press: Cambridge, UK, 2001.